HANDBOOK OF INFRARED SPECTROSCOPY OF ULTRATHIN FILMS

HANDBOOK OF INFRARED SPECTROSCOPY OF ULTRATHIN FILMS

Valeri P. Tolstoy
Irina V. Chernyshova
Valeri A. Skryshevsky

WILEY-INTERSCIENCE

A JOHN WILEY & SONS, INC., PUBLICATION

Library of Congress Cataloging-in-Publication Data:

Tolstoy, V. P. (Valeri P.)
    Handbook of infrared spectroscopy of ultrathin films / V. P.
Tolstoy, I. V. Chernyshova, V. A. Skryshevsky.
        p. cm.
    ISBN 0-471-35404-X (alk. paper)
    1. Thin films — Optical properties. 2. Infrared spectroscopy.
I. Chernyshova, I. V. (Irina V.) II. Skryshevsky, V. A. (Valeri
A.) III. Title.
    QC176.84.07 T65 2003
    621.3815′2 — dc21

2001007192

CONTENTS

PREFACE

In this book, we will designate ultrathin films, or, as they are also called in the literature, nanolayers, to mean layers ranging from submonolayers to several monolayers; these may be formed from a wide range of organic and inorganic substances or present adsorbed atoms, molecules, biological species, on a substrate or at the interface of two media. These films play an important role in many current areas of research in science and technology, such as submicroelectronics, optoelectronics, optics, bioscience, flotation, materials science of catalysts, sorbents, pigments, protective and passivating coatings, and sensors. It could even be argued that the rapid advances in thin-film technology has necessitated the development of special approaches in the synthesis and investigation of nanolayers and superlattices. Nowadays, these approaches are generally applicable in so-called nanotechnology, which includes the synthesis–deposition and characterization of ultrathin films with a prescribed composition, morphology–architecture and thicknesses on the order of 1 nm.

Common features in all studies in the field of nanotechnology arise from problems connected with the physicochemical investigation of ultrathin films, which originate in general from their extremely small thickness. To solve these problems, a number of technically complicated physical methods that operate under UHV conditions, such as AES, XPS, LEED, HREELS are used. Infrared (IR) spectroscopy and in particular Fourier transform IR (FTIR) spectroscopy — a method that enables the determination of molecular composition and structure — offers important advantages in that the measurements can be carried out for nanolayers located not only on a solid substrate but also at solid–gaseous, solid–liquid, liquid–gaseous, and solid–solid interfaces, including semiconductor–semiconductor, semiconductor–dielectric, or semiconductor–metal, with no destruction of either medium. Thus IR spectroscopy is one of a few physical methods that can be used for both in situ studies of various processes on surface and at interfaces and technological monitoring of thin-film structures in fields such as microelectronics or optoelectronics under serial production conditions. The versatility of modern FTIR spectroscopy provides means to characterize ultrathin coatings on both oversized objects (e.g., works of art) and small (10–20-μm) single particles, substrates with unusual shapes (e.g., electronic boards), and recessed areas (e.g., internal surfaces of tubes). It should be emphasized that IR spectroscopy can be highly sensitive to ultrathin films: Depending on the system, the sensitivity is $10^{-5}-10\%$ monolayer.

However, the various IR spectroscopy techniques must be adapted to measure spectra of very small amounts of substance in the form of ultrathin films. While for analyses of bulk materials it is possible to select the optimum mass of substance to record its spectrum, in the case of ultrathin films it is only possible to vary the conditions under which the spectra are recorded (measurement technique, polarization, angle of incidence, immersion media, number of radiation passages through the sample). For this purpose, it is necessary first to theoretically assess the effect of the recording conditions on the intensity of the absorption bands.

By understanding optical theory for stratified media, it is also possible to distinguish optical effects (artifacts), which present in each IR spectrum of an ultrathin film, and, hence, to avoid misinterpretation of the experimental data.

Although the optimum conditions for a number of simple systems are known (e.g., for ultrathin films on metals reflection–absorption (IRRAS) at grazing angles of incidence is commonly used, and for ultrathin films on transparent substrates, multiple internal reflection (MIR) is most suitable in many cases), the spectral contrast can be further enhanced by employing additional special technical approaches.

The material in this handbook is presented in such a way as to address these issues. Thus, the theoretical concepts associated with the interaction of IR radiation with matter and with thin films are considered and, to evaluate the optimum conditions routinely, simple algorithms for programming are given in Chapter 1. In Chapter 2, a theoretical evaluation of the optimum conditions for measuring nanolayer spectra is presented. In Chapter 3, a more detailed interpretation of the IR spectra of ultrathin films on flat and powdered substrates is discussed from the viewpoint of optical theory, including the authors' methods of determination of the optical constants of ultrathin films [1, 2] and molecular orientation [3]. In Chapter 4, technical approaches to measure good-quality IR spectra of ultrathin films are considered. Along with the techniques considered routine for studies of ultrathin films in many laboratories and original techniques described in the literature, techniques developed by the authors are included here, namely IR spectroscopy of single and multiple transmission in p-polarized radiation [4, 5]; IRRAS of the surface of semiconductors and dielectrics [6]; IRRAS of metals, semiconductors, and dielectrics in immersion media [7, 8]; diffuse transmission of disperse materials [9, 10]; and the special attenuated total reflection (ATR) technique for studying the semiconductor–solution interface [11]. An important feature of the optical accessories described is that they are placed in the sample compartment of a conventional continuous-scan or step-scan FTIR spectrometer, without any change in the spectrometer optical scheme. In addition, different attachments for specialized measurements are considered, including IR microscope objectives, in situ chambers, and spectroelectrochemical cells. Time-resolved spectroscopy and enhanced surface and structure sensitivity in IR spectroscopic techniques such as modulation spectroscopy and two-dimensional correlation analysis are described. Subsequent chapters illustrate some applications of these techniques in the study of thin-layer structures and semiconductor–electrolyte interfaces, which are currently of great practical significance in electronics, solar energy storage, sensors, catalysis, bioscience, flotation, and corrosion inhibition. Recommendations are given regarding application of these techniques for automated and on-line analysis of thin-film structures. PAS [12, 13] are not included here, because its sensitivity is insufficient for the measurements in question. Infrared emission studies of ultrathin films have recently been reviewed [14, 15]. Since this method, although rather surface sensitive under specific conditions, has not yet experienced extensive application, it is not considered here as well. Instead, fundamentals and techniques of transmission, diffuse transmission, IRRAS, ATR, and DRIFTS methods will be presented, with emphasis on their application to ultrathin films. Complementary information on how to use these methods and their history may be found in other monographs [16–21].

It should be noted that because the material is presented in such a manner, this monograph may serve as a handbook. It includes the theoretical foundations for the interaction of IR radiation with thin films, as well as the optimum conditions of measuring spectra of various systems, which are analyzed by computer experiments and illustrated by specific examples. Complementary to this, the basic literature devoted to the application of IR spectroscopy in the investigation of nanolayers of solids and interfaces is presented, and the necessary reference material for the interpretation of spectra is tabulated. Thus this book will be extremely useful for any laboratory employing IR spectroscopy, and for each industrial firm involved in the production of thin-film structures, as well as by final-year and postgraduate students specializing in the fields of optics, spectroscopy, or semiconductor technology.

Dr. Valeri Tolstoy (St. Petersburg State University, Russia) authored Chapter 2 and Sections 3.5, 3.10, 4.1.1, 4.1.2, 4.2.1, 4.2.2, 4.4, 4.5, 7.1–7.3. Dr. Irina Chernyshova (St. Petersburg State Polytechnical University, Russia) wrote Chapters 1, 3, 4, and 7 (except for the sections mentioned above) and coauthored Sections 2.3, 2.5, and 2.7. Prof. Valeri Skryshevsky (Kyiv National Taras Shevchenko University, Ukraine) presents Chapters 5 and 6. Tables in the Appendix were collected by Valeri Tolstoy and Irina Chernyshova. The language and style were edited by Dr. Roberta Silerova (University of Saskatchewan, Canada). Dr. Nadezhda Reutova (St. Petersburg State University, Russia) translated into English Chapters 2, 5, and 6 and Sections 3.3–3.5, 3.10, 4.1, and 4.2, and helped with translation of Chapter 1 and Sections 3.1 and 3.2.

Acknowledgements

We thank Dr. Silerova for editing the English language, improving the handbook style, and her great patience during the joint work several years long; Dr. Reutova for translating part of the text into English; and all the authors who have permitted citation of their results. VPT acknowledges a partial financial support granted by Interfanional Scientific Foundation (USA), Russian Foundation for Basic Research (RFBR), and Civilian Research and Development Foundation (CRDF) (USA). IVC thanks RFBR, the Swedish Institute, and Elena Chernetskaya. VAS thanks Ministry of Ukraine for Education and Science. We all thank friends and relatives who helped us with this work.

REFERENCES

1. V. P. Tolstoy, G. N. Kusnetsova, and I. I. Shaganov, *J. Appl. Spectrosc.* **40**, 978 (1984).
2. V. P. Tolstoy and S. N. Gruzinov, *Opt. Spektrosc.* **71**, 129 (1991).
3. I. V. Chernyshova and K. Hanumantha Rao, *J. Phys. Chem. B* **105**, 810 (2001).
4. V. P. Tolstoy, L. P. Bogdanova, and V. B. Aleskovski, *Doklady AN USSR*, **291**, 913 (1986) (in Russian).
5. V. P. Tolstoy, A. I. Somsikov, and V. B. Aleskovski, *USSR Inventor Certificate No. 1099255, Byul. Izobr.* **23**, 145 (1984).
6. S. N. Gruzinov and V. P. Tolstoy, *J. Appl. Spectrosc.* **46**, 480 (1987).
7. V. P. Tolstoy and V. N. Krylov, *Opt. Spectrosc.* **55**, 647 (1983).
8. V. P. Tolstoy and S. N. Gruzinov, *Opt. Spectrosc.* **63**, 489 (1987).
9. S. P. Shcherbakov, E. D. Kriveleva, V. P. Tolstoy, and A. I. Somsikov, *Russ. J. Equipment Tech. Exper.* **1**, 159 (1992).
10. V. P. Tolstoy and S. P. Shcherbakov, *J. Appl. Spectrosc.* **57**, 577 (1992).
11. I. V. Chernyshova and V. P. Tolstoy, *Appl. Spectrosc.* **49**, 665 (1995).
12. J. F. McClelland, R. W. Jones, S. Luo, and L. M. Seaverson, in P. B. Coleman (Ed.), *Practical Sampling Techniques for Infrared Analysis*, CRC Press, Boca Raton, FL, 1993, Chapter 5.
13. J. F. McClelland, S. J. Bajic, R. W. Jones, and L. M. Seaverson, in F. M. Mirabella (Ed.), *Modern Techniques in Applied Molecular Spectroscopy*, Wiley, New York, 1998, Chapter 6.
14. W. Suetaka, *Surface Infrared and Raman Spectroscopy. Methods and Applications*, Plenum, New York, 1995, Chapter 4.
15. S. Zhang, F. S. Franke, and T. M. Niemczyk, in F. M. Mirabella (Ed.), *Modern Techniques in Applied Molecular Spectroscopy*, Wiley, New York, 1998, Chapter 9.
16. F. M. Mirabella (Ed.), *Modern Techniques in Applied Molecular Spectroscopy*, Wiley, New York, 1998.
17. W. Suetaka, *Surface Infrared and Raman Spectroscopy. Methods and Applications*, Plenum, New York, 1995.
18. F. Mirabella (Ed.), *Internal Reflection Spectroscopy*, Marcel Dekker, New York, 1993.
19. P. B. Coleman, *Practical Sampling Techniques for Infrared Analysis*, CRC Press, Boca Raton, FL, 1993.

20. J. Workman, Jr. and A. W. Springsteen (Eds.), *Applied Spectroscopy: A Compact Reference for Practitioners*, Academic, San Diego, 1998.

21. J. Chalmers and P. Griffiths (Eds.), *Handbook of Vibrational Spectroscopy*, Vol. 1, Wiley, New York, 2002, Chapter 1.

VALERI TOLSTOY
IRINA CHERNYSHOVA
VALERI SKRYSHEVSKY

ACRONYMS AND SYMBOLS

A/D	analog–digital (convertor)
AES	Auger electron spectroscopy
AFM	atomic force microscopy
ARUPS	angle-resolved UV photoelectron spectroscopy
ATR	attenuated total reflection
AU	arbitrary units
AW	air–water (interface)
BF	Bruggeman formula
bi CMOS	bipolar and complementary MOS
BLB	Bouguer–Lambert–Beer (law)
BM	Bruggeman model
BML	buried metal layer
BP	bandpass
CMC	critical micelle concentration
CMLL	Clausius–Mossotti/Lorentz–Lorenz (model)
CMP	chemomechanical polishing
CS	compressed solid (L monolayer phase)
CVD	chemical vapor deposition
DA	Drude absorption, dispersion analysis
DAC	diamond anvil cell
DCT	dielectric continuum theory
DF	distribution function
DL	double layer
DMA	dynamic mechanical analysis
DR	diffuse reflection
DR	dichroic ratio
DRIFTS	diffuse reflectance infrared Fourier transform spectroscopy
DT	diffuse transmission
DTA	differential thermal analysis
DTGS	deuterated triglicine sulfate (detector)
DTIFTS	diffuse transmittance infrared Fourier transform spectroscopy

EDS	energy dispersive X-ray spectroscopy
EELS	electron energy loss spectroscopy
EFA	electric field analysis
EIRE	extended internal reflection element
EMIRS	electrochemically modulated infrared spectroscopy
EMT	effective medium theory
ERS	external reflection spectroscopy
EWAS	evanescent wave absorption spectroscopy
EXAFS	extended X-ray absorption fine-structure
FET	field-effect transistor
FPA	focal plane array
FTEMIRS	Fourier transform EMIRS
μ-FTIR	Fourier transform infrared microscopy
FTIR	Fourier transform infrared spectroscopy
FWHM	full width at half maximum
GIR	grazing internal reflection
GIXD	grazing-incidence X-ray diffraction
H/D	hydrogen/deuterium (exchange)
HATR	horizontal ATR
HL	Helmholtz layer
HM	hemimicelle model
HP	highpass
HPLC	high performance liquid chromatography
HREELS	high resolution electron energy loss spectroscopy
IC	integrated circuit
ILD	infrared linear dichroism
IOW	integrated optical waveguide
IP	in-phase
IR	infrared
IRE	internal reflection element
IRRAS	infrared reflection absorption spectroscopy
ITO	indium tin oxide
KK	Kramers–Krönig
KM	Kubelka–Munk
L monolayer	Langmuir monolayer
LB film	Langmuir–Blodgett film
LC	liquid condensed (L monolayer phase)
LE	liquid expanded (L monolayer phase)
LED	light-emitting data
LEED	low-energy electron diffraction
LIA	lock-in-amplifier
LO	longitudinal optical
LP	lowpass
LPD	liquid-phase deposition
LST	Lyddane–Sachs–Teller (law)

LT-OTTE	low-temperature OTTE
LUMO	lowest unoccupied molecular orbital
MAS NMR	magic angle spinning NMR
MBE	molecular beam epitaxy
MCT	mercury-cadmium-tellurium
MG	Maxwell–Garnett (dielectric function)
MGEMT	Maxwell–Garnett effective medium theory
MIR	multiple internal reflection
MIRE	multireflection internal reflection element
MIS	metal insulator semiconductor
MIT	multiple internal transition
ML	monolayer
MO	molecular orientation
MOATR	metal-overlayer ATR
MOS	metal–oxide–semiconductor
MOSFET	metal–oxide–semiconductor FET
MP	monolayer packing
MSEF	mean-square electric field
MTC	monothiocarbonate
NEXAFS	near-edge X-ray absorption fine structure
NIR	near IR
NMR	nuclear magnetic resonance
NMSEF	normalized mean-square electric field
OCP	open-circuit potential
ODT	order–disorder transition
OTE	optically transparent electrode
OTTLE	optically transparent thin-layer electrochemical (cell)
PAS	photoacoustic spectroscopy
PCA	principal component analysis
PDIR	potential-difference infrared (spectroscopy)
PECVD	plasma enhanced vapor deposition
PEDR	Perkin–Elmer diffuse reflectance
PEM	photoelestic modulator
PET	poly(ethylene teriphtalate)
PFPE	perfluoropolyesther
PM	polarization modulation
PMMA	poly(methyl methacrylate)
PSTM	photon STM
PTFE	poly(tetrafluoroethylene)
PVA	poly(vinyl acetate)
PVC	poly(vinyl chloride)
PVD	physical vapor deposition
Q	quadrature
QCM	quartz crystal microbalance
RA	reflectance-absorbance

RBS	Rutherford backscattering
RH	relative humidity
RTPM	real-time PM
S^2	step-scan
SAM	self-assembled monolayer
SAW	surface acoustic wave
SCE	saturated calomel electrode
SCR	space charge region
SE	semiconductor electrode
SEC	spectroelectrochemical
SEIRA	surface enhanced infrared absorption
SEM	scanning electron microscopy
SERS	surface enhanced Raman spectroscopy
SEW	surface electromagnetic waves
SFG	sum frequency generation
SHE	standard hydrogen electrode
SHG	second harmonic generation
SI	international system of units
SIA	sequential implantation and annealing
SIMOX	separation by implanted oxygen
SIMS	secondary ion mass spectrometry
SIPOS	semi-insulating polycrystalline silicon
SNIFTIRS	subtractively normalized interfacial FTIR spectroscopy
SNOM	scanning near field optical microscopy
SNR	signal-to-noise ratio
SOI	silicon-on-insulator
SPAIRS	single potential alternation IR spectroscopy
SPR	surface plasmon resonance
SSR	surface selection rule
STIRS	surface titration by internal reflectance spectroscopy
STM	scanning tunneling microscopy
STPD	stepwise thermo-programmed desorption
TDM	transitional dipole moment
TEM	transmission electron spectroscopy
TG	thermogravimetry
TIRF	total internal reflection fluorescence
TLC	thin layer chromatography
TMOS	tetraethylorthosilicate
TO	transverse optical
TPD	temperature programmed desorption
TR	time resolution
UHV	ultra high vacuum
ULSC	ultra large scale circuit
ULSI	ultra large scale integrated
UV	ultraviolet (radiation)

VLSI	very large scale integrated
VPE	vapor phase epitaxy
WAXS	wide-angle X-ray scattering
X, EX, BX, AX	xanthate (ethyl-, n-butyl-, amyl-)
XANES	X-ray absorption near-edge structure
XPS	X-ray photoelectron spectroscopy
2D IR	two dimensional correlation analysis of IR dynamic spectra

Symbols

$\hat{\alpha}$	Electric polarizability
α	Decay constant≡absorption coefficient
β	Restoring force
γ	Tilt angle, damping constant
δ	Bending mode
ε	Permittivity≡dielectric constant≡dielectric function
ε_∞	High-frequency dielectric constant≡screening factor
ε_m	Permittivity of metal
ε_{sm}	Permittivity of surrounding medium
ε_{st}	Static (low-frequency) dielectric constant
θ	Angle of bond
λ	Wavelength
μ	Permeability
ν	Wavenumber, Stretching mode
ρ	Mass volume density, resistivity, rocking mode
$\boldsymbol{\rho}$	Dynamic dipole moment
σ	Electrical conductivity
τ	Time
φ_1 or φ	Angle of incidence
ω	Wagging mode, angular velocity
ω_p	Plasma frequency
A	Absorbance
B	Magnetic induction
c	Velocity of light
C	Volume/mass concentration
d	Thickness
D	Electric displacement
d_p	Penetration depth
E	Electric field
E	Electrode potential, electric field
E_g	Energy gap
E	Integrated molar absorption coefficient
f	Filling fraction
g_k	Geometric factor

h	Plank constant
H	Magnetic field
I	Intensity of radiation
j	Current density
k	Extinction coefficient≡absorption index
k	Wave vector
m	Mass
M	Magnetic polarization
N	Surface or volume density of molecules (atoms), number of reflections
n	Refractive index
p	Dipole moment
P	Electric polarization
r	Reflection coefficient
R	Reflectance
S	Oscillator strength
t	Time/transmission coefficient
T	Temperature/transmittance
v	Velocity

INTRODUCTION

In the infrared (IR) spectroscopic range (200–4000 cm^{-1}), radiation is generally characterized by its wavenumber ν (cm^{-1}), related to the wavelength λ (μm), frequency $\tilde{\nu}$ (s^{-1}), and angular frequency ω (s^{-1}) as

$$\nu = \frac{1}{\lambda} = \frac{\tilde{\nu}}{c} = \frac{\omega}{2\pi c}, \tag{1}$$

where $c = 2.99793 \times 10^8$ m·s^{-1} is the velocity of electromagnetic radiation in a vacuum.

When IR radiation containing a broad range of frequencies passes through a sample, which can be represented as a system of oscillators with resonance frequencies $\nu_{0,i}$, then according to the Bohr rule,

$$\Delta E = h\nu \tag{2}$$

(where ΔE is the difference between the energy of the oscillator in the excited and ground states, ν is the frequency of photons, and $h = 6.626069 \times 10^{-34}$ J·s is Planck's constant), photons with frequencies $\nu = \nu_{0,i}$ will be absorbed. These photons will be eliminated from the initial composition of the radiation. Since all the elementary excitations have unique energy levels (*fingerprints*), measurements of the disappearing energy as a function of ν (*absorption spectrum of the sample*) enable these excitations to be identified, and microscopic information about the sample (e.g., molecular identity and conformation, intra- and intermolecular interactions, or crystal-field effects, etc.) may be obtained.[†]

[†] See M. Hollas, *Fundamental Aspects of Vibrational Spectroscopy. Modern Spectroscopy*, 3rd ed., Wiley, Chichester, 1996.

1

ABSORPTION AND REFLECTION OF INFRARED RADIATION BY ULTRATHIN FILMS

The primary characteristics that one identifies from infrared (IR) spectra of ultrathin films for further analysis are the resonance frequencies, oscillator strengths (extinction coefficients), and damping (bandwidths), related to different kinds of vibrational, translational, and frustrated rotational motion inside the thin-film material [1–7]. However, the microscopic processes inside or at the surface of a film (motion of atoms and electrons) give rise to the frequency dependence (the *dispersion*) not only of the extinction coefficient but also of the refractive index of the film. As a result, a real IR spectrum of an ultrathin film is, as a rule, distorted by so-called *optical effects*. Specifically, the spectrum strongly depends upon the conditions of the measurement, the film thickness, and the optical parameters of the surroundings and substrate impeding extraction of physically meaningful information from the spectrum. Thus after introduction of the nomenclature accepted in optical spectroscopy and a brief discussion of the physical mechanisms responsible for absorption by solids on a qualitative level, this introductory chapter will concentrate on the basic *macroscopic* or *phenomenological* theory of the optical response of an ultrathin film immobilized on a surface or at an interface.

The theoretical analysis of the IR spectra of ultrathin films on various substrates and at interfaces will involve two assumptions: (1) the problem is linear and (2) the system under investigation is macroscopic; that is, one can use the macroscopic Maxwell formulas containing the local permittivity. The first assumption is valid only for weak fields. The second assumption means that the volume considered for averaging, a (the volume in which the local permittivity is formed), is lower than the parameter of inhomogeneity of the medium, d (e.g., the effective thickness of the film, the size of islands, or an effective dimension of polariton), $a < d$. In this case, the response of the medium to the external electromagnetic field is essentially the response of a continuum. The

1

description of the medium properties using the macroscopic permittivity is called the dielectric continuum theory (DCT). One issue discussed repeatedly in the literature is the correctness of the DCT approximation for spectra of ultrathin films. In this context, semiphenomenological and microscopic models establishing the correlation between local field effects, molecular and atomic dynamics, and the IR spectrum of the film on the surface were proposed (for reviews, see Refs. [1, 6, 7]). These models greatly enhance the understanding of ultrathin films of submonolayer coverage. However, the corresponding relationships are cumbersome, deal with specific (as a rule, vacuum–metal) interfaces, require further refinements for most of the systems studied, and, hence, are unfeasible for routine analysis of the spectra. On the other hand, the macroscopic relationships can be used without substantial difficulties to solve many problems arising in the spectroscopy of ultrathin films. First, they allow one to choose the optimum conditions for recording IR spectra by comparing band intensities (Chapter 2), which is of prime technical importance. Second, by using spectral simulations based on the macroscopic formulas, it is possible to distinguish optical effects from physicochemical effects in the film, such as film inhomogeneity (porosity) and changes in the orientation of the film species (Chapter 3).

1.1. MACROSCOPIC THEORY OF PROPAGATION OF ELECTROMAGNETIC WAVES IN INFINITE MEDIUM

Optical thin-film theory is essentially based on the *Maxwell theory* (1864) [8], which summarizes all the empirical knowledge on electromagnetic phenomena. Light propagation, absorption, reflection, and emission by a film can be explained based on the concept of the *macroscopic* dielectric function of the film material. In this section, we will present the results of the Maxwell theory relating to an infinite medium and introduce the nomenclature used in the following sections dealing with absorption and reflection phenomena in layered media. The basic assertions of macroscopic electrodynamic theory can be found in numerous textbooks (see, e.g., Refs. [9–16]).

   1.1.1°. The optical properties of an infinite medium without any electric charge other than that due to polarization are described by Maxwell's equations [in International System (SI) of units]:

$$\nabla \times \mathbf{E} + \frac{\partial \mathbf{B}}{\partial t} = 0, \tag{1.1a}$$

$$\nabla \times \mathbf{H} - \frac{\partial \mathbf{D}}{\partial t} = \hat{\sigma}\mathbf{E}, \tag{1.1b}$$

$$\nabla \cdot \mathbf{D} = 0, \tag{1.1c}$$

$$\nabla \cdot \mathbf{B} = 0. \tag{1.1d}$$

Here, $\mathbf{E}$ and $\mathbf{H}$ are, respectively, the vectors of the macroscopic electric and the magnetic field; $\varepsilon_0 = 8.854 \times 10^{-12} \mathrm{C \cdot N^{-1} \cdot m^{-2}}$ and $\mu_0 = 4\pi \times 10^{-7} \mathrm{N \cdot A^{-2}}$ are the permittivity and permeability of free space, respectively; $\hat{\sigma}$ is the electrical *conductivity* (it appears as a coefficient in Ohm's law, $\mathbf{j} = \sigma \mathbf{E}$, where $\mathbf{j}$ is the current density), and t is time. The quantities $\mathbf{D}$ and $\mathbf{B}$ are the *electric displacement* and the *magnetic induction*, respectively. They characterize the action of the field on matter and are related to the field vectors by the so-called local material equations

$$\mathbf{D} = \varepsilon_0 \hat{\varepsilon} \mathbf{E} = \varepsilon_0 \mathbf{E} + \mathbf{P}, \tag{1.2a}$$

$$\mathbf{B} = \mu_0 \hat{\mu} \mathbf{H} = \mu_0 \mathbf{H} + \mathbf{M}, \tag{1.2b}$$

where $\mathbf{P}$ and $\mathbf{M}$ are the vectors of the electric and magnetic polarization, respectively; $\hat{\varepsilon}$ and $\hat{\mu}$ represent relative local *permittivity* (*dielectric constant*) and *magnetic permeability*, respectively. The *electric polarization* of the medium, $\mathbf{P}$, is defined as the sum of the dipole moments in a unit volume of the medium:

$$\mathbf{P} = \sum_i \mathbf{p}_i = N\mathbf{p}. \tag{1.3}$$

Here, $\mathbf{p}$ is the induced dipole moment of one atom and N is the atom density (1.3.1°).

1.1.2°. In general, the quantities $\hat{\mu}$, $\hat{\sigma}$, and $\hat{\varepsilon}$ are tensors. With a high degree of accuracy, the magnetic permeability $\hat{\mu}$ can be taken to be equal to unity in the optical range [11]. There is a straightforward correlation between the conductivity $\hat{\sigma}$ of the material and the imaginary part of the permittivity $\hat{\varepsilon}$ [see Eq. (1.16) below]. Thus, the parameter that most completely describes the interaction of the IR radiation with a medium is the permittivity $\hat{\varepsilon}$ of the medium. For an isotropic medium (or a cubic crystal), uniaxial crystals, and the biaxial crystals of an orthorhombic system, the permittivity $\hat{\varepsilon}$ is a symmetric tensor of the second rank with six distinct, complex (1.1.9°) elements, which can be transformed to a diagonal form by the proper choice of coordinate system [9, 14–16]. Depending on the symmetry of the film, the number of different principal values of the tensor $\hat{\varepsilon}$ can be 1, 2, or 3 (Section 3.11.2). In particular, for many materials (e.g., quartz, tourmaline, calcite) in the visible spectral region, where their absorption is small, two principal values of the refractive index [Eq. (1.9)] may be equal, $n_x = n_y \neq n_z$. As a result, along an arbitrary direction of propagation different from z (z being the *optical axis*) there exist two independent linearly polarized plane waves (1.1.5°) traveling with different phase velocities (1.1.4°). This phenomenon is named *double refraction*, or (for the real refractive index) *birefringence*. If an anisotropic material is absorbing [e.g., in the case of the minerals mentioned above, but in the region of their lattice vibrations (1.2.4°)], the extinction coefficient (1.1.10°) is also dependent on polarization. This phenomenon is referred to as *pleochroism*. A special case of pleochroism is *dichroism*, or the imaginary refractive index birefringence. For the biaxial crystals belonging to monoclinic and triclinic systems, the principal axes of the real and imaginary parts of the permittivity do not coincide and the tensor $\hat{\varepsilon}$ cannot be put in the diagonal form. In inhomogeneous or discontinuous media, the permittivity $\hat{\varepsilon}$ varies from region to region.

1.1.3°. For a homogeneous, isotropic, and nonabsorbing medium ($\hat{\sigma} = 0$), Eqs. (1.1a) and (1.1b) can be transformed into the standard wave equations,

$$\nabla^2 \mathbf{E} - \varepsilon_0 \hat{\varepsilon} \mu_0 \hat{\mu} \frac{\partial^2 \mathbf{E}}{\partial t^2} = 0, \tag{1.4a}$$

$$\nabla^2 \mathbf{H} - \varepsilon_0 \hat{\varepsilon} \mu_0 \hat{\mu} \frac{\partial^2 \mathbf{H}}{\partial t^2} = 0. \tag{1.4b}$$

Possible solutions of Eqs. (1.4a) and (1.4b) for the electric and magnetic field vectors are

$$\mathbf{E} = \mathbf{E}_0 e^{i(\omega t - \mathbf{k r})}, \tag{1.5a}$$

$$\mathbf{H} = \mathbf{H}_0 e^{i(\omega t - \mathbf{k r})}, \tag{1.5b}$$

which represents the transverse monochromatic plane wave propagating in the direction $\mathbf{k}$. Here, ω is the angular frequency of the wave, $\mathbf{r}$ is the location vector that depends on the coordinate system being used, $\mathbf{E}_0$ and $\mathbf{H}_0$ are the amplitudes of the electric and magnetic field, respectively, and $\mathbf{k}$ is the wave vector, which describes the propagation of the wave in the given medium. The terms *transverse* and *longitudinal* waves refer to the direction of the electric field vector $\mathbf{E}$ with respect to the direction of wave propagation, $\mathbf{k}$. Transversality ($\mathbf{E} \perp \mathbf{H} \perp \mathbf{k}$) of the electromagnetic wave can be obtained by substitution of Eqs. (1.5a) and (1.5b) into (1.1a) with allowance made for (1.1b) and (1.1d) (see also 1.3.7°).

1.1.4°. The equation $\omega t - \mathbf{kr} = \text{const}$ describes a plane normal to the wave vector $\mathbf{k}$. This plane is characterized by a constant phase $\Delta = \omega t - \text{const}$ and is called the *surface of constant phase* or the *wavefront*. The wavefront travels in a medium in the direction $\mathbf{k}$ with the velocity

$$\mathbf{v} = \frac{\omega}{\mathbf{k}}. \tag{1.6}$$

This is called the *phase velocity* of the electromagnetic wave in this medium.

1.1.5°. Along with the frequency of oscillation, ω, the amplitude $\mathbf{E}_0$, the phase Δ, and the direction of propagation $\mathbf{k}$, an electromagnetic wave is characterized by the direction of oscillation of the electric field vector $\mathbf{E}$. This is known as the *polarization state* of the electromagnetic wave. If during wave propagation the electric vector lies in a plane, the wave is said to be *linearly polarized*. The plane that contains the electric vectors and the direction of the wave propagation is called the *plane of polarization*. Superposition of two linearly polarized waves that are in phase results in a third linearly polarized wave.

1.1.6°. Substitution of Eq. (1.5a) into Eq. (1.4a) gives the following relationship for the wave vector:

$$\mathbf{k}^2 = \frac{\omega^2}{c^2} \hat{\varepsilon} \hat{\mu}. \tag{1.7}$$

Here, $c = 1/(\varepsilon_0\mu_0)^{1/2}$ is the velocity of electromagnetic radiation in vacuum. The dependence of ω on $\mathbf{k}$ (the energy–momentum relation) for an electromagnetic wave propagating through a crystal is called the *dispersion law*. Hence, Eq. (1.7) represents the dispersion law of a transverse electromagnetic wave in an infinite crystal [17].

1.1.7°. Substitution of Eq. (1.6) into Eq. (1.7) yields the following expression for the absolute value of the phase velocity of a wave $v = |\mathbf{v}|$ in the medium characterized by $\hat{\varepsilon}$:

$$v = \frac{c}{\sqrt{\hat{\varepsilon}\hat{\mu}}}. \qquad (1.8)$$

By definition, the *absolute refractive index* $\hat{n}$ of a medium is

$$\hat{n} = \frac{c}{v}. \qquad (1.9)$$

In terms of the permittivity and the magnetic permeability,

$$\hat{n} = \sqrt{\hat{\varepsilon}\hat{\mu}}. \qquad (1.10)$$

1.1.8°. In 1.1.3°–1.1.7°, we have assumed the medium to be nonabsorbing and, thus, the parameters $\hat{\varepsilon}$ and $\hat{n}$ to be constant and $\sigma = 0$. However, if the medium absorbs electromagnetic radiation, these quantities become dependent on the frequency of incident radiation, the function $\hat{\varepsilon}(\omega)$ termed the *dielectric function*. Below we will neglect the so-called spatial dispersion effects [18] connected with the dependence of the dielectric function on the wave vector $\varepsilon(\mathbf{k})$. This is permissible for the IR range (the limit $\mathbf{k} \to 0$).

In the case of *absorption of radiation*, the wave is damped and the wave equation (1.4a) transforms into

$$\nabla^2\mathbf{E} - \varepsilon_0\hat{\varepsilon}\mu_0\hat{\mu}\frac{\partial^2\mathbf{E}}{\partial t^2} - \mu_0\mu\sigma\frac{\partial\mathbf{E}}{\partial t} = 0. \qquad (1.11)$$

1.1.9°. The solution (1.5a) will satisfy both Eqs. (1.11) and (1.4a) if the wave vector, the permittivity, and the refractive index are allowed to be complex quantities,

$$\mathbf{k} = \mathbf{k}' - i\mathbf{k}'' = \hat{k}\mathbf{s} = (k' - ik'')\mathbf{s}, \qquad (1.12a)$$

$$\hat{\varepsilon} = \varepsilon' - i\varepsilon'', \qquad (1.12b)$$

$$\hat{n} = n - ik, \qquad (1.12c)$$

respectively. Here, $\mathbf{s}$ is the unit vector along the direction of propagation of the wave. Notice that Eq. (1.12a) is legitimate only for homogeneous plane waves, for which $\mathbf{k}'\|\mathbf{k}''$ [9, 10, 14–16]. The quantity $\hat{k} = k' - ik''$ is confusingly called the wavenumber, like the quantity $\nu = 1/\lambda$. Equations (1.7), (1.10), and (1.12a) give the following relationship between the wavenumber and the refractive index:

$$\hat{k} = \frac{\omega}{c}\hat{n} = \frac{2\pi}{\lambda}\hat{n},$$

or in view of (1.12a) and (1.12c)

$$k' = \frac{2\pi}{\lambda}n, \qquad k'' = \frac{2\pi}{\lambda}k \tag{1.13}$$

1.1.10°. By using Eqs. (1.12) and (1.13), one can rewrite Eq. (1.5a) in terms of the optical constants:

$$\mathbf{E} = \mathbf{E}_0 e^{i(\omega t - k'\mathbf{sr})} e^{-k''\mathbf{sr}} = \mathbf{E}_0 e^{i[\omega t - (2\pi n/\lambda)\mathbf{sr}]} e^{-(2\pi k/\lambda)\mathbf{sr}}. \tag{1.14}$$

From Eq. (1.14), we can see that the real parts of the wavenumber and the refractive index, k' and n, respectively, determine the phase velocity of the wave in the medium, whereas the imaginary parts, k'' and k, determine the attenuation of the electromagnetic field along the direction of propagation of the wave. By virtue of this, the imaginary part of the refractive index, k, is called the *extinction coefficient* or *absorption index*. Note that the symbol k is used to designate the wavenumber and the extinction coefficient. To distinguish between them, we label the wavenumber with a hat, as $\hat{k}$.

1.1.11°. In view of Eqs. (1.10), (1.12b), and (1.12c), substitution of Eq. (1.14) into Eq. (1.11a) gives

$$\varepsilon' = n^2 - k^2, \qquad \varepsilon'' = 2nk, \tag{1.15}$$

and

$$\varepsilon'' = \frac{\sigma}{\varepsilon_0 \omega}. \tag{1.16}$$

Equation (1.16) shows that absorbing media are characterized by nonzero optical conductivity.

1.1.12°. For practical purposes, it is convenient to make the inverse transformation of Eq. (1.15):

$$n = \left\{ \tfrac{1}{2}\left[(\varepsilon'^2 + \varepsilon''^2)^{1/2} + \varepsilon' \right] \right\}^{1/2},$$
$$k = \left\{ \tfrac{1}{2}\left[(\varepsilon'^2 + \varepsilon''^2)^{1/2} - \varepsilon' \right] \right\}^{1/2}. \tag{1.17}$$

The evident conclusion from relationships (1.15) and (1.17) is that the principal macroscopic parameters that characterize the interaction of an electromagnetic wave with an absorbing medium are n and k or ε' and ε''. As an example,

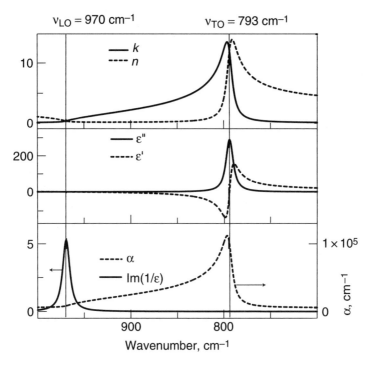

Figure 1.1. Frequency dependences of n, k, ε', ε'', and $\mathrm{Im}(1/\hat{\varepsilon})$ for β-SiC characterized by $S = 3.3$, $\varepsilon_\infty = 6.7$, $\gamma = 8.49$ cm^{-1}, and $\nu_0 = 793.6$ cm^{-1}. Also shown are optical transverse and longitudinal frequencies (ν_{TO} and ν_{LO}, respectively).

Fig. 1.1 shows these functions for β-SiC, which is a strong oscillator with weak damping (1.3.20°).

1.1.13°. The fundamental relations between the real and imaginary parts of the optical constants, which do not assume a model for calculating $\hat{\varepsilon}$ (Section 1.3), are rigorously described by the *Kramers–Krönig* (KK) relations, first introduced in 1926 [19, 20],

$$\varepsilon'(\omega) - \varepsilon_\infty = \frac{2}{\pi} P \int_0^\infty \frac{\varepsilon''(\omega')\omega'}{\omega'^2 - \omega^2}\, d\omega', \qquad (1.18)$$

$$\varepsilon''(\omega) = -\frac{2\omega}{\pi} P \int_0^\infty \frac{\varepsilon'(\omega') - \varepsilon_\infty}{\omega'^2 - \omega^2}\, d\omega', \qquad (1.19)$$

where P refers to the principal-value integral and ε_∞ is the offset value (also called the anchor point) that accounts for the background of the $\varepsilon'(\omega)$ function (1.3.6°).

The theoretical aspects of the KK relations are discussed in detail elsewhere [10, 21–23]. Using the KK relations, the optical parameters of a substance can be calculated from the measured spectra [24, 25] or the absorption spectrum can be calculated from the measured reflection spectrum [26]. The latter mathematical operation is called the *KK transform(ation)*. The efficient algorithms for this procedure are available for both normal and oblique incidence spectra [26–32] and are provided in most software packages of present-day Fourier transform infrared (FTIR) spectrometers.

1.1.14°. Electromagnetic theory shows [6–12] that the flux of the electromagnetic field energy normal to a unit surface containing the vectors of the electric and magnetic field is equal to the vector product

$$\mathbf{\Pi} = \mathbf{E} \times \mathbf{H},$$

called the *Poynting vector* or the "ray vector". Note that in the general case of an anisotropic crystal, the direction of the Poynting vector differs from that of the wave vector $\mathbf{k}$. The velocity of the energy propagation in the medium is named the ray (or energy) velocity. It can be shown [10, 14] that the phase velocity (1.1.4°) is the projection of the ray velocity onto the direction of the wave normal.

1.1.15°. The intensity of the wave, I, is the energy, time averaged over one period of oscillation, transferred by the wave across a unit area perpendicular to the direction of the Poynting vector per unit time. For a damped homogeneous wave of the type (1.14) propagating in the z-direction [6–13],

$$I = |\langle \mathbf{\Pi} \rangle| = \tfrac{1}{2} c n \varepsilon_0 |\mathbf{E}|^2 = \tfrac{1}{2} c n \varepsilon_0 |\mathbf{E}(0)|^2 e^{-4\pi k z/\lambda}, \qquad (1.20)$$

where $k = \mathrm{Im}(\hat{n})$ and $n = \mathrm{Re}(\hat{n})$ are the extinction coefficient and the refractive index of the medium, respectively, the angular brackets denote a time average over one period, and $\mathbf{E}(0)$ is the electric field at $z = 0$. Averaging the square of the vector of the electric field, we have $\langle E^2 \rangle = \tfrac{1}{2} |\mathbf{E}|^2$, which allows one to rewrite Eq. (1.20) as

$$I = c n \varepsilon_0 \langle E^2 \rangle. \qquad (1.21)$$

Thus the quantity I is proportional to the mean square of the electric field, $\langle E^2 \rangle$, and the refractive index of the medium (the speed of light in the medium).

1.1.16°. In spectroscopic practice, besides the extinction coefficient $k(1.1.10°)$, absorption of IR radiation is characterized by the decay constant (the absorption coefficient), the imaginary component of the permittivity, $\mathrm{Im}\,\hat{\varepsilon}$, and the imaginary component of the reciprocal of the permittivity, $\mathrm{Im}(1/\hat{\varepsilon})$. The conductivity σ is reserved to describe conductors in the frequency range where $\tilde{\nu} \ll \sigma/\varepsilon_0 \varepsilon$ [33].

1.1.17°. The decay constant (the absorption coefficient) α is defined as

$$\alpha \equiv \frac{1}{I} \frac{d}{dz} I. \qquad (1.22)$$

Here, I is the intensity of electromagnetic radiation and z is the distance that it has passed through the absorbing medium. The distance $1/\alpha$ measured in a direction normal to the surface plane is defined as the penetration depth of the radiation in the medium. When the electromagnetic wave has propagated through the penetration depth, its intensity decreases by a factor of e.

By integrating Eq. (1.22), we find that the intensity of light in an absorbing medium is attenuated according to

$$I(z) = I(0)e^{-\alpha z}, \tag{1.23}$$

where $I(0)$ is the intensity for $z = 0$ inside the medium. Equation (1.23) is the well-known *Bouguer–Lambert–Beer* (BLB) *law* [34], which describes the absorption of radiation (of "not high intensity") as it propagates through a medium. Comparing Eqs. (1.23) and (1.21) and taking into account Eq. (1.13) give the following relation between the decay constant α, the imaginary part of the wavenumber, k'', the dielectric function ε'', and the extinction coefficient k:

$$\alpha = 2k'' = \frac{4\pi}{\lambda}k = \frac{2\omega k}{c} = \frac{\varepsilon''\omega}{cn}. \tag{1.24}$$

Notice that the BLB law is valid only for "weak" absorption because it neglects the effect of the real part of the refractive index, n, in Eq. (1.21). The relationship between α and k is demonstrated in Fig. 1.1.

1.1.18°. According to the *Poynting theorem*, the real part of the divergence of the Poynting vector equals the energy dissipated from the electromagnetic field per unit volume per second. This quantity is related to the imaginary part of the permittivity of the medium, $\varepsilon''(\omega)$, the frequency of the field, ω, and the complex electric field vector **E** by

$$\mathrm{Re}(\nabla\cdot\langle\boldsymbol{\Pi}\rangle) = \tfrac{1}{2}[\varepsilon''(\omega)\varepsilon_0\omega]\mathbf{E}\cdot\mathbf{E}^* = \varepsilon''(\omega)\varepsilon_0\omega\langle E^2\rangle, \tag{1.25}$$

where $\langle E^2\rangle$ is the mean square of the electric field. Comparing Eqs. (1.25) and (1.21), we see that, as expected, the rate of absorption is proportional to the intensity of radiation and the imaginary part of the permittivity, $\varepsilon''(\omega)$. The frequency of the maximum of the function $\varepsilon''(\omega)$ is, by definition, the frequency of the *transverse optical* (TO) vibrations (1.3.7°) of the medium, ω_{TO}, and the function $\mathrm{Im}\,\hat{\varepsilon} = \varepsilon''(\omega)$ is called the TO *energy loss function*. For vibrations with weak damping ($\gamma \ll \omega_0$), the frequency of the TO vibrations is close to the resonance frequency, $\omega_{TO} \approx \omega_0$ (Section 1.3) [35].

1.1.19°. Light passing through an absorbing thin film can be attenuated in direct proportion to the quantity (Sections 1.5 and 1.6)

$$\mathrm{Im}\,\frac{1}{\hat{\varepsilon}(\omega)} = \frac{\varepsilon''(\omega)}{\varepsilon'^2(\omega) + \varepsilon''^2(\omega)} \tag{1.26}$$

known as the *longitudinal optical* (LO) *energy loss function*. By definition, the LO energy loss function has its maxima at the frequencies of the LO vibrations of the medium, ω_{LO} (see also 1.3.7°).

The energy loss functions for β-SiC are shown in Fig. 1.1 with the other optical parameters.

1.2. MODELING OPTICAL PROPERTIES OF A MATERIAL

In this section, the physical mechanisms and selection rules of IR absorption by bulk material due to vibrations, electronic excitations, and free carriers (electrons and holes) are briefly discussed on the qualitative level from the viewpoint of quantum mechanics. In general, this problem is highly specialized, and for fuller details several standard textbooks [21, 34, 36–50] are recommended. All of these mechanisms also apply to ultrathin films, but their appearance, which will be discussed in Chapters 3 and 5–7, is quite specific. Note that although we will discuss absorption by solids, the mechanisms to be considered are also applicable to liquids.

1.2.1°. The optical properties of solids that are electrically conducting (metals and semiconductors) differ considerably from the optical properties of dielectrics. These differences can be explained in the following manner. In a solid, there is an abundance of both electrons and energy levels (states) they can occupy. However, because of the periodicity of the crystal lattice, the energy states of electrons are confined to energy bands. The highest fully occupied energy band is called the *valence band*. The lowest partially filled or completely empty energy band is called the *conduction band*. The distance between the bottom of the conduction band and the top of the valence band is called the *band gap*. The energy of the band gap is symbolized by E_g (Fig. 1.2). The conductivity is determined not only by the amount of free electrons (holes) in the material, but also by the number of vacant places for their transfer. Therefore, the lowest conductivity will be observed when the valence band is completely filled and thus when the conduction band is empty. Such a material is called *dielectric*. In

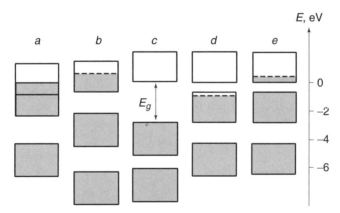

Figure 1.2. Schematic representation of band structure of (a, b) metals (overlapping bands and partially filled band); (c) insulators; (d, e) p- and n- degenerate semiconductors. Here, E_g is width of band gap and dash line indicates position of Fermi level.

dielectrics the valence electrons are "strongly" bound to the constituent atoms but capable of displacement by an electric field, that is, they can be polarized (1.3.2°).

1.2.2°. If the conduction band is half filled or the valence band overlaps in the energy space with the conduction band, the solids have the maximal conductivity. In such a material under the influence of an electric field, electrons can move to neighboring vacant states causing an electric current to flow. This phenomenon is characteristic for metals.

1.2.3°. In the intermediate case, when the upper band containing electrons has a small concentration of either empty or filled states and the temperature coefficient of the electrical resistance is negative at high temperatures, the substance is classified as a semiconductor.

1.2.4°. We shall restrict ourselves to possible mechanisms of IR absorption using as an example the transmission and reflection spectra of galena (natural PbS, semiconductor with $E_g \sim 0.4$ eV [46]) (Fig. 1.3). In these spectra, the regions corresponding to different excited states in solids are clearly discernable. In region I, falling in the range of $1800-200$ cm^{-1} for the majority of substances ($\nu < 500$ cm^{-1} for PbS), the interaction of light with TO vibrations of the crystal lattice — *phonons* — takes place. Here, the extinction coefficient k (1.1.10°) reaches values on the order of $10^{-1}-10^{1}$ [25, 51, 52]. The second region, II, dependent on the width of the band gap E_g is called the *transparency region* of the crystal. For dielectrics and semiconductors this region lies in the IR range. For many oxides its short-wavelength boundary ($\nu \sim 3000$ cm^{-1} for PbS) extends to vacuum ultraviolet (UV). However, because of the presence of defects and impurities in crystals and mutliphonon processes, the extinction coefficient does not reduce to zero everywhere over region II. At the maxima, the quantity k can reach values on the order of $10^{-3}-10^{-1}$ [25, 48, 51] depending on the

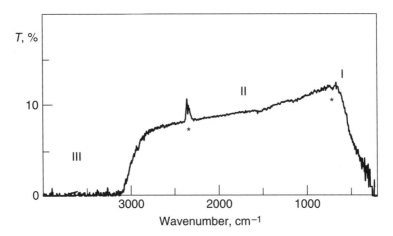

Figure 1.3. Absorption according to different mechanisms in transmission spectrum of a 2-mm-thick PbS plate. I — phonon absorption, II — transparency region, III — fundamental absorption. Asterisks indicate artifact due to atmosphere absorption.

concentration of impurities and defects. In regions I and II free-carrier absorption may manifest itself under specific experimental conditions (Section 3.7). For example, it is responsible for a monotonic background generated in the ATR spectra of semiconducting electrodes with changing potential (Figs. 3.44, 3.47, 7.34, 7.35, and 7.41). At wavenumbers $v > E_g/(h \cdot c)$ (region III) the electrons transfer from the valence band to the conduction band. This type of absorption is called *fundamental absorption* and the frequency $v = E_g/h$ is the *absorption edge*. Here, the extinction coefficient increases abruptly up to values on the order of $10^{-1}–10^0$ [25, 48, 51].

1.2.5°.   An electron excited into the conduction band and a hole left in the valence band can combine to form an electrically neutral species, called an *exciton*. Energy levels of excitons are located in the forbidden band gap and result in essential changes in the spectrum in the vicinity of the absorption edge.

1.2.6°.   The presence in metals of a great number ($\sim 10^{23}$ cm^{-3}) of free electrons gives rise to the high absorption ($k \sim 10^1–10^2$) observed in the IR range [25, 47]. In this spectral region, metals are characterized by high reflectivity, which, according to the microscopic theory of optical properties, is due to the absorption of light by conduction band electrons followed by rapid emission. As a result, the electromagnetic field behaves as if it were diffusing into the medium (1.3.11°).

1.2.7°.   The elementary excitations inside a specimen can cause absorption or emission of IR radiation only when special conditions — *selection rules* — are met. For example, all three vibrations of the H_2O molecule (v_{as}, v_s, and δ_s) and librations (the "frustrated" rotations of the H_2O molecules relative to each other) are active in the IR spectrum of water (Fig. 1.4), whereas the bands of the stretching vibrations of the C–C groups are absent in the IR spectra of a polyethylene film $(-CH_2-CH_2-)_n$. Below we shall formulate these selection rules. Readers interested in this topic should consult Refs. [53–64].

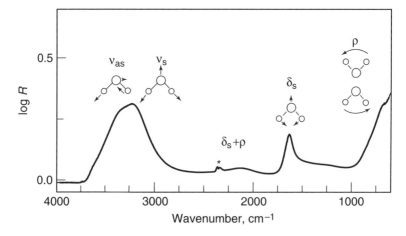

Figure 1.4. ATR spectrum of water. Asterisk indicates atmosphere CO_2 absorption.

1.2.8°. If a normal mode in a crystal, connected for example with a phonon or the photoionization of an impurity, gives rise to any change in the electric dipole moment **p**, then the *dynamic dipole moment* $\rho = \partial\mathbf{p}/\partial q_i$ is nonzero. Here q_i is the normal coordinate, which characterizes the corresponding normal mode and can be derived from normal coordinate analysis based on classical physics [55]. The value of **p** depends on the relative ionicity of the species and can be obtained only by quantum-chemical calculations (see Ref. [61] and the literature therein). In general, the more polar the bond, the larger the **p** term. The matrix element of the dynamic dipole moment, $|\langle j|\rho|i\rangle|$, is called the *transitional dipole moment* (TDM) of the corresponding normal mode.

1.2.9°. According to Fermi's golden rule [40, 42], the integral intensity A of the absorption band of the normal mode is proportional to the probability per unit time of a transition between an initial state i and a final state j. Within the framework of the first (dipole) approximation of time-dependent perturbation quantum theory [46, 65], this probability is proportional to the square of the matrix element of the Hamiltonian $\hat{H} = -\hat{\mathbf{E}}\cdot\hat{\mathbf{p}}$, where **E** is the electric field vector and **p** is the electric dipole moment, resulting in the absorption

$$A \propto E^2 |\langle j|\rho|i\rangle|^2 \cos^2\vartheta, \qquad (1.27)$$

where $|\langle j|\rho|i\rangle|$ is the TDM (1.2.8°) and ϑ is the angle between vectors **E** and ρ. Equation (1.27) represents the mathematical formulation of the selection rule for the activity of any excitation in the IR spectrum. From this equation it is clear that absorption of IR radiation is anisotropic. An excitation is active if (1) a change of the dipole moment takes place and (2) the projection of this change onto the direction of the electric field is nonzero. Maximum absorption is observed when **E** is parallel to the dynamic dipole.

1.2.10°. In the case of simple molecules, the question of whether a particular vibration is active in the IR spectrum can be answered by considering the forms of the normal modes [49–60, 66]. It can be seen in Fig. 1.4 that the dipole moment changes under all the active vibrations of the H_2O molecule. In contrast, the vibrations of homopolar molecules such as H_2 and N_2 do not produce a dipole moment and thus are inactive in the IR spectrum. When a polyatomic molecule contains a center of symmetry, the vibrations symmetrical about this center are active in the Raman spectrum but inactive in the IR spectrum of this molecule, and vice versa. This result is known as the alternative prohibition rule. In general, the activity of the excitation in the IR spectrum cannot be predicted from such a qualitative analysis, but rather must be determined using group theory [51–54, 62].

1.3. CLASSICAL DISPERSION MODELS OF ABSORPTION

To describe the interaction of radiation with a substance on the atomic level, resulting in absorption of light over a wide spectral range from the vacuum UV to the far-IR region, the quasi-classical approach is used [38–45]. It is based on the model proposed by Lorentz [67] in the beginning of the twentieth century, considering electrons and atoms (ions) in matter to be an ensemble of harmonic oscillators. In this section, the classical *Lorentz dispersion model* is discussed [10, 12, 22].

1.3.1°. A *harmonic oscillator* is a particle oscillating along the x-axis under the action of a quasi-elastic force $\mathbf{F} = -K\mathbf{x}$ (where K is the elasticity coefficient and $\mathbf{x}$ is the displacement of the particle from its equilibrium position) with a potential energy given by $V = \frac{1}{2}Kx^2 = \frac{1}{2}m\omega_0^2 x^2$. Here, $\omega_0^2 = K/m$ is the angular fundamental frequency and m is the particle mass. In the case of dipole oscillations, m is the reduced dipole mass. What is the response of such an oscillator to an external electric field?

1.3.2°. When one of these independent oscillators is exposed to an electromagnetic field, it becomes polarized. In the linear approximation, the induced dipole moment appearing in Eq. (1.3) for the electric polarization is proportional to the applied electric field

$$\mathbf{p} = \hat{\alpha}\varepsilon_0\mathbf{E}. \tag{1.28}$$

The coefficient $\hat{\alpha}$, which is related to the displacement of charged particles in a solid, is called the *electric polarizability* of the medium. This coefficient is the second-rank Hermitian tensor for an anisotropic particle, but it reduces to a scalar for isotropic particles. Note that to avoid confusion with the absorption coefficient α, the polarizability has been noted by the symbol $\hat{\alpha}$ (with a hat). From Eqs. (1.28) and (1.3) we obtain the following expression for the polarization $\mathbf{P} = \hat{\alpha}\varepsilon_0 N\mathbf{E}$. Thus, in view of Eq. (1.2a), the permittivity $\hat{\varepsilon}$ of an ensemble of particles can be written as

$$\hat{\varepsilon} = 1 + N\hat{\alpha}, \tag{1.29}$$

which assumes that all the particles respond to the exciting external electric field in phase, that is, coherently. This simple equation is the bridge between the macroscopic optical properties of the specimen, described in terms of the local dielectric function, and the microscopic parameter $\hat{\alpha}$ characterizing polarization of each specific particle under the action of the external electric field.

1.3.3°. Let an external electric field of the form $\mathbf{E} = \mathbf{E}_0 \exp^{i\omega t}$ be applied in the x direction to a harmonic oscillator representing either vibrations of electrons relative to the positively charged ions or phonons (1.2.4°). The electric field redistributes charges, inducing a dipole moment according to Eq. (1.28). The force bonding the ion (electron) within the lattice restricts this disturbance and produces a restoring force. Consider these forced vibrations when there is damping (from energy losses) in the system. The Newton equation for the motion of a harmonic oscillator is

$$m^*\frac{d^2\mathbf{x}}{dt^2} + m^*\gamma\frac{d\mathbf{x}}{dt} + \beta\mathbf{x} = -e^*\mathbf{E}, \tag{1.30}$$

where $\mathbf{x}$ is the displacement of the oscillator with respect to the equilibrium; β is the restoring force per unit displacement $\mathbf{x}$; γ is the damping force per unit mass and per unit velocity, or the damping constant; and e^* and m^* are, respectively, the effective charge and reduced mass of the dipole. After integration of Eq. (1.30), we obtain the following expression for the displacement:

$$\mathbf{x} = -\frac{e^*/m^*}{\omega_0^2 - \omega^2 + i\omega\gamma}\mathbf{E}, \tag{1.31}$$

where $\omega_0 = \sqrt{\beta/m^*}$ is the resonance angular frequency of the dipole.

1.3.4°. Taking into consideration Eqs. (1.28) and (1.31) and the fact that the dipole moment $\mathbf{p}$ induced by the external electric field is $\mathbf{p} = -e^*\mathbf{x}$, we can write the expression for the polarizability in the form

$$\hat{\alpha} = \frac{e^{*2}}{\varepsilon_0 m^*}\frac{1}{\omega_0^2 - \omega^2 + i\omega\gamma}. \tag{1.32}$$

Then, in view of Eqs. (1.29) and (1.32), the dielectric function proves to be a complex quantity written as

$$\hat{\varepsilon} = 1 + \frac{\omega_p^2}{\omega_0^2 - \omega^2 + i\omega\gamma}, \tag{1.33}$$

where

$$\omega_p = \sqrt{\frac{Ne^{*2}}{\varepsilon_0 m^*}} \tag{1.34}$$

is the so-called plasma frequency, which describes either the lattice vibrations or oscillations of electrons, and N is the number of the effective oscillators per unit volume. The value of ω_p determines the oscillator strength (1.3.15°). We have that high plasma frequencies result from large effective charges, small masses, and high densities. For phonons, the plasma frequency falls within the IR spectral range. For the plasma of free carriers (1.3.12°), this parameter varies from the IR range for semiconductors to the UV/visible range for metals.

1.3.5°. Notice that in the Lorentz model the difference between the local (actual) field $\mathbf{E}_{loc}$ and the macroscopic (applied) field $\mathbf{E}$ is neglected. Strictly speaking, this may be done only in the description of rare gaseous media, when $N\hat{\alpha} \ll 1$, and for delocalized (free) electrons [68]. In the general case of phonons and highly localized (bound) electrons, the local electric field that acts on the particle is the sum of both the external field and the electric field of the surrounding particles. It can be demonstrated [52] that from the *Clausius–Mossotti/Lorentz–Lorenz* (CMLL) law [12, 22, 33] the effect of the local field (the density of the material) results in the *red shift* of the resonance frequency from the value ω_0 to the value that can be calculated with the formula $\omega_0^2 = \beta/m^* - \frac{1}{3}\omega_p^2$.

1.3.6°. The Lorentz model becomes more useful and applicable over a wide frequency range if the effect of oscillators of different types is taken into account. In the simplest harmonic approximation, the oscillators representing the lattice and electron vibrations are assumed to be independent and the generalized electric polarizability $\hat{\alpha}$ is an additive quantity

$$\hat{\alpha} = \hat{\alpha}_e + \hat{\alpha}_v = \hat{\alpha}_e + \frac{e^{*2}/(\varepsilon_0 m^*)}{\omega_0^2 - \omega^2 + i\omega\gamma}, \tag{1.35}$$

where α_e is the electronic component of the polarizability (usually constant in the IR region) and $\hat{\alpha}_v$ is the vibrational part. In this case, ignoring orientational (dipolar) polarizability [22], Eq. (1.33) is transformed into the sum

$$\hat{\varepsilon} = 1 + \sum_\varsigma \frac{\omega_{p,\varsigma}^2}{\omega_{0,\varsigma}^2 - \omega^2 + i\omega\gamma_\varsigma} + \sum_l \frac{\omega_{p,l}^2}{\omega_{0,l}^2 - \omega^2 + i\omega\gamma_l}. \tag{1.36}$$

Here, $\omega_{0,\varsigma}$ and $\omega_{0,l}$ are the resonance frequencies of electrons bound within a ς-type lattice site and of the lattice vibrations of the l type, respectively; $\omega_{p,\varsigma}$, $\omega_{p,l}$, and γ_ς, γ_l are the plasma frequencies [Eq. (1.34)] and the phenomenological damping constants characterizing the electron and lattice vibrations, respectively.

When we deal with the IR spectral range, the first two terms in Eq. (1.36) can be replaced to a good approximation by the value of the dielectric function at the frequency ω in the transparency region (1.2.4°), where only the valence electrons are assumed to contribute to the polarization (atomic centers are rigidly fixed). This value is called the *high-frequency (optical) dielectric constant* or the *screening factor* and denoted as $\varepsilon_\infty [\varepsilon_\infty = \varepsilon(\omega)$ if $\omega_{0,l} \ll \omega \ll \omega_{0,\varsigma}]$. In particular, for a cubic diatomic ionic crystal modeled by one oscillator, the dielectric function Eq. (1.36) can be represented in the IR range as

$$\hat{\varepsilon} = \varepsilon_\infty + \frac{\omega_p^2}{\omega_0^2 - \omega^2 + i\omega\gamma}, \tag{1.37}$$

where ω_0 and ω_p are the resonance and plasma frequencies, respectively. In the absence of damping ($\gamma = 0$),

$$\varepsilon = \varepsilon_\infty + \frac{\omega_p^2}{\omega_0^2 - \omega^2}. \tag{1.38}$$

Passing to the limit $\omega \to 0$ in Eq. (1.38), one can obtain the following relationship:

$$\omega_p^2 = \omega_0^2(\varepsilon_{st} - \varepsilon_\infty), \tag{1.39}$$

where ε_{st} is the *static* (or *low-frequency*) *dielectric constant* defined as $\varepsilon_{st} = \lim_{\omega \to 0} \hat{\varepsilon}(\omega)$ or $\varepsilon_{st} = 1 + P_{st}/(\varepsilon_0 E_{st})$, where E_{st} and P_{st} are the static electric field and polarization, respectively.

Substituting Eq. (1.39) into Eq. (1.38), we have

$$\varepsilon = \varepsilon_\infty + \frac{(\varepsilon_{st} - \varepsilon_\infty)\omega_0^2}{\omega_0^2 - \omega^2}. \tag{1.40}$$

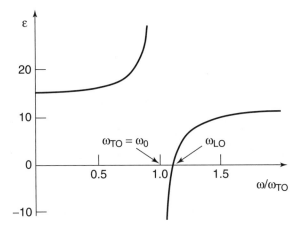

Figure 1.5. Function $\varepsilon(\omega)$ for model system without damping, covered by $S = 3$ and $\varepsilon_\infty = 12$.

1.3.7°. The behavior of the dielectric function described by Eq. (1.40) for a model undamped oscillator characterized by $\varepsilon_\infty = 12$, $\varepsilon_{st} - \varepsilon_\infty = 3$, and $\gamma = 0$ is shown in Fig. 1.5. It can be seen that in the absence of damping the dielectric function is negative over a certain frequency range. The high-frequency boundary of this range, where $\varepsilon(\omega) = 0$, is the frequency of the longitudinal optical wave and is designated as ω_{LO}. In fact, from Eq. (1.1b), when $\varepsilon(\omega) = 0$, we obtain $\nabla \times \mathbf{H} = 0$. This equation combined with Eq. (1.1d) leads to the conclusion that $\mathbf{B} = 0$ and, consequently, $\nabla \times \mathbf{E} = 0$. This means that the electric field is longitudinal (1.1.3°), while the magnetic field is absent at ω_{LO} [9, 11]. At the resonance frequency ω_0 the dielectric function has a pole, $\varepsilon(\omega_0) \neq 0$. From Eq. (1.1c) we obtain $\nabla \cdot \mathbf{E} = 0$, which represents the condition of the transversality of the electric field. Provided $\gamma \ll \omega_0$, $\omega_0 \approx \omega_{TO}$ (1.1.18°), and both the frequencies ω_0 and ω_{TO} are said to characterize the transverse excitation.

If $\varepsilon = 0$, Eq. (1.38) may be transformed to give the following relationship between the quantities ω_{LO} and ω_{TO} [69]:

$$\omega_{LO}^2 = \omega_{TO}^2 \frac{\varepsilon_{st}}{\varepsilon_\infty}. \tag{1.41}$$

This is known as the *Lyddane–Sachs–Teller* (LST) *law* [70]. In covalent monoatomic and organic crystals, with vanishing dipole moment, $\varepsilon_{st} \approx \varepsilon_\infty$, and hence $\omega_{TO} \approx \omega_{LO}$, which for these materials is usually denoted as ω_0. By using Eq. (1.41), we can rewrite Eq. (1.40) in the form

$$\varepsilon(\omega) = \frac{\omega_{LO}^2 - \omega^2}{\omega_{TO}^2 - \omega^2}. \tag{1.42}$$

1.3.8°. There are many excellent books [22, 33, 37, 46, 69, 71] in which the properties of the longitudinal and transverse optical excitations (LO and TO modes, respectively) are discussed in detail. Below, a short summary of these

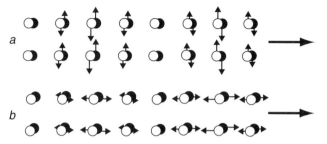

Figure 1.6. Displacements of ions in ionic cubic crystal in (a) transverse and (b) longitudinal waves.

discussions is given that will be used for the interpretation of IR spectra of thin films.

A qualitative, physical description of the longitudinal and transverse optical phonons is illustrated in Fig. 1.6. Kittel [22] phenomenologically explains the fact that $\omega_{LO} > \omega_{TO}$ as follows. The local electric field induces polarization of the surrounding atoms in the opposite direction to that of the longitudinal mode but in the same direction as the transverse mode. This polarization causes an increasing resistance to the longitudinal vibration relative to the transverse one (see also Ref. [33]).

In the case of strong oscillators, the range of frequencies from ω_{LO} to ω_{TO} is characterized by purely imaginary values of the refractive index (1.10) and the wavenumber (1.12a). This range is therefore forbidden for the propagation of electromagnetic waves. For the range $\omega_{LO} < \omega < \omega_{TO}$, the reflectance R at normal incidence [Eq. (1.70)] equals unity; that is, the incident wave is totally reflected. This causes the *reststrahlen* (residual ray) bands to arise in the specular reflection spectra of "strongly" absorbing substances.

From the general selection rule (1.27), it follows that, unlike the case of the transverse excitations ($\vartheta = 0°$), the longitudinal excitations ($\vartheta = 90°$) are nonradiative for any experimental geometry of experiment; that is, they do not interact with the transverse electromagnetic wave. For ultrathin films, absorption of p-polarized radiation at the frequency close to ω_{LO} takes place (Section 3.1). However, this absorption is due to the transverse surface mode produced by the so-called size effect (Section 3.2).

1.3.9°. It should be pointed out that ω_{LO} and ω_{TO} characterize not only optical phonons (1.2.4°) but also excitons (1.2.5°), plasmons, and so on. Consider the dielectric function of the plasmons.

Metals are denoted as free-electron metals if most of the electronic and optical properties are due to the conduction electrons alone. Examples are Al, Mg, and alkali metals. The dispersion of the optical constants of the free-electron metals is described by the *Drude model* [72], which can be regarded as a special case of the Lorentz model (1.3.3°) when the restoring (bounding) force β is equal to zero, and hence the resonance frequency of free carriers, $\omega_0 = \sqrt{\beta/m}$ (here, m is electron mass), is also equal to zero, $\omega_0 = 0$. The damping force γ results from collisions of electrons with the lattice. It is assumed that $\gamma = 1/\tau$, where τ is the average time between collisions (the *collision relaxation time*). Because the wave function of a free-electron gas is distributed uniformly throughout the metal, the field acting on a single electron is the average field and the local field correction is not necessary [50]. Thus, proceeding from Eq. (1.37), we can write the dielectric function for the electron plasma as

$$\hat{\varepsilon} = 1 + \frac{\omega_p^2}{i\gamma\omega - \omega^2}, \qquad (1.43a)$$

where ω_p is expressed by Eq. (1.34) in terms of the mass and charge of an electron. Formula (1.43a) is called the *Drude dielectric function*.

1.3.10°. If we further assume that $\gamma \ll \omega$, the dielectric function Eq. (1.43a) can be rewritten as

$$\varepsilon \approx \frac{\omega_p^2 - \omega^2}{-\omega^2}. \qquad (1.43b)$$

Comparing Eq. (1.43b) with Eq. (1.42), we see that $\omega_p = \omega_{LO}$. This means that plasma oscillations take place along the direction the electric field and therefore manifest themselves as longitudinal waves. Consequently, they do not interact with the transverse electromagnetic field (1.3.7°). In quantum mechanics, the particle with the energy $\omega_p \hbar$ is called the *volume plasmon*.

1.3.11°. Attenuation of the electric field by a metal is characterized by the skin depth δ. By definition, $\delta = 2/\alpha$, where α is the decay constant (1.1.17°). The electric field decreases exponentially by the factor of e over a distance of δ, whereas the light intensity [Poynting vector (1.1.15°)] decreases by the factor of e^2 over a distance of δ. Equation (1.24) gives

$$\delta = \frac{\lambda}{2\pi k} \approx 10^{-5} \text{ cm}. \qquad (1.44)$$

It should be kept in mind that Eq. (1.44) is valid only in the cases where the distances associated with spatial changes of the field are large compared to the free path of the conduction electrons (normal skin effect).

1.3.12°. For such metals as Al, Cu, Au, and Ag, the density of the electron plasma, N, involved in Eq. (1.34) is on the order of 10^{23} cm^{-3}. This means that $\omega_p \approx 2 \times 10^{16}$ s^{-1} ($\lambda_p \approx 100$ nm), so that in the IR spectral range $\omega \ll \omega_p$, and the dielectric function Eq. (1.43b) is negative ($\varepsilon < 0$), while the refractive index is purely imaginary (in practice, the refractive index of metals has the real component [25, 47, 52]). This leads to strong absorption and reflection of electromagnetic radiation by metals in the IR range. At $\omega > \omega_p$ metals become transparent and behave like dielectrics. In semiconductors, ω_p is shifted to the region of lower frequencies depending on the concentration N and the effective mass of free carriers, m^*. For example, for n-Ge with $N \approx 10^{19}$ cm^{-3}, $\lambda_p \approx$ 10 μm, which corresponds to the IR spectral range.

1.3.13°. The Drude model (1.3.9°) gives a correct description of optical characteristics of free-electron metals only. For the other metals and semiconductors, however, this model is unsatisfactory because of substantial contributions of interband transitions from lower lying bands into the conduction band (bound electrons) and lattice ions. In this case, the dielectric function can be described by the equation

$$\varepsilon(\omega) = \varepsilon_{\text{fe}}(\omega) + \delta\varepsilon_{\text{be}}(\omega) + \delta\varepsilon_L(\omega),$$

where $\varepsilon_{\text{fe}}(\omega)$ is described by Eq. (1.43a) and $\delta\varepsilon_{\text{be}}(\omega)$ and $\delta\varepsilon_L(\omega)$ are the second and third terms in Eq. (1.36) that are connected with bound electrons and the lattice vibrations, respectively.

1.3.14°. As a rule, the experimental (Section 3.2) frequency dependence of the free-carrier absorption (or the *Drude absorption*) disagrees with the ω^{-2} law predicted by the Drude model [Eq. (1.43a)]. The actual dependence of the decay constant α follows the ω^{-p} law [40, 42, 46], where p is a constant over the range $1 < p < 4$. The constant p depends on the semiconductor, the frequency range, the temperature, and the concentration of impurities and free carriers. The quantum-mechanical extension of the Drude theory [73] shows that $p = 3$ if free carriers are scattered on the optical phonons, $p = 3.5$ for ionized-impurity scattering, and $p = 1.5$ if the scattering centers are acoustic phonons. Therefore, the Drude absorption can be used for determining not only the concentration and mobility of free carriers but also the mechanism of the free-carrier scattering.

1.3.15°. To evaluate the intensity of the absorption band, the dimensionless constants S_j, which are called the oscillator strengths, are introduced with the formula

$$S_j = \frac{\omega_p^2}{\omega_{0j}^2}, \tag{1.45}$$

where ω_p and ω_{0j} are, respectively, the plasma [Eq. (1.34)] and resonance frequency of the jth oscillator. This allows one to rewrite Eq. (1.36) for a multiphonon system in the general form

$$\hat{\varepsilon}(\omega) = \varepsilon_\infty + \sum_{j=0}^{n} \frac{S_j \omega_{0j}^2}{\omega_{0j}^2 - \omega^2 + i\omega\gamma_j}, \tag{1.46}$$

which is convenient for parametrization of optical properties of a substance [24a, 74–76]. Here, the possibility of a free-carrier contribution [Eq. (1.43a)] is included in the $j = 0$ mode by setting $\omega_{00} = 0$ and γ_0 equal to the collision frequency. Equation (1.46) is known as the *Maxwell–Helmholtz–Drude dispersion formula*. In the case of randomly oriented weak oscillators of one type with weak damping ($\gamma \ll \omega_0$), substituting the value for the imaginary part of the dielectric function, ε'', from Eq. (1.46) [Eq. (1.53b)] into Eq. (1.24), one can derive the following relationship between the integral absorption over the absorption band $A(\nu)$ and the oscillator strength S [37]:

$$A(\nu) \equiv \int \alpha(\nu)d\nu = \frac{1}{6}N\pi\frac{\omega_0^2}{c}S, \tag{1.47}$$

where N is the concentration of the oscillators. This can be considered as a reformulation of the BLB law [Eq. (1.23)]: The integral absorption is proportional to the concentration of oscillators.

The quantity S is related to the concentration of oscillators N, their reduced mass m^*, the resonance frequency ω_0, and the phenomenological effective ionic charge e^* [46] by the equation

$$S = \frac{N e^{*2}}{m^* \varepsilon_0 \omega_0^2}. \tag{1.48}$$

The formula (1.48) or formulas based on another definition of the effective charge [40, 43, 46, 77] can be used for evaluating the degree of the bond polarity (the valency) from IR spectra of thin films if the density N of the electric dipoles is known from independent measurements [46, 78]. This parameter is of considerable importance, since it allows one to calculate the derivative of the surface potential with respect to the thickness of the adsorbate layer [79] and the interionic distance [40, 43, 46].

The oscillator strengths S_j satisfy the sum rule,

$$\varepsilon_{st} = \varepsilon_\infty + \sum_j S_j. \tag{1.49a}$$

For an ionic crystal with a two-atom cell, Eq. (1.49a) takes the simple form

$$\varepsilon_{st} = \varepsilon_\infty + S. \tag{1.49b}$$

Thus, the oscillator strength is proportional to the polarizability of the corresponding elementary excitations. Substituting Eq. (1.41) into Eq. (1.49b), we obtain

$$S = \frac{\varepsilon_\infty}{\omega_{TO}^2}(\omega_{LO}^2 - \omega_{TO}^2) \tag{1.50}$$

or

$$\omega_{LO}^2 = \omega_{TO}^2 \left(1 + \frac{S}{\varepsilon_\infty}\right). \tag{1.51}$$

Equation (1.51) demonstrates that the frequency separation of the longitudinal and transverse waves (TO–LO *splitting*) is proportional to the oscillator strength. For fixed values of the oscillator strength S and the high-frequency dielectric constant ε_∞, the TO–LO splitting will be less for higher frequency modes than for lower frequency modes. This is explained [80] by the fact that a low ω_{TO} implies a weak restoring force for the vibration, which will lead to an increased polarization and therefore a large TO–LO splitting.

1.3.16°. The quantum-mechanical description of the radiation–matter interaction uses the same classical expression (1.46) but assigns new meanings to the quantities involved [41, 44–65]. Thus, ω_{0j} is the frequency of the transition from the ground state 0 to the excited state j separated in energy by $\hbar\omega_{0j}$. The damping constants γ are connected with the transition probabilities from the state j to all the other states. The oscillator strength S_j is treated as the relative probability of a quantum-mechanical transition between two definite states 0 and j in times of the number of oscillators N available for the interaction:

$$S_j \propto N |\langle j | \boldsymbol{\rho} | i \rangle|^2, \tag{1.52}$$

where $|\langle j | \boldsymbol{\rho} | i \rangle|$ is the TDM of the excitation (1.2.8°).

1.3.17°.   Equation (1.46) can be split into real and imaginary parts as follows:

$$\varepsilon'(\omega) = \varepsilon_\infty + \sum_j \frac{S_j \omega_{0j}^2 (\omega_{0j}^2 - \omega^2)}{(\omega_{0j}^2 - \omega^2)^2 + \omega^2 \gamma_j^2}, \tag{1.53a}$$

$$\varepsilon''(\omega) = \sum_j \frac{\gamma_j S_j \omega_{0j}^2 \omega}{(\omega_{0j}^2 - \omega^2)^2 + \omega^2 \gamma_j^2}. \tag{1.53b}$$

Hence, the maximum of the TO energy loss function (ε'') for an elementary damped harmonic oscillator occurs at $\omega_{\max}$, where $\omega_{\max, j}^2 = \frac{1}{6}\{2\omega_{0j}^2 - \gamma_j^2 + [(2\omega_{0j}^2 - \gamma_j^2)^2 + 12\omega_{0j}^4]^{1/2}\}$, or, if $\gamma_j \ll \omega_{0j}$, $\omega_{\max, j}^2 \approx \omega_{0j}^2 - \frac{1}{4}\gamma_j^2$.

Consider the forms of the functions $\varepsilon'(\omega)$, $\varepsilon''(\omega)$, $n(\omega)$, $k(\omega)$, $|\hat{\varepsilon}(\omega)|$, and $\mathrm{Im}[1/\hat{\varepsilon}(\omega)]$ for the three most common cases:

1. a strong oscillator with weak damping, as for β-SiC (Fig. 1.1);

2. a model strong oscillator with strong damping (Fig. 1.7a); and

3. weak oscillators with weak damping, as for the νCH vibrations of polyethylene (Fig. 1.7b).

One can see in Fig. 1.1 that for a strong oscillator with weak damping the function $\varepsilon'(\omega) = \mathrm{Re}[\varepsilon(\omega)]$ lacks the discontinuity at the frequency $\omega = \omega_{\mathrm{TO}}$, which is observed when no damping occurs (Fig. 1.5), but has a pole. At ω_{LO}, the function ε' goes to zero ($n = k$). This is used in determining ω_{LO} according to Drude's method (Section 3.2.3). If either the damping constant increases or the oscillator strength decreases, the pole and the zero value of $\varepsilon'(\omega)$ disappear (Figs. 1.7a and b) so that for heavily damped strong oscillators and weak oscillators there is neither a zero value nor a pole for the function $\varepsilon'(\omega)$. As a consequence, the Drude rule becomes inapplicable [35]. Notice (Figs. 1.1 and 1.7a) that the TO energy loss curve, $\varepsilon''(\omega)$, is symmetric and centered practically at ω_{TO}. It can also be seen that for β-SiC and model inorganic substance (Figs. 1.1 and 1.7a) the maximum of the extinction coefficient k shifts toward higher frequencies with respect to the frequency ω_{TO}, and its contour is asymmetrical. This behavior is typical of a strong oscillator. It follows that, in contrast to the case of a "weak" oscillator (Fig. 1.7b), the transverse frequency of a strong oscillator cannot be found from a measured absorption spectrum alone.

1.3.18°.   The Maxwell–Helmholtz–Drude dispersion formula (1.46) provides a good model of the dielectric function in the case of moderate to weak TO

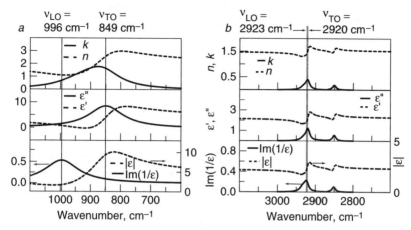

Figure 1.7. Frequency dependences of n, k, ε', ε'', and $\mathrm{Im}(1/\hat{\varepsilon})$ for (a) model inorganic substance characterized by $S = 1.37$, $\varepsilon_\infty = 3.68$, $\gamma = 138$ cm^{-1}, and $\nu_0 = 852$ cm^{-1} and (b) polyethylene in region of the νCH band, whose elementary oscillators are characterized by $S = 0.002, 0.0043, 0.0021$; $\gamma = 10, 11.5, 19$ cm^{-1}; $\nu_0 = 2850, 2920, 2930.5$ cm^{-1}; $\varepsilon_\infty = 2.22$ from Ref. [74].

mode damping. However, for a number of heavily damped strong oscillators, the dielectric function is often represented as the product [81–83]

$$\varepsilon(\omega) = \varepsilon_\infty \prod_j \frac{|\hat{\omega}_{\mathrm{LO}j}|^2 - \omega^2 - i\omega\gamma_{\mathrm{LO}j}}{\omega_{0j}^2 - \omega^2 - i\omega\gamma_j}, \tag{1.54a}$$

which allows the introduction of the LO mode parameters [the frequency $\hat{\omega}_{\mathrm{LO}j}$ and the damping coefficient $\gamma_{\mathrm{LO}j} = 2\,\mathrm{Im}(\hat{\omega}_{\mathrm{LO}j})$] in a straightforward way. This form of the dielectric function is known as the semiquantum four-parameter model. Assuming $\gamma = 0$ and letting $\omega \to 0$, the general LST relationship can be derived from Eq. (1.54a), namely [84],

$$\left(\frac{\varepsilon_{\mathrm{st}}}{\varepsilon_\infty}\right)^{1/2} = \prod_j \frac{\omega_{\mathrm{LO}j}}{\omega_{\mathrm{TO}j}}. \tag{1.54b}$$

1.3.19°. The functions $n(\omega)$ and $k(\omega)$ cannot be represented as a sum or product as for the dielectric function [Eqs. (1.46) and (1.54a)] and should be calculated from Eq. (1.17) by using Eqs. (1.53).

1.3.20°. The absorption spectrum containing many bands with partially overlapped contours is typical for the majority of materials [4]. The magnitudes of damping of the corresponding IR-active modes can be determined by nonlinear energy transfer processes from the given vibration to other vibrations. The formulas for the dielectric functions in the case of coupled modes were obtained by Barker and Hopfield [85]. The interaction of a (discrete) phonon with a continuous electronic excitation can result in specific band distortions [86] named the Fano resonances.

1.4. PROPAGATION OF IR RADIATION THROUGH PLANAR INTERFACE BETWEEN TWO ISOTROPIC MEDIA

Let an electromagnetic wave propagating in medium 1 with refractive index $\hat{n}_1$ encounter at $z = 0$ (Fig. 1.8) an ideal planar surface of medium 2 characterized by $\hat{n}_2 \neq \hat{n}_1$. At the interface, the incident wave splits into the reflected and transmitted (across the interface) components (Fig. 1.8). What are the directions and intensities of these components? Consider first the case when both contacting media are transparent ($k_1 = k_2 = 0$).

1.4.1°. *Law of reflection.* The incident ray, the reflected ray, and the normal to the interface at the point of incidence all lie in the same plane (the *plane of incidence*). The angle of incidence φ_1 is equal to the angle of reflection φ_1' (Fig. 1.8):

$$\varphi_1 = \varphi_1'. \tag{1.55}$$

All the angles are measured with respect to the surface normal (Fig. 1.8).

The reflection from the boundary when medium 1 is optically rarer than medium 2 ($n_1 < n_2$) is called the *external* or *specular*[†] *reflection* (Fig. 1.8). When the radiation travels from an optically denser to optically rarer medium, that is, when $n_1 > n_2$, it is said that *internal reflection* takes place (see Fig. 1.10 below).

1.4.2°. At a nonzero angle of incidence, $\varphi_1 \neq 0$, the direction of the transmitted beam differs from the direction of the incident beam. This process, the deflection of a beam by a medium, is called refraction.

Law of refraction. The incident ray, the refracted ray, and the normal to the interface at the point of incidence all lie in the same plane (the plane of incidence). The angle of incidence φ_1 is related to the angle of refraction φ_2 (Fig. 1.8) by

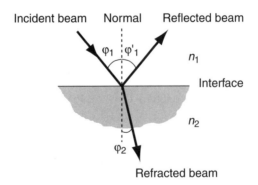

Figure 1.8. External reflection and refraction of light at interface of two phases, $n_2 > n_1$.

[†] From the Latin term *speculum*, which means "mirror."

Snell's law, which was established empirically in 1621:

$$\frac{\sin \varphi_1}{\sin \varphi_2} = \frac{n_2}{n_1},$$

(1.56)

where n_1 and n_2 are the refractive indices of the input and output media, respectively.

1.4.3°. To find the intensities of the reflected and transmitted beams, we need first to consider the relationship between the amplitudes of these beams. From the viewpoint of the Maxwell theory (1.1.1°), at an interface, the electric field vector of every monochromatic linearly polarized electromagnetic wave has the form described by Eq. (1.5a):

$$\mathbf{E}_{in}(\mathbf{r}, t) = \mathbf{E}_{in}(0)e^{i(\omega t - \mathbf{k}_{in}\mathbf{r})},$$

$$\mathbf{E}_r(\mathbf{r}, t) = \mathbf{E}_r(0)e^{i(\omega t - \mathbf{k}_r \mathbf{r})},$$

(1.57)

$$\mathbf{E}_t(\mathbf{r}, t) = \mathbf{E}_t(0)e^{i(\omega t - \mathbf{k}_t \mathbf{r})}.$$

Here, $\mathbf{k}_{in}$, $\mathbf{k}_r$, $\mathbf{k}_t$ and $\mathbf{E}_{in}(0)$, $\mathbf{E}_r(0)$, $\mathbf{E}_t(0)$ are, respectively, the wave vectors and amplitudes of the incident, reflected, and transmitted waves.

1.4.4°. For isotropic bounding media, a beam of IR radiation incident onto a surface can be resolved into two orthogonal and, therefore, independent components (1.1.5°) [6, 9] (Fig. 1.9): the *s*-component (*s-polarization*) and the *p*-component (*p-polarization*). In Fig. 1.9, these components are schematically denoted by the arrows lying in the plane of incidence coinciding with the plane of drawing (*p*) and by circles pointing out of the plane of drawing (*s*). The *s*-polarized component has the electric field vector $\mathbf{E}_s$ oriented perpendicular to the plane of incidence (*xz* in Fig. 1.9). The *p*-polarized component has the electric field vector $\mathbf{E}_p$ parallel to this plane. The vector $\mathbf{E}_s$ lies in the plane of the surface

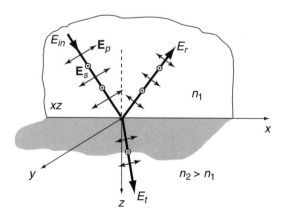

Figure 1.9. Electric field vectors in two-phase system.

at any angle of incidence φ_1, whereas $\mathbf{E}_p$ will only lie in the plane of the surface when $\varphi_1 = 0$.

1.4.5°. The amplitudes of the waves for p- and s-polarized incident radiation (1.57) can be interconnected in a formal way as

$$
\begin{aligned}
E_r^s(0) &= r^s E_{\text{in}}(0), \\
E_t^s(0) &= t^s E_{\text{in}}(0), \\
E_r^p(0) &= r^p E_{\text{in}}(0), \\
E_t^p(0) &= t^p E_{\text{in}}(0),
\end{aligned}
\tag{1.58}
$$

where r^s, t^s, r^p, and t^p are the *Fresnel amplitude coefficients* for s- and p-components of the reflected and transmitted wave.

1.4.6°. The quantities observable spectroscopically are the *reflectance* R_{12} and the *transmittance* T_{12} of the interface. They are defined as the ratios of the z-components (normal to the interface) of the time-averaged Poynting vectors (1.1.15°) for the reflected and transmitted waves to that of the incident wave [8]:

$$
R_{12} = \frac{\langle \Pi \rangle_{r,z}}{\langle \Pi \rangle_{\text{in},z}}, \qquad T_{12} = \frac{\langle \Pi \rangle_{t,z}}{\langle \Pi \rangle_{\text{in},z}}.
\tag{1.59}
$$

1.4.7°. The boundary conditions for nonferromagnetic materials require continuity of the electric and magnetic field vectors across the boundary [6–13]:

$$
\begin{aligned}
&D_{\text{normal}} \quad \text{continuous (no surface free charge),} \\
&E_{\text{tangential}} \text{ continuous,} \\
&\mathbf{B} = \mathbf{H} \quad \text{continuous (nonmagnetic media).}
\end{aligned}
\tag{1.60}
$$

Using the boundary conditions, one can determine the amplitudes [Eq. (1.58)] and intensities [Eq. (1.59)] of the reflected and transmitted waves as well as their directions of propagation [9, 14, 52, 87, 88].

1.4.1. Transparent Media

1.4.8°. For nonabsorbing materials, the boundary conditions (1.4.7°) lead to the *Fresnel formulas* for the amplitude of reflection and transmission coefficients (1.4.5°):

$$
\begin{aligned}
r_{12}^p &= \frac{\tan(\varphi_1 - \varphi_2)}{\tan(\varphi_1 + \varphi_2)}, \\
r_{12}^s &= -\frac{\sin(\varphi_1 - \varphi_2)}{\sin(\varphi_1 + \varphi_2)}, \\
t_{12}^p &= \frac{2n_1/\cos\varphi_2}{n_2/\cos\varphi_2 + n_1/\cos\varphi_1}, \\
t_{12}^s &= \frac{2\sin\varphi_2\cos\varphi_1}{\sin(\varphi_1 + \varphi_2)}.
\end{aligned}
\tag{1.61}
$$

These expressions were derived in a slightly less general form by Fresnel in 1823 on the basis of his elastic theory of light [9]. From Eqs. (1.59) and (1.20) we obtain the energetic coefficients

$$R_{12}^{s,p} = |r_{12}^{s,p}|^2,$$
$$T_{12}^{s,p} = |t_{12}^{s,p}|^2 \frac{n_2 \cos \varphi_2}{n_1 \cos \varphi_1}, \tag{1.62}$$

which, along with Eq. (1.61), completely express the relationship between the reflecting and transmitting properties of an interface and the refractive indices of the substances on both sides of that interface. The factor $(n_2 \cos \varphi_2)/(n_1 \cos \varphi_1)$ in the relationship for the transmittance [Eq. (1.62)] arises because the rate of the energy flow is proportional to the refractive index of the medium [Eq. (1.20)] and the cross-sectional area in medium 1 is $S \cos \varphi_1$, and in medium 2 is $S \cos \varphi_2$ (here, S is the area illuminated by the incident beam on the surface). It can be shown that $T_{12}^{s,p} + R_{12}^{s,p} = 1$, which implies that the boundary does not absorb or create energy.

1.4.9°. It can be shown from Eq. (1.61) that for the case of the transparent bounding media $r_{12}^p = 0$ when $\varphi_1 + \varphi_2 = \frac{1}{2}\pi$. This condition is fulfilled at the angle of incidence

$$\varphi_B = \arctan \left(\frac{n_2}{n_1} \right), \tag{1.63}$$

called the *Brewster angle*. At φ_B, the p-polarized beam passing through the interface and undergoing refraction is oriented at an angle of $\frac{1}{2}\pi$ with respect to the reflected (s-polarized) beam. That is, separation of the s- and p-polarized beams in space is observed. This effect is the basis for the operation of spatial polarizers, which can extract the desired polarization from nonpolarized light [89].

1.4.10°. In the case of internal reflection at the interface of two transparent media (Fig. 1.10), if the angle of incidence exceeds the angle

$$\varphi_c = \arcsin \left(\frac{n_2}{n_1} \right), \tag{1.64}$$

named the *critical angle, total internal reflection* occurs. This phenomenon was first described by Kepler in 1611. As can be seen from Snell's law [Eq. (1.56)], at $\varphi_1 = \varphi_c$ the refracted beam propagates along the boundary and at $\varphi_c < \varphi_1 < \frac{1}{2}\pi$ the interface is a perfect mirror ($R = 1$). However, if one inspects the details of the optical field near the interface — for example, experimentally by a scanning near-field optical microscope (SNOM) (Section 4.4) or theoretically by using the boundary conditions (1.4.7°) [9, 11] — one finds that at $\varphi_1 > \varphi_c$, tails of the electromagnetic field penetrate into the rarer medium in the form of an inhomogeneous wave [14, 90, 91]:

$$E \propto \exp[i(\omega t - k_x x + k_z z)], \tag{1.65}$$

where

$$k_x = \frac{\omega}{c} n_1 \sin \varphi_1,$$

$$k_z = i \frac{\omega}{c} n_2 \left[\left(\frac{n_1}{n_2} \right)^2 \sin^2 \varphi_1 - 1 \right]^{1/2} \tag{1.66}$$

are the parallel (x) and perpendicular (z) wave vectors to the interface. Since at $\varphi_1 > \varphi_c$ the quantity k_z is imaginary, the electric field vector decays exponentially with increasing z in the optically rarer medium, as illustrated schematically in Fig. 1.10. Such a surface wave is called evanescent.[†] The important peculiarity of the evanescent wave is that its phase velocity $v_x = c/(n_1 \sin \varphi_1)$ is larger than that of the bulk wave inside medium 1. This is exploited for excitation of surface electromagnetic waves (SEWs) (Section 3.2.1) [3, 92, 93]. If the incident radiation is s-polarized, the evanescent wave is transverse and does not manifest particular properties. However, in the case of p-polarization, the evanescent wave has a component of the electric field vector directed along the wave propagation (x-axis) and, therefore, is not transverse in the usual sense [90, 91].

1.4.11°. Using Eqs. (1.61) and (1.62), it can be shown that at $\varphi_1 > \varphi_c$ the energy of radiation is totally reflected regardless of the polarization state (1.4.5°), that is, $R^{s,p} = 0$ and $T^{s,p} = 0$ [14]. However, the Fresnel amplitude transmission coefficients have nonvanishing values ($t^{s,p} \neq 0$) and the wave vectors of the components parallel to the interface [Eq. (1.66)] are real. This means that the energy of the evanescent wave flows parallel to the boundary surface until it is reflected, that is, there is lateral displacement of the reflected wave (Fig. 1.10). This effect is called the *Goos–Hänchen shift* after Goos and Hänchen [94], who

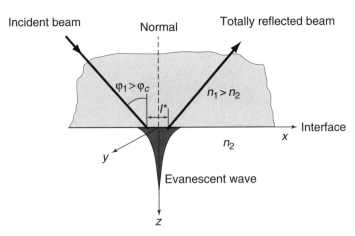

Figure 1.10. Schematic representation of total internal reflection of light at interface of two phases, $n_1 > n_2$. Here, l^* denotes the Goos–Hänchen shift.

[†] From the Latin root *evanscere*, which means "vanishing" or "passing away like a vapor."

detected it experimentally in 1947. Within the framework of Maxwell's theory, the Goos–Hänchen shifts for s- and p-polarization (ℓ_s^* and ℓ_p^*, respectively) are given by the expressions [95–98]

$$\ell_s^* = 2\frac{1}{|k_z|}\tan\varphi_1, \qquad \ell_p^* = \frac{2(1/|k_z|)\tan\varphi_1}{(\sin\varphi_1/\sin\varphi_c)^2 - \cos\varphi_1^2}. \tag{1.67}$$

It follows that the closer the angle of incidence is to either the critical angle or the grazing angle, the larger is the Goos–Hänchen shift.

The intensities and penetration depths of the x-, y-, and z-components of the electric field in the optically rare medium are examined in Section 1.8.3.

1.4.2. General Case

1.4.12°. Consider the general case in which contacting media can be absorbing. Electromagnetic theory shows [6–13] that the Fresnel formulas (1.61) can be adapted to this case by replacing in a formal way the real refractive indices with the complex quantities in Eqs. (1.12a–c). The resulting complex reflection and transmission coefficients are given by the following *generalized Fresnel formulas* [99–101]:

$$r_{12}^s = \frac{\xi_1 - \xi_2}{\xi_1 + \xi_2}, \qquad t_{12}^s = \frac{2\xi_1}{\xi_1 + \xi_2},$$

$$r_{12}^p = \frac{\hat{\varepsilon}_2\xi_1 - \hat{\varepsilon}_1\xi_2}{\hat{\varepsilon}_2\xi_1 + \hat{\varepsilon}_1\xi_2}, \qquad t_{12}^p = \frac{2\hat{n}_1\hat{n}_2\xi_1}{\hat{n}_2^2\xi_1 + \hat{n}_1^2\xi_2}, \tag{1.68}$$

where $\xi_i \equiv \hat{n}_i\cos\varphi_i$ ($i = 1, 2$) is the generalized complex index of refraction, which reduces to the complex index of refraction (1.12c) at normal incidence, and φ_i is the complex angle of refraction (1.4.7°). Using Snell's law (1.56), one can obtain

$$\xi_i \equiv \hat{n}_i\cos\varphi_i = \sqrt{\hat{n}_i^2 - \hat{n}_1^2\sin^2\varphi_1}. \tag{1.69}$$

As a result, the angles φ_i become complex. It is not expedient to undertake doubtful attempts to interpret the complex angles of incidence and refraction; they should simply be considered as mathematical representations. The real angle of refraction of the beam in an absorbing medium is calculated on the basis of the *Huygens-type construction* via the effective (real) refractive index of the medium, which is not equal to either n_i or $\hat{n}_i$ [9, 52].

1.4.13°. The reflectance and transmittance (1.4.6°) are related to the Fresnel coefficients (1.61) by the following [99, 100]:

$$R_{12}^{s,p} = |r_{12}^{s,p}|^2,$$

$$T_{12}^s = \frac{\text{Re }\xi_2}{\xi_1}|t_{12}^s|^2,$$

$$T_{12}^p = \frac{\text{Re}(\xi_2/\hat{n}_2^2)}{\xi_1}|\hat{n}_2 t_{12}^p|^2, \tag{1.70}$$

in which Re indicates the real part of the expression. Again, the factors Re ξ_2/ξ_1 and $\mathrm{Re}(\xi_2/\hat{n}_2^2)/\xi_1$ in the transmittance are due to the difference in the cross-sectional area of the incident and transmitted radiation.

1.4.14°. Equation (1.64) defines the critical angle for transparent media. However [99], when the contacting media are absorbing, so long as $k \ll 1$, the term "critical angle" is also used even though there is no total reflection at $\varphi_1 \geq \varphi_c$ and $\langle \Pi \rangle_{t,z} \neq 0$. This is called *attenuated total reflection* (ATR). This phenomenon in the visible optical region was observed by Isaac Newton in 1672 [102, 103]. It can be shown [9] that the intensity of the evanescent wave and consequently of the reflected radiation is attenuated at the wavelengths of the absorption of the optically rarer medium. This phenomenon provides the possibility of probing the layer next to the interface and is the basis of the ATR method.

1.4.15°. To provide a better understanding of the relationship between reflectance and the optical constants of the contacting media, some calculated values of reflectances at $\nu = 1000$ cm^{-1} for interfaces of air, Si, n-GaP, water, and Al are given in Fig. 1.11. The optical constants of these four media, which were used in the calculations, are tabulated in Table 1.1.

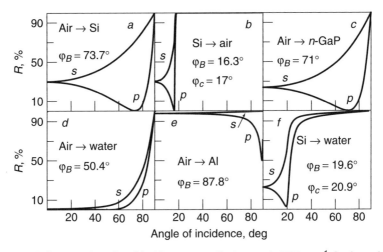

Figure 1.11. Influence of angle of incidence on reflectance at 1000 cm^{-1} for boundary of two phases: (*a*) air–Si; (*b*) Si–air; (*c*) air–*n*-GaP; (*d*) air–water; (*e*) air–Al; (*f*) Si–water. Optical constants are indicated in Table 1.1.

Table 1.1. Optical constants of media of Fig. 1.11

Material	n	k
Si	3.42	0
Water	1.218	0.0508
n-GaP	2.910	0.0269
Al	26	63

Source: From Refs. [25, 51].

It can be seen from Fig. 1.11 that in the case of two transparent contacting media (air and Si) the values of R_{12}^s for the external and internal reflection increase monotonically from the minimum value at $\varphi_1 = 0°$ to unity when $\varphi_1 = 90°$ and $\varphi_1 = \varphi_c$, respectively, whereas R_{12}^p goes through zero at the Brewster angle φ_B [Eq. (1.63)] before reaching unity. If medium 2 is absorbing, then even at φ_B there is a nonzero p-component in the reflected beam (Figs. 1.11e and f). Note that $R^p \approx 0$ at $\varphi_1 = \varphi_B$ for the external reflection from a medium with a relatively small extinction coefficient, as observed in the case of air and n-GaP (Fig. 1.11c). The larger the extinction coefficient of medium 2, the greater the intensity of the p-component reflected at $\varphi_1 = \varphi_B$. In this case, the angle of incidence defined by Eq. (1.63) corresponds to the minimum reflectance (and maximum transmittance) of the p-polarized component and is named the *pseudo-Brewster angle*. The (pseudo-) Brewster angles for the boundaries under examination are shown in Fig. 1.11.

1.4.16°. The Fresnel coefficients of reflection at the interface when the input medium is homogeneous, nonabsorbing, and isotropic and the final medium is homogeneous, nonabsorbing, and anisotropic, with the axes of symmetry normal and parallel to the interface, have been derived by Drude [72].

1.5. REFLECTION OF RADIATION AT PLANAR INTERFACE COVERED BY SINGLE LAYER

Several formulas have been introduced in the literature to describe the reflection from a film-covered planar interface in terms of the permittivities (1.1.1°) of the surrounding (immersion) medium, the film material and substrate, the thickness of the film, and the angle of incidence of the probe radiation. Born and Wolf [9] have presented a general theory for homogeneous, nonabsorbing, and isotropic media as developed by Abeles [104, 105]. Dluhy [106] used Abeles's formalism for computer simulations of external reflection spectra for monolayers adsorbed at the air–water (AW) interface. Below we give the explicit formulas of Hansen [99, 100] and Gruzinov and Tolstoy [107, 108], which are applicable to an isotropic film of an arbitrary thickness for both external and internal reflection at all angles of incidence. For ultrathin isotropic films ($d \leq 10$–20 nm), a quantitative analysis can also be performed within the thin-film approximation of McIntire and Aspnes [101, 109, 110] and through the use of the formulas derived by Teschner and Hubner [111] for the ATR geometry.

Anisotropy in the optical properties of a layer complicates the analytical expressions for reflectance since the complex dielectric function and the refractive index of the layer become tensors (1.1.2°). Determination of a film's anisotropy from its spectrum provides a wealth of information about the structure and molecular orientation in ultrathin films and therefore is of great importance in various areas of science and technology (Section 3.11). The theoretical approaches of Schopper [112] and Kuzmin et al. [113] (see the review in Ref. [114]) are applicable for calculation of the Fresnel amplitude reflection coefficients for external reflection from an absorbing, anisotropic, uniaxial film on an absorbing, isotropic, semi-infinite substrate in a transparent, isotropic, semi-infinite immersion medium, as is the case for a film at the AW interface. Although these models have different starting points, their results for monolayers differ by less than 0.25% [114]. Here, we give the thin-film approximation formulas of Yamamoto and Ishida [115], valid for both external and internal reflection, and formulas of Chabal [1] derived after Dignam and Moskovits [116], taking into account the anisotropy of the substrate.

1.5.1°. If an isotropic layer with a thickness d_2 is located at the planar interface of two semi-infinite media (Fig. 1.12), the incident wave gives rise to reflected and refracted waves in all the media except for the output half-space, where only the refracted wave exists. For such an optical configuration, the Fresnel coefficients (1.4.5°) can be rewritten in the Drude (exact) form [9] as

$$r_{123}^s = \frac{E_r^s(0)}{E_{in}^s(0)} = \frac{r_{12}^s + r_{23}^s e^{-2i\beta}}{1 + r_{12}^s r_{23}^s e^{-2i\beta}},$$

$$r_{123}^p = \frac{E_r^p(0)}{E_{in}^p(0)} = \frac{r_{12}^p + r_{23}^p e^{-2i\beta}}{1 + r_{12}^p r_{23}^p e^{-2i\beta}},$$

(1.71)

where

$$\beta \equiv \frac{2\pi d_2 \xi_2}{\lambda} = 2\pi \left(\frac{d_2}{\lambda}\right) \sqrt{\hat{\varepsilon}_2 - \hat{\varepsilon}_1 \sin^2 \varphi_1}$$

(1.72)

is the phase shift of the electromagnetic wave after one pass through the film, λ is the wavelength of light in vacuum, ξ_2 is the constant defined by Eq. (1.69) via the dielectric functions of the layer $\hat{\varepsilon}_2$ and the immersion medium $\hat{\varepsilon}_1$, $E_{in}^{s,p}(0)$ is the amplitude of the electric field of the incident wave in medium 1 at the

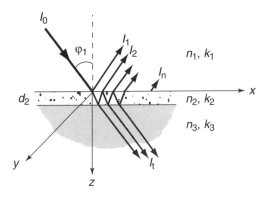

Figure 1.12. Scheme of beam propagation through stratified-layer system consisting of three phases. Angle of incidence of radiation is φ_1. Frequency-dependent optical constants of three media: $n_1, n_2, n_3, k_1, k_2, k_3$.

interface with the layer, and $E_r^{s,p}(0)$ is the amplitude of the electric field of the wave reflected into medium 1 at the interface with the layer (Fig. 1.12). Subscripts 1, 2, and 3 correspond to the ambient (immersion) medium, the layer, and the substrate, respectively. The formulas (1.71) remain the same for different conditions, including the cases of total internal reflection (1.4.11°) and strong absorption by phases 2 or 3.

1.5.2°. The exact expression for the reflectance of the three-media system, R_{123}, is obtained by multiplying Eqs. (1.71) by their complex conjugates [99]:

$$R_{123}^{s,p} = \frac{R_{12} + R_{23}e^{4\,\mathrm{Im}\,\beta} + R_{12}^{1/2}R_{23}^{1/2}e^{2\,\mathrm{Im}\,\beta}2\cos(\delta_{23}^r - \delta_{12}^r - 2\,\mathrm{Re}\,\beta)}{1 + R_{12}R_{23}e^{4\,\mathrm{Im}\,\beta} + R_{12}^{1/2}R_{23}^{1/2}e^{2\,\mathrm{Im}\,\beta}2\cos(\delta_{23}^r + \delta_{12}^r - 2\,\mathrm{Re}\,\beta)}, \quad (1.73)$$

where $R_{i,i+1}$ are the reflectances for the interface between the ith and $(i+1)$th medium in the three-phase system, computed by using Eq. (1.70); after substitutions $1 \to j$ and $2 \to (j+1)$ and assuming the layer is semi-infinite, β is given by Eq. (1.72) and

$$\delta^r = \arg r = \arctan\left[\frac{\mathrm{Im}(r)}{\mathrm{Re}(r)}\right] \quad (1.74)$$

represents the absolute phase change upon reflection for both s- and p-polarization. In Eq. (1.74) Im and Re indicate the imaginary and real parts of the expression, respectively.

Another form of the explicit expression for R_{123} was suggested in Refs. [107, 109],

$$R_{123}^{s,p} = \left|\frac{(q_1 - q_2)(q_2 + q_3)\exp(4\pi\chi_2 d_2/\lambda) + (q_1 + q_2)(q_2 - q_3)}{(q_1 + q_2)(q_2 + q_3)\exp(4\pi\chi_2 d_2/\lambda) + (q_1 - q_2)(q_2 - q_3)}\right|^2, \quad (1.75)$$

where $\chi_i = \sqrt{\varepsilon_1 \sin^2\varphi_1 - \hat{\varepsilon}_i}$, $\hat{\varepsilon}_i = n_i^2 - k_i^2 + 2in_ik_i^2$, and $q_i = \chi_i$ for s-polarization and $q_i = \hat{\varepsilon}_i/\chi_i$ for p-polarization.

1.5.3°. Since the reflectance for a film-free interface, which is described by Eqs. (1.64) and (1.68), is identical to the value of $R_{13}^{s,p}$, calculated by using Eqs. (1.73) or (1.75) with $d_2 = 0$, we can write

$$R_{13}^{s,p} = R^{s,p}(0), \qquad R_{123}^{s,p} = R^{s,p}(d_2). \quad (1.76)$$

In spectroscopic practice, as a rule, one measures not the absolute reflectances of a bare and coated surface, but rather the ratio $R^{s,p}(d_2)/R^{s,p}(0)$. However, the exact analytical expressions for this ratio are cumbersome and do not allow one to discern how the optical properties of the film, the ambient media, and the substrate affect the reflectivity. Convenient expressions for reflectances can be obtained within the framework of the thin-film approximation (1.5.5°) in terms of the so-called normalized reflectivity, or, simply, reflectivity, defined as

$$\frac{\Delta R}{R} = \frac{R_0 - R}{R_0} = 1 - \frac{R}{R_0}, \qquad (1.77)$$

where $R_0 = R(0)$ and $R = R(d_2)$ are the reflectances of the substrate alone and the substrate with the layer, respectively. If the reflectance R_0 of the substrate is close to unity (as in the case of metals), then $\Delta R/R \approx 1 - R$. The change in the reflectance of the interface due to the presence of a film, ΔR, is called, rather confusing, the absorption depth. The physical meaning of this quantity is clear from Fig. 1.13 (for more detail see 1.5.4°). Since the SNR is proportional to the

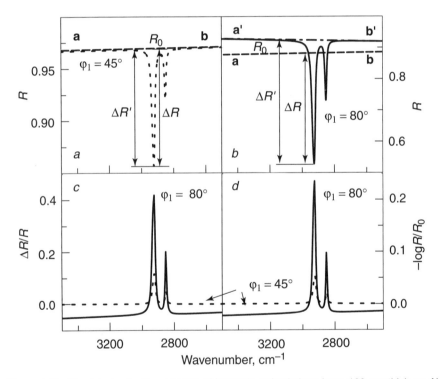

Figure 1.13. Simulated reflection spectra of isotropic polyethylene layer 100-nm thick on Al substrate in range of stretching vibrations of CH groups for angles of incidence $\varphi_1 = 45°$ (short-dashed line) and $\varphi_1 = 80°$ (solid line) and p-polarization, represented in different units: (a, b) reflectance R; (c) reflectivity $\Delta R/R$; (d) absorbance $A = -\log(R/R_0)$. Here, R_0 (dashed line) — reflectance of Al support without layer; R — reflectance of Al support with polyethylene layer; $\Delta R = R_0 - R$ — difference in reflectance of Al support without and with polyethylene layer (absorption depth); $\Delta R' = R'_0 - R$ — difference in reflectance of Al support with polyethylene layer at frequencies for which $k_2 \approx 0$ and $k_2 = $ max (band intensity). Also shown are baselines in measurements of reflectivity and absorption factor **ab** and **a'b'**, respectively. Optical constants of organic film and Al from Refs. [74] and [25], respectively.

absorption depth in a detector noise-limited spectrometer [117], the simulated spectra represented in this unit allow distinguishing the optimum conditions for their measurements (Chapter 2).

1.5.4°.   Instead of the reflectivity introduced by Eq. (1.77), Greenler [118] suggested characterizing the bands in reflection spectra using an *absorption factor* (or a *spectral sensitivity*), A_G, defined as

$$A_G \equiv \frac{R(d_2)_{k_2=0} - R(d_2)_{k_2>0}}{R(d_2)_{k_2=0}} = \frac{R(d_2)_{\varepsilon_\infty} - R(d_2)_{\varepsilon_2}}{R(d_2)_{\varepsilon_\infty}}.$$

Here, $R(d_2)_{k_2=0}$ is the reflectance of the film-covered substrate of the frequency at which the film material is transparent and $\Delta R' = R(d_2)_{k_2=0} - R(d_2)_{k_2>0}$ is the band intensity (Fig. 1.13*b*), $R(d_2)_{\varepsilon_2}$ denotes the reflectance of a thin film of the dielectric constant $\hat{\varepsilon}_2$, whereas $R(d_2)_{\varepsilon_\infty}$ denotes the reflectance of the so-called electromagnetically bleached film [119, 120], which is the same thin film but at the frequency where the film is transparent, that is, $\hat{\varepsilon}_2 = \varepsilon_\infty$ (1.3.6°). The band intensity unit permits quantifying the spectra of layers without preliminary recording of the substrate spectrum.

Notice that at $\varphi_1 = 80°$ (Fig. 1.13*b*) the reflectance values of the uncoated Al substrate, R_0, (**ab** baseline) appear to be significantly lower than the reflectance values of the **a′b′** baseline drawn through the band wings, which is distinct from the case of $\varphi_1 = 45°$ (Fig. 1.13*a*). This phenomenon, which is due to interference, was reported, for example, for CO adsorbed on glassy carbon [121]. Thus, the quantities A_G and the band intensity are more physically accurate than the reflectivity and the absorption depth, respectively, as they do not include the shift in the baseline due to the light interference in the film.

In practice, the most useful unit for representation of the reflection spectra is the *absorbance*, or *reflectance–absorbance* (RA), defined as

$$A = -\log \frac{R}{R_0}, \tag{1.78}$$

which characterizes the ratio of the reflectances of the surface with and without a thin film (Fig. 1.13*d*). This unit is particularly convenient as it is linearly proportional to the film thickness in a wider thickness range than is the reflectivity (Section 3.3). The quantity R/R_0 can be measured directly in the experiment and converted automatically into the absorbance by modern FTIR spectrometers. It can be demonstrated [99, 101] that the absorbance is proportional to the normalized reflectivity, $A \approx (\Delta R/R_0)/\ln 10$.

1.5.5°.   Since the thickness of an ultrathin film ($\sim 1-100$ nm) is much smaller than the IR wavelength ($\sim 10,000$ nm), $d_2 \ll \lambda$, the so-called *thin-film approximation* can be used in the spectral simulations. Using the approximation that $e^{-2i\beta} \approx$

$1 - 2i\beta$, the Fresnel amplitude coefficients [Eq. (1.71)] can be simplified to:

$$r_{123}^s \approx \frac{r_{12}^s + r_{23}^s(1 - 2i\beta)}{1 + r_{12}^s r_{23}^s(1 - 2i\beta)}, \qquad r_{123}^p \approx \frac{r_{12}^p + r_{23}^p(1 - 2i\beta)}{1 + r_{12}^p r_{23}^p(1 - 2i\beta)}. \qquad (1.79)$$

The use of Eq. (1.79) leads to the expressions derived by McIntyre and Aspnes [101, 109, 110] for a transparent immersion medium. Rewritten in terms of the reflectivity defined by Eq. (1.77) gives

$$\left(\frac{\Delta R}{R}\right)_s \approx \frac{8\pi d_2 n_1 \cos\varphi_1}{\lambda} \operatorname{Im}\left\{\left(\frac{\hat{\varepsilon}_3 - \hat{\varepsilon}_2}{\varepsilon_1 - \hat{\varepsilon}_3}\right)\right\},$$

$$\left(\frac{\Delta R}{R}\right)_p \approx \frac{8\pi d_2 n_1 \cos\varphi_1}{\lambda} \operatorname{Im}\left\{\left(\frac{\hat{\varepsilon}_3 - \hat{\varepsilon}_2}{\varepsilon_1 - \hat{\varepsilon}_3}\right)\left[\frac{1 - (\varepsilon_1/\hat{\varepsilon}_2\hat{\varepsilon}_3)(\hat{\varepsilon}_2 + \hat{\varepsilon}_3)\sin^2\varphi_1}{1 - (1/\hat{\varepsilon}_3)(\varepsilon_1 + \hat{\varepsilon}_3)\sin^2\varphi_1}\right]\right\}$$

$$(1.80)$$

for s- and p-polarized radiation, respectively. Here, φ_1 is the angle of incidence, $\hat{\varepsilon}_i = \varepsilon_i' - i\varepsilon_i''$ is the complex dielectric function of the ith medium, and λ is the wavelength in a vacuum. As seen from Eq. (1.80), the band intensity in the reflectivity spectrum of a thin film is linearly proportional to the film thickness d_2 within the framework of the thin-film approximation.

For the case of a transparent substrate, ($\varepsilon_3'' = 0$) Eq. (1.80) yields

$$\left(\frac{\Delta R}{R}\right)_s \approx -\frac{8\pi d_2 n_1}{\lambda \cos\varphi_1} C_y \operatorname{Im}\hat{\varepsilon}_2,$$

$$\left(\frac{\Delta R}{R}\right)_p \approx -\frac{8\pi d_2 n_1}{\lambda \cos\varphi_1}\left[C_x \operatorname{Im}\hat{\varepsilon}_2 + C_z\varepsilon_1^2 \operatorname{Im}\left(\frac{1}{\hat{\varepsilon}_2}\right)\right],$$

$$(1.81)$$

where

$$C_x = \frac{4\varepsilon_3 \cos^2\varphi_1}{\varepsilon_3 - \varepsilon_1}\frac{(\varepsilon_3/\varepsilon_1)\sin^2\varphi_1 - 1}{[(\varepsilon_3 + \varepsilon_1)/\varepsilon_1]\sin^2\varphi_1 - 1},$$

$$C_y = \frac{4\varepsilon_3 \cos^2\varphi_1}{\varepsilon_3 - \varepsilon_1},$$

$$C_z = \frac{4\varepsilon_3 \cos^2\varphi_1}{\varepsilon_3 - \varepsilon_1}\frac{(\varepsilon_3/\varepsilon_1)\sin^2\varphi_1}{[(\varepsilon_3 + \varepsilon_1)/\varepsilon_1]\sin^2\varphi_1 - 1}.$$

The degree of deviation of the thin-film approximation from the exact spectral simulations depends upon the refractive indices of the immersion medium n_1 and the substrate n_3, the oscillator strength of the film, and on the angle of incidence φ_1. The deviation can be more than 10% for a 50-nm film and several percent for a 1-nm film (Sections 2.2, 2.5, 3.3, and 3.10).

1.5.6°. As shown in Ref. [118], if $\hat{\varepsilon}_1 = 1$ and $\hat{\varepsilon}_2 \ll \hat{\varepsilon}_3$, which is the case for air surroundings and a metal substrate, the following simplification of Eq. (1.80) yields:

$$\left(\frac{\Delta R}{R}\right)_s \approx 1 - R_s \approx -8\pi d_2 \cos \varphi_1 \nu \operatorname{Im}(\hat{\varepsilon}_2),$$

$$\left(\frac{\Delta R}{R}\right)_p \approx 1 - R_p \approx -8\pi d_2 \sin \varphi_1 \tan \varphi_1 \nu \operatorname{Im}\left(\frac{1}{\hat{\varepsilon}_2}\right).$$
(1.82)

It can be seen from Eq. (1.82) that an IR spectrum of a thin film depends essentially on the TO and LO energy loss functions of the film substance (1.1.18°, 1.1.19°) for s- and p-polarization, respectively, the reflectivity values are positive, and the reflection spectrum is like the absorption spectrum. Moreover, the absorption of s-polarized radiation, $1 - R_s$, is greater at small angles of incidence, whereas the quantity $1 - R_p$ exhibits a maximum at grazing angles of incidence. This is not the case for a dielectric substrate [Eqs. (1.80) and (1.81)] (see Sections 2.2 and 2.3 for more detail).

1.5.7°. For ATR at a thin absorbing film whose thickness is far less than the penetration depth [Eqs. (1.110)], the angular condition for total reflection is [99, 122]

$$\varphi_c = \arcsin\left(\frac{n_3}{n_1}\right),$$
(1.83)

where n_1 and n_3 are the refractive indices of the input and output media, respectively. It should be emphasized that condition (1.83) is independent of the optical properties of the ultrathin-film material and corresponds to total reflection at the interface between the film and medium 3 if the film were absent [Eq. (1.65)]. The answers to a number of interesting questions concerning the band shape and intensity in ATR spectra may be estimated using the simple expressions obtained in Ref. [111], neglecting terms of order $\geq (2\pi d_2 / \lambda)^2$ in the series expansions,

$$\left(\frac{\Delta R}{R}\right)_s \approx 1 - R_s \approx -\frac{8\pi d_2 n_1 \cos \varphi_1}{\lambda(\varepsilon_1 - \hat{\varepsilon}_3)} \operatorname{Im} \hat{\varepsilon}_2,$$

$$\left(\frac{\Delta R}{R}\right)_p \approx 1 - R_p \approx -\frac{8\pi d_2 n_1 \cos \varphi_1 \varepsilon_1}{\lambda(A^2 + B^2)} \left(\xi_3^2 \operatorname{Im} \hat{\varepsilon}_2 - \hat{\varepsilon}_3^2 C \operatorname{Im} \frac{1}{\hat{\varepsilon}_2}\right),$$
(1.84)

where $A = \xi_1 \hat{\varepsilon}_3$, $B = \xi_3' \varepsilon_1$, $C = \varepsilon_0 \sin^2 \varphi_1$, $\xi_3' = i\xi_3$, and ξ_i is defined by Eq. (1.69).

1.5.8°. Yamamoto and Ishida [115] have obtained the following expressions for reflectances in the case of an anisotropic film on an isotropic substrate within the Abeles thin-film approximation (1.5.5°):

$$R^s \approx \left|\frac{n_1 \cos \varphi_1 - n_3 \cos \varphi_3}{n_1 \cos \varphi_1 + n_3 \cos \varphi_3}\right|^2 \left(1 - \frac{8\pi d_2 n_1 \cos \varphi_1}{(n_1 - n_3)(n_1 + n_3)} \nu \operatorname{Im}(\hat{\varepsilon}_{2y})\right),$$
(1.85)

$$R^p \approx \left| \frac{n_3 \cos \varphi_1 - n_1 \cos \varphi_3}{n_3 \cos \varphi_1 + n_1 \cos \varphi_3} \right|^2 \times \left| 1 - \frac{8\pi d_2 n_1 \cos \varphi_1}{(n_3 \cos \varphi_1 - n_1 \sin \varphi_1)^2 (n_1 - n_3)^2} \right.$$

$$\left. \times \left[(n_3 - n_1 \sin \varphi_1)(n_3 + n_1 \sin \varphi_1) v \operatorname{Im}(\hat{\varepsilon}_{2x}) - n_1^2 n_3^4 \sin^2 \varphi_1 v \operatorname{Im}\left(\frac{-1}{\hat{\varepsilon}_{2z}} \right) \right] \right|,$$

$$(1.86)$$

where $\hat{\varepsilon}_{2j} = \varepsilon'_{2j} + i\varepsilon''_{2j}$ is the principal component of the diagonal permittivity tensor of the film and j represents the coordinate axes x, y, or z, directed as shown in Fig. 1.12. Notice that Eqs. (1.85) and (1.86) are valid for both external and internal reflection (1.4.1°).

For the external reflection in a three-phase system, air–ultrathin anisotropic film–substrate, the components of absorbance $A_{s(y)}$, $A_{p(x)}$, and $A_{p(z)}$ can be calculated by Mielczarsky's approximate formulas, which are frequently used in practice [123, 124] due to their simple form:

$$A_{s(y)} \approx -\frac{16\pi}{\ln 10} \left[\frac{\cos \varphi_1}{n_3^2 - 1} \right] \frac{n_2 k_2 d_2}{\lambda},$$

$$A_{p(x)} \approx -\frac{16\pi}{\ln 10} \left[\frac{\cos \varphi_1}{\xi_3^2 / n_3^4 - \cos^2 \varphi_1} \right] \left[-\frac{\xi_3^2}{n_3^4} \right] \frac{n_2 k_2 d_2}{\lambda}, \qquad (1.87)$$

$$A_{p(z)} \approx -\frac{16\pi}{\ln 10} \left[\frac{\cos \varphi_1}{\xi_3^2 / n_3^4 - \cos^2 \varphi_1} \right] \frac{\sin^2 \varphi_1}{(n_2^2 + k_2^2)^2} \frac{n_2 k_2 d_2}{\lambda},$$

where ξ_3 is the generalized complex refractive index [Eq. (1.69)] of the substrate and n_2 and k_2 are, respectively, the refractive and absorption indices of the film. These equations were derived from the exact ones [100] after expanding them in terms of αd_2 and taking into account anisotropy of the film.

1.5.9°. If both the layer and the substrate are anisotropic (e.g., in the case of oriented molecules adsorbed onto an anisotropic crystal), Eqs. (1.80) of McIntyre and Aspnes take the form [1]

$$\left(\frac{\Delta R}{R} \right)_s \approx \frac{8\pi d_2 n_1 \cos \varphi_1}{\lambda} \operatorname{Im} \left[\frac{\hat{\varepsilon}_{3y} - \hat{\varepsilon}_{2y}}{\varepsilon_1 - \hat{\varepsilon}_{3y}} \right],$$

$$\left(\frac{\Delta R}{R} \right)_p \approx \frac{8\pi n_1 \cos \varphi_1}{\lambda}$$

$$\times \operatorname{Im} \left[\frac{\left[(\hat{\varepsilon}_{3x} / \hat{\varepsilon}_{2z}) d_{2z} - (\hat{\varepsilon}_{2x} / \hat{\varepsilon}_{3z}) d_{2x} \right] \varepsilon_1 \sin^2 \varphi_1 - (\hat{\varepsilon}_{3x} - \hat{\varepsilon}_{2x}) d_{2x}}{(\hat{\varepsilon}_{3x} - \varepsilon_1) - \left[(\hat{\varepsilon}_{3x} / \varepsilon_1) - (\varepsilon_1 / \hat{\varepsilon}_{3z}) \right] \varepsilon_1 \sin^2 \varphi_1} \right],$$

$$(1.88)$$

where $\hat{\varepsilon}_{2i}$ and $\hat{\varepsilon}_{3i} (i = x, y, z)$ are the principal values of the dielectric function tensors of the film and the substrate, respectively; d_{2i} is the formal parameter (the "effective" thickness), introduced by

$$(\hat{\varepsilon}_i - 1)d_{2i} = (N_s)_i \hat{\alpha}_i, \qquad i = x, y, z. \tag{1.89}$$

Here, $\hat{\alpha}_i$ is the principal value of the tensor of the generalized polarizability of one adsorbed molecule [Eq. (1.35)] and $(N_s)_i$ is the number of the adsorbed molecules per unit area that contribute to the ith component of the polarizability. The quantity on the right of Eq. (1.89) is called *surface susceptibility*.

1.5.10°. Chabal [1] has demonstrated how Eq. (1.88) can be simplified for the two cases of interest: the vibrational spectrum of an adsorbed monolayer (weak absorber) at an isotropic metallic substrate and the electronic spectrum of a surface state excitation (strong absorber).

In the case of an adsorbed anisotropic monolayer at an isotropic metallic substrate, the dependence of reflectivity [Eq. (1.77)] on the optical properties of the monolayer, substrate, and ambient medium and on the angle of incidence is given [1] by

$$\left(\frac{\Delta R}{R}\right)_s \approx \frac{8\pi d_{2y} n_1 \cos\varphi_1}{\lambda} \operatorname{Im}\left[\frac{\hat{\varepsilon}_{2y} - \hat{\varepsilon}_3}{\hat{\varepsilon}_3}\right],$$

$$\left(\frac{\Delta R}{R}\right)_p \approx \frac{8\pi n_1 \varepsilon_1 d_{2z}}{\lambda} \frac{\sin^2\varphi_1}{\cos\varphi_1} \operatorname{Im}\left[\frac{1/\hat{\varepsilon}_{2z}}{1 - (\varepsilon_1/\hat{\varepsilon}_3)\tan^2\varphi_1}\right]. \tag{1.90}$$

If the values of the dielectric function of an anisotropic thin film at an isotropic metal surface are large (as in the case of the electronic absorption of the *surface states*), Eqs. (1.87) can be rewritten as

$$\left(\frac{\Delta R}{R}\right)_s \approx \frac{8\pi d_{2y} n_1 \cos\varphi_1}{\lambda} \operatorname{Im}\left[\frac{-\hat{\varepsilon}_{2y}}{\hat{\varepsilon}_3}\right],$$

$$\left(\frac{\Delta R}{R}\right)_p \approx \frac{8\pi n_1 d_{2z}}{\lambda} \frac{1}{\cos\varphi_1} \operatorname{Im}\left[\frac{\hat{\varepsilon}_{2x}}{\hat{\varepsilon}_3}\right]. \tag{1.91}$$

The absence of the quantities $\hat{\varepsilon}_{2z}$ and d_{2z} in Eq. (1.91) means [1] that the radiation is so strongly refracted within the substrate that it only probes the tangential component of the dielectric function.

1.6. TRANSMISSION OF LAYER LOCATED AT INTERFACE BETWEEN TWO ISOTROPIC SEMI-INFINITE MEDIA

In this section, we list the exact formulas of Hansen [99] and Abeles [125] for isotropic films and the thin-film approximation of Yamamoto and Ishida [126] for the calculation of transmission spectra of anisotropic ultrathin films on planar isotropic supports. The spectral features and dependences predicted by these formulas are discussed in Sections 2.1 and 3.3.4.

1.6.1°. If an isotropic layer with a thickness d_2 is located at the interface between two isotropic semi-infinite media (Fig. 1.12), then the Fresnel coefficients (1.4.5°) for the transmitted wave are [63, 64]

$$t_{123}^s = \frac{E_t^s(0)}{E_{in}^s(0)} = \frac{t_{12}^s t_{23}^s e^{-i\beta}}{1 + r_{12}^s r_{23}^s e^{-2i\beta}},$$
$$t_{123}^p = \frac{E_t^p(0)}{E_{in}^p(0)} = \frac{t_{12}^p r_{23}^p e^{-i\beta}}{1 + r_{12}^p r_{23}^p e^{-2i\beta}},$$
(1.92)

where $t_{ij}^{s,p}$ and $r_{ij}^{s,p}$ are the two-phase Fresnel coefficients (1.68) for transmission and reflection, respectively, $E_t^{s,p}$ is the amplitude of the electric field of the wave transmitted through the layer into medium 3 at the interface with the layer, and the other coefficients are the same as in Eq. (1.71).

1.6.2°. Transmittance of radiation through a boundary with one isotropic layer can be expressed as [125]

$$T_{123}^{p,s} = \frac{16 n_1 n_3 (n_2^2 + k_2^2)}{bde^{2k_2\rho} + ace^{-2k_2\rho} + \cos n_2\rho + 2v \sin n_2\rho}.$$
(1.93)

Here, the coefficients are

$$a, b = (n_2 \pm n_1)^2 + k_2^2, \qquad c, d = (n_2 \pm n_3)^2 + (k_2 \pm k_3)^2, \qquad \rho = \frac{4\pi d}{\lambda},$$

$$v = 2k_2(n_3 + n_1)(n_2^2 + k_2^2 - n_1 n_3) - 2k_3[k_2 k_3 - 4(n_1^2 - n_2^2 - k_2^2)],$$

$$t = (n_1^2 + n_3^2 + k_3^2)(n_2^2 + k_2^2) - n_1^2(n_1^2 + k_3^2) + 4n_1 k_2(k_2 n_3 - n_1 k_3),$$

where subscripts 1, 2, and 3 refer to the external medium, layer, and substrate, respectively. At nonzero angles of incidence ($\varphi_1 \neq 0$), the values of n and k in formula (1.93) for s-polarized radiation should be replaced by p and q, respectively, as determined from the set of equations

$$p^2 - q^2 = n^2 - k^2 - n_1 \sin^2 \varphi_1, \qquad pq = nk.$$

For p-polarization, p' and q' should be used, where

$$p' = p\left[1 + \left(\frac{n_1^2 \sin^2 \varphi_1}{p^2 + q^2}\right)\right],$$

$$q' = q\left[1 - \left(\frac{n_1^2 \sin^2 \varphi_1}{p^2 + q^2}\right)\right].$$

Another form of the exact expression for $T_{123}^{s,p}$ was derived by Hansen [99]:

$$T_{123}^{s,p} = Q \frac{|t_{12}|^2 |t_{23}|^2 e^{2 \operatorname{Im} \beta}}{1 + R_{12} R_{23} e^{4 \operatorname{Im} \beta} + R_{12}^{1/2} R_{23}^{1/2} e^{2 \operatorname{Im} \beta} 2 \cos(\delta_{23}^r + \delta_{12}^r - 2 \operatorname{Re} \beta)}, \quad (1.94)$$

where

$$Q = \begin{cases} \operatorname{Re}\left(\dfrac{\xi_3}{\xi_1}\right) & \text{for } s\text{-polarization,} \\[2ex] \dfrac{(\hat{n}_3/n_1)^2 \operatorname{Re}(\xi_3/\hat{n}_3^2)}{(\xi_1/n_1^2)} & \text{for } p\text{-polarization;} \end{cases}$$

β and $\delta_{ij}^{s,r}$ are defined by Eqs. (1.72) and (1.74), respectively.

1.6.3°. Experimentally, transmittances of the bare T_0 and covered T surfaces are measured, and the absorbance

$$A \equiv -\log \frac{T}{T_0} \quad (1.95)$$

is calculated from these. This quantity is proportional to the film thickness within certain thickness limits and to the energy losses in the film (Section 3.3.4). *Transmissivity*, or transmission factor, is defined similar to reflectivity (1.77) as

$$\frac{\Delta T}{T} = \frac{T_0 - T}{T_0} = 1 - \frac{T}{T_0}. \quad (1.96)$$

If $T_0 \approx 1$, $\Delta T/T \approx 1 - T$. The physical meaning of the transmissivity is analogous to that of the reflectivity (Fig. 1.13). When transmission spectra are represented in absorbance units they are also referred to as absorption spectra.

1.6.4°. Applying the thin-film approximation (1.5.5°) to the transmissivity leads to the expressions [122]

$$\begin{aligned} T_{123}^s &\approx \frac{4\xi_1\xi_3}{B^2}\left(1 + \frac{4\pi d_2 \operatorname{Im}(\hat{\varepsilon}_2)}{\lambda}\right), \\ T_{123}^p &\approx \frac{4\xi_1\xi_3\varepsilon_1\varepsilon_3}{A^2}\left(1 + \frac{4\pi d_2[\xi_1\xi_3 \operatorname{Im}(\hat{\varepsilon}_2) + \varepsilon_1\varepsilon_3 C \operatorname{Im}(1/\hat{\varepsilon}_2)]}{\lambda A}\right), \end{aligned} \quad (1.97)$$

where $A = \xi_3\varepsilon_1 + \xi_1\varepsilon_3$, $B = \xi_1 + \xi_3$, and $C = \varepsilon_1 \sin^2 \varphi_1$. Further simplification for an ultrathin standing film in a transparent medium of the refractive index n_1 yields

$$\begin{aligned} \left(\frac{\Delta T}{T}\right)_s &\approx 1 - T_{123}^s \approx -\frac{4\pi d_2}{\lambda n_1 \cos \varphi_1} \operatorname{Im}(\hat{\varepsilon}_2), \\ \left(\frac{\Delta T}{T}\right)_p &\approx 1 - T_{123}^p \approx -\frac{4\pi d_2}{\lambda}\left[n_1^{-1} \operatorname{Im}(\hat{\varepsilon}_2) \cos \varphi_1 + \operatorname{Im}\left(\frac{1}{\hat{\varepsilon}_2}\right) \frac{n_1^3 \sin^2 \varphi_1}{\cos \varphi_1}\right]. \end{aligned}$$

$$(1.98)$$

It is noteworthy that for inorganic films the linear approximation (1.98) is valid only in the extremely narrow range of the layer thicknesses [126] (see discussion of this phenomenon in Section 3.3.4). To apply Eqs. (1.97) and (1.98) to anisotropic films, one should substitute $\hat{\varepsilon}_2$ by $\hat{\varepsilon}_{2y}$ in the formulas for s-component and by $\hat{\varepsilon}_{2x}$ and $\hat{\varepsilon}_{2z}$ in the first and second terms of the formulas for p-component, respectively.

The transmission spectral sensitivity is defined similarly to the case of reflection (1.5.4°):

$$C(\varphi_1) \equiv \frac{T_{\varepsilon_\infty}(\varphi_1) - T_{\varepsilon_2}(\varphi_1)}{T_{\varepsilon_\infty}(\varphi_1)}, \tag{1.99}$$

where $T_{\varepsilon_2}(\varphi_1)$ denotes the transmittance of a thin film of the dielectric constant $\hat{\varepsilon}_2$, whereas $T_{\varepsilon_\infty}(\varphi_1)$ denotes the transmittance of the same thin film at the angle of incidence φ_1 and polarization but at the frequency where $\hat{\varepsilon}_2 = \varepsilon_\infty$ (1.3.6°); then the sensitivity enhancement factor over a normal incidence measurement can be expressed from Eq. (1.98) for p-polarization as [127]

$$\frac{C(\varphi_1)}{C(0°)} \approx \frac{1}{\cos\varphi_1} \left[\left(\frac{\varepsilon_1}{|\hat{\varepsilon}_2|} \right)^2 \sin^2\varphi_1 + \cos^2\varphi_1 \right]. \tag{1.100}$$

Here, the geometric factor $1/\cos\varphi_1$ is due to the enlargement of the volume exposed to the beam. It is seen that a spectral contrast increases with increasing angle of incidence and the quantity $\varepsilon_1/|\hat{\varepsilon}_2|$. It means that the maximal spectrum enhancement will be achieved at the grazing angles of incidence, for a film of a low dielectric function, and using the immersion media (see Section 2.1 for details).

1.6.5°. Let us compare the thin-film approximation formulas for (1) the transmissivity (1.98); (2) the reflectivities for the external reflection from this film deposited onto a metallic substrate (1.82); (3) the internal reflection at $\varphi_1 \geq \varphi_c$ (1.84); and (4) the external reflection from this film deposited on a transparent substrate (dielectric or semiconducting) (1.81) (Table 1.2). In all cases s-polarized radiation is absorbed at the frequencies of the maxima of $\text{Im}(\hat{\varepsilon}_2)$, $\nu_{\text{TO}i}$ (1.1.18°), whereas the p-polarized external reflection spectrum of a layer on a metallic substrate is influenced only by the LO energy loss function $\text{Im}(1/\hat{\varepsilon}_2)$ (1.1.19°). The p-polarized internal and external reflection spectra of a layer on a transparent substrate has maxima at ν_{TO} as well as at ν_{LO}. Such a polarization-dependent behavior of an IR spectrum of a thin film is manifestation of the optical effect (Section 3.1).

1.6.6°. A very simple linear approximation for the transmittance of an anisotropic thin film deposited on a transparent isotropic substrate has been reported by Buffeteau et al. [128]:

$$T_p \approx T_p^{\text{sub}} \left\{ 1 - \frac{4\pi d_2 \nu}{\varepsilon_3 \cos\varphi_1 + (\varepsilon_3 - \sin^2\varphi_1)^{1/2}} \right.$$

$$\left. \times \left[\varepsilon_3 \sin^2\varphi_1 \, \text{Im}\left(-\frac{1}{\hat{\varepsilon}_{2z}} \right) + (\varepsilon_3 - \sin^2\varphi_1)^{1/2} \cos\varphi_1 \, \text{Im}(\hat{\varepsilon}_{2x}) \right] \right\}, \tag{1.101}$$

Table 1.2. Thin-film approximation formulas for reflectivity (IRRAS and ATR modes) and transmissivity of ultrathin film

External reflection, metallic substrate	Equation (1.82):
	$$\left(\frac{\Delta R}{R}\right)_s \approx -8\pi d_2 \cos\varphi_1 \nu \, \text{Im}(\hat{\varepsilon}_2)$$ $$\left(\frac{\Delta R}{R}\right)_p \approx -8\pi d_2 \sin\varphi_1 \tan\varphi_1 \nu \, \text{Im}\left(\frac{1}{\hat{\varepsilon}_2}\right)$$
External reflection, (semi)transparent substrate	Equation (1.81)
	$$\left(\frac{\Delta R}{R}\right)_s \approx -\frac{8\pi d_2 n_1}{\lambda \cos\varphi_1} C_y \, \text{Im}\,\hat{\varepsilon}_2$$ $$\left(\frac{\Delta R}{R}\right)_p \approx -\frac{8\pi d_2 n_1}{\lambda \cos\varphi_1}\left[C_x \, \text{Im}\,\hat{\varepsilon}_2 + C_z \varepsilon_1^2 \, \text{Im}\left(\frac{1}{\hat{\varepsilon}_2}\right)\right]$$
Internal reflection (ATR)	Equation (1.84)
	$$\left(\frac{\Delta R}{R}\right)_s \approx -\frac{8\pi d_2 n_1 \cos\varphi_1}{\lambda(\varepsilon_1 - \hat{\varepsilon}_3)} \, \text{Im}\,\hat{\varepsilon}_2$$ $$\left(\frac{\Delta R}{R}\right)_p \approx -\frac{8\pi d_2 n_1 \cos\varphi_1 \varepsilon_1}{\lambda(A^2 + B^2)}\left[\xi_3^2 \, \text{Im}\,\hat{\varepsilon}_2 - \hat{\varepsilon}_3^2 C \, \text{Im}\left(\frac{1}{\hat{\varepsilon}_2}\right)\right]$$
Transmission, standing film in air	Equation (1.98)
	$$\left(\frac{\Delta T}{T}\right)_s \approx -\frac{4\pi d_2}{\lambda \cos\varphi_1} \, \text{Im}(\hat{\varepsilon}_2)$$ $$\left(\frac{\Delta T}{T}\right)_p \approx -\frac{4\pi d_2}{\lambda}\left[\text{Im}(\hat{\varepsilon}_2)\cos\varphi_1 + \frac{\sin^2\varphi_1}{\cos\varphi_1} \, \text{Im}\left(\frac{1}{\hat{\varepsilon}_2}\right)\right]$$

where $\hat{\varepsilon}_{2x}$ and $\hat{\varepsilon}_{2z}$ are the permittivities of the layer along the x and z axes indicated in Fig. 1.14 and T_p^{sub} represents the transmittance of the bare substrate. As seen from Eq. (1.101), two types of bands can appear in the transmission spectrum at a nonzero angle of incidence. The frequencies and intensities of these bands correspond to the positions and maxima of the $\text{Im}(\hat{\varepsilon}_{2x})$ and $\text{Im}(1/\hat{\varepsilon}_{2z})$ functions. At the same time, the normal-incidence spectrum exhibits the only band that is described by $\text{Im}(\hat{\varepsilon}_{2x})$. This feature is used to determine the principal values of the permittivity tensor of anisotropic films (Section 3.11.3).

1.7. SYSTEM OF PLANE–PARALLEL LAYERS: MATRIX METHOD

To simulate the IR spectra of stratified media containing an arbitrary number of layers, there exist two approaches — the application of recursion relationships

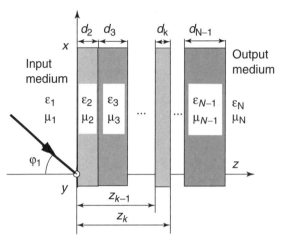

Figure 1.14. Scheme of stratified medium containing N phases and $N - 2$ layers. Shown are coordinates and nomenclature used throughout.

and the matrix method — both requiring computer programming. In the limiting cases, these approaches permit one to study wave propagation phenomena for a transition (inhomogeneous in depth) layer by "slicing" it into a large enough number of infinitely thin layers and, for a single layer, by equating this number to unity. Abeles [129, 130] was the first to suggest the use of 2×2 matrix transformations to simplify the calculations of the optical response in the case of isotropic layers. Abeles's formalism was developed [9, 52, 87, 131, 132] into a more convenient form for computation by introducing the Fresnel coefficients. Harbecke [133] and Ohta and Ishida [134] extended Abeles's approach to a system with phase incoherence, based on the concept of incoherent multiple reflections in real systems, in which there are neither plane nor parallel surfaces nor a monochromatic light source or some layers are sufficiently thick so that the period of interference is smaller than the resolution of the spectrometer.

The mathematical treatment of light propagation through an anisotropic stratified medium is not simple. Exact solutions can be obtained by the 4×4 matrix method [14, 135–139] generalizing Abeles's approach to anisotropic layers. The advantage of the Berreman 4×4 formalism [135] is in its applicability to materials with magnetic anisotropy and optically active materials. For the determination of orientation within the media, Parikh and Allara [138] found Yeh's treatment of the 4×4 matrix method [14] to be most flexible, extending it to a pile of absorbing films, each with any degree of anisotropy up to and including biaxial symmetry. However, according to Cojocaru [140], for practical purposes, the so-called 2×2 extended Jones matrix method [131] is adequate and easier to use. Hasegawa et al. [141] demonstrated that Hansen's 2×2 matrix method can be easily generalized for the case of anisotropy by introducing Drude's formulas developed originally for a two-anisotropic-phase system. They presented the matrix method for uniaxial media and tested it in the studies of the molecular orientation (MO) in the multilayered Langmuir–Blodgett films with uniaxial symmetry. Yamamoto and Ishida [115, 142] extended Hansen's method to the case of biaxial anisotropy. Based on the matrix method of Ohta and Ishida [134] and the generalization of Yamamoto and Ishida [115, 142], Buffeteau et al. [143] suggested the computation procedure applicable for an arbitrary succession of coherent anisotropic and incoherent layers.

In this section, we shall give the recursion relationships derived by Popov [144] and Leng et al. [145] and the explicit formulas and algorithm for programming the spectral simulations according to the Hansen matrix method [100], applicable to isotropic thin films. For the most part, the spectral simulations in the present handbook were carried out by using the program based on this algorithm. In Chapters 2 and 3 we will show that, using such a program or recursion relationship, any spectroscopist can simulate the IR spectra for the layered system of interest. The simulated spectra provide extensive information, assisting in, for example, the determination of the optimum conditions for recording the spectrum (Chapter 2) and the interpretation of the spectrum (Chapter 3).

1.7.1°. Let there be a general system of N isotropic layers with different optical properties (described by the complex refractive indices $\hat{n}_j$) and of arbitrary thicknesses d_j (Fig. 1.14). The recursion relationship of the corresponding reflectance that is conducive to computer programming has been obtained in [144]. In the case $\varphi_1 = 0$,

$$R^{s,p} = \left| h_1 + \cfrac{h_2^2}{-h_1 + \cfrac{1}{h_3 + \cfrac{h_4^2}{-h_3 + \cfrac{1}{h_5 \cdots + \cfrac{h_{2j-2}^2}{-h_{2j-3} + 1/h_{2j-1}}}}}} \right|^2 \tag{1.102}$$

with the intermediate variables defined as

$$h_{2j} = \frac{(1 - r_j^2)e^{-ik\hat{n}_j d_j}}{1 - r_j^2 e^{-2ik\hat{n}_j d_j}},$$

$$h_{2j-1} = r_j \frac{1 - e^{-2ik\hat{n}_j d_j}}{1 - r_j^2 e^{-2ik\hat{n}_j d_j}}, \tag{1.103}$$

$$r_j = -\frac{\hat{n}_j - 1}{\hat{n}_j + 1}, \tag{1.104}$$

where $k = 2\pi/\lambda$ and λ is the wavelength in a vacuum. For $\varphi_1 \neq 0$ and p-polarization, $\hat{n}_j$ is replaced by $\hat{n}_j \cos\varphi_j$ in Eq. (1.103) and by $|\hat{n}_j| \cos\varphi_j$ in Eq. (1.104), whereas for s-polarization, $\hat{n}_j$ is replaced by $\hat{n}_j \cos\varphi_j$ in Eq. (1.103) and by $(\cos \varphi_j)/\hat{n}_j$ in Eq. (1.104). The complex refractive index $\hat{n}_j$ and the complex angle of incidence φ_j follow Snell's law (1.56). Equation 1.102 is seldom used for spectral simulations, perhaps being as of yet relatively unknown.

For analysis of six-layer silicon on oxide films, Leng et al. [145] applied the following elegant regression formula:

$$R_n^{s,p} = \left(\frac{r_n^{s,p} + R_{n+1}^{s,p}}{1 + r_n^{s,p} R_{n+1}^{s,p}} \right) e^{2ik_{n,z}d_n}, \qquad (1.105)$$

where d_n is the thickness of the nth layer, $R_n^{s,p}$ is the reflectance of the s- or p-polarized radiation from layer n, and $r_n^{s,p}$ is the Fresnel coefficients for the interface between layers n and $n + 1$, calculated as

$$r_n^s = \frac{\mu_{n+1}k_{n,z} - \mu_n k_{n+1,z}}{\mu_{n+1}k_{n,z} + \mu_n k_{n+1,z}}, \qquad r_n^p = \frac{\hat{\varepsilon}_{n+1}k_{n,z} - \hat{\varepsilon}_n k_{n+1,z}}{\hat{\varepsilon}_{n+1}k_{n,z} + \hat{\varepsilon}_n k_{n+1,z}},$$

where $\hat{\varepsilon}_n$ and μ_n are the permittivity and permeability of the material in layer n, respectively, and $k_{n,z} = (2\pi/\lambda)(n_n + ik_n) \cos \varphi_n$ (φ_n is the angle of propagation in the nth layer). The Fresnel coefficients for each layer are calculated from the layer below.

1.7.2°. The theoretical background of the matrix method is covered in a great body of literature (see, e.g., Refs. [9, 14, 52, 87, 88, 131, 142]). The basic concept involves the construction of a characteristic transfer matrix for a pile of films, M, as the matrix product of the characteristic matrices of each film, M_j. In its turn, the characteristic matrix of a single film, M_j, is generated on the basis of the boundary conditions (1.4.7°). According to Abeles's approach [104, 105, 129, 130], such a matrix relates the tangential amplitudes of the electric and magnetic field vectors at the input and output film boundaries. The main characteristic matrix, $M = \prod_j M_j$, relates the tangential amplitudes of the electric and magnetic field vectors at the input, $z = 0$, and at the output boundary of the pile (Fig. 1.14). (The other type of matrix method uses amplitudes of the electric fields for directions of incidence and reflection [87, 132]. However, this method is not considered here since it cannot be generalized to anisotropy.) The Fresnel amplitude reflection and transmission coefficients of the system (1.4.5°) are expressed in terms of the matrix elements, which allows one to calculate the reflection and transmission spectra of the whole layered system.

1.7.3°. Let us now reproduce the algorithm of the Hansen method following the notations of Ref. [100]. The optical configuration used throughout the text is described in Fig. 1.14. Incoming parameters for the spectral simulations are as follows:

1. The number of media, N, of which 1st and Nth are the medium of incidence and the final medium, respectively. The number of layers is, therefore, $N - 2$.

2. The wavenumber dependences of the real refractive index $n(\nu)$ and the extinction coefficient $k(\nu)$ in the spectral range of interest. These dependences can (a) be calculated by using Eqs. (1.17) and (1.53) if the parameters of the corresponding oscillators are known, (b) be extracted from the reflection or absorption spectrum by the KK relations (1.18) (see description of the procedure, e.g., [142]), or (c) be reference data [16, 25, 47, 48, 51].

3. The magnetic permeability μ_j ($j = 1, 2, \ldots, N$) for each of the N media (1.1.2°).

4. The angle of incidence φ_1.

5. The thickness d_j for each of the $N - 2$ layers.

6. The wavenumber range and step.

1.7.4°.   The spectral simulation involves the stepwise calculation of the following quantities at each wavenumber, with the selected step over the range covered:

1. The complex vectors $\hat{n}_j = n_j + ik_j$ and $\hat{\varepsilon}_j = n_j^2 - k_j^2 + 2in_jk_j (j = 1, \ldots, N)$.

2. The cosines of the complex angle of refraction, $\cos\varphi_j = [1 - ((n_1 \sin\varphi_1)^2/\hat{n}_j^2)]^{1/2}$, as $\cos\varphi_j = |\,\mathrm{Re}(\cos\varphi_j)| + i|\,\mathrm{Im}(\cos\varphi_j)|$, $j = 2, \ldots, N$.

3. The generalized complex indices of refraction $\xi_j \equiv \hat{n}_j \cos\varphi_j (j = 1, \ldots, N)$, Eq. (1.69), where $\hat{n}_j$ is obtained at step 1 and $\cos\varphi_j$ is found at step 2.

4. $\beta_j \equiv 2\pi d_j\xi_j\nu_j (j = 1, \ldots, N)$, where ξ_j is the complex quantity obtained at step 3.

5. $p_j = (\hat{\varepsilon}_j/\mu_j)^{1/2} \cos\varphi_j$.

6. $q_j = (\mu_j/\hat{\varepsilon}_j)^{1/2} \cos\varphi_j$.

7. The *elementary characteristic matrix* M_j for each of the $N - 2$ constituent layers in the particular stratified medium is calculated as

   (a) $M_j^s = \begin{vmatrix} \cos\beta_j & \dfrac{-i}{p_j}\sin\beta_j \\ -ip_j\sin\beta_j & \cos\beta_j \end{vmatrix}$ for s-polarization,

   (b) $M_j^p = \begin{vmatrix} \cos\beta_j & \dfrac{-i}{q_j}\sin\beta_j \\ -iq_j\sin\beta_j & \cos\beta_j \end{vmatrix}$ for p-polarization.

   Here, $i = \sqrt{-1}$ and $j = 2, 3, \ldots, N - 1$.

8. The *characteristic matrix* of the whole multilayer structure is calculated as the product of the elementary matrices obtained at step 7:

$$M = M_2 M_3 \cdots M_{N-1} \equiv \begin{bmatrix} m_{11} & m_{12} \\ m_{21} & m_{22} \end{bmatrix}.$$

9. By using the elements of the M matrix, the Fresnel amplitude reflection and transmission coefficients (1.58) of the N-isotropic-phase medium are found from

$$r^s = \frac{(m_{11} + m_{12}p_N)p_1 - (m_{21} + m_{22}p_N)}{(m_{11} + m_{12}p_N)p_1 + (m_{21} + m_{22}p_N)},$$

$$r^p = \frac{(m_{11} + m_{12}q_N)q_1 - (m_{21} + m_{22}q_N)}{(m_{11} + m_{12}q_N)q_1 + (m_{21} + m_{22}q_N)},$$

$$t^s = \frac{2p_1}{(m_{11} + m_{12}p_N)p_1 + (m_{21} + m_{22}p_N)},$$

$$t^p = \frac{2q_1}{(m_{11} + m_{12}q_N)q_1 + (m_{21} + m_{22}q_N)}.$$

10. With the quantities computed at step 9, the reflectance and transmittance of the stratified medium are calculated as

(a) $R^{s,p} = r^{s,p} \cdot r^{s,p*} = |r^{s,p}|^2$,

(b) $T^s = \dfrac{\mu_1 \operatorname{Re}(\hat{n}_N \cos \varphi_N)}{\mu_N n_1 \cos \varphi_1} |t^s|^2$,

(c) $T^p = \dfrac{\mu_N \operatorname{Re}(\hat{n}_N \cos \varphi_N / \hat{n}_N^2)}{\mu_1 n_1 \cos \varphi_1 / n_1^2} |t^p|^2$.

1.7.5°.   Formulas may be derived on the basis of the matrix method for the reflectance and transmittance for a layer whose optical constants vary with the depth [9, 14, 87, 88, 109, 132, 146, 147]. For example, the following expressions for reflectance and transmittance were obtained in [147]:

$$R = \left| \frac{\begin{aligned}&\sqrt{\varsigma^2 + \chi^2}(r_{sub} + r_{f0}) + [\varsigma(r_{sub} - r_{f0})\\&+\chi(1 + r_{sub}r_{f0})]\tanh(\sqrt{\varsigma^2 + \chi^2})\end{aligned}}{\sqrt{\varsigma^2 + \chi^2}(1 + r_{sub}r_{f0}) + [\varsigma(1 - r_{sub}r_{f0}) + \chi(r_{sub} + r_{f0})]} \right|^2, \quad (1.106)$$

$$T = \left| \frac{-\displaystyle\int_0^d \partial\hat{n}_f(z)/\partial z \cdot \hat{n}_f(z)/2[\hat{n}_f^2(z) - \hat{n}_{f0}^2 \sin \varphi_1]}{\begin{aligned}&\cosh(\sqrt{\varsigma^2 + \chi^2})(1 + r_{sub}r_{f0}) + \sinh(\sqrt{\varsigma^2 + \chi^2})\\&\times[\chi(1 - r_{sub}r_{f0}) + \varsigma(r_{sub} + r_{f0})]\end{aligned}} \right|^2, \quad (1.107)$$

where r_{sub} is the Fresnel amplitude coefficient for the interface between the inhomogeneous film and the substrate, r_{f0} is the Fresnel amplitude coefficient for the interface between the surroundings and the film, $\hat{n}_f(z)$ is the refractive index as a function of the film depth, and $\hat{n}_{f0} = n_{f0} + ik_{f0}$ is the refractive index of the film at the interface with the surroundings,

$$\varsigma_s = \int_0^d dz \frac{\partial \hat{n}_f(z)}{\hat{n}_f(z)\partial z} \left(1 - \frac{\hat{n}_f^2(z)}{2[\hat{n}_f^2(z) - \hat{n}_{f0}^2 \sin \varphi_1]}\right),$$

$$\varsigma_p = \int_0^d dz \frac{\partial \hat{n}_f(z)}{\partial z} \left(\frac{\hat{n}_f(z)}{2[\hat{n}_f^2(z) - \hat{n}_{f0}^2 \sin \varphi_1]}\right),$$

$$\chi_s = \chi_p = \mathrm{Im} \int_0^d dz \sqrt{\hat{n}_f^2 - \hat{n}_{f0}^2 \sin \varphi_1},$$

and φ_1 is the angle of incidence.

1.7.6°. The Hansen formulas shown above can be adopted for anisotropic layers by redetermining the quantities β_j, p_j, and q_j, introduced in (1.7.4°), in the following way [126, 142]:

4*. $\beta_j^p \equiv 2\pi d_j v_j \hat{n}_{jx} \cos \varphi_j^p$, $\beta_j^s \equiv 2\pi d_j v_j \hat{n}_{jy} \cos \varphi_j^s$ ($j = 1, \ldots, N$, axes as shown in Fig. 1.14);

5*. $p_j = \cos \varphi_{jp}/\hat{n}_{jx}$; and

6*. $q_j = \hat{n}_{jy} \cos \varphi_{js}$.

Here, the angles φ_{jp} and φ_{js} are defined by the equations $\hat{n}_1 \sin \varphi_1 = \hat{n}_{jz} \sin \varphi_{jp}$ and $\hat{n}_1 \sin \varphi_1 = \hat{n}_{jy} \sin \varphi_{js}$, respectively, and $\hat{n}_{j\mathbf{k}} = n_{j\mathbf{k}} + ik_{j\mathbf{k}}(\mathbf{k} = x, y, z)$.

1.8. ENERGY ABSORPTION IN LAYERED MEDIA

According to Maxwell's theory, the rate at which radiation energy is absorbed is directly proportional to the *mean-square electric field* (MSEF), $\langle E^2 \rangle$, at the place where the absorption occurs (1.1.15°, 1.2.9°), which in turn is strongly dependent on the position within the layered medium, parameters of the experiment such as the polarization and the angle of incidence, and the material characteristics (the refractive indices of the layer, the substrate, and surroundings). Fry [148] proposed the use of the variation of the angle of incidence on the MSEF to determine optimal experimental conditions. This method, which is sometimes referred to as *electric field analysis* (EFA), has helped to understand the enhancement mechanisms in grazing-angle external reflection spectra of thin films on metals [101, 132, 149], *metal overlayer ATR* (MOATR) [2, 132, 150], and the ATR spectra of graphite-coated organic films [151]. Suzuki et al. [152] interpreted the phenomenon of spectral enhancement for ultrathin films on rough metal surfaces and islandlike metal underlayers using EFA. Sperline et al. [153, 154] proposed evaluation of surface excess of the adsorbed molecules at the solid–liquid interface in terms of the electric field intensities. Harbecke et al. [155] and Grosse and Offermann [2] interpreted the Berreman effect (Section 3.2) using the EFA expressions for the dissipated energy. By analyzing the electric field strengths,

Harrick [156] and Hansen [99] gained insight into the physics of the ATR spectrum of a layered structure and derived formulas for the penetration depth and the contribution of the surface layer to the net absorption. Electric field analysis has also been employed to explain the phenomena of SEWs [132, 157, 158] and the excitation of surface polaritons [159, 160]. In addition, EFA has been shown to be a basis for an approximate estimation of the molecular orientation (Section 3.11). However, as will be shown below, the MSEFs cannot be used to compare the spectral contrast for the same film in different optical systems.

In this section, the relationship between the MSEFs and the spectrum contrast will be considered to provide a proper understanding of the IR spectral features discussed in Chapters 2 and 3. The effect of experimental conditions for the three different spectroscopic methods — external reflection, transmission, and ATR — on the MSEFs in a model organic ultrathin film will be examined. The practical implications of these findings will be discussed.

On the basis of the Poynting theorem (1.1.18°), the energy loss of the beam with a cross-sectional area of unity in the thin film, ΔI, after integration of Eq. (1.25) over the volume of this film is given as [99, 101]

$$\Delta I = Cn_2\alpha_2\langle E_2^2\rangle\frac{d_2}{\cos\varphi_1}, \tag{1.108}$$

where C is a constant, $n_2 = \mathrm{Re}\,\hat{n}_2$ and α_2 are the real refractive index and the decay constant (1.22) of the film material, respectively, $\langle E_2^2\rangle$ is the MSEF in the film, d_2 is the film thickness, and φ_1 is the angle of incidence. The quantity $d_2/\cos\varphi_1$ represents the volume of the film irradiated by the beam. Thus, we can see that the spectral sensitivity is proportional to the MSEF in the ultrathin film. In what follows we shall analyze how the experimental conditions of the external reflection, transmission, and ATR influence the MSEFs inside a model organic ultrathin film at the boundary with the medium of incidence. The film parameters used in the EFA are as follows: $d_2 = 5$ nm, $n_2 = 1.5$, $k_2 = 0.1$, and $\nu = 1000$ cm^{-1}. It should be noted that the electric field across the 5-nm thin film is practically constant (data not shown). For the EFA, the equations given in the Appendix to this chapter have been programmed onto a personal computer in C^{++}.

1.8.1. External Reflection: Transparent Substrates

It can be seen in Figs. 1.15a and b that for the model ultrathin organic film at the air–Si and water–Si interfaces all components of the MSEFs within this film for all angles of incidence are weakened relative to the incident radiation ($\langle E_1^2\rangle = 1$). Comparison of the electric field intensities in the air and water environments reveals that the MSEF magnitudes within the film, $\langle E_2^2\rangle$, increase as the optical density of the surroundings ($\hat{n}_1$) increases. Deviation of the values of $\langle E_2^2\rangle$ from unity and their dependence on the optical parameters of the medium are due to the formation of interference patterns (standing waves) in

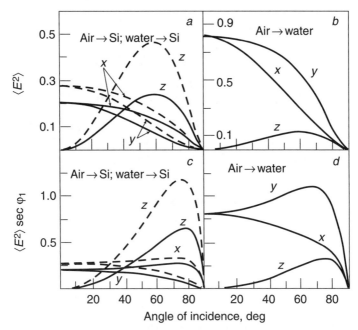

Figure 1.15. External reflection. Influence of angle of incidence on (a, b) mean-square electric fields $\langle E^2 \rangle$ and (c, d) normalized mean-square electric fields $\langle E^2 \rangle \sec \varphi_1$ at 1000 cm^{-1} inside model organic layer 5 nm thick ($n_2 = 1.5$ and $k_2 = 0.1$) located at boundaries: (a, c) air–Si (solid line) and water–Si (dashed line) and (b, d) air–water. Optical constants of water and Si indicated in Table 1.1.

the surroundings and the film. These patterns are composed of incoming and reflected beams, whose relative contributions are strongly influenced by the optical properties of the system components and the boundary conditions (1.4.7°). The interference distributions of the MSEFs with the coordinate z in such a standing wave formed in the air–model film–Si system at $\varphi_1 = 55°$ are illustrated in Fig. 1.16a. Notice that in Si we have a propagating wave (the MSEFs are constant with space). One can explain the observed increase in the MSEFs with increasing $\hat{n}_1$ by the decrease in the portion of the IR radiation reflected from the front surface of the film, based on the complex composition of the resultant standing wave. All conclusions made here are also true for the transmission geometry.

When the organic film is at the water surface (Fig. 1.15b), the maximum values of the normal MSEFs substantially decrease relative to those of the tangential component and, as for the Si substrate (Fig. 1.15a), all the absolute MSEF magnitudes over the whole angular range are less than for the incident beam. It can be shown that the field intensity in the film is less than for ordinary transmission ($\varphi_1 = 0°$, no polarization). The observed maximum of the tangential MSEFs at $\varphi_1 = 0°$ within the film (Fig. 1.15b) is due to the low reflectance of the interface at this angle (Fig. 1.11d).

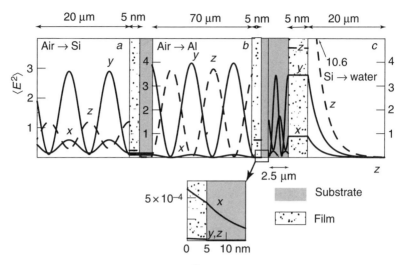

Figure 1.16. Mean-square electric field ($\langle E^2 \rangle$) as function of coordinate z in systems: (a) air–organic film–Si, $\varphi_1 = 55°$; (b) air–organic film–Al, $\varphi_1 = 80°$; and (c) Si–organic film–water, $\varphi_1 = 21°$. Inset shows decay of tangential MSEFs in Al. Frequency of incident radiation is 1000 cm^{-1}. Coordinate system is shown in Fig. 1.14. Parameters of film are as in Fig. 1.15. To gain a better demonstrativeness, segments of z-axis in medium of incidence, within film, and in output medium are represented in different scales.

An important observation from Fig. 1.15 is that the TDMs with all spatial components are active in the spectra: The y-components arise in the s-polarized spectra, while the x- and z-components arise in the p-polarized ones. Moreover, as seen from Fig. 3.89, the x- and z-components in the p-polarized spectra are characterized by the differently directed absorption bands. This regularity constitutes the *surface selection rule* (SSR) for dielectrics (see Section 3.11.4 for more detail).

1.8.2. External Reflection: Metallic Substrates

The angular dependences of the MSEFs in a film at the air–Al and water–Al interfaces shown in Fig. 1.17a are remarkable in three respects. First, independently of the immersion medium, the z-component of the electric field within the film is dominant, while the x- and y-components are almost zero at all angles of incidence φ_1. In other words, the tangential electric fields are quenched. This is observed for all metals. The second feature, which is common to all substrates in air (e.g., compare with Fig. 1.15a), is the attenuation of the perpendicular MSEF component, whose maximum value for Al is ~0.73. It can be shown [161] that such an attenuation of the $\langle E_z^2 \rangle$-component in the case of a metal substrate is observed for films with $n_2 \geq 1.4$, which includes the majority of films. Finally, it should be noted that if the radiation is incident from water onto a metal substrate, the perpendicular MSEFs within the film are enhanced by a factor of about 2, as for Si substrates (Fig. 1.15a).

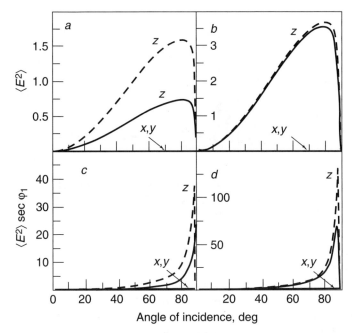

Figure 1.17. External reflection. Influence of angle of incidence on (*a, b*) mean-square electric fields $\langle E^2 \rangle$ and (*c, d*) normalized mean-square electric fields $\langle E^2 \rangle \sec \varphi_1$; (*a, c*) in model organic film ($n_2 = 1.5$, $k_2 = 0.1$, $\nu = 1000$ cm^{-1}) 5 nm thick at air–Al (solid line) and water–Al (dashed line) boundaries; (*b, d*) at bare air–Al (solid line) and water–Al (dashed line) boundaries. Optical constants of water and Al indicated in Table 1.1.

The distribution of the electric fields along z-axis in the air–model film–Al system is shown in Fig. 1.16*b*. The standing-wave patterns produced by the tangential electric field components exhibit nodes at a metal surface, while the normal component has an antinode. As seen from the insert in Fig. 1.16*b*, the tangential electric fields, which are continuous at interfaces (1.8.8°), decay dramatically after crossing the metal surface, typically at a distance similar to the depth of the skin layer (1.3.14°).

To interpret z-polarization of the electric field at a metal surface qualitatively, recall the basic laws of electrostatics, which state that at any point outside a conductor near its surface the electric field is always perpendicular to the surface, while at any point inside a conductor the net electric field is always zero. Therefore, in the case of the tangential external electric field, the strength of the induced electric field inside a metal equals that of the external field and their directions are antiparallel on both sides of the interface (Fig. 1.18*a*), which results in the vanishingly small strength of the total tangential electric field in the neighborhood of a metal surface. This is known as screening of the tangential electric field. If the external electric field is perpendicular to the surface, the induced dipole moment has an electric field parallel to the source field outside the metal (Fig. 1.18*b*). As a result, the net perpendicular electric field strength near

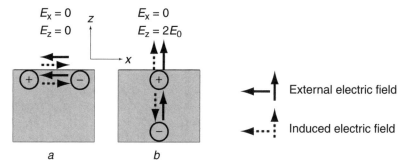

Figure 1.18. Sketch of instantaneous electric fields appearing at metal interface. Dashed arrows illustrate induced (image) field and solid arrows correspond to external field. Notice that in static limit, value of induced field is dictated by fact that net electric field inside a metal is nil.

a metal surface does not vanish. Since the intensity of an IR spectrum is defined by the interaction between the oscillating dipoles of the sample molecules and the electric field (1.2.9°), the result of these two effects is that only perpendicular vibrations are observed. This is known as the SSR for metal surfaces. This selection rule is rather universal, being valid not only for a plane–parallel but also a rough metal surface and metal particles (Section 3.9.4).

However, in the presence of a thin film at the air–metal interface, there is no enhancement of $\langle E_z^2 \rangle$ at the metal surface, as might be concluded based on the above consideration. The reason will be apparent if one analyzes the boundary conditions for the electromagnetic wave (1.4.7°). The continuity of the tangential components of the electric field and the perpendicular component of the electric displacement, $\mathbf{D} = \hat{\varepsilon}\mathbf{E}$, can be approximated for a three-phase system as [99, 101]

$$\langle E_{y1}^2 \rangle \approx \langle E_{y2}^2 \rangle \approx \langle E_{y3}^2 \rangle, \qquad \langle E_{x1}^2 \rangle \approx \langle E_{x2}^2 \rangle \approx \langle E_{x3}^2 \rangle,$$
$$|\hat{\varepsilon}_1|^2 \langle E_{z1}^2 \rangle \approx |\hat{\varepsilon}_2|^2 \langle E_{z2}^2 \rangle \approx |\hat{\varepsilon}_3|^2 \langle E_{z3}^2 \rangle, \tag{1.109}$$

where $\langle E_k^2 \rangle$ ($\mathbf{k} = x, y, z$) are the components of the MSEFs and $\hat{\varepsilon}$ are the permittivities. Subscripts 1, 3 and 2 refer to the input and output media and the film, respectively. Equations (1.109) imply that a boundary is transparent for the tangential electric field components, whereas the normal component, due to the polarization of the media, changes abruptly at a boundary, decreasing in the medium with the larger permittivity. This is illustrated by the plots of Fig. 1.16. When $|\hat{\varepsilon}_1|^2 = 1$ (input medium is air), it follows from Eqs. (1.109) that the ratio $\langle E_{z2}^2 \rangle / \langle E_{z1}^2 \rangle$ is less than unity and decreases inversely with respect to $|\hat{\varepsilon}_2|^2$. This conclusion is general for all substrates, and a metal substrate is a specific case.

Nonetheless, the spectral enhancement for films on metallic substrates exists and, as demonstrated by Greenler [122], can achieve values of 5000 and greater (!) relative to the normal-incidence transmission spectrum of the free film. Let us show that there is no contradiction with the conclusions listed above. As follows from Eq. (1.108), the spectral sensitivity depends not only on the MSEFs but also on the geometric factor $d_2/\cos\varphi_1$, which characterizes the volume of the film sampled by a radiation beam of unit cross-sectional area. Comparison of the angle-of-incidence dependences of the normalized MSEFs (NMSEF $\equiv$ $\langle E^2 \rangle \sec\varphi_1$) to those of the MSEFs (Fig. 1.17) reveals that the *normalization* by $\cos\varphi_1$ has a significant influence on the positions and values of the maxima. From Fig. 1.17, it can be inferred that the NMSEFs within the film at a metal substrate increase up to a factor of ~30 at grazing angles of incidence. Therefore, the enhancement of spectral sensitivity for the metal substrate is caused partly by the increase in the quantity of film material experiencing the field.

The other cause of the enhancement is that the spectra are represented in units of absorbance or reflectivity [Eqs. (1.78) and (1.77), respectively], which vary in inverse proportion with the light intensity reflected in the absence of the film. As seen from Fig. 1.11e, for p-polarization the reflectance of a metal, R_0, decreases significantly near the Brewster angle, causing $\Delta R/R_0$ to increase and thus the absorbance $A \approx (\Delta R/R_0)/\ln 10$ to increase. This factor gives an apparent enhancement in the spectrum at the expense of the SNR.

Both causes of spectral enhancement act in the case of external reflection from transparent or low-absorbing substrates. The "positive" geometric effect is demonstrated in Fig. 1.15. The "negative" effect due to the spectrum representation is the most pronounced in the p-polarized spectra measured at $\varphi_1 \approx \varphi_B$ because $R_0(\varphi_B) \approx 0$ (Figs. 1.11a, c, and d).

1.8.3. ATR

Figure 1.16c shows the dependence of the MSEFs on the z-coordinate for the Si–model film–water interface at $\varphi_1 \approx \varphi_c \approx 21°$. Periodic variation of the MSEF is observed in the medium of incidence, and as expected, the fields are almost constant within the film, decaying exponentially in water. Using the procedure outlined in 1.1.17° to determine penetration depth d_p and the MSEFs for a two-phase system, one can derive the general analytical expression for d_p as

$$d_p = -\frac{1}{2\pi\nu\,\text{Im}\,\xi_2}, \tag{1.110a}$$

which is independent of polarization. When the final phase is transparent, Eq. (1.110a) can be rewritten for $\varphi_1 \geq \varphi_c$ as

$$d_p = \frac{1}{2\pi\nu\sqrt{n_1^2\sin^2\varphi_1 - n_2^2}}. \tag{1.110b}$$

Notice that some authors [99, 163] have defined the penetration depth, replacing the factor 2 in the denominator in Eqs. (1.110) by a factor of 4. To gain some idea of the magnitudes of these penetration depths, Table 1.3 gives d_p values for several interfaces at 1000 and 3000 cm^{-1}.

Table 1.3. Typical penetration depths (μm) in ATR experiments for IREs from chalcogenide glass and Ge

Final Medium	Chalcogenide Glass, $n_1 = 2.37$, $\varphi_1 = 35°$	Germanium, $n_1 = 4.01$, $\varphi_1 = 45°$
Model polymer "in window of transparency": $n_2 = 1.5$, $k_2 = 0$, $\nu = 1000$ cm^{-1}	59.8($\varphi_1 = 40°$)	0.66
Water: $n_2 = 1.218$, $k_2 = 0.0508$, $\nu = 1000$ cm^{-1}	25.92	0.62
Water: $n_2 = 1.319$, $k_2 = 0.131$, $\nu = 1640$ cm^{-1}	1.96	0.38

Hirschfeld [163] attempted to correlate the Goos–Hänchen shift (1.4.11°) with the penetration depth and the effective thickness [Eq. (1.114)]. However, Epstein [164] reexamined the problem and revealed that there is no simple relation between these quantities, because the Goos–Hänchen shift involves non-homogeneous waves, whereas the penetration depth and the effective thickness are determined as the decay of a homogeneous wave.

As seen from Fig. 1.19 the largest values of $\langle E_{y2}^2 \rangle$ and $\langle E_{z2}^2 \rangle$ within the film always occur at the critical angle. In the case where the final medium is transparent (e.g., air) and $\varphi_1 = \varphi_c$, p-polarization gives only the electric field perpendicular to the interface, while at $\varphi_1 \gtrsim 2\varphi_c + 5°$, the tangential component dominates. This peculiarity, which presents the SSR for ATR, is useful in studying the orientation of adsorbed molecules.

When dealing with the ATR spectra of films at the boundary with an absorbing medium (Figs. 1.19b, d), it should be kept in mind that at $\varphi_1 = \varphi_c$ the p-polarized ATR spectrum contains parallel absorption bands as well as perpendicular absorption bands (caused by the dynamic dipole moments parallel and perpendicular to the interface, respectively). The surprising feature of Figs. 1.16c and 1.19a, b is that, contrary to the external reflection, the MSEFs within the ultrathin film are enhanced under ATR conditions at angles of incidence not far from the critical angle φ_c. This feature makes the ATR method more preferable for microsampling as compared to the external reflection [165].

It is noteworthy that the geometric factor (normalization) does not yield a remarkable increase in the NMSEF values as compared to the MSEF ones (compare Figs. 1.19a and b with c and d, respectively). Another feature to be mentioned is that the NMSEF values in the model organic film at the Ge–air and quartz–air interfaces are close to each other at $\varphi_1 > 50°$ (Fig. 1.20). By contrast, the absorption depths ΔR for the same systems are rather different (Fig. 2.23b). This means that analysis of the NMSEF plots can lead to incorrect conclusions concerning the SNR in the ATR spectra of thin films. Therefore, the simulations aiming to distinguish the optimum experimental conditions should be performed for ΔR instead of the NMSEFs. It seems that the mentioned inadequacy of the NMSEFs comes from a difference in the Goos–Hänchen shifts in different systems.

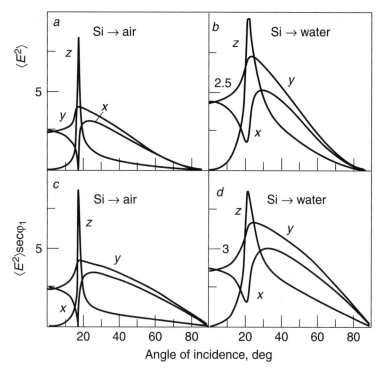

Figure 1.19. ATR. Influence of angle of incidence on (a, b) mean-square electric fields $\langle E^2 \rangle$ and (c, d) normalized mean-square electric fields $\langle E^2 \rangle \sec \varphi_1$; at 1000 cm^{-1} inside model organic layer 5 nm thick ($n_2 = 1.5$, $k_2 = 0.1$) located at boundary of two phases: (a, c) Si–air; (b, d) Si–water. Optical constants of water and Si indicated in Table 1.1.

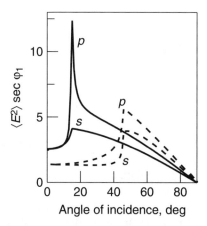

Figure 1.20. ATR. Normalized mean-square electric fields in model organic film 1 nm thick deposited at Ge (solid lines) and quartz (dashed lines) IREs. Optical parameters are as in Fig. 2.23.

The MSEF formalism was used for deriving the formulas for estimating the adsorption density at the internal reflection element (IRE)–solution interface [153, 154, 166–174]. Assuming that (1) species at the solid–solution interface are steplike distributed (with the maximum at the surface), (2) the refractive index of the adsorbed species is close to that of the solution, and (3) the absorption indices of the adsorbed and solvated species are close to each other, Tompkins [166] derived heuristically the following equation that describes the absorbance per reflection in the multiple internal reflection (MIR) spectrum of the species adsorbed at the IRE:

$$\frac{A}{N} = \frac{n_{31} E_0^2 \mathrm{E}}{\cos \varphi_1} \int_0^\infty C(z) e^{-2z/d_p} dz, \tag{1.111}$$

where A is the integrated absorbance (cm^{-1}), N is the number of internal reflections, E is the integrated molar absorption coefficient (cm$^{-2} \cdot$ mol^{-1}) for the species in the solution, which is obtained from the liquid-cell transmission measurements (e.g., for oleate at 20°C, E = 42,038 L $\cdot$ cm$^{-2} \cdot$ mol^{-1} [167]), z is the thickness, $C(z)$ is the concentration as a function of distance from the IRE (mol/L), $n_{31} = n_3/n_1$ is the refractive index of the solution relative to that of the IRE, and $E_0 = (|E_{0x}|^2 + |E_{0z}|^2)^{1/2}$ and $E_0 = E_{0y}$ are the mean electric fields in the solution at the interface for p- and s-polarization, respectively.

Defining the concentration profile as

$$C(z) = \begin{cases} C_i + C_b & \text{for } 0 < z < d_2, \\ C_b & \text{for } d_2 < z < \infty, \end{cases}$$

where C_b and C_i are the surfactant concentration in the bulk solution and at the interface, respectively, and d_2 is the adsorbed film thickness, and using the linear approximation $\exp(-2d_2/d_p) \approx 1 - 2d_2/d_p$ for $d_2 \ll d_p$, Sperline et al. [154] derived, from Eq. (1.111),

$$\frac{A}{N} = E C_b d_e + \mathrm{E}\left(\frac{2d_e}{d_p}\right)(C_i d_2). \tag{1.112}$$

Here,

$$d_e = \frac{n_{31} E_0^2 d_p}{2 \cos \varphi_1} \tag{1.113}$$

is the effective thickness of the material with n_3, which in the transmission spectrum at normal incidence gives the same absorbance as in the ATR spectrum. In practice, if the beam is not polarized, polarization of the light beam due to the properties of the spectrometer must be determined, and a value of d_e is calculated as a weighted average of the purely parallel and perpendicular values [168]. The value of N can be determined by using a reference solute [154] or by geometric consideration [156], which for a parallelepiped geometry is

$$N = \frac{L}{h} \cot \varphi_1,$$

where L and h are the IRE length and thickness, respectively. Recognizing that $C_i d_2/1000$ equals the Gibbs surface excess Γ (mol $\cdot$ cm^{-2}) and rearranging Eq. (1.112) give, for both s- and p-polarization,

$$\Gamma = \frac{(A/N) - EC_b d_e}{1000E(2d_e/d_p)} \qquad (1.114)$$

This so-called Sperline relationship has been verified in a series of in situ and ex situ studies of surfactant adsorption phenomena (see Refs. [169, 170] for examples and references). The corrected formula (1.114) is also applicable to ex situ ATR spectra [171] and adsorption onto a film-coated IRE [172]. The extension of Eq. (1.114) to anisotropic films is discussed by Fringelli [173]. Pitt and Cooper [174] represented Eq. (1.114) as

$$\Gamma = \frac{A_a C_b d_p}{2A_b}, \qquad (1.115)$$

where A_a is the absorbance of the adsorbed molecules at the interface and A_b is the absorbance of the bulk solution. The term $C_b d_p/2A_b$ is calculated from the ATR measurements. The formula of Pitt and Cooper gave the surface excess for adsorbed proteins that were in good agreement with the values obtained by another independent method. The advantage of Eq. (1.115) consists in no need for explicitly determining the extinction coefficient E.

In summary, due to the resonances or the formation of the evanescent wave, the electromagnetic field of the incident radiation is dramatically modified by the solid surface. As a result, the electric fields deviate from unity in an ultrathin film depending on the geometry of the experiment and the optical properties of the material.

If an ultrathin film is located at a metal substrate, the metal quenches the tangential components of the electric fields. Therefore, s-polarization is insensitive to ultrathin films, while for p-polarization only the perpendicular absorption bands are observed. This conclusion constitutes the SSR for metals.

Since at the boundary of a dielectric or semiconductor screening is much weaker, all the components of the TDMs are observed in the external reflection spectra of adsorbed molecules.

Independently of the substrate material, the MSEFs are attenuated in films if the input medium is air. An increase in $\langle E_{z2}^2 \rangle$ occurs if the support is immersed in a medium that is optically denser than the film. For all substrates, the spectral contrast is also increased due to (1) the increase in the quantity of the film material interacting with the incident radiation at the inclined angles of incidence and (2) the decrease in the reflectivity of the substrate. The latter factor, however, reduces the SNR and hence the quality of the external-reflection spectrum.

If one applies the ATR method at $\varphi_1 \approx \varphi_c$, where the spectral sensitivity is maximal, and the final medium is transparent, p-polarization will give only the spectrum of the excitations whose dipole moment is perpendicular to the interface. However, if the final medium is absorbing or $\varphi_1 \gtrsim \varphi_c + 5°$ the p-polarized spectra will contain both the perpendicular and the parallel bands. This is the SSR for ATR spectra.

To reveal the optimum conditions, the absorption depth (ΔR) or the band intensity $(\Delta R')$ should be used for the spectral simulations, instead of the MSEFs (NMSEFs) or absorbances.

1.9. EFFECTIVE MEDIUM THEORY

In the previous sections, we have dealt with the ideal case of homogeneous ultrathin films on flat substrates; on the microscopic level, all ultrathin films are inhomogeneous and surfaces are rough. Depending on the characteristic size, distribution, spacing, and orientation, these imperfections may contribute to the IR spectra, and to distinguish the spectral manifestation of these imperfections is rather complicated [175, 176]. It can be simplified if each particle (cell) is assumed to absorb and radiate coherently with the incident wave and the incident wave is assumed to be unable to resolve the individual particles. In this case, a discontinuous/composite film or the uppermost layer of a rough surface can be considered to be homogeneous, characterized by some effective dielectric constant and obeying the Fresnel formulas (Fig. 1.21). Such an approach may be legitimate provided that structural elements are small compared to the wavelength λ so that $r/\lambda < 10^{-2}$, where r is the characteristic dimension of the structural element [177]. In such a situation, the dielectric function of an inhomogeneous film can be derived within the framework of the *effective medium theory* (EMT).

The EMT was initially developed by Maxwell-Garnett in 1904 [178] to account for the colors of glasses containing microscopic metal spheres. It employs an analogy between ensembles of molecules and small particles in which the particles are regarded as giant molecules. Recall that when deriving the expression for the dielectric function of a dielectric continuum [Eq. (1.33)] within the framework of the Lorentz model, one assumes that a medium is composed of coupled dipoles (which model either phonons or electrons), each dipole being described by Eq. (1.32). Such an approximation permits summation of the individual polarizabilities using Eq. (1.29). The same idea is exploited in the case of an ensemble of small particles, but instead of the polarizability of a single dipole described by Eq. (1.32), the electrostatic polarizability of the particle is introduced in the summation. However, in contrast to the Lorentz model, the EMT

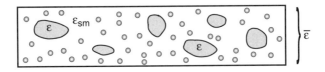

Figure 1.21. Sketch of effective medium.

takes into account the local-field effects: The embedded particles are polarized by the incident wave, causing distortions in the microscopic electromagnetic field around them. If the difference between the actual and external (applied) electric fields is described in accordance with the Clausius–Mossotti/Lorentz–Lorenz (CMLL) model [179–182], then the earliest variants of the EMT dielectric functions can be obtained. Note that the CMLL model is valid only for an isotropic arrangement of particles [12, 22, 33].

The derivation of the effective dielectric function of a medium composed of small spheres and a vacuum will be presented [183]. From elementary electrostatics [12], the polarization $\mathbf{P}$ of a sphere in a constant and uniform far field is uniform, and its polarizability $\hat{\alpha}$ $(1.3.2°)$ is given by

$$\hat{\alpha} = 4\pi r^3 \frac{\varepsilon - \varepsilon_{sm}}{\varepsilon + 2\varepsilon_{sm}}, \tag{1.116}$$

where r is the sphere radius and $\varepsilon(\omega)$ and ε_{sm} are the dielectric functions of the sphere material and the surrounding medium, respectively. It can be seen from Eq. (1.116) that the polarization of such a sphere will have a resonance when

$$|\varepsilon(\omega) + 2\varepsilon_{sm}| = \min. \tag{1.117}$$

The frequencies at which the condition (1.117) is met are referred to as *Frohlich frequencies* ω_F, and the corresponding modes of a sphere that is small compared to λ are referred to as the *Frohlich modes*, giving credit to the pioneering work of Frohlich [13], who determined theoretically that an ensemble of small spherical particles absorbs at ω_F (see the discussion in Section 3.9). When the imaginary part of $\varepsilon(\omega)$ vanishes, Eq. (1.117) is reduced to the *Mie condition*

$$\varepsilon(\omega) = -2\varepsilon_{sm}, \tag{1.118}$$

which follows from the Mie theory [184, 185].

Accounting for the CMLL local field converts Eq. (1.29) into the Clausius–Mossotti relation [9–13, 22, 43]

$$\hat{\alpha} = \frac{3}{N} \frac{\bar{\varepsilon} - 1}{\bar{\varepsilon} + 2}, \tag{1.119}$$

where N is the number of particles in a unit volume and $\bar{\varepsilon}$ is the dielectric function of the composite medium consisting of these particles and the vacuum. From Eqs. (1.116) and (1.119) one can obtain the following relationship in the case of $\varepsilon_{sm} = 1$:

$$\frac{\bar{\varepsilon} - 1}{\bar{\varepsilon} + 2} = f \frac{\varepsilon - 1}{\varepsilon + 2}. \tag{1.120}$$

Here, ε denotes the dielectric function of the sphere material and $f = NV$ is the volume fraction of these spheres, known as the filling fraction ($V = \frac{4}{3}\pi r^3$ is the volume of one particle). If, instead of the vacuum, the spheres are embedded in a matrix of a dielectric function ε_{sm}, then ε and $\bar{\varepsilon}$ are the dielectric constants relative to ε_{sm} and Eq. (1.120) can be rewritten as

$$\frac{\bar{\varepsilon} - \varepsilon_{sm}}{\bar{\varepsilon} + 2\varepsilon_{sm}} = f \frac{\varepsilon - \varepsilon_{sm}}{\varepsilon + 2\varepsilon_{sm}}. \qquad (1.121)$$

This formula is known as the Maxwell-Garnett effective medium (MGEM) expression. Note that it can be derived under various assumptions [175, 186, 187]. From Eq. (1.121), the Maxwell-Garnett (MG) dielectric function of the composite layer is expressed as

$$\bar{\varepsilon} = \varepsilon_{sm} \frac{\varepsilon (1 + 2f) + 2\varepsilon_{sm}(1 - f)}{\varepsilon(1 - f) + \varepsilon_{sm}(2 + f)}. \qquad (1.122)$$

One can see that the effective dielectric function of the effective medium differs from the simple average of the dielectric functions of the constituents. Moreover, Eq. (1.122) is inherently asymmetric in the treatment of the two constituents; the transformation of $\varepsilon \leftrightarrow \varepsilon_{sm}$ and $f \leftrightarrow 1 - f$ will result in different effective dielectric functions.

Using Eq. (1.122), it is straightforward to show for the undamped case [186] that the TO and LO energy loss functions of the layer have maxima at the frequencies ω'_{TO} and ω'_{LO}, respectively, obeying the conditions

$$\varepsilon(\omega'_{TO}) = -\varepsilon_{sm} \frac{2 + f}{1 - f}, \qquad \varepsilon(\omega'_{LO}) = -\varepsilon_{sm} \frac{2(1 - f)}{1 + 2f}.$$

It is seen that for $f \to 0$ both ω'_{TO} and ω'_{LO} approach the Frohlich frequency of the inclusions, Eq. (3.35), $\omega'_{TO} = \omega'_{LO} = \omega_F$. As f increases, the ω'_{TO} and ω'_{LO} bands move toward longer and shorter wavelengths, respectively, and the composite layer has a reflection band that extends from ω'_{TO} to ω'_{LO}. For $f \to 1$, ω'_{TO} and ω'_{LO} approach ω_{TO} and ω_{LO} of the bulk material, respectively. (Note that the maximum filling factor $f = 0.74$ in the cubic or hexagonal close-packed geometry for uniformly sized spheres [188].)

Assuming that the individual grains, representing both the particles and the surroundings, exist in some effective medium, in 1935 Bruggeman [189] derived the expression for the average dielectric function, which can be generalized to ellipsoids as [190]

$$\sum_{k=1}^{3} \left(f \frac{\varepsilon - \bar{\varepsilon}}{\bar{\varepsilon} + g_k(\varepsilon - \bar{\varepsilon})} + (1 - f) \frac{\varepsilon_{sm} - \bar{\varepsilon}}{\bar{\varepsilon} + g_k(\varepsilon_{sm} - \bar{\varepsilon})} \right) = 0. \qquad (1.123)$$

Here, ε, ε_{sm}, and $\bar{\varepsilon}$ are the dielectric functions of the particles, the surrounding, and the effective medium, respectively, and g_k ($k = 1, 2, 3$) is the geometric (shape) factor, which determines the self-polarizing effect of the ellipsoid and has a value between zero (needle) and unity (slab) so that $g_1 + g_2 + g_3 = 1$.

For a sphere $g_k = \frac{1}{3}$. Ellipsoids of rotation, or spheroids, which have two axes of equal length, are a special case. The prolate (column-shaped) spheroids, for which $g_2 = g_3$, are produced by rotating an ellipse about its major axis; the oblate (disk-shaped) spheroids, for which $g_1 = g_2$, are produced by rotating an ellipse about its minor axis. In the general case, the value of g_k can be obtained from the tables of Stoner [191] or by using the formulas developed by Osborn [192] (see also Ref. [175]).

As opposed to the MGEM expression (1.121), the Bruggeman formula is symmetrical relative to the particles and their surroundings. Although some authors [52] refer only to the Bruggeman model (BM) as the EMT, to differentiate from the MG-type theories. However, Aspnes [193] has shown that both theories are conceptually equivalent to the EMT since the Bruggeman formula can be derived from the MGEM if the CMLL model is used as the starting point. Nevertheless, the applicability range of these EMTs appears to be different. There is general consensus [52, 194] that the Bruggeman formula is more appropriate in the cases of composites containing two or more components at high filling factors.

In the literature, several EMTs have been reported. These EMTs originate from different assumptions of the shape, construction, number, and mutual orientations of the basic unit cells and various types of interactions between them (in addition to or instead of the *Lorentz field* — the long-range dipole–dipole field involved in the CMLL model). Rather than attempt to review this large body of theoretical work (see, e.g., Refs. [183, 195–197]), the results of only a few of them will be discussed.

Since the Bruggeman formula and MGEM expressions, which were established for essentially random structures, can be highly inaccurate for regular or partly ordered arrays, Radchik et al. [198] have derived a generalized EMT from first principles. In this version of the EMT, an exact expression for the effective filling factor f^* was derived from a conformal transformation that describes the periodic properties of an ordered array of cylindrical particles. The value of f^* is complex and wavelength dependent. The difference between the real filling factor and f^* is interpreted as an indication of screening.

The earlier MGEM-like dielectric functions of Gans [199], David [200], and Galeener [201, 202] of an anisotropic medium containing ellipsoids of identical shape and orientation have been further modified by Cohen et al. [203]. In the Cohen model, the shape of the fictitious Lorentz cavity ("near region"), which figures in the derivation of the local-field approximation, is postulated to be ellipsoidal instead of spherical. From a microscopic viewpoint, such a substitution reflects the presence of some structural disorder in the system of aligned particles or in their surroundings [204]. His expression for aligned ellipsoids in the form cited by Landauer [205] is given by the following:

$$\frac{\bar{\varepsilon} - \varepsilon_{\text{sm}}}{g_c\bar{\varepsilon} + (1 - g_c)\varepsilon_{\text{sm}}} = f\frac{\varepsilon - \varepsilon_{\text{sm}}}{g\varepsilon + (1 - g)\varepsilon_{\text{sm}}}. \tag{1.124}$$

The quantities g and g_c describe the shapes of the inclusion and the Lorentz cavity, respectively. For the general case of randomly oriented ellipsoids, Polder and von Santen [206] have derived the formula

$$\bar{\varepsilon} = \varepsilon_{sm} \frac{3 + 2f\tilde{\alpha}}{3 - f\tilde{\alpha}}, \tag{1.125}$$

which accounts for anisotropic dielectric properties of the particles. Here, $\tilde{\alpha}$ is the averaged polarizability of the randomly oriented ellipsoids, given by

$$\tilde{\alpha} = \frac{1}{3} \sum_{k=1}^{3} \hat{\alpha}_k, \tag{1.126}$$

where the summation is over the three principal axes of the ellipsoids and $\hat{\alpha}_k$ is the complex electric polarizability of the particle along the **k**-axis,

$$\hat{\alpha}_k = V \frac{\varepsilon - \varepsilon_{sm}}{\varepsilon_{sm} + (\varepsilon - \varepsilon_{sm})g_k}. \tag{1.127}$$

Here, $V = \frac{1}{3}4\pi abc$ is the ellipsoid volume ($a, b,$ and c are the radii along the principal axes) and g_k is the geometric factor. Equation (1.125) was further extended by Hayashi et al. [207] to particles that are nonuniform in size. Wiener [208] and Granqvist and Hunderi [209, 210] extended the EMT for the case of different shapes (circular cylinders, layers, rods, and spheres) and orientations of the cells. Schopper [211] introduced a Gaussian distribution of ellipsoidal shapes in the EMT, while Dobierzenska-Mozrzymas et al. [212] employed a log-normal type of distribution. As noted by Kreibig and Vollmer [177], a further generalization can be achieved if one applies the Sinzig n-shell formula [213] for the particle polarizability in calculating the average dielectric function. Ruppin and Yamaguchi [214] studied the effect of the substrate on the spectra of spherical particles, modeling such a system as spherical particles in a matrix with a dielectric function ε_{sm} that is an average of the vacuum value 1 and that of the support material ε_3 [$\varepsilon_{sm} = \frac{1}{2}(1 + \varepsilon_3)$]. It was found that interaction with a substrate leads to a red shift of the absorption bands. The Maxwell-Garnett theory has been improved by Stroud and Pan [215], Niklasson et al. [216, 217], Bohren and Huffman [175], and Bohren [218]. The EMT has been extended to multicomponent systems [219]. The effect of clustering has also been treated [220–222]. The difference in the application of the CMLL formula in the EMT to bulk and thin systems is discussed by Vinogradov [223].

The EMT can be considered as a basic tool for the analysis of IR spectra of inhomogeneous thin films, and it has been proven valid in many cases, provided that the working formula has been chosen properly (Section 3.9). However, the EMT is not sensitive to the positions or sizes of individual inclusions in the film. Moreover, in contrast to the Rayleigh law, it is inadequate in predicting the color and brightness of the sky, because it neglects incoherent scattering.

1.10. DIFFUSE REFLECTION AND TRANSMISSION

Reflected and transmitted radiation from a powder layer can be either specular or diffuse (Fig. 1.22). The *specular* (Fresnel) component I_{SR} reflected from the external boundary, which is comprised of all parts of the interface that have faces oriented in the direction of the "averaged" common interface. The magnitude of this component and its angular dependence can be determined by the Fresnel formulas (1.62). The specular (regular) transmission I_{RT} is the fraction of radiation that travels through the sample without any inclination. The other fractions of the radiation, the so-called diffuse reflection and transmission, I_{DR} and I_{DT}, respectively, are generated by the incoherent (independent) scattering and absorption by particles and do not satisfy the Fresnel formulas.

One can distinguish the surface and volume components in the diffuse transmission I_{DT} and the diffuse reflection I_{DR} (Fig. 1.22) [224–227]. The surface component, which is referred to as *Fresnel diffuse reflectance*, is the radiation undergoing mirrorlike reflection and still obeying the Fresnel reflection law but arising from randomly oriented faces. This phenomenon was first described by Lambert in 1760 [228] to account for the colors of opaque materials. The volume, or *Kubelka–Munk* (KM), component is the radiation transmitted through at least one particle or a bump on the surface (Fig. 1.22).

All the components mentioned interact with the powder and, therefore, contain information about its absorption coefficient. However, only the specular transmission and volume KM components give the absorptionlike spectra of the powder directly. The Fresnel components produce specular reflection (first derivative or inverted) spectra [which can be converted into the spectra of the absorption coefficient using the KK transformation (1.1.13°)]. Therefore, to obtain the absorption spectrum of a powder, the Fresnel components must be eliminated from the final spectrum. In practice, this can be achieved by immersion of the sample in a transparent matrix with a refractive index close to that of the powder, selection of appropriate powder size, or special construction of reflection accessories (Section 4.2).

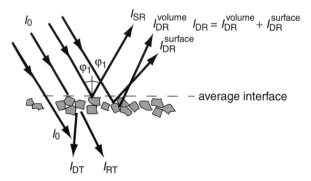

Figure 1.22. Illustration of components of radiation reflected from and transmitted through powder layer.

There is no general theory for diffuse reflection and transmission [176, 229–232]. The intensity of radiation traveling through a layer of particles is attenuated by incoherent scattering and absorption as well. Provided the particles are well separated so that multiple scattering may be neglected, this intensity can be approximated by [233]

$$I_t = I_0 e^{-N C_{ext} d}.$$ (1.128)

Here, N is the number of particles in the layer, I_0 and I_t are the intensities of the incident and transmitted radiation, respectively, and d is the layer thickness. The extinction cross section C_{ext} is the sum of an absorption cross section C_{abs} and a scattering cross section C_{sca}:

$$C_{ext} = C_{abs} + C_{sca}.$$ (1.129)

In 1871, using dimensional analysis, Rayleigh showed that for a single nonabsorbing particle that is small compared to the wavelength

$$C_{sca} \propto r^6 v^4,$$ (1.130)

where r is the particle size and v is the frequency of the radiation (see Ref. [234] for details.). In particular, the sky's blue color and its redness when one looks toward the sun at sunset are explained by the Rayleigh law [10]. Kerker et al. [235] have pointed out that the accuracy of the Rayleigh theory decreases as the $|n|$ of the particles increases. In 1908, Mie [184] derived a rigorous solution of the Maxwell equations with the appropriate boundary conditions in spherical coordinates for a homogeneous sphere of an arbitrary diameter located in a homogeneous medium. In his treatment, the spheres were assumed to be large enough for the dielectric continuum theory to be applicable, with no other limitation in their size. The input parameters were the sphere size and the dielectric functions of the sphere and the surrounding medium. In the general case, the solutions obtained in Mie's theory defy a simple analytical consideration. A detailed discussion of their properties can be found elsewhere [175, 230, 236]. In the limiting case when the radius of the sphere is much larger than the wavelength, the Mie solution gives the same result as diffraction theory and geometric optics. When the particle size is on the order of λ, the solution is complicated and varies from case to case. In particular, if the particle is not too highly absorbing, the Mie absorption and scattering coefficients tend to oscillate. In the limiting case of small particles, Mie's theory predicts that the intensity of scattering is proportional to v^4 (the same result as the Rayleigh theory), while the absorption varies as v.

For highly diluted randomly oriented ellipsoids, van de Hulst [233] obtained the following expressions for the absorption and scattering cross sections:

$$C_{sca} = \frac{\hat{k}^4}{6\pi} \frac{1}{3} \sum_{k=1}^{3} |\hat{\alpha}_k|^2,$$ (1.131)

$$C_{abs} = -\hat{k} \, \text{Im}(\tilde{\alpha}),$$ (1.132)

where $\hat{k} = 2\pi/\lambda$ is the wavenumber in vacuum (1.1.9°), k refers to a particular axis, and $\tilde{\alpha}$ is the average polarizability as determined by Eq. (1.126). A review of the results for coated spheres and cylinders is given by Bohren [176].

For a denser ensemble of scatterers, the theories can be classified into two groups. In the statistical theories (see, e.g., Ref. [237]), a scattering system is considered to be a series of individual particles or layers characterized by their size and concentration and the actual refractive indices of the parent materials and whose optical behavior is averaged. The continuum models consider the nonuniform medium to be a continuum and use phenomenological characteristics. The change in the intensity of an electromagnetic wave when it propagates through such a medium is found by solving the set of differential equations of first order (*equations of radiative transfer*). A solution of the problem was suggested by Gurevich in the early 1930s within the framework of the two-parametric two-flux approximation of the theory of radiation transfer in a scattering medium [238]. Analogous work was performed somewhat later by Kubelka and Munk [239] and by other authors (see review in Refs. [230, 240, 241]).

Separating the radiation flux in a disperse substance into two fluxes (one flux travels in the forward direction and one in the back direction), Kubelka and Munk obtained the following expression for the *KM reflectance* $F(R_\infty)$:

$$F(R_\infty) = \frac{(1 - R_\infty)^2}{2R_\infty} = \frac{K}{S}, \qquad (1.133)$$

where R_∞ is the reflectance of the sample relative to that of a nonabsorbing standard. The subscript ∞ means that the sample thickness is large enough ($d \rightarrow \infty$) to neglect the transmission of radiation through the sample and the reflection from the background boundary (for silica derivatives d_∞ is >2 mm and can be >5 mm if both K and S are small). The phenomenological parameters of the sample — absorption (K) and scattering (S) coefficients of an elementary volume — are called the *Kubelka–Munk absorption and scattering coefficients*. The KM absorption coefficient K is assumed to be equal to $\langle \alpha \rangle$, the true decay constant (1.22) of the compound of the particle, averaged over a unit volume, and the light intensity in a diffusely scattering sample is described by the BLB law as

$$I = I(0)e^{-Kd}. \qquad (1.134)$$

Equation (1.133) allows one to use the KM reflectance units to characterize a thin film on a powdered support.

The KM theory is valid if (1) the amount of the specular reflectance I_{SR} is negligibly small, (2) scattering is isotropic, (3) the sample thickness is large enough ($d \rightarrow \infty$) to allow contributions from transmission through the sample and reflection from the background boundary to be neglected, (4) the KM scattering coefficient S is independent of both the absorption and the wavelength, and (5) both S and K are constants within the sample volume. It was shown [232] that a quantitative determination of the KM function is restricted to the samples for which the scattering coefficient is much larger than the absorption coefficient $(K/S < 0.13)$ and $0.15 < R_\infty < 0.85$.

The KM theory has been adapted for the diffuse reflectance and transmittance measurements of adsorbed species on nondiluted powders [242, 243]. For example, to take into account the nonisotropic scattering directed preferentially forward, Boroumand et al. [243] introduced a third flux along with the upward and downward fluxes of the KM theory. The resulting expression connects the DT and DR signals with the KM parameters S and K and agrees well with the measured C≡N band intensities of a dye adsorbed on silica.

From a practical viewpoint, the main disadvantage of the KM theory is that the KM absorption and scattering coefficients do not relate to the actual optical parameters (the absorption and scattering coefficients) of the particle parent material in a straightforward manner [242, 243] and therefore the DR and DT spectra of a system cannot be calculated. In principle, the equation for radiative transfer can be generalized to include the dispersion effects and applied to a closed-packed medium of particles smaller than the wavelength, provided that the effective scatterers are considered as aggregates rather than the actual particles that make up the medium [230]. The size of such an effective particle should be on the order of a wavelength and its effective dielectric function is calculated within the framework of the EMT.

APPENDIX

Algorithm for Programming MSEF Calculations According to Hansen Matrix Formulas

The Hansen matrix formulas for the MSEFs introduced in Ref. [100] contain a number of misprints. Some of them have been corrected in Refs. [99, 101, 153], which also contain some misprints. Below, the Hansen formulas are corrected so as to produce results equivalent to those reported elsewhere [100, 101, 132, 244, 245].

To evaluate the MSEF (1.1.15°) at position z within the kth layer of an N-isotropic-phase system (Fig. 1.14), according to the Hansen treatment of the matrix method (Section 1.7), the following approach is recommended.

1. The column vectors $Q_1^{s,p}$ and $Q_{N-1}^{s,p}$, characterizing the amplitudes of the tangential components of the electric (s-polarization) and magnetic (p-polarization) fields at the first and final boundaries, respectively, are calculated as

$$Q_1^s \equiv \begin{bmatrix} E_{y1}^0 \\ H_{x1}^0 \end{bmatrix} = \begin{bmatrix} 1 + r^s \\ p_1(1 - r^s) \end{bmatrix},$$

$$Q_{N-1}^s \equiv \begin{bmatrix} E_{y(N-1)}^0 \\ H_{x(N-1)}^0 \end{bmatrix} = \begin{bmatrix} t^s \\ p_N t^s \end{bmatrix},$$

$$Q_1^p \equiv \begin{bmatrix} H_{y1}^0 \\ E_{x1}^0 \end{bmatrix} = \begin{bmatrix} 1 + r^p \\ q_1(1 - r^p) \end{bmatrix} n_1,$$

$$Q_{N-1}^p \equiv \begin{bmatrix} E_{y(N-1)}^0 \\ H_{x(N-1)}^0 \end{bmatrix} = \begin{bmatrix} t^p \\ q_N t^p \end{bmatrix} n_1.$$

Here, the terminology of 1.7.4° holds: $r^{s,p}$ and $t^{s,p}$ are the Fresnel amplitude reflection and absorption coefficients, respectively. The amplitude of the electric field vector of the incident beam in phase 1 is taken to be equal to unity.

2. For the jth layer ($j = 2, \ldots, N - 1$, $j \neq k$) of the N medium, the matrix $N_j^{s,p}$ is generated, which is the reciprocal of the characteristic matrix of this layer, $M_j^{s,p}$ (1.7.4°) ($N_j^{s,p} M_j^{s,p} = I$, where I is the unit matrix). To avoid calculating the inversion of the matrices, the matrix $N_j^{s,p}$ is obtained by changing the signs of the negative terms m_{21} and m_{21} of the matrices $M_j^{s,p}$ obtained in steps 7a and 7b in 1.7.4°.

3. For the kth phase ($1 \leq k \leq N$), within which the MSEF magnitude is to be evaluated at the position z ($z_{k-1} = \sum_{h=2}^{k-1} d_h \leq z \leq z_k = \sum_{h=2}^{k} d_h$), the matrix $N_k^{s,p}$ is constructed as

$$N_k^s(z) = \begin{bmatrix} \cos[2\pi \nu \xi_k(z - z_{k-1})] & \dfrac{i}{p_k} \sin[2\pi \nu \xi_k(z - z_{k-1})] \\ i p_k \sin[2\pi \nu \xi_k(z - z_{k-1})] & \cos[2\pi \nu \xi_k(z - z_{k-1})] \end{bmatrix},$$

$$N_k^p(z) = \begin{bmatrix} \cos[2\pi \nu \xi_k(z - z_{k-1})] & \dfrac{i}{q_k} \sin[2\pi \nu \xi_k(z - z_{k-1})] \\ i q_k \sin[2\pi \nu \xi_k(z - z_{k-1})] & \cos[2\pi \nu \xi_k(z - z_{k-1})] \end{bmatrix}.$$

4. The resultant vectors $Q_k^{s,p}(z)$ characterizing the field amplitudes at position z within the layered system are then found from the following relationships:

(a) at $-\infty < z \leq 0$, that is, in the input medium, where $k = 1$,

$$Q_1^{s,p}(z) = N_1^{s,p}(z) Q_1^{s,p};$$

(b) at $0 \leq z \leq d_1$, that is, in the first layer ($k = 2$),

$$Q_2^{s,p}(z) = N_2^{s,p}(z) Q_1^{s,p},$$

or, alternatively,

$$Q_2^{s,p}(z) = N_2^{s,p}(z) M_2^{s,p} M_3^{s,p} \cdots M_{N-1}^{s,p} Q_{N-1}^{s,p};$$

(c) at $\sum_{h=2}^{j-1} d_h \leq z \leq z_j = \sum_{h=2}^{j} d_h$, that is, within the layers with $(k-1) \geq 2$ except the final (Nth) phase,

$$Q_j^{s,p}(z) = N_j^{s,p}(z) \prod_{m=j-1}^{2} N_m^{s,p} Q_1^{s,p}$$

or

$$Q_j^{s,p}(z) = N_j^{s,p}(z) \prod_{m=j}^{N-1} M_m^{s,p} Q_{N-1}^{s,p};$$

(d) for the final phase ($k = N$), at $z \geq z_j = \sum_{h=2}^{N-1} d_h$, we get

$$Q_N^{s,p}(z) = N_N^{s,p}(z) Q_{N-1}^{s,p}.$$

5. If we designate the elements of the resultant vectors $Q_k^{s,p}(z)$ as

$$Q_k^{s,p}(z) \equiv \begin{bmatrix} U_k^{s,p}(z) \\ V_k^{s,p}(z) \end{bmatrix}$$

and introduce the quantity $W_k(z) \equiv n_1 \sin \varphi_1 U_k^p(z)/\hat{\varepsilon}_k$, where $\hat{\varepsilon}_k = \varepsilon_k' + i\varepsilon_k''$ is the dielectric function of the layer of interest, then calculation of the MSEFs at the depth z proceeds according to

$$\langle E_k^{s2}(z) \rangle = \langle E_{yk}^2(z) \rangle = |U_k^s(z)|^2,$$

$$\langle E_{xk}^2(z) \rangle = |V_k^p(z)|^2,$$

$$\langle E_{zk}^2(z) \rangle = |W_k^{(}z)|^2,$$

$$\langle E_k^{p2}(z) \rangle = \langle E_{xk}^2(z) \rangle + \langle E_{zk}^2(z) \rangle.$$

Note that the MSEF magnitudes are dimensionless quantities equal to the ratios of the MSEFs of interest to the MSEF of the incident beam.

REFERENCES

1. Y. J. Chabal, *Surf. Sci. Reports* **8**, 211 (1988).
2. P. Grosse and V. Offermann, *Vib. Spectrosc.* **8**, 121 (1995).
3. W. Suetaka, *Surface Infrared and Raman Spectroscopy: Methods and Applications*, Plenum Press, New York, 1995.
4. H. Kuzmany, in B. Schrader (Ed.), *Infrared and Raman Spectroscopy*, VCH, Weinheim, 1995, p. 372.

5. V. A. Skryshevsky and V. P. Tolstoy, *Infrared Spectroscopy of Semiconductor Structures*, Lybid', Kiev, 1991 (in Russian).

6. (a) H. Ueba, *Progress Surf. Sci.* **55**, 115 (1997); (b) P. Hollins and J. Pritchard, *Progress Surf. Sci.* **19**, 275 (1985).

7. (a) K. P. Lawley (Ed.), *Molecule Surface Interactions*, Vol. 76, Wiley, London, 1990; (b) B. E. Hayden, in J. T. Yates, Jr. and T. E. Madey (Eds.), *Vibrational Spectroscopy of Molecules on Surfaces,* Vol. 1, Plenum, New York, 1987, p. 267; (c) R. F. Willis, A. A. Lucas, and G. D. Mahan, in D. A. King and D. P. Woodruff (Eds.), *The Chemical Physics of Solid Surfaces and Heterogeneous Catalysis*, Vol. 2, Elsevier, Amsterdam, 1983, p. 59.

8. J. C. Maxwell, *Treatise on Electricity and Magnetism*, Dover, New York, 1954.

9. M. Born and E. Wolf, *Principles of Optics*, 5th ed., Pergamon, Oxford, 1975.

10. S. G. Lipson, H. Lipson, and D. S. Tannhauser, *Optical Physics*, 3rd ed., Cambridge University Press, New York, 1995.

11. L. Landau and E. Lifschitz, *Electrodynamics of Continuous Media*, Pergamon, New York, 1960.

12. J. D. Jackson, *Classical Electrodynamics*, 2nd ed., Wiley, New York, 1975.

13. H. Frohlich, *Theory of Dielectrics*, 2nd ed., Clarendon Press, Oxford, 1958.

14. P. Yeh, *Optical Waves in Layered Media*, Wiley, New York, 1988.

15. J. A. Kong, *Electromagnetic Wave Theory*, 2nd ed., Wiley, New York, 1990.

16. A. Miller, in M. Bass (Ed.), *Handbook of Optics*, Vol. 1, McGraw-Hill, New York, 1995, pp. 9.1–9.33.

17. E. A. Vinogradov, *Phys. Reports* **217**, 159 (1992).

18. V. M. Agranovich and V. A. Ginzburg, *Crystal Optics with Spatial Dispersion and Excitons*, Springer, Berlin, 1984.

19. H. A. Kramers, *Nature* **117**, 775 (1926).

20. R. de L. Krönig, *J. Opt. Soc. Am.* **12**, 547 (1926).

21. (a) F. Stern, in E. Seitz and D. Tirnbull (Eds.), *Solid State Physics*, Vol. 15, Academic, New York, 1963; (b) M. Cordona, in S. Nudelman and S. S. Mitra (Eds.), *Optical Properties of Solids*, Plenum, New York, 1969, p. 137; (c) H. M. Nussenzveig, *Causality and Dispersion Relations*, Academic, New York, 1972.

22. C. Kittel (Ed.), *Introduction to Solid State Physics*, 7th ed., Wiley, New York, 1996, p. 35.

23. H. Kuzmany, *Festkörperspektroskopie*, Springer, Berlin, 1990.

24. (a) D. Y. Smith, in E. D. Palik (Ed.), *Handbook of Optical Constants of Solids I*, Academic, New York, 1998, pp. 35–68; (b) S. S. Mitra, in E. D. Palik (Ed.), *Handbook of Optical Constants of Solids II*, Academic, New York, 1998, pp. 213–270.

25. E. D. Palik (Ed.), *Handbook of Optical Constants of Solids I, II, III*, Academic, New York, 1998.

26. K. Krishnan, *Applications of the Kramers-Kronig Dispersion Relations to the Analysis of FTIR Specular Reflectance Spectra*, FTIR/IR Notes 51, Biorad Digilab Division, Cambridge, MA, August 1987.

27. V. Hopfe, FSOS: Software Package for IR-Vis-UV Spectroscopy of Solids, Interfaces, and Layered Systems, Preprint No. 122, Technische Universitat Karl-Marx-Stadt, Chemnitz, 1989.

28. H. Abdullah and W. F. Sherman, *Vib. Spectrosc.* **13**, 133 (1997).

29. K. E. Peiponen, E. M. Vartiainen, and T. Asakura, *Opt. Rev.* **4**, 433 (1997).

30. (a) K. Yamamoto and A. Masui, *Appl. Spectrosc.* **49**, 639 (1995); (b) K. Ohta and H. Ishida, *Appl. Spectrosc.* **42**, 952 (1988); (c) J. P. Hauranek, P. Neelakentan, R. P. Young, and R. N. Jones, *Spectrochim. Acta A* **32**, 851 (1976).

31. (a) L. T. Tickanen, M. I. Tejedor-Tejedor, and M. A. Anderson, *Appl. Spectrosc.* **46**, 1849 (1992); (b) L. T. Tickanen, M. I. Tejedor-Tejedor, and M. A. Anderson, *Langmuir* **13**, 4829 (1997).

32. P. Grosse and V. Offermann, *Appl. Phys. A* **52**, 138 (1991).

33. N. W. Ashcroft and N. D. Mermin, *Solid State Physics*, Saunders College Publishing, Philadelphila, 1976.

34. H.-H. Perkampus, *Encyclopedia of Spectroscopy*, VCH, Weinheim, New York, 1995.

35. I. F. Chang, S. S. Mitra, J. N. Plendl, and L. C. Mansur, *Phys. Stat. Sol.* **28**, 663 (1968).

36. J. R. Hook, *Solid State Physics*, 2nd ed., Wiley, Chichester, 1991.

37. H. Ibach, *Solid-State Physics: An Introduction to Principles of Materials Science*, 2nd ed., Springer, Berlin, 1995.

38. S. R. Elliott, *The Physics and Chemistry of Solids*, Wiley, Chichester, 1998.

39. O. Madelung, *Introduction to Solid-State Theory*, Springer, Berlin, 1996.

40. K. Seeger, *Semiconductor Physics: An Introduction*, 5th ed., Springer, Berlin, 1991.

41. J. Callaway, *Quantum Theory of the Solid State*, 2nd ed., Academic, Boston, 1991.

42. J. I. Pankove, *Optical Processes in Semiconductors*, Prentice-Hall, Englewood Cliffs, NJ, 1971.

43. K. W. Böer, *Survey of Semiconductor Physics: Electrons and Other Particles in Bulk Semiconductors*, Van Nostrand Reinhold, New York, 1990.

44. B. Henderson and G. R. Imbusch, *Optical Spectroscopy of Inorganic Solids*, Clarendon, Oxford, 1989.

45. C. F. Klingshirn, *Semiconductor Optics*, Springer, Berlin, 1997.

46. P. Y. Yu and M. Cordona, *Fundamentals in Semiconductors*, Springer, Berlin, 1996.

47. R. A. Paquin, in M. Bass (Ed.), *Handbook of Optics*, 2nd ed., Vol. 2, McGraw-Hill, New York, 1995, pp. 35.1–35.78.

48. P. M. Amirtharaj and D. G. Seiler, in M. Bass (Ed.), *Handbook of Optics*, Vol. 2, McGraw-Hill, New York, 1995, pp. 36.1–36.96.

49. F. Wooten, *Absorption and Dispersion. Optical Properties of Solids*, Academic, New York, 1972.

50. T. S. Moss and M. Balkanski (Series Eds.), *Handbook on Semiconductors, Vol. 2: Optical Properties of Solids*, Elseiver, Amsterdam, 1994.

51. V. M. Zolotarev, V. N. Morozov, and E. V. Smirnova, *Optical Constants of Natural and Technical Media*, Khimia, Leningrad, 1984 (in Russian).

52. L. Ward, *The Optical Constants of Bulk Materials and Films,* 2nd ed., Institute of Physics Publishing, Bristol and Philadelphia, 1994.

53. M. Hollas, *Fundamental Aspects of Vibrational Spectroscopy. Modern Spectroscopy*, 3rd ed., Wiley, Chichester, 1996.

54. G. Herzberg, *Molecular Spectra and Molecular Structure II. Infrared and Raman Spectra of Polyatomic Molecules*, Van Nostrand, New York, 1945.

55. B. Wilson, Jr., J. C. Decius, and P. C. Cross, *Molecular Vibrations*, McGraw-Hill, New York, 1955.

56. R. L. Carter, *Molecular Symmetry and Group Theory*, Wiley, New York, 1998.

57. P. C. Painter, M. M. Coleman, and J. L. Koenig, *The Theory of Vibrational Spectroscopy and Its Applications to Polymeric Materials*, Wiley-Interscience, New York, 1982.

58. B. Smith, *Infrared Spectral Interpretation—A Systematic Approach*, CRC Press, Boca Raton, FL, 1998.

59. C. N. Banwell and E. M. McCash, *Fundamentals of Molecular Spectroscopy*, 4th ed., McGraw-Hill, London, 1994.

60. J. M. Brown, *Molecular Spectroscopy*, Oxford University Press, Oxford, 1998.

61. M. Diem, *Introduction to Modern Vibrational Spectroscopy*, Wiley, New York, 1993.

62. H. Haken and H. C. Wolf, *Molecular Physics and Elements of Quantum Chemistry: Introduction to Experiments and Theory*, Springer, Berlin, 1995.

63. K. Nakamoto, *Infrared and Raman Spectra of Inorganic and Co-ordination Compounds, Part A*, 5th ed., Wiley, New York, 1997.

64. D. Bougeard, in B. Schrader (Ed.), *Infrared and Raman Spectroscopy*, VCH, Weinheim, 1995, p. 445.

65. L. Landau and E. Lifschitz, *Quantum Electrodynamics*, 2nd ed., Pergamon, Oxford, 1982.

66. W. G. Fateley, F. R. Dollish, N. T. McDevitt, and F. F. Bentley, *Infrared and Raman Selection Rules for Molecular and Lattice Vibrations*, Wiley-Interscience, New York, 1972.

67. H. Lorentz, *The Theory of Electrons*, Dover, New York, 1952.

68. D. L. Lynch, in E. D. Palik (Ed.), *Handbook of Optical Constants of Solids I*, Academic, New York, 1998, pp. 189–212.

69. P. M. A. Sherwood, *Vibrational Spectroscopy of Solids*, Cambridge University Press, Cambridge, 1972.

70. R. H. Lyddane, R. G. Sachs, and E. Teller, *Phys. Rev.* **59**, 673 (1941).

71. H. Poulet and J.-P. Mathieu, *Spectres de vibration te symmetry des cristaux*, Gordon and Breach, Paris, 1970.

72. P. Drude, *Theory of Optics*, Dover, New York, 1959.

73. B. Jensen, in C. Kittel (Ed.), *Introduction to Solid State Physics*, 7th ed., Wiley, New York, 1996, pp. 169–188.

74. M. Milosevic, *Appl. Spectrosc.* **47**, 566 (1993).

75. J.-G. Zhang, X.-X. Bi, E. McRae, and P. C. Eklund, *Phys. Rev. B* **43**, 5389 (1991).

76. J.-G. Zhang, G. W. Lehman, and P. C. Eklund, *Phys. Rev. B* **45**, 4660 (1992).

77. B. Szigetti, *Proc. R. Soc. A* **258**, 377 (1960).

78. V. P. Tolstoy, Ph.D. Thesis, Synthesis and IR Spectroscopic Characterization of Ultrathin Films on Metal Surfaces, Leningrad State Technological University, Leningrad, 1980.

79. J. Appelbaum and D. A. Haman, *Phys. Rev. Lett.* **43**, 1839 (1979).

80. V. M. Da Costa and L. B. Coleman, *Phys. Rev. B* **43**, 1903 (1991).

81. D. W. Berreman and F. C. Unterwald, *Phys. Rev.*, **173**, 791 (1968).

82. F. Gervais and B. Pirou, *Phys. Rev. A* **135**, A1732 (1964).

83. D. W. Berreman and F. C. Unterwald, *Phys. Rev.* **173**, 791 (1968).

84. W. Cochran and R. A. Cowley, *J. Phys. Chem. Solids* **23**, 447 (1962).

85. A. S. Barker and J. J. Hopfield, *Phys. Rev. A* **135**, A1732 (1964).

86. U. Fano, *Phys. Rev.* **124**, 1866 (1961).

87. O. S. Heavens, *Optical Properties of Thin Solid Films*, Dover, New York, 1965.

88. A. Vašiček, *Optics of Thin Films*, North-Holland, Amsterdam, 1960.

89. J. M. Bennett and H. E. Bennett, in M. Bass (Ed.), *Handbook of Optics*, Vol. 1, McGraw-Hill, New York, 1995, pp. 10-1 – 10-164.

90. R. Belali and J. M. Vigoureux, *J. Opt. Soc. Am. B* **11**, 1197 (1994).

91. R. Belali, J. M. Vigoureux, and M. Camelot, *Spectrochim. Acta* **43A**, 1261 (1987).

92. D. L. Mills (Ed.), *Surface Polaritons. Electromagnetic Waves at Surfaces and Interfaces*, North-Holland, Amsterdam, 1982.

93. W. Knoll, *Annu. Rev. Phys. Chem.* **49**, 569 (1998).

94. F. Goos and H. Hänchen, *Ann. Phys.* **6**(1), 333 (1947).

95. K. Artmann, *Ann. Phys.* **6**(2), 87 (1948).

96. A. K. Ghatak and K. Thyagarajan, *Evanescent Waves and the Goos-Hänchen Effect. Contemporary Optics*, Plenum, New York, 1978, Chapter 11.

97. V. G. Fedoseyev, *J. Opt. Soc. Am. A* **3**, 826 (1986).

98. F. Falco and T. Tamur, *J. Opt. Soc. Am. A* **7**, 195 (1990).

99. W. N. Hansen, in P. Delahay and C. W. Tobias (Eds.), *Advances in Electrochemistry and Electrochemical Engineering*, Vol. 9, Wiley, New York, 1973, pp. 1–60.

100. W. N. Hansen, *J. Opt. Soc. Am.* **58**, 380 (1968).

101. J. D. E. McIntyre, in P. Delahay and C. W. Tobias (Eds.), *Advances in Electrochemistry and Electrochemical Engineering*, Vol. 9, Wiley, New York, 1973, pp. 61–166.

102. I. Newton, *Optics*, Book III, Part 1, Quaery 29, Dover, New York, 1952.

103. I. Newton, in A. Koyre and I. B. Cohen (Eds.), *Philosphiae naturalis principia mathematica*, Vol. I, 3rd ed., Harvard University Press, Cambridge, MA, 1972.

104. F. Abeles, *Ann. Phys.* **5**, 596 (1950).

105. F. Abeles, *Ann. Phys.* **5**, 706 (1950).

106. R. A. Dluhy, *J. Phys. Chem.* **90**, 1373 (1986).

107. V. P. Tolstoy and S. N. Gruzinov, *Opt. Spectrosc.* **63**, 489 (1987).

108. S. N. Gruzinov and V. P. Tolstoy, *J. Appl. Spectrosc. (Russ.)* **23**, 480 (1987).

109. J. D. E. McIntyre and D. E. Aspnes, *Surf. Sci.* **24**, 417 (1971).

110. J. D. E. McIntyre, in B. O. Seraphin (Ed.), *Optical Properties of Solids: New Developments*, North-Holland, Amsterdam, 1976, pp. 555–630.

111. U. Teschner and K. Hubner, *Phys. Stat. Sol. (b)* **159**, 917 (1990).

112. H. Schopper, *Z. Phys.* **132**, 146 (1952).

113. V. L. Kuzmin, V. P. Romanov, and A. V. Mikhailov, *Opt. Spectrosc.* **73**, 3 (1992).

114. R. Mendelsohn, J. W. Brauner, and A. Gericke, *Annu. Rev. Phys. Chem.* **46**, 305 (1995).

115. K. Yamamoto and H. Ishida, *Appl. Spectrosc.* **48**, 775 (1994).

116. M. J. Dignam and M. Moskovits, *J. Chem. Soc. Faraday Trans. II* **69**, 56 (1973).

117. P. R. Griffiths and J. A. de Haseth, *Fourier Transform Infrared Spectrometry*, Wiley, New York, 1986.

118. R. G. Greenler, *J. Chem. Phys.* **44**, 310 (1966).

119. W. N. Hansen and W. A. Abdou, *J. Opt. Soc. Am.* **67**, 1537 (1977).

120. K. Yamamoto and H. Ishida, *Vib. Spectrosc.* **15**, 27 (1999).

121. R. Ortiz, A. Cuesta, O. P. Marquez, J. Marquez, and C. Gutierrez, *J. Electroanal. Chem.* **465**, 234 (1999).

122. Y. Borensztein and F. Abeles, *Thin Solid Films* **125**, 129 (1985).

123. J. A. Mielczarski and R. H. Yoon, *J. Chem. Phys.* **93**, 2034 (1989).

124. T. Hasegawa, J. Umemura, and T. Takenaka, *J. Phys. Chem.* **97**, 9009 (1993).

125. F. Abeles, *Thin Solid Films* **34**, 291 (1976).

126. K. Yamamoto and H. Ishida, *Appl. Opt.* **34**, 4177 (1995).

127. R. Brendel, *J. Appl. Phys.* **72**, 794 (1992).

128. T. Buffeteau, B. Desbat, E. Pere, and J. M. Turlet, *Mikrochim. Acta* **14**, 631 (1997).

129. F. Abeles, *J. Opt. Soc. Am.* **47**, 473 (1957).

130. F. Abeles, in A. C. S. Van Heel (Ed.), *Advanced Optical Techniques*, North-Holland, Amsterdam, 1967, Chapter. 5.

131. R. M. A. Azzam and N. M. Bashara, *Ellipsometry and Polarized Light*, North-Holland, Amsterdam, 1977.

132. K. Ohta and H. Ishida, *Appl. Opt.* **29**, 1952 (1990).

133. B. Harbecke, *Appl. Phys. B* **39**, 165 (1986).

134. K. Ohta and H. Ishida, *Appl. Opt.* **29**, 2466 (1990).

135. D. W. Berreman, *J. Opt. Soc. Am.* **62**, 502 (1972).

136. A. I. Semenenko and F. S. Mironov, *Opt. Spectrosc.* **41**, 456 (1976).

137. S. Teitler and B. Henvis, *J. Opt. Soc. Am.* **60**, 830 (1970).

138. A. N. Parikh and D. L. Allara, *J. Chem. Phys.* **96**, 927 (1992).

139. H. Wohler, M. Fritsch, G. Haas, and D. A. Mlynski, *J. Opt. Soc. Am. A* **8**, 536 (1991).

140. E. Cojocaru, *Appl. Opt.* **36**, 2825 (1997).

141. T. Hasegawa, S. Takeda, A. Kawaguchi, and J. Umemura, *Langmuir* **11**, 1236 (1995).

142. K. Yamamoto and H. Ishida, *Vib. Spectrosc.* **8**, 1 (1994).

143. T. Buffeteau, D. Blaudez, E. Pere, and B. Desbat, *J. Phys. Chem. B* **103**, 5020 (1999).

144. U. A. Popov, *Zhurnal Prikladnoi Spestroskopii* **43**, 138 (1985).

145. J. M. Leng, J. J. Sidorowich, Y. D. Yoon, J. Opsal, B. H. Lee, G. Cha, J. Moon, and S. I. Lee, *J. Appl. Phys.* **81**, 3570 (1997).

146. Z. Knittl, *Optics of Thin Films*, Wiley, New York, 1976.

147. M. Milosevic, *Appl. Spectrosc.* **47**, 566 (1993).

148. T. C. Fry, *J. Opt. Soc. Am.* **22**, 307 (1932).

149. A. Francis and A. H. Ellison, *J. Opt. Soc. Am.* **49**, 131 (1959).

150. Y. Ishino and H. Ishida, *Appl. Spectrosc.* **42**, 1296 (1988).

151. C. Sellitti, J. L. Koehig, and H. Ishida, *Appl. Spectrosc.* **44**, 830 (1990).

152. Y. Suzuki, M. Osawa, A. Hatta, and W. Suetaka, *Appl. Surf. Sci.* **33/34**, 875 (1988).

153. R. P. Sperline, J. S. Jeon, and S. Raghavan, *Appl. Spectrosc.* **49**, 1178 (1995).

154. R. P. Sperline, S. Muralidharan, and H. Freiser, *Langmuir* **3**, 198 (1987).

155. B. Harbecke, B. Heinz, and P. Grosse, *Appl. Phys. A* **38**, 263 (1985).

156. N. J. Harrick, *Internal Reflection Spectroscopy*, Wiley, New York, 1967.

157. Y. Ishino and H. Ishida, *Surf. Sci.* **230**, 299 (1988).

158. Y. Ishino and H. Ishida, *Anal Chem.* **58**, 2448 (1986).

159. M. R. Philpott and J. D. Swalen, *J. Phys. Chem.* **69**, 2912 (1978).

160. A. Brillante, M. R. Philpott, and I. Pockland, *J. Phys. Chem.* **70**, 5739 (1979).

161. J. A. Mielczarski, *J. Phys. Chem.* **97**, 2649 (1993).

162. A. M. Bradshaw and E. Schweizer, in R. J. H. Clark and R. E. Hester (Eds.), *Spectroscopy of Surfaces*, Wiley, New York, 1988, pp. 413–483.

163. T. Hirschfeld, *Appl. Spectrosc.* **31**, 243 (1977).

164. D. J. Epstein, *Appl. Spectrosc.* **34**, 233 (1980).

165. N. J. Harrick, *Appl. Spectrosc.* **41**, 1 (1987).

166. H. G. Tompkins, *Appl. Spectrosc.* **28**, 335 (1974).

167. Y. Lu, J. Drelich, and J. D. Miller, *J. Colloid Interface Sci.* **202**, 462 (1998).

168. R. P. Sperline, *Appl. Spectrosc.* **45**, 677 (1991).

169. W. H. Jang and J. D. Miller, *Langmuir* **11**, 3159 (1993).

170. Y. Lu, J. Drelich, and J. D. Miller, *J. Colloid Interface Sci.* **202**, 462 (1998).

171. M. L. Free, W. H. Jang, and J. D. Miller, *Colloid Surf.* **93**, 127 (1994).

172. J. S. Jeon, R. S. Sperline, and S. Raghavan, *Appl. Spectrosc.* **46**, 1644 (1992).

173. U. P. Fringeli, in F. M. Mirabella, Jr. (Ed.), *Internal Reflection Spectroscopy, Theory and Application*, Marcel Dekker, New York, 1993, p. 255.

174. W. G. Pitt and S. L. Cooper, *Biomed. Mater. Res.* **22**, 359 (1988).

175. C. F. Bohren and D. R. Huffman, *Absorption and Scattering of Light by Small Particles*, Wiley, New York, 1998.

176. C. Bohren, in M. Bass (Ed.), *Handbook of Optics*, Vol. 1, McGraw-Hill, New York, 1995, pp. 6.1–6.21.

177. U. Kreibig and M. Vollmer, *Optical Properties of Metal Clusters*, Springer, Berlin, 1995.

178. J. C. Maxwell-Garnett, *Philos. Trans. R. Soc. (Lond.)A* **203**, 358 (1904).

179. R. Clausius, *Die mechanistische Wärmelehre II*, Vieweg, Braunschweig, 1879.

180. O. F. Mossotti, *Mem. Soc. Scient. Modena* **14**, 49 (1850).

181. H. A. Lorenz, *Wiedem. Ann.* **9**, 641 (1880).

182. L. Lorenz, *Wiedem. Ann.* **11**, 70 (1881).

183. F. Abeles, Y. Borensztein, and T. Lopez-Rios, in P. Grosse (Ed.), *Advances in Solid State Physics*, Vol. 34, Vieweg, Braunschweig, 1984, pp. 93–117.

184. G. Mie, *Ann. Phys.* **25**, 377 (1908).

185. P. Lilienfield, *Appl. Opt.* **30**, 4696 (1991).

186. L. Genzel and T. P. Martin, *Surf. Sci.* **34**, 33 (1973).

187. S. Barker, *Phys. Rev. B* **7**, 2507 (1972).

188. N. Sen, *Appl. Phys. Lett.* **39**, 667 (1981).

189. D. A. G. Bruggeman, *Ann. Phys. (Leipz.)* **5**, 636 (1935).

190. D. Podler and J. H. Van Santen, *Physica* **12**, 257 (1946).

191. E. C. Stoner, *Philos. Mag.* **36**, 803 (1945).

192. J. A. Osborn, *Phys. Rev.* **67**, 351 (1945).

193. D. E. Aspnes, *Thin Solid Films* **89**, 249 (1982).

194. R. A. Buhrman and H. G. Craighead, in L. E. Murr (Ed.), *Solar Materials Science*, Academic, New York, 1980, p. 277.

195. M. F. MacMillan, R. P. Devaty, and J. V. Mantese, *Phys. Rev. B* **43**, 13838 (1991).

196. J. C. Garland and D. B. Tanner (Eds.), *Electrical Transport and Optical Properties of Inhomogeneous Media (Ohio State University, 1977)*, Proceedings of the First Conference on the Electrical Transport and Optical Properties of Inhomogeneous Media, AIP Conf. Proc. No. 40, American Institute of Physics, New York, 1977.

197. A. M. Dykhne, A. N. Lagarkov, and A. K. Sarychev (Eds.), *Physica A* **241** (1997).

198. A. V. Radchik, P. Moses, I. L. Skryabin, and G. B. Smith, *Thin Solid Films* **317**, 446 (1998).

199. R. Gans, *Ann. Phys.* **37**, 881 (1912).

200. E. David, *Z. Phys.* **114**, 389 (1939).

201. F. L. Galleener, *Phys. Rev. Lett.* **27**, 421 (1971).

202. F. L. Galleener, *Phys. Rev. Lett.* **27**, 769 (1971).

203. R. W. Cohen, G. D. Cody, M. D. Coutts, and B. Abeles, *Phys. Rev. B* **8**, 3689 (1973).

204. R. Ossikovski and B. Drevillon, *Phys. Rev. B* **54**, 10530 (1996).

205. R. Landauer, in Ref. 196, p. 2.

206. D. Polder and J. H. von Santen, *Physika (Utr.)* **12**, 257 (1946).

207. S. Hayashi, N. Nakamori, and H. Kanamori, *J. Phys. Soc. Jpn.* **46**, 176 (1979).

208. O. Wiener, *Abn. Leipzig. Acad.* **32**, 509 (1912).

209. C. G. Granqvist and O. Hunderi, *Phys. Rev. B* **18**, 1554 (1978).

210. C. G. Granqvist and O. Hunderi, *Phys. Rev. B* **18**, 2897 (1978).

211. H. Schopper, *Z. Phys.* **131**, 215 (1952).

212. E. Dobierzenska-Mozrzymas, A. Radosz, and P. Bieganski, *Appl. Opt.* **24**, 727 (1985).

213. J. Sinzig, U. Radtke, M. Quinten, and U. Kreibig, *Z. Physik D* **26**, 242 (1993).

214. R. Ruppin and T. Yamaguchi, *Surf. Sci.* **127**, 108 (1983).

215. D. Stroud and F. P. Pan, *Phys. Rev. B* **17**, 1602 (1978).

216. G. A. Niklasson, C. G. Granqvuist, and O. Hunderi, *Appl. Opt.* **20**, 26 (1981).

217. G. A. Niklasson and C. G. Granqvuist, *J. Appl. Phys.* **55**, 3382 (1984).

218. C. Bohren, *J. Atmos. Sci.* **43**, 468 (1986).

219. R. Luo, *Appl. Opt.* **36**, 8153 (1997).

220. M. Gomez, L. F. Fonseca, L. Cruz, and V. Vargas, in L. Blum and F. B. Malik (Eds.), *Condensed Matter Theories*, *Proceedings of the 16th International Workshop*, Vol. 8, Plenum, New York, 1993, pp. 109–114.

221. R. Fuchs and F. Claro, *Phys. Rev. B* **35**, 3722 (1987).

222. N. Liver, A. Nitzan, and K. F. Feed, *J. Chem. Phys.* **82**, 3831 (1985).

223. A. P. Vinogradov, *Physica A* **241**, 216 (1997).

224. R. K. Vincent and G. R. Hunt, *Appl. Opt.* **7**, 53 (1968).

225. J. Brimmer, P. R. Griffiths, and N. J. Harrick, *Appl. Spectrosc.* **40**, 258 (1986).

226. J. Brimmer and P. R. Griffiths, *Appl. Spectrosc.* **41**, 791 (1987).

227. J. Brimmer and P. R.Griffiths, *Anal. Chem.* **58**, 2179 (1987).

228. J. H. Lambert, *Photometria*, Illuminating Engineering Society of North America, New York, 2001.

229. (a) G. Hecht, *J. Research Nat. Bur. Stand* **80A**, 567 (1976); (b) W. W. Wendlandt and H. G. Hecht, *Reflectance Spectroscopy*, Interscience, New York, 1966.

230. B. Hapke, *Theory of Reflectance and Emittance Spectroscopy*, Cambridge University Press, Cambridge, 1993.

231. J. E. Iglesias, M. Ocana, and C. J. Serna, *Appl. Spectrosc.* **44**, 418 (1990).

232. G. Kortum, *Reflectance Spectroscopy*, Springer-Verlag, New York, 1969.

233. H. C. van de Hulst, *Absorption and Scattering of Light by Small Particles*, Wiley, New York, 1957.

234. C. F. Bohren (Ed.), *Selected Papers on Scattering in the Atmosphere*, SPIE Vol. MS 7, SPIE Optical Engineering, Bellingham, WA, 1989.

235. M. Kerker, P. Scheiner, and D. D. Cooke, *J. Opt. Soc. Am.* **68**, 135 (1978).

236. E. J. McCartney, *Optics of the Atmosphere*, Wiley, New York, 1976.

237. D. J. Dahm and K. D. Dahm, *Appl. Spectrosc.* **53**, 647 (1999).

238. M. M. Gurevich, *Physik Ztshr.* **31**, 753 (1930).

239. P. Kubelka and F. Munk, *Ztshr. Techn. Phys.* **11a**, 593 (1931).

240. G. Hecht, *J. Res. Nat. Bur. Stand* **80A**, 567 (1976).

241. A. V. Bortkevich and M. M. Seredenko, *Opticheskii Zhurnal* **65**, N3, 3 (1998) (in Russian).

242. T. Burger, H. J. Ploss, J. Kuhn, S. Ebel, and J. Fricke, *Appl. Spectrosc.* **51**, 1323 (1997).

243. F. Boroumand, J. E. Moser, and H. van den Bergh, *Appl. Spectrosc.* **46**, 1874 (1992).

244. L. J. Fina and Y.-Sh. Tung, *Appl. Spectrosc.* **45**, 986 (1991).

245. S. Ekgasit, *Appl. Spectrosc.* **52**, 773 (1998).

2

OPTIMUM CONDITIONS FOR RECORDING INFRARED SPECTRA OF ULTRATHIN FILMS

One common problem when examining ultrathin films on various surfaces and at various interfaces by IR spectroscopy is that of selecting both the best IR method [transmission, IR reflection–absorption spectroscopy (IRRAS), ATR, DT, or DR] and the best experimental geometry (optical configuration) for this method. For films on plane substrates, this can be done using the optical theory introduced in Chapter 1 (Sections 2.1–2.6). In the case of powdered substrates, the optimum conditions are chosen based on the general theoretical and empirical regularities (Section 2.7).

Before proceeding to analysis of the optimum conditions for different IR spectroscopic techniques, it is useful to recall [1] that "it is easy to lose sight of the fact that these different techniques should all be compared to the benchmark transmission method. There is one overriding factor that should always be kept in mind: *If it can be done using transmission spectroscopy then do it!*" (p. 187). In fact, when selecting a technique, the experimental simplicity and lower cost of the transmission measurements is often forgotten.

Throughout this chapter, the nomenclature accepted in Figs. 1.12 and 1.14 for describing the optical parameters of layered systems is used.

2.1. IR TRANSMISSION SPECTRA OBTAINED IN POLARIZED RADIATION

If the substrate on which a nanolayer is located is transparent in the region of the nanolayer absorption, the spectrum of the nanolayer can be obtained by the transmission technique. According to this technique, a sample is inserted into the radiation beam of a spectrophotometer and the intensity of the transmitted radiation is recorded. In the case of zero angle of incidence (*normal-incidence transmission*), the main difficulties arise from insufficient spectral contrast and interference effects from the components of the radiation beam reflected inside the substrate. Thus, in practice, this technique is used mainly for the investigation of relatively thick layers (tens or hundreds of nanometers). The most typical of these studies is the investigation of the dielectric layers on semiconductors used in microelectronics [2–5] and Langmuir–Blodgett (LB) multilayers (Section 3.11). However, the potential applications of IR transmission spectroscopy can be extended by using polarized radiation and oblique angles of incidence (*oblique-incidence transmission*) [6–17]. As will be shown below, this effectively increases the optical path of the radiation in the layer under investigation, and the spectral contrast is enhanced proportionally.

The simplest optical schemes for recording transmission spectra of nanolayers are depicted in Figs. 2.1*a* and *b*. The advantage of these schemes is the absence of interference in the spectrum, as the substrate has only one planar surface. Calculations of the quantities T_0, ΔT, and $\Delta T/T$ as a function of the angle of incidence for the optical scheme presented in Fig. 2.1*a* are plotted in Fig. 2.2.

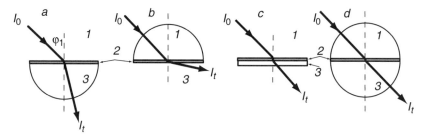

Figure 2.1. Optical schemes for recording oblique-incidence transmission spectra of nanolayers, in which layer is located (*a, b*) on surface of hemicylinder, (*c*) on surface of plane–parallel plate, (*d*) between two hemicylinders: (1) immersion medium with refractive index n_1, (2) layer under investigation of thickness d_2 and with optical constants n_2 and k_2, (3) transparent substrate with refractive index n_3.

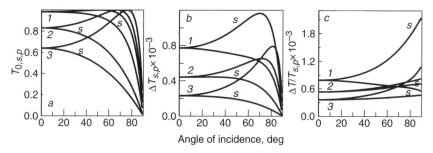

Angle of incidence, deg

Figure 2.2. (*a*) Dependence of *s*- and *p*-polarized transmittance $T_{0,s,p}$ of bare substrates, (*b*) absorption depth $\Delta T_{s,p}$, and (*c*) transmissivity $\Delta T/T_{s,p}$ on angle of incidence: $n_1 = 1$, $d_2 = 1$ nm, $n_2 = 1.564$, $k_2 = 0.384$, $\nu = 1200$ cm^{-1}; (1) $n_3 = 1.35$ (CaF$_2$), (2) 2.4 (ZnSe), (3) 4.0 (Ge).

The calculations were performed for substrates with n_3 equal to 1.35, 2.4, and 4.0 and a 1-nm layer; the oscillator parameters are $\varepsilon_\infty = 2.3$, $\nu_0 = 1200$ cm^{-1}, $\gamma = 15$ cm^{-1}, $S = 0.015$ (at the maximum of the $\nu = 1200$-cm^{-1} absorption band, $n_2 = 1.564$ and $k_2 = 0.384$), typical for a medium-strength IR band of an organic compound. The difference between the transmittance of the substrate without the layer and with it (absorption depth), $\Delta T = T_0 - T$, and the transmissivity $\Delta T/T$ [Eq. (1.96)] were used to estimate the degree of absorption of the radiation in the layer. One can see from Fig. 2.2a that, for all substrates, as the angle of incidence φ_1 increases, $T_{0,s}$ decreases smoothly from 0.978 (for CaF$_2$, $n_3 = 1.35$) and 0.64 (for Ge, $n_3 = 4.0$) at $\varphi_1 = 0°$ to zero at $\varphi_1 = 90°$. At the same time, $T_{0,p}$ first increases up to 1 as φ_1 increases from $0°$ to φ_B, but then decreases to zero at $\varphi_1 = 90°$. When the refractive index of the substrate is greater than that of the film, $n_3 > n_2$, the maximum values of ΔT and $\Delta T/T$ (Fig. 2.2b, c) are reached with p-polarization at grazing angles of incidence (these values are higher for Ge than for ZnSe). Otherwise, when $n_3 < n_2$ (curves 1), $T_p > T_s$, but $\Delta T_p < \Delta T_s$ and $\Delta T/T_p < \Delta T/T_s$, and the optimum angle is $\sim 70°$ (see also Fig. 3.24). This means that the maximal contrast in the transmission spectra can be achieved with either p- or s-polarization depending on whether the refractive index of the film is greater or less than that of the substrate. Since the refractive index n_2 varies substantially over one absorption band for strongly absorbing substances (Figs. 1.1 and 3.11), the transmission spectra of such films on substrates with low and high refractive indexes will have different shapes when recorded under the optimum conditions. This effect can be clearly seen in the dependence of ΔT_p and ΔT_s on φ_1 at the ν_{TO} and ν_{LO} frequencies that characterize the absorption band of the SiO$_2$ layer in the 1300–1000-cm^{-1} region (Fig. 2.3). For the film on a CaF$_2$ substrate, the absorption is maximal at the ν_{TO} frequency when measured with s-polarized radiation at an angle of incidence of $65°$–$70°$, while in the case of a Ge substrate the ΔT value reaches a maximum at ν_{LO} in the

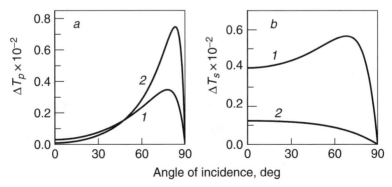

Figure 2.3. Angular dependence of absorption depth (a) ΔT_p at $\nu_{LO} = 1220$ cm^{-1} ($n_2 = 0.54$, $k_2 = 0.44$) and (b) ΔT_s at $\nu_{TO} = 1075$ cm^{-1} ($n_2 = 2.15$, $k_2 = 1.66$) for SiO$_2$ layer 1 nm thick on two substrates: (1) $n_3 = 1.35$ and (2) $n_3 = 4.0$; $n_1 = 1$. Optical constants of SiO$_2$ layer correspond to those of layer obtained by chemical deposition [2].

p-polarized spectra measured at grazing angles of incidence. Such a dependence of the spectrum shape on polarization is common for strongly absorbing films (see Sections 3.1–3.5 for more detail).

If the beam of IR radiation is incident on the layer from the medium of a larger refractive index (Fig. 2.1b), the transmission spectrum of the layer can be obtained only at $\varphi_1 < \varphi_c$, due to total internal reflection (1.4.10°). For example, with air as the input medium, the spectrum must be measured at $\varphi_1 < 47.78°$, 24.62°, 14.48° for substrates with refractive indices of 1.35, 2.4, and 4.0, respectively. The maximum values of ΔT and $\Delta T/T$ are reached as φ_1 approaches φ_c, when the optical path of IR radiation through the layer under investigation is maximized (Fig. 2.4). However, the reverse passage of the beam provides no gain in the spectrum contrast as compared to the geometry depicted in Fig. 2.1a. It also follows from Fig. 2.4 that if the refractive index of the substrate is greater than that of the film, $n_1 > n_2$, then $\Delta T_s > \Delta T_p$ and $\Delta T/T_s > \Delta T/T_p$. When $n_1 < n_2$, the interrelation of these quantities is opposite. Hence, as for the geometry depicted in Fig. 2.1a, the optimum polarization of IR radiation depends on whether the refractive index of the substrate is greater or less than that of the film.

In the case of a planar substrate supporting a nanolayer (Fig. 2.1c), the transmission spectrum of the layer contains interference signals (Figs. 2.5 and 2.6). Both the refractive index of the substrate (Fig. 2.5a–d) and its thickness (Fig. 2.5e–h) affect the interference pattern. Substrates with lower refractive indices and larger thicknesses exhibit less interference. As seen in Fig. 2.6, the polarization and the angle of incidence also exerts a definite influence on the spectra. A comparison of the spectra in Figs. 2.5 and 2.6 indicates that the interference is minimized with p-polarized radiation at an angle of incidence equal to the Brewster angle of the substrate (Fig. 2.6e). Under these conditions, light passes through the substrate with virtually no reflection from the interfaces (1.4.15°), which allows one to obtain the multiple transmission spectrum, with spectral contrast many times higher than that of a single transmission spectrum [7–9].

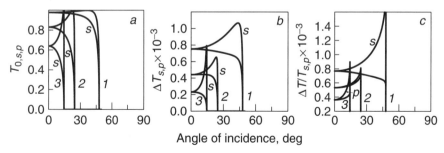

Figure 2.4. Dependence of s- and p-polarized (a) transmittance $T_{0,s,p}$, (b) absorption depth $\Delta T_{s,p}$, and (c) transmissivity $\Delta T/T_{s,p}$ on angle of incidence of IR radiation for different refractive indices of input medium (=substrate): (1) $n_1 = 1.35$, (2) $n_1 = 2.4$, and (3) $n_1 = 4.0$; $d_2 = 1$ nm, $n_2 = 1.564$, $k_2 = 0.384$, $\nu = 1200$ cm^{-1}, $n_3 = 1$.

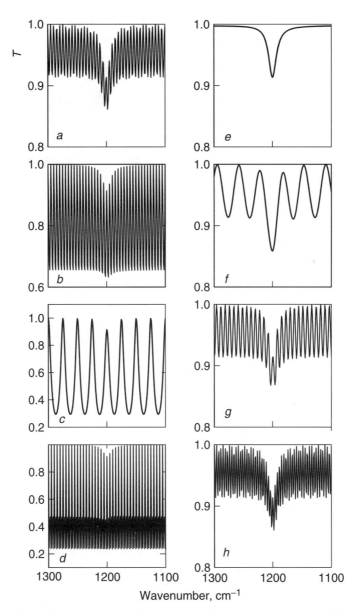

Figure 2.5. Dependence of normal-incidence ($\varphi_1 = 0°$) transmission spectra of 100-nm layer, whose optical parameters are $\varepsilon_\infty = 2.3$, $\nu_0 = 1200$ cm^{-1}, $\gamma = 15$ cm^{-1}, $S = 0.015$, on (a–d) refractive index and (e–h) thickness of substrate. Substrates are plane–parallel plates 3 mm thick with n_3 values of (a) 1.35, (b) 2.0, (c) 3.4, and (d) 4.0 and with $n_3 = 1.35$ and thicknesses d_3 of (e) 0, (f) 0.1, (g) 0.5, and (h) 1 mm.

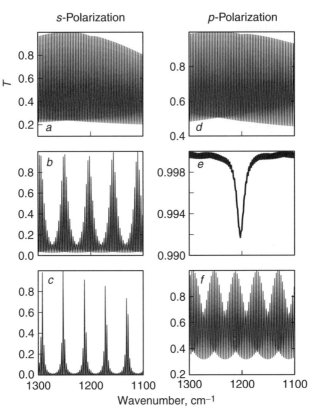

Figure 2.6. The s- and p-polarized oblique-incidence transmission spectra of 100-nm layer with optical parameters listed in Fig. 2.5. Layer is located on surface of 0.5-mm plane–parallel Si plate ($n_3 = 3.4$). Spectra are obtained at φ_1 values of (a, d) 40°, (b, e) 74°, and (c, f) 85°.

2.2. IRRAS SPECTRA OF LAYERS ON METALLIC SURFACES ("METALLIC" IRRAS)

The first attempts to apply IR spectroscopy to the investigation of ultrathin films on bulk metal samples were made in the late 1950s by Fransis and Ellison [18] and Pickering and Eckstrom [19] and in the early 1960s by Hannah [20] and Babushkin [21]. Already at that time, the significant influence of the angle of incidence and the type of polarization on the spectral sensitivity was recognized. Wider application of this method began when Greenler [22–24], using the model of an adsorbed layer on the surface of 19 metals, considered the interaction of IR radiation with a sample from the perspective of physical optics. To distinguish between measurement of the reflection spectra of a layer on a highly reflecting metal surface and that of a bulk sample, Greenler referred to the former measurement as *IR reflection–absorption spectroscopy* (IRRAS, IRAS, or RAIRS). This term and the terms *external reflection spectroscopy* (ERS) and *transflectance* are used in current scientific literature.

It should be emphasized that IRRAS is the only practical method for measuring IR spectra of layers on bulk metals and highly doped semiconductors. An alternative is the ATR in Otto's configuration (see Fig. 2.36d later).

Features of the IRRAS studies of ultrathin films on metals are discussed in Sections 3.1, 3.3.1, 3.5, 3.6, and 7.1.2. We now wish to consider the effect of the angle of incidence and the optical constants (n_3 and k_3) of a metallic substrate on the contrast in a spectrum measured by IRRAS. The angle-of-incidence dependence of the p-polarized absorption depth ΔR (1.77) at $v = 1200$ cm^{-1} for a 1-nm hypothetical layer with $n_2 = 1.564$ and $k_2 = 0.384$ on Al and Ti substrates is presented in Fig. 2.7. It is seen that for a strongly reflecting metal such as Al, this dependence has maximum at $\varphi_1 \approx 88°$. A similar dependence of ΔR on the angle of incidence for a strongly absorbing hypothetical layer simulating an oxide was obtained in Ref. [25]. For weakly reflecting metals such as Ti, the absolute values of absorption depth are several times smaller, and the dependence of ΔR is not so extreme. Calculations show that a lowering of the reflectance of the metal results in a decrease of $\Delta R/R_p$ to a lesser extent; this is because weakly reflecting metals have a smaller value of $R_{0,p}$ and thus a larger value of the ratio $\Delta R/R_p$.

It has been assumed above that the incident beam is parallel and the detector size is as large as all the radiation reflected from the sample detected. In practice, however, radiation is focused in an angular range of $5°-12°$, while the projection of the image of the source on the sample at grazing angles of incidence can be larger than the sample dimensions. Therefore angles of $80°-85°$ are more advantageous for single-reflection experiments.

The spectral contrast can be improved using multiple reflections between parallel metallic mirrors (see Fig. 4.7). However, increasing the number of reflections leads to a lowering of the reflectance. If one set the intensity of the radiation, reflected N times between metallic plates with no layer, equal to R_0^N and that with the layer equal to R^N, then the intensity of the absorption bands (the absorption depth, 1.5.3°) can be written as

$$\Delta R(N) = R_0^N - R^N. \tag{2.1}$$

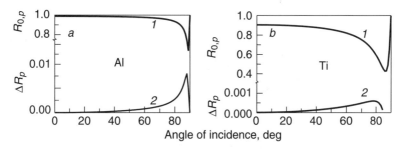

Figure 2.7. Dependence on angle of incidence of (1) reflectance $R_{0,p}$ of bare substrate and (2) absorption depth ΔR_p for 1-nm layer ($\hat{n}_3 = 1.564 - 0.384i$) at (a) Al ($\hat{n}_3 = 18.6 - 77i$) and (b) Ti ($\hat{n}_3 = 6.5 - 14.8i$) substrate; $v = 1200$ cm^{-1}; p-polarization, one reflection.

The quantity ΔR reaches its maximum value at N_{opt}, defined as

$$R^{N_{\text{opt}}} = \frac{1}{e} \approx 0.37. \qquad (2.2)$$

It follows from Ref. [24] that for strongly reflecting metals such as Au, Cu, Al, and Ag the maximum value of ΔR can be achieved by using many reflections before the background drops to 37%. On the other hand, only a few reflections (in most cases, one) should be used for weak reflectors. Another expression for $R_0^{N_{\text{opt}}}$,

$$R_0^{N_{\text{opt}}} \approx 0.1\text{--}0.2, \qquad (2.3)$$

holds when spectra are measured using a double-beam dispersive spectrometer [25].

Figure 2.7 shows that reflectance R varies with the angle of incidence. Consequently, an optimum number of reflections N_{opt} must exist for each angle. Figure 2.8a illustrates the dependence on the angle of incidence of the optimum values N_{opt} calculated with Eq. (2.2). It is seen that N_{opt} is smaller for large angles. The angular dependence of the absorption depths in the spectra obtained with the multiple-reflection technique is less pronounced than in the case when the single-reflection method is used. As an example, it can be seen in Fig. 2.8b that increasing the angle of incidence from $70°$ to $85°$ practically does not affect ΔR for the film on Al.

In practice, the number of reflections N is limited by the length of the sample. Moreover, a real beam of radiation will converge. An angular divergence of $\pm 6°$ reduces the ideal maximum $\Delta R/R$ value and shifts its position by $\sim 5°$ to less grazing angles [26]. As a results, the optimal angle of incidence is between $\varphi_1 = 75°$ and $\varphi_1 = 80°$.

Finally, it should be mentioned that the thickness dependence of the absorption depth of a strong absorber measured by metallic IRRAS at grazing angles of incidence is an approximately linear function at $d_2 < 20$ nm (Fig. 3.14 in Section 3.3.1). For weak-to-medium absorbers such as the νCH_2 and $\nu C{=}O$ modes, the linearity range extends up to ~ 60 nm [69] (see also Fig. 3.73 in Section 3.10).

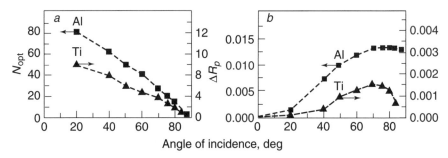

Figure 2.8. Dependence on angle of incidence of (a) optimum number of reflections N_{opt} and (b) absorption depth ΔR_p for N_{opt} reflections for 1-nm layer ($\hat{n}_3 = 1.564 - 0.384i$) at Al ($\hat{n}_3 = 18.6 - 77i$) and Ti ($\hat{n}_3 = 6.5 - 14.8i$) substrate; $\nu = 1200$ cm^{-1}; p-polarization.

Thus, in the case of weakly reflecting metal substrates, the maximum intensity of the IRRAS spectrum can be achieved with one reflection, whereas in the case of strongly reflecting metals, the optimum number of reflections, N_{opt}, should be calculated with Eqs. (2.2) or (2.3), depending on the type of spectrometer used. The optimum angles of incidence are $80-85°$ and $75°-80°$, respectively, for single and multiple reflection. In addition, surface sensitivity of the method can essentially be enhanced by using the "immersion medium" technique (Section 2.5.2).

2.3. IRRAS OF LAYERS ON SEMICONDUCTORS AND DIELECTRICS

The IRRAS method can be used to obtain information about ultrathin films not only at metals but also on semiconductor and dielectric (including liquid) substrates. This class of problem is applicable to many areas, including thin-film optics, electronic and electroluminescent devices [27] (Chapter 5), sensors and transducers [28], flotation technology [29] (Section 7.4.4), and biomedical problems [30, 31]. Although the sensitivity is much lower than when metallic substrates are used, the waiving of the metal selection rule allows both s- and p-polarized spectra to be measured and thus a more thorough investigation of molecular orientation within the layer.

2.3.1. Transparent and Weakly Absorbing Substrates ("Transparent" IRRAS)

For semiconductor and dielectric in the range of their transparency, the angular dependence of reflectance R substantially differs from the analogous dependence for metals (Fig. 1.11). A characteristic property of transparent media is the existence of the polarizing Brewster angle φ_B at which the intensity of the reflected component of p-polarized radiation is equal to zero (1.4.15°). The value of this angle [Eq. (1.63)] at $\nu = 3000$ cm^{-1} is equal to $55.6°$, $73.6°$, and $76°$ for quartz, Si, and Ge, respectively.

To determine the dependence of reflectance R and absorption depth ΔR (proportional to the SNR in a detector noise-limited spectrometer [32]) on the angle of incidence φ_1 for transparent, weakly and strongly absorbing layers, the optical modeling methods described in Section 1.7 may be used [33]. Analogous calculations have been done for Cu$_2$S and water [34], glass and indium tin oxide (ITO) [35], GaAs [36], water [37], glass and Si [33, 38], and glass and Ge [39]. Calculations for quartz and Si covered with a 1-nm hypothetical, isotropic layer with $n_2 = 1.3$ and the maximum value of $k_2 = 0.1$ in the absorption band are plotted in Figs. 2.9 and 2.10, respectively [33]. Experimental corroboration of these theoretical dependences can be found in Refs. [35–39]. The calculations were performed for the wavenumber $\nu = 3000$ cm^{-1}, at which these substrates are assumed to be totally transparent. The angular dependence of $\Delta R/R$ for p-polarized radiation exhibits the following specific features, which are distinct from systems on metal substrates. The spectra measured near φ_B in p-polarized radiation is characterized by the highest reflectivity, the lowest value of the absolute reflectance, and SNR close to zero for all substrates. The

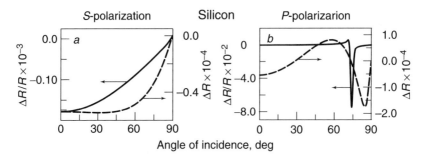

Figure 2.9. Reflectivity $\Delta R/R$ (solid line) and absorption depth ΔR (dashed line) of 1-nm layer ($n_2 = 1.3$, $k_2 = 0.1$) on Si ($n_3 = 3.433$) for (a) s-polarized and (b) p-polarized radiation versus angle of incidence φ_1 at $n_1 = 1$, $\nu = 3000$ cm^{-1}.

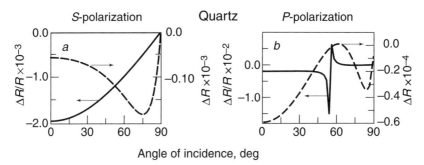

Figure 2.10. Reflectivity $\Delta R/R$ (solid line) and absorption depth ΔR (dashed line) of 1-nm layer ($n_2 = 1.3$, $k_2 = 0.1$) on quartz ($n_3 = 1.412$) for (a) s-polarized and (b) p-polarized radiation versus angle of incidence φ_1 at $n_1 = 1$, $\nu = 3000$ cm^{-1}.

$\Delta R/R$ function has a point of discontinuity of the second kind in the vicinity of φ_B, at which the absorption band changes its "direction". For a quartz substrate ($n_3 = 1.412$), negative reflectivity values $\Delta R/R$ are obtained at $\varphi_1 < \varphi_B$ and positive at $\varphi_1 > \varphi_B$. For a Si substrate ($n_3 = 3.433$), negative values of $\Delta R/R$ are observed at $\varphi_1 > \varphi_B$ and positive at $\varphi_1 < \varphi_B$.

The high noise level, along with the fact that the reflectivity values have the opposite sign on either side of the Brewster angle, makes it difficult to record a good spectrum at φ_B. This problem can be overcome in part if the IRRAS of layers on semiconductors is performed at angles of incidence close to but lower than φ_B ($\varphi_1 \approx \varphi_B - 3°$) in a pseudocollimated p-polarized radiation beam ($\Delta \varphi \sim 2°$) (technical details are described in Section 4.1.2) [33, 40]. Figure 2.9b clearly shows that if the spectral contrast (reflectivity) on a Si substrate is maximal in the vicinity of φ_B, the SNR is close to zero. Maxima of the SNR in the p-polarized IRRAS on Si are at the low and grazing ($\sim86°$) angles of incidence.

For quartz, the absolute values of ΔR for p-polarization in the vicinity of φ_B are substantially smaller than those at angles close to $0°$ (Fig. 2.10b), unlike for substrates with a high refractive index (e.g., Si). This means that using a pseudocollimated p-polarized radiation beam at $\varphi_1 \approx \varphi_B - 3°$ will not enhance the band intensity. On the other hand, measurement of the p-polarized spectra at angles of incidence close to zero is uninformative, as these spectra will be similar to the s-polarized spectra. It follows that for optimum SNR, p-polarized spectra of isotropic films on the quartz surface should be measured at $\sim 80°$, in the second maximum in the angle-of-incidence dependence of the band intensity.

For s-polarization and a Si substrate, both $\Delta R / R$ and ΔR decrease smoothly with increasing angle of incidence φ_1 (Fig. 2.9a). It follows that (i) the optimum angle of incidence for s-polarization is $0°-40°$; (ii) accuracy in setting the optimum angle of incidence in s-polarized radiation is not as important as it is when using p-polarized radiation; and (iii) to obtain the highest s-polarized spectral contrast, angles of incidence close to zero should be used. For quartz, only $\Delta R / R_s$ decreases smoothly with increasing φ_1, while ΔR has a maximum at $\varphi_1 \approx 70°-75°$ (Fig. 2.10b). Therefore, in this case, the maximum spectral contrast can be obtained at angles close to zero, while the maximum SNR is achieved at $\varphi_1 \approx 70°-75°$.

Comparing the maximum absorption depths (ΔR) for the same layer on Si and on quartz (Figs. 2.9 and 2.10), it can be seen that ΔR in the s-polarized spectra is higher for the substrate with a lower refractive index (quartz), while the p-polarized absorption bands have a higher intensity for the substrate with a higher refractive index (Si). It can be shown [33] that the best SNR for the detection of isotropic layers on transparent substrates can be achieved in p-polarized radiation for a substrate with a high refractive index at grazing angles of incidence.

Note (Section 3.11.5) that for anisotropic films the optimum conditions can be different.

Since the maximum spectral contrast for a given film may coincide with a low SNR, features of the apparatus such as the SNR of the spectrometer and differential technique capabilities must be carefully considered when choosing the optimum angle and state of polarization of radiation for each substrate.

For quantitative measurements of a film, the conditions to be optimized include the thickness range in which the band intensity is a linear function of the film thickness. The dependence of reflectivity on the thickness of an anisotropic inorganic layer on ZnSe and Ge is discussed in Section 3.3.2 (Fig. 3.21). For p-polarized IRRAS of an isotropic SiO_2 film on Si measured at the angle of incidence of $60°$ (Fig. 2.11), this dependence is virtually linear up to a thickness of approximately 20 nm (the divergence at the end of this interval is only 2%). In the linear range, an increase in the layer thickness of 1 nm causes a 1.3% increase in the reflectivity. If the reflectivity resolution is at least 1%, this means that under these conditions it is possible to detect a silicon oxide layer with a thickness between 0.8 and 20 nm with an accuracy of ± 0.8 nm. For s-polarized radiation (not shown), linearity is observed over almost the same interval (up to 30 nm). At the boundary of the linear range at $d \approx 30$ nm, $\Delta R / R$ is only 0.235%, and the reflectance is 54%, so in this linear region, the use of the multiple-reflection method at the angle of incidence will increase the s-polarized

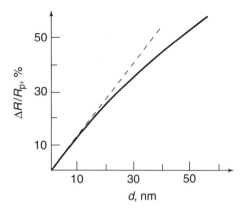

Figure 2.11. Dependence of reflectivity $\Delta R/R$ on thickness d of SiO_2 layer on Si for p-polarized radiation ($\nu = 1240$ cm^{-1}, $n_3 = 3.42$, $\varphi_1 = 60°$). Dashed line is tangent to reflectivity curve. Reprinted, by permission, from S. N. Gruzinov and V. P. Tolstoy, *Zhurnal prikladnoi spektroskopii* **46**, 775 (1987), p. 777, Fig. 2 Copyright © 1987 "Nauka i tekhnika".

spectral contrast. Thus, for the case considered, the maximum possible number of reflections when the value of reflectance drops to 1% is equal to 7 [33].

Additional enhancement of the spectral contrast for IRRAS can be achieved by recording differential spectra. These represent the spectral dependence of the quantity $\Delta R_{\varphi'/\varphi''}/R = 1 - R_{\varphi'}^0 R_{\varphi''}/R_{\varphi'} R_{\varphi''}^0$, where φ' is slightly smaller than the Brewster angle and φ'' is slightly larger. This technique exploits the fact that the reflectance will vary differently at angles of incidence close to but on opposite sides of the Brewster angle. This makes it possible to enhance the spectral contrast by a factor of approximately 2 [33].

To summarize the optimization of experimental conditions in the IRRAS of nanolayers located on the surface of transparent or weakly absorbing substrates, for each layer–substrate system, the optimum angle of incidence and the direction of polarization are specific and can be chosen properly only by way of calculation. Given the results from Si and quartz substrates, it can be noted that the maximum values of ΔR for an organic film can be achieved in IRRAS using p-polarized radiation at grazing angles of incidence and s-polarized radiation at an angle of incidence of $70°-75°$.

2.3.2. Absorbing Substrates

Doped semiconductors and carbon materials are characterized by the absorption index, which, as a rule, does not exceed 1–3 (Section 1.2). These values are intermediate between those characteristic of transparent substrates and those of metals. Correspondingly, the optical properties of such substrates have features inherent to both transparent substrates and metals. This is not the case for dielectrics in the region of the phonon bands, within which the optical constants vary significantly, as the refractive indices become less than or equal to 1.

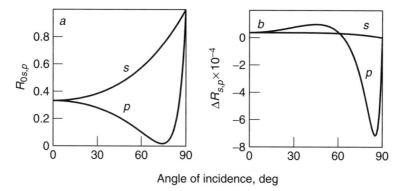

Figure 2.12. Dependences of (a) reflectance of bare substrate and (b) absorption depth of 1-nm model film ($n_2 = 1.3$, $k_2 = 0.1$) measured by transparent IRRAS with s- and p-polarization on angle of incidence φ_1: $n_1 = 1$, $n_3 = 3.4$, $k_3 = 1$, $\nu = 3000$ cm^{-1}.

Figure 2.12 refers to a substrate with the optical constants $n_3 = 3.4$ and $k_3 = 1$, which was chosen as a model substrate because its characteristics are close to those of doped silicon and one form of carbon [75]. Analysis of the curves in Fig. 2.12a (see also 1.4.15°) shows that the reflectance of p-polarized radiation from the clean substrate $R_{0,p}(\varphi_1)$ is close but not equal to 0 at angles of incidence φ_1 close to the pseudo-Brewster angle of the substrate, φ_B. The value $R_{0,p}(\varphi_B)$ increases with increasing k_3. Sign of the absorption depth ΔR_s of the layer on doped Si measured with s-polarization (Fig. 2.12b) is always positive, as opposed to transparent Si (Fig. 2.9). The form of the angle-of-incidence dependence for p-polarization differs greatly at $\varphi_1 < \varphi_B$, where ΔR_p is positive, but at $\varphi_1 > \varphi_B$ it changes sign, while its maximum is observed, as for transparent Si, in the spectra measured at angles of incidence larger than φ_B.

The dependences shown in Figs. 2.9 and 2.12, however, are unsuitable for comparing the spectrum contrast for the transparent and absorbing substrates, since a variation in the optical properties of the substrate shifts the spectrum baseline, which contributes to the absorption depth values. To elucidate how the absorption of a substrate affects the contrast of the spectra of an ultrathin film in IRRAS, p-polarized spectra of a model organic layer on a Si substrate were simulated as a function of the absorption index k_3. The angle of incidence was taken to be optimal, that is, 80°. Figure 3.29 (Section 3.4) shows that the spectral contrast practically does not change with increasing k_3 up to 2, while the band shape is essentially distorted.

The IRRAS spectra of ultrathin films can also be measured in the region of phonon absorption of the substrate. Figure 2.13 shows results of p-polarized ex situ IRRAS of calcite (CaCO$_3$) after adsorption of oleate for 5 min in a 3.3×10^{-5} M oleate solution at pH 10, measured by Mielczarski and Mielczarski [41] at different angles of incidence. It is important that although the absorption bands of adsorbed oleate at 1575, 1538, and 1472 cm^{-1} lie within the region of the *reststrahlen* band of calcite (1600–1400 cm^{-1}), they have sufficient

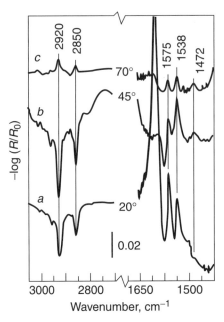

Figure 2.13. Reflection spectra of calcite after incubation for 5 min in 3.3×10^{-5} M oleate solution, pH 10, recorded for p-polarization at angles of incidence of (a) 20°, (b) 45°, and (c) 70°. Reprinted, by permission, from J. A. Mielczarski and E. Mielczarski, *J. Phys. Chem. B* **103**, 5852 (1999), p. 5856, Fig. 7. Copyright © 1999 American Chemical Society.

contrast to be used for further spectral analysis. The maximum intensity corresponds to small angles of incidence; this was confirmed by spectral calculations. (The quantity of adsorbed oleate estimated from these spectra correspond to approximately seven conventional monolayers, which implies satisfactory surface sensitivity.) On the basis of the spectral simulation, the intense band at 1640 cm^{-1} was assigned to the substrate.

To explain this technique using fused quartz as an example, several regions can be distinguished in the 1400–400-cm^{-1} spectral range of quartz absorption (Fig. 2.14a): (I) 1050–550 and 450–400 cm^{-1} (in which $n_3 > k_3$), (II) 1250–1050 and 500–480 cm^{-1} (in which $n_3 < k_3$), (III) 1400–1280 cm^{-1} (in which $k_3 \approx 0$, $n_3 \leq 1$), and (IV) close to 1100 and 470 cm^{-1} (in which $n_3 \approx k_3$). In each of these regions, a specific behavior of $\Delta R(\varphi_1)$ should be expected. As calculations show, in regions I and II the $\Delta R_p(\varphi_1)$ dependence for a film on quartz is similar to that for weakly absorbing dielectrics and weakly reflecting metals, respectively. In region III, the relationship between the optical constants of the material observed here is characteristic of dielectrics in the high-frequency region close to the phonon absorption. In this region, as seen from Fig. 2.14c, even if a transparent layer is deposited on the surface of such a specimen, IRRAS reveals a strong positive narrow absorption band (named the *substrate band* [41]) at the Christiansen frequency ($n = 1$) 1374 cm^{-1}. Although the physical origin of

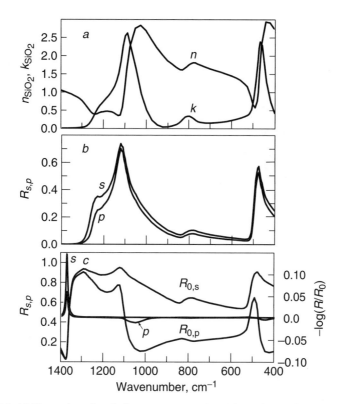

Figure 2.14. (a) Dispersion of optical constants n and k of fused quartz [75]; specular reflectance of quartz $R_{0,s}$ and $R_{0,p}$ obtained at (b) $\varphi_1 = 20°$ and (c) $\varphi_1 = 80°$, plus s- and p-polarized IRRAS spectra of transparent layer with optical constants $n_2 = 1.5$, $k_2 = 0$, and $d_2 = 10$ nm on quartz surface.

this absorption is not clear, this result could be explained qualitatively as follows. In the spectral region where the optical constants of a substrate are $k \approx 0$ and $n \approx 1$, the reflectance R of the bare substrate is close to zero (Fig. 2.14c). If a layer is deposited on such a substrate, it must cause a relatively large change in the total value of R. Apart from the strongest band at 1374 cm^{-1}, the substrate produces wide negative bands at ~1030 and 430 cm^{-1} in p-polarized spectra since at these frequencies reflectivity also approaches minimal values (Fig. 2.14c).

Mielczarski and Mielczarski [41], who first reported the effect of phonon absorption of a substrate on IRRAS, found that the substrate band absorption depth can vary in magnitude and sign, being very sensitive to small changes in factors such as angle of incidence and thickness of the deposited layer. For fused quartz, the angular dependence of the absorption depth of the "substrate" band $\Delta R(\varphi_1)$ for s-polarized radiation is depicted in Fig. 2.15a, which shows the maximum value of ΔR at $\varphi_1 \approx 80°$. Moreover (Fig. 2.15b), the absorption depth depends not only on the thickness of the layer but also on its refractive index.

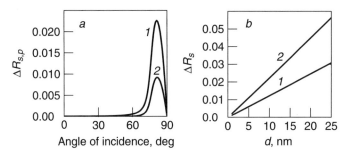

Figure 2.15. (a) Calculated change in substrate band intensities (1) ΔR_s and (2) ΔR_p in IRRAS of 10-nm transparent layer ($n_2 = 1.5$, $k_2 = 0$) on quartz substrate as function of angle of incidence at $\nu = 1360 \ \mathrm{cm}^{-1}$. (b) Calculated dependence of substrate absorption depth in s-polarized spectra obtained at $80°$ on layer thickness; $k_2 = 0$, $n_3 = 0.9702$, and $k_3 = 0.013$, $n_2 = (1) \ 1.3$ and $(2) \ n_2 = 1.5$.

The effect of the phonon absorption of a substrate in IRRAS of an adsorbed layer should be taken into account when ultrathin films are studied in this region.

2.3.3. Buried Metal Layer Substrates (BML-IRRAS)

Although IRRAS is a well-established method for studying monolayers on transparent substrates, its sensitivity is almost an order of magnitude lower than on metals. At the same time, "transparent" IRRAS offers an important advantage that p- and s-polarized spectra of the film can be measured, which is extremely valuable for orientational studies (Section 3.11.5). One can combine the advantages of metallic and transparent IRRAS by using a complex-substrate "transparent layer on a metal" (Fig. 2.16), rather than a single-substance substrate. The upper transparent layer, which imitates the surface chemistry of a bulk transparent substrate, is dubbed a *buffer* or *interference layer*. The technique that involves such a buffer layer–metal substrate is known as *buried metal layer* (BML)-IRRAS or *interference underlayer* IRRAS.

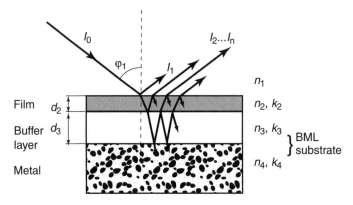

Figure 2.16. Schematic diagram of film on BML (interference) substrate.

There exist two different BML-IRRAS techniques. The first one that employs "chemically thick" but "optically thin" ($\sim$5–70 nm) buffer layers was used in studies of adsorption of CO_2 on $Cr_2O_3(0001)$ [42], ammonia on AlN and AlON on NiAl(111) [43], N_2O_5 on an ice–Au [44], methanol on an epitaxially grown Si(100) 2×1 surface [45], oxygen on GaAs [46], $Fe(CO)_5$ on SiO_2 [47], triethylgallium, ammonia, and hydrazine on MgO(100) [48], CF_3 species on Si [49], fluorine (XeF_2) adsorption and of oxygen–fluorine coadsorption on Si [50], high-pressure and high-temperature coadsorption of CO and H_2 on carbided Ni(100) [51], condensed layers of triethylaluminium and dimethylaluminium hydride on SiO_2–Al [52], and hydroxyl coatings on SiO_2 [53]. Theoretical studies of CO adsorbed at a semiconductor–metal superlattice consisting of very thin, alternating layers of metal and semiconductor has been carried out by Borg et al. [54, 55].

Optical modeling [45, 56, 57] showed that the optimum conditions for BML-IRRAS on optically thin buffer layers are similar to those of metallic IRRAS, that is, p-polarization and grazing angles of incidence, and the spectrum can be measured in the whole IR range with surface sensitivity above that in the transparent IRRAS. For example, Fukui et al. [53] obtained for the νOD band of adsorbed deuteroxyl species on SiO_2 the intensity enhancement by a factor of $\sim$3–4 and an increase in the SNR by an order of magnitude, as compared to transparent IRRAS on pure SiO_2. However, only modes perpendicular to the surface can be detected, and there is no gain in surface sensitivity as compared to metallic IRRAS (Fig. 2.21). Experimental evidence has been reported for ultra-thin (20–80 nm) poly(methyl methacrylate) films on a Si layer 20 nm thick on Cu relative to a pure Cu substrate [58].

The second BML-IRRAS technique employs buffer layers with the thickness on the order of the wavelength. A proper choice of the thickness and the angle of incidence allows measurements of modes both parallel and perpendicular to the substrate surface with a SNR above that in both transparent and metallic IRRAS, but in a narrow spectral range. The first experimental observation of this effect was reported by Vasil'ev et al. [59] who found that the intensity of the νOH band in the s-polarized spectrum of water adsorbed on the Al_2O_3 layer on Al is much higher than that in the p-polarized spectrum. Later the BML-IRRAS with thick buffer layers was applied to determe the structure of the Ge segregated surface on a Si–Ge–Si(001) BML substrate [60] and self-assembled monolayers (SAMs) on epitaxially grown Si [61]. The theoretical interpretation of this enhancement was first suggested by Tolstoy and Egorova [62]. This interpretation is considered below.

The radiation reflected from a two-layer system can be represented as a superposition of the components $I_1, I_2, \ldots, I_n$, which arise at the interface of each different layer (Fig. 2.16). These components are functions of the optical constants and thicknesses of the upper and lower layers (n_2, k_2, d_2 and n_3, k_3, d_3), the optical constants of the metal (n_4, k_4), and the angle of incidence of the light, φ_1. These functions can be calculated with the Fresnel formulas as described in Section 1.7. In the computations [62], values of the optical constants of the upper layer were chosen to be typical values for a weak-intensity absorption band of an organic material. The buffer layer has $n_3 = 1.46$, 1.7, which corresponds to SiO_2 or Al_2O_3, respectively, at $\lambda = 3$ μm; its thickness was varied from zero to 2 μm. The optical constants of the substrate were taken to be $n_4 = 3.2$ and $k_4 = 23$, characteristic of Al at $\lambda = 3$ μm. The computations were performed for the reflectivity $\Delta R/R' = 1 - R/R_0'$, where R is the reflectance of the two layer–substrate system and R_0' is that of the bare BML substrate. It follows from Fig. 2.17 that the reflectivity $\Delta R/R'(d_3)$ is a regular periodic function of the thickness of the buffer layer for both polarizations. For the angle of incidence $\varphi_1 = 60°$ this function reaches its highest values at the buffer layer thicknesses 0.5 and 1.5 μm in s-polarized radiation. This regularity can be interpreted as follows. Antinodes of the tangential fields in the standing wave formed at the air–metal interface (Fig. 1.16b) occur from the metal surface at distances $A\lambda(2k + 1)$, where λ is the wavelength, k is integer, and A is constant. To bring into coincidence such an antinode with the ultrathin film studied, one should insert an appropriately thick buffer layer in between the film and the metal surface.

Figure 2.18 demonstrates how the absorption depth $\Delta R = R_0' - R$ in s- and p-polarized BML-IRRAS of the upper organic film varies with the angle of incidence for different thicknesses of the buffer layer. This dependence is more complicated than that for a single transparent substrate (Figs. 2.9 and 2.10). As the angle of incidence φ_1 varies, the relative amplitudes of ΔR change significantly for s- and p-components and there is no general regularity. As calculations show, the interference phenomenon produces enhancement in surface sensitivity as compared to both transparent and metallic IRRAS. Thus, the maximum absorption depth for a film on a BML substrate can be greater by a factor of 50–100 relative to that for this film on the substrate without a buried metal and even twice (!) that for the film on the pure metal. The maximum values of the absorption depth are observed with s-polarization at the minimum values of the s-polarized reflectance of the buffer layer–substrate system, R_{0s}'.

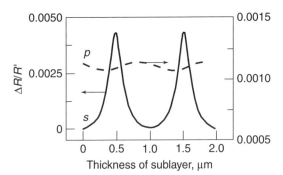

Figure 2.17. Calculated reflectivity of upper layer, $\Delta R/R'$, for s- and p-polarized radiation as function of thickness of lower layer, d_3, at $\varphi_1 = 60°$, $\lambda = 3$ μm, for $d_2 = 1$ nm, $n_2 = 1.3$, $k_2 = 0.1$, $n_3 = 1.7$, $k_3 = 0$, $n_4 = 3.2$, $k_4 = 23$. Reprinted, by permission, from V. P. Tolstoy and E. Yu. Egorova, *Opt. Spectrosc.* **68**, 663–647 (1990), p. 663, Fig. 1. Copyright © 1990 Optical Society of America.

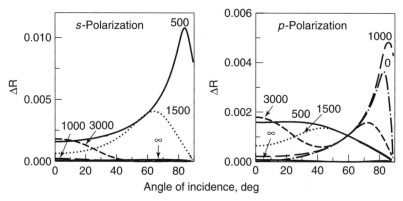

Figure 2.18. Calculated *s*- and *p*-polarized absorption depth of 1-nm upper layer on BML substrate, ΔR, as function of angle of incidence for different thicknesses (in nm) of buffer layer. Optical constants of system are as in Fig. 2.17. Curves for film on bare metal and transparent substrates are labeled with thickness 0 and ∞, respectively.

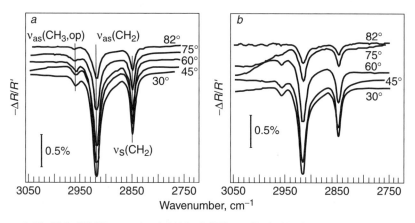

Figure 2.19. BML-IRRAS spectra of SAM of OTS on Si–CoSi$_2$–Si substrate with 200-nm buffer layer as function of angle of incidence: (*a*) *p*-polarization, (*b*) *s*-polarization. Reprinted, by permission, from Y. Kobayashi and T. Ogino, *Appl. Surf. Sci.* **100/101**, 407 (1996), p. 409, Fig. 2. Copyright © 1996 Elsevier Science B.V.

Figures 2.19 and 2.20 show experimental spectra measured in the νCH region by BML-IRRAS of an octadecyltrichlorosilane (OTS) SAM on the Si–CoSi$_2$–Si substrate with the Si buffer layer 200 and 70 nm thick, respectively [61]. A close inspection of these spectra reveals essential differences between them. First, for the 200-nm Si, the SNRs for *s*- and *p*-polarization are similar at the same angle of incidence, tending to increase at a lower angle of incidence. However, the SNR in the *p*-polarized spectra on a 70-nm Si is appreciably higher than that in the *s*-polarized spectra, and the *p*-polarized band intensities decrease with decreasing angle of incidence. Second, in contrast to the 200-nm buffer layer,

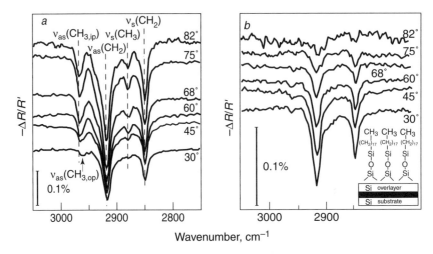

Figure 2.20. BML-IRRAS spectra of SAM of OTS on Si–CoSi$_2$–Si substrate with 70-nm buffer layer as function of angle of incidence: (a) p-polarization, (b) s-polarization. Reprinted, by permission, from Y. Kobayashi and T. Ogino, *Appl. Surf. Sci.* **100/101**, 407 (1996), p. 409, Fig. 1. Copyright © 1996 Elsevier Science B.V.

the 70-nm layer activates, apart from the $\nu_{as}CH_2$ and $\nu_s CH_2$ modes at $\sim$2920 and 2850 cm^{-1}, respectively, the bands at $\sim$2965 and 2880 cm^{-1} assigned to the $\nu_{as}^{ip}CH_3$ and $\nu_s CH_3$ modes, respectively. In the p-polarized spectra of the SAM on a pure Si (spectrum a in Fig. 3.90), the methyl stretching bands are also present but with negative intensity. It is worth noting here that the SNR provided by conventional transparent IRRAS is by an order of magnitude lower than that of BML-IRRAS.

The different composition in the p-polarized spectra measured by BML-IRRAS on buffer layers of different thicknesses and transparent IRRAS is explained by different surface selection rules (SSRs) for these methods. The vertical (z) component of the NMSEF at the angle of incidence of 80° is twice as large as the lateral (x) component in an organic film on pure Si (Fig. 1.15c) (see also Fig. 3.89a). However, the magnitude of z-NMSEF is negligible on the BML substrate with the 200-nm Si buffer layer (Fig. 2.21). As a result, the $\nu_{as}^{ip}CH_3$ and $\nu_s CH_3$ modes in the SAMs, which are almost perpendicular to the surface (Fig. 3.75), practically do not couple with IR radiation. The relationship between the values of x- and z-NMSEF changes with changing thickness of the buffer layer. In particular, the vertical NMSEF component becomes larger than the lateral one on the 70-nm Si, which makes the methyl stretching bands IR active in the p-polarized spectra.

It is also of interest to study the effect of the absorption coefficient k_3 of the buffer layer on $\Delta R/R'$. For this, calculations of R and $\Delta R/R'$ were performed for the $\nu_{as}CH_2$ band of a monolayer 2.5 nm thick of a long-chain surfactant described by $\nu_0 = 2920$ cm^{-1}, $\varepsilon_\infty = 2.22$, $S = 0.004$, and $\gamma = 10$ cm^{-1}. The absorption index k_3 of the buffer layer 750 nm thick with $n_3 = 1.4$ was varied

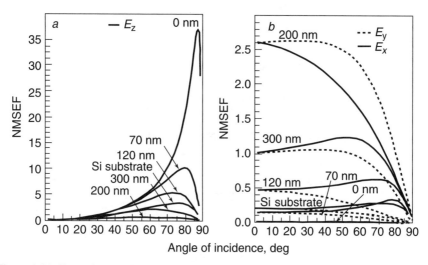

Figure 2.21. Dependence of NMSEF at surface of Si–CoSi$_2$ BML substrate on angle of incidence as function of Si buffer layer thickness. Result obtained from bulk Si substrate without BML is also shown for comparison. (a) Vertical components for p-polarization (E_z). (b) Lateral components for p- (E_x) and s-polarization (E_y). Here, $v = 2900$ cm^{-1}. Adapted, by permission, from Y. Kobayashi and T. Ogino, *Appl. Surf. Sci.* **100/101**, 407 (1996), p. 410, Fig. 3. Copyright © 1996 Elsevier Science B.V.

from zero to 0.08. The results obtained for the angle of incidence of 78° are shown in Fig. 2.22; it can be seen that in the presence of an absorption in a buffer layer the band intensity for s-polarization decreases, the band shape and the background becoming significantly distorted, while the p-polarized spectra are practically unaffected.

The limitation arising when the interference layer is applied should also be noted. This includes a narrowing of the spectral region with maximum spectral contrast, which results in different spectral regions being enhanced under different conditions. For example, for the system characterized at $v = 3000$ cm^{-1} by the parameters $n_1 = 1$, $n_2 = 1.3$, $k_2 = 0.1$, $n_4 = 3.2$, $k_4 = 23$, $d_2 = 1$ nm, $d_3 = 765$ nm, the basic condition for interference at $\varphi_1 = 85°$ is satisfied from 2950 to 3050 cm^{-1} only for s-polarization, [62]. This effect explains the background distortions in the BML-IRRAS spectra shown in Fig. 2.22. Nevertheless, even in such a narrow spectral range, the proposed method is useful when studying, for example, the vCH vibrations of adsorbed molecules. As a rule, the spectral contrast for such a system is fairly high, so that it is not necessary to increase the intensity by employing the multiple-reflection technique.

Several conclusions may be drawn from the results presented above in the case of a layer on a metal surface.

1. The use of a buffer (interference) layer between a film and the metallic substrate (the BML-IRRAS technique) allows one to measure the s-polarized spectra and to avoid the band distortions in the p-polarized spectra. At

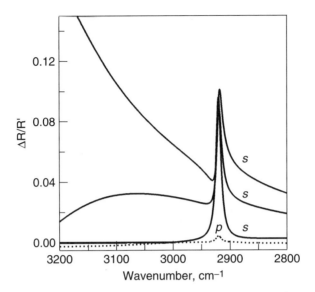

Figure 2.22. Reflectivity of film ($S = 0.004$, $\varepsilon_\infty = 2.22$, $\gamma = 10 \text{ cm}^{-1}$, $\nu = 2920 \text{ cm}^{-1}$, $d_2 = 2.5 \text{ nm}$) on BML substrate with buffer layer 750 nm thick characterized by $n_3 = 1.4$ and $k_3 = 0$ for p-polarization and, from bottom to top, $k_3 = 0$, 0.04, 0.08 for s-polarized radiation; $n_1 = 1$, $n_4 = 3.2$, $k_4 = 23$, $\varphi_1 = 85°$.

the optimum thickness of the buffer layer and angle of incidence, one can achieve the surface sensitivity enhancement by factors of 2–5 and 10–100 as compared to that in metallic and transparent IRRAS, respectively, but in a narrow spectral range.

2. By making the buffer layer "optically thin" ($< \sim 70$ nm thick) but "chemically thick," the absorption due to perpendicular modes of ultrathin films on dielectrics can be measured in the whole IR range with surface sensitivity above that in the transparent IRRAS but below that in metallic IRRAS.

2.4. ATR SPECTRA

Around 1960, Harrick [63] and Fahrenfort [64] demonstrated that ATR can be used for absorption measurements of thin films (the history of the method was well documented by Mirabella [65]). Since this time, the method has been extensively developed to study film on substrates with various optical properties (dielectrics, semiconductors, and metals) and shapes on bulk samples and on powders. The theory of ATR for thin layers is considered by Harrick [66] and Hansen [67] and has been reviewed in detail [68–72]. In this section, the experimental conditions necessary for the measurement of ATR in ultrathin films will be discussed; in particular the effects of the materials for the IRE substrate as well as of the angle of incidence will be considered. This will allow the capabilities of the ATR method for a particular system to be estimated and, to a certain

extent, appropriate conditions for the measurement to be chosen. The surface
enhancement effect due to an immersion medium is discussed in Section 2.5.4.

When considering the angle-of-incidence dependence of the reflectance (R)
and the absorption depth (ΔR) of the hypothetical model organic layer deposited
directly on an IRE, the limiting cases are when the IRE is made of a weakly or
strongly refracting substance (for Fig. 2.23 quartz and germanium, respectively,
were chosen). It can be seen in Fig. 2.23 that the maximal absorption band intensities ΔR (and hence the SNR maxima) are observed at angles close to the critical
angle ($\varphi_c \approx 14°$ for Ge and $\varphi_c \approx 48°$ for quartz). For the IRE with larger refractive index (Ge), the spectral contrast (ΔR) is higher, but only within a very narrow
range of angles ($\sim 1°$), and thus this superiority of the Ge element cannot actually
be observed experimentally since the divergence of a real beam is usually greater
than $1°$ and it is practically impossible to set the angle of incidence in an attachment with such accuracy. Hence, of the two materials for the IRE, the material
with the lower value of n (for which the angular dependences are considerably
wider and the precision of the adjustment will not strongly affect the spectral
contrast) is more preferable for studying thin films at the air–solid interface.

The sensitivity of ATR detection of ultrathin films can be increased by
using multiple reflections. Thus, if a certain band with a single reflection has
a reflectance $R \equiv I/I_0 = 1 - A$, where A represents reflection losses due to
the absorption of the evanescent wave, then with multiple reflections for small
A, $R_N = R^N = (1 - A)^N \approx 1 - NA$, where N is the number of reflections. In
other words, in the multiple-reflection spectra of nanolayers, the useful signal
increases proportionally to the number of reflections. However, the increase
in the absorption depth ($\Delta R = R_0 - R_N$) is limited due to the simultaneous
decrease in R_0 with increasing N (Section 2.2). The maximum value of ΔR for
the MIR method and the optimum number of reflections depend on the system
considered and the goals of the study. For example, to detect fractions of a SiH
monolayer on Si, the MIR geometry is most suitable [73]. However, this would
be inappropriate for studying films in the $1200–1000\text{-cm}^{-1}$ region, due to the
strong absorption by oxygen impurities inside the IRE. For detection of the νCH

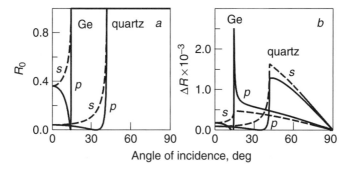

Figure 2.23. Angle-of-incidence dependence of (a) reflectance R_0 of quartz IRE–air and Ge
IRE–air interfaces and (b) absorption depth ΔR of layer with $n_2 = 1.3$, $k_2 = 0.1$, and $d_2 = 1$ nm
at these interfaces; $n_{quartz} = 1.402$, $n_{Ge} = 4.03$. In all cases, $n_3 = 1$, $\nu = 2800$ cm^{-1}.

bands of a phospholipid monolayer [74] deposited on a $45°$ Gc multiple-IRE in contact with air, 100 or more reflections would be desirable. However, for aqueous surroundings, the number of reflections should be decreased down to 20–40, due to the strong water absorption at these wavenumbers (Fig. 1.4). As a general rule, the parameters can be chosen properly *only by way of calculation*.

2.5. IR SPECTRA OF LAYERS LOCATED AT INTERFACE

The study of a layer buried at the interface between two different media is one of the most complex problems in applied spectroscopy. The difficulties arise from a masking of the analyzed material by the adjacent media and also from the fact that, as a rule, the thickness of such a layer is in the nanometer range. However, if at least one of the media is transparent at the IR absorption frequencies of the layer, then it is possible in principle to investigate the layer.

One frequently examined interface is the solid–liquid interface, where the solid phase may be a dielectric, a semiconductor, or a metal. Species located at these interfaces are of primary importance in electrochemistry and in chemistry of surface-active substances (surfactants). Another common type of interface is the solid–solid interface, specifically dielectric–dielectric, dielectric–semiconductor, dielectric–metal, semiconductor–semiconductor, semiconductor–metal, and metal–metal interfaces. These structures have an extremely important role in such areas as microelectronics and the chemistry of composites. Furthermore, positioning an ultrathin film at the interface of two media, one can substantially increase surface sensitivity of all IR spectroscopic methods.

When both bordering media are transparent, one can apply transmission spectroscopy in polarized radiation (Section 2.1) or, when there is a difference in the refractive indices of these media, the ATR method and IRRAS. For each type of solid–solid interface, except for the metal–metal interface, one can study the layers in the contact zone by IRRAS or ATR in the transparent spectral range of one of the media in the system. To choose the technique with which to investigate dielectric (semiconductor)–liquid, dielectric (semiconductor)–semiconductor, and dielectric–dielectric interfaces, several factors must be considered, including the region of transparency of the media under study and the relationship between their refractive indices. If the medium with the largest refractive index is the most transparent, one should use the ATR method; otherwise IRRAS is more appropriate.

2.5.1. Transmission

Figure 2.24 illustrates the effect of the relationship between the refractive indices of the input and output media (Fig. 2.1*d*) on the band intensities in the transmission spectra of the layers. One expects that increasing n_3 will lead to enhanced spectral contrast, because, as discussed in Section 1.8, the magnitude of the electric field within the film also increases. Comparing the results calculated for

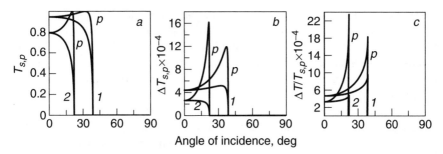

Figure 2.24. Dependence of s- and p-polarized (a) transmittance $T_{s,p}$, (b) absorption depth $\Delta T_{s,p}$, and (c) transmissivity $\Delta T/T_{s,p}$ on angle of incidence of IR radiation: (1) $n_1 = 2.4$, (2) $n_1 = 4.0$; $n_2 = 1.564$, $k_2 = 0.384$, $v = 1200$ cm^{-1}, $n_3 = 1.5$.

two different output media with $n_3 = 1.5$ (Fig. 2.24) and $n_3 = 1$ (Fig. 2.4), it is seen that when $n_3 = 1.5$ the maximum values of ΔT are higher by a factor of approximately 1.5.

When the values n_1 and n_3 are similar, the critical angle φ_c for the reflection at the interface between the input and output media can be as high as 80° and the permissible ranges of incident angles and, hence, geometric pathlengths of the beam inside the film are substantially extended. The dependence on the angle of incidence of ΔT_p for the optical schemes where one medium is chalcogenide glass IKS-35 [75] ($n = 2.372$) and the second medium is Irtran-2 ($n = 2.22$) or KRS-5 ($n = 2.374$) is shown in Fig. 2.25. The IKS-35 glass [76] has the composition AsSe$_x$I$_y$. Due to its low melting point (~90°C), it can be easily brought into optical contact with a substrate covered with a film under heating. For the set of the refractive indices $n_1/n_3 = 2.372/2.22$, $2.374/2.372$ the critical angle is $\varphi_c \approx 70°$ and $\varphi_c \approx 88°$, respectively. As can be seen from Fig. 2.25, when the radiation passes through the system from the medium with the higher refractive

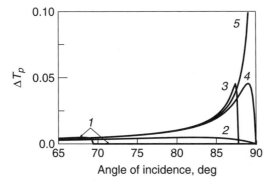

Figure 2.25. Dependence on angle of incidence of absorption depth ΔT_p at $v = 1200$ cm^{-1} in p-polarized transmission spectrum of 1-nm film of model organic substance ($n_2 = 1.564$, $k_2 = 0.384$) located at interface of two media with refractive indices n_1 and n_3: (1) 2.372 and 2.22, (2) 2.22 and 2.372, (3) 2.374 and 2.372, (4) 2.372 and 2.374, (5) $n_1 = n_3 = 2.372$.

index, the spectral contrast is greatest at the angles of incidence close to φ_c, the absolute values of ΔT being one order of magnitude higher than those measured for the optical systems characterized by Figs. 2.2, 2.4, and 2.24. The same spectral contrast is observed when the radiation passes in the reverse direction (from the medium with lower refractive index). The latter geometry is preferable from a technical viewpoint.

The refractive indices n_1 and n_3 can be equal when the layer is located inside a solid [77], or at the interface between a solid and a liquid with equal refractive indices, or when the oxidized sides of two semiconductor plates are tightly pressed together [14]. In this case, the use of oblique angles of incidence (up to 89°) provides a substantial increase in surface sensitivity (Fig. 2.25). Figure 2.26 shows the dependence of the intensity and the shape of the absorption band of a model organic film in different immersion media on the angle of incidence, the

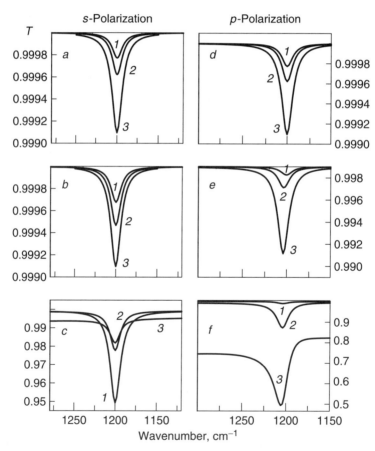

Figure 2.26. (a–c) The s- and (d–f) p-polarized transmission spectra of layer located at interface of two media with refractive indices (1) $n_1 = n_3 = 1.35$, (2) $n_1 = n_3 = 2.4$, (3) $n_1 = n_3 = 4.0$. Parameters of layer are as in Fig. 2.5; φ_1 : (a, d)0°, (b, e) 45°, (c, f) 89°.

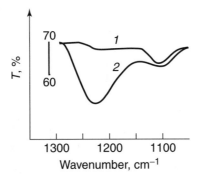

Figure 2.27. Experimental transmission spectrum of 1.4-nm SiO_2 layer on Si. Measurements were performed for p-polarization and angle of incidence $\varphi_1 = 60°$: (1) $n_1 = n_3 = 1.0$, (2) $n_1 = n_3 = 2.37$. Reprinted, by permission, from V. P. Tolstoy and S. V. Habibova, *Vestnilk Leningradskogo Universiteta. Phizika, Khimia* **2** 106 (1989), p. 109, Fig. 4 Copyright © 1989 St. Petersburg University Press.

refractive index of the medium, and the type of polarization. The maximum values of ΔT are reached at oblique angles of incidence for p-polarized radiation and for media characterized by maximum refractive indices (Fig. 2.26 f). However, under these conditions, the spectra are most distorted. A further enhancement can be achieved for modes perpendicular to the surface by directing p-polarized radiation onto the film/substrate cross section (formally, $\varphi_1 = 90°$) as illustrated in Fig. 6.17 [15, 16].

As an experimental demonstration of these dependences, Fig. 2.27 presents the transmission spectrum of a SiO_2 layer on a Si surface, obtained with two hemicylinders with refractive indices equal to 2.37. To obtain this spectrum, a special technique was used in which two KRS-5 ($n_1 = 2.374$) ATR hemicylinders were used as an immersion medium [6]. These hemicylinders were brought into optical contact with the SiO_2−Si system by lamination of the chalcogenide glass IKS-35, as shown in Fig. 2.1d. Using the immersion medium increases the intensity of the absorption bands of the SiO_2 layer by a factor of 7−8 relative to the spectrum obtained with no immersion medium, in agreement with theoretical predictions.

2.5.2. Metallic IRRAS

The effect on the band intensities of the angle of incidence of radiation, the radiation polarization, and the optical constants of the layer, immersion media, and the metal was analyzed in Refs. [40, 79, 80]. Figures 2.28−2.30 demonstrate how the refractive index of the input medium affects the band intensities in the IRRAS of ultrathin films of weakly and strongly absorbing material. The calculations show virtually no absorption of s-polarized radiation by the layer (for an explanation of this effect, see Section 3.2.2). For p-polarized radiation, the band intensity (1.5.4°) depends strongly on the angle of incidence and the

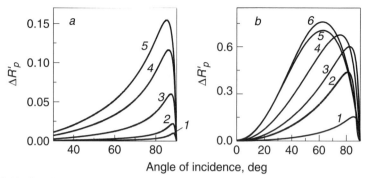

Figure 2.28. Angular dependence of band intensity $\Delta R'_p$ in IRRAS of layers characterized by oscillator parameters $\varepsilon_\infty = 2.3$, $\gamma = 15$ cm^{-1}, $\nu_0 = 1200$ cm^{-1}, $S = 0.015$ and located at immersion medium–Al interface: $n_3 = 19.52$, $k_3 = 57.7$, $\nu = 1200$ cm^{-1}; (a) $d_2 = 1$ nm, n_1: (1) 1, (2) 1.46, (3) 2.4, (4) 3.4, (5) 4.0; (b) $n_1 = 4.0$, d_2: (1) 1, (2) 5, (3) 10, (4) 20 (5), 40 and (6) 80 nm.

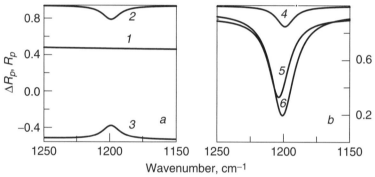

Figure 2.29. Spectra of (1) $R_{0,p}$, (2, 4–6) R_p, and (3) ΔR_p of 80-nm layer characterized by oscillator parameters $\varepsilon_\infty = 2.3$, $\gamma = 15$ cm^{-1}, $\nu_0 = 1200$ cm^{-1}, $S = 0.015$ and located at germanium–aluminum interface ($n_1 = 4.0$, n and k of Al from Ref. [75]); (a) $\varphi_1 = 85°$, (b) φ_1: (4) 85°, (5) 30°, and (6) 60°.

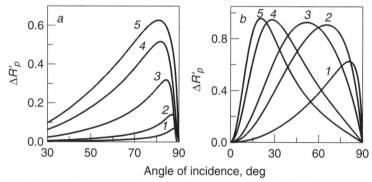

Figure 2.30. Angular dependence of band intensity $\Delta R'_p$ in IRRAS of SiO$_2$ layer at immersion medium–Al interface; $n_2 = 0.568$, $k_2 = 0.356$ [2], $n_3 = 18.55$, $k_3 = 56.26$ [75], $\nu = 1230$ cm^{-1} (ν_{LO}); (a) n_1: (1) 1, (2) 1.46, (3) 2.4, (4) 3.4, (5) 4.0, $d_2 = 1$ nm; (b) $n_1 = 4.0$, d_2: (1) 1, (2) 5 (3), 10, (4) 40, and (5) 80 nm.

refractive index of the output medium (Figs. 2.28a and 2.30a). For an angle of incidence of 75°, $\Delta R'_p$ in the spectrum of a weakly absorbing 1-nm layer increases by a factor of approximately 45 as n_1 increases from 1 to 4. However, for thicker layers, the degree of enhancement of the band intensity changes (Fig. 2.28b) and reaches its maximum at smaller angles; for example, a 40-nm layer with $n_1 = 4$ has a maximum at 60°. Therefore, as in the case of transmission, the band intensity increases with increasing refractive index of the input (immersion) medium. This phenomenon provides the basis for *immersion spectroscopy*. The first practical attempt to apply this technique in the studies of ultrathin films was made by Coleman et al. [78]. A sample was first conditioned with a solution, then dried and pressed to a KRS-5 IRE. Unfortunately, the spectra obtained showed a bad SNR. The sensitivity advantage was realized later in the nondestructive spectroscopic studies of ultrathin oxide films in the metal–oxide–semiconductor (MOS) structures [79] (see Chapter 6 for review).

As seen in Fig. 2.29a the reflectance of the interface increases in the presence of the film, although the spectrum retains its shape, and the spectral behavior of R is analogous with the transmission spectrum. This means that under the given conditions even a layer with $k_2 = 0$ affects the baseline of the spectrum (see also comments to Fig. 1.13). Fig. 2.29b shows how the band shape and intensity change with the angle of incidence. At angles $30° < \varphi_1 < 85°$, the shape of the band remains unchanged and its intensity reaches a maximum at 60°. The shift of the maximum band intensity for thicker films and optically denser immersion media to lower angles is due to "optical saturation" at grazing angles of incidence (Section 3.3.1). For strongly absorbing layers such as SiO_2, optical saturation becomes observable in thinner layers (5 nm in Fig. 2.30b). The optimum angle of incidence for a 40-nm layer immersed in a medium with $n_1 = 4$ is 20°. The most noticeable feature in the spectrum of a strongly absorbing layer (SiO_2, Fig. 2.31) is the significant change in the band shape that is observed with change in the angle of incidence and layer thickness. Thus, at 30° for 1- and 5-nm layers, there is a maximum at 1230 cm^{-1}; at 85°, it retains the same position for a 1-nm layer, but for thicker layers it shifts toward smaller wavenumbers down to 1020 cm^{-1}. Such band distortions were observed [78b] in experimental spectra of silicon oxide films in MOS structures measured by IRRAS.

The results presented in Figs. 2.28–2.30 lead to several conclusions regarding the spectral measurements of a layer located at the interface between a transparent medium (a semiconductor, a dielectric, or a liquid) and a metal. The application of p-polarized radiation is recommended as well as, for layers <1 nm, oblique angles of incidence ranging from 75° to 88°. For media with larger refractive indices, the angle corresponding to the maximum of ΔR_p shifts toward smaller values and for $n = 4$ it is 75°. For thicker layers, this value depends upon the optical properties of the layer and medium and on the layer thickness. For each structure being studied, these values may be chosen either by running the experiment or on the basis of calculations. It was established that under the most favorable experimental conditions and p-polarized radiation, when one of the

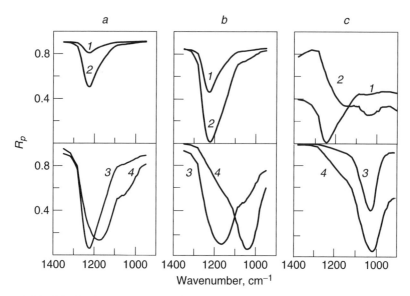

Figure 2.31. Calculated p-polarized IRRAS spectra of SiO_2 layer of thickness (1) 1, (2) 5, (3) 20, and (4) 100 nm located at Ge–Al interface. Spectra were obtained for φ_1 values of (a) 30°, (b) 60°, and (c) 85°. Optical constants of SiO_2 were specified in Ref. [2] and of Al in Ref. [75], $n_{Ge} = 4.0$.

media is germanium and the other a highly reflecting metal (Al), an increase of the band intensities of ~500 with respect to the normal-incidence transmission spectra of the same layer can be achieved. It is evident that multiple reflections in this structure provide the possibility of an additional increase of this value.

To observe experimentally the band intensity enhancement in spectra of layers at the metal surface, a special technique was developed [80], based on the low-melting-point chalcogenide glass with $n_1 = 2.37$ as an immersion medium. Spectra of the aluminum oxide, produced on the surface of Al by thermal oxidation in air at 550°C for 0.5 h (Fig. 2.32), were recorded using a special attachment whose optical scheme is depicted in Fig. 4.4. Comparison of the experimental spectra of the Al_2O_3 layer in the region of the most intense absorption band at 960 cm^{-1} (ν_{LO} of Al_2O_3) shows that the maxima in air and in contact with the immersion medium virtually coincide. However, in the latter case, the intensity is greater by a factor of ~5.

Measuring the spectrum of a layer located at the semiconductor (dielectric)–metal interface is more complicated. Such systems are widespread and are found, for example, in microelectronics, and in particular in integrated circuits. To record the spectra of such layers, a special method was proposed [79] in which the incoming beam is incident onto a transparent plate at the angle $\varphi_1 = \varphi_B = \arctan(n_2/n_1)$ (n_1 and n_2 are the refractive indices of the surroundings and the semiconductor plate, respectively) (Fig. 2.33). At such angle of incidence, the intensity of p-polarized radiation reflected from the front plane of

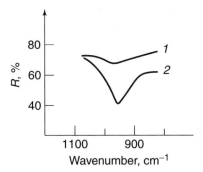

Figure 2.32. Experimental spectra of Al_2O_3 layer on Al mirror heated at $t = 550°C$ in air for 0.5 h, using p-polarized IRRAS at $\varphi_1 = 60°$: (1) sample in air, (2) sample in contact with immersion medium (KRS-5). Reprinted, by permission, from V. P. Tolstoy and S. N. Gruzinov, *Opt. Spectrosc.* **63**, 489–491 (1987), p. 490, Fig. 4. Copyright © 1988 Optical Society of America.

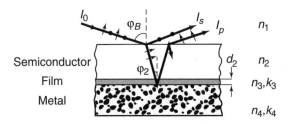

Figure 2.33. Diagram of propagation of beam within metal–oxide–semiconductor (MOS) structure under following conditions: $\varphi_B = \arctan(n_1/n_2)$, where n_1 and n_2 are refractive indices of surroundings and semiconductor, respectively, n_3 and k_3 relate to dielectric layer, and n_4 and k_4 to metal. Adapted, by permission, from V. P. Tolstoy and V. N. Krylov, *Opt. Spectrosc.* **55**, 647–649 (1983), p. 648, Fig. 2. Copyright © 1984 Optical Society of America.

the plate is equal to zero (Fig. 1.11*a*), so that the whole radiation flux will pass into the plate. Then it is reflected from the layer–substrate interface, passes out through the plate, and can be analyzed by a spectrophotometer.

When this scheme was used to investigate the layers in the metal–oxide–semiconductor (MOS) structure Si–SiO$_2$–Al, the angle of incidence onto the air–Si interface was set to $\varphi_1 = 73°$, which, according to Snell's law [Eq. (1.56)], corresponds to an angle of incidence onto the layer under study inside the plate of $\varphi_2 = 16°$. To enhance the contrast of the spectra recorded at this angle of incidence, one can use a multiple-reflection strategy similar to the technique described in Section 4.1.2 using, for example, two identical parallel Si plates.

Another way to enhance the contrast in the spectrum is to direct the radiation beam to the semiconductor plate from the side of the immersion medium rather than from air ($n_1 \neq 1$ in Fig. 2.33). For such a configuration, the angle of refraction of the radiation inside the semiconductor plate will be larger than the above-mentioned incident angle of 16°. In particular, when a KRS-5 prism or the low-melting-point chalcogenide glass IKS-35 is used as an immersion medium, the angle of incidence of radiation onto the immersion medium–Si plate interface is $\varphi_1 \approx 55°$. This determines the angle of incidence of radiation onto the nanolayer–metal interface to be approximately 30°. Calculations show that such increase in the angle of incidence leads to a five-fold increase in the absorbance.

2.5.3. Transparent IRRAS

As was the case for transmission and metallic IRRAS, there exists an additional way to increase the spectral contrast by increasing the MSEF within the layer studied by immersing a transparent substrate in a transparent medium with refractive index >1. The analysis of changes in IR spectra of the nanolayers located at the substrate–immersion medium interface was first performed in Refs. [79, 80]. Below, we consider examples of weakly absorbing layers located at a $CBrCl_3$–ZnSe interface and a chalcogenide glass $AsS_{1.5}Br_2$–ZnSe interface (Fig. 2.34).

Comparing curves 1–6 in Fig. 2.34a, one can see that at the medium–substrate interface, the reflectances R_{0s} and R_{0p} decrease relative to those at the air–substrate interface and reach a minimum for the interface systems with the least difference between n_1 and n_3. At the same time, the absorption depths ΔR_s and ΔR_p (SNR) increase (Fig. 2.34b), and their maximum value for p-polarized radiation is reached at $\varphi_1 > \varphi_B$ when the difference between n_1 and n_3 is at a minimum, as in the cases of transmission (Section 2.5.1) and ATR (Section 2.5.4). For example, for curve 11 of Fig. 2.34b, which represents the $AsS_{1.5}Br_2$–ZnSe interface, the maximum values of ΔR_p are reached at $\varphi_1 = 85°$, and ΔR_p is enhanced by a factor of >13 relative to the spectrum of the same layer located at the air–ZnSe interface. This value is not a limit, and calculations performed over a wide range of optical constants for adjacent media demonstrate the increase

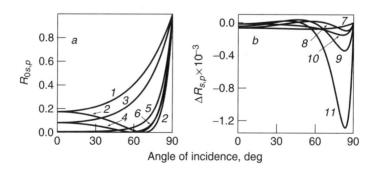

Figure 2.34. Angular dependences of (a) reflectance $R_{0s,p}$ of uncoated interfaces and (b) absorption depth ΔR_s (7, 10) and ΔR_p (8, 9, 11) in IRRAS spectra of 1-nm layer: (1, 2, 7, 8) air–ZnSe, (3, 4, 9) $CBrCl_3$–ZnSe, and (5, 6, 10, 11) $AsS_{1.5}Br_2$–ZnSe interfaces; $n_{CBrCl3} = 1.4773$, $n_{AsS1S1.5Br2} = 2.1$, $n_{ZnSe} = 2.43$, $n_2 = 1.3$, $k_2 = 0.1$, $v = 2900$ cm^{-1}.

in this factor with increasing n_3 and the ratio n_1/n_3. For a layer located at the interface of two semiconductors, for example amorphous silicon ($n_1 = 3.8$) and germanium ($n_3 = 4$), the calculated value of the enhancement factor is $\sim$100.

2.5.4. ATR

The effect of placing a sample into an immersion medium on the ATR spectrum is of interest, for example, when studying films in contact with a solvent or at the interface between two dielectrics with different refractive indices. Two immersion media, hexafluorobenzene (for the quartz IRE) and amorphous silicon (for the Ge IRE), were chosen for simulation purposes. In both cases, the ratio $n_3/n_1 \approx 0.95$ at 2800 cm^{-1}, and hence the critical angle is the same ($\sim$72°). Comparing Figs. 2.35 and 2.23, it can be seen that the absorption depth ΔR_p increases when the sample is placed into an immersion medium. For quartz, the maximum value of ΔR_p for the layer increases by approximately a factor of 3, and for germanium by a factor of $\sim$100. This means that for a constant ratio n_3/n_1, ΔR_p increases with increasing n_1 (notice this dependence can be analytically proven [66, 71]). Comparison of the maximal band intensity in the ATR spectrum of the layer at the Ge–amorphous Si interface with that at the quartz–air interface shows that in the former case the spectral contrast is $\sim$600(!) times higher. This advantage was employed in the IR spectroscopic studies of layers at the solid–solid interfaces [81]. In particular, Bruesch et al. [81a] studied a 0.7-nm silica layer formed at the interface between polycrystalline silicon ($n_1 = 3.40$) and crystalline silicon ($n_3 = 3.25$), which is of crucial importance for passivation of high-voltage devices. The ATR spectra were measured at the angle of incidence of 80°. The layer thickness was determined by a direct comparison with the ATR spectrum of a reference sample containing a 10-nm SiO$_2$ interface layer and spectral simulations. The SNR in the reported ATR spectra implies interface sensitivity of $\sim$0.1–0.2 nm. However, enhancing sensitivity to ultrathin films, an increase in the refractive index of the output medium substantially reduces the linear range of the intensity–film thickness dependence.

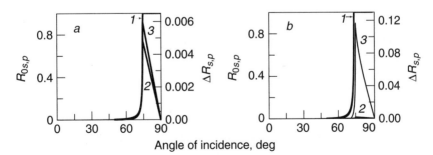

Figure 2.35. Dependence on angle of incidence of (1) reflecance R_0 for (a) quartz–hexafluorobenzene interface and (b) Ge–amorphous Si interface and of absorption depths (2) ΔR_s and (3) ΔR_p for layer at given interfaces; $n_2 = 1.3$, $k_2 = 0.1$, $d_2 = 1$ nm, $\nu = 2800$ cm^{-1}, (a) $n_1 = 1.402$, $n_3 = 1.351$, (b) $n_1 = 4.03$, $n_3 = 3.84$.

For example, for the SiO_2 films sandwiched between a Ge prism and a Si substrate (Fig. 3.3*a*), the linear approximation (1.5.5°) is valid at thicknesses less than 1.5 nm [81b]. The use of high-refractive-index IREs and high angles of incidence gives small penetration depths (Section 1.8). This effect has been exploited by Garton et al. [82] with a Ge IRE at 60° to probe the uppermost layer of a polymer. Alternatively, in the "barrier film" technique [83], the penetration depth is controlled by varying the thickness of the coating separating an IRE and polymer while ensuring that the ATR condition is met at the IRE–coating interface.

More complicated dependences are observed when two layers are located on the surface of the ATR element. The optical properties of a hemicylindrical IRE–thin ($d < 50$ nm) metal film–film system, called the *Kretschmann configuration* [84] (Fig. 2.36*a*), were actively investigated in the seventies and eighties (see, e.g., Ref. [85]) regarding the possibility of SEW excitation at the metal–outer layer interface. However, even without exploiting this and surface-enhanced infrared absorption (SEIRA) (Section 3.9.4) effects, optical enhancement may be achieved in the ATR spectrum of a layer deposited on metal. Because of this, the Kretschmann configuration has found wide application in the investigation of nanolayers located on the metal surfaces, especially at the metal–solution interface (Section 4.6.3).

Due to the strong absorption of IR radiation by metals, the intensity of the beam penetrated into the layer under investigation will decrease with increasing thickness of the metal layer. For highly reflecting metals, only metal layers less than 20–50 nm thick will be semitransparent in the IR region. Another important feature of the Kretschmann configuration is the abrupt decrease in the penetration depth of the radiation into the outer layer; this restricts the investigation to monolayers adjacent to the metal layer. The dependence of the layer spectrum on the thickness of the metal layer for a Ge–Ag–organic layer system is shown in Fig. 2.37. The calculations indicate a sharp decrease in ΔR_s when the Ag layer appears between the IRE and the layer under study, which is explained by the quenching of the tangential field by conduction band electrons in the metal (Section 1.8.2). Although, in general, ΔR_p decreases with increasing thickness of the Ag layer, there is a region between thicknesses of about 3 and 5 nm in which a 30% increase in the *p*-polarized band intensities is observed. This enhancement is attributed to the multiple reflection of the IR beam in the metal film, as confirmed experimentally by Nakao and Yamada [86]. Investigating the

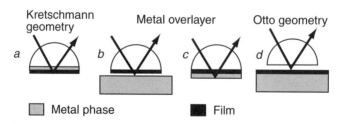

Figure 2.36. Basic ATR configurations for thin-film studies.

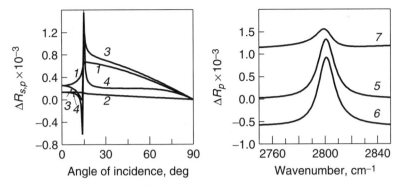

Figure 2.37. Angular dependence of quantities (1, 2) ΔR_s and (3, 4) ΔR_p for ATR spectrum of organic layer in Ge–Ag–organic layer structure (Kretschmann geometry) and (5–7) spectrum ΔR_p of this layer calculated for $\varphi_1 = 15°$: $n_1 = 4.0$; d_{Ag}: (1, 3, 5) 0, (6) 3, (2, 4, 7) 12 nm; $n_{Ag} = 1.94$, $k_{Ag} = 22.1$, $d_3 = 1$ nm, $n_3 = 1.52$, $k_3 = 0.12$, $n_4 = 1$, $\nu = 2800$ cm^{-1}.

ATR spectra of polymers deposited onto a metal-coated (Ag, Ni, Pd, or Pt) KRS-5 hemicylinder, they observed maximum intensities with 2.6-nm Ni films that were homogeneous and smooth, which calculations show to have the maximum transmittance.

Another important consideration is finding the optimal combination of refractive indices for the IRE and the metal film. This problem was studied theoretically and experimentally by Johnson et al. [87] using systems comprised of a Ge ($n = 4.0$), GaAs ($n = 3.316$), or BaF$_2$ ($n = 1.5$) IRE and a 20-nm layer of Pt ($n = 2.76$, $k = 9.72$) or Au ($n = 0.57$, $k = 20.14$). Calculations using the Fresnel formulas showed that the intensity of the absorption band at 3 μm in the ATR spectra of adsorbed water decreases by the following fractions: BaF$_2$–Au(3%) > Ge–Au(0.55%) > BaF$_2$–Pt(0.5%) > Ge–Pt(0.3%). [These spectra were obtained in p-polarized radiation at angles slightly above the critical angle for the ATR crystal–solution pair (20° and 75° for Ge and BaF$_2$, respectively).] In other words, the largest ATR signal is observed for the ATR crystal with the lowest refractive index (BaF$_2$) combined with the metal having the highest reflectivity. In this case, the intensity of the reflected radiation was more than 96%, which permits application of the MIR arrangement for further improvement to the spectral contrast (six reflections will amplify the water signal up to 10%). Although the origin of the enhancement has not been discussed by Johnson et al., it appears to be analogous to that for a film located directly on an IRE, namely the increase of multiple reflections in the metal film and the decrease in the penetration depth. Comparison of the theoretical predictions with experimental data shows a very good correlation in many cases, but discrepancies between experimental and calculated spectra do arise that are caused, in particular, by the formation of new phases at the interface between the ATR crystal and the metal. In general, it was found that experimental absorption peaks were somewhat larger than the calculated ones, probably due to surface enhancement by imperfections in the metal layer (Section 3.9.4).

A theoretical analysis of the system consisting of either the ZnSe or Ge IRE, an Fe or hematite substrate layer, an adsorbate layer, and a solution of methylene chloride has been performed by Loring and Land [88]. The system consisting of the ZnSe IRE, Al_2O_3 intermediate layer, the sputtered Si substrate layer, and water has been analyzed within the framework of the Fresnel formalism by Sperline et al. [89]. Calculations also reveal that within a narrow wavelength range and at a certain ratio of the optical refractive indices, enhancement of intensities is observed in the spectrum of a given layer in an arbitrary two-layer structure located on an IRE. The ATR spectra of such structures were considered in detail in Ref. [66]. This enhancement is attributed to interference of the radiation, as is the enhancement of the reflectivity in BML-IRRAS (Section 2.3.3).

An alternative ATR technique for recording IR spectra of nanolayers located at semiconductor (dielectric)–metal interfaces is the *metal overlayer ATR* (MOATR) [90], which is also referred to as the *grazing internal reflection* (GIR) [91]. There are two approaches to do this. In the first one, a flat metal plate [2, 92] or metal-coated elastomer film [93, 94] is pressed onto an IRE that already has a film in place (Fig. 2.36*b*). The air gap in such a configuration essentially determines the spectral contrast. The angular dependences of the reflectance without the film and the change in the reflectance caused by the film in a Ge–film–air gap–Ag composite are shown in Fig. 2.38. Figures 2.38 and 2.23 show a significant increase in the *p*-polarized spectral contrast when a metal is in the vicinity of the IRE and a sharp decrease as the air gap increases; the maximum enhancement is observed when there is optical contact between the metal and the film, which is already the IRRAS configuration. The ΔR_s values are not shown because they are many times smaller than those of ΔR_p. The absorption depth ΔR_p of the organic layer was found to decrease, and the maximum shifted to higher angles of incidence, with a decrease in the refractive index of the IRE, n_1, and a constant air gap. For smaller values of n_1, the dependence of the absorption depth on the magnitude of the air gap decreased (spectra not shown).

The second approach involves sputtering a metal film directly onto the film to be studied (Fig. 2.36*c*). It is evident that depending on the thickness of the metal

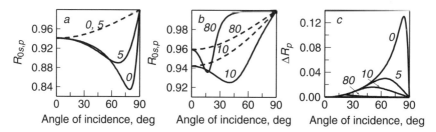

Figure 2.38. Dependence on angle of incidence of (*a, b*) *p*- (solid lines) and *s*- (dashed lines) polarized reflectance of Ge–air gap–Al MOATR configuration and (*c*) absorption depth of film on Ge for different air gaps; Ge: $n_1 = 4.0$; film: $\hat{n}_2 = 1.52 - 0.12i$, $d_2 = 1$ nm; air gap: $n_3 = 1.0$, $k_3 = 0$, values of d_3 in nanometers are indicated on curves; Al: $\hat{n}_4 = 1.94 - 22.10i$; $v = 2800$ nm^{-1}.

overlayer, one deals with either the ATR or IRRAS method, which have different optimum conditions for the spectrum recording. In the case when the metal overlayer thickness is less than the thickness of the metal skin layer (1.3.11°), it is in fact the ATR technique, since the film is probed by the evanescent wave. When the thickness of the metal exceeds the skin layer thickness, it is the immersion IRRAS method, the IRE playing the role of the immersion medium. The optimum conditions for this case are the subject of Section 2.5.2.

Enhancement of the absorption by thin films on metal surface can be achieved if the reflection spectra are measured with *Otto's configuration*. In it, the film is deposited on a metal substrate that is positioned near the reflecting surface of the IRE so that there is a gap (air or liquid) between the film and the IRE, as sketched in Fig. 2.36*d*. In the case of an air gap, this geometry is used for excitation of SEWs (Section 3.2.1) [95–97]. However, the optimal configuration for this measurement is not necessarily the ATR configuration. For example, Suzuki et al. [98] observed the maximum spectral contrast for a film on a gold substrate with an air gap of 20 μm between a ZnSe hemicylinder and the substrate when *p*-polarization and angles of incidence 1°–1.5° smaller than the critical angle for total reflection from the IRE–air interface were used. This effect was attributed to the Fabry–Perot interference (reflection cavity effect) that takes place inside this gap (see below for more detail).

Besides exploiting the interference effect due to the presence of the thin metal film over- and underlayers, it is possible to obtain an additional enhancement of the band intensities in all the above-mentioned configurations by using multiple reflections. The maximum value of ΔR for the multiple-reflection method and the optimum number of reflections depend on the system considered (Section 2.4). However, when a relatively thick (3–20 nm) metal layer participates in the optial scheme, multiple reflections may not improve the quality of the spectrum because of the high absorption by the metal film. The spectrum contrast can be enhanced if the metal layer has islandlike structure, due to an increase in the effective surface and the SEIRA effect (Section 3.9.4).

We now wish to consider the thin-layer optical configuration that is used in spectroelectrochemical (SEC) measurements. Its main components are a window, a thin layer of a liquid, and a flat-sided polished substrate covered with an ultra-thin film (Fig. 4.46). To select optimally the window material and the angle of incidence, the spectrum simulations were performed by Faguy and Fawcett [99a] for the window–acetonitrile layer–adsorbed acetonitrile–gold system with four different windows (CaF$_2$, ZnSe, Si, and Ge). The effect of a poorly defined electrolyte layer (a wedge-shaped layer or a layer of varying thickness) was taken into account by representing the result as an unweighted average of five electrolyte layers with the integral thicknesses in the 4–8-μm range. The simulations showed that the maximum spectral contrast is observed at the angles of incidence a few degrees larger than the critical angle for the window–bulk liquid interface. This contrast occurs to be higher than for the same system but without the input window. The largest enhancement (by a factor of 4.7) is provided by the window from CaF$_2$ (the material with the closest matched refractive index to the electrolyte layer), while for Ge the enhancement is by a factor of 2.1. For CaF_2 the optimal angle of incidence approaches 90°. Such a large angle is impractical, since the SNR is low due to a low reflectivity of a metal support at grazing angles of incidence (Section 1.8) and a larger effective thickness of the "thin layer". In addition, it is technically difficult to realize grazing incident angles in the in situ measurements [99b]. The calculations [100] also revealed the electric field enhancements in a thin-layer cell comprised of a ZnSe or Ge hemishpherical window, an Au electrode, and a water layer.

The optimum conditions in the case of nonmetalic (cuprous sulfide) substrate covered by an organic ultrathin film were studied theoretically by Mielczarski et al. [101, 102]. It was shown that, independently of the angle of incidence and the water interlayer thickness, the band positions in the s-polarized spectra are almost identical to those obtained by the normal-incidence transmission. However, the band intensities, which are negative, are much lower than in the p-polarized spectra (Fig. 7.40b). The p-polarized bands have derivativelike shape, in agreement with the SSR for dielectrics (Sections 3.3.2 and 3.11.5). Their intensity changes sign at a certain angle of incidence depending on the water layer thickness. As in the case of a metal substrate, spectral contrast for p-polarization is maximal for the window with the lowest refractive index (CaF_2). However, such spectra are significantly distorted, being affected by a small change in either the thickness of the water layer or the angle of incidence. In the other limiting case (Ge), the problem is that the highest contrast is achieved in a narrow ($\sim 2°$–$5°$) incident angle range. Due to a wider conventional beam divergence, the spectra measured using a Ge window at the optimum angle of incidence are a superposition of the positive and negative absorption, which complicates the interpretation. Hence, windows with medium (2–3) values of the refractive index are more appropriate in this case.

The important parameter affecting spectral contrast is the thickness of the liquid thin layer. This can be understood through spectral simulations for a 0.2-nm hypothetical organic film in the ZnSe–water layer–film–Pt system. The optical constants of the film chosen were similar to those of a monolayer of CO adsorbed on a Pt(111) electrode. Increasing thickness of the water layer from 0.25 to 5 μm leads to attenuation of the p-polarized reflection spectra measured at the optimum angle of incidence by a factor of 2 (Fig. 2.39). In this thickness range, the optimum angle changes, decreasing from 43° at 0.25 μm to 31° at 5 μm (Fig. 2.40), that is, from the value above φ_c to the value below φ_c (33° for the ZnSe–water interface). The critical angle is crossed at the water layer thickness of ~ 2 μm. When the thickness exceeds this value, the angular dependence of the absorption depth changes sign in the vicinity of the optimum angle (curve 5 in Fig. 2.40), which makes these conditions troublesome from the viewpoint of the spectrum interpretation. If the water layer is increased from 1 to 5 μm at the constant angle of incidence of 33°, the spectrum intensity decreases by a factor of ~ 20. Similar results were obtained for a nonmetalic (Cu_2S) substrate [101, 102].

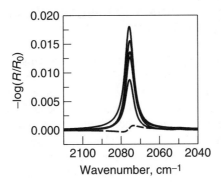

Figure 2.39. (Solid lines, from top to bottom) p-polarized reflection spectra of 0.2-nm film with optical constants similar to those of complete monolayer of adsorbed CO on Pt(111) in ZnSe–water layer–film–Pt system for different thicknesses d_2 of water layer and optimum angle of incidence φ_1 : $d_2 = 0.25\ \mu m$, $\varphi_1 = 43°$; $d_2 = 0.5\ \mu m$, $\varphi_1 = 36°$; $d_2 = 1\ \mu m$, $\varphi_1 = 33°$; $d_2 = 2\ \mu m$, $\varphi_1 = 31°$; and $d_2 = 5\ \mu m$, $\varphi_1 = 31°$. Dashed line: $\varphi_1 = 34°$, $d_2 = 5\ \mu m$. Here, $n_1 = 2.42$ for ZnSe, $\hat{n}_2 = 1.319 - 0.015i$ for water, and $\hat{n}_4 = 5.02 - 20.43i$ for Pt. Optical constants of film are $\nu_0 = 2050\ cm^{-1}$, $\gamma = 5\ cm^{-1}$, $S = 0.05$, $\varepsilon_\infty = 2$.

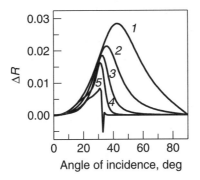

Figure 2.40. Angle-of-incidence dependence of absorption depth ΔR at 2075 cm^{-1} in p-polarized IRRAS spectra for 0.2-nm film with optical constants as for Fig. 2.39 in ZnSe–water layer–film–Pt system for water layer thicknesses: (1) 0.25, (2) 0.5, (3) 1, (4) 2, and (5) 5 μm.

Thus, for both metallic and transparent substrates, p-polarization, angles of incidence above φ_c, and water layers thinner than 1–2 μm are preferable. Under these conditions ($\varphi_1 > \varphi_c$ and $d_2 < d_p$, where d_p is the penetration depth), the cell windows act as IREs. In other words, the in situ IRRAS geometry under the optimum conditions is in fact the ATR geometry in Otto's configuration (Fig. 2.36d).

In general, the application of IR spectroscopy to investigate nanolayers located at various interfaces must involve the determination of optimum experimental conditions for each system to be analyzed. These conditions include angle of incidence, polarization of radiation, and the number of reflections of radiation from the layer–substrate interface.

2.6. CHOOSING APPROPRIATE IR SPECTROSCOPIC METHOD FOR LAYER ON FLAT SURFACE

Successful IR spectroscopy of ultrathin films is very sensitive to the choice of the method and the optical geometry of the experimental set-up, maximizing spectral contrast and the amount of information obtained about the film. These choices should be made on the basis of a comparison of band intensities in film spectra calculated for different experimental conditions. In this section, this approach will be demonstrated using a 1-nm weakly absorbing hypothetical layer that models an isotropic organic monolayer with optical constants $n_2 = 1.3$ and $k_2 = 0.1$ in the region of the νCH vibrations ($\nu = 2800$ cm^{-1}). The layer is assumed to be located on a Ge or Al substrate. The spectra were calculated for p-polarized reflection IRRAS and ATR and single transmission.

Simulated band intensities for the hypothetical layer under the chosen optical conditions are shown in Fig. 2.41, where it can be seen that in air the highest spectral contrast will be achieved in IRRAS on Al and the lowest one in IRRAS on Ge and transmission of the Ge–air interface (with consideration for the beam divergence of $\pm 6°$). In the ATR spectrum recorded near φ_c, the absorption depth will be comparable with that in the best transmission spectra. However, as can be seen in Fig. 2.42, for a 100-nm layer, the maximum intensities in the spectra from metallic IRRAS and the ATR method are practically equal and only 30% higher than those in the transmission spectra. This illustrates the point that when selecting the proper experimental method it must be remembered that the best method for a given layer may be different for different thicknesses, because for each method the dependence of the band intensity on the layer thickness is specific.

A layer located at the interface of two solids must be considered separately (Fig. 2.43). For such a layer, the angle-of-incidence dependence of the p-polarized spectra recorded by transmission, IRRAS, and ATR have been discussed in Section 2.5. A comparison of Figs. 2.41 and 2.43 shows that the immersion of the

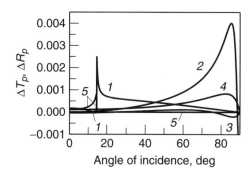

Figure 2.41. Dependence on angle of incidence of absorption of 1-nm model weakly absorbing layer in p-polarized (1–3) reflection ΔR_p and (4, 5) transmission ΔT_p spectra. Layer is located at (1, 3–5) Ge and (2) Al. Spectroscopic methods: (1) ATR, (2, 3) IRRAS, (4) transmission at air–Ge interface, (5) transmission at Ge–air interface; $n_2 = 1.3$, $k_2 = 0.1$, $\nu = 2800$ cm^{-1}, $n_{Ge} = 4.0$, $n_{Al} = 4.22$, $k_{Al} = 27.66$.

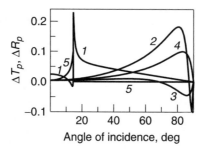

Figure 2.42. Dependence on angle of incidence of absorption of 100-nm model weakly absorbing layer in p-polarized (1–3) reflection ΔR_p and (4, 5) transmission ΔT_p spectra. Layer is located at (1, 3–5) Ge and (2) Al. Spectroscopic methods: (1) ATR, (2, 3) IRRAS, (4) transmission at air–Ge interface, (5) transmission at Ge–air interface; $n_1 = 1$, $n_2 = 1.3$, $k_2 = 0.1$, $\nu = 2800\ \mathrm{cm^{-1}}$, $n_{Ge} = 4.0$, $n_{Al} = 4.22$, $k_{Al} = 27.66$.

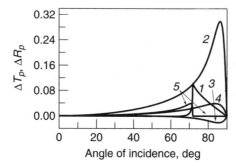

Figure 2.43. Dependence on angle of incidence of absorption of 1-nm model weakly absorbing layer in p-polarized (1–3) reflection ΔR_p and (4, 5) transmission ΔT_p spectra. Layer is located at interfaces (1, 5) Ge–a-Si, (3, 4) a-Si–Ge, and (2) Ge–Al. Spectroscopic methods: (1) ATR, (2, 3) IRRAS, (4) transmission at aSi–Ge interface, (5) transmission at Ge–aSi interface; $n_{aSi} = 3.8$, $n_2 = 1.3$, $k_2 = 0.1$, $\nu = 2800\ \mathrm{cm^{-1}}$, $n_{Ge} = 4.0$, $n_{Al} = 4.22$, $k_{Al} = 27.66$.

substrate into an optically dense medium results in large (40–75-fold) increases in the band intensities. For the 1-nm model organic layer, as for the layer in air, IRRAS yields spectra with the highest contrast at the Ge−Al interface and the lowest contrast at the a-Si−Ge interface. A comparison of the angle-of-incidence dependence of the band intensities for 1- and 10-nm layers of the same material (Figs. 2.43 and 2.44, respectively) shows again that the best method for one layer will be the worst choice for the other layer of the same material. Moreover, even such a large increase in the layer thickness makes little difference to the maximum spectral contrast in the spectra of the layer on the Al substrate collected by IRRAS, while in the ATR spectrum the signal increases by only a factor of 2; in the transmittance and IRRAS methods at the a-Si−Ge interface, the spectral contrast increases 10-fold, in proportion to the layer thickness.

Choosing the technique to obtain spectra, one should take into account that the low spectral intensity may be compensated by using multiple reflection or

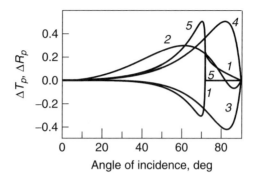

Figure 2.44. Dependence on angle of incidence of absorption of 10-nm model weakly absorbing layer in p-polarized (1–3) reflection ΔR_p and (4, 5) transmission ΔT_p spectra. Layer is located at interfaces (1, 5) Ge–a-Si, (3, 4) a-Si–Ge, and (2) Ge–Al. Spectroscopic methods: (1) ATR, (2, 3) IRRAS, (4) transmission at a-Si–Ge interface, (5) transmission at Ge–a-Si interface; $n_{aSi} = 3.8$, $n_2 = 1.3$, $k_2 = 0.1$, $\nu = 2800$ cm^{-1}, $n_{Ge} = 4$, $n_{Al} = 4.22$, $k_{Al} = 27.66$.

transmission. For example, in ATR spectra, 100 or more reflections may be used, although in practice, there are problems connected with the manufacture of IREs having comparatively large dimensions (for e.g., $100 \times 10 \times 2$ mm) and sufficient precision. Transmission and IRRAS methods (with or without an immersion medium) can also be applied under the multiple passages of the radiation through the film (up to 20–25) (Sections 2.1 and 2.2). However, using modern FTIR spectrophotometers with a SNR between 1000 : 1 and 10,000 : 1 for a 1–4-min scan time and a spectral resolution of 2–4 cm^{-1} enables one, under the optimum experimental conditions, to obtain good spectra of nanolayers without the use of multiple transmissions.

2.7. COATINGS ON POWDERS, FIBERS, AND MATTE SURFACES

Measurement of the IR spectra of an ultrathin film on a powder sample may be carried out using transmission, diffuse transmittance (DT), diffuse reflectance (DR), or ATR techniques. As mentioned in Section 1.10, calculations to model the IR spectra of ultrathin films on powders under a different set of experimental conditions have not yet been realized. Compared to the stratified systems considered in Sections 2.1–2.6, optimization of the measurements on powders is significantly more complicated. Moreover, this problem has not yet been studied in a systematic fashion. Below current knowledge concerning the optimization of such measurements will be presented, with emphasis on the requirements of the sample. The technical aspects (the production of IR spectra of powders) are discussed in Section 4.2.

2.7.1. Transmission

Infrared transmission is the oldest and most highly developed IR spectroscopic method used to study ultrathin films on powdered solids [103, 104]. Buswell et al. [105] were among the first to demonstrate that adsorbed molecules on powders can be detected by IR spectroscopy, by measuring the IR spectra of water adsorbed on montmorillonite. Transmission spectroscopy has been applied to characterize the adsorption of molecules onto catalysts, sorbents, metal particles, and minerals since the 1940s, following the work of Terenin [106, 107] (see review in Refs. [108, 109]). Surface sensitivity of the method as high as $10^{-5}\%$ of CO_2 on Cab-O-Sil was reported by Parkyns and Bradshaw [110].

In the transmission method, the I_{RT} component of radiation that is regularly transmitted through the sample is detected (Fig. 1.22). Measurements of this component do not require any special device — the pellet pressed from the powder is directly positioned in the holder in the sample compartment so as to be in the beam focus. This means that studies of adsorption and catalysis in situ are quite straightforward, based on reactive chambers with a fairly simple optical scheme [111]. Theoretical and practical aspects of the IR transmission spectroscopy of molecules adsorbed on powders have been treated in excellent monographs [108, 109] and will not be repeated here, except to list the basic requirements of the sample.

To increase the radiation traversing a sample in a straight line, I_{RT}, and to minimize scattering, I_{DT} (Fig. 1.22), a self-supporting disk is made by pressing a powder containing particles $<1-2$ μm in diameter. If the particle size is not small enough, the scattering takes place, increasing, according to the Rayleigh law [Eq. (1.130)], with increasing frequency. As a result, at frequencies higher than 2000 cm^{-1}, the spectrum baseline slops toward higher absorbance, distorting the spectrum. Obviously, a higher surface area (smaller and/or porous particles) increases the surface sensitivity. By pressing, a single solid is formed, characterized by a low level of scattering by individual particles. Since the scattering is less for powder substances with low refractive indices, the transmission method is preferable for materials with low refractive indices and gives good results for various large effective surface area forms of silica, such as Aerosil [108], which is comprised of 10–80-nm silica spheres with a refractive index close to that of fused quartz ($\approx$1.43).

Unlike ordinary transmission spectroscopy of powders, which is improved by the use of immersion of the analyte into a homogeneous IR transparent cold-sintering solid (alkali halide) or liquid (oil) of compatible refractive index, this approach, as a rule, is not applied in transmission spectroscopy of adsorbed species. In fact, the purpose of such a dilution is to get rid of the Fresnel reflectance, which distorts intensive absorption bands, and to reduce the intensity of these bands [112]. However, high surface sensitivity can be obtained only in the region of the adsorbent transparency because in the spectral regions where the powder itself strongly absorbs IR radiation (the regions of the phonon absorption) it is impossible to accurately subtract the background spectrum from the resulting spectrum, due to the influence of a shell on the Frohlich frequencies of a small particle and vice versa (Section 3.9.2). In the region of the powder transparency, the immersion medium for the adsorbed species is the powder material. Therefore, a dilution, which decreases detection threshold, is unnecessary. However, this procedure can be justified in the case of relatively thick films [e.g., the oxidation products on PbS powder (see Fig. 2.51 below, curve 5)], in the transmission studies of adsorption onto small metal particles with large effective surface areas [113, 114] and coarse particles with high refractive indices, and in the regions where the adsorbent has weak absorption bands (e.g., the 1500–1650-cm^{-1} region for silica and silica gel [115]) (see also Section 2.7.2). In these cases, it should be kept in mind that the immersion media may modify the adlayer and block reactive sites [108, 109].

To obtain a good-quality transmission spectrum of an adsorbed species, the disk thickness is chosen so that the maximum absorbance of the analyte in the resulting spectrum is within the 0.3–0.8 range, in which spectrophotometric error is minimal. Thus, the disk thickness may be different for different functional groups. For example, a thicker disk is required to study the hydrogen-terminated functional groups (which are relatively weak) on the surface of silica derivatives in the spectral region of the silica transparency (4000–2300 cm^{-1}) than for the adsorbate backbone vibrations within the region where silica has weak absorption bands (overtones) (2300–1300 cm^{-1}). For the former, the recommended thickness, expressed as weight per surface area, is 20–40 $mg·cm^{-2}$, while for the latter, the recommended thickness is less (2–5 $mg·cm^{-2}$). When choosing the pellet thickness, the increase in scattering with increasing frequency should be kept in mind.

2.7.2. Diffuse Transmittance and Diffuse Reflectance

The method of diffuse transmittance (DT) is based on measurement of the radiation component I_{DT} (Fig. 1.22) that passes diffusely through an inhomogeneous layer. This method was first applied to the IR spectroscopic analysis of thin films on samples in powder form by Tolstoy in 1985 [116, 117], who obtained DT spectra of water adsorbed onto silica gel. When used in conjunction with a FTIR spectrometer, the method is called diffuse-transmittance infrared Fourier transform spectroscopy (DTIFTS). DTIFTS is the most recently developed IR spectroscopic methods for studying powder surfaces and has already found application in high-performance liquid chromatography (HPLC) and thin-layer chromatography (TLC) [118, 119]. Of increasing popularity are DTIFTS measurements of powders that use an IR microscope to collect radiation [112, 119] (Section 4.3).

Diffuse-reflectance infrared Fourier transform spectroscopy (DRIFTS) in the mid-IR region has been widely used for studying surfaces since the beginning of the 1980s, when FTIR spectrometers became routine research tools, found in many laboratories, and optical accessories were invented for collecting IR radiation (Section 4.2.3). In particular, DRIFTS has been applied in the characterization of surfaces of catalysts (Section 7.1.1), Ag [120] and Au [121] particles. Diffuse reflectance has been used to study the light-induced yellowing of paper [122, 123], adhesion of coatings to roughened steel sheets [124–126], lubrication of magnetic disks [127], and ultrathin films on polymer films [128–130], polymer [131–133] and glass fibers [134, 135], activated carbon, graphitized carbon fibers and synthetic diamond powder [136, 137], zeolites [138–142], and wood [143, 144] as well as for identification of spots in HPLC and TLC [119, 145–149] and adsorption of surfactants on minerals (Section 7.4).

Because of the internal scattering, the mean pathlength of the IR radiation is increased many fold in both the DRIFTS and DTIFTS measurements, which increases the detectable level of the adsorbed species as compared with the standard one-pass transmission method [150]. The highest contrast in DRIFTS is observed for species adsorbed onto supported catalysts and metal particles, due to higher scattering and multiscattering by the powders. A high SNR with detection limits approaching 10^{-6} monolayer coverage of CO on metal-supported catalysts has been reported by Every and Griffiths [187]. This detection limit is comparable to that of some UHV methods often used in heterogeneous catalyst characterization. A high SNR in DRIFTS can be obtained by exploiting the surface enhancement (SEIRA) phenomenon (Section 3.9.4). This has been shown by Makino et al. [128] using KBr particles covered by 6-nm Ag films. Kim and co-workers [168, 151] studied adsorption on fine (2–3.5-μm) Ag particles and found that the SNR of DRIFTS is 20–30 times higher than in IRRAS.

Figure 2.45 shows spectra from DRIFTS and DTIFTS of adsorbed monolayers of two reagents, 5-cyano-3,3-dimethylpentyl-(dimethylamino)-dimethylsilane (DMP.CN) and 3,3-(dimethylbutyl)-(dimethylamino)-dimethylsilane (DMB), on Cab-O-Sil (a silica derivative), reported by Boroumand et al. [152–154]. Both types of spectrum were measured on each sample, the DTIFTS signal being detected with a pyroelectric detector/holder positioned at the bottom of the

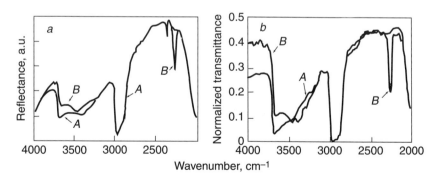

Figure 2.45. (a) Diffuse reflectance and (b) normalized diffuse transmittance of 1.3-mm-thick samples of Cab-O-Sil silica powders entirely covered by (A) DMB and (B) DMP.CN siloxy substituents. (a) Number of coadded scans: 250, resolution: 4 cm^{-1}; interferometer scanning speed: 0.5 cm·s^{-1}. (b) Photopyroelectric signals normalized to spectrum of empty sample holder/detector. Number of coadded scans: 75, resolution: 4 cm^{-1}; interferometer scanning speed: 0.03 cm·s^{-1}. Adapted, by permission, from F. Boroumand, J. E. Moser, and H. Vandenbergh, *Appl. Spectrosc.* **46**, 1874 (1992), pp. 1883 (Fig. 7) and 1885 (Fig. 10). Copyright © 1992 Society for Applied Spectroscopy.

DRIFTS sample cup (note that this geometry actually measures the sum of transmission and DT but is referred to as DT in the cited work). The DMB has structure similar to that of DMP.CN, but with no cyano groups. The spectra in Fig. 2.45 display broad absorption peaks below 2100 cm^{-1} and above 2700 cm^{-1} due to the OH groups on the silica surface and intensive absorption bands around 2900 and 2247 cm^{-1} from the C−H and C≡N groups, respectively. Comparing the DRIFTS and DTIFTS spectra, one can see that under the selected measurement conditions DRIFTS provides a higher spectral resolution: in the DRIFTS spectrum, the C≡N band is narrower and there is an additional narrow component at a higher frequency, which is not interpreted by the authors. However, the spectrum contrast is somewhat higher in DTIFTS.

Figure 2.46 allows one to compare the effect of the layer thickness on the absolute DR and DT at the absorption peak of the CN groups for two different-morphology silicas treated up to surface saturation by DMP.CN. These are Cab-O-Sil (*nonporous* silica with a primary particle diameter of 1–1.5 nm, aggregated in 10-nm grains with a specific surface area of 191 m$^2 \cdot$g^{-1} and a density of 150 g L^{-1}) and LiChrosorb Si 100 (precipitated *porous* silica characterized by a particle diameter of 5–10 μm, specific surface area of 321 m$^2 \cdot$g^{-1}, and density of 350 g L^{-1}). For 1–3-mm porous LiChrosorb silica layers, DR does not vary and DT is absent. Therefore, these layers can be considered as pseudoinfinite with $R = R_\infty = 0.38 \pm 0.02$. For the Cab-O-Sil sample, which has 50% less effective surface area and a generally looser structure, the asymptotic value of the reflectance is much lower ($R = R_\infty = 0.044 \pm 0.002$). Even though the

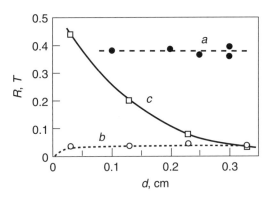

Figure 2.46. Effect of depth of layer *d* on absolute diffuse reflectance and transmittance of derivatized LiChrosorb and Cab-O-Sil silica powder samples. Reported values are measured at maximum of absorption peak of CN group at $v = 2247$ cm^{-1}. (*a*) Diffuse reflectance of various LiChrosorb silica layers treated up to surface saturation by DMP.CN; (*b*) diffuse reflectance of various Cab-O-Sil powder layers treated by DMP.CN; (*c*) diffuse transmittance of Cab-O-Sil powder under same conditions. Reprinted, by permission, from F. Boroumand, H. Vandenbergh, and J. E. Moser, *Anal. Chem.* **66**, 2260 (1994), p. 2263, Fig. 2. Copyright © 1994 by American Chemical Society.

scattering power of Cab-O-Sil is smaller than that of LiChrosorb, the saturation is observed already for layers as thin as 0.3 mm. In addition, there is a principal difference in the thickness dependence of transmittance at the νCN band maximum for Cab-O-Sil layers: The transmittance value decreases with the thickness increasing from 0.3 to 3 mm. As the reflectance is not affected, such a variation is due to the absorption of C≡N groups. Thus, DTIFTS is more surface sensitive of the two techniques. A similar conclusion was drawn from the computer model simulations [154].

When studying adsorption processes on powders by DRIFTS and DTIFTS, one of the key problems is the correct choice of particle size. For the "classic" DR of solids in powder form, the ideal particle size depends on the complex refractive indices of both the powder and the matrix and the packing properties (density) of the powder material [155–158]. Optimal particle sizes were found to be 5–10 and 6 μm for organic substances [157, 159] and diamond powder [160], respectively. For smaller particles, scattering effects are greater and, as seen from Eqs. (1.128) and (1.129), the effective penetration depth is shorter, which was confirmed experimentally [161, 162]. The effect of particle size on the diffuse-reflectance spectra of inorganic (ionic) compounds has been studied by Vincent and Hunt [163, 164] and Chalmers et al. [165]. It was found that the Fresnel reflection cannot be eliminated for the phonon (strong) bands, even with particles as small as 2 μm. On the other hand, for particles up to 10 μm, the distortions of "very weak" bands are absent and the band intensities increase with particle size. For particles of intermediate size, the contribution of the volume component dominates and the Fresnel reflection is suppressed for the particle size below some critical value. When the particle size is comparable to the IR wavelength, spectra from DR and DT measurements [166] are distorted, as predicted by Mie's theory (Section 1.10). For very small particles (<1 μm), the DR signal decreases due to the particle extensive aggregation driven by van der Waals forces.

To date, influence of the powder particle size on the band intensities in DRIFTS and DTIFTS of adsorbed species has not been studied systematically. From general consideration, one can expect that the optimal particle size for both the DR and DT measurements on the adsorbed corresponds to the maximum value of the product $d_{eff}S_{eff}$, where d_{eff} is the mean pathlength of the radiation through the scattering medium and S_{eff} is the effective surface of the powder. With reducing the particle size, the quantity S_{eff} increases, while parameter d_{eff} changes in a complex manner, depending on the refractive index, particle size and surface morphology, and packing properties (powder density) of the powder material [155, 162]. For example, absorption at the νCN frequency in DRIFTS is higher for Cab-O-Sil — the silica derivative with smaller particles, smaller surface area, and looser packing — as shown in Fig. 2.46. For this sample the penetration depth is longer. Cab-O-Sil is also more preferable for the DT measurements.

For transparent particles of CaF_2, KBr, ZnSe, and Si (Table A.2) having the size distribution of $50 < d < 90$ μm, the effective penetration depths measured from the DRIFTS are 530, 1005, 447, and 350 μm (the maximum is observed for the KBr powder). The scattering coefficient varies in the opposite way and presents a minimum for KBr [162]. In another work [167], the penetration depth for such high-refracting substances as Ge and Si was found to be only a few micrometers below the powder sample surface.

The effect of the particle size on the DRIFTS intensity for a monolayer of dodecyl amine adsorbed on quartz particles is shown in Figs. 2.47a, b. The spectral contrast is higher for the <5-μm fraction than for the 38–150-μm one. This is akin to the data for adsorption of a photosensitizer on silica gel [115]. The results of Kim and co-workers [168] indicate that contrast of the DRIFTS of 4-dimethylaminobenzoic acid adsorbed on 2–3.5-μm Ag particles, for which the SEIRA phenomenon is expected, and 30-nm TiO_2 powder is practically identical. For both DRIFTS and DTIFTS, the particle size should be smaller than the shortest wavelength in the spectral range, as particles of a size comparable to the

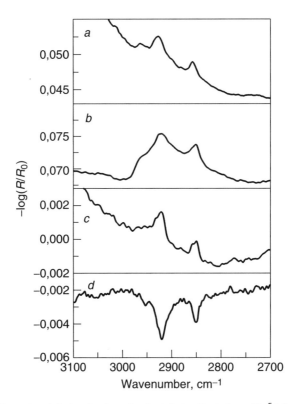

Figure 2.47. IR spectra of dodecylamine adsorbed during 5 min from 10^{-5} M aqueous solution on (a–c) powdered quartz and (d) polished quartz surface. DRIFTS of *undiluted* quartz powder: (a) $-150 + 38$ μm and (b) -5 μm size (200 scans); (c) DRIFTS of -5 μm size but *diluted* in the 1 : 5 ratio by KBr (200 scans); (d) s-polarized IRRAS spectrum of polished surface after treatment with amine, measured at $\varphi_1 = 70°$ (500 scans). Background spectrum is (a) and (b) initial quartz powder; (c) KBr; (d) freshly polished quartz surface.

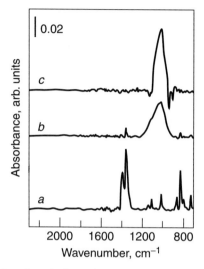

Figure 2.48. DRIFT spectra of *p*-nitrobenzoic acid adsorbed on Ag powders with particle sizes of (*a*) 2, (*b*) 5, and (*c*) 11 μm. Reprinted, by permission, from H. S. Han, C. H. Kim, and K. Kim, *Appl. Spectrosc.* **52**, 1047 (1998), p. 1051, Fig. 5. Copyright © 1998 Society for Applied Spectroscopy.

wavelength of irradiation exhibit complex scattering phenomena (Section 1.10). This is demonstrated in Fig. 2.48, which shows the spectra from DRIFTS of *p*-nitrobenzoic acid on Ag particles of different diameter [169]. It is seen that for 5- and 11-μm particles an artifact band is present in the spectra at 1100 cm^{-1}, instead of the absorption bands of the adsorbate.

The spectra *b* and *c* of Fig. 2.47 indicate that dilution of quartz powder with particles <5 μm by KBr decreases the intensities of the DR νCH bands of adsorbed dodecylamine. However, this does not mean that matrices transparent in the IR cannot be used for scattering measurements. For particles larger than the wavelength whose size cannot be reduced (as is the case for fibers), mixing with KBr increases the multiple reflections between the particles and, therefore, the spectral contrast. To obtain the IR spectra of species adsorbed onto fibers, McKenzie et al. [170] suggested the salt-overlayer technique (Section 4.2.2). Chatzi et al. found that this approach increases the sensitivity of DRIFTS of fiber surfaces, allowing the study of water adsorbed onto polyamide (Kevlar) fibers [171]. This phenomenon was explained to be a result of increasing the multiple reflections between the fibers in the presence of a highly scattering immersion matrix. Deposition of a layer of a transparent powder can also improve the quality of spectra of a film on a rough surface. In the case of substances of a higher refractive index (e.g., Ge, Si, semiconducting sulfides), mixing the powder with KBr increases the penetration depth and, therefore, the spectrum contrast. In fact, the reflection from the front surface, which is responsible for most of the energy losses, decreases. Moreover, the medium in contact with the powder particles is no longer air but an optically denser medium. A dilution in KBr or KCl can be recommended to minimize spectral distortions in the case of (1) thick films (e.g., oleate adsorbed on Ca minerals at alkaline pH [172, 173]) and (2) weakly absorbing substrates [115] (Section 2.7.1).

Figure 2.47 allows comparison of the spectra from DRIFTS and IRRAS of an organic monolayer (dodecylamine) on crystalline quartz in the region of the νCH vibrations. Optimum conditions for IRRAS (70° angle of incidence and s-polarized radiation) were chosen as described in Section 2.3.1. The resolution is higher in the spectrum collected by IRRAS, since the polished substrate surface is more uniform. However, the SNR is substantially higher in DRIFTS, which can be explained, in addition to the optical effects, by a higher surface density of the surfactant adsorbed on a fine powder than on a polished surface (Section 7.4.3). As can be concluded from the spectra reported in Ref. [174], the sensitivity threshold for the νCH bands in the DRIFTS is about 0.1 monolayer (ML).

2.7.3. ATR

ATR (as a rule, in the MIR mode) spectroscopy was first applied to powder surfaces by Harrick [66]. Since the penetration depth d_p is proportional to the wavelength [Eqs. (1.110)], the ATR method is more effective at longer wavelengths. A routine procedure consists in placing fine particles or a suspension on the top of a horizontal IRE (Fig. 4.32) or in a CIRCLE cell (Fig. 4.14). One can maximize d_p by working at the angle of incidence equal to the critical angle φ_c, using an IRE with a low refractive index (Section 2.4), and minimizing scattering. The latter condition requires that the particle size be as small as possible. Pressing the powder layer onto the IRE with a special press (Fig. 4.32) also increases the penetration depth and the powder density and, as a result, the spectral contrast.

The data presented in Sections 1.8 and 2.5.4 show that immersion of the powder into a transparent liquid with a refractive index close to that of the IRE can substantially increase the SNR of the ATR spectrum of the interfacial layer due to the increase of the MSEF at the interface. This approach also reduces scattering and so is particularly effective in studying in situ adsorption on powders. If the $r/\lambda < 10^{-2}$ condition is met, the refractive index of the composite layer (powder plus the immersion medium) may be evaluated within the framework of the EMT (Section 1.9), which allows one to calculate the value of φ_c and simulate the ATR spectrum [175, 176].

To demonstrate the enhancement effect of immersion, Fig. 2.49 shows horizontal ATR (HATR) s-polarized spectra of a pure limonite ($Fe_2O_3 \cdot nH_2O$, orange form/China) powder with 1–5-μm particles and of this powder after the addition of paraffin oil. Figure 2.50 illustrates the SNR in the in situ ATR spectra of sulfate (a strong oscillator) coordinated on a hematite (Fe_2O_3) particle film at a coverage increasing up to about a complete monolayer [177]. The particle size was 10–25 nm. The spectra were obtained using the HATR accessory (Fig. 4.32) with the ZnSe crystal in contact with a 0.01 M KCl solution (pH 3). One can see that the minimal reliably detectable amount of adsorbed sulfate is ~0.05 ML.

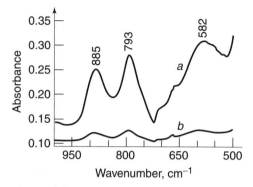

Figure 2.49. ATR spectra of limonite ($Fe_2O_3 \cdot nH_2O$, orange form/China, 5 mg): (*a*) with Nujol film as immersion layer; (*b*) measurement with dry powder. Reprinted, by permission, from U. Kunzelmann, H. Neugebauer, and A. Neckel, *Langmuir* **10**, 2444 (1994), p. 2448, Fig. 6. Copyright © 1994 American Chemical Society.

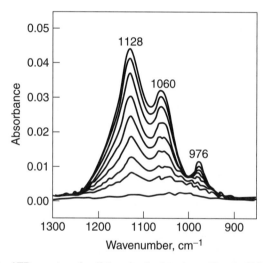

Figure 2.50. In situ ATR spectra of sulfate adsorbed on hematite at pH 3. Spectra were run on Bio-Rad FTS-45 instrument equipped with MCT detector and horizontal ATR unit with 45° ZnSe IRE (thickness, 6 mm; dimension of upper face, 10×70 mm^2; five internal reflections). For each measurement 1000 spectra were coadded. Spectral resolution: 4 cm^{-1}. Reprinted, by permission, from S. J. Hug, *J. Colloid Interface Sci.* **188**, 415 (1997), p. 418, Fig. 4b. Copyright © 1997 Academic Press.

The same sensitivity can be achieved in the in situ ATR spectra of the $v_{as}CH_2$ bands of surfactants with the chains 12–16 carbon lengths adsorbed on colloid oxide films [174, 178]. For coarser particles of a higher refractive index, the sensitivity decreased. For example, for xanthate adsorbed on chalcopyrite ($CuFeS_2$) particles <30 μm size (Fig. 7.26), surface sensitivity is about 0.3 ML [179]. For such substrates, the SNR is comparable to that from the in situ ATR spectra measured at the polished flat surface with one reflection at the critical angle of incidence but higher than in the in situ IRRAS (Table 7.4). The in situ ATR on powders provides about a two- to fivefold higher SNR as compared to that in the in situ MIR, as follows from comparison of the νCH band intensities for ~1 ML of CTAB (hexadecyltrimethylammonium bromide) adsorbed on a sol–gel TiO_2 film [178] and on a SiO_2–Si IRE (25 reflections, $\varphi_1 = 45°$) [180] and a LB monolayer of oleic acid (C_{17}) deposited on the TiO_2 film–Ge IRE (13 reflections, $\varphi_1 = 50°$) [181].

ATR in Otto's configuration (Fig. 2.36d) can be applied to characterizing surfaces of carbon fibers. To obtain the spectra, a fiber cloth is pressed to an IRE. The optimum conditions for such a system were studied by Ohwaki and Ishida [182]. It was shown that a multiple reflection Ge IRE, s-polarization, and an angle of incidence of $30°$ provide the maximum spectral contrast. However, in terms of SNR, the use of unpolarized radiation at an angle of incidence of $35°–40°$ is more advantageous.

2.7.4. Comparison of IR Spectroscopic Methods for Studying Ultrathin Films on Powders

The important benefits of the transmission method are the applicability of the BLB law for quantitative analysis of the spectra and a very high sensitivity to the species adsorbed onto powders of high surface area. Disadvantages include the need to press pellets for sample preparation whose thickness must fall in some predetermined range. Fuller and Griffiths [150] have shown that when small quantities of IR-absorbing materials are present in a nonabsorbing matrix, the intensity of bands in a transmission spectrum is lower than the intensity of the corresponding bands in the DR spectra of the same amount of the sample, since the effective absorption pathlength is enhanced by multiple scattering events when the measurement is performed in the DR mode.

The advantages of the DT method over ordinary transmission include simpler sampling procedures for qualitative analysis and higher sensitivity to ultrathin films especially on highly scattering powders of carbides, semiconductors such as Si, Ge, sulfides, and GaAs. The disadvantages are the special devices required for collecting radiation (Section 4.2.2) and the high sensitivity of the band intensities to any inhomogeneity in the sample (surface or bulk). The latter necessitates special skills to make a sample for quantitative measurements.

The main advantages and disadvantages of DR are the same as those of DT. The sampling procedure is simple, resulting in rapid qualitative analysis, but special optics are required for collecting radiation (Section 4.2.3) and the theory is rather sophisticated (Section 1.10). If one compares DR with DT, in situ measurements in different gaseous environments may be made with the former using commercially available thermostabilized cells. The disadvantages of DR over DT are lower sensitivity to changes in adsorbate absorption (Section 2.7.2) and the high cost of the integrating-sphere accessories relative to the simple DTIFTS optics described in Section 4.2.2. A comparative analysis of surface-modified polyethylene fibers by Taboudoucht [183] using both DTIFTS and DRIFTS showed that orientation of the sample has little effect on the DTIFTS signal relative to that from DRIFTS. Another advantage of DR is the possibility of accessing the spectral region below 1000 cm^{-1} in studying the surface of silica, in which DT and transmission methods are inapplicable due to the strong absorption by the silica phonon bands [186].

In general, the DRIFTS and DTIFTS band intensity varies nonlinearly with the adsorbent quantity [153, 184]. As contrast to the transmission spectrometry, where the BLB law is met, the SNR in the DR spectra of analytes at low concentration is proportional to $c^{1/2}$ [185], which increases sensitivity of the DR method in detecting adsorbate at low levels. However, for analytical applications it is important to recognize that if the DR is properly measured [115] and represented (Section 4.2.3), a linear correlation between the DR band intensity and surface coverage can be achieved over a certain coverage range. For example, Every and Griffiths [187] found that the band intensities of gases adsorbed on catalysts, represented in the KM units [Eq. (1.133)], vary linearly with surface coverage for low coverages and show a negative deviation from linearity at high coverages (see also Refs. [968, 980] in Chapter 7).

The problem is worse when a powder weakly absorbs (as a rule, due to the overtones or adsorbed water) in the spectral range where the adsorbate bands are considered. This takes place for silica in the 1500–1700-cm^{-1} range. In this case, the support absorption can be eliminated in both transmission and DRIFTS studies, provided that silica is diluted in KBr, the spectra are obtained against the pure KBr reference, and a background correcting procedure is used [115]. The correction consists in adjusting the spectra of the different samples on a selected band of the standard made of pure silica and KBr in the same ratio as the samples. The integrated intensities (in units of KM· cm^{-1}) of the DRIFTS bands due to adsorbate, which were obtained after the correction procedure, proved to be linear functions of the adsorbate coverage in the 0.04–1-ML range for both ground and nonground samples.

The main advantage of the ATR method is the convenient and rapid sample preparation and linearity of the ATR band intensity–surface coverage dependence. In addition, it is the best method to study in situ the powder–liquid interface, especially in the spectral region where the liquid absorbs.

Spectra collected by these four methods of an oxidized layer on 50–70-μm galena particles (natural PbS, $n \approx 4$) using the same number of scans (100) and resolution (4 cm^{-1}) and a Perkin-Elmer 1760X FTIR spectrometer equipped with a mercury–cadmium–tellurium detector are shown in Fig. 2.51. Because of strong backscattering by the PbS particles and, as a result, a small penetration depth of radiation, the transmission spectrum obtained from the powder squeezed between two plane–parallel KBr plates represents mainly the component I_0 that has passed by the particles (Fig. 1.22) and, hence, bears no information on the sample absorption. The DRIFTS and DTIFTS spectra of the PbS powder and the transmission spectrum of a mixture of PbS and KBr spectra are more informative. The distinct absorption bands of surface oxidation products at 1440, 1400, and 1200–1100 cm^{-1} are assigned to lead carbonate, hydroxide, and sulfoxide [109],

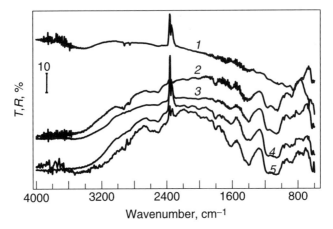

Figure 2.51. Comparison of IR spectra of oxidized surface compounds on galena (PbS) powder 50–70 μm size: (1) transmission of powder layer deposited from acetone on KBr plate after evaporation of acetone; (2) DTIFTS of powder layer deposited from acetone on ZnSe wedge (Section 4.2.2); (3) DRIFTS; (4) DRIFTS of 1:3 mixture with KBr; (5) transmission of 1:3 mixture of galena and KBr squeezed between two KBr windows.

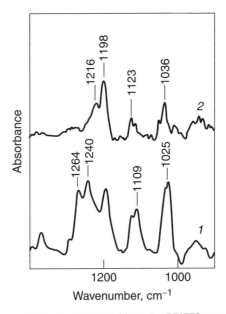

Figure 2.52. Comparison of (1) in situ ATR and (2) ex situ DRIFTS spectra of same chalcopyrite sample <30 μm size contacted with ethyl xanthate solution of pH 9.5 of 7.6×10^{-4} *M*. Baseline correction was done for both reflection spectra. Adaptated, by permission, from J. A. Mielczarski, J. M. Cases, and O. Barres, *J. Colloid Interface Sci.* **178**, 740 (1996), p. 744, Fig. 5. Copyright © 1996 Academic Press.

respectively. Comparing spectra 2 (DT), 3 and 4 (DRIFTS), and 5 (transmission of the mixture with KBr), the DT spectrum (2) is characterized by a somewhat higher noise level.

Figure 2.52 allows one to compare the SNRs in the in situ ATR spectrum and DRIFTS of ultrathin films on highly scattering powders. Both spectra were measured on the same chalcopyrite powder <30 μm size covered by about one statistical monolayer of adsorbed xanthate. The DRIFTS was measured after air drying and gentle mixing of the powder in the 1 : 7 proportion with KBr, which was also used as the reference. First, one can see that the SNR in the ATR spectrum is higher than that in the DRIFTS. The same is true for the in situ ATR and DRIFTS spectra of carbonate absorbed on γ-Al_2O_3 colloid particles [188]. Second, these spectra are qualitatively different. In addition to the bands at ~1200, 1110, and 1030 cm^{-1} assigned to the C$-$O$-$C and C$=$S groups of the surface copper$-$xanthate complex [179], the ATR spectrum exhibits the bands at 1263, 1239, 1110, and 1026 cm^{-1} of physically adsorbed dixanthogen (see Section 7.4.4 for more detail). The absence of the latter bands in the DRIFTS is explained by decomposed/evaporated dixanthogen during mixing with KBr. (The 1216-cm^{-1} band in the DRIFTS is due to the iron hydroxy$-$xanthate complex, which is hidden by the dixanthogen absorption in the ATR spectrum.) Hence, the in situ ATR method is more appropriate for the given system from the viewpoints of the SNR and safety of the adsorbed layer.

REFERENCES

1. J. P. Blitz, in F. M. Mirabella (Ed.), *Modern Techniques in Applied Molecular Spectroscopy*, Wiley, New York, 1998, pp. 185–220.

2. A. V. Rakov, *Spectrophotometry of Semiconducting Structures*, Sovetskoe Radio, Moscow, 1975 (in Russian).

3. V. Gopal and M. Gopal Rao, *Phys. Stat. Solidi (a)* **11**, 783 (1972).

4. K. Takehiko and K. Mototaka, *Jpn. J. Appl. Phys.* **11**, 15 (1972).

5. M. K. Gunde and B. Aleksandrov, *Appl. Spectrosc.* **44**, 970 (1990).

6. V. P. Tolstoy and S. V. Habibova, *Vestnik leningradskogo universiteta. Phyz. Khim.* **2**, 106 (1989).

7. V. P. Tolstoy, A. I. Somsikov, and V. B. Aleskovski, *USSR Inventor Certificate No. 1099256, Izobretenia i otkrytia* **23**, 145 (1984).

8. V. P. Tolstoy, L. P. Bogdanova, and V. B. Aleskovski, *Doklady AN USSR* **291**, 913 (1986).

9. B. Harbecke, B. Heinz, and P. Grosse, *Appl. Phys. A.* **38**, 263 (1985).

10. P. Grosse and V. Offerman, *Vib. Spectrosc.* **8**, 121 (1995).

11. K. Yamamoto and H. Ishida, *Appl. Opt.* **34**, 4177 (1995).

12. K. Sakamoto, R. Arafune, and S. Ushioda, *Appl. Spectrosc.* **51**, 541 (1997).

13. J. Heidberg, N. Y. Gushanskaya, O. Schonekas, and R. Schwarte, *Mikrochim. Acta Suppl.* **14**, 643 (1997).

14. R. Brendel, *J. Appl. Phys.* **72**, 794–796 (1992).

15. A. Sassella, A. Borghesi, and B. Pivac, *Mikrochim. Acta Suppl.* **14**, 343 (1997).

16. A. Borghesi and A. Sassella, *Phys. Rev. B* **50**, 17756 (1994).

17. (a) T. Buffeteau, B. Desbat, E. Pere, and J. M. Turlet, *Mikrochim. Acta Suppl*. **14**, 631 (1997); (b) I. Pelletier, H. Bourque, T. Buffeteau, D. Blaudez, B. Desbat, and M. Pezolet, *j. Phys. Chem. B* **106**, 1968 (2002); (c) J. Wang, *Thin Solid Films* **379**, 224 (2000).

18. S. A. Fransis and A. H. Ellison, *J. Opt. Soc. Am.* **49**, 131 (1959).

19. H. L. Pickering and H. C. Eckstrom, *J. Phys. Chem.* **63**, 512 (1959).

20. R. W. Hannah, *Appl. Spectrosc.* **17**, 23 (1963).

21. A. A. Babushkin, *Zhurnal fizicheskoi khimii* **38**, 1843 (1964).

22. R. G. Greenler, *J. Chem. Phys.* **44**, 310 (1966).

23. R. G. Greenler, *J. Chem. Phys.* **50**, 163 (1969).

24. R. G. Greenler, *J. Vac. Sci. Technol.* **12**, 1410 (1975).

25. V. P. Tolstoy, G. N. Kuznetsova, S. I. Koltsov, and V. B. Aleskovski, *J. Appl. Spectrosc.* **32**, 167 (1980).

26. P. W. Faguy and W. R. Fawcett, *Appl. Spectrosc.* **44**, 1309 (1990).

27. L. Zugang, Z. Weiming, J. Rongbing, Z. Zhilin, J. Xueyin, X. Minzhao, and F. Bon, *J. Phys. Condens. Matter* **8**, 3221 (1996).

28. T. R. E. Simpson, D. A. Russel, I. Chambrier, M. J. Horn, and S. C. Thorpe, *Sensors and Actuators B* **29**, 353 (1995).

29. E. J. Suoninen, K. S. E. Forssberg, and A. N. Buckley (Eds.), *Application of Surface Science to Advancing Flotation Technology*, Elsevier, Amsterdam, 1997.

30. M. W. Urban, *Vibrational Spectroscopy of Molecules and Macromolecules on Surfaces*, Wiley, New York, 1993.

31. J. D. Swalen, D. L. Allara, J. D. Andrade, E. A. Chandross, S. Garoff, J. Israelachvili, T. J. McCarthy, R. Murray, R. F. Pease, J. F. Rabolt, K. J. Wynne, and H. Yu, *Langmuir* **3**, 932 (1987).

32. P. R. Griffiths and J. A. de Haseth, *Fourier Transform Infrared Spectrometry*, Wiley, New York, 1986.

33. S. N. Gruzinov and V. P. Tolstoy, *Zhurnal prikladnoi spektroskopii* **46**, 775 (1987).

34. J. A. Mielczarski, *J. Phys. Chem.* **97**, 2649 (1993).

35. A. Udagawa, T. Matsui, and S. Tanaka, *Appl. Spectrosc.* **40**, 794 (1986).

36. T. Hasegawa, J. Umemura, and T. Takenaka, *J. Phys. Chem.* **97**, 9009 (1993).

37. (a) A. Gericke, A. V. Michailov, and H. Huhnerfuss, *Vib. Spectrosc.* **4**, 335 (1993); (b) C. R. Flach, A. Gericke, and R. Mendelsohn, *J. Phys. Chem. B* **101**, 58 (1997); (c) Y. Ren, C. W. Meuse, S. L. Hsu, and H. D. Stidham, *J. Phys. Chem.* **98**, 8424 (1994).

38. (a) H. Brunner, U. Mayer, and H. Hoffmann, *Appl. Spectrosc.* **51**, 209 (1997); (b) D. Blaudez, T. Buffeteau, B. Desbat, P. Fournier, A.-M. Ritcey, and M. Pezolet, *J. Phys. Chem. B* **102**, 99 (1998).

39. P. G. H. Kosters and R. P. H. Kooyman, *Thin Solid Films* **327–329**, 283 (1998).

40. V. P. Tolstoy, *UV-Vis and IR Spectroscopy of Nanolayers*, St. Petersburg University Press, St. Petersburg, 1998 (in Russian).

41. J. A. Mielczarski and E. Mielczarski, *J. Phys. Chem. B* **103**, 5852 (1999).

42. O. Seiferth, K. Wolter, B. Dillmann, G. Klivenyi, H.-J. Freund, D. Scarano, and A. Zecchina, *Surf. Sci.* **421**, 176 (1999).

43. V. M. Bermudez, *Thin Solid Films* **347**, 195 (1999).

44. A. B. Horn, T. G. Koch, M. A. Chesters, A. R. S. McCousta, and J. R. Sodeua, *J. Phys. Chem.* **98**, 946 (1994).

45. W. Erley, R. Butz, and S. Mantle, *Surf. Sci.* **248**, 193 (1991).

46. V. M. Bermudez and S. M. Prokes, *Surf. Sci.* **248**, 201 (1991).

47. S. Sato, S. Minoura, T. Urisu, and Y. Takasu, *Appl. Surf. Sci.* **90**, 29 (1995).

48. V. M. Bermudez, *J. Phys. Chem.* **98**, 2469 (1994).

49. V. M. Bermudez, *Appl. Phys. Lett.* **62**, 3297 (1993).

50. V. M. Bermudez, *J. Vac. Sci. Technol. A* **10**, 3478 (1992).

51. A. S. Glass and V. M. Bermudez, *J. Vac. Sci. Technol. A* **8**, 2622 (1990).

52. Y. Imaizumi, Y. Zhang, Y. Tsusaka, T. Urisu, and S. Sato, *J. Mol. Structure* **352–353**, 447 (1995).

53. K. Fukui, H. Miyauchi, and Y. Iwasawa, *Chem. Phys. Lett.* **274**, 133 (1997).

54. A. Borg, P. Apall, and O. Hunderi, *Superlattices Microstructures* **3**, 103 (1987).

55. A. Borg, P. Apall, and O. Hunderi, *Phys. Scripta* **35**, 868 (1987).

56. V. M. Bermudez, *J. Vac. Sci. Technol. A* **10**, 152 (1992).

57. A. Horn, in R. J. H. Clark and R. E. Hester (Eds.), *Spectroscopy for Surface Science*, Vol. 26, Wiley, New York, 1998, pp. 273–339.

58. S. J. Finke and G. L. Schrader, *Spectrochim. Acta* **46A**, 91 (1990).

59. A. F. Vasiliev, I. Yu. Gumanskaia, G. N. Zhizhin, and V. A. Iakovlev, *Opt. Spectrosc.* **64**, 700 (1988) (in Russian).

60. Y. Kobayashi, K. Sumitomo, and T. Ogino, *Surf. Sci.* **427–428**, 229 (1999).

61. Y. Kobayashi and T. Ogino, *Appl. Surf. Sci.* **100/101**, 407 (1996).

62. V. P. Tolstoy and E. Yu. Egorova, *Opt. Spectrosc.* **68**, 663 (1990).

63. N. J. Harrick, *J. Phys. Chem.* **64**, 1100 (1960).

64. J. Fahrenfort, *Spectrochim. Acta* **17**, 698 (1961).

65. F. Mirabella, in F. Mirabella (Ed.), *Internal Reflection Spectroscopy*, Dekker, New York, 1993, pp. 1–16.

66. N. J. Harrick, *Internal Reflection Spectroscopy*, Wiley, New York, 1967.

67. W. N. Hansen, in P. Delahay and C. W. Tobias (Eds.), *Advances in Electrochemistry and Electrochemical Engineering*, Vol. 9, Wiley, New York, 1973, pp. 1–60.

68. F. Mirabella, in F. Mirabella (Ed.), *Internal Reflection Spectroscopy*, Marcel Dekker, New York, 1993, pp. 17–52.

69. K. Yamamoto and H. Ishida, *Vib. Spectrosc.* **8**, 1 (1994).

70. J. W. Strojek, J. Mielczarski, and P. Nowak, *Adv. Colloid Interface Sci.* **19**, 309 (1983).

71. W. Suetaka, *Surface Infrared and Raman Spectroscopy: Methods and Applications*, Plenum, New York, 1995, pp. 117–161.

72. Y. J. Chabal, *Surf. Sci. Reports* **8**, 211 (1988).

73. J. Rappich and H. J. Lewerenz, *Electrochim. Acta* **41**, 675 (1996).

74. U. P. Fringeli, in F. Mirabella (Ed.), *Internal Reflection Spectroscopy*, Marcel Dekker, New York, 1993, pp. 255–324.

75. V. M. Zolotarev, V. N. Morosov, and E. V. Smirnova, *Optical Constants of Natural and Technical Media*, K̇himia, Leningrad, 1984 (in Russian).

76. V. M. Zolotarev, *J. Opt. Technal.* **67**, 309 (2000).

77. M. K. Weldon, V. E. Marsico, Y. J. Chabal, D. R. Hamann, S. B. Christman, and E. E. Chaban, *Surf. Sci.* **368**, 163 (1996).

78. (a) R. E. Coleman, H. E. Powell, and A. A. Chochran, *Trans. AIME* **238**, 408 (1967); (b) J. Izumitani, M. Okuyama, and Y. Hamakawa, *Appl. Spectrosc.* **47**, 1503(1993).

79. V. P. Tolstoy and V. N. Krylov, *Opt. Spektrosc.* **55**, 647 (1983).

80. V. P. Tolstoy and S. N. Gruzinov, *Opt. Spektrosc.* **63**, 489 (1988).

81. (a) P. Bruesch, T. Stockmeier, F. Stucki, P. A. Buffat, and J. K. N. Lindner, *J. Appl. Phys.* **73**, 7701 (1993); (b) C. H. Bjorkman, T. Yamazaki, S. Miyazaki, and M. Hirose, *J. Appl. Phys.* **77**, 313 (1995).

82. A. Garton, K. Ha, and M. Adams, *Proc. ACS Div. Polym. Mater. Sci. Eng.* **64**, 36 (1991).

83. F. M. Mirabella, in F. M. Mirabella (Ed.), *Modern Techniques in Applied Molecular Spectroscopy*, Wiley, New York, 1998, pp. 127–184.

84. E. Kretschmann, *Z. Phys.* **241**, 313 (1971).

85. V. M. Agranovich and D. L. Mills (Eds.), *Surface Polaritons: Electromagnetic Waves at Surfaces and Interfaces*, North-Holland, Amsterdam, 1982.

86. Y. Nakao and H. Yamada, *J. Electron Spectrosc. Related Phenom.* **44**, 121 (1987).

87. (a) B. W. Johnson, B. Pettinger, and K. Doblhofer, *Ber. Bunsen. Phys. Chem.* **97**, 412 (1993); (b) B. W. Johnson and K. Doblhofer, *Electrochim. Acta* **38**, 695 (1993).

88. J. S. Loring and D. P. Land, *Appl. Opt.* **37**, 3515 (1998).

89. R. P. Sperline, J. S. Jeon, and S. Raghavan, *Appl. Spectrosc.* **49**, 1178 (1995).

90. Y. Ishida and H. Ishida, *Appl. Spectrosc.* **42**, 1296 (1988).

91. J. Izumitani, M. Okuyama, and Y. Hamakawa, *Appl. Spectrosc.* **47**, 1503 (1993).

92. F. Muller, N. Schwartz, V. Petrova-Koch, and F. Koch, *Appl. Surf. Sci.* **39**, 127 (1989).

93. C. G. Khoo and H. Ishida, *Appl. Spectrosc.* **44**, 512 (1990).

94. A. Watanabe, *Appl. Spectrosc.* **47**, 156 (1993).

95. A. Otto, *Z. Phys.* **216**, 398 (1968).

96. A. Hjortsberg, *Opt. Commun.* **25**, 65 (1978).

97. Y. Ishino and H. Ishida, *Anal. Chem.* **58**, 1448 (1986).

98. Y. Suzuki, S. Shimada, A. Hatta, and W. Suetaka, *Surf. Sci.* **219**, L595 (1989).

99. (a) P. W. Faguy and W. R. Fawcett, *Appl. Spectrosc.* **44**, 1309 (1990); (b) P. W. Faguy and N. S. Marinkovic, *Appl. Spectrosc.* **50**, 394 (1996).

100. P. A. Brooksby and W. R. Fawcett, *Anal. Chem.* **73**, 1155 (2001).

101. J. A. Mielczarski, E. Mielczarski, J. Zachwieja, and J. M. Cases, *Langmuir* **11**, 2787 (1995).

102. J. A. Mielczarski, Z. Xu, and J. M. Cases, *J. Phys. Chem.* **100**, 7181 (1996).

103. R. P. Eischens, *Science* **146**, 486 (1964).

104. R. W. Duerst, M. D. Duerst, and W. L. Stebbings, in F. Mirabella (Ed.), *Modern Techniques in Applied Molecular Spectroscopy*, Wiley, New York, 1998, Chapter. 1.

105. A. M. Buswell, K. Krebs, and W. H. Rodebush, *J. Am. Chem. Soc.* **59**, 2603 (1937).

106. A. N. Terenin, *J. Phys. Chem.* **14**, 1362 (1940).

107. N. G. Yaroslavski and A. N. Terenin, *Doklady AN USSR* **66**, 885 (1949).

108. A. V. Kiselev and V. I. Lygin, *Infra-red Spectra of Surface Compounds*, Wiley, New York, 1975.

109. L. H. Little, *Infra-red Spectra of Adsorbed Species*, Academic, London, 1966.

110. N. D. Parkyns and D. I. Bradshaw, in H. A. Willis, J. H. van der Maas, and R. G. J. Miller (Eds.), *Laboratory Methods in Vibrational Spectroscopy*, 3rd ed., Wiley, Chichester, 1991, pp. 363–410.

111. C. H. Rochester, *Progr. Colloid Polymer Sci.* **67**, 7 (1980).

112. A. M. Hofmeister, in H. J. Humecki (Ed.), *Practical Guide to Infrared Microscopy*, 2nd ed., Marcel Dekker, New York, 1995, pp. 377–416.

113. R. P. Eischens, S. A. Francis, and W. A. Pliskin, *J. Phys. Chem.* **60**, 194 (1956).

114. W. A. Pliskin and R. P. Pliskin, *Adv. Catal.* **10**, 1 (1958).

115. E. Péré, H. Cardy, O. Cairon, M. Simon, and S. Lacombe, *Vib. Spectrosc.* **25**, 163 (2001).

116. V. P. Tolstoy and A. I. Somsikov, *USSR Inventor Certificate No. 1245898, Izobretenia i otkrytia* **27**, 131 (1986).

117. V. P. Tolstoy, *Pribory i tekhnika experimenta* **32**(3, Part 2), 734 (1989).

118. A. Fong and G. Heiftje, *Appl. Spectrosc.* **48**, 394 (1994).

119. F. Rouessac and A. Rouessac, *Chemical Analysis: Modern Instrumental Methods and Techniques*, Wiley, Chichester, 2000, Chapter 10.

120. A. C. Ontko and R. J. Angelici, *Langmuir* **14**, 1684 (1998).

121. K. Shih and R. J. Angelici, *Langmuir* **11**, 2539 (1995).

122. A. J. Mitchell, C. P. Garland, and P. J. Nelson, *J. Wood Chem. Technol.* **9**, 85 (1989).

123. A. J. Mitchell, P. J. Nelson, and C. P. Garland, *Appl. Spectrosc.* **43**, 1482 (1989).

124. M. Thibault, J.-C. Bavay, J. Hernandez, and J.-M. Leroy, *Surface Eng.* **14**, 256 (1998).

125. K. C. Cole, A. Pilon, D. Noel, J.-J. Hechler, A. Chouliotis, and K. C. Overbury, *Appl. Spectrosc.* **42**, 761 (1988).

126. S. Bistac, M. F. Vallat, and J. Schultz, *Appl. Spectrosc.* **51**, 1823 (1997).

127. M. Yanagisawa, *Tribol. Trans.* **36**, 484 (1993).

128. N. Makino, K. Mukai, and Y. Kataoka, *Appl. Spectrosc.* **51**, 1460 (1997).

129. S. R. Culler, M. T. McKenzie, L. J. Fina, H. Ishida, and J. L. Koenig, *Appl. Spectrosc.* **38**, 791 (1984).

130. N. Dumont and C. Depecker, *Vib. Spectrosc.* **20**, 5 (1999).

131. E. G. Chatzi, H. Ishida, and J. L. Koenig, *Appl. Spectrosc.* **40**, 847 (1986).

132. M. T. McKenzie, S. R. Culler, and J. L. Koenig, *Appl. Spectrosc.* **38**, 786 (1984).

133. R. T. Graf, J. L. Koenig, and H. Ishida, in H. Ishida (Ed.), *Fourier Transform Characterization of Polymers*, Plenum, New York, 1987, p. 397.

134. R. T. Graf, J. L. Koenig, and H. Ishida, *Anal. Chem.* **56**, 773 (1984).

135. M. T. McKenzie, S. R. Culler, and J. L. Koenig, *Appl. Spectrosc.* **38**, 786 (1984).

136. A. Dandekar, R. T. K. Baker, and M. A. Vannice, *Carbon* **36**, 1821 (1998).

137. Y. El-Sayed and T. J. Bandosz, *J. Colloid Interface Sci.* **242**, 44 (2001).

138. T. V. Voskoboinikov, B. Coq, F. Fajula, R. Brown, G. McDougall, and J. L. Couturier, *Microporous Mesoporous Mater.* **24**, 89 (1998).

139. M. Suvanto and T. A. Pakkanen, *J. Mol. Catal. A: Chem.* **138**, 211 (1999).

140. S. Myllyoja and T. A. Pakkanen, *J. Mol. Catal. A: Chem.* **136**, 153 (1998).

141. M. Kurhinen and T. A. Pakkanen, *Langmuir* **14**, 6907 (1998).

142. J. J. Benitez, I. Carrizosa, and J. A. Ordizola, *Appl. Surf. Sci.* **84**, 391 (1995).

143. E. Zavarin, S. J. Jones, and L. G. Cool, *J. Wood Chem. Technol.* **10**, 495 (1990).

144. J. F. Manville and J. R. Nault, *Appl. Spectrosc.* **51**, 721 (1997).

145. M. Kawahara, H. Nakamura, and T. Nakajima, *Anal. Sci.* **5**, 485 (1989).

146. M. Kawahara, H. Nakamura, and T. Nakajima, *J. Chromatogr.* **515**, 149 (1990).

147. N. D. Danielson, J. E. Katon, S. P. Bouffard, and Z. Zhu, *Anal. Chem.* **64**, 2183 (1992).

148. R. White, *Chromatography/Fourier Transform Infrared Spectroscopy and Its Applications*, Marcel Dekker, New York, 1990.

149. G. W. Somsen and T. Visser, Liquid Chromatography/Infrared Spectroscopy, in R. A. Meyers (Ed.), *Encyclopedia of Analytical Chemistry*, Wiley, Chichester, 2000, p. 8.

150. M. P. Fuller and P. R. Griffiths, *Appl. Spectrosc.* **43**, 533 (1980).

151. S. J. Lee and K. Kim, *Vib. Spectrosc.* **18**, 187 (1998).

152. F. Boroumand, H. Vandenbergh, and J. E. Moser, *Anal. Chem.* **66**, 2260 (1994).

153. F. Boroumand, J. E. Moser, and H. Vandenbergh, *Appl. Spectrosc.* **46**, 1874 (1992).

154. A. Mandelis, F. Boroumand, and H. Vandenbergh, *Spectrochim. Acta A* **47**, 943 (1991).

155. K. Moradi, C. Depecker, and J. Corset, *Appl. Spectrosc.* **48**, 1491 (1994).

156. D. J. J. Fraser and P. R. Griffiths, *Appl. Spectrosc.* **44**, 193 (1990).

157. M. P. Fuller and P. R. Griffiths, *Anal. Chem.* **50**, 1906 (1978).

158. M. L. E. Tevrucht and P. R. Griffiths, *Talanta* **38**, 839 (1991).

159. J. K. Drennen, E. G. Kraemer, and R. A. Lodder, *Crit. Rev. Anal. Chem.* **22**, 443 (1991).

160. J. M. Brackett, L. V. Azarraga, M. A. Castles, and L. B. Rogers, *Anal. Chem.* **56**, 2007 (1984).

161. P. R. Griffiths, in R. Hester (Ed.), *Advances in Infrared and Raman Spectroscopy*, Vol. 9, Wiley, New York, 1983, p. 64.

162. K. Moradi, C. Depecker, J. Barbillat, and J. Corset, *Spectrochim. Acta A* **55**, 43 (1999).

163. R. K. Vincent and G. R. Hunt, *Appl. Opt.* **7**, 53 (1968).

164. G. R. Hunt and R. K. Vincent, *J. Geophys. Res.* **73**, 6039 (1968).

165. J. M. Chalmers, M. W. Chalmers, and M. W. MacKenzie, in M. W. MacKenzie (Ed.), *Advances in Applied Fourier Transform Infrared Spectroscopy*, Wiley, Chichester, 1988, pp. 145–150.

166. W. Maddams, *Internet J. Vib. Spectrosc.*, available on-line: http://www.ijvs.com/volume1/edition1/section1.html.

167. D. J. J. Fraser and P. R. Griffiths, *Appl. Spectrosc.* **44**, 193 (1990).

168. S. J. Lee, S. W. Han, M. Yoon, and K. Kim, *Vib. Spectrosc.* **24**, 265 (2000).

169. H. S. Han, C. H. Kim, and K. Kim, *Appl. Spectrosc.* **52**, 1047 (1998).

170. M. T. McKenzie, S. R. Culler, and J. L. Koenig, *Appl. Spectrosc.* **38**, 786 (1984).

171. E. G. Chatzi, S. L. Tidrick, and J. L. Koenig, *J. Polym. Sci. B* **26**, 1585 (1988).

172. K. Hanumantha Rao, J. M. Cases, and K. S. E. Forssberg, *J. Colloid Interface Sci.* **145**, 330 (1991).

173. K. Hanumantha Rao, J. M. Cases, P. De Donato, and K. S. E. Forssberg, *J. Colloid Interface Sci.* **145**, 314 (1991).

174. E. Mielczarski, Ph. de Donato, J. A. Mielczarski, J. M. Cases, O. Barres, and E. Bouquet, *J. Colloid Interface Sci.* **226**, 269 (2000).

175. P. Grosse and V. Offermann, *Vib. Spectrosc.* **8**, 121 (1995).

176. T. Kume, T. Kitagawa, S. Hayashi, and Y. Yamamoto, *Surf. Sci.* **395**, 23 (1998).

177. S. J. Hug, *J. Colloid Interface Sci.* **188**, 415 (1997).

178. K. D. Dobson, A. D. Roddick-Lanzilotta, and A. J. McQuillan, *Vib. Spectrosc.* **24**, 287 (2000).

179. J. A. Mielczarski, J. M. Cases, and O. Barres, *J. Colloid Interface Sci.* **178**, 740 (1996).

180. D. J. Neivandt, M. L. Gee, M. L. Hair, and C. P. Tripp, *J. Phys. Chem. B* **102**, 5107 (1998).

181. J. Drelich, Y. Lu, L. Chen, J. D. Miller, and S. Guruswamy, *Appl. Surf. Sci.* **125**, 236 (1998).

182. T. Ohwaki and H. Ishida, *Appl. Spectrosc.* **49**, 341 (1995).

183. A. Taboudoucht, *Diss. Abstr. Int.* **50**(9), 105 (March 1990).

184. T. Burger, H. J. Ploss, J. Kuhn, S. Ebel, and J. Fricke, *Appl. Spectrosc.* **51**, 1323 (1997).

185. M. P. Fuller and P. R. Griffiths, *Appl. Spectrosc.* **34**, 533 (1980).

186. J. P. Blitz, *Colloid Surf.* **63**, 11 (1992).

187. K. W. van Every and P. R. Griffiths, *Appl. Spectrosc.* **45**, 347 (1991).

188. H. Wijnja and C. P. Schulthess, *Spectrochim. Acta A* **55**, 861 (1999).

3

INTERPRETATION
OF IR SPECTRA
OF ULTRATHIN FILMS

Generally speaking, IR spectral measurements of ultrathin films are undertaken to determine (i) the chemical identity of adsorbed species (including dissociation fragments); (ii) the geometric or structural arrangement (orientation) of these species and their positions with respect to the surface atoms of the substrate; (iii) their vibrational, rotational, and translational motion on the surface; (iv) the charge distribution and energy level structure of the valence electrons in both adsorbate and substrate; and (v) the effects of external perturbations, such as the electric and magnetic fields, photons, electrons, heating, and surface pressure on factors (i)–(iv) [1]. However, such information cannot be extracted directly from the IR spectra of ultrathin films due to the strong dependence of the shape and relative intensity of the bands on the geometry of the experiment, the optical properties of the substrate, the surroundings, the gradient of the optical properties at the film–substrate interface and in the film itself, as well as on the film thickness, and the particle size in the case of powder or islandlike supports. These effects and their physical background are discussed in Sections 3.1–3.5 and 3.9. In addition, there can be an intensity transfer between modes of coadsorbed species and a change in the band position with surface coverage. This effect is used for determining the mode of the surface filling and the degree of intermixing of different species at the surface (Section 3.6). If the film under study is located at the electrode–solution interface, the resulting spectrum is made up of contributions of all species in the path of radiation that are affected by the electrode potential, which further complicates the interpretation. Apart from technical means (Chapter 4), there exist analytical and mathematical approaches to resolve contributions of different species in such a complex spectrum (Sections 3.7 and 3.8). Another feature of the IR spectra of ultrathin films that is perhaps surprising and unexpected is their sensitivity to the volume fraction, shape, and orientation of inhomogeneities (e.g., pores or inclusions) in the film. Moreover, if the characteristic size of the particles is small compared to the wavelength, there is a mutual effect of the film on the IR spectrum of the powdered substrate and, vice versa, of the substrate on the absorption bands of the film (Section 3.9). Finally, IR spectra allow for characterization of ultrathin films in terms of optical constants, molecular order and orientation, and homogeneity of the molecular packing (Sections 3.10 and 3.11).

3.1. DEPENDENCE OF TRANSMISSION, ATR, AND IRRAS SPECTRA OF ULTRATHIN FILMS ON POLARIZATION (BERREMAN EFFECT)

As shown in Chapter 2, to optimize the contrast in the IR spectrum of an ultrathin film, it is necessary in many cases to use *nonnormal angles of incidence* and *p-polarized radiation*, which creates specific difficulties in the interpretation of the spectrum. The problems stem from the appearance of additional bands in the spectra of samples that are small relative to the wavelength; these bands are due to the surface charges resulting from the polarization of the samples. The dependence of the transverse vibrational frequency of a polar crystal on the crystal size, called the *size effect*, was discovered by Frohlich [2]. A convincing explanation of this effect in the IR spectra of thin films was presented in 1963 by Berreman [3] while studying the transmission of 325–348-nm LiF layers. Consequently, this size effect in the IR spectra of ultrathin films became known as the *Berreman effect* by Harbecke et al. [4].

In this section the manifestation of the Berreman effect in experimental work will be illustrated by some typical examples and interpreted qualitatively. In the following section this effect will be considered from the viewpoint of Maxwell's

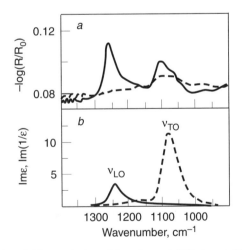

Figure 3.1. (a) Experimental transmission of two thermal SiO_2 layers 26 nm thick each produced on both sides of Si plate by heating in air at 800°C for 20 min: s- (dashed line) and p-polarization (solid line); $\varphi_1 = 74°$. (b) The LO (solid line) and TO (dashed line) energy loss functions calculated on basis of optical constants of thermal Si dioxide grown at 1150°C [52].

theory. The spectral "anomalies" in the case of powdered supports and rough surfaces are discussed in Section 3.9.

Figure 3.1 shows the s-polarized transmission spectrum obtained at the Brewster angle of incidence of two 26-nm thermal SiO_2 layers, one on either side of a Si plate. Besides the absorption around 1102 cm^{-1} by oxygen contained in the plate, this spectrum exhibits a single asymmetric absorption band at 1075 cm^{-1}. A similar spectrum is observed with s-polarized radiation for all angles of incidence (the spectra are not shown). However, the p-polarized transmission spectrum obtained at nonnormal incidence has an additional absorption band at 1253 cm^{-1}. In grazing-angle p-polarized IRRAS of a 1.3-nm sputtered SiO_2 layer on Al, only this "additional" band is observed (Fig. 3.2) (the shoulder near 1150 cm^{-1} is attributed to nonstoichiometric SiO oxide), but in the s-polarized spectrum, this film is undetectable. The polarization dependence of ATR spectra is analogous to that for the transmission spectrum. Figure 3.3a (solid line) shows the p-polarized ATR spectrum of a 10-nm thermal SiO_2 layer measured at $\varphi_1 = 60°$ with a Ge prism that exhibits an intense band near 1240 cm^{-1} and a weak band at 1080 cm^{-1}. In the ATR spectra of a 10-nm WO_3 film sputtered on a Ge substrate ($\varphi_1 = 45°$), there is a pronounced absorption band at 970 cm^{-1} in the p-polarized spectrum that is absent in the s-polarized spectrum (Fig. 3.3b).

A similar band doubling is observed in p-polarized spectra of all polar inorganic films and organic films incorporating highly polar bonds such as C≡O, C=O, and S=O, characterized by $k \sim 0.3-1$, especially at the oblique angles of incidence [5]. However, this effect is not observed for the bands, whose extinction coefficient k is lower than 0.3, such as the bands originating from C−H and Si−H. In this case, a change in the polarization leads only to slight band distortions and shifts (1−2 cm^{-1}).

Such a strong dependence on polarization in the IR spectra of ultrathin films may suggest a molecular-level, structural interpretation or, because of the contrast with ordinary transmission spectra, orientational effects. However, this is not the

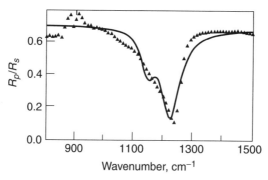

Figure 3.2. IRRAS spectra of MOS Si–photo-CVD SiO_2 2 nm thick–Al system (CVD = Chemical vapor deposition): experiment (triangles) and two-oscillator simulation (solid line); p-polarization, $\varphi_1 = 80°$. Reprinted, by permission, from R. Brendel, *Appl. Phys.* **A50**, 587 (1990), p. 591, Fig. 3a. Copyright © 1990 Springer-Verlag.

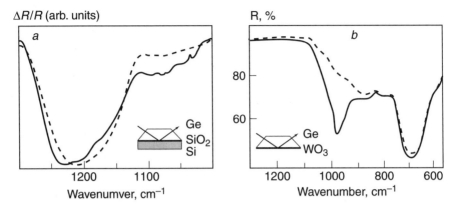

Figure 3.3. (a) Experimental (solid line) and calculated (dashed line) p-polarized ATR spectra of 10-nm SiO_2 film on Si obtained using 60° Ge prism (see insert). Reprinted, by permission, from C. H. Bjorkman, T. Yamazaki, S. Miyazaki, and M. Hirose, *J. Appl. Phys.* **77**, 313 (1995), p. 316, Fig. 6. Copyright © 1995 American Institute of Physics. (b) Experimental ATR spectra of 10-nm WO_3 layer on 45° Ge prizm: s-polarization (dashed line) and p-polarization (solid line). Adapted, by permission, from T. A. Taylor and H. H. Patterson, *Appl. Spectrosc.* **48**, 674 (1994), p. 677, Fig. 6. Copyright © 1994 Society for Applied Spectroscopy.

case; the band doubling in question is a macroscopic phenomenon that can be described within the framework of classical electromagnetic theory, considered in Chapter 1. In fact, as can be seen immediately from Fig. 3.1, the 1075-cm^{-1}-peak position in the s- and p-polarized spectra is near the maximum of the TO energy loss function and therefore is near ν_{TO} (1.1.18°). Furthermore, the position of the additional peak at ~1240–1253 cm^{-1} in the p-polarized spectrum is close to the maximum of the LO energy loss function, near ν_{LO}, in agreement with the thin-film approximation formulas (Table 1.2). To distinguish more accurately possible band doubling in p-polarized spectra, the explicit Fresnel formulas (Sections 1.5–1.7) can be invoked. To illustrate this, Figs. 3.2 and 3.3a and Figs. 3.9 and 3.10 in the next section show the results of a spectral simulation using exact formulas. It can be seen in these figures that both the form and the position of the spectral bands are reproduced rather well.

A qualitative understanding of the appearance of the absorption band near ν_{LO} in p-polarized spectra can be reached by considering the electric polarization of a slab-shaped cubic ionic crystal. If the external electric field $\mathbf{E}_0$ is perpendicular to the slab, the positive ions are displaced to one side of the slab and the negative ions to the other (Fig. 3.4). The excess charge on the surfaces results in a polarization $\mathbf{P}$ perpendicular to the slab plane and given by

$$\mathbf{P} = \varepsilon_0(\varepsilon - 1)\mathbf{E}. \tag{3.1}$$

Here, ε is the dielectric constant of the slab and $\mathbf{E}$ is the electric field inside the slab, which is the sum of the external field $\mathbf{E}_0$ and the depolarization field

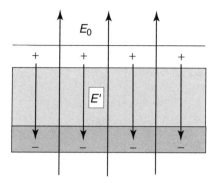

Figure 3.4. Polarization of slab in external electrostatic electric field $\mathbf{E}_0$ perpendicular to slab.

$\mathbf{E}' = -\mathbf{P}/\varepsilon_0$ (Fig. 3.4):

$$\mathbf{E} = \mathbf{E}_0 + \mathbf{E}'. \tag{3.2}$$

Inserting Eq. (3.2) into Eq. (3.1) gives the boundary condition of the continuity of the normal component of the electric displacement D_{normal} (1.60):

$$\mathbf{E} = \frac{\mathbf{E}_0}{\varepsilon}. \tag{3.3}$$

It follows that the polarization and the internal electric field show resonance behavior at the frequency of the LO mode of the film, ν_{LO}, where $\varepsilon(\nu) = 0$ (1.3.7°). Since the dielectric function of a cubic ionic crystal is a scalar, the polarization is parallel to the external electric field and, according to the general selection rule in Eq. (1.27), strong absorption occurs at ν_{LO}. When the external electric field is parallel to the film surface (s-polarization and $\mathbf{E}_x$ from p-polarization, Fig. 1.9), the additional polarization due to the surface charges is absent, $\mathbf{E} = \mathbf{E}_0$, and the absorption resonance position coincides with the frequency of the transverse optical mode of the cubic ionic crystal.

Although the treatment of Maxwell's theory outlined in Chapter 1 will reveal band doubling through spectral simulations or the approximation formulas, it says nothing about the physical nature of the bands. This may result in another misconception in that the appearance of the "additional" blue-shifted band in the p-polarized spectra is simply an optical effect, which needs only be kept in mind when interpreting the IR spectra of ultrathin films (see, e.g., Refs. [5–9]). Thus, many points remain unclear, including which elementary excitations are responsible for the bands in the s- and p-polarized spectra of ultrathin films and why, in the IR spectrum of an ultrathin film on a metal substrate, only the band near ν_{LO} is detected.

To clarify the origin of the absorption bands in the IR spectra of ultrathin films, consider a plane wave impinging on a cubic ionic crystal film whose thickness is small compared to the wavelength but significantly larger than the value of the lattice constant. If the polar vibrations in an infinite crystal result in the appearance of the electric polarization **P** (1.1.1°), these vibrations represent the elastic waves, called the *normal polarization waves*, or normal polarization modes [10, 11]. The polarization **P** and the electric field **E** arising from them change periodically with space and time, obeying the law of the atom displacements. As shown in Section 1.3, the polarization waves can be either *longitudinal* or *transverse* in their form, but only the latter type can interact with transverse waves of the external electromagnetic radiation.

If a transverse polarization wave with its wave vector parallel to the film surface propagates along this film [11], positive ions will be displaced to one surface of the film and negative ions to the other by the electric field of this wave. The excess charge arising on the surfaces creates an electric field perpendicular to the film plane and virtually uniform at a distance exceeding the lattice constant (Fig. 3.5). This field will vary with the wave frequency and at each instant will act on the lattice ions in the direction opposite to the direction of their displacement from the equilibrium position, thus contributing to the restoring force. This force is similar to that acting perpendicular to the front of a longitudinal wave propagating in an infinite crystal (Fig. 1.6). Therefore, the frequency of the transverse wave in the film considered here will be close to the frequency of the longitudinal wave in the infinite crystal, ν_{LO}. As the film thickness increases (and the ratio of surface to volume decreases), the influence of the surface charge density on the internal field decreases, and the frequency of the transverse vibrations with the wave vector parallel to the surface will decrease, approaching ν_{TO}.

When the transverse polarization wave propagates normal to the film surface (and thus the electric field is parallel to the surface), an additional field does not arise because the film is regarded as an infinite plane, and the frequency of the wave coincides with that of the transverse vibration, ν_{TO}, in the infinite crystal. Such a wave interacts only with the tangential components of the electric field.

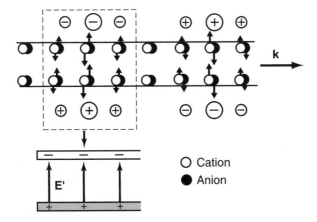

Figure 3.5. Displacements of ions inside ultrathin film of ionic cubic crystal in transverse polarization wave propagating along film. Below is shown polarization field due to ion displacement.

Therefore, the s-spectrum of the film will consist of the single band near ν_{TO}, independent of the angle of incidence.

Since the p-polarized radiation at $\varphi_1 \neq 0°$ has tangential and normal electric field components (1.4.4°), these components will interact with those dipoles within the film that are oriented parallel and perpendicular to the surface, respectively, according to the selection rule (1.27). Furthermore, under action of the external electric field normal to the surface, the dipoles experience an additional electric field induced by the polarized chemical bonds within the layer. It follows that in the p-polarized spectra, the band near the longitudinal optical mode ν_{LO} will be observed, in addition to the band near ν_{TO} resulting from the dipoles parallel to the film. In light of the discussion in Section 1.8.2, it is evident that because the dipole oscillations responsible for the ν_{TO} band are directed along the surface, these oscillations will be quenched by a metal surface and, hence, be absent in p-polarized spectra of films on metal substrates, as demonstrated in Fig. 3.2.

More rigorous treatment of the problem is given within the framework of the Maxwell macroscopic theory, which is the subject of the following section.

3.2. THEORY OF BERREMAN EFFECT

It is now generally accepted that the IR spectra of ultrathin films taken with different polarizations exhibit quite different elementary excitations. The most rigorous interpretation of this phenomenon is given by the *polariton theory* [12–17]. According to this theory, which is based conceptually on Maxwell's macroscopic electrodynamics, the eigenstates of the wave in a crystal are the coupled photon–polarization excitation modes [15, 16, 18], or, using quantum-mechanical terminology, composite particles (quanta) of photons and dipole-active (polar) excitations inside the crystal. The latter arise from lattice vibrations (phonons), oscillations of free carriers (plasmons), and electron–hole excitations (excitons). The coupled excitations are called *polaritons*. In general, the polariton energy differs from the energy of the corresponding uncoupled polar mode, excited, for example, by electron bombardment and analyzed by high-resolution electron energy loss spectroscopy (HREELS) [19–21] and inelastic tunneling vibrational spectroscopies [22]. When the frequency of a polariton is close to a particular electric dipole excitation such as optical phonon or plasmon, it is customary to refer to this polariton as an optical *phonon–polariton*, or as a *plasmon–polariton*, respectively, to indicate that the polariton is largely optical phonon or plasmon in character. In polar-doped semiconductors, the characteristic frequencies of lattice vibrations and the bulk plasmon frequency are close to each other. In this case, the polariton is called a *phonon–plasmon*.

The use of special terminology from condensed-matter optics and the emphasis on the dispersion relations of the different polaritons inside a layered structure rather than on the physical origin of the corresponding modes may discourage the use of the polariton theory by spectroscopists. Fortunately, it was found [12, 23–25] that for ultrathin films ($d \ll \lambda$) and small particles, the problem can be simplified if the interaction between the radiation and the excitations is

ignored and the modes within the film are considered separately, in the so-called
quasi-static regime.

In this section, the reader will first be introduced to the general properties
of polaritons at the interface of two media and within a layer. Then, following
Vinogradov [26] and Ruppin [23], the results of a quasi-static treatment will be
discussed to gain insight into the physical nature of the absorption bands observed
in the IR spectra of ultrathin films. For further detailed discussion of both the
theoretical and experimental aspects of the polariton theory and its approximations, several reviews [12–14] and monographs [15, 16, 24] are recommended.
Modifications of polaritons in superlattices are considered in Refs. [27–29].

3.2.1. Surface Modes

It was shown in Sections 1.4 and 1.8 that at an interface there can be propagating reflected and refracted waves or a wave that is propagating on the side
of the input medium and evanescent (decaying) on the other side. However,
this picture is incomplete, and an additional mode that is completely confined
by a surface can arise under certain conditions, when the incident electromagnetic wave is p-polarized. Such a mode, called a surface mode, has been known
since 1909 from the work of Sommerfeld in radio wave propagation and was
investigated for optical waves by Fano [30] in 1941. The surface modes were
studied extensively in the 1960s and 1970s following the pioneering work of Teng
and Stern [31], Ritchie et al. [32], Otto [33, 34], and Kretschmann [35, 36], who
developed experimental methods for their observation.

For simplicity's sake, consider p-polarized radiation incident from a non-
absorbing medium with $\varepsilon_1(\omega) = \text{const} \geq 1$ onto a second medium with $\varepsilon_2(\omega)$.
Assume that the dielectric function of the second medium is real and expressed
by Eq. (1.40). The coordinate axes are defined with respect to the interface as
shown in Fig. 1.9.

The wave equations (1.5) for the electric and magnetic fields at the interface
have the forms

$$\mathbf{A}_1 = \mathbf{A}_{10} e^{i(\omega t - \mathbf{k}_{x1}\mathbf{x} - \mathbf{k}_{z1}\mathbf{z})} \quad \text{at } z < 0, \tag{3.4a}$$

$$\mathbf{A}_2 = \mathbf{A}_{20} e^{i(\omega t - \mathbf{k}_{x2}\mathbf{x} - \mathbf{k}_{z2}\mathbf{z})} \quad \text{at } z > 0, \tag{3.4b}$$

where $\mathbf{A}$ stands for $\mathbf{E}$ and $\mathbf{H}$; $\mathbf{k}_{x1}$ and $\mathbf{k}_{x2}$ are the wave vectors in the x-direction,
$\mathbf{k}_{z1}$ and $\mathbf{k}_{z2}$ are those in the z-direction, and ω is the angular frequency.

From the boundary conditions (1.60), it follows that the tangential components
of $\mathbf{E}$ and $\mathbf{H}$ must be equal at the interface, where $z = 0$,

$$E_{x1} = E_{x2}, \tag{3.5a}$$

$$H_{y1} = H_{y2}, \tag{3.5b}$$

or, taking into account Eqs. (3.4), $k_{x1} = k_{x2} = k_x$. At the same time, Eqs. (3.4) and (1.1b) give

$$k_{z1} H_{y1} = \omega \varepsilon_0 \varepsilon_1 E_{x1}, \tag{3.6a}$$

$$k_{z2} H_{y2} = -\omega \varepsilon_0 \varepsilon_2(\omega) E_{x2}. \tag{3.6b}$$

From Eqs. (3.5) and (3.6), it follows that a nontrivial solution only exists if

$$\frac{k_{z1}}{k_{z2}} = -\frac{\varepsilon_1}{\varepsilon_2(\omega)} \tag{3.7}$$

when the permittivity of the second medium is negative [$\varepsilon_2(\omega) < 0$] and its refractive index n_2 is imaginary, implying that this mode is evanescent inside the second medium. In the IR region, this condition is met for metals (1.3.9°), doped semiconductors, and inorganic compounds represented by weakly damped strong oscillators within the frequency range $\omega_{TO} < \omega < \omega_{LO}$ (Fig. 1.5). From Eqs. (1.1a), (1.1b), and (3.6) and recalling that $c = (\varepsilon_0 \mu_0)^{-1/2}$, one can obtain

$$k_x^2 + k_{zj}^2 = \left(\frac{\omega}{c}\right)^2 \varepsilon_j \tag{3.8}$$

or

$$k_{zj} = \sqrt{\left(\frac{\omega}{c}\right)^2 \varepsilon_j - k_x^2}, \qquad j = 1, 2. \tag{3.9}$$

On the basis of Eqs. (3.7) and (3.9), the dispersion relation for electromagnetic waves at the interface of two isotropic nonconducting semi-infinite media can be written as

$$\frac{k_x^2 c^2}{\omega^2} = \frac{\varepsilon_1 \varepsilon_2(\omega)}{\varepsilon_1 + \varepsilon_2(\omega)}. \tag{3.10}$$

The solutions of Eq. (3.10) are the normal modes of the given system. Since we have assumed that $\varepsilon_1 \geq 1$ and the energy does not dissipate in the system [$\varepsilon_2(\omega)$ is real], then the wave vector $\mathbf{k}_x$, representing propagation of the given normal mode along the interface, is also real and $k_x^2 > 0$. Because $\varepsilon_2(\omega) < 0$, the denominator of Eq. (3.10) will be negative. Hence, a solution to Eq. (3.7) exists only if the permittivities of the contacting media satisfy the inequality

$$-\infty < \varepsilon_2(\omega) < -\varepsilon_1. \tag{3.11}$$

A medium with a dielectric function $\varepsilon_1 > 0$ is called inactive. A medium whose dielectric function satisfies Eq. (3.10) is called active.

Figure 3.6 shows the dispersion curve Eq. (3.10) of a surface polariton, with $\varepsilon_2(\omega)$ given by Eq. (1.40), and the dispersion curve of the incident light in the medium of incidence. It is seen that over the whole frequency range, the wave vector of the incident photon is smaller than k_x of the mode under consideration,

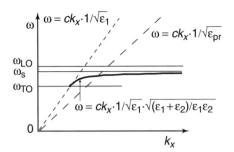

Figure 3.6. Dispersion curves of surface polariton (solid line) at interface of inactive and active media characterized by ε_1 and ε_2, respectively. Short-dashed line: dispersion line of incident photon in inactive medium with ε_1; long-dashed line: in ATR prism with $\varepsilon_{pr} > \varepsilon_1$.

provided condition (3.11) is met. Hence, the expression under the square root in Eq. (3.9) is always negative, and k_{zj} of the mode is always pure imaginary. It follows that the corresponding wave propagates along the interface in the x-direction and its amplitude decays exponentially in the z-direction with increasing distance from the interface into each medium, and thus the wave is confined by the interface (Fig. 3.7). To differentiate between this mode and those existing in the bulk crystal, it can be called a *surface mode*, a *surface electromagnetic wave* (SEW), or a *surface polariton* (surface phonon–polaritons, surface plasmon–polaritons, etc.)

Depending on whether or not the dispersion curve of a particular mode falls to the left or the right of the dispersion line of the incident radiation in the medium of incidence (Fig. 3.6), the mode is classified as *radiative* or *nonradiative*, respectively. Physically, this means that the corresponding fields in the input

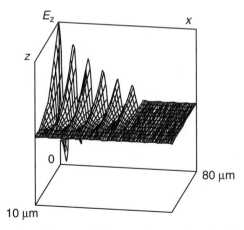

Figure 3.7. Sketch of surface phonon–polariton at air–quartz interface in xy-plane propagating as damped wave in x-direction; $\nu = 1110 \text{ cm}^{-1}$.

and output media are either periodic or evanescent, respectively. Because both z-components are pure imaginary, surface modes are nonradiative and cannot be detected with "ordinary" optical configurations.

One can see from Eq. (3.10) that as $\varepsilon_2(\omega)$ approaches $-\varepsilon_1$, the value of k_x approaches infinity. The resonance condition

$$\varepsilon_2 = -\varepsilon_1 \tag{3.12}$$

is satisfied at a characteristic frequency given by

$$\omega_S = \omega_{TO}\sqrt{\frac{\varepsilon_{st} + \varepsilon_1}{\varepsilon_\infty + \varepsilon_1}}, \tag{3.13}$$

which is called the *surface phonon–polariton frequency*. Here, ε_∞ and ε_{st} are the values of the dielectric function of the active medium, $\varepsilon_2(\omega)$, at $\omega \gg \omega_{TO}$ and $\omega \ll \omega_{TO}$, respectively (1.3.6°). It follows from Eq. (3.13) that

$$\omega_{TO} < \omega_S < \omega_{LO}. \tag{3.14}$$

Substituting the dielectric function defined in Eq. (1.43b) instead of that defined in Eq. (1.40) into Eq. (3.10), we obtain the expression for the frequency of surface plasmon–polaritons, ω_{SP}, in the form

$$\omega_{SP} = \frac{\omega_p}{\sqrt{1 + \varepsilon_1}}, \tag{3.15}$$

which gives $\omega_{SP} = \omega_p/\sqrt{2}$ when the ambient medium is air. Here, ω_p is the frequency of the volume plasmon, defined by Eq. (1.34). It follows from the typical values of ω_p that surface plasmons arising at the dielectric–metal interface do not absorb in the IR spectral range, unlike for doped semiconductors.

In real systems, both ε_2 and k_x are complex. As a consequence, the resonance condition (3.12) becomes $|\varepsilon_1 + \varepsilon_2| = \min$. Moreover, the surface mode propagating along an interface has a finite length $L_x = 1/k_x''$ (of the order of $10–50$ μm for dielectrics). To illustrate this, a surface phonon at the air–quartz surface is depicted in Fig. 3.7. In both media, the penetration depth of the surface polariton depends on the permittivity; for example, at the air–quartz interface at 1110 cm^{-1}, the penetration depths are $d_{z1} \approx 3.5$ μm and $d_{z2} \approx 0.56$ μm in air and quartz, respectively.

Although nonradiative, the surface mode can nevertheless participate in absorption and emission of electromagnetic radiation under special "artificial" conditions. These include a rough surface (Section 3.9.5) or ATR conditions, under which photons are not coupled directly to the active medium–dielectric interface but via the evanescent tail of the radiation, which is totally reflected internally at the base of a high-index prism (with $\varepsilon_{pr} > \varepsilon_1$) [16, 36]. In the latter case, the radiation is characterized by a larger momentum (Fig. 3.6, long-dashed line) and can therefore excite the surface mode. The electric field analysis [37, 38] showed that any excitation of a surface mode is accompanied by the enhancement by several orders of magnitude of the electric field component perpendicular to the interface, giving rise to substantial intensity enhancements in the spectra of ultrathin films deposited onto an active medium. It can be shown [39] that the appearance of an ultrathin film on the surface that sustains the surface mode disturbs the dispersion relation of this surface mode at the free surface. The corresponding resonances of the surface polariton refractive index are near the film resonance frequencies. This phenomenon has led to the development of such novel techniques as surface plasmon resonance (SPR) spectroscopy in the visible/near-IR (Vis/NIR) spectral region [40] and SEW spectroscopy in the IR region [41, 42]. However, SPR and SEW are beyond the scope of the present handbook; for further information Refs. [40–42] are recommended.

3.2.2. Modes in Ultrathin Films

The electromagnetic field of the surface mode at the interface of two semi-infinite media is "quantized" by the second boundary, which causes the transformation of the nonradiative surface polariton into a set of new polariton states, the surface and the bulk (interference and waveguide) polaritons of the film [13, 26]. Of those, only the interference polaritons are radiative, and they determine the IR absorption and emission spectra of the film.

When the film thickness is small compared to the wavelength of light, the macroscopic electric field can be assumed to be constant across the film, and the solution can be found in the *quasi-static regime* [24, 25]. Formally, the quasi-static regime is reached by setting the velocity of light to be infinite ($c \to \infty$). As a result, instead of the Maxwell equations, the electrostatic laws

$$\nabla \times \mathbf{E} = 0, \tag{3.16a}$$

$$\nabla \cdot \mathbf{D} = 0. \tag{3.16b}$$

are used to describe the interaction between the light and the film. This is justified by comparing the results in the limit of ultrathin films and the exact results of the polariton theory. To elucidate the resonance frequencies and the electric fields of the normal modes of the film, Eqs. (3.16) with the material relations (1.2) are solved separately for each medium in the vacuum–film–substrate system, and the corresponding fields are matched at the interfaces, according to the boundary conditions (1.60) [26]. The electric fields are obtained in the form $\mathbf{E} = \mathbf{E}_q(z)e^{iqx}$, where $\mathbf{q}$ is the two-dimensional wave vector lying in the plane of the film (xy, Fig. 3.8). Casting the solution in this form allows Eqs. (3.16) to be split into a pair of coupled equations for the E_x and E_z field components for p-polarization and the equation for the E_y-component for s-polarization. At the second stage, the interaction of these normal modes with the transverse electromagnetic field is taken into account and the corresponding band intensity is

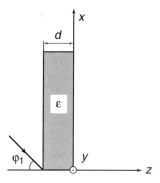

Figure 3.8. Coordinate system and optical constants in Vinogradov model.

calculated via the interaction Hamiltonian according to the equation [analogous to Eq. (1.27)]

$$A \propto \left| \int \mathbf{P}_q^*(z) \cdot \mathbf{E}_q^\perp(z)\, dz \right|^2 , \tag{3.17}$$

where $\mathbf{P}_q(z)$ is the polarization corresponding to the given mode and the vector $\mathbf{E}_q^\perp(z)$ describes the transverse external electromagnetic field:

$$\mathbf{E}_q^\perp(z) \propto \begin{cases} (0, \sin[k_z(z+d)], 0) & \text{for } s\text{-polarization;} \tag{3.18a} \\ (i\cos\varphi_1 \sin[k_z(z+d)], & \\ 0, i\sin\varphi_1 \cos[k_z(z+d)].) & \text{for } p\text{-polarization.} \tag{3.18b} \end{cases}$$

Here, d is the film thickness, $k_z = (\omega/c)\cos\varphi_1$, and φ_1 is the angle of incidence (Fig. 3.8).

Omitting the intermediate calculations, let us consider only the results summarized in Table 3.1. For the surrounding–film–substrate system, the phonon theory predicts five types of eigenstates of the electromagnetic field inside an ultrathin film (*normal modes*); one mode has a y-component of electric polarization and therefore is named the s-polarized mode, and the other four modes

Table 3.1. Modes in thin films on metal substrates

Mode	Polarization	Frequency	Intensity	Charge Surface	Charge Volume	Transformation at $d \to \infty$
1	s	ω_{TO}	$\propto (kd)^3 \cos^2\varphi_1$	0	0	TO
2	p	$\omega \approx \omega_{LO}$	$\propto kd \sin^2\varphi_1$	+	0	S_+
3	p	$\omega \approx \omega_{TO}$	$\propto (kd)^3$	+	0	S_-
4	p	ω_{TO}	$\propto (kd)^3$	0	0	TO
5	p	ω_{LO}	0	+	+	LO

Source: Adapted, by permission, from E. A. Vinogradov, *Phys. Rep.* **217**, 159 (1992). Copyright © 1992 Elsevier.

have x- and/or z-components and thus are p-polarized according to accepted terminology (1.4.4°).

The frequency of the s-polarized mode (mode 1) is equal to ω_{TO}. The corresponding electric field vanishes, but the electric polarization inside the film is finite and can be represented as a linear combination of the linearly independent functions

$$P_y(z) = \begin{cases} A_n \sin\left[\left(\dfrac{\pi l}{d}\right)(z+d)\right], & l = 1, 2, \ldots\,, & (3.19a) \\[4mm] B_n \cos\left[\left(\dfrac{\pi l}{d}\right)(z+d)\right], & l = 0, 1, 2, \ldots\,. & (3.19b) \end{cases}$$

From Eqs. (3.19), it is apparent that this mode has harmonically oscillating electric fields that occupy the whole film volume. The properties of this mode are similar to those of the TO modes in an infinite crystal; specifically, the vibrations in the film do not result in bulk or surface charge. From Eqs. (3.17)–(3.19), it can be shown that mode 1 interacts with s-polarized radiation (is *radiative*), and the integral intensity A of the corresponding absorption peak in the IR spectrum is proportional to the third power of the film thickness (Table 3.1).

Although in the case of an ultrathin film the energy of all the five modes is distributed within the film volume, by convention the four p-polarized modes are classified as either surface (modes 2 and 3) or volume (modes 4 and 5) in order to emphasize their different origins. Below, p-polarized modes are examined in more detail.

If the film is located on a metal substrate, the electric fields of the p-polarized surface mode (mode 2) are expressed as

$$E_x(z) = E \sinh[q(z+d)], \tag{3.20a}$$

$$E_z(z) = -iE \cosh[q(z+d)]. \tag{3.20b}$$

Equations (3.20) show that the electric field is not harmonic and, in the case of thick films, is localized near the vacuum–film interface. The frequency of mode 2 is described by the equation

$$\omega_2 = \omega_{TO}\sqrt{\frac{\varepsilon_{st} + \tanh(dq)}{\varepsilon_\infty + \tanh(dq)}}, \tag{3.21}$$

where ε_{st} and ε_∞ are the parameters entering into the real dielectric function of the film [Eq. (1.40)]. Hence, as $d \to 0$, the frequency of mode 2 approaches the frequency of the LO phonon, ω_{LO} [Eq. (1.41)]. In the other limiting case, when $d \to \infty$, mode 2 transforms into a nonradiative surface mode of the surroundings–film interface with a frequency of ω_s [Eq. (3.13)]. For this reason, this mode is classified as a surface mode, and its frequency and intensity are strongly dependent on the dielectric function of the surrounding medium, ε_{sm} (an increase in ε_{sm} has the same effect as an increase in the film thickness), but are relatively unaffected by the substrate properties. Inside an ultrathin film, the electric field of mode 2 [Eqs. (3.20)] is almost perpendicular to the film plane and virtually constant throughout the film. Moreover, the vibrations do not lead to bulk charge but give rise to surface charge at the film boundaries. From Eqs. (3.17), (3.18), (3.20), and (1.2), it follows that mode 2 is radiative and its integral absorption is proportional to the film thickness (Table 3.1).

The polarization components where the electric field is zero for the p-polarized surface mode 3 are

$$P_x(z) = P \cosh(qz), \tag{3.22a}$$

$$P_z(z) = -i P \sinh(qz). \tag{3.22b}$$

In thick films, mode 3 is localized near the interface with the substrate, and its amplitude decays exponentially with increasing depth into the film. In ultrathin films, the polarization vector $\mathbf{P}$ of this mode is almost parallel to the film plane and approximately constant across the film. For a film on a metal substrate with the dielectric function ε_m, the mode 3 frequency is given by:

$$\omega_3 = \omega_{TO} \left(1 + \frac{\varepsilon_{st} - \varepsilon_\infty}{2\varepsilon_m}\right) \tanh(qd). \tag{3.23}$$

From Eq. (3.23), for an ultrathin film at an ideally conducting substrate ($|\varepsilon_m| \to \infty$) the mode 3 frequency ω_3 is close to ω_{TO} and increases with increasing $\mathrm{Re}(1/\varepsilon_m)$. From Eqs. (3.22), the integral intensity of the absorption band of mode 3 is proportional to the third power of the film thickness, indicating that this mode is radiative. However, because of the quenching of the tangential electric field by metals (Section 1.8.2), the band near ω_{TO} arises only in the spectra of relatively thick films on a metal substrate. It was also found [12, 23, 26] that the properties of mode 3 are independent of the surroundings. This can be qualitatively understood by recognizing that, with increasing film thickness, mode 3 is transformed into a nonradiative surface mode (3.13) of the film–substrate interface.

Mode 4 corresponds to p-polarized transverse vibrations. Its frequency is ω_{TO} and its polarization is the linear superposition of $P_x(z)$ and $P_z(z)$, where

$$P_x(z) = \sum_l C_{l,q} \cos\left(\frac{\pi l(z+d)}{d}\right), \qquad l = 1, 2, \ldots, \tag{3.24a}$$

$$P_z(z) = -i\frac{qd}{\pi} \sum_l C_{l,q} \sin\left(\frac{\pi l(z+d)}{d}\right), \qquad l = 1, 2, \ldots . \tag{3.24b}$$

As for mode 1, the corresponding vibrations are not accompanied by either bulk or surface charges and the frequencies of these modes are independent of the substrate material. On the basis of Eqs. (3.24), (3.17), and (3.18), the integral absorption coefficient for mode 4 was found to be proportional to the third power of the film thickness d.

Finally, mode 5 represents a p-polarized longitudinal mode. Satisfying the equation $\varepsilon(\omega) = 0$, this mode is similar to a longitudinal mode in the bulk (1.3.7°), and their frequencies and properties coincide. The electric fields in mode 5 are harmonic and expressed as

$$E_x(z) = C_l \sin\left(\frac{\pi l(z+d)}{d}\right), \qquad l = 1, 2, \ldots, \qquad (3.25a)$$

$$E_z(z) = -iC_l \frac{\pi l}{d} \cos\left(\frac{\pi l(z+d)}{d}\right), \qquad l = 1, 2, \ldots. \qquad (3.25b)$$

The matrix element of the interaction Hamiltonian in Eq. (3.17) is zero, and so, unlike the four previous modes, mode 5 of the film does not interact with electromagnetic radiation, as for the LO modes of a bulk crystal, and thus it is nonradiative. As the film thickness increases, modes 4 and 5 transform into the bulk TO and LO modes, respectively.

It should be stressed that due to the limitations of the quasi-static regime, the film thickness dependence of the band positions in the IR spectra is correctly described only within the framework of the polariton theory. Therefore, Eqs. (3.21) and (3.23), which relate to pure surface modes, give an inadequate thickness dependence of the band positions in the spectra. The general dispersion relation has the following simple form for the vacuum–film–metal interface and p-polarization [13]:

$$\tan(\beta d) = -i\frac{\beta_0 \varepsilon(\nu)}{\beta}, \qquad (3.26)$$

where $\beta_0 = \nu\sqrt{1 - \sin^2\varphi_1}$, $\beta = \nu\sqrt{\varepsilon(\nu) - \sin^2\varphi_1}$, and φ_1 is the angle of incidence. Equation (3.26) indicates that increasing film thickness shifts the absorption band near ω_{TO} toward the red, whereas the band near ω_{LO} exhibits a blue shift (Section 3.3.1).

Numerous experimental IR spectra of ultrathin films confirm the above theoretical interpretation (see Ref. [26] and literature therein). The p-polarized emission spectra of a ZnSe film on the Si−Al BML substrate (Fig. 3.9) are given as an example [26]. As the Si underlayer thickness increases, the intensity of the ZnSe band near ν_{TO} (206 cm^{-1}) increases. In fact, according to the quasi-static interpretation, the dipole moment of mode 3 (mainly responsible for the band near ν_{TO}) is directed along the x-axis, parallel to the interface. At small thicknesses, mode 3 will be quenched by the image in the metal (Section 1.8.2), but as the film thickness increases, this effect diminishes. Mode 2, whose band is near ν_{LO} (252 cm^{-1}) with polarization parallel to the surface, is unaffected by the presence of the underlayer. It is noteworthy that the experimental spectra agree so well with the spectral simulations from the Fresnel formulas. Another result, which agrees with the theoretical treatment, is the frequency dependence of the band near ν_{TO} in spectra of the ZnSe layers on metals with different permittivities $\hat{\varepsilon}_m$

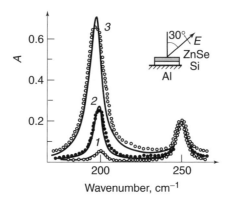

Figure 3.9. The *p*-polarized thermostimulated spectra of 0.6-μm ZnSe film on buried metal layer (BML) substrates Si–Al with Si buffer layer thickness: (1) 0, (2) 2, and (3) 3 μm. Points are experimental data; solid curves are calculated. Experimental points are shifted downward for spectra 1 and 2 by 0.06 and for spectrum 3 by 0.13 (reduction of background radiation). Reprinted, by permission, from E. A. Vinogradov, *Phys. Rep.* **217**, 159 (1992), p. 201, Fig. 26. Copyright © 1992 Elsevier.

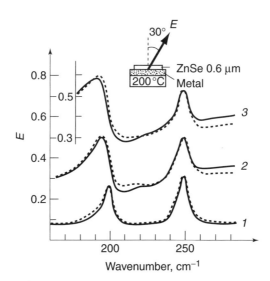

Figure 3.10. The *p*-polarized thermostimulated radiation spectra of 0.6-μm ZnSe film on (1) Al, (2) Cr, and (3) Ti. Points are experimental data; solid curves are calculated. Reprinted, by permission, from E. A. Vinogradov, *Phys. Rep.* **217**, 159 (1992), p. 200, Fig. 25. Copyright © 1992 Elsevier.

(Fig. 3.10). The main observation from Fig. 3.10 is that as ε_m decreases in the series Al, Cr, and Ti, the band near ν_{TO} shifts to lower frequencies [in agreement with (Eq. 3.23)], whereas the band near ν_{LO} is unaffected (the latter conclusion also follows from Fig. 3.9). Similar results were obtained for IRRAS of LiF [3] and CdS [43] films on metallic and dielectric substrates.

Thus, the different components of the electric vector interact with the different normal modes of the film. The s-polarized light excites mode 1, resulting in one absorption peak at ν_{TO} (e.g., for the SiO_2 films, it is the 1075-cm^{-1} peak in Fig. 3.1). This mode is suppressed in an ultrathin film at a metal. The p-polarized light interacts with three modes of a thin film (Table 3.1), including mode 2, located near ν_{LO} of the film substance. For an ultrathin film on a metal, this mode produces a single absorption peak in the p-polarized spectrum (the 1240-cm^{-1} band in IRRAS of the SiO_2 film in Fig. 3.2). The properties of mode 2 are independent of the substrate but are functions of the surroundings. The other two p-polarized modes (modes 3 and 4) absorb near ν_{TO} and the absorption intensities depend in the same manner on the film thickness. However, only the position and intensity of mode 3 are sensitive to changes in the optical characteristics of the substrate. For ultrathin films, the absorption at ν_{TO} is primarily due to mode 3 (a surface mode), whereas in the case of thicker films, the contribution of mode 4 (a bulk mode) becomes dominant. As will be shown in Section 3.9.3 and Chapter 5, the difference in positions of the absorption bands at ν_{TO} and ν_{LO} is very sensitive to the distance between vibrators, which gives information on such properties of the film as crystallinity and the presence of voids or other inclusions, even if these species do not absorb IR radiation.

It should be noted that the band near ν_{LO} due to surface mode 2 is sometimes associated with the LO phonon of the film substance, mode 5. In this context, it is important to note that the theoretical interpretation of thin-film spectra given above is not entirely consistent with the nomenclature currently in use. Some researchers prefer a macroscopic terminology, ignoring its physical sense. With this terminology, the peaks observed near ν_{TO} and ν_{LO} are simply referred to as the TO and LO bands, respectively (e.g., see Refs. [7–9, 44]). However, these names deprived of a physical sense are unacceptable. In fact, all the modes excited in an ultrathin film are transverse, while the LO modes are nonradiative. Since the terminology concerning the modes under consideration is still being developed, we can avoid the ambiguity of the formal macroscopic nomenclature, referring simply to the ν_{TO} and ν_{LO} bands instead of the TO and LO bands, respectively. It is hoped that this will not cause undue confusion for readers more familiar with macroscopic nomenclature.

3.2.3. Identification of Berreman Effect in IR Spectra of Ultrathin Films

The first practical problem in the interpretation of experimental IR spectra of a layered structure [particularly for nonmetallic supports and relatively thick films ($d > 100$ nm) on metals] is to distinguish between the ν_{TO} and ν_{LO} bands. In the following, practical solutions to this problem will be presented.

The most rigorous approach is to generate spectral simulations using the exact Fresnel formulas (Sections 1.5–1.7); however, this requires a knowledge of the optical constants of the film over the spectral range of interest. Special methods may be used to measure the optical parameters of thin films in their final form (Sections 3.10 and 3.11). To simplify the problem, differences between the optical parameters of thin films and those of the parent bulk materials can be neglected; the latter can be simply obtained using KK transformations (1.1.13°) or from the literature [45, 46].

If the optical constants of the film or the film material are already known, ν_{TO} and ν_{LO} can be approximated by the peak positions of the TO and LO energy loss functions, respectively (Fig. 3.11b), without resorting to spectral simulations. If the film material is a weakly damped strong oscillator, ν_{LO} may also be evaluated using *Drude's method*, determining the frequency at which the refractive index of the film material is equal to the extinction coefficient, $n = k$, or, in other words, at which the real part of the dielectric function $\varepsilon'(\nu)$ is zero [Eq. (1.15)]. Figure 3.11c is a graphical representation of such an approach using a SiO_2 layer on Ge as an example. As mentioned (1.3.17°), Drude's method is inapplicable when $\varepsilon'(\nu)$ is nonzero at all frequencies, such as is the case for multiresonance oscillators (especially when the oscillator strengths differ considerably) and heavily damped oscillators. For the latter case, the use of the *minima* (LO) and *maxima* (TO) of the modulus of the complex dielectric function, $|\hat{\varepsilon}(\nu)|$, was suggested by Chang et al. [47].

The following experimental techniques can be employed for the evaluation of ν_{LO} of the film, depending on the accuracy required and the resources available [45, 47]:

1. Direct determination from the band maxima in IRRAS of ultrathin films ($d < 50$ nm) on metals.

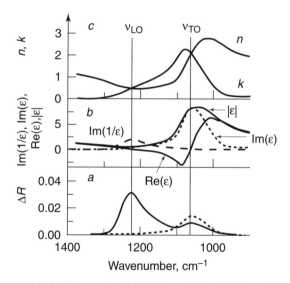

Figure 3.11. (a) Calculated ATR spectra of SiO_2 layer 2.5 nm thick on Ge for s-polarization (short-dashed line) and p-polarizarion (solid line) at $\varphi_1 = 15°$; (b) dispersion of TO (short-dashed line) and LO (long-dashed line) energy loss functions and $|\varepsilon|$ and $Re(\varepsilon)$ (solid line) of SiO_2; (c) dispersion of refractive index n and extinction coefficient k of CVD SiO_2 after Rakov [52].

2. Calculation using the *LST law* for systems with narrow absorption bands. In this case, then, ν_{TO} can be determined from the s-polarized transmission or reflection spectra, and the values ε_{st} and ε_∞ can be measured at frequencies much lower and much higher that ν_{TO}, respectively.

3. Use of the alternative methods whose selection rules admit of spectral activity of the LO modes. One such method is polarized Raman spectroscopy, which is applicable to substances with cubic zinc blend and hexagonal wurtzite structures such as ZnS, ZnSe, CdS, ZnO, ZnTe, and the III–V compounds (see Refs. [47–49] and literature cited therein). The most direct method to measure ν_{LO} is inelastic neutron scattering (INS) since there are no selection rules for INS spectroscopy and as a result all modes are allowed [50, 51].

The simplest way to interpret p-polarized spectra is to use literature ν_{TO} and ν_{LO} values. In Table A.1 (Appendix), frequencies for a wide range of inorganic substances are listed. The values presented in this table were obtained using one of the approaches described above, as indicated in the table.

3.3. OPTICAL EFFECT: FILM THICKNESS, ANGLE OF INCIDENCE, AND IMMERSION

The dependence of the positions, intensities, and shapes of IR absorption bands on film thickness, angle of incidence, and the optical properties of the substrate and surroundings is usually referred to collectively as an "optical effect". In this section, optical effects in the IR spectra of specific thin films will be considered, and conditions under which the spectra are relatively unaffected will be discussed.

3.3.1. Effect in "Metallic" IRRAS

In this section, oxide films (strong TO–LO splitting) will be considered exclusively. Figures 3.12 and 3.13 illustrate how an increase in the film thickness influences the band position and shape of SiO_2 and α-Fe_2O_3 layers on Al and Fe, respectively. Simulation was done only for p-polarization, because according to the SSR, only p-polarized radiation can couple to an ultrathin film on a metal (Section 1.8.2). The optical constants of a SiO_2 film formed by chemical vapor deposition (CVD) were taken from Ref. [52]. The parameters of the dielectric function of isotropic α-Fe_2O_3 are listed in Table 3.2. The optical constants of the metals were specified according to Palik [45]. Figures 3.12 and 3.13 show that for small thicknesses (<10 nm) each spectrum displays only one ν_{LO} band: 1238 and 662 cm^{-1} for SiO_2 and α-Fe_2O_3, respectively. An increase in thickness is accompanied by an increase in the intensity of these bands and, in the case of the α-Fe_2O_3 film, by the appearance of additional ν_{LO} bands at 478 and 384 cm^{-1}. The corresponding ν_{TO} bands are observed at 1060 and 525 cm^{-1} for

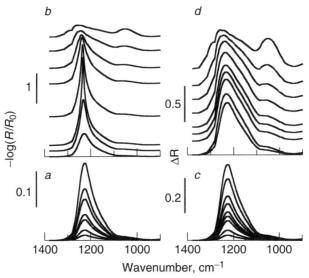

Figure 3.12. Simulated p-polarized IRRAS spectra of SiO_2 layers on Al as function of layer thickness, $\varphi_1 = 80°$. Spectra are represented in (a, b) absorbance and (c, d) absorption depth ΔR units. Layer thickness increases from lower to upper curve in series: (a) 1, 2, 4, 5, 8, 10, and 15 nm; (b) 50, 90, 110, 130, 160, 200, 300, and 500 nm; (c) 1, 2, 4, 5, 6, 8, 10, 15, and 20 nm; (d) 50, 90, 100, 110, 130, 160, 200, 300, and 500 nm.

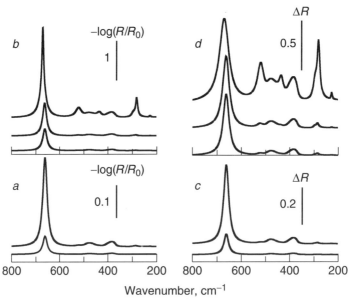

Figure 3.13. Simulated p-polarized IRRAS spectra of α-Fe_2O_3 layers on Fe in (a, b) the absorbance and (c, d) ΔR units, $\varphi_1 = 75°$. Layer thickness increases from lower to upper curve in series: (a, c) 10 and 50 nm and (b, d) 100, 150, and 500 nm. Parameters of Maxwell–Helmholtz–Drude formula (1.46) for α-Fe_2O_3 listed in Table 3.2.

Table 3.2. Oscillator parameters of Maxwell–Helmholtz–Drude dispersion formula (1.46) for α-Fe$_2$O$_3$

	ν_{TOj} (cm^{-1})	ν_{LOj} (cm^{-1})	S_j	γ_j (cm^{-1})
$(E\|c)^a$ $\varepsilon_\infty = 6.7$	299	414	11.5	15
$\varepsilon_{st} = 20.6$	526	662	2.2	30
$E\perp c$	227	230	1.1	4
$\varepsilon_\infty = 7.0$	286	368	12.0	8
$\varepsilon_{st} = 24.1$	437	494	2.9	20
	524	662	1.1	25

^{a} E is the electric field vector; c is the optical axis.

Source: From Ref. S. Thibault, *Mat. Chem.* **1**, 71 (1976).

SiO$_2$ and α-Fe$_2$O$_3$, respectively, only if the layer thickness exceeds $\sim$100 nm (see Section 3.2.2 for an explanation).

Once the reflectance at the band maximum is reduced to zero ($R \sim 0$), the band depth drops and the band shape becomes distorted, narrowing if the spectrum is represented in absorbance [$-\log(R/R_0)$] units and broadening in ΔR (or $\Delta R/R$) units. To designate the onset of significant distortions in p-polarized spectra, the term *Berreman thickness* d_B is sometimes applied [4]. At the air–metal interface, the Berreman thickness can be approximated by [4]

$$d_B = \frac{\lambda}{2\pi} \frac{\cos\varphi_1}{\sin^2\varphi_1} \left[\text{Im}\left(\frac{1}{\hat{\varepsilon}_2}\right) \right]^{-1}_{max}, \tag{3.27}$$

where $\hat{\varepsilon}_2$ is the complex dielectric function of the film, φ_1 is the angle of incidence, and λ is the wavelength. It follows that d_B increases when either the angle of incidence or the LO energy loss function decreases. Thus, for the SiO$_2$ layer on Al, the value of d_B is equal to $\sim$100 and $\sim$500 nm for $\varphi_1 = 80°$ and $\varphi_1 = 60°$, respectively, and greater than 1200 nm for $\varphi_1 = 40°$ (Fig. 3.14). Such values of d_B are typical for oxides on metals. Experimental confirmation of these calculations have been reported by Scherubl and Thomas [53] for Cr$_2$O$_3$ films on Ni.

Another important observation from Fig. 3.14 is that at small thicknesses the intensity of the ν_{LO} band depends linearly on the film thickness, in agreement with the thin-film approximation [Eq. (1.82)]. This linear region extends to approximately half the Berreman thickness and depends upon the angle of incidence in the same way as d_B. The linear range of the band intensity in absorbance is somewhat broader than that in reflectance, as illustrated by comparing curves 1 and 4. This implies that absorbance units should be used in any quantitative analysis of thin films, rather than reflectivity units.

Figure 3.15 shows the theoretical and experimental shifts of the ν_{LO} band position for Cr$_2$O$_3$ film on Ni as a function of film thickness and angle of incidence [53]. The ν_{LO} band shows a blue shift as film thickness increases. The same effect was reported for the Cu$_2$O–Cu [54] and TiO$_2$–Al [55] systems. This shift

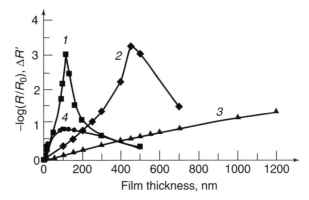

Figure 3.14. Band intensity measured in (1–3) absorbance and (4) reflectance $\Delta R'$ for ν_{LO} band of SiO_2 films on Al as function of film thickness in simulated IRRAS spectra. Angle of incidence is (1, 4) 80°, (2) 60°, and (3) 40°.

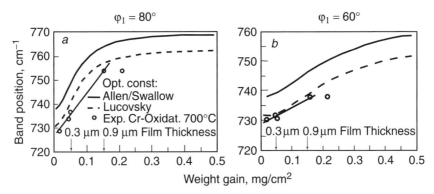

Figure 3.15. Position of ν_{LO} band in experimental (circles) and calculated (lines) IRRAS of spectra Cr_2O_3 films on Ni as function of film thickness. Angle of incidence is (a) 80° and (b) 60°. Theoretical curves were generated using Eq. (3.26) and different optical constants of Cr_2O_3. Reprinted, by permission, from Th. Scherubl and L. K. Thomas, *Appl. Spectrosc.* **51**, 844 (1997), p. 847, Fig. 3. Copyright © 1997 Society for Applied Spectroscopy.

becomes more pronounced as the film thickness exceeds the Berreman thickness, $d > d_B$ (see also Figs. 3.12 and 3.13). It is interesting to note that over a certain range of $d > d_B$, the ν_{LO} band shift is approximately proportional to the film thickness; this can be exploited to determine film thicknesses.

The experimental spectra of Al_2O_3 [56], RbI [57], and GaAs [58a] ultrathin films on metals exhibit an additional decrease in the ν_{LO} frequency, which cannot be described adequately using the dielectric function of the corresponding bulk materials. For example, as the GaAs film thickness decreased from 10 to 0.5 nm, the ν_{LO} band shifted to the red by approximately 40 cm^{-1}. This effect was attributed to lower force constants for the atoms at the surface relative to those in the bulk [58b], implying a difference between the "microscopic" dielectric function of the superficial layer and that of the bulk phase.

There are two important considerations when using the IRRAS immersion technique (Section 2.5.2). First, the Berreman thickness drops as the refractive index of the surroundings, n_1, increases. As an example, from Izumitani et al. [60], in the MOS structure $Si-SiO_2-Al$, the d_B value is $\sim$70 nm at $\varphi_1 = 20°$, $\sim$10 nm at $\varphi_1 = 40°$, $\sim$7 nm at $\varphi_1 = 60°$, and only $\sim$2 nm at $\varphi_1 = 80°$. Second, the immersion also affects the band shapes and positions (Fig. 2.31). In the case of SiO_2, one can see from Fig. 3.16 that as the refractive index n_1 is increased from 1 to 3.42, the ν_{LO} band broadens slightly toward the red, particularly when the film thickness is larger than d_B (Fig. 3.16a, curve 2). This phenomenon is attributed to a decrease in surface charge due to partial screening by the immersion medium, resulting in a smaller restoring force, which in turn decreases the resonance frequency. The opposite occurs for the $\alpha\text{-}Fe_2O_3$ film on Fe: The main absorption band broadens toward the blue with increasing n_1 (Fig. 3.16b). This difference originates from the fact that $\alpha\text{-}Fe_2O_3$ is a multimode material with specific relationships between oscillator strengths, which allows mode coupling (Section 3.6). The same phenomenon is the basis of an increase in the relative intensity of the low-frequency ν_{LO} bands with increasing n_1 (Fig. 3.16b).

The practical implications of these results are the following:

1. The condition $d = d_B$ provides the optimum conditions for recording the spectrum, guaranteeing the greatest spectral contrast with negligible distortions.

2. The Berreman thickness drops abruptly if either the angle of incidence or the refractive index of the immersion medium is increased. If for a grazing angle of incidence, $d > d_B$, the angle should be decreased so that $d = d_B$.

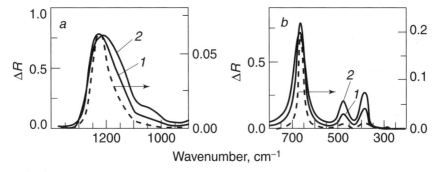

Figure 3.16. Simulated p-polarized IRRAS spectra of (a) SiO_2 films (1) 2 nm and (2) 4 nm thick located at Si–Al interface (solid lines) and 2 nm thick at air Al interface (dashed line) at $\varphi_1 = 80°$ and (b) $\alpha\text{-}Fe_2O_3$ films (1) 20 nm and (2) 40 nm thick at IKS-35–Fe interface (solid lines) and 20 nm at air Fe interface (dashed line) at $\varphi_1 = 75°$.

3. For single- and multiple-mode film materials, immersion causes red and blue band shifts, respectively.

3.3.2. Effect in "Transparent" IRRAS

In Fig. 3.17, the experimental spectra of an amorphous SiN_x layer deposited on a silicon wafer, measured by Yamamoto and Ishida [7] at various angles of incidence, are shown as an example of IRRAS of a strongly absorbing layer on a transparent substrate. A silicon wafer with no film was used as a reference. These spectra demonstrate two principal differences between IRRAS of a film on a metal substrate and a film on a transparent substrate. First, the v_{TO} band of SiN_x at 840 cm^{-1} is present in the p-polarized spectra, along with the v_{LO} band, unlike with a metal substrate. Second, the signs of the v_{TO} and v_{LO} bands in the p-polarized spectra are opposite and change at a certain angle of incidence.

These effects can be easily interpreted using the Fresnel formulas. Consider the predictions of optical theory for an anisotropic inorganic film in which not only TO–LO splitting but also anisotropy of the film are manifested in IRRAS. Figure 3.18 shows the angle-of-incidence dependence of the band intensities for a model inorganic (Table 3.3) 10-nm film, calculated using the exact matrix method [8]. The simulation was done for substrates with refractive indices of 1.5 (BaF$_2$), 2.4 (ZnSe), 3.4 (Si), and 4 (Ge). The intensity of the v_{TO} mode polarized in the y-direction (see the axes definitions in Fig. 1.12) is negative throughout the angular range, whereas ΔR of the v_{TO} mode polarized in the x-direction, v_{TOx}, is always opposite to ΔR of the v_{LO} mode polarized in the z-direction, v_{LOz}. The sign of the v_{TOx} and v_{LOz} intensities changes at the Brewster angle (56.3° for BaF$_2$, 67.4° for ZnSe, 73.6° for Si, and 76° for Ge). The dependence of the band polarity for the x- and z- components in p-polarized transparent IRRAS on the orientation of the corresponding TDM in the film is referred to as the *surface selection rule* (SSR) for dielectrics (see also Section 3.11.4).

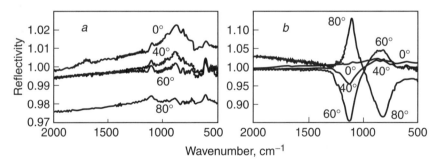

Figure 3.17. IRRAS spectra of SiN_x thin film on silicon wafer with (a) s-polarization and (b) p-polarization. Angles of incidence are shown. Silicon wafer without film is used as reference. Angle of incidence, $\varphi_1 = 0°$, represents measurement at $\varphi_1 = 11°$ without any polarizer. Reprinted, by permission, from K. Yamamoto and H. Ishida, *Vibrational Spectrosc.* **8**, 1 (1994), p. 22, Fig. 21. Copyright © 1994 Elsevier Science B.V.

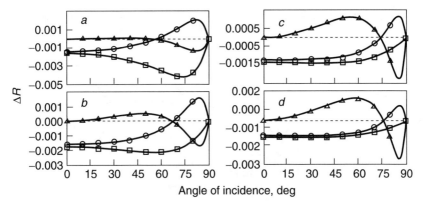

Figure 3.18. Simulated ΔR of ($\square$) ν_{TOy}, ($\circ$) ν_{TOx}, and ($\triangle$) ν_{LOz} bands in IRRAS spectra of 10-nm inorganic anisotropic film on (a) BaF$_2$ ($n_3 = 1.5$), (b) ZnSe ($n_3 = 2.4$), (c) Si ($n_3 = 3.4$), and (d) Ge ($n_3 = 4.0$) as function of angle of incidence. Optical parameters of film material are shown in Table 3.3; $n_1 = 1.0$. Adapted, by permission, from K. Yamamoto and H. Ishida, *Appl. Spectrosc.* **48**, 775 (1994), p. 779, Fig. 5. Copyright © 1994 Society for Applied Spectroscopy.

Table 3.3. Dispersion parameters for anisotropic and inorganic model

Parameter	x-Axis	y-Axis	z-Axis
ε_∞	2.89	2.89	2.89
S	0.2	0.2	0.2
γ, cm^{-1}	9	10	11
ν_0, cm^{-1}	900	1000	1100

Another peculiarity of spectra from IRRAS can be seen in the calculated spectra of SiO$_2$ layers of different thicknesses on Si at $\varphi_1 = 60°$ (Fig. 3.19): The ν_{TO} peak in the p-polarized spectra changes sign with an increase in layer thickness. This effect, which is explained by the redistribution of contributions of different modes contributing to the ν_{TO} band (Table 3.1), makes the ν_{TO} band in the p-polarized spectra unacceptable for quantitative measurements.

In general, the spectral distortions arising in isotropic inorganic, anisotropic organic, and isotropic organic (weakly absorbing) films can be easily deduced. A cautionary example is the case of a weakly absorbing film where TO–LO splitting is low. The calculated dependencies of the spectra of a 1-nm-thick isotropic water layer on quartz and the dispersion of the optical constants of water are shown in Fig. 3.20. It is seen that the p-polarized spectra change their shape depending on the angle of incidence. This effect is explained by the change in sign of the ν_{LO} band intensity at φ_B (vide supra). In contrast, the s-polarized spectra maintain their shape, making them superior for quantitative analysis. In addition, s-polarized spectra have a higher SNR (Section 2.3.1).

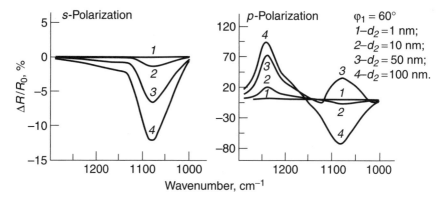

Figure 3.19. Simulated s- and p-polarized IRRAS spectra of SiO_2 layers on Si as function of layer thickness at $\varphi_1 = 60°$. Optical constants of thermal SiO_2 are specified according to Ref. [52]; $n_3 = 3.42$.

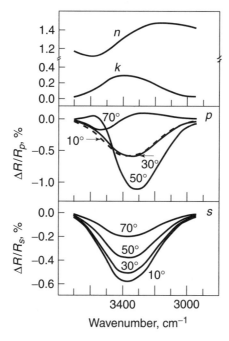

Figure 3.20. Simulated s- and p-polarized IRRAS spectra and optical constants k_2 and n_2 of 1-nm thick water layer on quartz as function of angle of incidence.

Figure 3.21 allows one to deduce the applicability of the thin-film approximation to IRRAS of an anisotropic inorganic layer at the ZnSe and Ge substrates. The deviation from the results of the exact formulas is greatest (up to 50% for a 200-nm-thick film) for the ν_{LO} band and the Ge substrate.

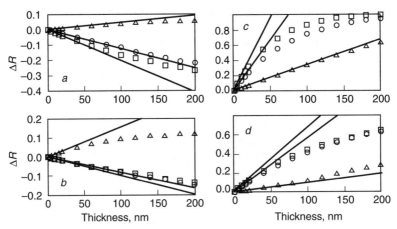

Figure 3.21. Simulated ΔR for (○) ν_{TOy}, (□) ν_{TOx}, and (△) ν_{LOz} bands in (a, b) IRRAS and (c, d) ATR spectra of inorganic anisotropic film on (a, c) ZnSe ($n_3 = 2.4$) at $\varphi_1 = 40°$ and (b, d) Ge ($n_3 = 4.0$) at $\varphi_1 = 60°$ as function of film thickness. Solid lines are calculated according to thin-film approximation Eq. (1.85) and (1.86). Optical parameters of film material are shown in Table 3.3; $n_1 = 1.0$. Reprinted, by permission, from K. Yamamoto and H. Ishida, *Appl. Spectrosc.* **48**, 775 (1994), p. 780, Fig. 7. Copyright © 1994 Society for Applied Spectroscopy.

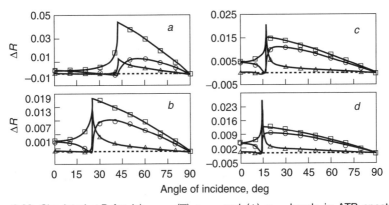

Figure 3.22. Simulated ΔR for (○) ν_{TOy}, (□) ν_{TOx}, and (△) ν_{LOz} bands in ATR spectra of 10-nm inorganic anisotropic film on (a) BaF$_2$ ($n_3 = 1.5$), (b) ZnSe ($n_3 = 2.4$), (c) Si ($n_3 = 3.4$), and (d) Ge ($n_3 = 4.0$) as function of angle of incidence. Optical parameters of film material are shown in Table 3.3; $n_1 = 1.0$. Reprinted, by permission, from K. Yamamoto and H. Ishida, *Appl. Spectrosc.* **48**, 775 (1994), p. 779, Fig. 6. Copyright © 1994 Society for Applied Spectroscopy.

3.3.3. Effect in ATR Spectra

As opposite to IRRAS, the band intensity in ATR spectra is positive, independent of the optical properties of the substrate and orientation of the dipole moment of the mode[†], which simplifies interpretation of ATR spectra. As seen from Fig. 3.22 [8],

---

[†] This is true only for ultrathin films in a three-phase system "IRE–film–surrounding" at $\varphi_1 \geq \varphi_c$. In multilayer systems, the ATR bands can be negative (Figs. 3.67 and 7.40b).

the intensity ΔR of the $\nu_{\mathrm{LO}z}$ band is much smaller than that of the $\nu_{\mathrm{TO}x}$ band, except at the angles of incidence close to φ_c. This makes the ATR method less suitable for orientation measurement than IRRAS, for which, over a certain range of the angle of incidence, the band intensity of the $\nu_{\mathrm{LO}z}$ band is comparable to that of the $\nu_{\mathrm{TO}x}$ band (Fig. 3.18). At the same time, the low intensity of the absorption component perpendicular to the surface makes the ATR spectrum shape practically independent of the angle of incidence, which facilitates both qualitative and quantitative analysis of the film.

A comparison between band intensities obtained with the exact formulas and those obtained by the thin-film approximation (Fig. 3.21c, d) reveals that the highest deviation from linearity is observed for the modes polarized in the film plane.

Significant shape distortions (negative-going peaks at the high-energy side of the band and an increased background at the low-energy side) can be observed for the bands in the in situ ATR spectra of ultrathin films on metal surfaces. Figure 3.23a shows ATR spectra of CO adsorbed from a solvent (CH_2Cl_2) on a 1-nm Pt film evaporated on a 45° Ge IRE (Kretschmann's configuration) measured in situ for two different treatments of the sample. Both the spectra are due to linearly (2050 cm^{-1}) and bridged CO (1825 cm^{-1}). In spectrum 2 the band associated with linearly bound CO is clearly dispersive. Spectral calculations [61a] for the IRE–smooth Pt layer–adsorbed film–water system using

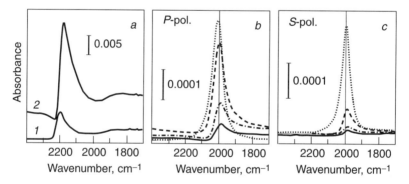

Figure 3.23. (a) Unpolarized in situ ATR spectra (Kretschmann's configuration) of CO adsorbed from CH_2Cl_2 on 1-nm Pt film evaporated on Ge. Spectra were recorded with flow-through liquid cell. N_2 saturated CH_2Cl_2 was first pumped (reference spectrum). Then CO was adsorbed by pumping CH_2Cl_2 saturated with 0.5% CO in Ar through the cell. After 30 min spectrum 1 was recorded. Thereafter H_2 saturated solvent was pumped through the cell, and spectrum 2 was recorded 5 min after switching to H_2. (b, c) Calculated p- (b) and s- (c) polarized ATR spectra for monolayer adsorbate on Pt film in contact with liquid phase. Parameters of monolayer correspond to CO adsorbed on Pt: $\nu_0 = 2000$ cm^{-1}, $\varepsilon_\infty = 2.56$, $S = 0.07$, $\gamma = 60$ cm^{-1}, $d = 0.3$ nm. Parameters of configuration: $n_{\mathrm{Ge}} = 4.01$, $\hat{n}_{\mathrm{Pt}} = 4.75 - 19.48i$, $n_{\mathrm{solvent}} = 1.4$, $\varphi_1 = 45°$. Thickness of metal layer is 0, 10, 20, and 30 nm (from top to bottom). Adapted, by permission of the Royal Society of Chemistry on behalf of the PCCP Owner Societies, from T. Bürgi, *Phys. Chem. Chem. Phys. (PCCP)* **3**, 2124 (2001), p. 2128, Figs. 8 and 9. Copyright © 2001 The Owner Societies.

optical constants of bulk Pt and a monolayer of CO adsorbed on Pt (Fig. 3.23*b*, *c*) reproduce such a band shape and show that the distortion is enhanced as the underlayer thickness increases. However, the calculations predict that the band distortion becomes noticeable for Pt films 20–30 nm thick, whereas it is experimentally observed for a 1-nm Pt film. This discrepancy and the difference in the experimental spectra for different treatments of the sample (Fig. 3.23*a*) stem from the difference in the optical constants of the metal underlayer (see Section 3.9.4 for more detail).

Spectrum distortions in the in situ ATR spectra measured in Otto's configuration with a nonmetalic substrate are discussed in Ref. [61b].

3.3.4. Effect in Transmission Spectra

As in the case of IRRAS and ATR, transmission spectra of ultrathin films, when measured at an inclined angle of incidence and with *p*-polarized radiation, are subject to TO–LO splitting (Fig. 3.1). However, as for ATR, interpretation is simplified by the fact that the band intensities ΔT are always positive (Figs. 2.2–2.4). One source of spectral distortion in the transmittance is interference fringes (Figs. 2.5 and 2.6). To reduce this effect, spectra may be represented as the ratio of the transmittance of the film–substrate system to that of the substrate itself [9]. Analysis of Fig. 2.6 shows that the least distortion is exhibited in *p*-polarized spectra measured at the Brewster angle. It should be noted that in practice the interference effect from the substrate is usually not observed in the form shown in Figs. 2.5 and 2.6 because of poor flatness of the substrate and beam divergence in the light source. However, it may still manifest itself as averaged smoothed distortions of the band shapes and intensities and, therefore, should be taken into account when the spectra are interpreted. Another source of spectral distortions in the transmittance spectra may be interference within the film itself. This occurs when the effective thickness of the film is such that the transmittance in the band maximum goes to zero. For an anisotropic layer, whose optical parameters are represented in Table 3.3, it becomes pronounced in the *s*-polarized spectra of a 100-nm film at $\varphi_1 = 80°$ (Fig. 3.24). This thickness is analogous to the Berreman thickness (Section 3.3.1), but instead of the ν_{LOz} band, it relates to the ν_{TOy} band.

According to spectral simulations [9], increasing the film thickness up to 100 nm causes the band positions in the *s*- and *p*-polarized transmission spectra to change by no more than 0.5 cm^{-1}. A more important characteristic is the linear range of the band intensity dependence on the film thickness. As evident from Fig. 3.25, this dependence already deviates significantly from linearity at small thicknesses (up to about 10 nm), which means that the quantity of strongly absorbing species cannot be estimated correctly by applying the BLB law to transmission spectra.

Variation in IR spectra of a thin poly(methyl methacrylate) PMMA film measured with different methods are shown in Fig. 3.26 [7]. The $\nu C = O$ band at ~ 1730 cm^{-1} is the most distorted due to its relatively high oscillator strength.

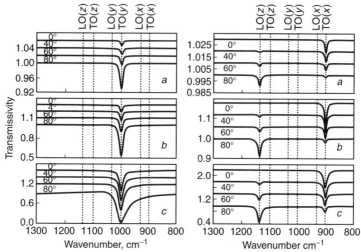

Figure 3.24. Simulated transmission spectra of standing film of anisotropic inorganic material (Table 3.3) at several angles of incidence. Film thickness: (*a*) 1 nm, (*b*) 10 nm, (*c*) 100 nm. The TO and LO frequencies of model material are shown. Results for *s*-polarization are shown on left and those for *p*-polarization are shown on right. Reprinted, by permission, from K. Yamamoto and H. Ishida, *Appl. Optics* **34**, 4177 (1995), p. 4180, Figs. 3 and 4. Copyright © 1995 Optical Society of America.

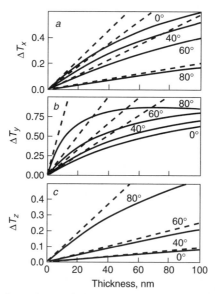

Figure 3.25. Thickness dependence of absorption depth ($\Delta T = 1 - T$) for (*a*) ν_{TOx}, (*b*) ν_{TOy}, and (*c*) ν_{LOz} bands in transmission spectra of standing film of anisotropic inorganic material (Table 3.3) at several angles of incidence: (*a, c*) for *p*-polarization; (*b*) for *s*-polarization. Dashed lines are calculated according to thin-film approximation Equation (1.98). Reprinted, by permission, from K. Yamamoto and H. Ishida, *Appl. Opt.* **34**, 4177 (1995), p. 4181, Fig. 6. Copyright © 1995 Optical Society of America.

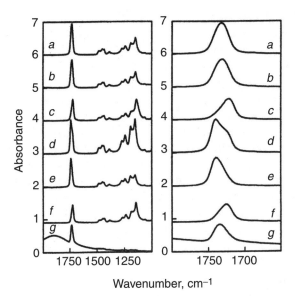

Figure 3.26. Comparison of simulated spectra for poly(methyl methacrylate) (PMMA) film measured with different methods; (a) decay constant, $\alpha/1000$, (b) transmission ($d_2 = 2\ \mu m$, $\varphi_1 = 0°$), (c) ATR ($n_1 = 2.4$, $\varphi_1 = 45°$, s-polarization), (d) IRRAS on Au ($\varphi_1 = 0°$), (e) IRRAS on Au ($d_2 = 20$ nm, $\varphi_1 = 80°$, p-polarization), (f) MOATR ($d_2 = 20$ nm, $n_1 = 4.0$, $\varphi_1 = 80°$, p-polarization), (g) surface electromagnetic wave (SEW) spectroscopy (Otto's configuration, $d_2 = 20$ nm, $n_1 = 2.4$, $\varphi_1 = 24.58°$, air gap (d_3) $= 14\ \mu m$, p-polarization). Spectra are vertically shifted and (e) is multiplied by factor of 15. Reprinted, by permission, from K. Yamamoto and H. Ishida, *Vib. Spectrosc.* **8**, 1 (1994), p. 11, Fig. 8. Copyright © 1994 Elsevier Science B. V.

3.4. OPTICAL EFFECT: BAND SHAPES IN IRRAS AS FUNCTION OF OPTICAL PROPERTIES OF SUBSTRATE

Before looking at the effect of the substrate in IRRAS, it is convenient to divide all substrates into the following five types, depending on their optical properties: (1) metals, which strongly and nonselectively absorb and reflect IR radiation; (2) dielectrics in the region of phonon absorption, which strongly and selectively absorb and reflect IR radiation; (3) transparent substrates with high refractive indices ($n_2 > 2.0$), including semiconductors and some kinds of optical glasses; (4) transparent substrates with low refractive indices ($n_3 = 1.4-2.0$), including oxides and transparent chalcogenides; and (5) weakly absorbing substrates, including doped semiconductors and dielectrics in the overtone spectral region or within the "wings" of the phonon bands. Spectra calculated for a wide range of layers on various substrates and experimental spectra measured under a variety of conditions were analyzed in Refs. [7, 62–64]. Below, the main results are outlined.

1. *Metals.* The effect of the optical properties of the metal substrate in IRRAS of ultrathin films was discussed by Tobin [64]. The band shape is practically Lorentzian in the IRRAS of CO on Pt measured at 87°. As the optical conductivity

of the substrate decreases from Pt to Fe, the band intensity is attenuated and its shape becomes asymmetrical; an apparent onset on the low-frequency side and a tail at higher frequencies is observed. At a lesser angle of incidence, the asymmetry practically disappears even for Fe.

In practice, adlayers are not perfectly homogeneous, which results in inhomogeneous broadening of the absorption bands. In the case of CO sparsely adsorbed in an atop position and at a low surface coverage, this mechanism produces a low-frequency tail (Fig. 3.36 in Section 3.6), which counterbalances the optical broadening. For highly compressed adlayers, such as the $(2 \times 2)-3CO$ structure on Pt(111) [65], the low-frequency tail is quenched due to strong band intensity transfer (Section 3.6), and the high-frequency tail becomes distinct. The effect of optical properties of metal substrates on SEIRA spectra is discussed in Section 3.9.4.

2. The strongest and most complicated distortions in spectra from IRRAS are expected in the region of the phonon absorption of the substrate. As discussed in Section 2.3.2, these distortions are different for each substrate and can be distinguished only using spectral simulations. Here, monolayers on water are considered, which are of interest from many practical and scientific viewpoints [66, 67]. The simulated spectra of a 10-nm anisotropic inorganic film (Fig. 3.27) at $\varphi_1 = 60°$ [8] will help to interpret the experimental data. It is evident that the

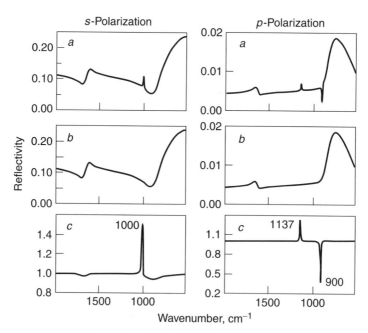

Figure 3.27. (*a*) Simulated *s*- and *p*-polarized IRRAS spectra of 10-nm film of anisotropic inorganic material (Table 3.3) at air–water (AW) interface, (*b*) reflection spectrum of water at $\varphi_1 = 60°$, and (*c*) ratio *a/b*. Adapted, by permission, from K. Yamamoto and H. Ishida, *Appl. Spectrosc.* **48**, 775 (1994), p. 784, Fig. 12. Copyright © 1994 Society for Applied Spectroscopy.

ratio of the p-polarized ratio spectra is practically distortion free while that of the s-polarized spectra exhibits "extra" bands in the regions of water absorption (3000–3600, 1600–1700, and 800–900 cm^{-1}). These bands are caused by changes in the reflectivity of the water surface in these regions in the presence of a film.

3. It is of interest to elucidate how the absorption of a substrate affects the band shape in p-polarized IRRAS of ultrathin films on transparent substrate. A representative example is the IRRAS spectra of a 1-nm hypothetical organic layer whose dielectric function is characterized by $S = 0.001$, $\gamma = 10$ cm^{-1}, $\nu_0 = 2800$ cm^{-1}, and $\varepsilon_\infty = 1.7$ (Fig. 3.28). The simulations are for $\varphi_1 = 83°$ corresponding to the maximum SNR (ΔR) (Section 2.3.1). As seen immediately from Fig. 3.28, for substrates with a low refractive index $(n_3 < 2)$ and for substrates with a high refractive index $(n_3 = 2.0\text{–}4.0)$, the absorption bands are oppositely directed while in the intermediate case the band is distorted assuming a derivative-like shape.

Figure 3.29 shows simulated p-polarized spectra of a model organic layer on a Si substrate as a function of the absorption index k_3, which can be modulated by changing the level of doping of Si. The angle of incidence was taken to be 80°. It is seen that the dependence of sign and intensity of the absorption band on the absorption index of the substrate is similar to the dependence on n_3 (Fig. 3.28); as k_3 increases, the negative absorption band becomes positive. At intermediate values of k_3 (around 1.5–2.5), when reflectivity at the resonance frequency (2800 cm^{-1}) is minimal, the band has derivative-like shape. Such spectral distortions were observed experimentally in the p-polarized spectra of organic films on glassy carbon ($\hat{n} = 3.5 - 1.5i$) [68] and chalcocite ($\hat{n} = 5.1 - 0.18i$) [61b].

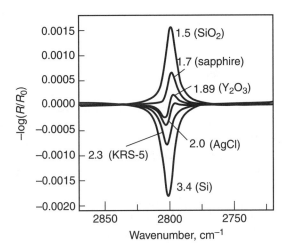

Figure 3.28. Band shape in p-polarized IRRAS spectra of hypothetical organic layer 1 nm thick as function of refractive index n_3 of transparent substrate (indicated in figure). Dielectric function of film was specified by $S = 0.001$, $\gamma = 10$ cm^{-1}, $\nu_0 = 2800$ cm^{-1}, and $\varepsilon_\infty = 1.7$; $\varphi_1 = 83°$.

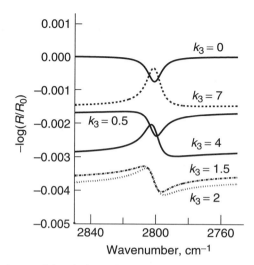

Figure 3.29. Dependence of band shape in p-polarized IRRAS spectrum of hypothetical organic layer 1 nm thick on absorption index of substrate with $n_3 = 3.4$. Absorption indices are indicated. Dielectric function of film was specified by $S = 0.001$, $\gamma = 10$ cm^{-1}, $\nu_0 = 2800$ cm^{-1}, and $\varepsilon_\infty = 1.7$; $\varphi_1 = 80°$.

The band distortions are explained by the redistribution of contributions of the vertical and lateral components of the mode in the p-polarized spectrum as either the real or imaginary part of the refractive index of the substrate is changed. In spite of this, however, the SSR for dielectrics (Sections 3.3.2 and 3.11.4) is independent of the optical properties of the substrate. The phenomenon outlined above can also be considered from the viewpoint of geometric optics, rather than invoking the complex origin of the absorption bands in the p-polarized spectra (Section 3.2). The intensity of the radiation reflected from the film–substrate system can be represented as the sum of the intensities of the radiation reflected from the front film–substrate interface, I_1, and the radiation multiply reflected in and emerging from the film, $I_2, \ldots, I_n$ (see Fig. 1.12). Clearly, in IRRAS, the spectrum is the sum of $I_1 + I_2 + \cdots + I_n$. For a layer on a transparent substrate with a small n_3, the $I_2, \ldots, I_n$ values are small relative to I_1, and so the signal consists primarily of I_1, which is similar to the specular reflection of the layer substance (the bands are negative). In the case of a large n_3, the contribution of the components $I_2, \ldots, I_n$ to the resulting reflected radiation increases, and the spectrum approaches the spectrum of the radiation passed through the layer, so that the reflection spectrum is similar to the absorption spectrum of the layer. The ratio of the I_1 and $I_2, \ldots, I_n$ components is also a function of the optical constants n_2 and k_2 of the layer and therefore changes within the limits of the band because of the dispersion of the optical constants. As a result, with a certain combination of optical constants, reflectivity can be positive and negative at different frequencies within the band limits, giving the derivative-like shape of the absorption band (Figs. 3.28 and 3.29).

It is noteworthy here that some bands of anisotropic films on transparent substrates can vanish from the p-polarized IRRAS [69, 70] or polarization modulation (PM) IRRAS [71, 72] spectra. This phenomenon is a consequence of the SSR for dielectrics (see Section 3.11.4 for more detail).

4. The effect of optical anisotropy of the substrate in IRRAS has been analyzed by Carson and Granick [73] for organic self-assembled monolayers on mica (muscovite). This crystal has biaxial optical properties and produces large interference fringes in transmission or reflection spectra because of reflections off the front and back of a sheet. To reduce the intensities of the interference fringes, it has been suggested to measure spectra at φ_B of the mica with p-polarized radiation, incident along one of its principal optical axes

The following is a summary of the current section.

1. *Layers on Metals*. The bands in IRRAS are positive, independent of the angle of incidence, as in the absorption spectrum. The band shape is sensitive to the dielectric properties of the metal substrate and the angle of incidence: The band shape becomes asymmetrical with decreasing optical conductivity of the metal and at grazing angles.

2. In the region of the phonon absorption of the substrate material, artifact (substrate) bands are present in IRRAS of any ultrathin film.

3. The band shape is complex in the case of weakly absorbing layers. With a certain combination of the angle of incidence and the optical constants of both the layer and the substrate, the absorption band may exhibit a derivative-like shape.

3.5. OPTICAL PROPERTY GRADIENTS AT SUBSTRATE–LAYER INTERFACE: EFFECT ON BAND INTENSITIES IN IRRAS

When calculating band intensities in IR spectra of layers measured by either transmission or IRRAS, the models most often used to describe the reflection, refraction, and absorption assume an abrupt change in the optical properties at the layer–substrate interface. An exception is microscopic models that take into account the existence of a transition layer, but the relationships derived based on these model are too cumbersome to find wide application in the analysis of optical layers [41, 74, 75].

Nevertheless, a comparison of the band intensities calculated on the basis of the typical model (assuming a sharp interface) and experimental results [76] show that in the experimental spectra of layers, the band intensities are several times lower than predicted. Therefore, absorption indices of layers determined from experimental spectra using the relationships derived with the sharp interface model (Sections 1.4–1.7) include a systematic error. The root of this problem lies not only in the difference between the optical properties of the thin-film materials and solid materials and the error introduced by an uncertainty in the angle of incidence, imperfection of the polarizer, and surface roughness but also in the obvious inadequacy of the sharp interface model.

It is evident that in each case the contribution of the optical property gradient will depend upon the optical properties of the samples under investigation, the gradient depth, and the conditions under which the spectra are recorded and must be analyzed individually for each system. It can be instructive to pinpoint general trends observed in the spectra by comparing the spectra calculated using two different models, one with a sharp interface and one with an optical property gradient. Tolstoy and Gruzinov [76] performed such a comparative study for weakly and strongly absorbing layers (from substances such as organic compounds containing C−H groups) and silicon oxide layers, as well as transparent, weakly and strongly absorbing substrates, such as Si, doped Si, and Al. The optical property gradient was specified at the interface with the relationship (1.102), treating the interfacial region of the substrate as a set of m layers with a thickness on the order of several nanometers and allowing the optical properties to vary exponentially. To simplify the calculations, the optical property gradient of the layer was specified using a simple three-zone model [77] in which the layer is divided into three zones: (1) adjacent to the substrate, (2) next to the surroundings, and (3) a central one, whose optical properties correspond to those of the given bulk substance (Fig. 3.30).

Figure 3.31 shows the p-polarized IRRAS spectrum of a SiO_2 layer on Si simulated for $\varphi_1 \approx \varphi_B - 3°$ using the sharp interface model. In the spectra calculated with the optical property gradient model for the surfaces of both the layer and the substrates, the intensity of the 1240-cm^{-1} band at ν_{LO} is less, and that of the 1090-cm^{-1} band at ν_{TO} is greater, than in the spectrum shown in Fig. 3.31. This is demonstrated in Fig. 3.32 for the three-zone SiO_2 layer on Si (the lower 1-nm SiO zone with optical constants taken from Ref. [78], the upper 1-nm zone with $n_2 = 1.3$ and $k_2 = 0$ and the middle 0–10-nm SiO_2 zone with optical constants taken from Ref. [78]; $\varphi_1 = 71°$, $n_3 = 3.42$, $k_3 = 0$). The calculations were also performed with an optical property gradient only in the layer being analyzed. The gradient was incorporated into the classical three-phase stratified-layer model of the $Si-SiO_2$ system in which the formation of a SiO layer at the $Si-SiO_2$ interface

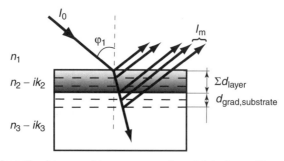

Figure 3.30. Schematic diagram of beam propagation at interface with gradient of optical properties.

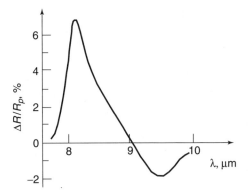

Figure 3.31. Simulated p-polarized IRRAS spectrum of 1-nm-thick layer of SiO_2 on Si, measured at $\varphi_1 = 71°$. Reprinted, by permission, from V. P. Tolstoy and S. N. Gruzinov, *Opt. Spectrosc.* **71**, 77–80 (1991), p. 78, Fig. 1. Copyright © 1991 Optical Society of America.

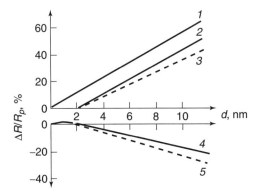

Figure 3.32. Calculated reflectivity $\Delta R/R_p$ for bands at 1240 cm^{-1} (1–3) and 1090 cm^{-1} (4, 5) in p-polarized IRRAS spectrum ($\varphi_1 = 71°$) of SiO_2 layer on Si surface versus thickness of SiO_2 layer in assumption of (1) sharp surfaces, (2, 4) gradient in optical properties of SiO_2 layer, and (3, 5) gradient in both SiO_2 layer and substrate (depth of transition layer in substrate with exponential variation of n_3 is equal to 50 nm). Reprinted, by permission, from V. P. Tolstoy and S. N. Gruzinov, *Opt. Spectrosc.* **71**, 77–80 (1991), p. 78, Fig. 2. Copyright © 1991 Optical Society of America.

and the appearance of a 1-nm water layer on the SiO_2–surroundings interface are included. The calculated band intensities (Fig. 3.32, curves 2 and 4) reveal that the divergence of the curves characterizing the layer thickness dependence of the reflectivity is determined by the depth of the optical property gradient of the substrate. The difference in the band intensities increases if the gradient depth is increased. On the other hand, the deviation of these dependencies from linearity, especially in the initial part of the curves, is caused by the optical property gradient in the layer under investigation.

The fact that the optical constant gradients of the layer and of the substrate have different effects on the band intensities can be used to determine the gradient depth directly from experimental spectra. First, a dependence of the intensities of the ν_{LO} and ν_{TO} bands on the gradient depth is calculated and is then used to graphically determine the gradient depth for the given layer–substrate system. The accuracy can be improved if the ratio $\Delta R_{LO}/\Delta R_{TO}$ of the absolute intensities of the bands at 1240 and 1090 cm^{-1} (ΔR_{LO} and ΔR_{TO}, respectively) is used in the calculations, rather than the absolute value of the band intensities, since the ratio is more sensitive to the presence of the gradient. The corresponding curves for the Si–SiO$_2$ system are shown in Fig. 3.33.

One might suggest that this technique is limited because of the necessity to determine dependencies on gradient depth not only for each layer–substrate system but also for layers synthesized by different methods; optical constants of

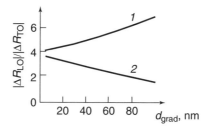

Figure 3.33. Variation in ratio $|\Delta R_{LO}|/|\Delta R_{TO}|$ (p-polarization, $\varphi_1 = 71°$) for 1-nm SiO$_2$ layer on Si with thickness of transition layer; (1) $\varphi_1 = 80°$ and (2) $\varphi = 70°$. Curves correspond to exponential variation of refractive index of Si substrate in region of transition layer; $n_3 = 3.42$, $k_3 = 0$. Reprinted, by permission, from V. P. Tolstoy and S. N. Gruzinov, *Opt. Spectrosc.* **71**, 77–80 (1991), p. 79, Fig. 3. Copyright © 1991 Optical Society of America.

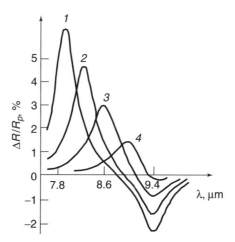

Figure 3.34. Simulated p-polarized IRRAS of layer 1 nm thick with optical properties specified by oscillator parameters $\nu_0 = 1090$ cm^{-1}, $\gamma' = \gamma/\nu_0 = 0.05$, $\varepsilon_\infty = 2.5$, $S = (1)0.8$, (2) 0.6, (3) 0.4, (4) 0.2; $n_3 = 3.42$, $\varphi_1 = 71°$. Reprinted, by permission, from V. P. Tolstoy, *Methods of UV-Vis and IR Spectroscopy of Nanolayers*, St. Petersburg Univ. Press, St. Petersburg, 1998, p. 179, Fig. 1.15. Copyright © 1998 St. Petersburg University Press.

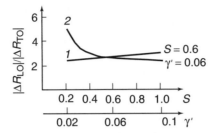

Figure 3.35. Variation in ratio $|\Delta R_{LO}|/|\Delta R_{TO}|$ in p-polarized IRRAS spectra of hypothetical strongly absorbing 1-nm layer with optical properties described by oscillator parameters $v_0 = 1064$ cm^{-1}, $\varepsilon_\infty = 2.5$, (1) $S = 0.6$ at different $\gamma' = \gamma/v_0$, (2) $\gamma' = \gamma/v_0 = 0.06$ at different S; $\varphi_1 = 71°$, $n_3 = 3.42$. Reprinted, by permission, from V. P. Tolstoy and S. N. Gruzinov, *Opt. Spectrosc.* **71**, 77 (1991), p. 79, Fig. 4. Copyright © 1991 Optical Society of America.

such layers have been shown to differ considerably. Therefore, the effect of the effective oscillator parameters, which describe the optical properties of the layer, on $\Delta R_{LO}/\Delta R_{TO}$ was analyzed [76]. The optical constants of the layer and the oscillator parameters S and γ were related by Eq. (1.46) for $j = 1$. The calculated spectra of a hypothetical layer on a silicon surface are depicted in Fig. 3.34, and the variation in $\Delta R_{LO}/\Delta R_{TO}$ versus the parameters S and $\gamma' = \gamma/v_0$ is presented in Fig. 3.35. As the calculations show, for $S = 0.4$–0.8 and $\gamma' = 0.04$–0.08, representing strong absorption bands, the change in $\Delta R_{LO}/\Delta R_{TO}$ is less than 10–15%, which is considerably smaller than the change in $\Delta R_{LO}/\Delta R_{TO}$ caused by a gradient in the optical properties of the substrate. Thus, this technique can also be recommended for the determination of the gradient from experimental spectra of layers with slightly different structure.

The spectrum simulations taking into account the optical property gradient of the substrate give a decreased intensity by up to 50% for the case of weak absorbers on transparent or weakly absorbing substrates but practically the same intensity for weak absorbers on strongly absorbing substrates (e.g., metal). This is largely because the optical property gradient depends on an additional contribution I_m (Fig. 3.30) to the intensity of the radiation reflected from the layer–substrate system. The relative value of this component is comparatively high for weakly reflecting substrates and small for strongly reflecting ones, whose total level of reflected radiation reaches 70–100%.

Thus, the difference between the band intensities in the experimental spectra measured by IRRAS and the calculated ones based on the sharp interface model can be connected with the existence of the optical property gradient in real optical systems, which is most clearly manifested in IRRAS of strong absorbers on the surface of transparent and weakly reflecting substrates.

3.6. DIPOLE–DIPOLE COUPLING

As shown in Section 3.3, the dependences of IR spectra on the amount of a substance are different for the substance in the form of an ultrathin film and bulk phase. In the present section, we shall consider an additional optical effect that is responsible for intensity and energy transfer between modes and dependence of the band position and intensity in the IR spectra of the adsorbed molecules on coverage.

As reported already in the first IR spectroscopic studies of adsorption on Pt catalysts [79, 80], as the surface coverage of dipoles aligned perpendicular to the surface increases up to a complete monolayer, the absorption band can shift upward. Figure 3.36 illustrates this phenomenon using the CO–Ir(111) system in ultra high vacuum (UHV) at 300 K as an example [81]. The νCO band assigned to atop (terminal) CO [82, 83] shifts from $\sim$2030 cm^{-1} at a coverage of 0.002 ML upward by $\sim$50 cm^{-1} with increasing surface coverage. This effect retains when the substrate is placed in an aqueous environment (Fig. 3.37a) [84] (see also Ref. [85]). Although the coverage-dependent band shift is rather typical — for example, a blue shift of up to 36 cm^{-1} was observed for the νCO band of atop CO in the CO–Pt(110) system at 300 K [86], up to 85 cm^{-1} for atop CO in the CO–Ir(110) system at room temperature [87], and up to 100 cm^{-1} in the CO–Pd(100) system at 100 K [88] — there are systems, for example NO–Ir(111) (Fig. 3.38a, open circles) and CO–Cu [89] (Section 7.1.2), that are indifferent to surface coverage in terms of the band position.

The band shift with surface coverage has been studied intensively theoretically on the microscopic level [90–95] (for review, see Refs. [96–99]) and attributed mainly to dipole–dipole coupling of adsorbed molecules. Before introducing

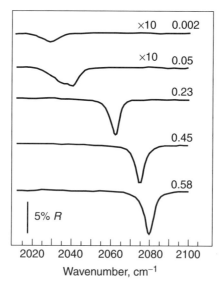

Figure 3.36. IRRAS spectra of various fractional coverages of CO on Ir(111) measured at 300 K. Coverages in monolayers are denoted beside each spectrum. Reprinted, with permission, from J. Lauterbach, R. W. Boyle, M. Schick, W. J. Mitchell, B. Meng, and W. H. Weinberg, *Surf. Sci.* **366**, 228 (1996), p. 228, Fig. 2. Copyright © 1996 Elsevier Science Ltd.

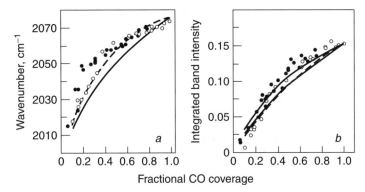

Figure 3.37. Plots of (a) peak ν_{CO} frequency and (b) integrated ν_{CO} intensity versus fractional coverage for CO adsorbed on Ir(111) electrode surface in 0.1 *M* HClO₄ in (○) absence and (●) presence of coadsorbed NO [at +0.4 and +0.45 V (saturated hydrogen electrode, SHE), respectively]. Spectra were measured in situ by IRRAS. Dashed and solid traces are corresponding predicted plots extracted from dipole-coupling simulations. Adapted, by permission, from C. Tang, S. Zou, M. W. Severson, and M. J. Weaver, *J. Phys. Chem. B* **102**, 8546 (1998), pp. 8550 (Fig. 5) and 8551 (Fig. 7). Copyright © 1998 American Chemical Society.

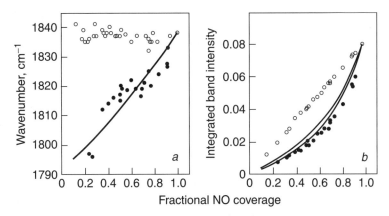

Figure 3.38. Plots of (a) peak ν_{NO} frequency and (b) integrated ν_{NO} intensity versus fractional coverage for NO adsorbed on Ir(111) electrode surface in 0.1 *M* HClO₄ at 0.45 V (SHE) in absence (○) and presence (●) of coadsorbed CO. Spectra were measured in situ by IRRAS. Dashed and solid traces are corresponding predicted plots extracted from dipole-coupling simulations. Adapted, by permission, from C. Tang, S. Zou, M. W. Severson, and M. J. Weaver, *J. Phys. Chem. B* **102**, 8546 (1998), pp. 8550 (Fig. 6) and 8551 (Fig. 8). Copyright © 1998 American Chemical Society.

these results, it is instructive, however, to understand the effect from a macroscopic point of view, by considering the dielectric function (Section 1.3).

The adsorption–desorption process or any physicochemical treatment of an adlayer, which changes density of dipoles, N, and/or the effective charge, changes, in agreement with Eq. (1.48), the oscillator strength of this adlayer.

The influence of the oscillator strength on the spectrum of an adlayer on a metal measured by IRRAS can be interpreted in the following manner. Combining the Lorentzian oscillator model equation (1.46) for a single mode with the thin-film approximation equation (1.82) gives the reflectivity spectrum [100]

$$\left(\frac{\Delta R}{R}\right)_p (v) = 8\pi d_2 \sin\varphi_1 \tan\varphi_1 \frac{Sv^2\gamma/\varepsilon_\infty^2}{(v_{LO}^2 - v^2)^2 + \gamma^2 v^2} \tag{3.28a}$$

and the maximum value of reflectivity

$$A_{max} = 8\pi d_2 \sin\varphi_1 \tan\varphi_1 \frac{S}{\gamma\varepsilon_\infty^2}, \tag{3.28b}$$

which is observed at $v = v_{LO}$. From this, it can be seen that the v_{LO} band intensity is directly proportional to S, as in conventional spectroscopy [see Eq. (1.47)]. At the same time, contrary to conventional spectroscopy, it follows from Eq. (1.51) that increasing S, for example, due to an increase in the oscillator surface density N_s, results in a blue shift of the v_{LO} band. This phenomenon can be explained by *long-range dipole–dipole interactions* in which each vibrating molecule gives rise to a long-range dipole field that is felt by the other molecules in the ensemble, and thus all the molecules interact coherently with the incident radiation. As mentioned earlier (1.3.2°), this effect is incorporated into the macroscopic dielectric function through the sum of the polarizabilities of the individual molecules.

However, the preceding macroscopic treatment fails to describe quantitatively the effect of surface coverage on the absorption band frequency. Moreover, contrary to what one might expect from Eq. (3.28b), even taking into account that molecules can tilt at higher coverages due to the strong repulsive interaction between neighboring CO molecules [101, 102], the relationship between the spectral intensity and the number of adsorbed molecules is not linear. As seen from Figs. 3.36 and 3.37b, the vCO band growth is significant at low surface coverage and slows down at higher coverages, implying a decrease in absorption per CO. These inconsistencies mainly arise from the fact that the high-frequency dielectric constant ε_∞ depends not only on the number of dipoles per unit area (surface coverage), N_s, but also on the effective electronic charge. The latter quantity is highly sensitive to the dielectric properties of the substrate and the microstructure within the layer (the surroundings of the adsorbed molecules). Parameterizing an adsorbate layer by the generalized polarizability equation (1.35), instead of the dielectric function in Eq. (1.46), and assuming that the surface is randomly filled, Persson and Ryberg [103] (see also Refs. [65, 104]) derived the following formulas for the band position and intensity:

$$v^2 = v_0^2 \left(1 + \frac{\hat{\alpha}_v \theta U_0}{1 + \hat{\alpha}_e \theta U_0}\right) \tag{3.29}$$

and

$$\int \frac{I - I_0}{I_0} dv = \frac{\hat{\alpha}_v \theta U_0}{(1 + \hat{\alpha}_e \theta U_0)^2}, \tag{3.30}$$

respectively. Here, $\theta = N_s/N_s^0$ is the coverage, where N_s is the number density of the species in the adsorbate layer and N_s^0 is that in the outermost layer of the substrate, ν is the frequency of the dipoles oriented perpendicular to the metal surface, $\hat{\alpha}_v$ and $\hat{\alpha}_e$ are the vibrational and electronic polarizability of the adsorbate, respectively, ν_0 is the so-called *singleton frequency*, and U_0 is the *lattice sum*. The singleton frequency is defined as the frequency of the isolated molecule on the surface, which interacts only with this surface. Its value can be obtained by extrapolating to zero coverage the band frequency corrected for chemical shifts (see below). For example, ν_0 of CO adsorbed on Ir(111) in an electrochemical environment (Fig. 3.37a) is 2000 cm^{-1}, while in UHV (Fig. 3.36) it is 2028 cm^{-1} [81] (this difference will be interpreted below). Comparison of Eqs. (3.29) and (1.51) shows that the singleton frequency is a phenomenological analog of the resonance frequency. The lattice sum U_0 takes into account the contributions of surrounding dipoles, their image dipoles, and the self-image dipole in the actual adsorbate structure at complete coverage of the two-dimensional surface lattice. The vibrational polarizability $\hat{\alpha}_v$[†] is expressed in terms of the effective charge e^*, the reduced mass m^*, and the resonance frequency ν_0 as

$$\hat{\alpha}_v = \frac{e^{*2}}{\varepsilon_0 m^* \nu_0^2}. \tag{3.31}$$

Using Eqs. (1.35) and (1.89), the electronic polarizability $\hat{\alpha}_e$[‡] can be related to the high-frequency dielectric constant

$$\varepsilon_\infty = 1 + \frac{N_s}{d_2} \hat{\alpha}_e. \tag{3.32}$$

It follows that $\hat{\alpha}_e$ in Eqs. (3.29) and (3.30), much like ε_∞ in Eqs. (1.51) and (3.28), takes into account screening of the external electric field by the polarizable electronic cloud of the local environment of the dipole which results in an attenuation of the dynamic dipole moment (depolarization) and, hence, a decrease of the normalized band intensity per single dipole. Originated from the contribution of the depolarization field to the local field, this effect is termed *dielectric screening*. As seen from Eq. (3.30), dielectric screening is negligible at very low θ but increases progressively toward higher θ, thereby yielding nonlinear intensity–coverage dependences.

The Persson–Ryberg model is adequate for the systems with random occupation of adsorption sites, for example, for CO adsorbed at 300 K on Ru(001) [103] and at 90 K on Ir(111) [81]. However, as mentioned, the band intensity of adsorbed NO increases linearly with coverage (Fig. 3.38), while the position of the absorption band is practically unchanged. This suggests that on Ir(111) the NO adlayer forms close-packed, nanoscale (or larger) clusters, so that the "local" (microscopic) coverage remains high. The result is strong

---

[†] Typical values for CO adsorbed on transition metals are 0.2–0.4 Å^3 [65, 81, 84, 85, 93, 103].
[‡] Typical values for adsorbed CO are 1.5–3 Å.

dipole coupling that is practically constant at all coverages examined. The same interpretation was suggested for CO adsorbed on Cu (Section 7.1.2). When clusters or a superstructure is formed during adsorption, instead of weak next-nearest-neighbor interactions, nearest-neighbor interactions become more important, yielding a greater contribution to the dipole sum. This effect can be included in the Persson–Ryberg model, which was accomplished by Lauterbach et al. [81] for quantifying the IRRAS data for CO on Ir(111). Pfnur et al. [105] used the dependence of the band frequency and width on the cluster sizes for estimating the cluster size from the IR spectra. Based on the features of the dipole coupling, models of island formation have been proposed for partial oxidation of the CO adlayer on silica gel-supported Pt catalysts [106] and low-index Pt-group electrodes [107, 108].

Coadsorption of highly polarizable species (e.g., solvent molecules or metal adatoms) can cause a red shift of the absorption band of the adsorbate by up to 50–80 cm^{-1} [65, 84, 85, 109, 110] and an attenuation of the band intensity by a factor of 2–20 [110]. When an oscillator is surrounded by such species, the local electric field acting on it decreases due to dielectric screening. This is described by substituting the electronic polarizability of the coadsorbate $\hat{\alpha}_e^{coads}$ ($\hat{\alpha}_e^{coads} > \hat{\alpha}_e$) instead of that of the oscillator $\hat{\alpha}_e$ in Eqs. (3.29) and (3.30). If the surrounding coadsorbate is further replaced progressively by the oscillators (θ increases), $\hat{\alpha}_e^{coads}$ is replaced by $\hat{\alpha}_e$. This model allowed Weaver and co-workers [84, 85] to reproduce the spectral dependences for CO adsorbed at the Ir(111)–water interface. The parameters used in this simulation were $\hat{\alpha}_e(CO) = 2.5$ Å^3, $\hat{\alpha}_v(CO) = 0.41$ Å^3, $\hat{\alpha}_v(H_2O) = 0$ Å^3, $\hat{\alpha}_e(H_2O) = 1.5$ Å^3, and $v_0CO = 2000$ cm^{-1}. To extend the model to an electrochemical environment (an immersion medium), the method of Greenler and co-workers [111, 112] was used. As seen from Figs. 3.37 and 3.38, the calculated curves describe reasonably the experimental data. An attempt has also been undertaken [113–116] to explain the coverage dependence of the band frequency and intensity in terms of a combination of the Stark effect and backdonation (Section 3.7), that is, by the change in the surface potential by the solvent dipolar charge. However, this hypothesis was not supported by the vCO–electric field dependences obtained by Lambert et al. [117–119] when the interfacial electrostatic field was adjusted in the absence of coadsorbates by applying voltages to the substrate. In the general case, studies of the dipole–dipole interaction between adsorbate and coadsorbed species can be hampered by coadsorption-induced alternations in the adsorbate coordination and spatial structure (Section 3.7.6).

Dielectric screening is the specific case of another particularly interesting phenomenon arising from long-range dipole–dipole interactions — the so-called intensity and energy transfer. An illustration of this is the decrease in intensity and frequency of the NO absorption band when NO is coadsorbed with CO (Fig. 3.38). The oscillator strengths of the individual modes of α-Fe$_2$O$_3$ listed in Table 3.2 and the relative intensities in the spectra of the isotropic α-Fe$_2$O$_3$ films on Fe obtained by IRRAS (Fig. 3.13) indicate that the intensity transfer occurs from the strong low-frequency modes at 368 and 414 cm^{-1} to the weak high-frequency

mode at 662 cm^{-1}. As a result, the most intense band in the spectrum corresponds
to the high frequency mode with the lowest oscillator strength.

To illustrate the intensity and energy transfer in more detail, Fig. 3.39a shows
the simulated spectrum obtained by IRRAS for an ultrathin film modeled as the
sum in Eq. (1.46) of two strong oscillators ($S_1 = S_2 = 0.2$) with resonant frequencies 1000 and 1100 cm^{-1}. The intensity of the low-frequency band of the "two-
oscillator" film is quenched relative to the spectrum of the "single-oscillator"
film while the intensity of the high-frequency band increases significantly. Moreover, the frequency difference between the bands increases from that when the
modes are uncoupled; that is, band repulsion takes place. This implies that when
coupling [the summation in Eq. (1.46)] is allowed, the modes repel one another
and band intensity transfer occurs from the lower to the higher frequency mode.
This effect is absent if the oscillator strength of the lower frequency mode is

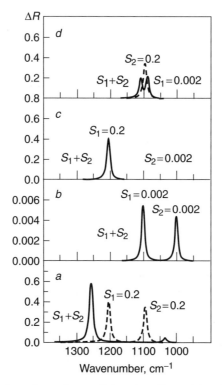

Figure 3.39. Redistribution of seeming oscillator strengths and band "repulsion" in IRRAS
spectra of two-oscillator-model films 1 nm thick on Au ($\hat{n}_3 = 10.84 - 51.6i$) for $\varphi_1 = 85°$. Film
materials are represented as sum of two oscillators (solid lines) with resonance frequencies
$\nu_1 = 1100$ cm^{-1} and $\nu_2 = 1000$ cm^{-1}, dampings $\gamma_1 = \gamma_2 = 10$ cm^{-1}, and oscillator strengths
(a) $S_1 = S_2 = 0.2$, (b) $S_1 = S_2 = 0.002$, (c) $S_1 = 0.2$ and $S_2 = 0.002$, and (d) $S_1 = 0.002$ and
$S_2 = 0.2$. For comparison dashed lines show spectra in case of corresponding spatially
separated oscillators. In (b) and (c), solid and dashed lines coincide.

weak, regardless of whether the higher frequency mode is weak (Fig. 3.39*b*) or strong (Fig. 3.39*c*). When a negligibly weak higher frequency mode and a strong lower frequency mode are involved, the higher frequency mode is significantly enhanced (Fig. 3.39*d*). It was shown by Da Costa and Coleman [100b] that the band repulsion and the intensity transfer increase as the frequencies and the frequently difference of the coupled modes decrease. Thus, in the spectra of a multioscillator film obtained by IRRAS, the relationship between the band intensities may not reflect the relationship between the corresponding oscillator strengths.

The physical origin of the suppression of the lower frequency band intensity is more efficient screening of this oscillator by the higher frequency dynamic dipole. The macroscopic depolarization field arising from the accumulation of charge at the film surface mediates any interaction of these oscillators, which no longer vibrate independently: Each oscillator experiences a macroscopic depolarization field arising from the other. As a result, due to larger polarizabilities and weaker restoring forces, the low-frequency modes are screened by the high-frequency ones but not vice versa.[†]

Intensity transfer to the high-frequency mode can cause the surface concentration of a coadsorbate or a surface structure to appear lower than it actually is. In particular, when more than one binding site is occupied such as atop, bridging, or threefold hollow sites, multiple bands are expected in the spectra. Since the frequency of the chemisorbate decreases with increasing coordination number, the intensity from molecules chemisorbed in the bridging or threefold hollow position can appear less than it should [65, 120], and even the corresponding peaks can be missing in some structures [82, 90, 121]. At the same time, if the spectral regularities caused by dipole coupling are known, intensity transfer effects can be utilized for studying different structural sites at the surface [113, 122] and determining the degree of microscopic intermixing of coadsorbed different molecules [84, 123]. For example, the decrease in intensity of the NO absorption band when NO is coadsorbed with CO (Fig. 3.38*b*) is consistent with molecular intermixing, although the observed nonlinear νCO$-\theta$ dependence (Fig. 3.37*a*) also suggests [84] the formation of locally enriched CO regions at intermediate compositions. One can see that an "intensity transfer" can occur even when the vibrational frequencies are substantially different ($\sim$200–300 cm^{-1}) provided that the oscillating partners are closely juxtaposed.

To discriminate between dipole–dipole coupling and changes in the electronic environment due to chemical effect, *Hammaker's method* of isotopic dilution

---

[†] This can be understood using a simple two-oscillator model outlined in Refs. [65, 89b, 111, 142]. If two oscillators at the surface are coupled, they generate two modes, one in which the two oscillate in phase and one in which they oscillate out of phase. For a positive coupling constant, the in-phase mode has the higher frequency. If the two oscillators are identical (the singleton frequencies are equal), the in-phase mode intensity is twice that of the single oscillator, while the out-of-phase mode has zero intensity. If the two oscillators have different singleton frequencies, the cancellation in the out-of-phase mode is partial, resulting in a nonzero intensity, yet smaller than that of the in-phase mode [65].

is often used [124]. This method involves the comparison of the absorption of two isotopes of the same molecule. For a given coverage, the chemical environment will be the same and, in the absence of long-range dipole–dipole interactions, the spectrum will be independent of the ratio of these isotopes. Otherwise, the spectrum will demonstrate some intensity transfer and changes in the band frequencies, which will depend upon the isotope ratio. For example, typical data collected from the adsorption of CO on Pt (111) [125] show a shift in the absorption band from 2065 cm^{-1} at low coverage to 2101 cm^{-1} at saturation. Varying the $^{13}CO-^{12}CO$ ratio at constant total coverage, in accordance with Hammaker's method of isotopic dilution, Crossley and King [126] demonstrated that the blue shift is caused mainly by the long-range dipole coupling, rather than by chemical bonding effects (see also Section 7.1.2).

Dipole coupling also affects the shape of bands [93, 96–98]. Typically, narrow bands are considered as proof of high order of the adlayer structure, while broad bands are associated with disorder. However, even in the cases of disordered structures characterized by broad absorption bands, dipole coupling can produce narrow resultant bands [127]. Therefore, in the absence of other arguments, interpretation of a narrow band as proof of a well-ordered structure should be careful.

The dipoles of adsorbed molecules can couple through metal substrate electrons, partly contributing to the observed frequency shifts with the surface coverage [92, 128]. Finally, plasma oscillations of the metal electrons in the electric field of the IR beam (Section 1.3) can couple by an electron–hole pair creation mechanism with the adlayer modes whose dipole moment vector is parallel to the surface [129–132]. Being forbidden (Section 1.8.2), these modes manifest themselves in metallic IRRAS as the so-called antiabsorption peaks (increase of reflectance). Examples are frustrated rotations of CO on Cu(100) and frustrated translations of H on W(100) and Mo(100) [129, 130]. This effect, however, should not be considered as the exception from the SSR for metals.

3.7. SPECIFIC FEATURES IN POTENTIAL-DIFFERENCE IR SPECTRA OF ELECTRODE–ELECTROLYTE INTERFACES

According to electrochemical theory, the kinetics of an electrochemical reaction is controlled by the potential drop between the solid and solution phases [133–136]. A dynamic zone extending in both directions from the electrified interface over which this drop exists is called the double layer (DL) of charge. The DL in the solution is made up of adsorbed and solvated ions (molecules) and solvent. Its dense part, which is referred to as the *Helmholtz layer* (HL), plays the major role in the interfacial processes. At low ion concentration, there is also a diffuse layer (*Gouy layer*) in the solution. The countercharged part of the DL in a metal electrode is comprised of a "skin" layer with an excess or a deficit of electrons. The DL in a semiconductor electrode is called the space charge layer. It consists of an accumulation, depletion, or inversion layer with an excess or a deficit of electrons or holes and ionized donor or acceptor states, depending on the semiconductor properties and the position of the Fermi level at the surface relative to the band edges [137–139]. All of these components of the DL are sensitive to the potential drop and can interact with IR radiation. In addition, the IR spectra measured in situ of the electrode–electrolyte interface exhibit a number of features associated with artifacts due to disadvantages of the techniques employed. Correct band assignment in such a complex spectrum, while not trivial, is possible since the spectral features of different nature depend on potential in different ways and, in some cases, have specific appearance. IRRAS has been treated thoroughly in recent reviews [113, 140–145], while ATR has been discussed in Refs. [146, 147]. The results of these will be summarized below and supplemented by additional data.

Typically, a potential-difference spectrum of the electrode–electrolyte interface presents negative, positive, and bipolar bands and smooth background absorption, which are due to all species in the path of IR radiation that are affected by the electrode potential (Figs. 3.40, 3.43a, 4.47a, 4.50a, 7.45, and 7.47). In addition to the absorption bands that reflect change in the population in the HL of reagents/products for the reaction under study, these may include the bands due to (1) the electrolyte species in the diffuse layer, (2) the electrolyte species in the HL, (3) the reagent/product species whose absorption (parameters of the elementary oscillators) is modulated by potential and coadsorption of electrolyte species, (4) delocalized and localized charge carries, and (5) optical effects of various nature. Consider these effects in more detail.

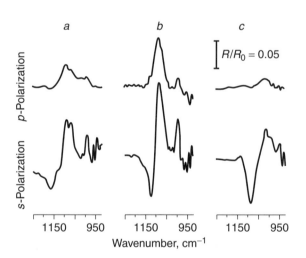

Figure 3.40. The s- and p-polarized SNIFTIRS of adsorbed phosphate species on Au(111) electrode taken for different solutions: (a) 0.005 M KOH, 1.5×10^{-2} M KH$_2$PO$_4$; (b) 0.01 M KOH, 1 M KF, 1.5×10^{-2} M KH$_2$PO$_4$; (c) 0.05 M KOH, 1.5×10^{-2} M KH$_2$PO$_4$; reference potential: $E_0 = 0.03$ V versus Pd/H$_2$; sample potentials: (a, b) $E = +1.1$ V versus Pd/H$_2$; (c) $E = +1.2$ V versus Pd/H$_2$. Spectra were measured with Bomem DA-8 FTIR spectrometer equipped with MCT detector. Each spectrum is average of 1000×15 scans. Reprinted, by permission, from M. Weber and F. C. Nart, *Electrochim. Acta* **41**, 653 (1996) p. 655, Fig. 1. Copyright © 1996 Elsevier Science Ltd.

3.7.1. Absorption Due to Bulk Electrolyte

One of the most serious problems in interpreting spectra of the solid–liquid interface is the problem of distinguishing the absorption that is due to bulk electrolyte from the total absorption. The importance of this has recently been demonstrated, for example, for bisulfate adsorbed on Pt [141, 148–150]. This problem does not arise in the ATR-SEIRA technique, whose sensitivity is on the order of one to two monolayers, but is especially critical for thin-layer cells that are characterized by rather large time constants (Section 4.6.2). For example, a competitive adsorption of hydroxyls/protons [151, 152] or Faradaic reactions such as reduction of oxygen [153] causes protons to be consumed (released) at the interface, which may change the hydrolysis state of solvated (buffer) species, giving rise to strong bands overlapping with the bands of the adsorbate. Figure 3.40 shows s- and p-polarized subtractively normalized interfacial FITR spectroscopy (SNIFTIRS) spectra of the Au(111)–(0.015 M KH$_2$PO$_4$) interface in a series of electrolytes [(a) 0.005 M KOH, (b) 0.01 M KOH and 1 M KF, and (c) 0.05 M KOH] during a potential step from 0.03 V (where no phosphate adsorption is expected) to 1.2 V (SHE). The spectra are represented as the ratio between the sample (R) and the reference (R_0) spectra. Thus a loss of any species results in a positive spectral feature, and a gain of any species results in negative bands. The s-polarized spectra show positive bands that correspond to the solution species. Based on what would be expected from an increase in the pH in the series (a), (b), and (c), the spectra are assigned to (a) H$_2$PO$_4^-$ (1157 and 1078 cm^{-1}) and HPO$_4^{2-}$ (1076 and 988 cm^{-1}), (b) HPO$_4^{2-}$, and (c) PO$_4^{3-}$ (1010 cm^{-1}). The p-polarized spectra for (a) and (b) are complex, containing components from both adsorbate and solution. In case (a), there are positive bands at 1077 and 988 cm^{-1} due to HPO$_4^{2-}$ (whose concentration decreases in the diffuse layer at the sample potential), and negative bands at 1157 and 1078 cm^{-1}. By comparing these spectra with the s-polarized spectrum, these bands were attributed to changes in the solution arising from a shift in the equilibrium between HPO$_4^{2-}$ and H$_2$PO$_4^-$ during the potential step [151]. In case (b), the solution is responsible for a weak negative band at 1078 cm^{-1} in the p-polarized spectrum, in addition to the positive bands. In case (c), the p-polarized spectrum is practically free from interference with the solution spectrum, since the high concentration of hydroxyls in the solution provides fast replenishment of hydroxyls that are adsorbed.

The above example demonstrates that interference with the solution spectrum can be reduced considerably by use of an appropriate buffer that has high concentration of coadsorbed species and no absorption bands in the spectral region under study (see also Refs. [154, 155]). The solution bands can be distinguished in p-polarized in situ IRRAS by comparison with the s-polarized spectra in which only the absorption bands of the species in solution are present (Section 1.8.2). However, since the SSR holds at a distance of $\sim$1 μm out into the solution from the metal surface, this approach can be inefficient. In this case, apart from using other means (vide infra), one can resolve the contribution of bulk solution comparing the dependence of s- and p-polarized spectra on the buffer concentration [156]. Comparison with the reference spectra of the electrolyte components can be useful, in any case (e.g., see the spectrum analysis procedure in Refs. [157–159]). The bulk solution spectrum may be experimentally disengaged by employing flow cell tactics (Section 4.6.2) or a modulation method (Sections 4.7 and 4.9.2), and correlation analysis (Section 3.8). In other approaches, the potential dependence of the band center is analyzed: Contrary to bulk solution species, the bands of an adsorbed species may be specifically subjected to the Stark effect at a fixed surface coverage (see below), dipole coupling (Section 3.6) at changing coverage, or a donating/backdonating effect (see below).

Due to the diffusion, production, and consumption of IR-active species, results from the potential-difference experiments may be difficult to interpret when Faradaic reactions are studied. Using oxidation of glucose on Pt(100) in perchloric acid as an example, Faguy and Marinkovic [160] showed that the spectra measured upon a single and multiple alternation of potential can be rather different. Because of the large time constant of thin-layer cells, SNIFTIRS probes the diffuse layer when its composition still changes at both the reference and sample potential, which leads to a suppressed intensity of the bands due to accumulation of ClO_4^- and depletion of glucose in the diffuse layer.

If absorption due to electrolyte in the diffuse layer does not interfere with that of the adsorbate, analysis of the electrolyte spectrum may be useful, for example, in determining the actual pH at the interface [161, 162] (with an accuracy of 10^{-4} pH [163]) and studying redox reaction in electrolyte. This has been used, for example, to probe products of the successive reduction of the heteropolyanions $SiW_{12}O_{40}^{4-}$ or $PW_{12}O_{40}^{3-}$ at a Ge electrode [164], the oxidation of ferrocyanide on n-GaAs [165], and the anodic decomposition of water and various organic solvents on TiO_2 [166–169]. Provided that diffusion in and out of the thin layer is slow on the time scale of the potential change, the intensities and signs of the electrolyte bands in the s- [170] and p- [65, 84, 85, 154] polarized spectra can be used for quantifying the potential-dependent surface coverage (the surface excess) of the adsorbate. Analysis of the absorption due to electrolyte species allows gaining insight into the DL structure (Section 7.6) and even detecting IR nonactive species (Section 3.7.2).

3.7.2. (Re)organization of Electrolyte in DL

The IR spectra measured in situ are sensitive to (re)organization of electrolyte (both solute and solvent) species in the DL. On the one hand, this can provide unique information about the DL structure and, on the other hand, it can complicate determination of the spectrum baseline. The spectral changes associated with reorientation of an electrolyte species can be distinguished by using the corresponding SSR (Sections 1.8 and 3.11.4), as shown in the IR studies of perchlorate [157–159, 171], $Cr(DMSO)_6^{3+}$ [157], sulfate [172], bisulfate [173], nitrate [158] ions at Au, tetraethylammonium ions at Pt [151], perchlorate ions at Ge [174, 175], and ferri/ferrocyanide at Si [176]. The situation is complicated when an electrolyte species with degenerate IR-active modes appears in the DL [115, 174, 177] or forms complexes with products of the electrochemical reactions [178]. In this case, the degeneration can be canceled, resulting in band splitting.

A second-order effect that contributes to the in situ IR spectra consists of change in the interfacial structure of solvent, including the coordination to the surface, solute, and self-organization. Correlation between the structure of solvent and the HL ionic composition and the electrode surface properties is a considerable objective not only in electrochemistry but also in other numerous areas of science and technology dealing with surface modification. A number of systems have been studied to date, including acetonitrile [110, 115, 159, 179], acetone [110, 179], methanol [110, 180], and benzene [110] at a Pt electrode. However, particularly interesting but yet little understood is the most common solvent, water.

IR spectra of the water molecules at electrified interfaces are sensitive to the applied potential which controls the manner of the water adsorption and the strength of the water–surface bonding [181, 182]. In addition, it is now well understood theoretically [182–185] and confirmed using vibrational spectroscopy [186–189] that water forms an extensive three-dimensional network, and small localized perturbations in the chemical potential of water induce compensating changes in its properties that can propagate through the network over considerable distances from the point of origin of the perturbation. Water near a nonpolar surface has a high interfacial energy, and the chemical potential is equilibrated by an increase in the partial molar volume (a decrease in the water density). Water accomplishes this by forming a more complete icelike three-dimensional network of self-associated molecules near the surface than in the bulk water. The resulting energetic effect causes H bonds near a hydrophobic surface to break and may even move water away from the surface, leaving only a thin vapor interlayer (essentially the effect of drying) [190, 191]. Conversely, the H bonded network of water near a surface bearing Lewis sites that can compete with the self-association of water molecules collapses, causing an increase in the chemical potential that accommodates the decreased interfacial energy. This results in an increased density of water. Thus, the degree of self-association and the local density of water scales with the wettability of the surface. However, the influence of the solid extends over three water layers [192].

A pioneering IR spectroscopic study of interfacial water at a platinum electrode was reported by Bewick and co-workers [193]. Habib and Bockris [194] studied the potential dependence of the OH stretching band intensity at a platinum electrode using PM-IRRAS. Nuzzo et al. [195a] reported IRRAS of artificial quasi-boundary water on hydrophilic and hydrophobic surfaces of SAM-functionalized gold substrates. In both cases, the spectra testified that at temperatures <120 K water is in amorphous solid state, while at >130 K it is polycrystalline ice. The band of "free" hydroxy groups at 3700 cm^{-1} was observed for both substrates. However, its intensity was substantially higher for water on the hydrophilic surface. The same strategy was used by Bensebaa and Ellis [195b] and Engquist et al. [195c–195e]. A decrease in the number of H bonds to hydrophobic (covered by SAMs) substrates was detected by IRRAS when the D_2O overlayer was annealed from amorphous ice at 100 K to polycrystalline-like ice at 140 K [195d]. The spectral changes accompanying the structural transition were consistent with a change from a mainly flat overlayer to condensed three-dimensional clusters. These microscopic ice clusters qualitatively mimicked the shape of macroscopic water drops on the same SAMs. Faguy and Richmond [196] demonstrated that a combination of in situ real-time polarization modulation (RTPM) (Section 4.7) and conventional potential-difference spectroscopy (SPAIRS) allows better discrimination between surface and bulk water absorption. In the case of a copper electrode with adsorbed imidazole, these authors observed a distinct difference in the RTPM spectra of the interfacial water as compared to those for adsorbed glucose and thiocyonate. It was suggested that this difference can be rationalized in terms of different hydrophilicity of the electrode surface. Specific changes in the δH_2O band intensity and position in the in situ ATR spectra of the PbS–electrolyte interface upon adsorption of a hydrophobizing reagent are discussed in Section 7.5.2.

One can expect that water coordinates to a metal electrode through hydrogen or oxygen at potentials below and above the potential of zero charge (PZC), respectively, and forms icelike clusters near the PZC. To differentiate between these structures, the νOH band position can be used: the band shifts downward in the order O-down (>3400 cm^{-1}) > icelike structure ($\sim$3200 cm^{-1}) > H-down (2900–3100 cm^{-1}) (Table 3.4). Applying this correlation to the spectra of water adsorbed on the Pt(111)–0.1 M HClO$_4$ interface measured by in situ IRRAS, Iwasita and Xia [197] found that at $+0.35$ V (SHE) water orientation changes from H-down to O-down, indicating that the PZC of Pt(111) is close to this value. Similar results were earlier obtained by Toney et al. [192] for an Ag(111) electrode but with in situ X-ray scattering techniques. In the case of O-down coordination, the red shift of the δH_2O band was found [181] to scale with strengthening of the water–metal interaction unless the intermolecular H bonding is very strong. The largest downward shift (to 1520–1560 cm^{-1}) was observed for the water molecules adsorbed at the Ru(001) surface in UHV [181a]. Since the δH_2O band position is at 1600 cm^{-1} and 1610 cm^{-1} for Hg [156c] and Au [171] electrodes, respectively, it can be concluded [156c] that the water–metal bond weakens in the order Ru > Hg > Au. The red shift of the δH_2O band can be attributed to donation of electron density from the partly nonbonding $3a_1$-orbital of water (lone pair electrons) to the metal [181]. This orbital partly occupies a position between the two H atoms, producing a screening effect on the proton charges. As a result, donation of $3a_1$-electrons causes an increase in the H–O–H angle and a consequent decrease in the δH_2O frequency. The donation is also consistent with the red shifts of both the stretching and bending modes that were observed for the adsorbed water on the Pt(111) electrode as the electrode potential was increased from the PZC up to the onset of dissociation [197].

Ataka and co-workers [171] have presented the following interpretation of the absorption bands of water in the ATR-SEIRA spectra of the Au–0.5 M HClO$_4$ interface measured in the DL region from $+0.1$ to $+1.3$ V (SHE) (Fig. 3.41).

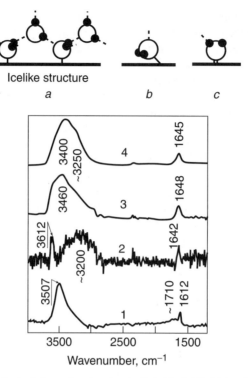

Figure 3.41. Proposed orientations of water at potentials (a) slightly higher than PZC, (b) below PZC, and (c) much higher than PZC. ATR-SEIRA of water at Au thin-film electrode in 0.5 M HClO₄ at *average* potential of (1) +0.12, (2) +0.77, and (3) +1.22 V (SHE). Spectra were acquired using Bio-Rad FTS-60A/896 FTIR spectrometer with resolution of 4 cm⁻¹ during one potential sweep from 0.1 to 1.3 V at rate of 5 mV · s⁻¹. Each spectrum was obtained during 15 s by coadding 100 interferograms. Therefore, indicated potentials are averages of every 75-mV interval. Reference spectrum obtained at +0.1 V. Spectrum (4) is transmission spectrum of 0.5 M HClO₄. All spectra are shown by scaling strongest bands to be equal. Adapted, by permission, from K. Ataka, T. Yotsuyanagi, and M. Osawa, *J. Phys. Chem.* **100**, 10664 (1996), p. 10667, Fig. 5. Copyright © 1996 American Chemical Society.

The bands at 3200 and 1642 cm⁻¹ are observed at a potential slightly higher than +0.55 V (PZC) (Fig. 3.41, curve 2). Because these peaks dominate in the spectra of ice, they have been attributed [171] to the coupled symmetric OH stretch mode of tetrahedrally H coordinated (highly ordered) water molecules (Fig. 3.41, structure *a*). The sharp peak at 3612 cm⁻¹ has been assigned to the dangling (free) bonds in the first monolayer of an icelike structure at the positively charged electrode surface. The positions and narrow widths of the bands at ~3500 and 1612 cm⁻¹ in the spectrum measured at a potential below the PZC (Fig. 3.41, curve 1) correspond to weak intermolecular H bonding when water interacts with the metal surface via a lone pair on oxygen (Fig. 3.41, structure *b*). These bands also indicate disorder in the molecular arrangement of water. A broad weak band at ~1710 cm⁻¹ is due to the doubly degenerate δ_{as}OH mode (ν_{4a}) of hydronium ion (H_3O^+). Asymmetrically H bonded water molecules [when only one OH moiety is H bonded to another water molecule (Fig. 3.41, structure c)] associated with electrolyte anions at the interface give absorption bands at 3300–3500 and 1648–1650 cm^{-1} (Fig. 3.41, curve 3).

To selectively resolve the boundary water near a nonmetal–water interface, Hasegawa et al. [199] employed the polarized ATR at different angles of incidence in conjunction with principal component analysis (PCA). The surface of a hemicylindrical Si IRE was cleaned by use of ozone cleaner, which yielded Si–OH species at the surface. The νOH band of the interfacial water was similar to that in Fig. 3.41, spectrum 2 which implies that such nanometer-scale information can be obtained on a nonmetallic surface without using the SEIRA effect. Based on the polarization dependence, this spectrum was assigned to H-down water that forms strong symmetric double H bonds with the hydrophilic surface. The surface-perturbed water layer was found to be several monolayers thick, in agreement with the X-ray scattering data [192].

The position of the absorption bands of water in the hydration spheres depends on the origin of the ions [200] (Table 3.4, Part A). Certain correlations were also observed between ability of adsorbed ions to structure the water of hydration and IR spectra of interfacial water (Table 3.4, Part B). For a series of aqueous

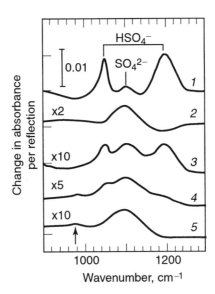

Figure 3.42. Detection of IR-non-active species: Ga^{3+} ions produced by GaAs dissolution in 0.5 M H_2SO_4 + 0.1 M H_2O_2. Change in ATR absorbance, in sulfate region, produced by (1) addition of 0.5 M H_2SO_4 to pure water, (2) pH increase from 0.3 to 0.5 at fixed total sulfate concentration, (3) addition of 20 mM Na_2SO_4 to 0.5 M H_2SO_4, (4) addition of 10 mM $Ga_2(SO_4)_3$ to 0.5 M H_2SO_4, and (5) 1 min GaAs dissolution at open-circuit potential. Adapted, by permission, from B. H. Erne, F. Ozanam, M. Shchakovsky, D. Vanmaekelbergh, and J.-N. Chazalviel, *J. Phys. Chem B* **104**, 5961 (2000), p. 5963, Fig. 4. Copyright © 2000, American Chemical Society.

solutions of NaX (X = F, Cl, Br, I, PF_6, BF_4, and ClO_4) a sharp absorption band between 3600 and 3500 cm^{-1} has been reported [156, 188], whose frequency decreases in the order $PF_6^- > BF_4^- > I^- > Br^- > Cl^-$, pointing to strengthening of the H bond of the hydrating water. A specific case is that of hydronium ion accumulated in the DL, which is characterized by a very broad absorption band centered at about 1710 cm^{-1} (Fig. 3.41, curve 1) and a weak feature at 1150 cm^{-1} [225, 200]. Coadsorption of metal atoms (ions) also red shifts the νOH frequencies of adsorbed water. The shift is weakest for Cs^+ (<30 cm^{-1} for νOH [201]) and strongest for K^+ ($\sim$500 cm^{-1} for νOH [202, 203] and 250–350 cm^{-1} for νOD [113, 115, 116]), which is consistent with the strongest Stark effect in the latter case due to the formation of the $K^+ \ldots e^-$ "dipole" [113, 115, 116]. One interesting phenomenon is the adsorbate-induced hydrophilicity reported by Kizhakevariam et al. [203], in which adsorbed chlorine atoms (probably as $Cl^{\delta-}$) cause the thermal stabilization of large numbers (4–13 per Cl) of water molecules on Ag(100). Some evidence for mutual solvation of $Cl^{\delta-}$–K^+ pairs was obtained non situ for the Pt(111)–water interface. Supplemental data on absorption of interfacial water can be found in Table 3.4. Comparing IR band positions collected in Table 3.4, it should be taken into account that the data obtained under different measurement conditions can differ due to optical effect. In particular, the δH_2O band of water measured by ATR in the IRE–Pt layer–water system (Kretschmann's configuration) is distorted [61a], as in the case of adsorbates (Figs. 3.23 and 3.67). The baseline is higher at the lower wavenumber side of the band and there is a steep offset at the higher wavenumber side. Also, the δH_2O band maximum shifts from 1641 cm^{-1} without Pt to 1617 cm^{-1} with 20 nm Pt on a 45° Ge IRE. Calculations showed that the band shape strongly depends on the optical constants of the IRE and metal film and the metal film thickness. The distortion increases with increasing thickness and the refractive index of the thin metal film and with decreasing the refractive index of the IRE, getting more pronounced when approaching the critical angle.

An example of misinterpretation of the absorption of interfacial water has been discussed by Chazalviel and Ozanam [204]. The in situ ATR spectra of the silicon surface rinsed in HF solutions of increasing concentration exhibit a decrease of the νSiH band at $\sim$2100 cm^{-1} and a new band at $\sim$2230 cm^{-1}. The latter was initially assigned [205, 206] to the SiH_2F_2 surface groups. However, the latter band is more likely to be due to the combination ($\delta + \rho$) mode of the interfacial water, as it is present in the ATR spectra of the Ge/HF–solution interface as well. This band appears to be very sensitive to the degree of organization of the water. It is located at around 2125 cm^{-1} in pure water (Fig. 1.4) and shifts to 2270 cm^{-1} in ice, with a corresponding increase in intensity. Chazalviel and Ozanam have shown that if the background correction is made using the ATR spectra of the Ge/HF–solution interfacial water, then the hydrogen coating of Si is unaffected by the increasing HF concentration.

Careful analysis of the electrolyte absorption allows one to detect IR-non-active species in the DL. An example that demonstrates the utility of this property is the IR study by Chazalviel and co-workers [146, 147, 178] of the Ga^{3+} ions produced

Table 3.4. IR frequencies of water in different states

νOH (cm^{-1})	δHOH (cm^{-1})	Sample, Method, and Reference
	A. Bulk Water	
3756 (ν_{as} OH) 3657 (ν_s OH)	1595	Gas [632]
3633–3604 (ν_{free}) 3524–3527 (ν_3) 3364–3382^b ($2\nu_2$) 3260–3209^b (ν_1) br (FWHM$^a \approx 500$), max at ~3350 cm^{-1}, with at least four components (see also Fig. 1.4)	1636 (ν_2) FWHM ≈ 100 2125 ($\delta + \rho$) [632]	Liquid pure water (ATR, ZnSe IRE) [633]
3627 (ν_{free}) 3552 (ν_3) 3427^b ($2\nu_2$) 3269^b (ν_1) br (FWHM ≈ 500), max at ~3375 cm^{-1}, with shoulder at ~3240 cm^{-1}		Liquid pure water (ATR, Si IRE) [200]
3400 3250	1645	Transmission of 0.5 M HClO$_4$ [171]
3350, 2900 2900	1170 1760 1710 [δ(O...H...O)], 1370 [δ(O...H...O)], 1170 [ν_{as}(O...H...O)]	ν_{as}(O...H...O) of H$_5$O$_2^+$ [634] H$_5$O$_2^+$ ion; ATR of 1.0 M HCl, ZnSe [635] H$_5$O$_2^+$ ion; ATR of 37.4% HCl [634]
3560 (νOH$^-$), 2900	1900 (δH$_2$O) 1550 (δOH$^-$), 1130 (δH$_2$O)	H$_3$O$_2^-$ ion; ATR of 48.1% KOH [634]
3560 (ν_1), 3510 (ν_3) 3385 (ν_1), 3470 (ν_3)	1600 (ν_4), 1095 (ν_2) 1700 (ν_4), 1635 (ν_4) 1635 (NaCl), 1633 (KCl) 1615 (NaI), 1617 (KI)	(H$_3$O)$^+$ SbCl$_6^-$ (solution) [636] (H$_3$O)$^+$ in liquid SO$_2$ [636] ATR of solutions [637]
3406, 3000–2900 3450, 3000–2900	1622 (NH$_4$Cl) 1610 (NH$_4$I)	ATR of solutions (ZnSe IRE) [633a]

Wavenumber (cm⁻¹)	Assignment
3600 (w), 3470 (w), 3000–2900; 3600–3450	Water solvating COO⁻, F⁻, SO₃⁻ groups (ATR, ZnSe IRE [633b])
3645	Water solvating amino groups and the neighboring aliphatic tail (ATR, ZnSe IRE [633b])
3680	Water–methanol mixture (ATR, ZnSe IRE [633b])
3450–3400 (sh) (H bond disorder); 3200–3220 (max) (complete tetrahedral coordination)	Bulk ice [187, 632, 636, 638, 639]
1703, 1665 (NH_4F)	
1650 (δ); 2125 ($\delta + \rho$) [632]	
1630–1670	Strong H bonds formed by two H atoms in $CaHAsO_4 \cdot 2H_2O$ [640]
1620–1580	Asymmetric system where one H atom forms a strong H bond, whereas the second one is not involved in H bonding in $ZnSO_4 \cdot 4H_2O$ [640] or coordination of oxygen to cation [181a]
1800–1500, br (FWHM $\approx$ 350–400) and irregularly structured band with peaks at ~1700 and ~1560 for $K_2Fe(SO_4)_2 \cdot 6H_2O$ Doublets: 1685.6, 1621.5, 1689.3, 1628.4, 1717.2, 1637.8	Anharmonic coupling structure in Tuton salts $M^I M^{II}(XY_4)_2 \cdot 6H_2O$, where M^I = K, NH_4, Rb, Cs; M^{II} = Mg, Ni, Co, Fe, Cu, Zn; X = S, Se, Be: Y = O when X = S or Se and F when X = Be; transmission of KBr pellets [641]
1658, 1720	$CaSO_4 \cdot 2H_2O$, $CaSeO_4 \cdot 2H_2O$, $YPO_4 \cdot 2H_2O$; transmission of KBr pellets [642]
Vbr (FWHM $\approx$ 800), max at 3300–3100	$CaHPO_4 \cdot 2H_2O$; transmission of KBr pellets [643]

Table 3.4. (Continued)

νOH (cm^{-1})	δHOH (cm^{-1})	Sample, Method, and Reference
Structured band with narrow (FWHM $\approx$ 100) components and center of gravity from ~3000 (MnAsO$_4$ · H$_2$O) to ~3250 (MnSO$_4$ · H$_2$O)	Multifold narrow (FWHM $\approx$ 20) peaks spread in region from ~1480 to 1740	Kieserite family M^{II}RO$_4$ · H$_2$O (M^{II} = Mg, Mn, Fe, Co, Ni or Zn; R = S or As); transmission of KBr pellets [644]
~3350 (FWHM$\approx$150)	1640 (w), 1510 (w)	CuSO$_4$ · H$_2$O [645]
	1660 (s), ~1525 (w), ~1430 (s)	MKPO$_4$ · H$_2$O: M = Mn [646],
~3350 (FWHM$\approx$80)	~1700 (w), ~1600 (w), ~1470 (s)	M = Co, Ni, Mg [646, 647] Band splitting and low δ H$_2$O frequency component attributed to Fermi resonance; transmission of KBr pellets
νOH^{Ic} (br, st): 3337 (ν_{TO})	δH$_2$O^{I,III} (m):	NaAl(SO$_4$)$_2$ · 12H$_2$O(γ)^d,
νOH^{III} (vbr, s): 2995 (ν_{LO}), 2937 (ν_{TO})	1625 (δ_{LO}), 1615 (δ_{TO})	
νOH^{I} (br, s): 3387 (ν_{TO})	δH$_2$O^{I,III} (m):	NaAl(SO$_4$)$_2$ · 12H$_2$O(α)^d,
νOH^{III} (vbr, s): 2990 (ν_{LO}), 2941 (ν_{TO})	1645 (δ_{LO}), 1635 (δ_{TO})	
νOH^{I} (br, s): 3383 (ν_{LO}), 3353 (ν_{TO})	δH$_2$O^{I,III}(m):	NaAl(SO$_4$)$_2$ · 12H$_2$O(β)^d; KK transform of specular reflection of single crystals [648]
νOH^{III} (vbr, s): 3017 (ν_{LO}), 2972 (ν_{TO})	1650 (δ_{LO}),	
νOH^{III} (sh): ~2800 (ν_{LO}), ~2800 (ν_{TO})	1650 (δ_{TO})	
3285 (ν_1), 3100 (ν_3)	1577 (ν_4), 1175 (ν_2)	[OH$_3$]$^+$ ClO$_4^-$ (solid) [636]
2780 (ν_1), 2780 (ν_3)	1680 (ν_4), 1175 (ν_2)	[OH$_3$]$^+$ NO$_3^-$ (solid) [636]
	B. Interfacial Water	
$\approx$3700$^{a)}$	1595	Water monomer, SHG of air–sulfate solution interface [649]
3700		"Free" OH of ice near the carboxylic acid or methyl terminated Au surface at 80–145 K; non situ IRRAS [195a]

Water dimer at a Pt electrode in 1 M HCl [650]	1600, 1620	3570–3580
Water in diffuse part of DL [651][e]	1650	3400–3150
Water adsorbed directly at Au electrode [651][e]		3500
Ex situ IRRAS [652]:		
Fe(110)	1630	3380
Ru(001)	1520	3400
Ni(110)	1610	3350
Pd(100)	1605	3405
Pd(100)	1645	3460
Pd(100)	1597	—
Pt(100)	1630	3380
Pt(111)	1625	3400
Cu(100)	1589	—
Ag(110)	1660	3410
O/Ag(110)	1590	3230
Si(111) × (2 × 1)	1575	3385
Al(100)	1655	3510
Al(111)	1655	3445
Water co-adsorbed with anions on Au from [651][e]:		
LiCl		3698
LiBr		3610
KCl		3582
KBr		3595
KI solution		3588
OH...M bonded; ex situ IRRAS [198, 181, 653] in situ IRRAS [197]	1600–1610	~2900 / ~3000
O-down orientation on Pt(111); ex situ IRRAS [632]	1625	3400

199

Table 3.4. (Continued)

νOH (cm^{-1})	δHOH (cm^{-1})	Sample, Method, and Reference
3150–3180		O-down orientation to Pt(111); in situ IRRAS [197]
3270		Co-adsorbed OH$^-$ ions Pt(111); in situ IRRAS [197]
3442–3388	1758–1788	H$_3$O$^+$ on Pt(111) in 0.5 M H$_2$SO$_4$; in situ IRRAS [654]
	1710	H$_3$O$^+$ paired with ClO$_4^-$ on Au(111) [171]^e
3570	1625	In situ IRRAS [156d] of Au(111)–0.5 M HF +
3555		0.05 M NaI,
3525		0.05 M NaBr,
		0.05 M NaCl
		In situ IRRAS [156a] of Au(111)–
3633		0.1 M KPF$_6$
3626		0.5 M NaBF$_4$
3612		0.5 M NaClO$_4$
3620, 3492	1690	Hg in 0.1 M HClO$_4$ at +0.3 V (AgCl); in situ IRRAS [156b]
3507	1612	Weakly H bonded water at negatively charged Au [171]^e
3612 (nar), 3200 (br)	1642	Ice-like structure on Au(111) in 0.5 M HClO$_4$ at a potential slightly above PZC [171]^e
3612 (nar), 3217 (br)		Strong H–O–H$\cdots$O bonds with a Si surface; ATR [199]
3650, 3150	1650	HOO$^-$ (H$_2$O$_2$) on TiO$_2$; in situ IRRAS [655]
3100–3500 (vbr, structured)		Water in primary hydration sphere of Li$^+$ in DL; in situ IRRAS [156b]

3100–3300 (vbr, structured)		Water in primary hydration sphere of Mg^{2+} in DL; in situ IRRAS [156b]
	1660 (δ_{as})	H_3O^+ paired with NO_3^- on Au(111); in situ IRRAS [158]
	1150 (δ_s), 1705 (δ_{as})	H_3O^+ on Pt(111) [225, 656, 657]
3430	1633	H_2O co-adsorbed with H_3O^+ on Pt(111); in situ IRRAS [225]
~3350 (FWHM ≈ 250) ~3250 with shoulder at ~3400		Ice near the carboxylic acid or methyl terminated Au surface; non situ IRRAS [195a]; amorphous solid below 120 K, polycrystalline at 130–145 K
3475 (FWHM ≈ 250)	1652 (FWHM ≈ 70)	Ordered structure of H_2O coadsorbed with Cu and SO_4^{2-} on Au(111) [410b]
3386–3396	1652	Ice bilayer on sulfate ion on Pt; non situ IRRAS [225]
3697 (vnar), 3612 (nar), 3386 (br)	1652	Ice bilayer on (4 × 2)–3CO on Pt(111); non situ IRRAS [227]
3662–3400	1666	Water on (2 × 2)–3CO on Pt(111) in 0.1 M H_2SO_4 at +0.4 V (SHE); non situ IRRAS [227]

[a] Full width at half maximum in cm^{-1}.
[b] Fermi doublet of ν_1 and $2\nu_2$.
[c] Superscripts I and III refer to water molecules coordinated to M^I and M^{III}, respectively.
[d] Crystallographic class.
[e] ATR-SEIRA.

by the dissolution of GaAs in an acidic sulfate medium. Curves 1–4 in Fig. 3.42 present the ATR spectra of reference solutions. Addition of Na_2SO_4 (curve 3 in Fig. 3.42) leads to an identical increase of the SO_4^{2-} and HSO_4^- concentrations, whereas addition of $Ga_2(SO_4)_3$ (curve 4) leads to an increase of SO_4^{2-}, due to preferred association of Ga^{3+} with the SO_4^{2-} anions. The latter effect is also manifested by the small peak at 980 cm^{-1}, attributable to the breathing mode of SO_4^{2-}, which becomes weakly IR active due to symmetry breaking. Spectrum 5 in Fig. 3.42 measured in situ at the GaAs–electrolyte interface is a combination of spectrum 4 and the spectrum of sulfate ion (curve 2), which provides evidence for an increase in pH and Ga^{3+} formation. It is possible to apply this method quantitatively provided the IR signals in solution are preliminarily calibrated.

Thus, the IR spectra measured in situ are affected by a change in composition, concentration and reorganization of electrolyte ions in the path of the IR radiation, and reorganization of the solvent that solvates these ions and the surface.

3.7.3. Donation/Backdonation of Electrons

Upon adsorption onto a metal, lone-pair electrons of the adsorbate may be donated to empty metal orbitals of appropriate symmetry (*σ-type overlap*). Alternatively, the metal may backdonate electrons from filled *d*-orbitals to the lowest unoccupied molecular orbitals (LUMO) on the adsorbate. When a molecule is adsorbed onto a clean, uncharged metal surface, its vibrational frequency may decrease from the frequency of the free molecule if the LUMO is antibonding and the backdonation is dominant. When the charge on the metal is made negative, the bond is weakened due to further backdonation and the band frequency further shifts to lower wavenumbers. Backdonation to the empty π^*-orbitals is typical for CO adsorbed on Pt electrodes [107, 202, 207], where it favors filling bringing rather than terminal sites. If the donation interaction is dominant, the mode frequency changes depending on the bonding character (localization) of the lone pair and the vibration form. Since the upper filled 5σ-orbital of CO is slightly antibonding, the νCO frequency increases upon the donation [104], while both the νOH and δH_2O bands of the O-down adsorbed water shift toward lower wavenumbers (Section 3.7.2). The donation effect is observed with an increase in potential, because the Fermi level of the electrode is downshifted as the potential is increased with respect to the molecular levels. However, the situation is not so simple [208]: The electrode potential influences many factors that determine the absorption frequency of the CO stretching mode, such as the electronic interactions (donation and backdonation), Stark effect (see below), the steric repulsion, and adsorption geometries, which affect in different manners the frequencies of CO adsorbed in different configurations. As a result, it is almost impossible to decompose the observed frequency shift into each contribution experimentally without theoretical consideration.

3.7.4. Stark Effect

The potential drop across the Helmholtz layer (the dense part of the DL) typically ranges from 0.1 to 1 V, and because this occurs across a very short distance (nanometers) the field strength may be up to 10^8–10^9 $V \cdot m^{-1}$. Such fields may be strong enough to perturb vibrating dipoles [118, 119, 202, 209]. This so-called vibrational Stark effect can manifest itself in the in situ IR spectra of adlayers at fixed coverage as a linear blue shift[†] of the band with increasing electrode potential. Typical values of the frequency shifts are 30 $cm^{-1} \cdot V^{-1}$ for CO at polycrystalline Pt [210], 10–20 $cm^{-1} \cdot V^{-1}$ for CN^- and SCN^- on poly-crystalline Pt and Pd [211], and 190 $cm^{-1} \cdot V^{-1}$ for carbonate ions [212]. This effect can be accompanied by a decrease in the band intensity, as observed in the spectra of CN adsorbed on Pt(111) at potentials higher than the potential of the transition from N- to C-bound species (in the latter the dipole moment is oriented against the electric field [213]). Application of a strong static electric field can distort polarizable electrons in a molecule, resulting in an induced dipole moment perpendicular to the metal surface (exciting of a forbidden mode), such as occurs in pyrene [214, 215] and tetracyanoethylene [216] on a Pt electrode. The Stark effect can be first or second order for a molecule with a static dipole moment or inversion symmetry (e.g., SO_4^{2-} [217]), respectively [218]. Unfortunately, in practice, it is difficult to ascribe a priori the observed potential-dependent frequency shift to either the Stark effect or backdonation [140, 202]. Moreover, ab initio and semiempirical calculations based on these hypotheses are unable to describe the experiment, as neither model can account for the relatively large shifts observed [219]. The "chemical model" does not explain the effects observed when the electrolyte is varied, and the "pure" Stark effect model is inconsistent with the data for CO-saturated Pt [220]. To distinguish the electric field effects, model calculations that take into account the structure of the double layer have been developed [113].

3.7.5. Bipolar Bands

Thus, changing the electrode potential can change not only the population of a species in the DL but also parameters of the elementary oscillators that characterize its absorption. In addition to the dipole–dipole coupling effects (Section 3.6), a mode perpendicular to the surface at the fixed coverage can be perturbed by the donation/backdonation and Stark effects, which results in the bipolar form of its band. To illustrate the manifestation of these effects in the IR spectra measured in situ, Fig. 3.43a shows a set of spectra for Ru(0001) in 0.05 M H_2SO_4, taken at 0.1-V intervals, starting from -0.01 and ranging up to $+0.79$ V (SHE) [150]. A well-defined, bipolar peak at 1250–1280 cm^{-1} that blueshifts with increasing potential is observed. As previously proposed for Pt(111) [221, 222], this band can be assigned to adsorbed bisulfate HSO_4^-(ad) or a sulfate–hydronium ion pair. At potentials above $+0.4$ V, where the Ru(0001) oxidation starts, a positive-going band at 1051 cm^{-1} due to the totally symmetric stretching ν_1(sol) band of

---

[†] The exception is CO on some metal surfaces, where potential-induced perturbations of the surface binding geometry result in abrupt changes in vibrational frequencies [S.-L. Yau, X. Gao, S.-C. Chang, B. C. Schardt, and M. J. Weaver, *J. Am. Chem. Soc.* **113**, 6049 (1991); I. Villegas and M. J. Weaver, *J. Chem. Phys.* **101**, 1648 (1994).]

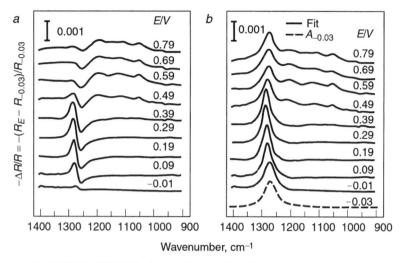

Figure 3.43. (a) IRRAS-SNIFTIRS obtained from Ru(0001) electrode in 0.05 M H$_2$SO$_4$. Reference spectrum obtained at -0.03 V versus SHE and sample spectra taken every 0.1 V, from -0.01 to $+0.79$ V; 4096 scans were coadded in 16 cycles, 256 scans each; resolution was 8 cm^{-1}. Spectra are offset for clarity. (b) Same spectra after adding spectrum at reference potential of -0.03 V. Reprinted, by permission, from N. S. Marinkonic, J. X. Wang, H. Zajonz, and R. R. Adzic, *J. Electroanal. Chem.* **500**, 388–394 (2001), pp. 391, (Fig. 3) and 392 (Fig. 5). Copyright © 2001 Elsevier Science B.V.

the HSO$_4^-$ species in solution appears in the spectra, while the negative-going part of the bipolar band centered at 1248 cm^{-1} becomes most pronounced.

To interpret these features, one needs to recall that the in situ spectra are differential (i.e., are related to the spectrum taken at some potential) (Section 4.6.4). A unipolar, potential-dependent band is obtained in a potential different spectrum if adsorbed species are present at a sample potential E_s and absent at the reference E_r. If the adsorbed species is present at both potentials and the frequency of its mode is potential dependent, this mode is characterized by a bipolar band. The absence of a band indicates the lack of a measurable change in the population or the parameters of the elementary oscillators of the mode associated with this band. Since the spectra are represented in $-\Delta R/R$ units, features pointing up are due to species gained at the sample potential (Section 4.6.4).

The bipolar shape of the band in the 1250–1280-cm^{-1} range due to adsorbed bisulfate implies that a substantial coverage of bisulfate is attained already at the reference potential of -0.03 V and does not change up to $+0.39$ V. The presence of the adsorbed bisulfate is manifested only by the Stark and/or donation tuning; otherwise the absorption would be invisible. The surface oxidation at potentials higher than $+0.4$ V causes desorption of bisulfate and the corresponding increase in concentration of bisulfate anion in the diffuse part of the DL. This is responsible for the observed increase in the negative lobe of the bipolar band at 1248 cm^{-1} and the appearance of the positive band at 1051 cm^{-1} at potentials equal to or higher than 0.49 V.

Since bipolar bands are difficult to analyze quantitatively, Marinkovic et al. [150] suggested a numerical procedure to transfer such a band into the unipolar one. The procedure is based on the fact that the reference spectrum is the same for a set of potential-difference spectra. Therefore, the three parameters of the elementary oscillator for a band in the virtual absolute spectrum taken at the reference potential can be obtained by fitting the potential-difference spectra of the adsorbate at almost the same coverage. Figure 3.43b shows the sum of the measured potential difference spectra and the reconstructed "absolute" absorption band of the bisulfate adsorbed at the reference potential $A_{-0.03}$. In this way, the unipolar band near 1280 cm^{-1} represents the absorbance of the adsorbed bisulfate. It is now obvious that the bisulfate adsorption occurs over the whole potential region investigated. The weak intensity of the bipolar band at -0.01 V in Fig. 3.43a results from a small band shift in the potential range from -0.03 to -0.01 V rather than from the onset of adsorption of bisulfate. Analogously, the lack of a positive-going band near 1280 cm^{-1} at potentials above $+0.49$ V results from a red shift of the center frequency in the $-\Delta R/R$ spectra, due to a replacement of the adsorbed bisulfate by hydroxyl species (dipole–dipole decoupling), not from a complete desorption of bisulfate.

3.7.6. Effect of Coadsorption

Comparing spectra measured in situ and ex situ, one should bear in mind that coadsorption of a solvent species can change the spectrum of the adsorbate in terms of the band position, intensity, and width and cause the band splitting. The effect of solvent on IRRAS of adlayers has been studied using the so-called non situ strategy, or UHV double-layer modeling, developed by Sass and co-workers during the 1980s [223, 224] and Weaver and co-workers (see review in Ref. [113]) and Ito and co-workers [225–227] during the 1990s. The approach consists of modeling the DL in UHV by dosing onto a clean metal surface of the DL components (solvent, reagent, and electrolyte ions). It was found [110, 115, 116, 120] that solvation of a saturated CO adlayer with methanol, acetonitrile, acetone, benzene, and ammonia results in attenuation, broadening (by 5–20 cm^{-1}), and red shift (by 15 cm^{-1} for water and 50 cm^{-1} for ammonia) of the atop and bridging νCO bands from the case of UHV surroundings. As for bulk solvation, the broadening can be associated in part with vibrational and orientational relaxation [228]. The red shift and intensity attenuation can be explained by dipole–dipole coupling (Section 3.6) and/or a combination of the electrostatic Stark effect with a solvent-induced increase in the $d\pi \rightarrow 2\pi^*$ backdonation [110, 113, 115, 116, 120]. Significantly higher saturation coverage of CO adsorbed on Pt(111) at low potentials ($\theta = 0.75$) as compared to the case of UHV ($\theta = 0.5$) implies stabilization of a compression CO structure by water coadsorption [227, 229]. At the same time, this effect is not observed for Ir(111) [85]. Solvation of CO [110, 113, 115, 116, 118, 120] or NO [114] adlayers at low coverages can cause displacement of atop configuration by bridging and/or threefold hollow configuration. This effect was explained [110] by a more favorable electrostatic interaction between the multidentate configuration and adjacent oriented solvent species than in the case of the less polar atop geometry. The absence of solvent-induced site transfer for saturated CO adlayers was attributed [113] to the fact that the solvent molecules are constrained to lie on top of such an adlayer, rather than be coadsorbed alongside the admolecules. Coadsorption of potassium with CO and a solvent on Pt(111) induces a red shift of the atop and bridging νCO bands [116]. This shift was shown [115, 116] to be consistent with the Stark effect, which is modified by charging the surface negatively via formation of a K^+–electron pair. In the case of ammonia, K^+ cations cause reorientation [230].

3.7.7. Electronic Absorption

Infrared radiation interacts not only with the vibrational excitations of the material but also with the free carriers (Section 1.2), creating phenomena such as free-carrier absorption, excitation across the energy gap, exciton transitions, or light scattering by free electrons [231]. Analysis of these phenomena can clarify the mechanism of charge transport across the semiconductor–electrolyte interface, which can take place either through changes in the space charge and the associated band bending or through changes in the population of electronic interface states [146, 147].

If the concentration of free carriers in the space charge layer of a semiconductor electrode changes upon scanning of the potential, the background absorption also changes [232–234]. Such an absorption is referred to as Drude absorption (DA) (1.3.22°). The ATR method was developed by Harrick in connection with measurements of DA [235]. The in situ ATR spectra of the interface between a low-doped n-type Ge electrode and 1 M $HClO_4$ during negative and positive scans are shown in Fig. 3.44a [175, 174]. Besides weak bands at 1100 and $\sim$1630 cm^{-1} from the solution, a positive band at $\sim$2000 cm^{-1} (hydrogen vibrations), and a negative band at 3200 cm^{-1} (vibrations of surface hydroxyls and coadsorbed water), a smooth increase in the background toward lower wavenumbers associated with DA is observed.

More accurate conclusions can be drawn if DA is evaluated as a deviation of the background absorption at a specific wavenumber [236, 237]. The same is true if DA is simply measured through an optical filter without scanning the interferometer [238], and afterward calibrated in terms of the changes in free carrier concentration, if the doping level of the semiconductor and free-carrier IR cross section are known [234]. In general, at increasing potential, the DA will be positive if holes are added to the space charge layer (onset of the accumulation regime in the case of a p-type semiconductor or onset of the inversion regime for a narrow-band n-type semiconductor) and negative when electrons are subtracted from the space charge layer of a n-type surface. In the absence of any surface effect, a steep increase of free-electron concentration is expected during the negative potential scan for an n-type semiconductor electrode at the flat-band potential, which is typically more negative by E_g/e than the onset of the inversion regime during the positive scan [238, 239].

The difference of the spectra of holes and electrons in Ge, which is caused by the presence of intravalence band optical transitions in the former

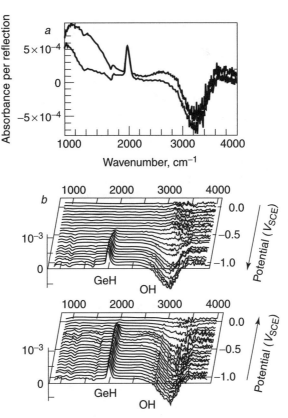

Figure 3.44. (a) ATR-SNIFTIRS (p-polarization) for (100) Ge surface in 1 M $HClO_4$ electrolyte at -1.025 V (Saturated calomel electrode, SCE) during negative scan (lower trace) and at -0.225 V (SCE) during positive scan (upper trace). Smooth background at low frequencies arises from free-electron and free-hole absorption, respectively. (b) ATR spectra (p-polarization) taken using Bomem MB 100 FTIR spectrometer at 4 cm^{-1} resolution during potential scan. Reference spectrum was obtained at positive potential bound. Spectra were recorded every 5 s by averaging three interferograms. Full potential scan was repeated typically 30 times, and spectra of various scans recorded at same potentials were coadded to improve SNR. Smooth background corresponding to free-carrier absorption present in (a) was subtracted from spectra. Adapted, by permission, from F. Maroun, F. Ozanam, and J.-N. Chazalviel, *Surf. Sci.* **427–428**, 184 (1999), p. 185, Fig. 1 and p. 188, Fig. 3. Copyright © 1999 Elsevier Science B.V.

case (Fig. 3.44a), allows one to differentiate the DA of holes and electrons. Figure 3.45b shows DA recalculated into the surface concentration of free electrons and holes as a function of potential. During the negative scan, the free-electron concentration increases smoothly, with maxima at the onset of the α, β, and γ peaks in the voltammogram (Fig. 3.45a), which were attributed to the formation of the GeH_2 and GeH species and breaking of the Ge–Ge bond by penetrated hydrogen, respectively. A steep increase in absorption by free electrons is observed at the onset of hydrogen evolution. During the positive

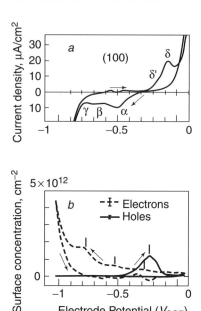

Figure 3.45. (a) Current–voltage curve of (100) Ge electrode in 1 M HClO$_4$ electrolyte taken when spectra shown in Fig. 3.44 were measured. Scan rate 10 mV · s^{-1}. (b) Free-electron and free-hole absorption during potential scan under same conditions as in Fig. 3.43. Adapted, by permission, from F. Maroun, F. Ozanam, and J.-N. Chazalviel, *Surf. Sci.* **427–428**, 184 (1999), p. 185, Fig. 1 and p. 188, Fig. 3. Copyright © 1999 Elsevier Science B.V.

scan, free-electron concentration drops to zero when no hydrogen is evolved, and a free-hole peak is observed at the anodic peak onset. The shift in the band edge that is observed has been attributed predominantly to a change in the surface charge balanced by an ionic charge in the electrolyte, rather than to a dipolar effect associated with the GeH-to-GeOH substitution favored in the literature.

Erne et al. [240] have recently shown that the potential dependence of the DA under the depletion conditions is similar to the Mott–Schottky line, which allows determination of the flat-band potential of the electrode. Other examples of employing DA in spectroelectrochemical studies are analysis of the relaxation process of free carriers in semiconductor electrodes [241] and elucidating the mechanisms of interfacial processes (Section 7.5).

Because of another structure of the energy bands, charge carrier absorption in conducting polymers is characterized by a more complex absorption. For illustration, Fig. 3.46 shows IRRAS of conducting polythiophene films 70–100 nm thick on Pt in acetonitrile [242]. The spectra are represented in $\Delta R/R$ units, with the negative-going bands representing species gained at the sample potential. The main negative-going features of these spectra are a broad absorption at wavenumbers above 1500 cm^{-1} and several bands in the 1500–700-cm^{-1} region. These features are assigned to *(bi)polaron*-type charge carriers: The broad absorption in the high-wavenumber region is due to transition from the

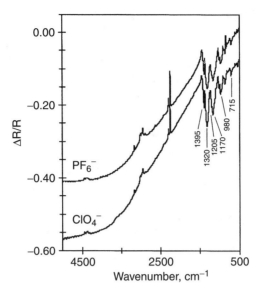

Figure 3.46. In situ IRRAS of poly-(3-methyl thiophene)(ClO_4^-) film in Bu_4NClO_4 in acetonitrile at +0.35 V (with offset of −0.1 reflectance units) and poly-(3-methyl thiophene)(PF_6^-) film in Bu_4NPF_6 in acetonitrile at +0.65 V. Reference spectrum recorded at −0.35 V. Spectra were recorded using Bruker IFS-113v FTIR spectrometer with MCT detector. Spectral resolution was 4 cm^{-1} and number of scans for each spectrum was 128. Reprinted, by permission, from E. Lankinen, G. Sundholm, P. Talonen, T. Laitinen, and T. Saario, *J. Electroanal. Chem.* **447**, 135–145 (1998), p. 141, Fig. 6. Copyright © 1998 Elsevier Science B.V.

valence band to the lowest subgap state, and the latter are caused by the coupling of this electronic excitation to the lattice vibrations of the polymer. Christensen et al. used the absorption above 1500 cm^{-1} due to interband transitions to estimate the conductivity of polybithiophene films [243].

Of particular interest for electrochemists is the possibility of monitoring surface states that are located within the band gap. Being of varied origins, these states play an important role in the charge transfer process [137, 138]. Surface state absorption generally originates from transitions of electrons between surface state levels in the gap and either of the two bands. The low-frequency threshold of the surface state absorption corresponds to the energy separation between the surface state level and the relevant band (Section 6.5). When the potential of a semiconductor electrode is chosen to be near the flat-band potential, DA is minimized, and absorption associated with surface states (localized charge carriers) appears. This absorption can be distinguished by the differential technique, in which the base spectrum is obtained so that the surface states are a priori known to be completely empty or populated [233, 244]. For example, surface state absorption of the silicon–electrolyte interface manifests itself as a background absorption at wavenumbers larger than 4000 cm^{-1} (Fig. 3.47) [237, 245, 246]. Since the shift of this part of the background spectrum has been found to correlate with changes in the absorption of the SiOH groups upon varying the

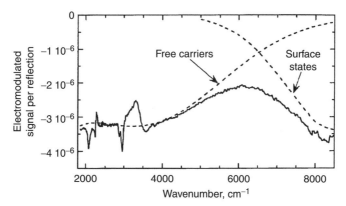

Figure 3.47. Typical Fourier transform electrochemically modulated IR spectroscopy (FTEMIRS) ATR spectrum of *n*-Si in acetonitrile–0.1 *M* tetrabutylammonium perchlorate. Baseline can be analyzed as sum of free-electron and surface state contribution. These two contributions appear in different proportions when potential modulation range is changed, giving direct information on splitting of modulated charge between space charge and surface states. Adapted, by permission, from A. Venkateswara Rao, J.-N. Chazalviel, and F. Ozanam, *J. Appl. Phys.* **60**, 686 (1986), p. 699, Fig. 3. Copyright © 1986 American Physical Society.

electrode potential, this background absorption has been attributed to populating the SiOH surface states [245, 246].

Thus, a potential-dependent smooth background in IR spectra can be due to modulation of concentration/population of free carries and surface states. In contrast to common capacitance measurements, IR spectra of free carrier can provide information on the space charge capacitance of the semiconductor electrode under accumulation as well as under depletion conditions, without interference from the surface state capacitance. The DA–potential plots under depletion conditions allow measurements of the flat-band potential, while absorption of charge carriers in conducting polymer films can be used for estimating the film conductivity. It is also possible to follow the filling of surface states and relate these energy levels to the chemical composition of the interface.

3.7.8. Optical Effects

In all the studies mentioned, the optical effect caused by the accumulated products of any electrode reactions (formally, by the appearance of an ultrathin film at the interface) was ignored. This effect produces the baseline distortions in the regions of the absorption bands of the solvent (Fig. 3.27). Using the air–water interface as an example, Buffeteau et al. [247] demonstrated that to reproduce these distortions correctly, the optical constants of interfacial rather than bulk water should be used in the spectrum simulations. As underlined by Chazalviel et al. [146, 147] for ATR spectra of a film at the semiconductor–water interface, the uncompensated absorption of the solvent due to increasing the film thickness can be either negative or positive. If the film is compact and has a low refractive index (this is typically the case for an oxide film), total reflection takes place at the semiconductor–film interface, and the electric field of the evanescent wave in the electrolyte is decreased. This effect leads to decreased electrolyte absorption, as observed upon the growth of anodic SiO_2 on a Si electrode (Fig. 7.30). The opposite behavior may occur if the film is porous and made of a high-index material. In particular, the formation of porous Si on Si (Fig. 7.32b) [248–250] or porous arsenic hydride on GaAs [178] yields increased electrolyte absorption, which has been ascribed to enhancement of the IR electric field in the electrolyte-filled pores.

The optical effect can also arise from potential/adsorption-induced changes in the electrode optical properties. In particular, electrochemical processes can modify morphology and, hence, optical properties of SEIRA electrodes, resulting in a significant change in the band intensity and shape in ATR- (Fig. 3.23a and Section 3.9.4) and IRRAS-SEIRA spectra not only of adsorbates (Sections 3.9.4 and 3.9.5) but also of solution (Sections 3.7.2). It was revealed [61a] that this effect in ATR-SEIRA is more pronounced for stronger absorbers (both the solution and the film) and at the angles of incidence closer to the critical angle. In the case of a metal electrode, the background reflectivity changes with electrode potential, as the concentration of free electrons in the skin layer changes [251]. The background shift resulting from diffusive scattering of conducting electrons by adsorbates has been treated theoretically by Persson [252]. Reversible changes in the background reflectance level were observed by in situ ATR-SEIRA spectroscopy [253, 254]. Since these changes were parallel to surface coverage of a chemisorbate, they were attributed [254] to the chemisorption-induced changes in the optical properties of the Au-film electrode. An upward slope of the high-wavenumber background was observed in the ATR-SEIRA spectra upon adsorption of different species on Au-film electrodes (see Fig. 3.65 later). This slope can be assigned to a variation in the optical properties of the islandlike electrode film–adlayer effective medium generated by replacement of interfacial water by an adsorbate or a change in the adsorbate–substrate bonding (for more detail, see Section 3.9.4).

Background slopes toward higher wavenumbers were also observed in the spectra of the semiconductor–electrolyte interface, in particular, in ATR for a GaAs electrode at the reduction potential [255]. By analogy with the ATR spectra of discontinuous Ag deposits on Si (see Figs. 3.58 and 3.59 later), this background behavior was explained in terms of SPR generated by the islandlike metallic Ga deposit. In contrast, based on the dependence on the electrode potential, a similar background observed in the ATR spectra of the PbS–solution interface (Figs. 7.35, 7.41, and 7.43) was assigned to the surface layer defects created during polishing of the electrode [256].

Dissolution/deposition of the electrode material can cause roughening of the electrode surface and, as a consequence, an increase in the diffuse-reflectance component in the reflected radiation. As illustrated by Figs. 3.48 and 7.32a, similar to optical effect and surface and space charge layer states, surface roughness generates an increase of the spectrum background toward larger wavenumbers. Although no quantitative analysis of this feature has been performed yet, there

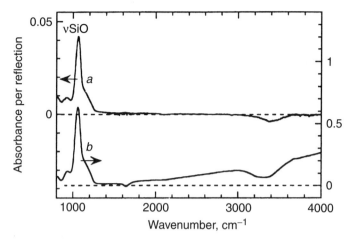

Figure 3.48. Surface roughening of p-Si during anodic dissolution in transpassivation regime Electrolyte is 1 M NH_4Cl + 0.025 M HF + 0.025 M NH_4F. Potential (a) +5 V (slow electropolishing) and (b) +40 V (transpassivation). Reprinted, by permission, from J. N. Chazalviel, B. H. Erne, F. Maroun, and F. Ozanam, *J. Electroanal. Chem.* **502**, 180–190 (2001), p. 188, Fig. 9. Copyright © 2001 Elsevier Science B.V.

is an opinion [146, 147] that it must be feasible. In addition, roughening can yield surface enhancement. For example, Maroun et al. [174] observed a three-fold increase of the surface-averaged sensitivity of ATR on a ripped Ge surface with roughness much less than IR wavelength as compared to that for a flat surface at the same amount of adsorbate. Calculation of the electric field map at a microfaceted Ge surface–electrolyte interface revealed that the local electric field is enhanced in the vicinity of surface pits, and significant enhancements in sensitivity can be expected from (111) microfacets of a (100) surface.

3.8. INTERPRETATION OF DYNAMIC IR SPECTRA: TWO-DIMENSIONAL CORRELATION ANALYSIS

Spectra measured using phase-sensitive detection when a periodic perturbation is applied (a *frequency domain*) as well as spectra measured as a function of time (a *time domain*) or any other state parameter of the system (e.g., a temperature, pressure, strain, distance, concentration domain) are referred to, by convention, as *dynamic spectra* to distinguish them from ordinary static (average) spectra. Dynamic spectra can offer an advanced opportunity for separating contributions of different subsystems to spectra of complex systems and quantifying the characteristic half-lives of these subsystems. This opportunity is especially valuable when complex spectra such as the spectra of electrode–electrolyte interfaces and biological films are interpreted. Below we will consider interpretation of dynamic IR spectra, while the technical side of the problem is discussed in Section 4.9.

Frequency-Domain Spectra. The principle behind extraction of dynamic information from absorption modulation spectra is based on the fact that when a system is periodically perturbed to generate a specific species, concentration of this species is modulated with the same frequency, whereas all inert molecules remain unaffected. Since the phase delay of the signal due to the active species depends upon its half-life, different species affected by the stimulus are characterized by different phase delays and frequency dependences. At the limiting frequency corresponding to its half-life, a reaction intermediate becomes inactive in the modulation spectrum, which may be especially useful in studies of reaction mechanisms through the identification of intermediates.

The field of application of the absorption modulation method strategy is applicable when the external periodic perturbation is of any type, including mechanical, electrical, thermal, or chemical [257, 258]. Chazalviel and co-workers [245, 246, 259–261] and then Griffiths and co-workers [262–264], and Ataka and co-workers [265] have demonstrated by spectroelectrochemistry that the absorption modulation method can be used not only for increasing the SNR and canceling out the solution background (Section 4.9.2) but also for measuring the kinetic and dynamic characteristics of the process. Due to the conceptual similarity to electrochemical impedance spectroscopy [266], the method of analyzing potential-modulated spectra to differentiate and quantify various contributions to an electrode process was termed the generalized electro-optical impedance method by Chazalviel et al. [146, 147].

Modulation spectroscopy provides three techniques for discriminating bands in the in situ spectra of the electrode–electrolyte interface: correlation analysis, phase rotation, and the changing modulation frequency. To illustrate the correlation analysis technique, Fig. 3.49a shows the absorption spectrum for a phase setting of $40°$ measured by Fourier transform electrochemically modulated IR spectroscopy (FTEMIRS) in the ATR geometry for the n-Ge–(1 M $H_2SO_4 + 10^{-2} M$ α-$H_4SiW_{12}O_{40}$) interface [267]. The modulation frequency was 125 Hz, while the modulation range was from -0.15 to -0.35 V (SCE). This spectrum exhibits bipolar bands due to the νWO modes of the redox species $(SiW_{12}O_{40}{}^{4-} \Rightarrow H_4SiW_{12}O_{40}{}^{5-})$ and the νSO modes of the DL species, which are generated/consumed upon the potential change. To separate the contributions of the Faradaic and non-Faradaic processes to the redox process, the two-dimensional phase map, showing isolevel lines of the spectrum as a function of wavenumber and phase settings (levels are chosen as $3 \times 10^{-6} \times 2^n$, with $n = 0, 1, 2, \ldots$), was constructed (see Fig. 3.49b). One can see that concentration of the DL species is maximal at the phase delay of $200°$, which corresponds to the RC time constant of the cell, while concentration of the Faradaic species is maximal at a phase of $240°$. In a more general form, correlation analysis can be performed using two-dimensional synchronous and asynchronous plots (see below).

However, application of the above technique can be problematic when strong solution-phase absorption obscures weak bands of a surface species. To overcome this limitation, the phase rotation approach [263] can be used. Phase-sensitive detection such as with a lock-in amplifier (LIA) provides two signals: the signal that is in phase (IP) and the signal that is out of phase (the quadrature, Q) with the external perturbation [264]. These quantities can be represented at each

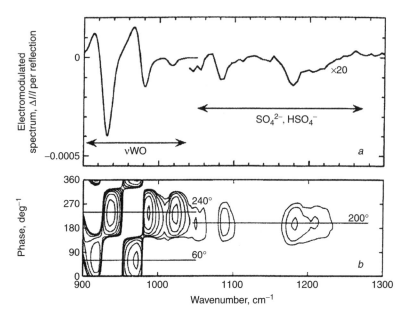

Figure 3.49. Differentiating between Faradaic and DL species by using phase effects in FTEMIRS. Spectra were measured by ATR for n-Ge/(1 M H_2SO_4 + 10^{-2} M α-$H_4SiW_{12}O_{40}$) interface. Modulation frequency was 125 Hz, while modulation range was from -0.15 to -0.35 V (SCE). (*a*) Spectrum for phase setting of 40°. (*b*) Two-dimensional phase map, showing isolevel lines of spectrum as function of wavenumber and phase settings (levels are chosen as $3 \times 10^{-6} \times 2^n$, $n = 0, 1, 2, \ldots$). Reprinted, by permission, from J. N. Chazalviel, B. H. Erne, F. Maroun, and F. Ozanam, *J. Electroanal. Chem.* **502**, 180–190 (2001), p. 187, Fig. 7. Copyright © 2001 Elsevier Science B.V.

wavenumber by the magnitude (M) and the phase (ϑ) as

$$M = \sqrt{(IP)^2 + Q^2}, \tag{3.33a}$$

$$\vartheta = \arctan\left(\frac{Q}{IP}\right). \tag{3.33b}$$

The magnitude of the signal represents the total optical response of the system to the modulation frequency Φ, while the phase represents the phase delay. Provided the phase delay across the solution band is constant (the species is in a homogeneous environment), the value of ϑ is measured for a solution species, and then the entire complex spectrum is rotated into the quadrature channel, as shown in Fig. 3.50. As a result, only the absorption due to surface species with different phase delays remains in the in-phase channel. This technique was used for disentangling strong bands due to the ferro/ferricyanide Faradaic species that were superimposed on weak bands of these species in the adsorbed state on a Pt electrode [263].

The third technique consists in measuring spectra as a function of the modulation frequency. Due to mass transport limitations, the characteristic times of

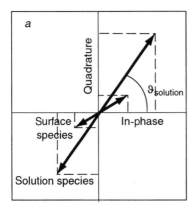
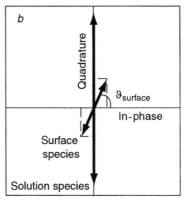

Figure 3.50. Projection of solution and surface bands into in-phase and quadrature channels (*a*) before and (*b*) after phase rotation.

solution species are longer than those for surface species and thus one can expect increasing surface sensitivity as the modulation frequency increases [263].

Analyzing the phase delay between the stimulus and the resulting infrared signal as a function of the modulation frequency, the kinetics of the modulated process may be deduced, provided that the rate constants of the forward and reverse processes are equal [268]. Using the ATR-SEIRA method, Ataka and co-workers [265] measured the rate constant as large as $\sim 10^5$ s^{-1} for the charge transfer between an adsorbate and an electrode (Section 7.6).

Thus the use of phase-sensitive detection allows discrimination between the spectra of solution and surface species and extraction of dynamic information about the products of a reaction.

Two-Dimensional Correlation Analysis. Two-dimensional correlation analysis of dynamic IR spectra (2DIR) allows further, extensive examination of ultrathin films by extracting information from dynamic IR spectra that is obscured in static spectra. Together with partial least-squares curve-fitting, spectral derivatives, and deconvolution (multivariative analysis) techniques [269–273], this technique is a powerful tool for enhanced structural resolution. The basis of 2DIR, developed by Noda in 1986 [274–276], is that different components of an anisotropic or inhomogeneous system respond to an externally applied perturbation differently according to their structural characteristics and environments. In the generalized approach [277–280], the perturbation can be chemical or physical, and of any form, including step, exponential decay, periodic, and noise. The first application of this method was the reo-optical study of atactic polystyrene films [274, 281]. To obtain detailed information about local submolecular motions of polystyrene, the dynamic IR transmission spectra of the film under periodic (~ 20 Hz) mechanical strain were collected. Later, this technique was extended to studies of the double-layer structure (the perturbation being the electrode potential) [282], conformations in protein films [the perturbation being hydrogen–deuterium (H/D) exchange] [283], phase transitions in phospholipid monolayers (the perturbation being the surface pressure) [284], electro-optical properties of films (the perturbation being an electric field) [285], and molecular association in solution (the perturbation being temperature) [286].

To increase interpretability, the dynamic IR spectra are subjected to mathematical cross-correlation to produce two different types of 2DIR correlation spectra, or two-dimensional correlation maps. These maps, in which the x- and y- axes are independent wavenumber axes (ν_1, ν_2), show the relative proportions of in-phase (synchronous) and out-of-phase (asynchronous) response (Figs. 3.51 and 3.52). Initially, the mathematical formalism was based on the complex Fourier transformation of dynamic spectra [277]. To simplify the computational difficulties, the Hilbert transform approach was developed [280], which produces two-dimensional correlation maps from a set of dynamic spectra as follows. First, the average spectrum $\bar{y}(\nu)$ is subtracted from each spectrum in the set, $\tilde{y}(\nu, P_j) = y(\nu, P_j) - \bar{y}(\nu)$, where P_j is the dynamic parameter. Then, the synchronous spectrum, $S(\nu_1, \nu_2)$, and the asynchronous spectrum, $A(\nu_1, \nu_2)$, are calculated as

$$S(\nu_1, \nu_2) = \frac{1}{n-1} \sum_{j=1}^{n} \tilde{y}(\nu_1, P_j)\tilde{y}(\nu_2, P_j),$$

$$A(\nu_1, \nu_2) = \frac{1}{n-1} \sum_{j=1}^{n} \tilde{y}(\nu_1, P_j) \sum_{k=1}^{n} M_{jk}\tilde{y}(\nu_2, P_k).$$

(3.34)

In Eqs. (3.34), ν_1 and ν_2 represent two independent wavenumbers, n is the number of spectra used in the calculations, and M_{jk} is the Hilbert transform

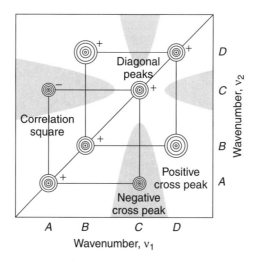

Figure 3.51. Schematic contour map of synchronous 2DIR correlation spectrum. Shaded areas represent negative-intensity regions. Adapted, by permission, from I. Noda, A. E. Dowrey, and C. Marcott, *Appl. Spectrosc.* **47**, 1317 (1993), p. 1318, Fig. 2. Copyright © 1993 Society for Applied Spectroscopy.

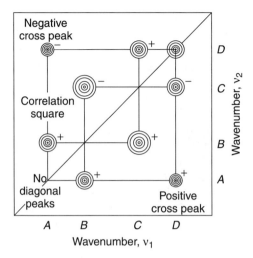

Figure 3.52. Schematic contour map of asynchronous 2DIR correlation spectrum. Shaded areas represent negative-intensity regions. Reprinted, by permission, from I. Noda, A. E. Dowrey, and C. Marcott, *Appl. Spectrosc.* **47**, 1317 (1993), p. 1318, Fig. 3. Copyright © 1993 Society for Applied Spectroscopy.

matrix, defined as

$$M_{jk} = \begin{cases} 0 & \text{if } j = k, \\ \dfrac{1}{\pi(k-j)} & \text{otherwise.} \end{cases}$$

Modern instruments usually include the necessary software.

The *synchronous* correlation map shows the coherence of dynamic changes in IR signals at two different wavenumbers. Strong correlation shows that the changes at those wavenumbers occur synchronously (in phase) with each other. A positive correlation indicates that they change in the same direction and a negative correlation indicates that they change in the opposite direction. The two-dimensional synchronous spectra are symmetrical with respect to the diagonal (Fig. 3.51). The intensity maxima appearing along the diagonal are called *autopeaks* and are always positive. Autopeaks correspond to the autocorrelation of perturbation-induced modes. The intensity maxima located at off-diagonal positions are called *cross peaks*. A pair of cross peaks may be positive or negative. In Fig. 3.51 signals *A* and *C* are changing simultaneously but in opposite directions. Signals *B* and *D*, on the other hand, are changing together in the same direction.

The *asynchronous* correlation map allows the elementary components of overlapping bands or spectra measured in situ to be distinguished, showing the uncorrelated or independent changes of IR signals at two different wavenumbers. Strong correlation peaks at (*B*, *C*) and (*C*, *B*) in the asynchronous correlation map (Fig. 3.52) show that the changes at the wavenumbers *B* and *C* are out of phase or occur at different rates with respect to one another. Only cross peaks located at off-diagonal positions appear in two-dimensional asynchronous spectra, and these spectra are antisymmetric with respect to the diagonal (Fig. 3.52): A pair of symmetrically positioned cross peaks are necessarily of opposite intensity. The sign of asynchronous peaks provides information about the temporal relationship between the responses of the modes at ν_1 and ν_2 to the external perturbation. The sign is positive if the absorption change induced by the perturbation at ν_1 occurs before that at ν_2.

An example of the possibilities of 2DIR is provided by looking at the area where 2DIR has a significant role to play — determination of the structure of protein (polymer) films. Nabet and Pezolet applied two-dimensional correlation analysis to resolve secondary structure of a myoglobulin film deposited on a Ge IRE [283]. They constructed the two-dimensional maps from the dynamic spectra upon H/D exchange. Figure 3.53 shows synchronous and asynchronous correlation maps for myoglobulin collected for a short time after the beginning of H/D exchange. There are three main autocorrelation peaks in the amide I region at 1615, 1640, and 1675 cm^{-1}, which can be assigned to β-sheets, a random coil, and β-turns, respectively (Table 7.9). It follows that the amide groups in these conformations are exchanged first. The strongest peak corresponds to the amide II band at 1530 cm^{-1}, in agreement with the dominant N–H bond contribution. This peak correlates with all three amide I components, while a

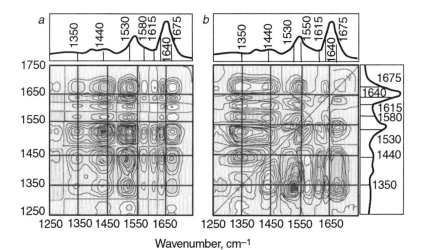

Figure 3.53. (a) Synchronous and (b) asynchronous two-dimensional ATR spectra of myoglobin film on Ge IRE (50 × 20 × 2 mm, 45°) calculated from first 10 spectra recorded during hydrogen–deuterium exchange process. Clear and dark peaks relative to gray background are positive and negative, respectively. One-dimensional spectrum shown is average of 10 spectra. Spectra were measured with Nicolet Magna 550 spectrophotometer equipped with narrow-band MCT detector. Each spectrum is average of 16 scans (10 s). Background spectrum was obtained at beginning of experiment by recording 250 scans. About 250 16-scan spectra, separated by a time delay of 30 s, were needed to complete deuteration of protein within 10–15%. Reprinted, by permission, from A. Nabet and M. Pezolet, *Appl. Spectrosc.* **51**, 466 (1997), p. 468. Copyright © 1997 Society for Applied Spectroscopy.

weak peak at 1580 cm^{-1} correlates only with the bands at 1640 and 1675 cm^{-1}. In the amide II$'$ region, two peaks at 1350 and 1440 cm^{-1} correlate with all the amide I and amide II components. As opposite to the amide I region, the synchronous map in the amide II and amide II$'$ regions does not display clearly resolved spectral components. In the synchronous map constructed from the spectra measured 1 h later, only one autopeak at 1655 cm^{-1} was found in the amide I region, assigned to α-helices.

There are two out-of-phase cross peaks at 1615–1640 cm^{-1} and 1675–1640 cm^{-1} in the asynchronous map of myoglobulin (Fig. 3.53b), which confirms the fact that the peaks at 1615, 1640, and 1675 cm^{-1} in the synchronous map are due to three different conformations. In the amide I–amide II cross-correlation region, a new component that is correlated out of phase with the 1615-, 1640-, and 1675-cm^{-1} amide I bands is observed at 1550 cm^{-1}, revealing a fourth conformation with a different H/D exchange rate from those in the intermolecular β-sheets, the random coil, and the β-turns. A strong cross peak is also seen in the amide II region at 1530–1550 cm^{-1}. As discussed above, the 1530-cm^{-1} component that is unresolved in the synchronous map could be assigned to all three rapid exchanging conformations. Thus, two-dimensional correlation analysis showed that myoglobin in the film form exhibits at least four conformations (α-helices, β-sheets, a random coil, and β-turns).

The interpretation of two-dimensional correlation maps can be complicated by noise and baseline fluctuations [287a] or by perturbation-induced changes in the band frequency, FWHM, and oscillator strength [288]. Specific complications arise when applying 2DIR correlation analysis to the IRRAS of monolayers at the air–water interface [284].

3.9. IR SPECTRA OF INHOMOGENEOUS FILMS AND FILMS ON POWDERS AND ROUGH SURFACES. SURFACE ENHANCEMENT

In Sections 3.1 and 3.2 the effect of size on IR spectra was discussed solely in the context of ultrathin films with plane–parallel boundaries. However, this size effect can be seen for all particles whose size is small relative to the wavelength and can lead to additional, abnormal absorption by both the particles and ultrathin films coating such particles. This phenomenon is well known for metals and causes metallic ultrathin films to have different colors than bulk metals. In 1857, Faraday proposed that such a color transformation is associated with the intrinsic aggregating nature of metallic films. His hypothesis has since been confirmed and understood based on Maxwell electrodynamics, and these effects have subsequently been found in the IR range for metals, dielectrics, and semiconductors. Moreover, it has been established that the particle shape also affects the IR spectrum of an ultrathin film in the closest vicinity of a system of particles that are small compared to the wavelength of irradiation. The abnormal absorption of inhomogeneous films remains the subject of intense theoretical investigations, due to the wide practical implications. However, the purpose of this section is not to review this theory in depth but rather to concentrate on the practical aspects of this effect in IR spectroscopy of ultrathin films. For further details of the theory, textbooks [289–292] and reviews [293–300] can be consulted.

Before considering the IR spectra of inhomogeneous films and ultrathin films on powders and rough surfaces, the effect of the shape of an isolated particle on the IR spectra will be discussed.

3.9.1. Manifestation of Particle Shape in IR Spectra

The main advantage in interpreting the optical response of small particles is that the retardation can be neglected, and, consequently, the electric field can be simply computed within the framework of electrostatics. As for ultrathin films with plane–parallel boundaries (Section 3.2.2), this approach uses the electrostatic polarizability of a body to describe the IR absorption by a macroscopically small, but microscopically large, particle of the same shape. The major assumption is that the applied electric field is uniform across the particle.

Sphere. A complete description of the coupling of an electromagnetic wave and the eigenmodes of an isolated sphere of any size, given by *polariton theory* based on Mie's formalism (Section 1.10), indicates that all modes of a sphere-shaped crystal are radiative [293, 298]. These modes are called *surface modes* since their origin lies in the finite size of the sample [297]. For very small spheres, there is only the lowest order surface mode (the *Frohlich mode*), which is neither transverse nor longitudinal [293]. Its frequency (the *Frohlich frequency*) is given by

$$\omega_F^2 = \omega_{TO}^2 \frac{\varepsilon_{st} + 2\varepsilon_{sm}}{\varepsilon_\infty + 2\varepsilon_{sm}}, \qquad (3.35)$$

where ε_{st} and ε_∞ are the static and high-frequency limits of the permittivity of the particle, respectively, and ω_{TO} is the TO mode frequency of the particle material. This frequency corresponds to the polarization resonance in a sphere and can be obtained by inserting Eq. (1.40) into Eq. (1.118). Due to the small particle dimensions, the Frohlich mode has a constant amplitude over the entire sphere.

As the radius of the sphere is increased, the ω_{TO} band of the bulk mode appears in the spectrum. For example, for ZnS particles this occurs at $r > 2$ μm. As the radius increases further, (i) both the surface- and bulk-mode absorption bands broaden and split due to the appearance of higher order surface modes, (ii) the band maxima shift toward lower frequencies, and (iii) the ratio of the intensities of the surface to bulk modes decreases [293]. As can be deduced from Eq. (3.35), an increase in the dielectric constant of the surrounding medium will also cause the surface modes to shift toward the red, as does an increase in the particle dimensions. The explanation is the same as for plane–parallel films (Section 3.3.1). The presence of a dispersion in particle size causes the two absorption bands to broaden, the fine structure to disappear, and the ω_{TO} band to shift to lower frequencies. An analogous effect can be observed if the damping constant γ of the sphere material increases.

Figures 3.54 and 3.55 show the transmission spectra of 0.1-μm α-Fe_2O_3 (hematite) spheres and 0.5-μm amorphous SiO_2 spheres, respectively, pressed

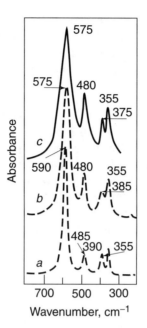

Figure 3.54. Comparison between (a, b) calculated and (c) experimental spectra of spherical α-Fe$_2$O$_3$ microcrystals in KBr ($\varepsilon_m = 2.25$): (a) calculated for isolated ($f = 0.001$) spherical ($g_1 = 0.33$) particles; (b) calculated for aggregated ($f = 0.3$) spherical ($g_1 = 0.33$) particles; (c) experimental spectrum. Reprinted, by permission, from J. E. Iglesias, M. Ocana, and C. J. Serna, *Appl. Spectrosc.* **44**, 418 (1990), p. 421, Fig. 4. Copyright © 1990 Society for Applied Spectroscopy.

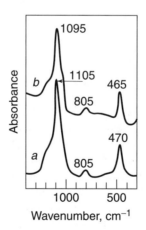

Figure 3.55. Experimental transmission spectra of spherical particles of amorphous SiO$_2$ dispersed in (a) KBr ($\varepsilon_m = 2.25$) and (b) TlCl ($\varepsilon_m = 5.1$). Reprinted, by permission, from J. E. Iglesias, M. Ocana, and C. J. Serna, *Appl. Spectrosc.* **44**, 418 (1990), p. 424, Fig. 9. Copyright © 1990 Society for Applied Spectroscopy.

in KBr. Comparing these to the spectra of the TO energy loss functions of the corresponding oxides reveals a difference in the band positions. The band frequencies in the powder spectra are by $20-50$ cm^{-1} higher than ν_{TO}. A red shift of the absorption bands of particles is observed as the refractive index of the immersion medium is increased, as illustrated in Fig. 3.55.

The dashed lines in Fig. 3.54 show the absorption of the α-Fe$_2$O$_3$ microspheres in KBr, simulated using a MGEM dielectric function for the actual ($f = 0.001$) and overestimated ($f = 0.3$) filling factors, together with the experimental transmission spectrum. The experimental spectrum is better represented by the spectrum calculated for $f = 0.3$ than that for $f = 0.001$. Iglesias et al. [301a] have attributed the observed discrepancy between the actual and theoretical filling factors to aggregation (clustering) of the particles; this is supported by the scanning electron microscopy (SEM) analysis of the KBr pellets. It was concluded that the clustering effect can be incorporated into the MGEM theory by using a priori overestimated values of f.

Ellipsoids, Cylinders, and Disks. For ellipsoids, the solution for a sphere is simply generalized. Ellipsoids are the only class of bodies that are polarized uniformly in a uniform external field. This class includes cylinders, disks, spheres, and plates as limiting shapes. If the principal axis $\mathbf{k}$ of an ellipsoid is parallel to the external field $\mathbf{E}_0$, the corresponding polarization $\mathbf{P}_k$ is described by

$$\mathbf{P}_k = \varepsilon_0 \hat{\alpha}_k \mathbf{E}_0, \qquad (3.36)$$

where $\hat{\alpha}_k$ is given by Eq. (1.127). From Eqs. (3.36) and (1.127), it is evident that when the incident electric field is not parallel to any of the ellipsoid axes but has components along two or three of them, the anisotropy of the particle splits the threefold degenerate absorption of the Frohlich mode into two or three absorption peaks. The frequencies of these peaks satisfy the conditions

$$|\varepsilon_{sm} - (\varepsilon - \varepsilon_{sm})g_k| = \min \quad k = 1, 2, 3, \qquad (3.37)$$

which can be considered as a generalized Mie condition (1.118). Inserting Eq. (1.40) into Eq. (3.37) gives the frequencies of the surface modes of an isolated small ellipsoid. It was shown [294] that increasing the aspect ratio $\eta = a/c$ of the oblate ellipsoid (where a and c are the radii along the long and short axes of the ellipsoid, respectively; see Fig. 3.62 later) results in a red shift and increased intensity of the low-frequency surface mode in which the induced dipole is perpendicular to the major short (c) axis. For a single cylinder of infinite length, the explicit solution was obtained within the framework of polariton theory, with no limits to the cylinder radius [293]. If the radius is small, two absorption bands — one at ω_S defined by Eq. (3.10) and the other near ω_{TO} — can be observed in the IR spectra. One difference between scattering by spheres and scattering by cylinders is that cross sections for cylinders depend on the polarization state of the incident radiation [300].

Morales et al. [301b] discussed the applicability of IR spectroscopy to determine the orientation of the crystallographic axes in anisometric particles. Two samples of monocrystalline hematite powder consisting of uniform particles of similar size ($\sim$300 nm in length) and shape (aspect ratio of $\sim$5) obtained by different preparation methods were used in this study. It was shown that in spite of the similar composition, crystalline structure, and morphological characteristics of the samples, their IR spectra present important variations in the position and relative intensity of the absorption bands due to the different orientations of the crystallographic axes. Such behavior can be accounted for by the EMT using the optical constants of hematite. These results were corroborated by electron diffraction carried out on individual particles. Therefore, from the analysis of the IR spectrum of a powder, it is possible to identify the orientation of the crystallographic axes of the constituent anisometric particles.

Other Shapes. For other shapes of particles and finite cylinders, mathematical difficulties arise due to the presence of the edges and corners [302]. The surface modes in small *cubes* were calculated by Fuchs [303] and Napper [304].

3.9.2. Coated Particles

The formulas for the polarizability of an isolated sphere and/or ellipsoid with a concentrical shell were given by Aden and Kerker [305], Van de Hulst [290], Guttler [306], Morriss and Collins [307], and Bohren and Huffman [289]. Eagen [308] derived the following expression for the polarizability of a metal oblate ellipsoid surrounded by a dielectric shell:

$$\hat{\alpha}_{\|,\perp} = \frac{(\varepsilon - \varepsilon_{sm})[\varepsilon_m g_1 + \varepsilon(1 - g_1)] + Q(\varepsilon_m - \varepsilon)[\varepsilon(1 - g_2) + \varepsilon_{sm} g_2]}{[\varepsilon g_2 + \varepsilon_{sm}(1 - g_2)][\varepsilon_m g_1 + \varepsilon(1 - g_1)] + Q(\varepsilon_m - \varepsilon)(\varepsilon - \varepsilon_{sm})g_2(1 - g_2)},$$
$$(3.38)$$

where ε_{sm}, ε_m, and ε are the dielectric constants of the host medium, the core, and the shell, respectively. The subscripts of the geometric factor g (1 and 2) refer to the value of core and coated ellipsoids, respectively. Here, Q is the ratio of the core ellipsoid volume to the coated ellipsoid volume ($Q = V_1/V_2$). For the general case, the n-shell problem for a sphere has been solved by Bhandari [309] and for an ellipsoid by Sinzig et al. [310, 311]; this allows the IR spectrum of a particle to be modeled with a radial gradient of the dielectric function. Interaction of radiation with coated cylinders is discussed by Barabas [312] and Sharma and Balakrishnan [313].

Using the appropriate expression for the dielectric function of a coated particle, the influence of a mantle on both the mantle and core modes can be analyzed. For example, Fig. 3.56 shows IR spectra of a coated sphere when both the core and the mantle are polar dielectrics (CdTe and CdO, respectively). If the mantle thickness is less than 10 nm and the radius of the core is of the order of 5 μm, one bulk mode absorption peak appears slightly below ω_{TOa} and three surface mode peaks appear in the IR spectra, one between ω_{TOa} and ω_{LOa} and the other two lying very close to ω_{TOb} and ω_{LOb}, respectively (here the subscripts a and b

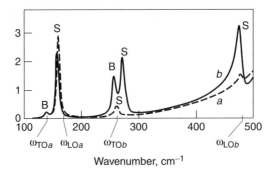

Figure 3.56. Calculated extinction cross section of CdTe sphere of radius 3 μm with CdO coating of (a) 10 nm and (b) 100 nm thick; S and B denote surface and bulk mode absorption peaks, respectively. Reprinted, by permission, from R. Ruppin, *Surf. Sci.* **51**, 140 (1975), p. 145, Fig. 3. Copyright © 1975 North-Holland Publishing Company.

correspond to the core and mantle, respectively). As the thickness of the coating increases, the spectrum becomes more complicated, with first the bulk ω_{TOb} mode and then the $l = 2$ surface mode of the mantle appearing and the separation between the two high-frequency surface mode bands decreasing.

For oxidized metal spheres, there is only one absorption band, near the ν_{LO} frequency of the coating material, as shown by the example of a Mg sphere with a MgO coating [298]. This is similar to the case of an ultrathin film on a metal plane substrate (Section 3.2). It follows that the usual SSR (Section 1.8.2) may also be applied to the surfaces of metal powders (see also the discussion in Section 3.9.4). Barnickel and Wokaun [314] reported that a dielectric coating shifts the resonance frequency of a metallic particle toward the red. Applying the MGEM dielectric function, Martin [315] calculated the reflection spectrum of an ensemble of 5-μm Zn spheres coated with ZnO of variable thickness. He obtained the same qualitative result as Ruppin for the isolated particle [298], that in the limiting case of the absence of the core, the absorption maximum is at the Frohlich frequency. As the thickness of the coating decreases relative to the core radius, the surface-mode frequency increases and approaches ν_{LO} monotonically.

Chen et al. [316] found a fit between the EMT and the IR absorption spectra of CVD β-SiC spherical particles. These particles were either solid or had a Si core and/or were hollow. As seen from Fig. 3.57, the EMT predicts that the whole β-SiC particle is represented by a single absorption peak near the Frohlich frequency of β-SiC. In the case of a Si core with an ultrathin β-SiC coating, the absorption spectra exhibit two peaks, close to 976 and 794 cm^{-1} (the ν_{LO} and ν_{TO} frequencies of β-SiC, respectively). The peak splitting decreases (the peak at the LO side shifts to a lower frequency and that on the TO side to a higher frequency) as the Si core decreases. These spectral features are similar to those shown in Fig. 3.56 for an isolated coated sphere. When the β-SiC spheres are hollow, with no Si core, an absorption peak near ν_{TO} with a shoulder at ν_{LO} appears. When there are both holes and cores, again peaks appear at ν_{LO} and ν_{TO}, but the intensity of the ν_{LO} band increases as the volume of the hole decreases.

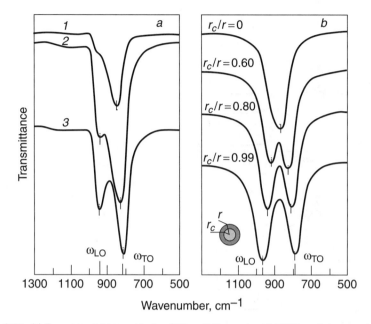

Figure 3.57. (a) Experimental transmission (KBr pellet) spectra of β-SiC particles consisting of β-SiC shell, silicon core, and hollow within core. Particles were prepared from $SiH_4-CH_4-H_2$ system at reaction temperature (1) $T = 1673$ K, (2) $I = 1623$ K, and (3) $I = 1573$ K. At $T = 1673$ K particles consist of β-SiC shell and hollow but not of Si core. (b) Effect of silicon core radius r_c on absorption spectra of β-SiC particles of radius $r = 50$ nm consisting of β-SiC shell and silicon core (no inner hollow within core), simulated using EMT. Reprinted, by permission, from L. Chen, T. Goto, and T. Hirai, *J. Mat. Sci.* **25**, 4273 (1990), p. 4276, Figs. 7 and 8. Copyright © 1990 Plenum.

3.9.3. Composite, Porous, and Discontinuous Films

Composite, porous, and discontinuous thin films have increasingly wide application and practical significance. Since many of the physicochemical properties of heterogeneous films are structure sensitive, an interesting and practical application of IR spectroscopy is the characterization of their microstructure (density, porosity, polycrystallinity, surface roughness) and inhomogeneities (specifically distribution and morphology of the latter). This characterization is based on EMT (Section 1.9). The main problem is that, although there exist several expressions for the dielectric functions of composite media, none of them are general, and thus each particular system must be investigated separately to determine the appropriate expression. The EMT approach for extracting microstructural information from IR spectra is illustrated below.

Figures 3.58*a,b* show the *p*- and *s*-polarized IR transmission spectra ($\varphi_1 = 45°$) of 0.5-μm SiO_2 films deposited by the radio-frequency (RF) cosputtering method, with 6–15-nm Ge microcrystals embedded in Si wafers at different values of the filling factor f [317]. As f increases, the intensity of the ν_{LO} band at 1240 cm^{-1} decreases significantly and the band FWHM increases (Fig. 3.58*c*),

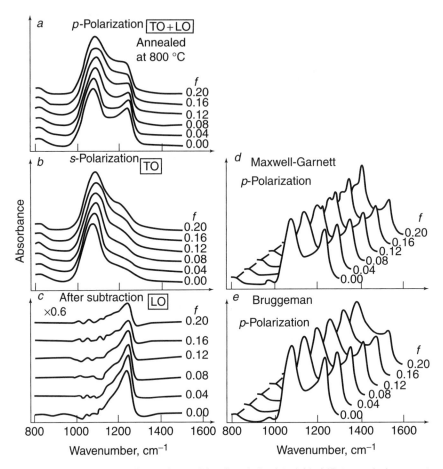

Figure 3.58. Dependence of experimental (*a–c*) and simulated (*d, e*) IR transmission spectra ($\varphi_1 = 45°$) of Ge–SiO$_2$ composite film on Si wafer on filling factor *f* of Ge microcrystals (6–15 nm size). Obtained with (a) *p*- and (b) *s*-polarized incident light; (c) result of substraction. Spectra in (b) and (c) correspond to adsorption at ν_{TO} and ν_{LO}, respectively. Reprinted, by permission, from M. Fujii, M. Wada, S. Hayashi, and K. Yamamoto, *Phys. Rev B* **46**, 15930 (1992), pp. 15932, (Fig. 2) and 15933, (Fig. 6). Copyright © 1992 American Physical Society.

while the peak intensity of the ν_{TO} band at 1080 cm^{-1} remains practically unchanged and the peak position is slightly blue shifted. The results of spectral simulations based on the Fresnel formulas are shown in Figs. 3.58*d,e*. The film was assumed to be a SiO$_2$–Ge composite, and optical constants of the film were calculated using both the MGEM and Bruggeman Model (BM) expressions. Compared to the experimental spectra, the spectra from IRRAS simulated with the MGEM dielectric function exhibit an extra peak at 1115 cm^{-1}, which grows with increasing f. This peak is assigned to the cavity mode that arises when an inhomogeneity such as a void with a positive dielectric function is embedded in a matrix with a negative dielectric function. Since the cavity mode peak is absent in the experimental spectra, the MGEM model is inadequate for describing the optical properties of the system under consideration. On the other hand, the shape of absorption spectra calculated with the BM (Fig. 3.58e) is very similar to that of the experimental spectra and demonstrates that the f-dependence is the same; specifically, the intensities of the ν_{LO} band and the high-frequency shoulder of the ν_{TO} band increase with increasing f. Thus, for this system the BM is more appropriate.

Another feature that may influence the IR spectra is the film *porosity*. This has been investigated by Wackelgard for anodic alumina (Al_2O_3) films [318]. In her study, 0.5-μm alumina films were grown by anodic oxidation of Al in 2.5 M phosphoric acid at 15 V for 15 min at 19°C. Analysis by SEM, transmission electron microscopy (TEM), and X-ray diffraction showed that pores are fibrous and oriented perpendicular to the surface; their volume fraction was about 30%, the average diameter of the pores 30 nm, and the alumina host amorphous. As would be expected for any oxide, the spectra measured by IRRAS (Fig. 3.59) exhibit a strong Berreman effect (Section 3.1). The ν_{LO} absorption band at 11.4 μm with the long-wavelength shoulder caused by the ν_{TO} band is observed in the p-polarized spectra and the ν_{TO} band at about 15.9 μm in the s-polarized spectra. The TO and LO energy loss functions and the real part of the dielectric function, calculated from the reflection spectra of the porous alumina films, are shown in Fig. 3.60. The value of $\lambda_{TO} = 1/\nu_{TO} = 15.9$ μm is the same for both the porous and nonporous films, suggesting that the Al−O bonds in these films are identical. The value of $\lambda_{LO} = 1/\nu_{LO}$ for the porous alumina is 11.4 μm, compared to 10.6 μm for the nonporous alumina, implying that the difference in the values of λ_O is caused by the porosity. To describe this effect, both the MG and Bruggeman expressions were used, and the pores were treated as cylinders perpendicular to the surface. The results proved to be independent

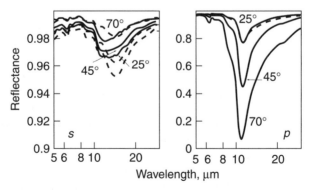

Figure 3.59. Experimental (solid lines) and calculated (dashed lines) IRRAS spectra of porous anodic alumina film 500 nm thick on Al, measured at three different angles of incidence with s- and p-polarized radiation. Calculation was performed using experimentally derived refractive index. Reprinted, by permission, from E. Wackelgard, *J. Phys. Condens. Matter* **8**, 4289 (1996), p. 4294, Fig. 2. Copyright © 1996 IOP Publishing Ltd.

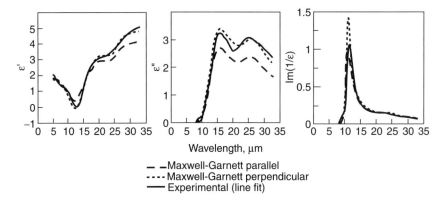

Figure 3.60. The TO and LO energy loss functions and real part of effective dielectric function, obtained using Maxwell-Garnett effective medium theory, are compared with corresponding experimental functions. Filling factor is 0.3 and pores are regarded as cylinders perpendicular to surface. Reprinted, by permission, from E. Wackelgard, *J. Phys. Condens. Matter* **8**, 4289 (1996), p. 4297, Fig. 4. Copyright © 1996 IOP Publishing Ltd.

of the type of EMT, and therefore, the Cohen formula (1.124) was used in the final analysis. For $f = 0.3$ and shape factors g of 0 and 0.5 for the directions perpendicular (z) and parallel (x) to the substrate, respectively, Eq. (1.124) gives

$$\varepsilon_z = 0.3\varepsilon_{\text{pore}} + 0.7\varepsilon_{\text{alumina}},$$

$$\varepsilon_x = \frac{\varepsilon_{\text{alumina}}(0.65\varepsilon_{\text{pore}} + 0.35\varepsilon_{\text{alumina}})}{(0.35\varepsilon_{\text{pore}} + 0.65\varepsilon_{\text{alumina}})}. \tag{3.39}$$

In Fig. 3.60, the optical parameter functions calculated according to Eq. (3.39) are compared with those from the spectra obtained by IRRAS, and satisfactory agreement between the theory and experiment, especially in the case of the ε_z spectrum, is apparent. Therefore, a decrease in frequency of the maximum of the LO energy loss function of a film and hence in the frequency of the ν_{LO} band in the IR spectra due to the presence of pores may be used to characterize the film porosity (volume fraction or shape of pores). It should be noted that the maximum of the LO energy loss function calculated with the EMT is 10.9 μm, while the experimental value is 11.4 μm. To generate the larger shift, a pore fraction of as high as 0.45 must be used, which would not agree with the TEM data. These observations are similar to those discussed above for the α-Fe$_2$O$_3$ particles (Fig. 3.54) and also appear to originate from the effect of clustering.

Atsumi and Miyagi [319] proposed a two-step EMT to predict the optical properties of anodized alumina films based on the assumption that spherical aluminum particles are distributed in columnar structures within an anodized alumina film. Using this model, theoretical parameters such as the volume ratio and distribution range of alumina particles were determined, their dependencies on the anodization conditions were evaluated, and the optical constants of the alumina films were calculated. Roos et al. [320] determined the optical properties of tin oxide coated Al and anodized Al according to the BM. McPhedran and Nicorovici [321] applied the Rayleigh method to derive a formulation yielding the MG effective dielectric tensor of a periodic composite of an array of elliptical cylinders placed in a matrix with a dielectric constant of unity. The MG approach was developed for calculating optical properties of aligned carbon nanotube films by Garcia-Vidal et al. [322].

A technique for evaluating the film density and volume fraction of the pores from the IR spectra is described for SiO_2 films in Ref. [323, 324]. It is well known for silicon oxide films that variation of the frequency of the ν_{TO} band is a measure of the density of the silica network (Sections 5.1 and 5.2). If one defines the difference between the ν_{TO} frequency of fused silica (1075 cm^{-1}) and that measured on SiO_2 films as $\Delta\nu_{TO}$, and $\Delta\rho$ is the difference between the measured mass density and that of bulk fused silica ($\rho = 2.2$ g $\cdot$ cm^{-3}), then for fused silica densified by applying hydrostatic pressure $\Delta\nu_{TO}/(\Delta\rho/\rho) = -3.6$ cm^{-1}/% [325a] and for silica films grown by thermal oxidation at high temperature $\Delta\nu_{TO}/(\Delta\rho/\rho) = -3.2$ cm^{-1}/% [325b]. In contrast, variations in ν_{LO} are mainly due to changes in the pore volume fraction f, as was demonstrated by applying the EMT to the experimental ν_{LO} values measured for a series of SiO_2 films [324b]. Approximately linear dependence between the $\Delta\nu_{LO}$ and f values was obtained for f up to 0.5, for example, $f = 15 \pm 2\%$ and $f = 28 \pm 2\%$ for $\Delta\nu_{LO} = 35$ cm^{-1} and $\Delta\nu_{LO} = 70$ cm^{-1}, respectively. The f values evaluated from the ν_{LO} frequencies agreed with those converted from the νOH water band near 3300 cm^{-1}. It was found [324] that absorption of water brings about a reduction in the porosity of silicon oxide films. Because the pore filling either void ($n = 1$) or water ($n = 1.33$) did not have too much influence on the determination of f from $\Delta\nu_{LO}$, it was concluded [324b] that this method can be used for silicon oxide films with different types of porosity: isolated pores, partially isolated and connected pores, and fully connected pores. Ohwaki et al. [326] attributed deviations between the measured IR spectra of native silicon oxide films on Si and the spectra calculated assuming that the native oxide was pure silicon dioxide with voids.

The effect of voids on the IR spectra of thin films produced by cosputtering of $Mo_{25}Si_{75}$ and $Mo_{36}Si_{64}$ onto Si substrates followed by a rapid thermal annealing has been analyzed by Srinivas and Vankar [327]. To access the microstructural state of these films, the BM was used in a multiphase–multilayer mode. Alterovitz et al. [328] studied plasma-enhanced chemical vapor deposition (PECVD) SiN films that had undergone rapid thermal annealing. Structural information was obtained using models that assumed an effective medium approximation for (1) Si_3N_4 and voids; (2) Si_3N_4 and a-Si; and (3) Si_3N_4, voids, and a-Si. Fiske and Coleman [329] described the spectra from far IRRAS of porous 0.5–1-μm CsBr, RbI, RbBr, and KI films, produced by thermal evaporation on Al, using the Cohen formula (1.124). in situ measurements of the porosity of thin oxide films by IR emission spectroscopy were performed by Pigeat et al. [330] on the basis of the BM. The optical properties of sintered SiC thermoelectric semiconductors consisting of crystalline grains, metallic inclusions, pores, and an intergranular material were determined by Okamoto et al. [331].

Discontinuous metal and ceramic–metal films exhibit an insulator–conductor (IC) transition in their optical, electrical, and other physical properties at some critical filling factor of conducting grains, which is called the *percolation threshold* f_c [332]. The value of f_c depends on the film constituents and their microstructure. For example, for Ni–MgO composites $f_c = 0.32$ [333], for Pt–Al$_2$O$_3$ thin films $f_c \approx 0.55$ [334], and for VO$_2$ thin films $f_c \approx 0.7$ [335]. As mentioned above, the Bruggeman dielectric function (1.123) can be used to qualitatively evaluate the percolation thtovrreshold, to give $f_c = g_k$ for a Drude metal (for spherical inclusions, $g_k = \frac{1}{3}, \frac{1}{2}$ for three- and two-dimensional systems, respectively) [336]. By contrast, the MGEM asymmetric dielectric functions give no access to this phenomenon [333, 337]. Sheng [338] suggested that a relatively high value of f_c indicates a coated-grain topology. By assuming the basic cells to be dielectric-coated metal and metal-coated dielectric spheres, he obtained a symmetric generalization of the MGEM dielectric function. The original Sheng theory for three-dimensional films gives $f_c \approx 0.455$ and after modification $f_c = 0.577$, which agrees well with the values obtained from the spectra of 110–210 nm Pt–Al$_2$O$_3$ cermet films [339]. Another problem is that at high filling factors, the cluster size is usually larger than the wavelength. At the same time, the quasi-static approximation is restricted to particles that are small compared to λ. This means that the EMT is no longer strictly valid for large filling factors, and the optical properties can no longer be described by an effective dielectric function past the percolation threshold [340]. Recently, some models have been proposed to resolve this problem based on various scaling laws [341–343]. Buhrman and Craighead [344] correlated EMT results with the microstructure of various composites and found that the MGEM expression adequately describes composites of metallic grains well dispersed in a dielectric, but composites of intermixed microcrystals of metal and dielectric have optical properties that are better described by the BM. Heilmann et al. [345] applied variations of EMT to interpret the absorption spectra of polymer–Ag composite films prepared by plasma polymerization and metal evaporation, effected either simultaneously or alternately. The best results were obtained with the extension of the MGEM theory for parallel ellipses.

The manifestation of discontinuity in the IR spectra of an ultrathin film can be predicted within the framework of the EMT, treating the film as an effective medium consisting of particles and air. Many EMT studies (see, e.g., Ref. [346]) have been devoted to metal films because of their applications in solar energy conversion and surface enhancement spectroscopies (see below) and as radiation filters. Some results for ultrathin metallic films will be discussed below. In principle, these should apply to ionic crystal clusters in the spectral range where $n < k$ as well as metallic clusters, because there is no physical difference in the interpretation of absorption spectra of metallic and ionic crystal clusters.

The optical properties of Ag islandlike films have been studied most extensively since their utilization in surface-enhanced Raman scattering and

IR spectroscopy. The discussion below is limited to several structure-related results. McKenna and Ward [347] employed David's shape factors, modified by an interaction term, to show how the Ag island shapes can be deduced from the wavelengths of anomalous absorption peaks in the spectra of Ag films. In all the cases investigated, the shapes predicted were in good agreement with those determined from profile electron micrographs. The shape factors did not show a significant variation with particle size and indicated roughly hemispherical particles in all cases. Using the modified long-wavelength approximation (MLWA) developed by Schatz and co-workers [348, 349], Jensen et al. [350] modeled extinction of a single oblate ellipsoid on mica, Ge, and Si substrates. The simulated spectra were found to be similar to the experimental absorption ones, exhibiting the peaks in the mid-IR range, which red shift and increase in magnitude as the dielectric constant of the substrate is increased.

The optical properties of discontinuous Cu films on dielectric substrates were investigated by Dobierzewska-Mozrzymas et al. [351]. The transmittance spectra of Cu films with low volume fractions were interpreted using the MGEM theory, which incorporates the size effect, the island shape, and the surface layer. To explain the experimental results for films with higher volume fractions, a new method, the renormalization approach, was developed. This method was applied to digitized transmission electron micrographs to calculate the transmittance spectra of discontinuous metal films. The calculated transmittance spectra were in good agreement with the experimental data for both the resonant absorption and the percolation threshold. The EMT analyses of angular selectivity of IR transmittance through obliquely sputter-deposited metallic and partially oxidized films of Cr, Al, Ti, and W have been reviewed by Mbise et al. [352]. Several types of samples were subjected to elaborate theoretical analysis based on EMTs. The angular and polarization dependences of the spectra of obliquely deposited films can be described if the parameters used to specify the film microstructures have been properly chosen. The optical properties of obliquely evaporated Ni films were studied by Yang et al. [353]. The anisotropic optical properties of these films were elucidated from the angle-of-incidence dependences of the SPR at 3.391 μm, measured in Otto's geometry at different polarizations. The data were fitted using the Fresnel theory, and an anisotropic permittivity tensor was evaluated. This tensor permitted the surface profile, depolarization factor, and volume composition to be determined. Yagil et al. [354] studied the optical reflectance and transmittance of percolating Au films close to the metal–insulator transition over an extended wavelength range of 2.5–500 μm. It was shown that the inhomogeneities of films with a typical grain size of 10 nm dictate the optical properties at wavelengths up to 500 μm. To describe the percolation threshold, a scaling model was developed that yielded agreement over the wavelength range studied, in contrast to the EMT. The optical properties of discontinuous Pd films were studied by Sullivan and Parsons [355].

To summarize, IR spectroscopy is able to provide structural information about inhomogeneous films in terms of volume fractions and the form of organization of the inclusions and the percolating degree, requiring an assumption/knowledge of the inclusion shape. The procedure is an iterative fitting of the spectra simulated using an EMT equation into the experimental ones.

3.9.4. Interpretation of IR Surface-Enhanced Spectra

Enhancement of absorption bands in the IR spectra of ultrathin films in the presence of discontinuous (islandlike) under- and overnanolayers of Ag and Au was discovered by Hartstein et al. [356] in the early 1980s. Although these researchers believed that they observed an increase in the νCH band intensities for p-nitrobenzoic acid (p-NBA), benzoic acid, and 4-pyridine-COOH films, it was recently shown [350] that the spectra reported are in actual fact due to fully saturated hydrocarbons (possibly vacuum pump oil). In any case, this discovery has stimulated various research activities and led to the development of *surface-enhanced IR absorption* (SEIRA) spectroscopy. To date, the SEIRA phenomenon has been exploited in chemical [357] and biochemical IR sensors (see [357–360] and literature therein), in studying electrode–electrolyte interfaces [171, 361–365], and in LB films and SAMs [364, 366–370]. Other metals that demonstrate this effect are In [371] and Cu, Pd, Sn, and Pt [372–375]. The metal films can be prepared by conventional metal deposition procedures such as condensation of small amounts of metal vapor on the substrate, spin coating of a colloidal solution, electrochemical [388], or reactive deposition [299] (see also Section 4.10.2).

Depending on the molecular system, the morphology of the films involved, substrate, and optical geometry of the measurement, the enhancement factor for Ag may range from 40 to 500 relative to the metal film-free case. For the other metals the effect is weaker (e.g., a factor of 10 was achieved for the Au needle substrates [376]). The SEIRA was measured using transmission [377–383], IRRAS [359, 360, 384], ATR [265, 350, 357, 358, 361–364, 376, 385], and DRIFTS [370, 386] for analytes with polar as well as apolar functional groups [376] and various basic molecular structures (e.g., aromatics [377, 387], aliphatics [376, 380, 385], inorganics [380, 388], polymers [385], pesticides [357, 376], and biomaterials [358, 359, 389]). However, the observed enhancement was reported to be not equivalent for all absorption bands, being significantly greater for polar groups having large dipole moment gradients [390] and molecules that bind to metals [387]. In addition, the effect extends to a distance of $\sim$4–5 nm from the metal surface [403, 366].

The SEIRA phenomenon can be illustrated by the p-NBA–Ag system, which has been extensively studied [382, 384, 390–393] since the work of Hartstein et al. [356]. Curve a in Fig. 3.61 shows the SEIRA spectrum of p-NBA adsorbed from an acetone solution onto a CaF$_2$ window coated by an 8-nm Ag islandlike layer, after removal of physisorbed p-NBA by washing with acetone. This spectrum was recorded in the transmission configuration at $\varphi_1 = 0°$. For comparison, the p-polarized spectrum obtained by IRRAS ($\varphi_1 = 80°$) of p-NBA adsorbed on a thick Ag layer and the transmission spectrum of p-nitrobenzoate potassium salt in a KBr pellet are also shown. The intensity of the SEIRA spectrum is enhanced about 10-fold relative to the spectrum obtained by IRRAS; of this enhancement, a factor of 3 was attributed to the increase in the total surface area. Compared to the transmission spectrum at normal incidence, the enhancement is $\sim$200.

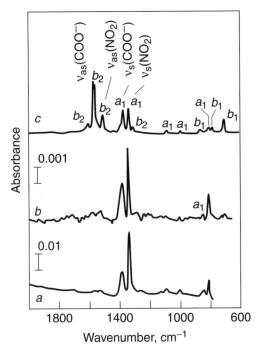

Figure 3.61. (a) Transmission-SEIRA and (b) metallic IRRAS spectra of p-nitrobenzoic acid on Ag. (c) Transmission spectrum of potassium salt of p-nitrobenzoate (KBr pellet). Silver film used in transmission-SEIRA measurements was vacuum evaporated on CaF₂. Assignment of absorption bands shown. Reprinted, by permission, from M. Osawa, K. Ataka, K. Yoshii, and Y. Nishikawa, *Appl. Spectrosc.* **47**, 1497 (1993), p. 1498, Fig. 1. Copyright © 1993 Society for Applied Spectroscopy.

It is seen from Fig. 3.61 that the SEIRA spectrum is identical to the one obtained by IRRAS. Only the symmetric (a_1) modes (1352 and 1413 cm^{-1} for CO$_2$ and NO$_2$ groups, respectively) appear, while the antisymmetric (b_1) modes (1528 and 1592 cm^{-1}, respectively) are practically absent in both spectra. This can be understood assuming that p-NBA is adsorbed at the Ag surface as the p-nitrobenzoate ion with its C_2 axis normal to the metal surface, as sketched in Fig. 3.62, provided charge transfer does not occur [390]. If this is the case, the dynamic dipole moment of the symmetric and antisymmetric CO$_2$ and NO$_2$ stretches is directed perpendicular and parallel to the metal surface, respectively. From this observation, Osawa and Yoshii [361] concluded that the surface selection rule (SSR) for metal surfaces is also valid for SEIRA, which was confirmed by Zhang et al. [367]. In fact, analysis of the polarizability tensor of a molecule adsorbed on a metal particle [392, 394] has confirmed the dominance of the α_{zz} component, where the z-axis is normal to the surface at the adsorption site. Greenler et al. [395] have performed both classical and quantum-mechanical calculations for the interaction of the electromagnetic field with particles of varying sizes and arrived at the conclusion that the SSR should only be applied to particles larger than about 1.5 nm. Nevertheless, the applicability of the SSR to SEIRA systems has been disputed by Merklin and Griffiths [383, 396]. Kwan Kim et al. [370, 386] have attributed the apparent deviations from the SSR to particular structural features of the adsorbed molecule.

There have been extensive efforts over the past 20 years to explain SEIRA (see Ref. [374] for a review). This effect is not yet perfectly understood, although it has been accepted by many researchers that at least two mechanisms are responsible for it. As follows from the general selection rule (1.2.9°), the band intensity can be increased by increasing (1) the dynamic dipole moment (chemical mechanism) and (2) the electric field along the given dynamic dipole moment (electromagnetic mechanism). The first mechanism is associated with a redistribution of the electron density due to chemical attachment of the analyte molecule to the metal film. It has been invoked to explain the high dependence of SEIRA on the molecular structure of the adsorbed molecules [374, 383, 390] and the band shape anomaly, especially for p-polarization [397]. On the other hand, it was shown [385, 391, 398, 399] that chemical attachment to an islandlike metal film is not required for surface enhancement. Furthermore, an increase in the ATR spectra of polymer films was observed in contact with smooth Ni nanolayers evaporated onto a KRS-5 prism [372, 373], which implies predominance of the electromagnetic mechanism. Thus, there is a general consensus that the electromagnetic effect causes the majority of the enhancement.

As shown in Chapter 2, the electromagnetic enhancement can be generated within a film that has a smooth metal under- or overlayer when there is a specific combination of the refractive indices of the various components of the system. This effect is described by the Fresnel formulas (Sections 1.5–1.7) when the surfaces of the metal film are assumed to be smooth [400, 401].

In addition, the magnitude of the electric field within a molecule and, hence, the absorption of IR radiation [Eq. (1.27)] are increased when the molecule is near a curved surface, as is the case for a cavity[†] [402], a rough surface [174], or an inhomogeneous (islandlike) film [299]. This effect can be understood modeling surface inhomogeneities by oblate ellipsoids. (Wetting effects cause metal clusters sputtered or evaporated on a dielectric substrate to form more or less regular oblate ellipsoids with varying axial ratios and the long axes oriented parallel to the substrate surface [24, 403].) The local electric field around an ellipsoid is the sum of the polarizing electric field $\mathbf{E}_0$ and the electric field induced by the virtual dipole with the dipole moment $\mathbf{p}_k = \mathbf{P}_k$ at the origin of the ellipsoid [polarization $\mathbf{P}_k$ is described by Eqs. (3.36) and (1.127)]. As can be seen from Eqs. (1.28) and (1.127) and illustrated by Fig. 3.62 the lower the geometric factor (or the larger the radius) of an axis, the larger the electric field directed along this

---

[†] Yakovlev and coworkers [402a, 402b] reported the IR absorption enhancement by more than one order of magnitude for hydrocarbons adsorbed inside porous silicon, and assigned this effect to photon confinement in the microcavity acting like a multipass (Fabry–Perot type) cell. Recently, Jiang et al. [402c] observed a 50 times enhancement in the in situ IRRAS of CO adsorbed on Pd nanoparticles synthesized in cavities of Y-zeolite, as compared to the cases when the supports were ultrathin Pd films deposited directly on the zeolite or on amorphous alumosilicate layer.

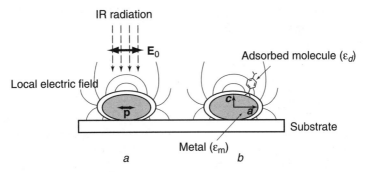

Figure 3.62. Polarization of metal particle in incident electric field. Thin lines represent electric field produced by induced dipole **p** in particle. Metal particle is modeled by prolate ellipsoid. Adsorbed molecule is modeled by thin layer covering metal core.

axis. All the enhancement effects will be substantially suppressed if the external electric field is directed along the short axis.

When the applied electric field is directed along the long axis of the ellipsoid, a further enhancement of the electric field in the space between the ellipsoids and a red shift of the resonance frequency take place due to the strong electric field coupling that cannot occur when the electric field is directed along the short axis. This coupling, which is called *surface plasmon resonance* (SPR), broadens the surface plasmon modes at IR wavelengths [294, 349, 350, 404] and promotes the SEIRA effect.

This model explains why SEIRA is observed in both *s*- and *p*- polarized IRRAS [384] and ATR [391, 405] spectra and in normal-incidence transmission spectra [377] and why the enhancement is not uniformly spread over each metal island but occurs mainly on the lateral faces of the metal islands [378, 384, 385]. The quasi-static interpretation of the SEIRA also defines the material parameters necessary for excitation and observation of SPR: (1) The resonance frequency determined from the general Mie condition must be as low as possible and (2) Im $\varepsilon(\omega_{\mathrm{res}})$ must be as small as possible. The maximum enhancement effect should be observed for the absorption bands near the Mie (resonance) frequency of the particle. As mentioned in Section 3.9.1, the resonance frequencies of metal particles lie in the visual or near-IR range. However, they can be shifted into the mid-IR range by (1) increasing the aspect ratio of the ellipsoids, (2) adding the support to an immersion medium, (3) coating the particles by a dielectric shell [24, 406], or (4) varying the optical properties of the support [24, 349, 350, 384]. As emphasized by Metiu [299], the surface enhancement effect is not restricted to metals but can also be observed for such semiconductors as SiC and InSb.

To estimate the contribution of SPR in the SEIRA, Osawa and co-workers [382, 384, 407] modeled the electric field coupling between the particles within the framework of the EMT. The optical constants of the effective medium, which consists of the Ag oblate ellipsoids with *p*-nitrobenzoate shells and the host medium (air), were calculated using both the MGEM formula (1.125) and the Bruggeman formula in the form

$$\bar{\varepsilon} = \varepsilon_{sm} \frac{3(1 - f) + f\alpha}{3(1 - f) - 2\alpha}$$

with the polarizability α of the coated ellipsoid described by Eq. (3.38). The transmission-SEIRA spectra were simulated using Fresnel theory for the three-layer system external medium–effective medium layer–substrate. It was found that the BM describes the experimental data of Osawa and co-workers much better than the MGEM theory, predicting an enhancement factor of 140 for the spectrum shown in Fig. 3.61a compared to the value of 200 found experimentally. The BM was found to provide good fit into the angle-of-incidence dependences of the ATR-SEIRA spectra in Refs. [391, 405]. Maroun et al. [388] studied applicability of the EMT models for Ag islandlike films of different morphology and packing. For nonpercolating grainlike Ag deposits with interisland distances of $\sim$300 nm, which were obtained by electrodeposition from a solution of $AgNO_3$ in 0.1 M $HClO_4$, the ATR-SEIRA spectra were well fitted by the Bruggeman formula (Fig. 3.63). However, for closely packed columnar patterns with a small tendency to percolation, which were obtained from a solution of $AgNO_3$ in 0.1 M ethylenediaminetetraacetic acid (EDTA) (pH 4), the MGEM Bergman formula [408] was found to be more suitable (Fig. 3.64). The latter film generated the SEIRA effect four times stronger than the deposits from $HClO_4$.

Consideration must be given to optical effects in the SEIRA spectra. As seen from Figs. 3.63 and 3.64, increasing thickness of the Ag film results in the baseline shift and increase toward higher wavenumbers. This effect is more pronounced for the more SEIRA-efficient films (Fig. 3.64). Similar background shifts and slopes were observed in the ATR-SEIRA spectra upon adsorption of anions [409a], combined adsorption of anions and metal deposition on Au(111) [409b], and adsorption of organic molecules [403] (Fig. 3.65), but attributed [409b] to changing the dielectric screening ε_∞ of an adlayer due to a potential-induced change in the adlayer density and electronic state (electronic density) [Eq. (3.32)] which modulates the SPR. This assignment was supported by spectral simulations for the Si/10-nm Au film/0.2-nm adlayer/solution. Assuming that the adlayer is transparent, while its dielectric screening varies from 1.0 to 1.85, the permittivity of the Au film was obtained with the Bruggeman equation (1.132), and the spectra were calculated using the Fresnel formulas. As seen from Fig. 3.66, a slight change (from 1.70 to 1.85) in ε_∞ of the adlayer yields an increase in the IR background absorption by $\sim$0.05 at 4000 cm^{-1} and 0.004 at 2000 cm^{-1} in the absorbance scale. However, since SEIRA is strongly dependent on the metal film morphology [378] and the metal particle size [407], it is not inconceivable, while unchecked yet, that the background effects result from potential-induced changes in the morphology of the thin-film electrodes.

According to the EMT, the effective optical constants of an islandlike film depend on the packing density of the particles, the optical constants of the surroundings, and the polarizability of the mutually interacting metal islands

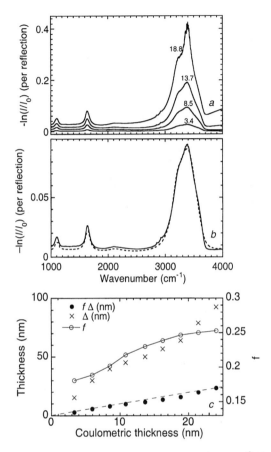

Figure 3.63. Silver electrodeposition on silicon from 1 M $HClO_4 + 10^{-3}$ M $AgNO_3$ electrolyte. (a) Typical absorbance spectra for different integrated charges: 3.4, 8.5, 13.7, and 18.8 $mC \cdot cm^{-2}$. (b) Bruggeman fit of spectrum with 8.5 $mC \cdot cm^{-2}$ integrated charge. (c) Layer thickness Δ obtained from fits with Bruggeman's theory, its volume fraction f, and $f\Delta$ plotted as a function of the coulometric thickness. Notice that $f\Delta$ falls close to coulometric thickness, which provides good check of fitting procedure. Reprinted, by permission, from F. Maroun, F. Ozanam, J. N. Chazalviel, and W. Theiss, *Vib. Spectrosc.* **19**, 193–198 (1999), p. 196, Fig. 1. Copyright © 1999 Elsevier Science B.V.

(Section 3.9.3). The polarizability itself is a complicated function of the shape of the metal particles. Using IR ellipsometry, Roseler and co-workers [410] found that for SEIRA films of a nominal thickness of 6 nm, k varies from 0 to 3, while n is from 5 to 8, both indices being actually independent of the wavelength throughout the IR range. This behavior differs from the usual dispersion of the optical constants of bulk metals in the IR range. For comparison, the refractive index of bulk gold increases from about 1.2 at 4000 cm^{-1} to 12 at 1000 cm^{-1}, while the absorption index increases from 15 to 55. The vanishing absorption of the SEIRA films was attributed to the inhibited lateral conductivity of the film

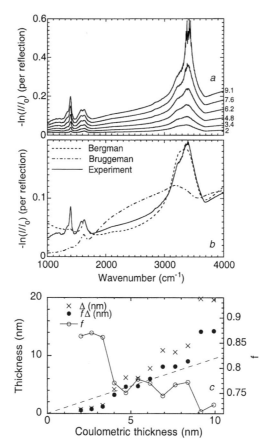

Figure 3.64. Silver electrodeposition on silicon from 0.1 M EDTA, pH 4 $+10^{-2}$ M AgNO$_3$ electrolyte. (a) Typical absorbance spectra for different integrated charges: 2, 3.4, 4.8, 6.2, 7.6, and 9.1 mC $\cdot$ cm^{-2}. (b) Bruggeman and Bergman fits of spectrum with 5.5 mC $\cdot$ cm^{-2} integrated charge. (c) Layer thickness Δ obtained from fits with Bergman's theory, its volume fraction f, and $f\Delta$ plotted as function of coulometric thickness. Notice that, as in Fig. 3.63c, $f\Delta$ falls close to coulometric thickness, which provides a good check of fitting procedure. Reprinted, by permission, from F. Maroun, F. Ozanam, J. N. Chazalviel, and W. Theiss, *Vib. Spectrosc.* **19**, 193–198 (1999), p. 196, Fig. 2. Copyright © 1999 Elsevier Science B.V.

when the metal islands are mutually separated and insulated by the dielectric substrate. In agreement with model calculations [410d], higher enhancement factors were found for islandlike films with lower ($\sim$0.5) absorption indices, which also agrees with the theoretical data of Brouers et al. [411], who have shown that the intensity of the local electric field in metal–dielectric granular films may exhibit giant fluctuations in the IR range when the dissipation in metallic grains is low.

Bands in SEIRA spectra can be anomalously shifted and distorted, as reported for CO adsorbed on thin Pt films in contact with aqueous electrolyte [61a, 412], CO adsorbed from the gas phase and aqueous solutions on Pt and Pd

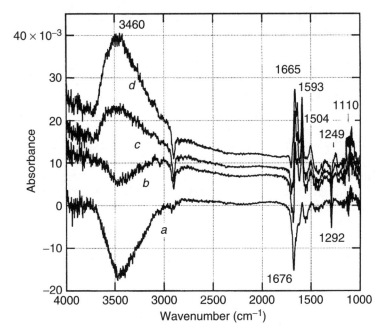

Figure 3.65. ATR-SEIRA spectra from uracil on Au electrode at various potentials in 10 m*M* uracil +0.1 *M* LiClO$_4$ solution: (*a*) +0.3 V, (*b*) +0.6 V, (*c*) +0.7 V, (*d*) +0.9 V. Reference potential is +0.20 V (Ag/AgCl). Positive peaks correspond to increase in absorption at sample potential. Reprinted, by permission, from M. Futamata and D. Diesing, *Vib. Spectrosc.* **19**, 187–192 (1999), p. 190, Fig. 3. Copyright © 1999 Elsevier Science B.V.

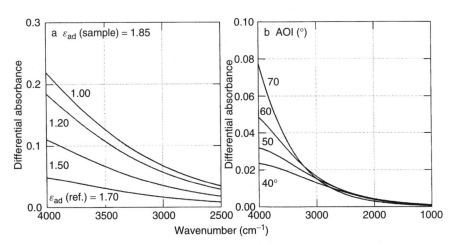

Figure 3.66. Theoretical IR background shift in ATR-SEIRA spectra as function of (*a*) dielectric constant of adsorbate layer without chemisorption [ε_{ads}(ref)] at incident angle of 60° and (*b*) angle of incidence (AOI). Dielectric constant of adsorbate layer under chemisorption ε_{ads} (sample) is 1.85. Reprinted, by permission, from M. Futamata, *Chem. Phys. Lett.* **333**, 337–343 (2001), p. 342, Fig. 5a. Copyright © 1999 Elsevier Science B.V.

thin films [413], CO adsorbed on thin metal films [414], and p-NBA [383] and water [415] on Cu. Figure 3.23a [61a] demonstrates distortions for linearly bound CO adsorbed on a Pt electrode. As argued for CO adsorbed on Pt [412] and transmission spectra of adsorbed CO [416] these spectral features can be generated by the Fano resonance (1.3.20°). An alternative explanation is optical effect [61a]. In fact, spectrum 2 in Fig. 3.23a that was measured after a hydrogen treatment of the sample is more distorted, while its intensity is enhanced with respect to spectrum 1 by a factor of ~4. According to the scanning tunneling microscopy (STM) data, the hydrogen treatment leads to sintering of the Pt islands and, hence, changing the optical properties of the Pt electrode. Figure 3.67 shows how optical properties of a metal film electrode affect the νCO band measured by both s- and p-polarized ATR in Kretschmann's geometry. Although an increase in both n and k influences the band shape, the effect of n is more pronounced. An increase in k (more absorbing metal) also leads to a decreased absorbance for the adsorbate. It should be noticed that the s-polarized band is inverted for $= 10-10i$. Spectrum simulations also revealed that the derivative-like bands are more typical for strong absorbers (both solution and adsorbate).

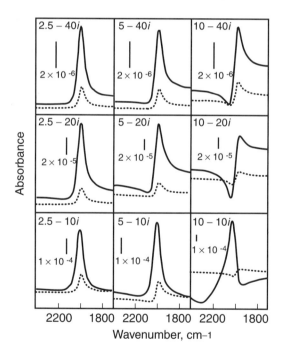

Figure 3.67. Calculated p- (solid lines) and s-(dotted lines) ATR spectra for monolayer adsorbate on 20-nm metal film in contact with solvent as function of complex refractive index of metal film. Except for metal, the system is as in Fig. 3.23b, c. The respective complex refractive index of metal film is given at top part of each spectrum. Reprinted, by permission of the Royal Society of Chemistry on behalf of the PCCP Owner Societies, from T. Bürgi, *Phys. Chem. Chem. Phys. (PCCP)* **3**, 2124 (2001), p. 2128, Fig. 10. Copyright © 2001 The Owner Societies.

The studies of optimum conditions for SEIRA measurements [356, 357, 368, 376, 378, 393, 410b] showed that the optimum thickness of a metal islandlike film is in the 4–25-nm range, depending on the system and the metal film morphology. For thicker films, the enhancement is reduced and eventually disappears because of absorption from the metal itself. The greatest enhancements were observed in metal nonpercolating deposits with sharply defined islets [357, 388] and for particles with a small curvature (e.g., needle-shaped) [357]. Ordered arrays of metal particles produce enhancement comparable to that on disordered vapor-deposited island films [350]. It is important for analytical purposes that the SEIRA band intensities are linear functions of the film thickness for the first two to three adsorbed monolayers only [357].

Optimum conditions in the case of Ag films on the Si or ZnSe IREs and the ATR geometry were found [403] to be the film thickness of 9 nm and the angle of incidence of 30°–40°. As in ordinary spectra obtained by transparent IRRAS (Section 2.3.1), the s-polarized IRRAS-SEIRA bands are negative, independent of the angle of incidence, while the p-polarized bands are negative at $\varphi_1 < \varphi_B$ and positive at $\varphi_1 > \varphi_B$, where φ_B is the Brewster angle of the substrate [384]. The maximum band intensity is observed at small and oblique angles of incidence in s- and p-polarization, respectively [384]. However, the maximum absorption of s-polarization is always smaller than that for p-polarization, independent of the metal film thickness and the optical properties of the substrate [359, 360, 384]. The enhancement factor increases as the refractive index of the substrate decreases [350, 384].

The effects mentioned above are also observed in the case of metal overlayers [356, 384]. When the ATR geometry is used, the overlayers have been found to be less suitable, most likely because of the larger distance to the ATR interface [358]. It is interesting that the SEIRA spectrum of a film can be different for metal underlayers and metal overlayers due to different cordination of molecules to the Ag surface [417].

3.9.5. Rough Surfaces

Apart from scattering, which increases the DR spectral contrast of a film, and the intensity enhancement, a rough surface can give a specific contribution to the IR spectra of an ultrathin film deposited on this surface. As pointed out in Section 3.2.1, under ordinary experimental conditions, a surface polariton on a flat surface is nonradiative. However, when the surface of an active medium is grated with a spacing a, the x-component of the incident-photon wave vector (Fig. 1.9) becomes [418]

$$k_x = \frac{\omega}{c} \sin \varphi_1 \pm N \frac{2\pi}{a},$$

where N is the diffraction order. Tuning the angle of incidence φ_1, one can achieve the resonance condition (3.10), which is used for SEW excitation. A randomly rough surface can be treated as a superposition of gratings described by a set of wave vectors k_{xa}. Each grating gives rise to its own diffracted order. For a number of these gratings, the condition (3.10) can be met, allowing the excitation of the nonradiative surface polariton. Thus, a rough surface reflects radiation differently than a flat one, producing an extra band near ω_S [Eq. (3.10)] in each *reststrahlen* band if the correlation length of the roughness is of the same order as or larger than the wavelength of the incident radiation.

Although many groups have studied electromagnetic scattering by rough surfaces and rough films, the exact solution to this class of problems has not yet been achieved except in a few oversimplified cases [419, 420]. By considering the surface roughness to be in the form of minute hemispherical pits and bumps well separated from one another, Berreman [421] has shown that such surface imperfections influence the reflection near the *reststrahlen* bands, resulting in the appearance of secondary maxima below ω_{LO}. This phenomenon is attributed to localized resonances of the ionic vibrations in the vicinity of extremely small pits or bumps, which induce additional electric fields around the curved surfaces and, therefore, shifts the bulk-mode frequency. In conventional terminology, this amounts to excitation of the surface vibrational modes of tiny particles situated at a surface. A theoretical study of the optical response of a BeO surface covered by oblate and prolate ellipsoids, truncated spheres, and holes by Andersson and Niklasson [422] revealed that such a system is characterized by two types of absorption peaks: (1) at ν_{LO} and ν_{TO} of the substrate and (2) at the Fröhlich frequencies of the imperfections. This phenomenon can be used to assess the quality of polishing of a polar crystal surface [423]. In general, the resonance frequencies have been found to be higher for pits than for bumps. Abnormal light reflection is also observed from the surfaces of rough and corrugated thin films, which is currently under extensive investigation [424]. The EMT treats a rough surface as a geometrically inhomogeneous medium composed of bumps made of the substrate material and the embedding medium. Using such an approach, Dignam and Moskovits [425] have described an optical model that takes into account the effect of the deposition of an ultrathin film on a slightly rough metal surface. They found that the reflection spectra of methanol and oxygen adsorbed on a rough Ag surface are very strongly dependent on the microscopic roughness of the surface. Bjerke and co-workers [426a] reported that electrochemical platinization of a Pt electrode provides up to 20 times the enhancements of the νCO band measured by in situ IRRAS as compared to smooth Pt. At the same time, as the degree of platinization was increased, the band shape became asymmetrical, then bipolar, and finally appeared as a reflection maximum. The S-shape of the adsorbate band was more pronounced near the percolation threshold of the Pt underlayer. The spectrum simulations based on the Bergman dielectric function [408] showed that the dielectric constants of the Pt-matrix composite layer also change drastically near percolation. This effect can be due to the fact that scattering of electrons at the interface becomes an important relaxation mechanism in the case of microscopically rough surfaces [426b]. At the percolation threshold, when the metal islands contact each other, the conductivity of the films strongly changes and with it the effective optical constants.

3.10. DETERMINATION OF OPTICAL CONSTANTS OF ISOTROPIC ULTRATHIN FILMS: EXPERIMENTAL ERRORS IN REFLECTIVITY MEASUREMENTS

Optical constants of ultrathin films are relevant in technologies such as those involved in solar absorption, temperature control, radiative cooling, and optical coatings and are of interest for IR spectroscopic research including determination of the optimum conditions for the spectrum measurements (Chapter 2), isolation of optical effects (Sections 3.1–3.5), and molecular orientation investigations (Section 3.11). It should be emphasized that the whole problem is treated as the inverse problem of mathematical physics and is thus inaccurate [427]. Therefore, in each case it is necessary to use all of the a priori information available about the system under investigation, to choose the best model of the layer, and the optimal method of solving the inverse problem. Methods for determining the optical constants (refractive and absorption indices) and thicknesses of ultrathin films have been well described for the UV/Vis region [45, 346]. They involve measurement of transmission and phase changes [428, 429], reflection and transmission at certain thicknesses [430, 431], reflection and interference fringes [432], and reflection at Brewster angle [433], as well as methods such as ellipsometry [434–436], polarimetry [437, 438], and SPR [439–441].

In the IR range, the most accurate and direct methods of determining the optical constants of ultrathin films appear to be the interferometric SEW method [39, 442, 443] and SPR [444], which both require special equipment, including an IR laser. Other methods are conventionally divided into *single-wavelength* and *multiwavelength* approaches. In the single-wavelength approach, n and k are calculated interactively at each frequency from the transmittance and reflectance measured at normal and/or oblique angles of incidence using the Fresnel formulas [445–451]. However, the solutions $n-ik$ are not unique for a given pair of measured R and T values, due to the highly nonlinear relationships among the variables involved and the singularity of the solutions. To achieve efficient convergence on the true solution, this approach requires good initial estimates of n and k [247, 452]. In the multiwavelength approach, the experimental spectra are fitted by simulated ones, and the spectral simulations are performed using optical constants from Kramers–Krönig (KK) relations [452–454] or dispersion formulas [455–463]. The use of ATR spectra is reviewed in Refs. [44, 456, 457].

In this section, only the optical constants of isotropic films determined by the multiwavelength approach in IRRAS will be discussed. The optical constants are assumed to be independent of the film thickness, and any gradient in the optical properties of the substrate (Section 3.5) is ignored. This undoubtedly lowers accuracy of the results. Anisotropic optical constants of a film are more closely related to real-world ultrathin films. At this point, it is worth noting that approaches to measuring isotropic and anisotropic optical constants are conceptually identical: An anisotropic material shows a completely identical metallic IRRAS spectrum to the isotropic one if the complex refractive index along the z-direction for the anisotropic material is equal to that for the isotropic one [44]. However, to extract anisotropic constants, IR measurements along three mutually orthogonal directions with respect to the film have to be performed (Section 3.11.3).

Dispersion analysis seems to be the simplest technique for extracting the complex refractive indices of ultrathin films. It involves fitting the experimental reflection spectra with simulated ones. For these simulations, either the Fresnel formulas (Section 1.5) or the matrix method (Section 1.7) is used, and the complex dielectric function is treated as a superposition of a specified number of oscillators. The fitting parameters are S_j, γ_j, ν_{0j}, and ε_∞ in Eq. (1.46) or ν_{LOj}, ν_{0j}, γ_{LOj}, γ_j, and ε_∞ in Eq. (1.54a). Success of the curve-fitting method depends on the adequacy of the analytical dispersion law adopted, which depends upon the proper choice of the number of oscillators. The zero-approximation values of these parameters can be drawn directly from the experimental spectra. Specifically, the resonance frequencies ν_{0j} are given by the peak positions in the *s*-polarized transmission or reflection spectra [in which the band position coincides with the maximum of the TO energy loss function (Figs. 3.1 and 3.11)] and the damping constants by the band FWHM. The value of ε_∞ can be measured independently, for example by ellipsometry [433]. Subsequently it is sufficient to vary only the oscillator strength S_j to attain good agreement of experimental and calculated spectra. For example, it was found [458] that in the case of polycrystalline and quasi-amorphous films of MgO (a single oscillator), Eq. (1.54a) gives better results than Eq. (1.46). Dispersion analysis using Eq. (1.54a) has been performed for 85–95-nm MgO films deposited by vacuum evaporation on Al [458]. To simplify the calculations of the optical constants, the ν_{LO} frequency was taken directly from the IRRAS spectrum of the MgO layer on an Al substrate and the ν_{TO} frequency from the transmission spectrum of a thick MgO film deposited on a transparent substrate. The oscillator parameters that were obtained as a result of fitting the spectra obtained by IRRAS are shown in Table 3.5. The parameters for the polycrystalline film were quite close to those obtained by a dispersion analysis of the specular reflection from a single MgO crystal ($\varepsilon_\infty = 3.01$, $\nu_{LO} = 725$ cm^{-1}, $\nu_{TO} = 401$ cm^{-1}, and $\gamma = 7.6$ cm^{-1}) [459]. For the quasi-amorphous MgO film, satisfactory agreement between the theoretical and measured spectra was obtained using a considerably higher value of γ. The dispersion analysis of the spectra obtained by IRRAS of strongly absorbing AlN and SiC layers prepared by laser ablation on Si (111) at 800°C [460] also gave damping constants much larger than those of the crystalline materials, indicating a low degree of crystallinity in the samples (Table 3.5). The ε_∞ had an imaginary part, which was significant in the case of SiC. This effect has been ascribed to the carbon admixture found in the Raman spectra. For a SiC/double layer on Si, a model comprised of characteristics from each film (SiC and AlN) has been found to be unsatisfactory and requires inclusion of the Drude term [Eq. (1.43)], which describes the presence of free carriers. The dispersion analyses employing the Drude dielectric function and the multioscillator model [Eq. (1.46)] have been applied to determining the optical constants of thin metal films in the IR region [461] and weakly absorbing (*polymeric*) films [455b, 462, 463].

Table 3.5. Parameters of dielectric function of MgO, SiC, and AlN thin films obtained by dispersion analysis

System	ε_∞	Oscillator		
		v_0	S	γ
Polycrystalline MgO (85–95 nm thick) on Al [458]	3.00	400	0.525	7.6
Quasi-amorphous MgO (85–95 nm thick) on Al [458]	2.95	400	0.525	40
SiC (108 nm thick) on Si [460]	6.38 + 0.36 i	799	2.82	39
AlN (813 nm thick) on Si [460]	4.25 + 0.02 i	662	3.39	17

The KK method of determining optical constants in the IR range is based on the fact that the optical constants are not independent of one another but are related through KK relations [Eq. (1.18)]. These relations yield an additional equation between the real and imaginary parts of the complex function and make it possible to reduce the number of unknown parameters required to determine the optical constants of a layer. The application of these relationships to determine the optical constants of ultrathin films on metals was studied theoretically by Konovalova et al. [464]. The model film material was represented by the dispersion parameters $v_0 = 1000$ cm^{-1}, $S = 0.1$, $\gamma = 0.1$, and $\varepsilon_\infty = 3$ (typical for a medium-intensity band found in the absorption of a dielectric), and the substrate was Al. The optical constants $n_2(v)$ and $k_2(v)$ were calculated by Eq. (1.17), and the p-polarized metallic IRRAS spectra $R(v)$ were simulated within the film thickness range of 0.01–0.3 μm using the exact formulas (1.75). Subsequently, the reverse procedure was followed: The simulated spectra were approximated by Eq. (1.82) (the thin-film approximation), allowing the spectrum of the LO energy loss function $\mathrm{Im}(1/\hat{\varepsilon}_2)$ to be obtained directly. The function $\mathrm{Re}(1/\hat{\varepsilon}_2)$ was then extracted using the KK relation of the form

$$\mathrm{Re}\,\frac{1}{\hat{\varepsilon}_2(v)} = \frac{2}{\pi} \int_{v_1}^{v_2} \frac{v'\,\mathrm{Im}[1/\hat{\varepsilon}_2(v')]\,dv'}{v'^2 - v^2} + C(v), \qquad (3.40)$$

where v_1 and v_2 are the integration boundaries and $C(v)$ is a weakly varying function that describes the contribution of the spectrum beyond the frequency interval under consideration (between 0 and v_1 and between v_2 and ∞). [In a first approximation, $C(v)$ may be considered to be constant and determined by the refractive index in the visible range, $C = 1/n_\infty$.] The real and imaginary parts of the refractive index, $n_2(v)$ and $k_2(v)$, were retrieved using Eq. (1.17) for the model layer. Good agreement between these values of $n_2(v)$ and $k_2(v)$ and those specified initially was achieved for films of thickness $d_2 < 0.01\lambda$. This limitation arises because of the restricted applicability of the linear approximation.

The results of this approach to extracting $n_2(v)$ and $k_2(v)$ from the experimental spectra of MgO layers on the surface of Al mirrors are discussed below.

These layers were deposited by magnetron sputtering on Al substrates maintained at temperatures of $25°$ and $250°C$. The spectrum obtained by IRRAS for the film deposited at $25°C$ (Fig. 3.68a) shows an intense ν_{LO} absorption band at 725 cm^{-1}. In the spectrum of the film deposited at $250°C$, the maximum of this band is shifted to higher frequencies, and its FWHM is smaller than that in the spectrum of the film obtained at $25°C$. The resulting spectral dependences of $n_2(\nu)$ and $k_2(\nu)$ for the MgO layers are presented in Fig. 3.68b. The optical constants of the layer sputtered on the $250°C$ substrate are close to those of a MgO crystal. The lower frequency and larger bandwidth of the $k_2(\nu)$ band suggest an amorphous phase in the layer.

However, the thin-film approximation may yield significant errors even for films a few tenths of nanometers thick [7]. Yamamoto and Ishida [44], based on the KK technique of Hansen and Abdou [465] for a multi-interface system, and Buffeteau and Desbat [452], based on the Abeles matrix method and the fast Fourier transform for the KK relations, suggested procedures for measuring the optical constants from IRRAS without employing the thin-film approximation. These techniques can yield the optical constants of both organic and inorganic ultrathin films with high accuracy if the film thickness and ε_∞ are known.

As was shown by Selci et al. [466], the KK method allows the complex dielectric function associated with the semiconductor surface states to be calculated. The surface is treated as an absorbing layer of thickness $d \ll \lambda$ located between a substrate and an external medium. The differential quantity $\Delta R/R_d = (R_{c\ell} - R_{ox})/R_{ox}$ was used as a measure of the surface reflectivity, where R_{ox} is

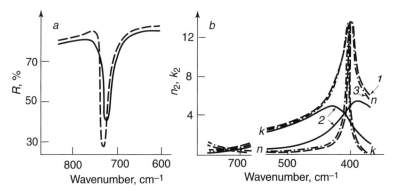

Figure 3.68. (a) IRRAS spectra of MgO layer on Al substrate, recorded in p-polarized radiation at $\varphi_1 = 60°$, $d_2 = 100$ nm at temperatures $t = 25°C$ (solid line) and $t = 250°C$ (dashed line). Reprinted, by permission, from I. I. Shaganov, O. P. Konovalova, and O. Y. Rusakova, *Sov. J. Opt. Technol.* **55**, 402 (1988); p. 403, Fig. 1, Copyright © 1988 Optical Society of America. (b) Dispersion of optical constants k and n of MgO: (1) polycrystalline film deposited at $t = 250°C$ (dashed lines); (2) amorphous film deposited at $t = 25°C$ (solid lines); (3) MgO monocrystal (dashed-dotted lines). Reprinted, by permission, from O. P. Konovalova, O. Y. Rusakova, and I. I. Shaganov, in N. G. Bakhshiev (Ed.), *Spectrochemistry of Inter- and Intramolecular Interactions*, issue 4, Leningrad State University Press, Leningrad, 1988, p. 206; p. 209, Fig. 2. Copyright © 1988 St. Petersburg University Press.

the saturation value of the reflectance after prolonged exposure to oxygen (surface states are removed) and $R_{c\ell}$ is the reflectance of the cleaved semiconductor surface. The thickness of the oxide, d_2, was assumed to be equal to d and the equation for $\Delta R/R_d$ was derived from a simple linear approximation for a three-phase model (1.81). In the general case $d_2 \neq d$, and the effective dielectric function for the oxide, $\varepsilon_{2,\text{eff}}$, must be considered and is defined as $\varepsilon_{2,\text{eff}} = (\varepsilon_2 + 1)d_2/d - 1$. Since the oxide is usually transparent in the region of the surface state absorption ($\varepsilon_2'' = 0$), for $\varphi_1 = 0°$ and $n_1 = 1$, Eq. (1.81) reduces to

$$\frac{\Delta R}{R_d} = d[A\varepsilon_s'' - B(\varepsilon_s' - \varepsilon_2')], \tag{3.41}$$

where

$$A = \frac{8(\pi/\lambda)(\varepsilon_3' - 1)}{(1 - \varepsilon_3')^2 + (\varepsilon_3'')^2},$$

$$B = \frac{8(\pi/\lambda)\varepsilon_3''}{(1 - \varepsilon_3')^2 + (\varepsilon_3'')^2}.$$

It follows that to distinguish the contribution of the surface to the reflectivity, the dielectric function of the bulk substrate should be known. However, in the case of a transparent substrate at energies less than E_g, $\varepsilon_s'' = 0$ and the reflectivity becomes $dA\varepsilon_s''$. For transitions with energies $h\nu > E_g$, the second term in Eq. (3.41) cannot in principle be neglected. However, it was found that $B \approx 0$ at energies lower than 3.2 eV for Si, 2 eV for Ge, 2.8 eV for GaAs, and 3.5 eV for GaP. When $B \neq 0$, the IRRAS spectrum is related to both ε_s' and ε_s'' (see also Section 6.5).

Errors in measuring the reflectivities $\Delta R/R$ for calculation of the optical constants of layers can be divided into two groups [467]: (1) errors connected with the photometric accuracy with which R is determined and (2) errors inherent in the method of obtaining the spectra, including inaccuracy in the angle of incidence of radiation, convergence of the radiation beam and its influence on the accuracy, and the ideality of the polarizer.

An error in the photometric measurements of ±0.005 [456] corresponds to an accuracy of ±0.01 for $\Delta R/R$ for a single reflection. Since the quantity being measured is the ratio R/R_0, the error introduced by radiation loss from the IRRAS accessory and reflection from the sample is partially corrected.

The error in the reflectivity that arises from the accuracy limits of the angle of incidence was estimated theoretically [458] for the spectrum obtained by IRRAS of an ultrathin film on a highly reflecting metal ($n_3 = 15$, $k_3 = 60$) for which the dependence of $\Delta R/R$ on φ_1 is substantial. The calculations revealed that, for a *single reflection*, the error in $\Delta R/R$ introduced by the angle of incidence over the angular range of $75° \pm 30'$ when the angular aperture of the beam is $\pm5°$ is 3–4% $\Delta R/R$ or less. As seen from Fig. 3.69, for *multiple reflections*, the rays incident onto the plate at larger angles undergo a greater absorption in the layer (due to the geometric factor) and fewer reflections between the plates than the

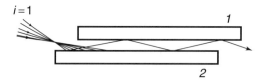

Figure 3.69. Schematic diagram of beam propagation between two reflecting plates (1 and 2).

rays incident at smaller angles. Hence, under multiple-reflection conditions, the dependence of $\Delta R/R$ on the angle of incidence φ_1 is reduced and thus leads to a smaller error. Deviations in the reflectivity were calculated for the optimum multiple-reflection conditions from highly reflecting metals. The reflectance R_0 of a metal was found from Eqs. (1.68)–(1.70). It was established that out of the $\varphi_B \pm 10°$ range when φ_1 can be set to within $30'$ and the angular aperture of the beam is set to within $\pm 5°$, the error in determining $\Delta R/R$ does not exceed 1–2%. However, the uncertainty in the reflectivity can be much greater in the vicinity of the Brewster angle φ_B [468].

It is also of importance to know the effect of the beam convergence on the accuracy of $\Delta R/R$. If the beam is represented as a sum of equal-intensity beams incident onto a sample at different angles, then the reflectivity of multiple reflections from a film–metal system can be written as

$$\left(\frac{\Delta R}{R}\right)^N = \frac{\sum_{i=1}^{r} (R_i^0)^N - \sum_{i=1}^{r} (R_i)^N}{\sum_{i=1}^{r} (R_i^0)^N}, \qquad (3.42)$$

where r is the number of beam components, R_i^0 and R_i are the reflectances of the ith component for the bare metal and the metal with a layer, respectively, and N is the number of light reflections from the sample. This accounts for the beam convergence in the plane of incidence only. The beam convergence perpendicular to the plane of incidence has little effect on the accuracy of the reflectances [458]. As follows from Eq. (3.42), the components of a beam with an angular aperture of $\pm 5°$, incident onto the sample at an angle φ_1, make opposite contributions to the change in $\Delta R/R$ and thus change its value at a given angle φ_1 very little.

For transparent substrates, the spread of incident angles around the average value for the central incoming ray affects the p-polarized absorbance as shown in Fig. 3.70 for spreads of 0, $\pm 8°$, and $\pm 16°$. When the angle of incidence is near the Brewster angle, the error in the measured absorbance increases abruptly due to the uncertainty in the beam convergence.

Measurement of the reflectivity $\Delta R/R$ from experimental spectra of thin dielectric layers on metals is complicated by the fact that if the absorption depth ΔR is taken as the band intensity then the experimental values of reflectivity, $(\Delta R/R)_{\exp}$, will differ from the calculated values of $\Delta R/R$. The reasons are the following. First, the band wings are distorted by the presence of the layer, effectively causing a shift of the spectral baseline relative to the reflectance of the bare

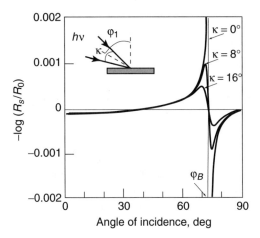

Figure 3.70. Calculated absorbances for p-polarized radiation in IRRAS spectra of hypothetical absorbate vibration at 3000 cm^{-1} on silicon ($n_3 = 3.42$) as function of angle of incidence φ_1 for different cone angles κ of incident radiation. Film parameters: $d_2 = 1$ nm, $n_2 = 1.5$, $k_2 = 0.1$. Reprinted, by permission, from H. Brunner, U. Mayer, and H. Hoffmann, *Appl. Spectrosc.* **51**, 209 (1997), p. 215, Fig. 7. Copyright © 1997 Society for Applied Spectroscopy.

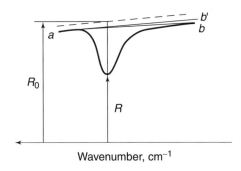

Figure 3.71. Scheme of drawing base line in IRRAS spectrum of film on metal surface: a–b', recommended baseline.

metal (1.5.4°). Hence, for calculating n_2 and k_2 it is advisable to derive the relationship between n_2, k_2, n_3, k_3, d_2, φ_1, and the absorption factor $A_G = (R'_0 - R)/R'_0$, where R'_0 is the reflectance of metal with the bleached layer. Second, in the case of the ν_{LO} band, there can be a contribution to the absorption from the ν_{TO} band of the layer substance on the low-frequency side of the band maximum (Fig. 3.12). Thus, the difference between $(\Delta R/R)_{exp}$ and A_G for ultrathin layers is somewhat larger on the low-frequency side. To decrease this difference, it is therefore recommended to draw the baseline in the spectrum (as shown in Fig. 3.71) by extrapolating the background at high frequencies toward lower frequencies. The values of $(\Delta R/R)_{exp}$ calculated from spectra treated in this manner are in better agreement with the values of $\Delta R/R$ determined for known R_0 [458].

Another error in the optical constants n_2 and k_2 of the layer, calculated from experimental values of $\Delta R/R$, is caused by an uncertainty in the optical constants n_3 and k_3 of the substrate. The partial derivatives $\partial(\Delta R/R)/\partial n_3|_{k_3=\text{const}}$; $\partial(\Delta R/R)/\partial k_3|_{n_3=\text{const}}$ were calculated for $\varphi_1 = 75°$ and n_2, with k_2 in the range from 0 to 3 [458]. It was found that the relative error in $\Delta R/R$ generated from n_3 and k_3 (whose accuracy is 20%) is less than 0.5% and can be neglected. Moreover, if the optical constants of an ultrathin film are measured using the technique of Yamamoto and Ishida [7], the optical constants of the metal substrate do not matter, and those of any real metal can be employed.

To estimate the possible error in n_2 and k_2 caused by an inaccuracy in the layer thickness d, $\Delta R/R$ was calculated as a function of d using Eq. (1.75). The results, presented in Fig. 3.72, show a linear dependence of $\Delta R/R$ on d at small layer thickness and an exponential dependence as d increases. The limit of the linear dependence varies, depending on the maximum value A_{max} of the LO energy loss function $\text{Im}(1/\hat{\varepsilon}_2)$ and the angle of incidence (Fig. 3.73). Thus, for $\varphi_1 = 75°$, the linear dependence occurs for thicknesses that satisfy $d/\lambda \ll 1 \times 10^{-3}$–$2 \times 10^{-3}$ for a wide range of values of A_{max}. The analysis of the change in $\Delta R/R$ as a function of the thickness d allows one to assume that for ultrathin layers the relative error $\Delta A_{\text{max}}/A_{\text{max}}$ caused by inaccuracy in specifying d is approximately equal to $\Delta d/d$. This result means that the inaccuracy in d makes a major contribution to the accuracy of experimental values of the optical constants.

The effect of nonideality in the polarizer on the accuracy of the optical constants has been analyzed in Refs. [468–471]. If one assumes that a fraction x of s-polarized radiation passes through the polarizer when it is set to transmit p-polarized light, then the effective p-polarized reflectance is given by [470]

$$R_p^{\text{eff}} = (1-x)R_p + xR_s, \qquad (3.43)$$

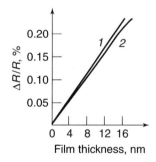

Figure 3.72. Dependence of reflectivity in IRRAS spectrum of film on metal on film thickness, calculated with (1) linear approximation and (2) exact Fresnel formulas $\varphi_1 = 75°$, $\hat{n}_2 = 0.5 - 0.12i$, $\hat{n}_3 = 15 - 60i$, $\nu = 1000 \text{ cm}^{-1}$. Reprinted, by permission, from V. P. Tolstoy, *Methods of UV-Vis and IR Spectroscopy of Nanolayers*, St. Petersburg University Press, St. Petersburg, 1998, p. 189, Fig. 5.20. Copyright © St. Petersburg University Press.

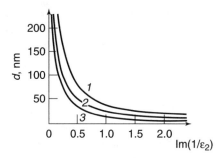

Figure 3.73. Variation in limit of linear approximation [Eq. (1.82)] depending on maximum value A_{max} of LO energy loss function $Im(1/\hat{\varepsilon}_2)$ and angle of incidence of radiation, $\varphi_1 = (1)\ 60°$, (2) $\varphi = 75°$, (3) $\varphi = 80°$; $n_3 = 15$, $k_3 = 60$, $\nu = 1000\ cm^{-1}$. Reprinted, by permission, from V. P. Tolstoy, *Methods of UV-Vis and IR Spectroscopy of Nanolayers*, St. Petersburg University Press, St. Petersburg, 1998, p. 189, Fig. 5.21. Copyright © St. Petersburg University Press.

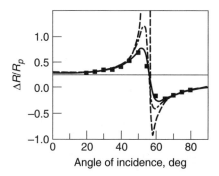

Figure 3.74. Angle-of-incidence dependence of reflectivity of $\nu_{as}CH_2$ band of 11-monolayer CdAr LB films on glass measured from IRRAS with ideal polarizer (dashed line), with polarizer with 98.5% degree of polarization (solid line), and for range of angles of incidence of $\pm5°$ around nominal value (dotted-dashed line). Solid squares are experimental peak intensity calculated from $\nu_{as}CH_2$ band of reflectivity spectra. Optical constants used in simulations are $n_{2x} = n_{2y} = 1.49$, $n_{2z} = 1.55$, $k_{2x} = k_{2y} = 0.5$, $k_{2z} = 0.04$, $d_2 = 29\ nm$, $n_3 = 1.47$. Reprinted, by permission, from D. Blaudez, T. Buffeteau, B. Desbat, P. Fournier, A.-M. Ritcey, and M. Pezolet, *J. Phys. Chem. B* **102**, 99 (1998), p. 102, Fig. 5. Copyright © 1998 American Chemical Society.

where R_p and R_s are the ideal reflectances for p- and s-polarization, respectively. Figure 3.74 shows that the experimental results can be well fitted with $(\Delta R_p/R_p)^{eff}$ if both the nonideality of the polarizer ($x = 1.5\%$)[†] and the beam convergence ($\pm5°$) are taken into consideration. One can deduce from Fig. 3.74 that the error due to the imperfection of the polarizer increases steeply around the Brewster angle. Because this angular region has a poor SNR in the absorbance spectra (due to a decrease in the denominator in $-\log R/R_0$; Fig. 1.11) and a greater error due to beam convergence, it is unsuitable for quantitative measurements.

[†] This value of the polarizer leakage is given in the commercial specifications of the polarizer [468].

To sum up, if the absorption band intensities are measured with a single-reflection IRRAS technique, the error in $\Delta R/R$ arises largely due to limited accuracy in the angle of incidence and the photometric error of the spectrometer. When using the multiple-reflection IRRAS technique, the error is determined by the photometric inaccuracies. Independently of the IRRAS technique, the error in calculating the optical constants is strongly dependent on the accuracy to which the layer thickness d can be determined. These conclusions are also valid for transmission and ATR spectra. For spectra obtained by IRRAS of ultrathin films on transparent substrates, the measurements performed near φ_B are more sensitive to errors due from leakage of the polarizer and beam convergence.

3.11. DETERMINATION OF MOLECULAR PACKING AND ORIENTATION IN ULTRATHIN FILMS: ANISOTROPIC OPTICAL CONSTANTS OF ULTRATHIN FILMS

It is well known that molecules tend to be adsorbed on surfaces and at interfaces with a preferential orientation. Such preferred orientations ultimately reflect the orientational dependence of the potential energy surface describing the adsorbent–adsorbate interaction [472]. This orientational dependence is a multiparameter function of head group size, distance between surface sites, surface pressure, chain branching, the strength of intermolecular and adsorbent–adsorbate bonds, temperature, kinetics of deposition, and the surroundings. Characterization of the molecular orientation (MO) and molecular packing (MP) in systems as diverse as biomimetic membranes, biological sensors, electronic and optical organic devices, novel solid lubricants, corrosion inhibitors, and hydrophobic and hydrophilic coatings has gained considerable attention [473–477]. Apart from IR spectroscopy, methods for quantifying MP and MO in ultrathin films include ellipsometry [478], X-ray photoelectron spectroscopy (XPS) [479], scanning tunneling microscopy (STM) [480], second-harmonic generation [481], near-edge X-ray absorption fine structure (NEXAFS) [482, 483], polarized ultraviolet (UV) spectroscopy [484], angle-resolved photoelectron spectroscopy (ARUPS) [485], Brewster angle microscopy [486], near-infrared (NIR) Fourier transform surface-enhanced Raman spectroscopy [487], on- and off-specular high-resolution electron energy loss spectroscopy (HREELS) [488], nuclear magnetic resonance (NMR) spectroscopy [489], and grazing-incidence X-ray diffraction (GIXD) [490] (see also reviews [491, 492]). The advantage of FTIR spectroscopy in the determination of MO and MP is that it provides simultaneously information about the whole adsorbed species. It can be applied nondestructively in situ to a wide variety of systems. However, IR spectroscopy can only measure the first coefficient in the orientational distribution function, and other techniques must be used (excluding birefringence) to extract the higher coefficients. Generally speaking, laborious calculations are necessary to achieve a high degree of accuracy in the MO data, and, unlike AFM, STM, and GIXD, IR spectroscopy yields only qualitative information about MP. Below, characterization of the MP and MO in ultrathin films by IR spectroscopy will be demonstrated, and then selection of the proper IR spectroscopic method and experimental conditions will be discussed for the measurement of the anisotropic optical constants and MO in ultrathin films. Particular attention will be paid to three types of ultrathin films of *long-chain molecules*: (i) densely packed molecular monolayers specifically adsorbed from solution onto solid substrates [self-assembled monolayers (SAMs)], (ii) monolayers of insoluble amphiphiles deposited at the air–water (AW) interface [Langmuir (L) monolayers], and (iii) films obtained when L monolayers are transferred to solids via the Langmuir–Blodgett (LB) technique (LB films).

3.11.1. Order–Disorder Transition

Infrared spectra allow comparison of the molecular order in different mesophases of the film. As pointed out by Tredgold [477], although the correct definition of the term order is a thermodynamic concept, it is of little use when discussing the properties of ultrathin films which are far from thermodynamic equilibrium. It has been suggested to regard the most ordered state as the one which corresponds most closely to some preconceived structure which one wishes to bring about. Monitoring of order–disorder transitions (ODTs) in ultrathin films is important when constructing superlattices with prescribed physical or chemical properties and a known thermal stability and in the modeling of phase transitions in biologically relevant membranes. Pioneering work in this field has been done by Naselli et al. [493, 494], who reported a two-step melting process in LB films of cadmium arachidate (CdAr) and suggested that these transitions may involve *trans–gauche* isomerization within the chains, a change in the average tilt angle, rotator mesophases, chain diffusion, and eventually desorption.

Currently, the most thoroughly studied systems are those comprised of long alkyl chain molecules for which specific spectral features of the C–H vibrations accompany the ODT. However, studies concerning absorption of head groups, such as carboxylic and carboxylate groups of fatty acids [495–497], amide [498–502] and phosphate [503] groups of biophysical LB films (Section 7.7), and the CN groups of chromophoric molecules [504], have also been reported.

As seen from Fig. 3.75, the TDMs of the $\nu_s CH_2$ and $\nu_{as} CH_2$ vibrations (at ~ 2850 cm^{-1} and ~ 2920 cm^{-1}, respectively) of hydrocarbon chains are orthogonal and perpendicular to the carbon chain axis. The TDM of the $\nu_s CH_2$ vibration lies in the plane of the molecular CCC backbone, while the TDM of the $\nu_{as} CH_2$ is perpendicular to it. The terminal methyl group has three modes. The TDM of the symmetric mode ($\nu_s^{ip} CH_3 \approx 2870$ cm^{-1}) lies in the plane of the backbone and is inclined by 35.5° from the molecular axis. One asymmetric stretch (near 2970 cm^{-1}), denoted by $\nu_{as}^{ip} CH_3$, is inclined 54.5° from the molecular axis in the plane of the molecular backbone. The other asymmetric stretch (near 2955 cm^{-1}), denoted as $\nu_{as}^{op} CH_3$, is perpendicular to the plane of the molecular backbone. The TDMs of the scissoring ($\delta_{scis} CH_2$) at ~ 1470 cm^{-1} and wagging ($\delta_{wag} CH_2$) at ~ 1350 cm^{-1} bending vibrations are directed perpendicular and parallel to the carbon chain axis, respectively. The characteristics of the νCH bands are collected in Table 3.6.

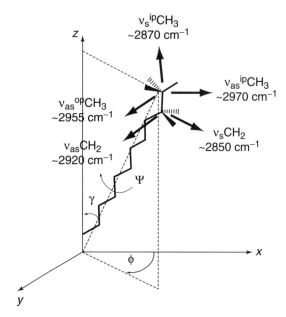

Figure 3.75. Definition of tilt γ, azimuth ϕ, and twist ψ angles for an all-trans hydrocarbon chain in laboratory coordinate system.

Table 3.6. Isotropic absorption indexes (k_{iso}), FWHM, and TDM orientations for CH stretching bands of dioctadecyl ($C_{18}H_{37}$) group of dioctadecyl disulfide [a]

Vibration	Peak Frequency (cm^{-1})	k_{iso}	FWHM (cm^{-1})	TDM Unit Vector $\{a, b, c\}$[b]
$\nu_{as}^{ip}CH_3$	2962	0.030	12	$\sin 55°, 0, \cos 55°$
$\nu_{as}^{op}CH_3$	2950	0.023	10	$0, 1, 0$
$\nu_{as}^{FR}CH_3$	2937	0.008	12	$-\cos 55°, 0, \sin 55°$
$\nu_{as}\alpha\text{-}CH_2^c$	2927	0.070	10	$0, 1, 0$
$\nu_{as}CH_2$	2919	0.305	12	$0, 1, 0$
$\nu_{as}^{FR}CH_2$	2906	0.028	11	$1, 0, 0$
$\nu_{as}^{FR}CH_2$	2892	0.030	17	$1, 0, 0$
$\nu_s CH_3$	2878	0.018	10	$-\cos 55°, 0, \sin 55°$
$\nu_s\alpha\text{-}CH_2^c$	2860	0.015	16	$1, 0, 0$
$\nu_{sym}\beta\text{-}CH_2^c$	2853	0.021	10	$1, 0, 0$
$\nu_{sym}CH_2$	2851	0.174	12	$1, 0, 0$

[a] Derived from absorption index (k) spectrum of bulk dioctadecyl disulfide.
[b] Unit vector coordinates in molecular coordinate system defined by hydrocarbon chain axis (c-axis) and C-atom backbone plane (a, c plane).
[c] Absorptions of CH_2 groups in α- and β-position to disulfide group.
Abbreviations: ip, in plane; op, out of plane; FR, Fermi resonance.

Source: Reprinted, by permission, from H. Hoffmann, U. Mayer, and A. Krischanitz, *Langmuir* **11**, 1304 (1995), p. 1308. Copyright © 1995 by American Chemical Society.

The frequencies of the CH_2 stretching modes of hydrocarbon chains are extremely sensitive to the conformational ordering of the chains in a layer. Although these frequency shifts are small (within 5–7 cm^{-1}), they can be measured routinely with modern FTIR spectrometers. When the chains in the monolayer are in all-trans zigzag conformation and highly ordered, the narrow absorption bands $\nu_{as}CH_2$ and $\nu_s CH_2$ appear at around 2918 ± 1 and 2850 ± 1 cm^{-1}, respectively, or at even lower wavenumbers (vide infra). With the formation of gauche rotomers, the bands shift upward to 2926 and 2856 cm^{-1}. This is due to a coupling between the carbon atoms and a methylene hydrogen, which, due to conversion around the C–C bond, is positioned in the plane defined by the carbon atoms, resulting in an increased force constant for that C–H bond. In contrast, for the all-trans conformation, all methylene hydrogens are out of plane [505]. In general, the order of the hydrocarbon chains decreases with decreasing chain length [495].

For the ODT analysis of films on transparent substrates, the νCH_2 band frequencies measured with the tangential component of the electric field (i.e., in the s-polarized reflection or normal-incidence transmission spectra) are used, since the p-polarized bands have complex composition (Section 3.4) and origin (Section 3.2). In addition, TO–LO splitting for the νCH_2 modes is up to 6 cm^{-1}.[†] The ODT data obtained from the $\nu_s CH_2$ band are more reliable because of fewer overlapping bands around the peak relative to the $\nu_{as}CH_2$ band [507, 508]. Moreover, in the case of crystalline films, the $\nu_{as}CH_2$ band can exhibit splitting due to the crystal field effect [504].

Gauche defects are expected to be concentrated near the free ends of the chains, and this has been experimentally confirmed by Nuzzo et al. [509, 510]. Gericke et al. [511, 512] reported the lowest frequencies for the $\nu_{as}CH_2$ and $\nu_s CH_2$ modes 2911.8 ±0.2 and 2848.9 ±0.1 cm^{-1}, respectively, for methylene groups adjacent to the head group of the half-deuterated hexadecanoic acid L monolayer in the liquid condensed–solid (LS) phase at the AW interface in the presence of Zn^{2+} subphase cations. These authors have measured slightly higher values for the same monolayer on a Pb^{2+} subphase ($\nu_{as}CH_2 = 2914.2 \pm 0.2$ and $\nu_s CH_2 = 2848.1 \pm 0.2$ cm^{-1}) and on pure water ($\nu_{as}CH_2 = 2915$ and $\nu_s CH_2 = 2849$ cm^{-1}). For comparison, the methylene groups of the whole chain of the hexadecanoic acid L monolayer in the S-phase absorb at 2917.1 and 2850 cm^{-1}, and those in 1-alkanol L monolayers at the AW interface at 5°C are at 2915.9 and 2848.3 cm^{-1} [513, 514]. The increase in order in the presence of cations in the subphase has been attributed to the bridging character of the bidentate chelate formed between the cation and the carboxylate head group [511].

The νCH_2 temperature dependence has been used in determining the melting point of a wide range of LB films [503, 515–519] and SAMs [370, 520]. The νCH_2 molecular area dependences have been used to study the mesophase transformations in L monolayers [508, 521–524], while the νCH_2 time dependences have been helpful in studies of the kinetics of aggregation (self-assembly)

[†] Being negligibly small for near-normal orientation of the chains [506], the TO–LO shift will arise in p-polarized spectra with increasing the tilt angle.

of adsorbed molecules [515, 525–528]. To illustrate the sensitivity of the νCH_2 frequencies to film melting, Fig. 3.76a shows the temperature dependence of the $\nu_{as}CH_2$ band frequency in the normal-incidence transmission spectra of a five-monolayer LB film of N-octadenoyl-L-alanine (a chiral molecule used for modeling enzymes, shown in Fig. 3.76b) on a CaF_2 substrate [529]. At around 115°C, the frequency changes steeply from values characteristic for a highly ordered state to those for a highly disordered state, indicating a clear first-order phase transition. The temperature dependence of the νCH_2 and νCD_2 frequencies was used to monitor separation of lipid phases (for review, see Ref. [530]). Miscibility of components can be measured by the extent to which the melting temperature and cooperativity of the individual components are related in the mixture. Another approach to distinguish phase separation is to compare MP and MO in the single-component and mixed films. An example of ODT found by the νCH_2 shift is discussed in Section 7.4.3.

The bandwidths (FWHM) of the νCH_2, δCH_2, and ρCH_2 bands are proportional to the degree of rotational mobility and flexibility within the layer, and increase with decreasing order [531], such that the FWHM can also be used for diagnostics of the degree of aggregation of the molecules in the film along with the νCH_2 frequencies [504, 519, 524]. However, in some cases, increased packing of the molecules may not yield a decrease in FWHM, as it was found in the coexistence regions of L monolayers, where the FWHM is independent from [521, 522] or even increases with [495] increasing surface pressure, presumably due to the increasing number of conformers.

The ODT can also be monitored by plotting the *peak height* or *integrated peak area* of the methylene stretching bands as a function of the external conditions

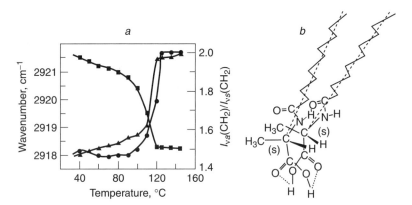

Figure 3.76. (a) Temperature dependence of (▲) wavenumber and (■) intensity (in arbitrary units) of antisymmetric CH_2 stretching bands and (●) intensity ratio of antisymmetric CH_2 stretching band to symmetric one in normal-incidence transmission spectra of five-monolayer LB film of N-octadecanoyl-L-alanine fabricated onto CaF_2 substrate by vertical dipping method at surface pressure 40 mN · m⁻¹. (b) Pair of N-octadecanoyl-L-alanine molecules in tightly aggregated state. Reprinted, by permission, from X. Du, B. Shi, and Y. Liang, *Langmuir* **14**, 3631 (1998), p. 3635, Figs. 6 and 8. Copyright © 1998 American Chemical Society.

(temperature, surface pressure, or time) [521, 522, 524]. The advantage of using the peak values instead of the band area is that the results are less dependent on interference from adjacent, partially overlapped satellite bands. Changes in these features can be explained for the most part by changes in the average tilt angle, the relative amounts of trans and gauche conformers [532], surface density, and the oscillator strengths (TDMs) of the νCH_2 vibrations [533] (Section 3.11.3). If reorientation dominates, an increase in the chain disorder results in a decrease in the band intensities of the methylene stretching vibrations in the s-polarized reflection and normal-incidence transmission spectra (see an example in Fig. 3.76). At the same time, in IRRAS of LB films on metal substrates, the opposite trend is observed: With increasing disorder the intensity of the methylene stretching bands decreases [520, 534].

Another spectral parameter that depends on the order of the molecules within the film is the intensity ratios and the dichroic ratio for the modes whose TDMs have different orientations relative to the molecular frame. The origin of these dependences, which is connected with an increase in concentration of cis-isomers and/or the appearance of rotational freedom upon the transition from biaxial to uniaxial symmetry, is discussed in Section 3.11.3.

As disorder increases, the number of different gauche conformers increases, which causes changes in the absorption from methyl stretching modes (see, e.g., Ref. [509]). In addition, the wagging ($\delta_{wag}CH_2$) bands between 1300 and 1400 cm^{-1} are useful indicators of gauche defects. These bands are generally quite pronounced in the p-polarized spectra of adsorbed long-chain molecules, as the TDM of the wagging modes is parallel to the hydrocarbon chain [512]. There are several types of gauche defects, both single and double, which may be located either near the chain ends or near the chain center. For highly organized molecules with chains mostly in all-trans conformation, the CH_2 wagging modes couple and split, producing a series (progression) of bands between 1180 and 1350 cm^{-1}. The number of these bands is equal to twice the total number of carbon atoms in the chain [535]. As demonstrated by Snyder [536], in disordered films the end-gauche conformers are characterized by absorption at 1341 cm^{-1}, gauche–gauche at 1353 cm^{-1}, and wagging modes from the sum of gauche–trans-gauche and gauche–trans-gauche (kink) sequences appear at 1368 cm^{-1}. The wagging vibration bands in the DRIFTS-SEIRA spectra of stearic acid self-assembled on 2-μm silver particles are shown in Fig. 3.77, and the band assignments are summarized in Table 3.7.

3.11.2. Packing and Symmetry of Ultrathin Films

Films can be divided into three classes — isotropic, uniaxial, and biaxial — depending on the form of their permittivity tensor $\hat{\varepsilon}$ (1.1.2°). For isotropic films, the three principal values of the permittivity tensor are equal, $\hat{\varepsilon}_1 = \hat{\varepsilon}_2 = \hat{\varepsilon}_3$, and there is no preferred orientation within the film. The permittivity tensor of a *uniaxial* film has two different principal values, one that describes the propagation of radiation in the film plane and another that is perpendicular to the film plane, so that $\hat{\varepsilon}_x = \hat{\varepsilon}_y \neq \hat{\varepsilon}_z$. For a *biaxial* film all of the principal values of the permittivity tensor are different, so that $\hat{\varepsilon}_x \neq \hat{\varepsilon}_y \neq \hat{\varepsilon}_z$. For crystalline films with triclinic

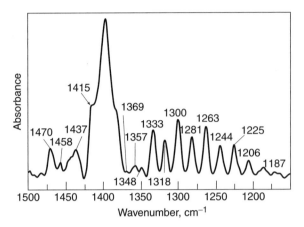

Figure 3.77. DRIFTS spectra of stearic acid (C_{18}) self-assembled on silver 2-μm particles in region of 1100–1500 cm^{-1}. Reprinted, by permission, from S. J. Lee and K. Kim, *Vib. Spectrosc.* **18**, 187 (1998), p. 192, Fig. 2. Copyright © 1998 Elsevier Science B.V.

Table 3.7. Assignment of IR peaks associated with scissoring and wagging modes of methylene groups of stearic acid self-assembled from ethanol on 2-μm silver particles[a]

Peak Frequency (cm^{-1})	Assignment
1470w	Scissor of all-trans CH_2 chain
1458vw	Scissor of CH_2 group next to a gauche bond
1437w	CH_2 wag + scissor for an end-gauche defect
1416sh	Scissor of CH_2 group adjacent to COO^-
1369vvw	CH_2 wag for internal kink defect
1357vw	CH_2 wag for double-gauche defect
1348vw	CH_2 wag for an end-gauche defect
1333m	Twist-rock + wag progression band
1318m	Twist-rock + wag progression band
1300m	Twist-rock + wag progression band
1281m	Twist-rock + wag progression band
1263m	Twist-rock + wag progression band
1244m	Twist-rock + wag progression band
1225m	Twist-rock + wag progression band
1206w	Twist-rock + wag progression band
1187w	Twist-rock + wag progression band

[a]DRIFTS, see Fig. 3.77.
Abbreviations: v, very; w, weak; sh, shoulder; m, medium.

Source: Reprinted, by permission, from S. J. Lee and K. Kim, *Vibrational Spectrosc.* **18**, 187 (1998), p. 192, Table 2. Copyright © 1998 Elsevier Science B.V.

and monoclinic crystal symmetry (Table 3.8), the permittivity tensor cannot be diagonalized (1.1.2°). However, the difference between the principal axes of the real and imaginary parts of the refractive index is generally ignored, so that this reduces to a biaxial class of film.

As seen from Eq. (1.29), the macroscopic optical anisotropy of the film (the number of different principal values of $\hat{\varepsilon}$) stems from the anisotropy of the molecular polarizability $\hat{\alpha}$ and the arrangement of the molecules in the film (their orientation and packing). If the molecules are randomly oriented, their distribution is characterized by three rotational degrees of freedom. The crystalline-like phase is characterized by a long-range periodic positional/translational order and the macroscopic symmetry of the permittivity tensor depends upon the point group of the lattice unit cell [537]. Depending on the number and order of rotational symmetry axes, there are seven categories of unit cells: *cubic*, *hexagonal*, *tetragonal*, *trigonal*, *orthorhombic*, *monoclinic*, and *triclinic*. The elementary primitive unit (*Bravais*) lattice cells and the determining symmetry elements for each category are shown in Table 3.8, which also demonstrates how the seven point group categories are distributed among the three symmetry classes of $\hat{\varepsilon}$. Note that the nodes in the cells can symbolize molecules. In the case of ultrathin long-chain molecular films, additional cells can form, which does not occur for three-dimensional structures [538].

As originally shown by Kitaigorodskii [539], the packing (subcell structure) of long-chain molecules in bulk crystalline phases is determined by geometric compatibility of the chains. Figure 3.78 shows the views of the aliphatic molecules packed in orthorhombic (tilt angle $0°$) and triclinic (tilt angle $\approx 30°$) cells. The hydrogen atoms on the chain are shown as circles, with the open circles representing hydrogen atoms attached to odd carbon atoms and the shaded circles representing those attached to even carbon atoms. The most densely packed is the *triclinic* cell, in which all molecules are identically oriented, such that there is one

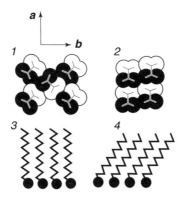

Figure 3.78. Schematic depictions of alkyl chains viewed along hydrocarbon backbone showing (1) orthorhombic and (2) triclinic subcell packings. Lines represent bonds and circles represent electron radii of hydrogen atoms. (3) Cartoon of aliphatic amphiphile displaying how zigzag carbon backbones of adjacent molecules can nest correctly if molecules are untilted with respect to surface normal. (4) Nesting of all-trans chains is undisturbed if all molecules are tilted by specific angle such that adjacent molecules are translated by one repeat distance along chain direction. Intermediate tilt angles do not preserve chain nesting; *a* and *b* are axes in molecular system. Adapted, by permission, from D. K. Schwartz, *Surf. Sci. Reports* **27**, 241 (1997), p. 290, Fig. 15. Copyright © 1997 Elsevier Science B.V.

Table 3.8. Primitive unit cells of Bravais lattices, which characterize symmetry of film†

Film Type	Crystal System	Axial Relationship	Point Group Symbols[a] International	Point Group Symbols[a] Schönflies	
Biaxial, $\hat{\varepsilon}_1 \neq \hat{\varepsilon}_2 \neq \hat{\varepsilon}_3$	Triclinic	$a \neq b \neq c$ $\alpha \neq \beta \neq \gamma$	1 $\bar{1}$	C_1 S_2	
	Monoclinic	$a \neq b \neq c$ $\alpha = \gamma = 90°$ $\beta \neq 90, 120°$	2 m $2/m$	C_2 C_v C_{2h}	
	Orthorhombic	$a \neq b \neq c$ $\alpha = \beta = \gamma = 90°$	$mm2$ 222 mmm	C_{2v} D_2 D_{2h}	
Uniaxial, $\hat{\varepsilon}_1 = \hat{\varepsilon}_2 \neq \hat{\varepsilon}_3$	Trigonal (rhombohedral)	$a = b = c$ $\alpha = \beta = \gamma \neq 90°$	3 $\bar{3}$ $3m$ 32 $\bar{3}/m$	C_3 S_6 C_{3v} D_3 D_{3d}	

System	Lattice parameters	International	Schoenflies
Tetragonal	$a = b \neq c$ $\alpha = \beta = \gamma = 90°$	4	C_4
		$\bar{4}$	S_4
		$4/m$	C_{4h}
		$4mm$	C_{4v}
		422	D_{2d}
		$4mm$	D_4
		$4/mmm$	D_{4h}
		$\bar{4}2m$	D_{2d}
Hexagonal	$a = b \neq c$ $\alpha = \beta = 90°$ $\gamma = 120°$	6	C_6
		$\bar{6}$	C_{3h}
		$6/m$	C_{6h}
		$6mm$	C_{6v}
		622	D_6
		$\bar{6}2m$	D_{3h}
		$6/mmm$	D_{6h}
Cubic	—	23	T
		$m\bar{3}$	T_h
		$\bar{4}3m$	T_d
		432	O
		$m\bar{3}m$	O_h
Isotropic, $\hat{\varepsilon}_1 = \hat{\varepsilon}_2 = \hat{\varepsilon}_3$			

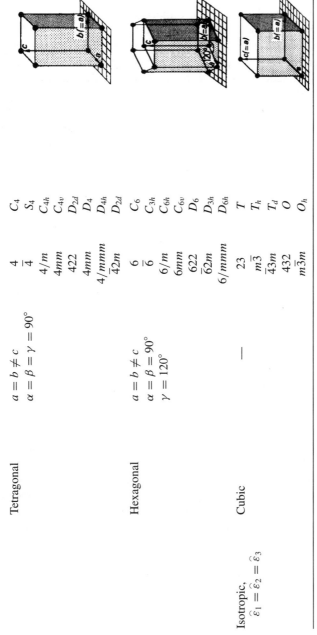

†To characterize arrangement of the adsorbate head groups on the substrate surface the IUPAC nomenclature for two dimensional Bravais cells is used [A. M. Bradshaw and N. V. Richardson, *Pure Appl. Chem.* **68**, 457 (1996)].

molecule per unit cell. The tangential contact of open and shaded circles implies that chains are tilted by ≈30°. The *orthorhombic* cell (the so-called *herringbone packing*) consists of two molecules with different orientations. Open and shaded circles contact only their own type in this arrangement, allowing the possibility of an untitled structure. In the general case, both the aforementioned types of packing result in biaxial symmetry within the film, but, if the planes of the chains with the CCC backbone in the orthorhombic subcell are orthogonal to each other, the resulting symmetry is uniaxial [468]. In the case of adsorbed monolayers, additional factors such as the relationship between the size of the head and end groups, the distance between and symmetry of the adsorption sites, and the conformational disorder in the tail part of the chain can distort the crystallinity as well [540].

Besides isotropic (liquidlike) and crystalline-like states, molecular films may go through a series of intermediate *mesophases* in which the molecules have some rotational mobility. The best-known three-dimensional mesophases are liquid crystals [541]. If there is long-range correlation in the orientation of the molecules in these mesophases, they are categorized in terms of the rotational degrees of freedom of the constituent molecules. The orientation of a long-chain molecule in the laboratory cartesian frame (the z-axis being defined perpendicular to the film and the y-axis along the direction of the s-polarization) is defined by three Euler angles (Fig. 3.75): the *tilt* angle γ (between the chain and the z-axis), the *azimuth* angle ϕ (between the projection of the chain axis onto the xy-plane and the x-axis), and the *twisting* angle ψ (between the CCC backbone plane and the plane formed by the z-axis with the chain axis vector). Films in which the molecular chains have the same angle γ with the surface normal and no preferred angles ϕ and ψ have uniaxial symmetry and are formally described as distributions with two rotational degrees of freedom. The long-range order of the adsorbed molecules with preferred tilt and azimuth angles has biaxial symmetry and one rotational degree of freedom.

Two-dimensional condensed mesophases in L monolayers have been thoroughly studied by various methods [474, 491, 542], as the method of monolayer formation allows for successive transitions from a less organized assembly to a more organized assembly by decreasing the molecular area. The dependence of the surface pressure of the L monolayer on the molecular area is called the $\pi - A$ *isotherm* (Fig. 3.79) and illustrates the aggregating process on the water surface. At high molecular areas, the isotherm exhibits the so-called gaseous (G)/liquid expanded (LE) state. Upon compression, a pure LE is achieved, followed in succession by a LE/liquid condensed (LC) phase, and a liquid condensed/solid (LS) phase (see an alternative terminology in Refs. [543, 544, 491]). These mesophases are called *rotator mesophases* since there is no long-range order with respect to the rotational degree of freedom of the molecules about their long axis and the chains can be considered as rods. When the monolayer is compressed beyond the LC phase, a kink appears in the isotherm, indicative of the transition to the liquid solid (LS) mesophase, and after that the relatively incompressible collapsed solid (CS) phase forms. Recently, it has been shown [538] that the untilted CS,

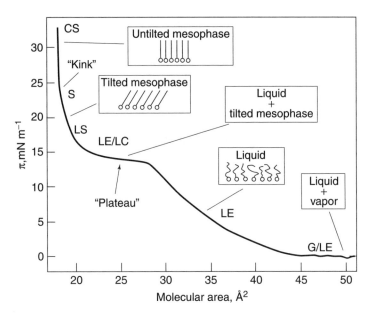

Figure 3.79. Surface pressure (π) versus molecular area isotherm of pentadecanoic (C_{15}) acid L monolayer on pure water at about 30°C. Various features of isotherm can be identified with different structural phases. Adapted, by permission, from D. K. Schwartz, *Surf. Sci. Reports* **27**, 241 (1997), p. 248, Fig. 1. Copyright © 1997 Elsevier Science B.V.

S, and LS phases of *n*-alkanes correspond to a herringbone crystal, a distorted rotator with no long-range herringbone order, and a hexagonal rotator, respectively. The tilted phases can be correlated with the same three categories with respect to distortion and herringbone order that characterize the untilted phases if the distortion is measured perpendicular to the chain axis. In addition, many ultrathin films exhibit a domainlike structure (examples include L monolayers in transition between phases [491] and simple fatty acid LB monolayers deposited from the LE phase [545, 546]). Therefore, the macroscopic symmetry depends on the domain symmetry and distribution. If the ultrathin film consists of planar molecules, such as oligothiophenes [547], the terminology accepted for liquid crystals is used to classify its symmetry.

Details of chain packing can be revealed by analyzing the shape and position of the methylene scissoring ($\delta_{scis}CH_2$) band (1460–1474 cm^{-1}) [495, 548–554]. It is known [551, 554] that the hydrocarbon chains in the first monolayer of LB films of fatty acids and their salts have significant positional, rotational, and conformational disorder but are, on an average, normal to the surface and hexagonally packed even after deposition of subsequent layers. The subsequent layers themselves are highly organized in the orthorhombic subcell [491]. The evolution of the $\delta_{scis}CH_2$ band of stearic acid deposited on Ge as the number of LB layers increases is shown in Fig. 3.80. The ATR spectrum of the first monolayer displays a single broad (FWHM $\approx$10–11 cm^{-1}) band at ~1466 cm^{-1} attributed to

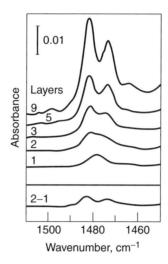

Figure 3.80. ATR spectra in CH_2 scissoring region of one-, two-, three-, five-, and nine-layer LB films of stearic acid on germanium. Adapted, by permission, from F. Kimura, J. Umemura, and T. Takenaka, *Langmuir* **2**, 96 (1986), p. 98, Fig. 5. Copyright © 1986 American Chemical Society.

a relatively disordered hexagonal subcell packing where the hydrocarbon chain is free to rotate around its long axis. (In a totally disordered structure, the scissoring motion of a methylene group next to a gauche bond is characterized by a similar broad band at ∼1466 cm^{-1}.) This band is split into the doublet with narrow (3–5-cm^{-1}) components at 1473 and 1463 cm^{-1} in thicker films, due to intermolecular interactions between two adjacent molecules in an orthorhombic perpendicular subcell. The same splitting was observed for a monoclinic (∼30° tilt) subcell [468, 548, 549]. Therefore, to distinguish between these two cases, additional information on the tilt of the chains is required. In the case of an orthorhombic inclined (∼30°) subcell, the splitting is 5–8 cm^{-1} [536, 556]. A further increase in packing density has been observed for a long-chain carboxylic acid on a Zn^{2+} subphase [511]. In this case a sharp, narrow (FWHM ∼3–6-cm^{-1}) singlet band is observed at 1471 cm^{-1}, indicative of a triclinic subcell packing (the most dense) where the hydrocarbon chains are parallel [555].

The $\delta_{scis}CH_2$ band splitting can be used for estimating the degree of chain segregation in mixtures, in which one component is hydrogenated and the other deuterated [549]. The method depends on the fact that the intermolecular coupling produces splitting only in the spectra of isotopically identical chains, if the chains are isotopically different, the coupling is negligible and the spectrum is not affected. As a consequence, clusters of like chains are vibrationally isolated from surrounding chains. The band splitting increases with the size of the domain.

The TDM of the higher frequency component (1472–1473 cm^{-1}) of the $\delta_{scis}CH_2$ mode of hydrocarbon chains packed in the orthorhombic subcell is directed along the a-axis of the subcell, while the lower frequency component

$(1462–1463 \text{ cm}^{-1})$ is directed along the b-axis [536, 556] (Fig. 3.78). The low-frequency component is significantly more sensitive to structural changes than the high-frequency one [556]. The setting angle θ (the angle between the chain direction and the a-axis) can be obtained from the intensity ratio of the scissor-band component intensities as [556]

$$\cot \theta = \left(\frac{A_{1460}}{A_{1470}} \right)^{1/2}.$$

Therefore, when the chain is perpendicular to the surface, the intensities of these components are equal. The same holds for the polarized spectra if the subcell is oriented 45° with respect to the x, y laboratory axes. Otherwise the intensity ratio depends upon the orientation of the subcell. Flach et al. [468] were able to determine the orientation of behenic acid ester in L monolayers ($\pi \approx 14 \text{ mN·m}^{-1}$) on pure D_2O (which was found to be 36° with respect to the y-axis) from this orientational dependence observed in the s- and p-polarized spectra obtained by IRRAS (40° angle of incidence) (Fig. 3.81). The absence of $\delta_{scis}CH_2$ bands in the spectra obtained by IRRAS of a thin film on a metal indicates an almost perpendicular orientation of chains in this film [557].

When the layer is inhomogeneous (domainlike), the use of this structure–frequency correlation strategy can be difficult. For example, in the IR spectra of 1-hexadecanol [526] L monolayers on pure water as the surface area decreases from 0.266 to 0.192 nm^2/molecule, the $\delta_{scis}CH_2$ band broadens only slightly and shifts from 1465.0 ± 0.2 to $1466.50 \pm 0.2 \text{ cm}^{-1}$. This has been ascribed to the

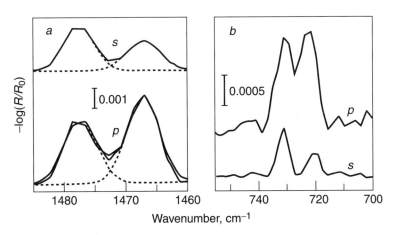

Figure 3.81. Split of methylene (a) scissors and (b) rocking mode in s- and p-polarized IRRAS of behenic acid ester L monolayer ($\pi \approx 14 \text{ mN · m}^{-1}$) on pure (a) D_2O and (b) H_2O. Spectra were measured at 40° angle of incidence. Adapted, by permission, from C. R. Flach, A. Gericke, and R. Mendelsohn, *J. Phys. Chem. B* **101**, 58 (1997), pp. 61 (Fig. 6) and 64 (Fig. 12). Copyright © 1995 American Chemical Society.

fact that under the given conditions the film is a mixture of orthorhombic (slightly tilted) and hexagonal (untilted) packed subcells.

The presence of two molecules per unit cell can also be deduced by the splitting of the *rocking* (ρCH_2) band into a doublet at 729 and 720 cm^{-1} (Fig. 3.81*b*) [468, 505, 548, 549].

Although the effect of the mutual arrangement of the molecules in a subcell on the CH *stretching* bands has been found to be negligible [507, 527], the ratio of the band intensities can be used for determining the film symmetry [558] (see Section 3.11.3 for details).

3.11.3. Orientation

The MO measurements provide information about the angular distribution of molecules in the x, y, and z film coordinates. To extract MO data from IR spectra, the general selection rule equation (1.27) is invoked, which states that the absorption of linearly polarized radiation depends upon the orientation of the TDM of the given mode relative to the local electric field vector. If the TDM vector is distributed anisotropically in the sample, the macroscopic result is selective absorption of linearly polarized radiation propagating in different directions, as described by an anisotropic permittivity tensor $\hat{\varepsilon}$. Thus, it is the anisotropic optical constants of the ultrathin film (or their ratios) that are measured and then correlated with the MO parameters. Unlike for "thick" samples, this problem is complicated by optical effects in the IR spectra of ultrathin films, so that optical theory (Sections 1.5–1.7) must be considered, in addition to the statistical formulas that establish the connection between the principal values of the permittivity tensor $\hat{\varepsilon}$ and the MO parameters. In fact, a thorough study of the MO in ultrathin films requires judicious selection not only of the theoretical model for extracting MO data from the IR spectra (this section) but also of the optimum experimental technique and conditions [angle(s) of incidence] for these measurements (Section 3.11.5).

Theoretical and experimental IRRAS results of studying MO in monolayers at the AW interface have been reviewed by Mendelsohn et al. [67] and Horn [532]. Horn [532] has also discussed the IRRAS experimental data for SAMs, LB films, and small molecules on metals. The application of ATR to MO measurements has been considered by Mirabella [559] from a theoretical and practical point of view. Although polarized transmission spectroscopy at inclined angles of incidence is as sensitive as the ATR and IRRAS methods (Section 2.1), the former method has been used in very few MO studies [560–563]. Normal-incidence transmission measurements combined with metallic IRRAS has been exploited for the analysis of MO, first by Greenler [564] and then by others [247, 551, 565–568]. Yarwood et al. have proposed to use ATR in combination with metallic IRRAS [569] or the normal-incidence transmission method [570]. Ishino and Ishida [571] combined normal-incidence transmission and MOATR. Below, these problems are discussed, mainly for long-chain molecules.

By neglecting dispersion in the real part of the refractive index for matter in which all the dynamic dipole moments of the given mode are perfectly aligned,

Eqs. (1.15), (1.52), and (1.53a,b) give an absorption index k_{max} that is proportional to the number of oscillators per unit area or volume, N, and the TDM value $|\langle j|\mathbf{\rho}|i\rangle|$, so that

$$k_{max} \propto N|\langle j|\mathbf{\rho}|i\rangle|^2. \qquad (3.44)$$

It follows that when the dipoles have a particular orientation in the film, the principal values of the absorption index k_i, where $i = x$, y, z, are related to k_{max} as

$$k_i = k_{max} \cos^2(\mathbf{i}, \mathbf{k}_{max}), \qquad (3.45)$$

where $\mathbf{k}_{max}$ ($|\mathbf{k}_{max}| = k_{max}$) is the vector directed along the given TDM and $(\mathbf{i}, \mathbf{k}_{max})$ are the angles between $\mathbf{k}_{max}$ and the laboratory axes (Fig. 3.82). After elementary trigonometric manipulations, Eq. (3.45) yields [507]

$$
\begin{aligned}
k_x &= k_{max} \sin^2 \theta \cos^2 \phi, \\
k_y &= k_{max} \sin^2 \theta \sin^2 \phi, \\
k_z &= k_{max} \cos^2 \theta,
\end{aligned}
\qquad (3.46)
$$

where the Euler angles θ and ϕ are defined in Fig. 3.82. Since the chain axis and the TDMs of the $\nu_{as}CH_2$ and ν_sCH_2 stretching vibrations in the all-trans hydrocarbon chains are mutually orthogonal, the tilt angle of the chain, γ_{chain}, can be deduced from the measured tilts of the methylene stretching vibrations (θ_{as} and θ_s) from the following relationship:

$$\cos^2 \gamma_{chain} + \cos^2 \theta_{as} + \cos^2 \theta_s = 1. \qquad (3.47)$$

The correlation between the absorption index and the MO and MP must be considered when estimating the surface concentration of adsorbed molecules from band intensities using a method such as that suggested by Sperline et al. [572, 573].

If the molecules are uniaxially or biaxially distributed, characterization of the MO becomes more complicated, and order parameters are used, which are

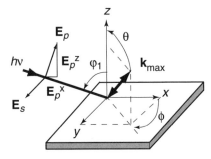

Figure 3.82. Definition of Euler angles describing orientation and principal components of maximum absorption index vector, $\mathbf{k}_{max}$.

some functions of averaged directional cosines $\langle \cos^2(\mathbf{i}, \mathbf{k_{max}}) \rangle$, where averaging is over all molecules in the film. The probability that the molecules adopt a specific orientation (γ, ϕ, ψ) in the film (laboratory) axis frame (Fig. 3.75) is called the (orientational) *distribution function* (DF), $N(\gamma, \phi, \psi)$. The theory of DFs can be found in the work of Zannoni [574] and Jarvis et al. [575]. In brief, the DF is based on a Legendre polynomial expansion P_i [576], the form of which depends upon the symmetry of the molecule and its distribution in the film. To characterize the DF completely, P_i should be measured up to $n = \infty$, which is impossible.

For a uniaxial distribution of rodlike molecules, uneven terms in the DF vanish upon averaging, and the main contribution is from the second-order term s (sometimes referred to as the long-axis order parameter [577], first introduced by Tsvetkov [578]):

$$s \equiv \langle P_2(\cos \gamma) \rangle = \tfrac{1}{2}\langle 3 \cos^2 \gamma - 1 \rangle. \tag{3.48}$$

Here, γ is the tilt angle of the long axis and the brackets indicate averaging over all molecules in the ensemble. By using IR spectroscopy, one can measure this quantity exclusively; it ranges from unity, characterizing perfect alignment along the z-axis ($\gamma = 0°$), to -0.5 for perfect alignment perpendicular to z ($\gamma = 90°$). However, several different DFs could give rise to the same average values of s [577, 579]. For the present case, the principal values of the absorption index are given by Fraser's equations [580]:

$$\begin{aligned}
k_x &= k_y = k_\perp = k_{max}[\tfrac{1}{2}s(\sin^2 \alpha) + \tfrac{1}{3}(1 - s)], \\
k_z &= k_\parallel = k_{max}[s \cos^2 \alpha + \tfrac{1}{3}(1 - s)],
\end{aligned} \tag{3.49}$$

where α is the angle between the given TDM and the long molecular axis. When the TDM is directed perpendicular the the long molecular axis, so that $\alpha = 90°$, Eq. (3.49) transforms into

$$k_\perp = \tfrac{1}{4}k_{max}(\langle \cos^2 \gamma \rangle + 1), \qquad k_\parallel = \tfrac{1}{2}k_{max}\langle \sin^2 \gamma \rangle, \tag{3.50}$$

using the classical averages $\langle \cos^2 \varphi \rangle = \langle \sin^2 \varphi \rangle = \tfrac{1}{2}$ in Eq. (3.46). One can see immediately from Eq. (3.50) that when $\langle \sin^2 \gamma \rangle = \tfrac{2}{3}$ ($s = 0$), the order does not affect the absorption, and this case is formally indistinguishable from the case of random orientation. The angle $\gamma \approx 54.7°$ at which this occurs is called the *magic* or *isotropic angle*. One can see from Eq. (3.50) that if the average tilt of a hydrocarbon chain increases from $0°$, the absorbance in the normal-incidence transmission spectrum decreases, but increases in the metallic IRRAS, which is typically monitored in the ODT studies (see, e.g., Refs. [500, 504, 534, 581, 582]).

If the film is macroscopically uniaxial but the molecules have biaxial symmetry, the short molecular axes are not aligned identically relative to the film's z-axis. The formulas for the principal values of the absorption index in this case have been given by Korte [583, 584]:

$$k_\perp = \tfrac{1}{3}k_{max}[1 - s(1 - \tfrac{3}{2}\sin^2\alpha) - \tfrac{1}{2}d\sin^2\alpha\cos 2\psi],$$
$$k_\parallel = \tfrac{1}{3}k_{max}[1 + 2s(1 - \tfrac{3}{2}\sin^2\alpha) + d\sin^2\alpha\cos 2\psi],$$

(3.51)

where ψ is the molecular twist angle (Fig. 3.75), $d \equiv \tfrac{3}{2}\langle\sin^2\gamma\cos 2\psi\rangle$ is an additional (transverse-axes) order parameter, and α is the angle between the given TDM and the long molecular axis.

The general expressions for the DF and the order parameters of a biaxial film comprised of molecules with cylindrical symmetry are rather cumbersome [577]. For the TDMs perpendicular to the long molecular axis such as those for $\nu_{as}CH_2$ and ν_sCH_2 stretching vibrations [517],

$$k_x = \tfrac{1}{2}k_{max}(1 - \sin^2\gamma\cos^2\phi),$$
$$k_y = \tfrac{1}{2}k_{max}(1 - \sin^2\gamma\sin^2\phi),$$
$$k_z = \tfrac{1}{2}k_{max}\sin^2\gamma,$$

(3.52)

where γ and ϕ are the tilt and azimuth angles of the carbon–hydrogen chain, respectively. At ϕ of $45°$ and $135°$ (the magic angles), biaxial symmetry is indistinguishable from the uniaxial one.

The principal components of the real part of the refractive index, n_i, can also be expressed in terms of the order parameters. General formulas for the ordinary and extraordinary refractive indices have been developed that depend upon the molecular orientation in the film [585]. For an ensemble of uniaxially distributed hydrocarbon chains, the principal components are

$$n_x = n_y = n_\perp = n_{ext}\sin^2\gamma + n_{ord}\cos^2\gamma,$$
$$n_z = n_\parallel = n_{ext}\cos^2\gamma + n_{ord}\sin^2\gamma,$$

(3.53)

where the ordinary (n_{ord}) and extraordinary (n_{ext}) refractive indices are perpendicular and parallel to the chain axis, respectively. Another way to obtain the real part of the refractive index is by applying the KK relations (vide infra).

It is evident from Eqs. (3.46) and (3.50)–(3.53) that the MO may be determined by fitting (1) the principal components of the absorption index or (2) the ratios of the principal components of the absorption index. These two approaches will be considered in more detail.

The first ("spectrum fitting") approach involves producing simulated spectra to fit the experimental ones in an iterative manner. In the method of Parikh and Allara [507], k_{max} is obtained from the relationship (easily determined from the spatial averaging [507])

$$k_{max} = 3.0k_{iso},$$

(3.54)

where k_{iso} is the optical constant of material in which the TDMs are distributed isotropically. Assuming that the oscillator strengths do not change appreciably upon chemisorption of the molecules, the value of k_{iso} can be determined from the transmission spectrum of the molecular species in a KBr pellet and the KK transformation. Typical values of k_{iso} and FWHMs for the CH stretching bands are shown in Table 3.6. After selecting the appropriate formulas for the absorption indices [Eqs. (3.46), (3.49)–(3.52)] which depend on the molecule and film symmetry, the real part of the refractive indices is calculated with the KK relations, and the film spectra are simulated using the matrix method (Section 1.7). The procedure is repeated by a coordinate transformation (rotation of the molecular axes in the laboratory coordinate system) [586] until the calculated spectrum matches the experimental one. The advantage of this method is that films with anisotropy up to and including full biaxial symmetry may be studied. However, as follows from Eq. (3.45), k_{iso} and, hence, k_{max} are related to the molecular density and packing. This explains in part the wide diversity in the k_{max} values reported in the literature. For example, for the $\nu_{as}CH_2$ bands of a LB film of CdAr, the values of 1.04 [471] and 0.6 [588] have been reported, and for the L monolayer of octadecanol in the solid state, the values of 1.31 [589] and 0.51 [470] were obtained. An illustration of this can be seen in the variation with the temperature of k_{iso} for the methylene and carboxylate stretching vibrations of bulk cadmium stearate (Fig. 3.83). These effects can be eliminated by varying *both* the orientation angles and k_{max} [506] or deriving k_{iso} (or k_{max}) from a suitable reference phase in which the molecules are in the same phase state as in the film [590, 591]. It has been reported [592] that k_{max} can be adjusted so that the derived tilt angle is independent of the film thickness, smaller thicknesses requiring larger values of k_{max}.[†] To determine the MO in thin ethylene-vinyl acetate copolymer layers deposited on Al mirrors, Brogly et al. [587] obtained the absorption index spectra from the specular reflection spectrum of a single copolymer layer measured at near-normal angle of incidence and the KK transforms.

A more accurate spectrum fitting approach [247, 566–568, 593] is to determine the anisotropic optical constants of the film from the metallic IRRAS and transmission spectra. It can be seen from Eqs. (1.90) that, within the linear approximation, the spectra of an ultrathin film on a metal substrate obtained by IRRAS give the LO energy loss function (which characterizes the energy dissipation in the direction perpendicular to the film surface) $Im(1/\hat{\varepsilon}_\parallel)$, where $\hat{\varepsilon}_\parallel = \hat{\varepsilon}_{2z}$, and the normal-incidence transmission spectrum [Eq. (1.98)] gives the TO energy loss function $Im\,\hat{\varepsilon}_\perp$ ($\hat{\varepsilon}_x = \hat{\varepsilon}_y = \hat{\varepsilon}_\perp$). Alternatively, the value of $Im(1/\hat{\varepsilon}_\parallel)$ can be obtained from the incline-angle p-polarized transmission or reflection spectrum using the value of $Im\,\hat{\varepsilon}_\perp$ calculated from the normal-incidence transmission or s-polarized reflection spectrum, respectively. If the imaginary parts are known, the real parts of $\hat{\varepsilon}_\parallel$ and $\hat{\varepsilon}_\perp$ can be calculated by KK relations. Then, the imaginary part of the permittivity is perturbed until the difference between the simulated and experimental spectra is minimized. In this stage, the spectrum fitting is performed using a matrix method. Anisotropic absorption indices for uniaxial CdAr LB

[†] The difference between the optical constants of a material in the bulk and ultrathin-film form was discussed in R. Arnold, A. Terfort, and C. Woll, *Langmuir* **17**, 4980 (2001).

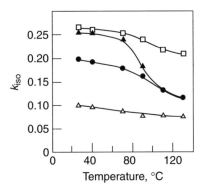

Figure 3.83. Temperature dependence of absorption index (k_{iso}) in maximum of ($\square$) $\nu_{as}CH_2$, ($\bullet$) ν_sCH_2, ($\blacktriangle$) $\nu_{as}COO^-$, and ($\triangle$) ν_sCOO^- bands derived from transmission (KBr pellet) spectrum of bulk CdSt. Reprinted, by permission, from T. Hasegawa, S. Takeda, A. Kawaguchi, and J. Umemura, *Langmuir* **11**, 1236 (1995), p. 1241, Fig. 8. Copyright © 1995 American Chemical Society.

films [247] are shown in Fig. 3.84 and Tables 3.9 and 3.10. These indices correspond to a tilt angle of ~12° that is in good agreement with the values available in literature. Notice that the k_{iso} for the $\nu_{as}CH_2$ and ν_sCH_2 modes were 0.32 and 0.22, respectively, where $k_{iso} = \frac{1}{3}(k_\parallel + 2k_\perp)$. These values are close to those obtained by Popenoe et al. [594] from the transmission spectra of crystalline octadecathiol in KBr, which testifies to the validity of using the crystalline phase as a reference for LB films in this case. The same strategy has been applied to a uniaxial monolayer LB film of bacteriorhodopsin (bR) (the only protein found in purple membranes of the halophilic bacteria *H. halobium*) (Fig. 3.85) deposited on CaF$_2$ (for the transmission measurement) and Au (for the IRRAS measurement) [568a]. As seen from Figure 3.85, k_{max} of the amide I band is ~1.0, which will result in a TO–LO splitting of ~20 cm^{-1} when the mode TDM is perpendicular to the surface. This uncertainty in the band position should be taken into account when the protein secondary structure–frequency correlation (Table 7.9 in Chapter 7) is applied to IR spectra of ultrathin biofilms.

The method of Buffeteau et al. [247, 567] was extended to measure the optical constants of biaxial film [568b, 593a]. In this case, at least two measurements with the electric field parallel to the film surface are required, where for the second measurement the sample is rotated by 90° around the z-axis or the normal-incidence transmission is measured with s- and p-polarization. Considering the molecular symmetry and transforming the complex permittivity tensor to the molecular coordinate system, the direction of a specific TDM relative to the molecular axis (molecular conformation) can be determined [568b]. This advantage is valuable for characterizing the secondary structure of absorbed proteins (Section 7.8.1). When the optical constants in the molecular system are known, spectra for different orientations of the molecule under different experimental conditions can be simulated. This approach was used for quantifying the MOs in monolayers from the PM-IRRAS [566, 568b] and ATR [593].

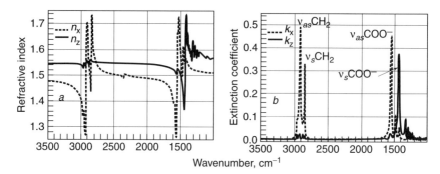

Figure 3.84. (a) Anisotropic refractive indices ($n_x = n_y = n_\perp$ and $n_z = n_\parallel$) and (b) absorption indices ($k_x = k_y = k_\perp$ and $k_z = k_\parallel$) of 10 CdAr monolayers deposited on solid substrates. In-plane and out-of-plane complex refractive indices were calculated from transmittance and IRRAS spectra, respectively, using refractive indices in visible region of $n_{\infty,\perp} = 1.49$ and $n_{\infty,\parallel} = 1.55$ and a monolayer thickness of 26.4 Å. Reprinted, by permission, from T. Buffeteau, D. Blaudez, E. Pere, and B. Desbat, *J. Phys. Chem. B* **103**, 5020 (1999), p. 5025, Fig. 3. Copyright © 1999 American Chemical Society.

Table 3.9. Maximum in-plane absorption indices ($k_{x,\mathrm{max}}$) for several IR modes of CdAr LB films of absorption index spectrum in Fig. 3.84

Layer Number	Substrate	$\nu_{as}CH_2$, 2919 cm^{-1}	$\nu_s CH_2$, 2850 cm^{-1}	$\nu_{as}COO^-$, 1545 cm^{-1}	$\nu_s COO^-$, 1422 cm^{-1}
8	CaF$_2$	0.458	0.320	0.426	0.053
	CaF$_2$	0.483	0.338	0.418	0.068
	CaF$_2$	0.470	0.331	0.417	0.067
10	ZnSe	0.490	0.336	0.448	0.055
	Si	0.464	0.327	0.443	0.055

Source: Reprinted, by permission, from T. Buffeteau, D. Blaudez, E. Pere, and B. Desbat, *J. Phys. Chem. B* **103**, 5020 (1999), p. 5025. Copyright © 1999 American Chemical Society.

Table 3.10. Maximum out-of-plane absorption indices ($k_{z,\mathrm{max}}$) for several IR modes of CdAr LB films of absorption index spectrum in Fig. 3.84

Layer Number	Substrate	Angle of Incidence	$\nu_{as}CH_2$, 2919 cm^{-1}	$\nu_s CH_2$, 2850 cm^{-1}	$\nu_{as}COO^-$ 1548 cm^{-1}	$\nu_s COO^-$ 1432 cm^{-1}
8	Gold	75°	0.024	0.014	0.024	0.370
	Gold	80°	0.023	0.014	0.025	0.371
	Gold	85°	0.022	0.013	0.025	0.363

Source: Reprinted, by permission, from T. Buffeteau, D. Blaudez, E. Pere, and B. Desbat, *J. Phys. Chem. B* **103**, 5020 (1999), p. 5025. Copyright © 1999 American Chemical Society.

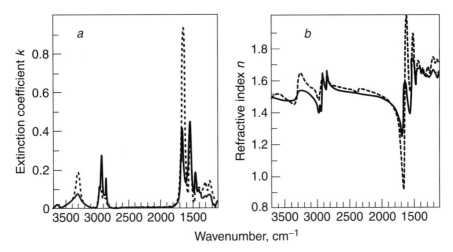

Figure 3.85. (*a*) Anisotropic absorption indices $k_\perp$ (solid line) and $k_\parallel$ (dashed line) and (*b*) refractive indices $n_\perp$ (solid line) and $n_\parallel$ (dashed line) of monolayer of bacteriorhodopsin (bR). In-plane and out-of-plane complex refractive indices were calculated from transmittance and metallic IRRAS spectra, respectively, using refractive indices in visible region of $n_\perp = 1.52$ and $n_\parallel = 1.55$ and monolayer thickness of 38 ± 3 Å. Reprinted, by permission, from D. Blaudez, F. Boucher, T. Buffeteau, B. Desbat, M. Grandbois, and C. Salesse, *Appl. Spectrosc.* **53**, 1299 (1999), p. 1301, Fig. 2. Copyright © 1999 Society for Applied Spectroscopy.

Since the substrate may influence the anisotropic optical properties of the overlying film [595], the method of Buffeteau et al. [247, 566–568, 593] is conceptually more reliable when the MO is studied on solid transparent substrates, whereas the initial anisotropic optical constants are extracted from normal- and oblique-incidence transmission or polarized reflection of the same film on the same substrate. In the case when different substrates participate into the measurements (e.g., when MO in monolayers at the AW interface is studied), the comparison of the simulated and experimental spectra can be used for distinguishing chemical effects generated by specific film–substrate interactions [568b]. In particular, the k_{max} values derived from spectra of monolayers at the AW interface obtained by IRRAS are usually larger than those obtained by ellipsometric measurements of thin films on solid supports [247]. This difference has been attributed to a gradient in the optical properties of the interfacial water [71].

To simplify the calculations for L monolayers, Hasegawa et al. [581] have used a modified form of Eq. (3.54) to describe the dependence of the absorption index on the maximum of the LB film on compression:

$$k_{max} = 3.0 k_{iso} \frac{d_{st}}{d_{LB}^*} \frac{A_{st}}{A_{LB}}. \tag{3.55}$$

Here, d_{LB}^* and d_{st} are the thicknesses of an LB film of tilted molecules and a standard LB film of stearic acid, and A_{LB} and A_{st} are the molecular areas of the former and the latter, respectively. Mendelsohn et al. [542] used a fitting procedure, in which the anomalous dispersion of the real and imaginary parts of the absorption indices modeled by a Lorentzian or Gaussian distribution are given by (for the Lorentzian case)

$$
k_{\perp(\parallel)}(\nu) = \frac{k_{\perp(\parallel),\max} g^2}{4\Delta^2 + g^2},
$$
$$
n_{\perp(\parallel)}(\nu) = \frac{n_{\perp(\parallel)} - 2\Delta k_{\perp(\parallel),\max} g}{4\Delta^2 + g^2},
$$

(3.56)

where $g = 2\pi c(\text{FWHM})$, $\Delta = 2\pi c(\nu - \nu_0)$, c is the speed of light, ν_0 is the center frequency, and ν is the frequency for which the calibration is being made. The anisotropic optical constants are specified by Eqs. (3.49) and (3.53) with $n_{\text{ord}} = 1.46$ and $n_{\text{ext}} = 1.58$. Flach et al. used the same procedure [468] but neglected the anisotropy of the real part of the refractive index. Instead, the isotropic values of 1.41 and 1.40 were used for the $\nu_{as}\text{CH}_2$ and $\nu\text{C=O}$ modes, respectively, based on the fact that changes in n within the 1.4–1.6 range have little effect on the absorbance of p-polarized radiation at angles far from φ_B [468]. This can be further simplified by fitting only the band peak intensity. Such simulations have been performed for uniaxial films of long-chain molecules using the classical three-layer Fresnel formulas (Section 1.5) [506, 523, 581, 589, 596, 597] or the thin-layer approximation (1.87) [588, 598] for spectra measured by IRRAS in either one or both polarization states. The real parts of the refractive indices can be assumed to be equal to their asymptotic values in the high-frequency region [468]. Another approach to the determination of anisotropic refractive indices for organic thin films uses Schopper's theory [599] and can be found in works of Tomar and Srivastava [600–602] and Elsharkawi and Kao [603].

Experimental parameters such as the angle of incidence and film thickness are canceled out if the orientation is calculated from the ratio of the band intensities in the metallic IRRAS of the film and in the transmission (KBr) spectrum of the isotropic bulk sample. For a crystalline-like film, the ratio of the intensities of the $\nu_{as}\text{CH}_2$ and $\nu_s\text{CH}_2$ bands (A_{as} and A_s, respectively) gives the twist angle directly [604–606]:

$$
\psi = \arctan\left[\left(\frac{A_{as}}{A_s}\right)_{\text{film}} \left(\frac{A_s}{A_{as}}\right)_{\text{KBr}}\right]^{1/2}.
$$

(3.57)

Servant et al. [503, 607] have shown that the spectra of the imaginary parts of the surface susceptibility tensor [Eq. (1.89)], which are easily obtained from the polarized IR spectra with no measurement of the film thickness, take into account the domainlike composition typical of LB films. However, this approach demands a priori knowledge of the polarizability of the adsorbed molecule — a parameter that can be calculated assuming a certain microscopic model of the molecular packing only.

The MO in the film can be determined by fitting the dichroic ratio (DR) of a selected band in ATR [517, 518, 559, 608–617] or IRRAS [558, 618–620] spectra or the ratio of the band intensities in the metallic IRRAS and normal-incidence transmission spectra [551, 621], which will be referred to below as the "DR fitting" approach. In the spectroscopy of ultrathin films, the DR is defined as the ratio of the peak (or integrated) intensity of the absorption band in the s-polarized spectrum, A^s, to that in the p-polarized spectrum, A^p:

$$DR = \frac{A^s}{A^p}. \tag{3.58}$$

Similar to the surface susceptibility method [503, 507], DR fitting allows one to exclude the film thickness and optical constants as input parameters in spectral simulations, a very important advantage in the case of adlayers when there is no way to determine the film thickness accurately [622]. As opposed to spectrum fitting, DR fitting is applicable to the cases when (1) properties of the film depend on the substrate (the general case), (2) preliminary information about the film is poor or absent, and (3) when the film can not be perfectly reproduced in a series of experiments. Moreover, as will be shown below, DR fitting is more sensitive to MO.

A typical procedure of converting the DR measured from ATR spectra or A_{IRRAS}/A_{tr} into angles of orientation requires the MSEF components within the layer.[†] For example, the DR measured from the ATR spectrum of an uniaxial film for a mode perpendicular to a long-chain molecule is expressed as [517]

$$DR = \frac{E_y^2(2 - \sin^2 \gamma)}{E_x^2(2 - \sin^2 \gamma) + 2E_z^2 \sin^2 \gamma}, \tag{3.59}$$

where E_x^2, E_y^2, and E_z^2 are the MSEF components along the laboratory axes, calculated as described in Section 1.8, assuming that the film is transparent ($k_2 = 0$). As can be seen from Fig. 3.86a, for an Si ATR element and $\varphi_1 = 45°$, the change in the absorption index of the film from 0 to 0.6 substantially affects the DR value for modes perpendicular to the molecule axis (e.g., the CH_2 stretching vibrations) at tilt angles larger than 20°. For modes directed along the molecule axis, error in the DR is essential for angles from 10° to 80° [517a]. This systematic error might explain the inadequacy noticed by Schwartz [491]

[†] By analogy with Eq. (3.44), the absorbance A_i of the layer, measured with the IR radiation polarized along the ith laboratory axis, is written as:

$$A_i = \int_\gamma \int_\phi \int_\psi (M_i E_i)^2 N(\gamma, \phi, \psi) \, d\gamma \, d\phi \, d\psi,$$

where M_i and E_i are the respective projections of the TDM and the electric field vector on the ith axis. $N(\gamma, \phi, \psi)$ is the orientation distribution function, which is expressed by the common statistical formula.

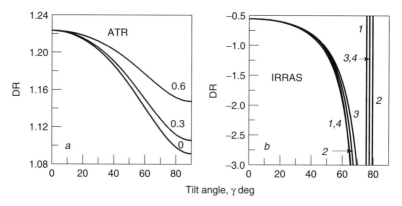

Figure 3.86. Effect of absorption index on dichroic ratio (DR) of the $\nu_{as}CH_2$ band of carbon–hydrogen chain as function of chain tilt angle in (a) ATR and (b) IRRAS spectra of uniaxial monolayer 2 nm thick on silicon and quartz, respectively: (a) $n_1 = 3.42$, $\varphi_1 = 45°$, $n_{2\parallel} = n_{2\perp} = 1.5$, absorption indices $k_{2\,max}$ are shown; (b) $\varphi_1 = 73°$, $n_3 = 1.49$; (1) $n_{2\parallel} = n_{2\perp} = 1.5$, $k_{2\,max} = 0.7$; (2) $n_{2\parallel} = n_{2\perp} = 1.5$, $k_{2\,max} = 0.9$; (3) $n_{2\parallel} = 1.55$, $n_{2\perp} = 1.49$, $k_{2\,max} = 1.04$; (4) $n_{2\parallel} = n_{2\perp} = 1.5$, $k_{2\,max} = 1.04$.

in the MO data obtained by Ahn et al. [517a] for the LB films of stearates of Ca, Cd, and Pb deposited on Ge, etched Si, and oxidized Si and the difference in orientation of α-helix of the bacteriorhodopsin (bR) LB films, obtained by DR fitting [502] and spectrum fitting [568a].

Some investigators [499, 613] evaluated the MSEFs using the so-called two-phase approximation that neglects the film at the interface. This model was found [608] to give more accurate MO results than the three-phase (substrate–film–environment) approximation, which was explained by a significant change in the optical properties as the film thickness decreases from the thick-film regime to the ultrathin regime. Finally, the MSEFs can be calculated using a matrix method (the thickness- and absorption-dependent formalism) based on the optical parameters of the film or Harrick's thin-film approximation formulas [237]. To test the validity of the different formalisms, s- and p-polarized ATR spectra of a lipid bilayer were simulated for different acyl chain tilt angles [593b]. The results show that for dry bilayers, the acyl chain tilt angle varies with the formalism used, while no significant variations are observed for the hydrated bilayers. The thickness- and absorption-dependent formalism using the mean values of the electric fields over the film thickness gives the most accurate values of acyl chain tilt angle in dry lipid films. However, for dry ultrathin ($d < 4$ nm) lipid monolayers or bilayers, the tilt angle can be determined with an acceptable accuracy using the Harrick thin-film approximation. This study also shows that the uncertainty on the determination of the tilt angle comes mostly from the experimental error on the dichroic ratio and from the knowledge of the film refractive index.

Chernyshova and Rao [558] suggested to characterize the MO and MP from IRRAS of ultrathin films on transparent substrates by fitting the DR calculated using either a matrix method or the linear approximation expressions derived by Mielczarski [Eq. (1.87)]. For a uniaxial film Eqs. (3.50) and (1.87) give

$$
\mathrm{DR} = \frac{A^y}{A^x + A^z}
$$

$$
= \frac{(2 - \sin^2 \gamma)/4(n_3^2 - 1)}{-\dfrac{\xi_3^2(2 - \sin^2 \gamma)}{4(\xi_3^2 - n_3^4 \cos^2 \varphi_1)} + \dfrac{\sin^2 \varphi_1 \sin^2 \gamma}{2(\xi_3^2/n_3^4 - \cos^2 \varphi_1)[n_2^2 + k_{2\,\mathrm{max}}^2 (\sin^2 \gamma/2)^2]^2}},
$$

(3.60)

where φ_1 is the angle of incidence, $\xi_3 \equiv \sqrt{\hat{n}_3^2 - \sin^2 \varphi_1}$ is the generalized complex refractive index of the substrate, n_2 and k_2 are, respectively, the refractive and absorption indices of the film, and n_3 is the refractive index of the substrate. As seen from Eq. (3.60), the quantity k_{max}^2 multiplied by the factor $(\frac{1}{2}\sin^2 \gamma)^2$, which is much less than unity, is present only in the second term in the denominator (responsible for A^z). Therefore, one can expect that the effect of k_{max} on the DR calculated with Eq. (3.60) is much weaker than on the absorbance [see Eqs. (1.87)]. This is confirmed by the dependences in Fig. 3.86b calculated for the $\nu_{\mathrm{as}}\mathrm{CH}_2$ band of a monolayer of a long-chain molecule adsorbed on quartz, measured by IRRAS at the optimal ($73°$) angle of incidence (Section 3.11.5). Figure 3.86 also allows comparison of the sensitivity of the DR fitting MO measurements from the ATR and IRRAS spectra. Changing γ from $0°$ to $30°$ produces the decrease in the DR by ~1% in the ATR spectra and by ~20% in the IRRAS spectra. Therefore, the DR fitting approach coupled with the IRRAS spectra measured under the optimum conditions is much more effective in MO measurements than that based on the ("standard") ATR spectra.

The standard criterion of biaxial symmetry of a sample is nonidentity of the polarized normal-incidence transmission spectra. One can also distinguish symmetry of the film comprised of long-chain molecules by analyzing either the DRs or the ratio of the intensities for the $\nu_{\mathrm{as}}\mathrm{CH}_2$ and $\nu_{\mathrm{s}}\mathrm{CH}_2$ bands, $A_{\mathrm{as}}/A_{\mathrm{s}}$, as follows.

The DR for biaxial distribution of the hydrocarbon chains can be calculated substituting Eqs. (3.52) in Eqs. (1.87). Figure 3.87 shows the DR of such a film plotted against the tilt angle γ and the azimuth angle ϕ. The function $\mathrm{DR}(\phi)$ has a period of $180°$, the maximum at $\phi = 90°$, and the regions of discontinuity when $|A^x| = |A^z|$. Obviously, since the DR surface is symmetrical relative to $\phi = 90°$, averaging the two DR values, which correspond to the arguments shifted by $90°$, gives the DR value at the magic angles ($45°$ or $135°$). This fact is of important practical significance, permitting extracting both the angles γ and ϕ from the DR values for the $\nu_{\mathrm{as}}\mathrm{CH}_2$ and $\nu_{\mathrm{s}}\mathrm{CH}_2$ modes ($\mathrm{DR}_{\mathrm{as}}$ and DR_{s}, respectively) whose TDMs and the chain axis are mutually orthogonal (Fig. 3.75). Namely, using the averaged value of the DRs ($\mathrm{DR}_{\mathrm{av}}$),

$$
\mathrm{DR}_{\mathrm{av}} = \tfrac{1}{2}(\mathrm{DR}_{\mathrm{as}} + \mathrm{DR}_{\mathrm{s}}),
$$

(3.61)

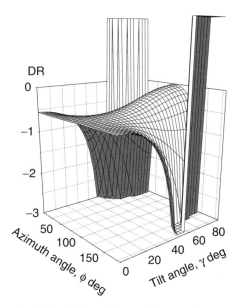

Figure 3.87. Dichroic ratio (DR) calculated for $\nu_{as}CH_2$ band in IRRAS spectra of biaxial monolayer of hexadecylamine on quartz as function of tilt and azimuth angles. Monolayer was characterized by $d_2 = 2.13$ nm, $n_{2\parallel} = 1.55$, $n_{2\perp} = 1.49$, $k_{2\,max} = 1.04$; refractive index of quartz $= 1.49$; angle of incidence $= 73°$. Reprinted, by permission, from I. V. Chernyshova and K. Hanumantha Rao, *J. Phys. Chem. B* **105**, 810 (2001), p. 816, Fig. 7. Copyright © 2001 American Chemical Society.

one can find the tilt angle from the DR(γ) plots for the uniaxial film characterized by the same set of optical parameters. After that, the azimuth angle can be determined from the DR(ϕ) plots constructed for the found tilt angle.

Nuzzo et al. [509] observed that with increasing temperature from 80 to 400 K, the intensity of the $\nu_{as}CH_2$ band of alkyl thiol SAMs on Au(111) decreases while that of the $\nu_s CH_2$ band is almost the same, which was attributed to the intrinsic temperature dependences of these modes. However, this effect can also be due to reducing the film symmetry. As follows from Eqs. (3.50) and (3.52), the A_{as}/A_s value obtained from the s-polarized spectrum of a uniaxial film is the same as for the isotropically distributed film material (if intrinsic spectral perturbations, such as the splitting due to the crystal field effect [504], do not redistribute the intensities). On the other hand, biaxial symmetry should change this ratio.

Using Eqs. (3.50) and (3.52), one can easily quantify this effect in the s-polarized IRRAS (or the normal-incidence transmission) spectra. Taking into account the shift of $90°$ in the ϕ values for the $\nu_{as}CH_2$ and $\nu_s CH_2$ modes, Eq. (3.50) for k_y gives

$$\left(\frac{A_{as}}{A_s}\right)^s = \frac{k_{max,as}}{k_{max,s}} \frac{1 - \sin^2 \gamma \sin^2 \phi}{1 - \sin^2 \gamma \cos^2 \phi}, \qquad (3.62)$$

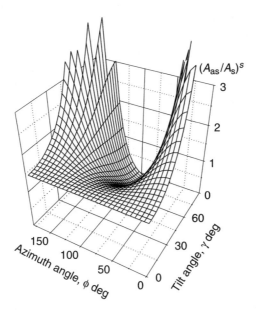

Figure 3.88. Dependence of ratio of peak intensities of $\nu_{as}CH_2$ and $\nu_s CH_2$ bands in s-polarized IRRAS spectra, $(A_{as}/A_s)^s$, of biaxial monolayer on quartz on tilt and azimuth angles, $k_{max,as}/k_{max,s} = 1.5$. Adapted, by permission, from I. V. Chernyshova and K. Hanumantha Rao, *J. Phys. Chem. B* **105**, 810 (2001), p. 816, Fig. 9. Copyright © 2001 American Chemical Society.

where A_{as} and A_s are, respectively, the peak intensities of the $\nu_{as}CH_2$ and $\nu_s CH_2$ bands in the s-polarized spectrum and $k_{max,as}$ and $k_{max,s}$ are, respectively, the maximum absorption indices for the $\nu_{as}CH_2$ and $\nu_s CH_2$ modes. It is seen that the loss of the azimuthal degree of freedom results in the dependence of the $(A_{as}/A_s)^s$ value on both the tilt and azimuth angles. The orientation dependence of the quantity $(A_{as}/A_s)^s$ for a biaxial monolayer is shown in Fig. 3.88. This dependence can be used for evaluating the azimuth angle [558]. However, to realize this approach, the value of $k_{max,as}/k_{max,s}$ is necessary. For ultrathin films in the crystalline state, one can assume that this quantity is equal to A_{as}/A_s[†] in the IR spectrum of the isotropically distributed crystals and can be measured, for example, from the transmission spectrum of the crystalline substance in a KBr pellet. If only the s-polarized or normal-incidence transmission spectra are available, the difference between the values of $(A_{as}/A_s)_{exp}$ and $k_{max,as}/k_{max,s}$ can be used for discriminating between uniaxial and biaxial symmetry of the film, in addition to the features of the $\delta_{cis}CH_2$ band (Section 3.11.2).

Another approach to distinguishing symmetry of the film is to compare the A_{as}/A_s values in the normal transmission (or s-polarized IRRAS) and metallic IRRAS spectra of the same film [623, 624]. This approach is valid provided that the change of the substrate does not affect the MO in the film, which rarely

---

[†] This assumption is invalid for the transformation from the crystalline to liquid form (Fig. 3.83).

occurs in practice. Binder et al. [625] presented the basic equations for the quantitative analysis of DR of biaxial lamellar structures measured by ATR in terms of transverse and longitudinal molecular order parameters. This formalism was applied to characterizing symmetry of a biaxial diene lipid bilayer, based on the DR of selected in-plane and out-of-plane vibrations of the diene groups. In the IR spectra of ultrathin films on metals Lee and Kim [370] have suggested the use of the ratios $A_{as}CH_2/A_{as}CH_3$ and $A_sCH_2/A_{as}CH_3$ as an indicator of the ODT, since the direction of the TDM vector of the $\nu_{as}^{in}CH_3$ was found to be almost independent of the molecular tilt with respect to the surface normal [626, 627].

By analogy with thick polymer films, a mixed (domainlike) distribution of the TDMs within the film may be envisaged in which a fraction of the molecules is perfectly aligned and the rest are randomly oriented (the Fraser–Beer model). Alternatively, the film may be considered as a collection of differently aligned species [579]. To take into account the difference in the packing and orientation of adsorbed molecules near the interface and in the end-group regions, a gradient in the anisotropic optical constants in the z-direction can be introduced in the calculations [470]. An ultrathin film may be considered as a stratified structure, and the distance from the geometric interface over which the orientation persists (the persistence length of molecular orientation) may be defined [628].

3.11.4. Surface Selection Rule for Dielectrics

The calculated angle-of-incidence dependences of the absorbance in the s- and p-polarized IRRAS of a 1-nm hypothetical isotropic layer on silicon and on glass (Figs. 3.89, 2.09, and 2.10) show that

$$A_p(\varphi_1) > 0 \text{ (for Si) and } A_p(\varphi_1) < 0 \text{ (for SiO}_2) \text{ at } \varphi_1 < \varphi_B,$$
$$A_p(\varphi_1) < 0 \text{ (for Si) and } A_p(\varphi_1) > 0 \text{ (for SiO}_2) \text{ at } \varphi_1 > \varphi_B. \tag{3.63}$$

The net absorbance in the p-polarized spectra A_p^x shown is further resolved into two components, A_p^x and A_p^z, which describe the film's interaction with the $\mathbf{E}_x$ and $\mathbf{E}_z$ components of the electric field, respectively (Fig. 3.82). The sign of these components differs from that of the sum. For both substrates,

$$A_p^x(\varphi_1) < 0 \quad \text{and} \quad A_p^z(\varphi_1) > 0 \quad \text{at } \varphi_1 < \varphi_B,$$

while $\tag{3.64}$

$$A_p^x(\varphi_1) > 0 \quad \text{and} \quad A_p^z(\varphi_1) < 0 \quad \text{at } \varphi_1 > \varphi_B.$$

Relationships (3.64) constitute the *surface selection rule* (SSR) for dielectrics (see also Fig. 3.18).

The SSR for dielectrics allows one to obtain a quick, qualitative impression about molecular orientations in ultrathin films. For example (Fig. 3.90), the spectrum of a SAM, obtained by IRRAS at $\varphi_1 = 80 \pm 3°$, exhibits positive and negative absorbance due to the methylene and methyl stretching modes,

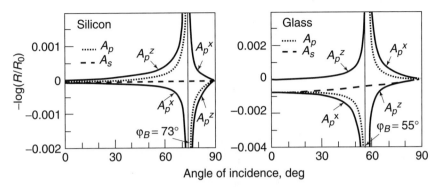

Figure 3.89. Calculated absorbances $-\log(R/R_0)$ in IRRAS spectra of hypothetical adsorbate vibration at 3000 cm^{-1} on silicon and glass surface as function of light incidence angle φ_1 for p-polarized radiation (A_p, dotted lines) and s-polarized radiation (A_s, dashed lines). Solid lines denoted with A_p^x and A_p^z represent parallel and perpendicular components, respectively, of total absorbance A_p; φ_B is Brewster angle. Optical constants; $d_2 = 1$ nm, $n_2 = 1.5$, $k_2 = 0.1$, $n_{glass} = 1.5$, $n_{Si} = 3.42$. Reprinted, by permission, from H. Brunner, U. Mayer, and H. Hoffmann, *Appl. Spectrosc.* **51**, 209 (1997), p. 211, Fig. 2. Copyright © 1997 Society for Applied Spectroscopy.

respectively. As follows from Eq. (3.64), the methylene and methyl stretching modes are mainly oriented in and out of the film plane, and, hence, the aliphatic chains are almost perpendicular to the surface (Fig. 3.75) (the numerical analysis gave the average chain tilt of 11° [598]). At the same time, all the bands are negative in the p-polarized IRRAS of a film of paraffin oil on Si, which testifies [Eq. (3.63)] that the film is isotropic (liquidlike). If a film is heterogeneous, consisting of both organized and isotropic domains, the p-polarized spectrum obtained by IRRAS represents a weighed sum of contributions from each domain (Fig. 3.90, curve b). Thus, the difference in the sign of the absorbance in the p-polarized spectra of isotropic and anisotropic films can be exploited not only for analyzing the degree of order in the film but also for elucidating the domainlike structure of the film [598].

Information about molecular orientations in ultrathin films on transparent substrates can also be gained from the polarity of the PM-IRRAS signals [Eq. (4.8)]. When the spectra are measured at the optimum angle of incidence, which is above the Brewster angle [629], a TDM lying in the surface plane gives rise to a positive PM-IRRAS signal, while a TDM perpendicular to the surface gives rise to a negative signal. This property is the SSR for PM-IRRAS. In the intermediate case, the PM-IRRAS band is a weighted sum of these two signals. As a result, for a particular ("vanishing") tilt angle, the two contributions balance and the absorption band disappears from the spectrum [71, 72]. The value of the vanishing angle depends on the refractive index of the substrate and the angle of incidence, increasing as n_3 increases. Observation of a vanished band in PM-IRRAS allows determination of the title angle of the corresponding mode and, hence, the title angle of the molecule. For example [72], the intensity of the νC−O band at

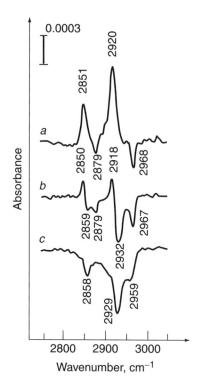

Figure 3.90. IRRAS spectra of (a) SAM of tetradecyltrichlorosilane (TTS) prepared in dry inert-gas atmosphere, (b) partly disordered film of TTS prepared in ambient atmosphere, and (c) liquid film of paraffin oil ($d \approx 10\text{Å}$) on silicon substrates (p-polarization, $80° \pm 3°$ incidence). Reprinted, by permission, from H. Hoffmann, U. Mayer, and A. Krischanitz, *Langmuir* **11**, 1304 (1995), p. 1311, Fig. 7. Copyright © 1995 American Chemical Society.

1060 cm^{-1} of a long-chain 1-alcohol monolayer on water is either zero or very slightly negative in the PM-IRRAS measured at the angle of incidence of $\sim76°$. Under these optical conditions, this fact means that the C–O mode is titled by $\sim36°$. Taking into account the sp^3 tetrahedral hybridization of the carbon and oxygen atoms and the fact that the OH polar group points into the water, one can expect that the aliphatic chains are perpendicular to the surface, which agrees with the high intensity of the νCH_2 bands.

3.11.5. Optimum Conditions for MO Studies

Of paramount importance when investigating MO is the appropriate choice of experimental method and conditions under which the spectra will have the highest SNR and sensitivity to the MO. The latter implies that three principal components of the absorption index can be measured with comparable SNRs. Except for films on metals, for which only the modes perpendicular to the surface can be detected (Section 1.8.2) and the optimum conditions are known (*p*-polarized radiation and grazing angles of incidence, Section 2.2), this is a multifaceted problem. Hence, a compromise must be found that will vary greatly from system to system, depending on features such as the preferential direction of the analyzed modes and the optical properties of the stratified system.

Selection of the IR spectroscopic method for the spectral measurements is nontrivial when the substrate is transparent in the IR region: One can study the MO in the film by analyzing the spectra from transmission, ATR, IRRAS, or BML-IRRAS spectra measured at one or several angles of incidence. From these, ATR is the least suitable method to the MO measurements on the νCH_2 bands of hydrocarbon chains with tilt angles $<30°$ (see also [593b] and the discussion of Fig. 3.86). As seen from Figs. 1.20 and 3.22, at angles of incidence far from φ_c, the z-component of the absorbance is negligible, which means that the ATR method is insensitive to the mode component directed along the surface normal. In the vicinity of φ_c, the vertical component becomes comparable to the lateral components for the substrates with high refractive indices (Si and Ge). However, this takes place in a narrow range of angles of incidence, which is hard to realize in practice, requiring well-collimated radiation. However, this method is rather adequate for the amide I band of α-helices [593a]. A higher accuracy in the tilt angle of protein species can be achieved when the ratio of the amide I and amide II band intensities rather than the DR is fitted [593a]. It also follows from the data reported in Ref. [593a] that DR fitting is more sensitive to the azimuth angle than spectrum fitting for α-helices oriented parallel to the surface ($\gamma = 90°$). The optimum conditions for the BML-IRRAS depend on the thickness of the buffer layer (Section 2.3.3). Below we will consider in more detail the optimum conditions for MO measurements from IRRAS spectra.

In the case of uniaxial LB films (tilt angle of 16°) on glass, the angle-of-incidence dependences of the SNR and the reflectivity for the $\nu_{as}CH_2$ band in IRRAS are shown in Fig. 3.91. One can see that for this system the optimum angle of incidence from the SNR viewpoint is $70°-73°$ (for the $\nu_{as}COO^-$ bands this angle is $77°$). For the AW interface, the highest SNR in IRRAS has been reported at $\varphi_1 \approx 30°$ in both s- and p-polarized IRRAS [589] and at $\varphi_1 \approx 71°-76°$ in PM-IRRAS [496, 71]. For all transparent substrates, the SNR in the p-polarized spectra is minimal and extremely poor at $\varphi_1 = \varphi_B$ (Section 2.3.1).

The highest sensitivity to changes in the MO of long-chain molecules measured by the spectrum fitting approach has been found in p-polarized spectra at $\sim80°$ (which is somewhat higher than $73°$) if ΔR units are used [471, 506]. If absorbance or reflectivity units are chosen [514, 630], the sensitivity is maximal near φ_B, significantly decreasing for tilt angles below $\sim30°$ in uniaxial monolayers on substrates with a low refractive index (1.4–1.6), regardless of the angle of incidence [542]. However, the poor SNR and exaggerated errors (Section 3.10) undo this advantage for measurements near φ_B.

The DR fitting measurements of MO in ultrathin films on substrates with low refractive indices (quartz) at angles of incidence higher than $\sim70°$ are optimum from the viewpoints of both maximum MO sensitivity and minimum inherent method error (for tilt angles $<40°$) [558]. At the same time, a marked increase

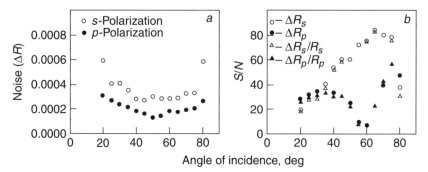

Figure 3.91. (a) Peak-to-peak noise in difference ΔR IRRAS spectra, obtained from experimental spectra of 11-monolayer CdAr LB films on glass (peak-to-peak noise was determined by using OMNIC software) in 3000–3100-cm^{-1} spectral region for s- and p-polarization. (b) SNR ratio for the $\nu_{asym}CH_2$ band in difference ΔR and reflectivity $\Delta R/R$ spectra for s- and p-polarization (solid and open symbols, respectively). Reprinted, by permission, from D. Blaudez, T. Buffeteau, B. Desbat, P. Fournier, A.-M. Ritcey, and M. Pezolet, *J. Phys. Chem. B* **102**, 99 (1998), p. 103, Figs. 6a and 7. Copyright © 1998 American Chemical Society.

in the error due to uncertainty in the monolayer optical parameters limits the optimal angle range from above by ∼75°. Hence, the angle of 73° optimal from the SNR viewpoint is also optimal for the IRRAS-DR fitting method [558].

Finally, it is important to know *absolute accuracy* in the IR spectroscopic measurements of MO. According to the simulations performed for IRRAS of a uniaxial LB film on glass [471], the spectral noise makes the accuracy in the tilt angle measured by the spectrum fitting method not better than 10°. The systematic error in the MO data determined using the Parikh–Allara method [507] from IRRAS of crystalline SAMs on Si was claimed [527] to be 1°–3° for the tilt angle. Another important factor influencing the accuracy of MO data is the surface roughness not only of the front [589] but also of the back [631] surface of the substrate.

REFERENCES

1. J. D. E. McIntyre, in B. O. Seraphin (Ed.), *Optical Properties of Solids: New Developments*, North-Holland, Amsterdam, 1976, p. 555.

2. H. Frohlich, *Theory of Dielectrics*, 2nd ed., Clarendon Press, Oxford, 1958.

3. D. W. Berreman, *Phys. Rev.* **130**, 2193 (1963).

4. B. Harbecke, B. Heinz, and P. Grosse, *Appl. Phys. A* **38**, 263 (1985).

5. K. Yamamoto and A. Masui, *Appl. Spectrosc.* **50**, 759 (1996).

6. K. Yamamoto, A. Masiu, and H. Ishida, *Vib. Spectrosc.* **13**, 119 (1997).

7. K. Yamamoto and H. Ishida, *Vib. Spectrosc.* **8**, 1 (1994).

8. K. Yamamoto and H. Ishida, *Appl. Spectrosc.* **48**, 775 (1994).

9. K. Yamamoto and H. Ishida, *Appl. Opt.* **34**, 4177 (1995).

10. H. Poulet and J.-P. Mathieu, *Spectres de vibration et symmetry des cristaux*, Gordon and Breach, Paris, 1970.

11. A. N. Lazarev, A. P. Mirgorodski, and I. S. Ignat'ev, *Vibrational Spectra of Complex Oxides*, Nauka, Leningrad, 1975 (in Russian).

12. R. Ruppin and R. Englman, *Rep. Prog. Phys.* **33**, 149 (1970).

13. K. L. Kliewer and R. Fuchs, in I. Prigogine (Ed.), *Advances in Chemical Physics*, Vol. 27, Wiley, New York, 1974, pp. 355–542.

14. E. Burstein, *Phys. Reports* **194**, 253 (1990).

15. E. Burstein and F. de Martini (Eds.), *Polaritons*, Pergamon, London, 1974.

16. V. M. Agranovich and D. L. Mills (Eds.), *Surface Polaritons: Electromagnetic Waves at Surfaces and Interfaces*, North-Holland, Amsterdam, 1982.

17. B. K. Ridley, O. Al-Dossary, N. C. Constantinou, and M. Babiker, *Phys. Rev. B* **50**, 11701 (1994).

18. R. S. Knox, *Theory of Excitons*, Academic, New York, 1963.

19. H. Ibach and D. L. Mills, *Electron Energy Loss Spectroscopy and Surface Vibrations*, Academic, New York, 1982.

20. M. de Crescenzi and M. N. Piancastelli, *Electron Scattering and Related Spectroscopies*, World Scientific, Singapore, 1996.

21. M. Rocca, H. Ibach, S. Lehwald, and T. S. Rahman, in W. Schommers and P. von Blanckenhagen (Eds.), *Surface and Dynamics of Surfaces*, Vol. 1, Springer-Verlag, Berlin, 1986.

22. P. K. Hansma (Ed.), *Tunneling Spectroscopy*, Plenum, New York, 1982.

23. R. Ruppin, *Surf. Sci.* **34**, 20 (1973).

24. U. Kreibig and M. Vollmer, *Optical Properties of Metal Clusters*, Springer, Berlin, 1995.

25. L. Genzel and T. P. Martin, *Surf. Sci.* **34**, 33 (1973).

26. E. A. Vinogradov, *Phys. Rep.* **217**, 159 (1992).

27. D. L. Mills, in M. Cordona and G. Guntherodt (Eds.), *Light Scattering in Solids*, Vol. 5, Springer-Verlag, Berlin, 1989, p. 13.

28. T. Dumelow, T. J. Parker, S. R. P. Smith, and D. R. Tilley, *Surf. Sci. Reports* **17**, 151 (1993).

29. P. Apell and O. Hunderi, in E. D. Palik (Ed.), *Handbook of Optical Constants*, Vol. 2, Academic, San Diego, 1998, Chapter 5.

30. U. Fano, *J. Opt. Soc. Am.* **31**, 213 (1941).

31. Y. Y. Teng and E. A. Stern, *Phys. Rev. Lett.* **19**, 511 (1967).

32. R. H. Ritchie, E. T. Arakawa, J. J. Cawan, and R. N. Hamn, *Phys. Rev. Lett.* **21**, 1530 (1968).

33. A. Otto, in P. Grosse (Ed.), *Advances in Solid State Physics*, Vol. 14, Vieweg, Braunschweig, 1974.

34. A. Otto, in B. O. Seraphin (Ed.), *Optical Properties of Solids: New Developments*, North-Holland, Amsterdam, 1976, Chapter 13.

35. E. Kretschmann and H. Raether, *Z. Naturforscchung* **22a**, 1623 (1976).

36. E. Kretschmann, *Z. Physik* **241**, 313 (1971).

37. Y. Ishino and H. Ishida, *Surf. Sci.* **230**, 299 (1990).

38. Y. Ishino and H. Ishida, *Appl. Spectrosc.* **42**, 1296 (1988).

39. E. A. Alieva, G. Beitel, L. A. Kuzik, A. A. Sigaev, V. A. Yakovlev, G. N. Zhizhin, A. F. G. van der Meer, and M. J. van der Wiel, *Appl. Spectrosc.* **51**, 584 (1997).

40. W. Knoll, *Annu. Rev. Phys. Chem.* **49**, 569 (1998).

41. Y. J. Chabal, *Surf. Sci. Reports* **8**, 211 (1988).

42. W. Suetaka, *Surface Infrared and Raman Spectroscopy: Methods and Applications*, Plenum, New York, 1995.

43. M. Balkanski and R. Le Toullec (Eds.), *Optical Properties of Dielectric Films*, in Proceedings of the Symposium held in Boston, Mass., on May 6–7, 1968, Electrochemical Society, New York, 1968.

44. K. Yamamoto and H. Ishida, *Vib. Spectrosc.* **15**, 27 (1997).

45. E. D. Palik, *Handbook of Optical Constants of Solids I, II, III*, Academic, New York, 1998.

46. *Reference Database of Optical Constants*, www.astro.spbu.ru/JPDOC/1-dbase.html.

47. (a) I. F. Chang, S. S. Mitra, J. N. Plendl, and L. C. Mansur, *Phys. Stat. Solidi* **28**, 663 (1968); (b) S. S. Mitra, in S. Nudelman and S. S. Mitra (Eds.), *Optical Properties of Solids*, Plenum Press, New York, 1969, Chapter 14.

48. W. J. Tropf, M. E. Thomas, and T. J. Harris, in M. Bass (Ed.), *Handbook of Optics*, Vol. 2, McGraw-Hill, New York, pp. 33.1–33.101.

49. G. R. Wilkenson, in A. Anderson (Ed.), *The Raman Effect*, Vol. 2: *Applications*, Marcel Dekker, New York, 1973.

50. B. Dorner, *Coherent Inelastic Neutron Scattering in Lattice Dynamics*, Springer-Verlag, Berlin, 1982.

51. R. K. Singh, *Phys. Reports* **85**, 259 (1982).

52. A. V. Rakov, *Spectrophotometry of Semiconducting Structures*, Sovetskoe Radio, Moscow, 1975 (in Russian).

53. Th. Scherubl and L. K. Thomas, *Appl. Spectrosc.* **51**, 847 (1997).

54. A. F. Vasil'ev, I. Y. Guschanskaya, G. N. Zhizhin, and B. A. Yakovlev, *Zhurnal prkladnoi spektroskopii* **48**, 405 (1988).

55. B. C. Trasferetti, C. U. Davanzo, N. C. da Cruz, and M. A. B. de Moraes, *Appl. Spectrosc.* **54**, 687 (2000).

56. P. Bruesch, R. Kotz, H. Neff, and L. Pietronero, *Phys. Rev. B* **29**, 4691 (1983).

57. T. G. Fiske and L. B. Coleman, *Phys. Rev. B* **45**, 1414–1424 (1992).

58. G. Kanellis, J. F. Morhange, and M. Balkanski, *Phys. Rev. B* **28**, 3390 (1983).

59. P. Bruesch, R. Kotz, H. Neff, and L. Pietronero, *Phys. Rev. B* **29**, 4691 (1984).

60. J. Izumitani, M. Okuyama, and Y. Hamakawa, *Appl. Spectrosc.* **47**, 1503 (1993).

61. (a) T. Burgi, *Phys. Chem Chem Phys. (PCCP)* **3**, 2124 (2001); (b) J. A. Mielczarski, E. Mielczarski, J. Zachwieja, and J. M. Cases, *Langmuir* **11**, 2787 (1995).

62. V. P. Tolstoy and S. N. Gruzinov, *Zhurnal prikladnoi spektroskopii* **54**, 162 (1991).

63. A. Udagava, T. Matsui, and S. Tanaka, *Appl. Spectrosc.* **40**, 794 (1986).

64. R. G. Tobin, *Phys. Rev. B* **45**, 12110 (1992).

65. M. W. Severson, C. Stuhlmann, I. Villegas, and M. J. Weaver, *J. Chem. Phys.* **103**, 9832 (1995).

66. R. A. Dluhy, S. M. Stephens, S. Widayati, and A. D. Willams, *Spectrochim. Acta A* **51**, 1413 (1995).

67. (a) R. Mendelsohn, J. W. Brauner, and A. Gericke, *Annu. Rev. Phys. Chem.* **46**, 305 (1995); (b) R. Mendelsohn and C. R. Flach, in J. Chalmers and P. Griffiths (Eds.), *Handbook of Vibrational Spectroscopy*, Vol. 2, Wiley, New York, 2002, p. 1028.

68. M. D. Porter, T. B. Bright, D. L. Allara, and T. Kuwana, *Anal. Chem.* **58**, 2461 (1986); (b) C. Selliti, J. L. Koenig, and H. Ishida, *Appl. Spectrosc.* **44**, 830 (1990); (c) T. Ohwaki and H. Ishida, *Appl. Spectroc.* **49**, 341 (1995).

69. H. Brunner, U. Mayer, and H. Hoffmann, *Appl. Spectrosc.* **51**, 209 (1997).

70. H. Brunner, T. Vallant, U. Mayer, and H. Hoffmann, *Surf. Sci.* **368**, 279 (1996).

71. D. Blaudez, J. M. Turlet, J. Dufourcq, D. Bard, T. Buffeteau, and B. Desbat, *J. Chem. Soc. Faraday Trans.* **94**, 4 (1996).

72. C. Alonso, D. Blaudez, and A. Renault, *Chem. Phys. Lett.* **284**, 446 (1998).

73. G. Carson, and S. Granick, *Appl. Spectrosc.* **43**, 473 (1989).

74. A. Bagchi and A. Rajagand, *Solid State Commun.* **31**, 127 (1979).

75. V. L. Kuzmin, *Opt. Spektrosc.* **38**, 745 (1975).

76. V. P. Tolstoy and S. N. Gruzinov, *Opt. Spektrosc.* **71**, 77 (1991).

77. V. P. Tolstoy, L. P. Bogdanova, and V. B. Aleskovski, *Doklady USSR* **291**, 913 (1986).

78. V. M. Zolotarev, V. N. Morosov, and E. V. Smirnova, *Optical Constants of Natural and Technical Media*, Khimia, Leningrad, 1984 (in Russian).

79. R. P. Eichens, S. A. Fransis, and W. A. Pliskin, *J. Phys. Chem.* **60**, 194 (1956).

80. R. P. Eischens and W. A. Pliskin, *Adv. Catal.* **10**, 1 (1958).

81. J. Lauterbach, R. W. Boyle, M. Schick, W. J. Mitchell, B. Meng, and W. H. Weinberg, *Surf. Sci.* **350**, 32 (1996).

82. E. Schweizer, B. N. J. Persson, M. Tushaus, D. Hoge, and A. M. Bradshaw, *Surf. Sci.* **213**, 49 (1989).

83. B. N. J. Persson, M. Tushaus, and A. M. Bradshaw, *J. Chem. Phys.* **92**, 5034 (1990).

84. C. Tang, S. Zou, M. W. Severson, and M. J. Weaver, *J. Phys. Chem. B* **102**, 8546 (1998).

85. C. Tang, S. Zou, M. W. Severson, and M. J. Weaver, *J. Phys. Chem. B* **102**, 8796 (1998).

86. R. K. Sharma, W. A. Brown, and D. A. King, *Surf. Sci.* **414**, 68 (1998).

87. K. Lyons, J. Xie, W. J. Mitchell, and W. H. Weinberg, *Surf. Sci.* **325**, 85 (1995).

88. A. Ortega, F. M. Hoffmann, and A. M. Brawdshaw, *Surf. Sci.* **119**, 79 (1982).

89. (a) J. P. Hollins and I. Pritchard, in R. F. Willis (Ed.), *Vibrational. Spectroscopy of Adsorbates*, Springer-Verlag, Berlin, 1980, p. 171; (b) P. Hollins and J. Pritchard, *Prog. Surf. Sci.* **19**, 275 (1985).

90. M. Scheffler, *Surf. Sci.* **81**, 562 (1979).

91. G. D. Mahan and A. A. Lucas, *J. Chem. Phys.* **68**, 1344 (1978).

92. M. Moskovitz and J. E. Hulse, *Surf. Sci.* **78**, 397 (1978).

93. R. Ryberg, in K. P. Lawley (Ed.), *Advances in Chemical Physics*, Wiley, Chichester, 1989, pp. 1–44.

94. H. Ueba, *Progress Surf. Sci.* **55**, 115 (1997).

95. V. P. Zhdanov, *Elementary Physicochemical Processes on Solid Surfaces*, Plenum, New York, 1991.

96. D. A. King, in R. F. Willis (Ed.), *Vibrational Spectroscopy of Adsorbates*, Springer-Verlag, Berlin, 1980.

97. Y. J. Chabal, *Surf. Sci. Reports* **8**, 211 (1988).

98. A. M. Bradshaw and E. Schweizer, in R. J. H. Clark and R. E. Hester (Eds.), *Spectroscopy of Surfaces*, Wiley, New York, 1988, pp. 413–483.

99. M. Urban, *Vibrational Spectroscopy of Molecules and Macromolecules on Surfaces*, Wiley, New York, 1993.

100. (a) H. Ibach, *Surf. Sci.* **66**, 56 (1977); (b) V. M. Da Costa and L. B. Coleman, *Phys. Rev. B* **43**, 1903 (1991).

101. P. Hofmann, S. R. Bare, and D. A. King, *Surf. Sci.* **117**, 245 (1982).

102. P. Hofmann, S. R. Bare, and D. A. King, *Surf. Sci.* **144**, 347 (1984).

103. B. N. J. Persson and R. Ryberg, *Phys. Rev. B* **24**, 6954 (1981).

104. H. Ueba, *Surf. Sci.* **188**, 421 (1987).

105. H. Pfnur, D. Menzel, F. M. Hoffmann, A. Ortega, and A. M. Brawdshaw, *Surf. Sci.* **93**, 431 (1980).

106. P. T. Fanson, W. N. Delgass, and J. Lauterbach, *J. Catal.* **204**, 35 (2001).

107. S. Chang and M. J. Weaver, *J. Chem. Phys.* **92**, 4582 (1990).

108. M. W. Severson and M. J. Weaver, *Langmuir* **14**, 5603 (1998).

109. Z. Xu, M. G. Sherman, J. T. Yates, and R. P. Antoniewicz, *Surf. Sci.* **276**, 249 (1992).

110. N. Kizhakevariam, I. Villegas, and M. J. Weaver, *Langmuir* **11**, 2777 (1995).

111. F. M. Leibsle, R. S. Sorbello, and R. G. Greenler, *Surf. Sci.* **179**, 101 (1987).

112. R. K. Brandt, R. S. Sorbello, and R. G. Greenler, *Surf. Sci.* **271**, 605 (1992).

113. M. J. Weaver and S. Zou, in R. J. H. Clark and R. E. Hester (Eds.), *Spectroscopy for Surface Science*, Vol. 26, Wiley, New York, 1998, pp. 219–272.

114. I. Villegas, R. Gomez, and M. J. Weaver, *J. Phys. Chem. B* **99**, 14832 (1997).

115. I. Villegas and M. J. Weaver, *J. Phys. Chem. B* **101**, 5842 (1997).

116. I. Villegas and M. J. Weaver, *J. Phys. Chem. B* **101**, 10166 (1997).

117. J. S. Luo, R. G. Tobin, and D. K. Lambert, *Chem. Phys. Lett.* **204**, 445 (1993).

118. D. K. Lambert, *Electrochim. Acta* **41**, 623 (1996).

119. D. K. Lambert, *J. Chem. Phys.* **89**, 3847 (1988).

120. I. Villegas and M. J. Weaver, *J. Chem. Phys.* **101**, 1648 (1994).

121. E. Borguet and H.-L. Dai, *Chem. Phys. Lett.* **194**, 57 (1992).

122. C.-S. Kim and C. Korzeniewski, *Anal. Chem.* **69**, 2949 (1997).

123. C. Tang, S. Zou, and M. J. Weaver, *Surf. Sci.* **412/413**, 344 (1998).

124. R. A. Hammaker, S. A. Francis, and R. P. Eichens, *Spectrochim. Acta* **21**, 1295 (1965).

125. R. A. Shigeishi and D. A. King, *Surf. Sci.* **58**, 379 (1976).

126. A. Crossley and D. A. King, *Surf. Sci.* **68**, 528 (1977).

127. B. N. J. Persson and R. Ryberg, *Chem. Phys. Lett.* **174**, 443 (1990).

128. P. Gao and M. J. Weaver, *J. Phys. Chem.* **93**, 6205 (1989).

129. J. E. Reutt, Y. J. Chabal, and S. B. Christman, *Phys. Rev. B* **38**, 3112 (1988).

130. C. J. Hirschmugl, G. P. Willams, F. M. Hoffmann, and Y. J. Chabal, *Phys. Rev. Lett.* **65**, 80 (1990).

131. B. N. J. Persson, *Chem. Phys. Lett.* **185**, 292 (1991).

132. B. N. J. Persson, *Surf. Sci.* **269**, 103 (1992).

133. J. O'M Bockris and A. N. Reddy, *Modern Electrochemistry*, Plenum, New York, 1967.

134. B. E. Conway, *Theory and Principles of Electrode Processes*, Ronald, New York, 1965.

135. B. B. Damaskin and O. A. Petriy, *Introduction to Electorchemical Kinetics*, Vysshaya Shkola, Moscow, 1975 (in Russian).

136. K. B. Oldman and J. C. Myland, *Fundamentals of Electrochemical Science*, Academic, San Diego, 1994.

137. S. R. Morrison, *The Chemical Physics of Surfaces*, 2nd ed., Plenum, New York, 1990.

138. H. O. Finklea, *Semiconductor Electrodes*, Elsevier, Amsterdam, 1988.

139. A. J. Nozik and R. Memming, *J. Phys. Chem.* **100**, 13061 (1996).

140. J. Lipkowski and P. N. Ross (Eds.), *Structure of Electrified Interfaces*, VCH, New York, 1993.

141. T. Iwasita and F. C. Nart, *Progress Surf. Sci.* **55**, 271 (1997).

142. C. Korzeniewski, *Crit. Rev. Anal. Chem.* **27**, 81 (1997).

143. C. Korzeniewski and M. W. Severson, *Spectrochim. Acta* **51A**, 499 (1995).

144. S. M. Stole, D. D. Popenoe, and M. D. Porter, in H. D. Abruna (Ed.), *Electrochemical Interfaces: Modern Techniques for In Situ Interface Characterization*, VCH, New York, 1991.

145. R. J. Nichols, in J. Lipkowski and P. H. Ross (Eds.), *Adsorption of Molecules at Metal Electrodes*, VCH, Weinheim, 1992, pp. 347–389.

146. J. N. Chazalviel, B. H. Erne, F. Maroun, and F. Ozanam, *J. Electroanal. Chem.* **502**, 180 (2001).

147. J. N. Chazalviel, B. H. Erne, F. Maroun, and F. Ozanam, *J. Electroanal. Chem.* **509**, 108 (2001).

148. V. B. Paulissen and C. Korzeniewski, *J. Electroanal. Chem.* **351**, 329 (1993).

149. P. W. Faguy, N. Marinkovic, and R. R. Adzic, *Langmuir* **12**, 243 (1996).

150. N. S. Marinkonic, J. X. Wang, H. Zajonz, and R. R. Adzic, *J. Electroanal. Chem.* **500**, 388 (2001).

151. M. Weber and F. C. Nart, *Electrochim. Acta* **41**, 653 (1996).

152. F. C. Nart and T. Iwasita, *J. Electroanal. Chem.* **322**, 289 (1992).

153. I. V. Chernyshova, *Langmuir* **18**, 6962 (2002).

154. D. S. Corrigan and M. J. Weaver, *J. Electroanal. Chem.* **239**, 55 (1988).

155. (a) Z. Ping, H. Neugebauer, and A. Neckel, *Electrochim. Acta* **41**, 767 (1996); (b) M. A. Almanza-Workman, S. Raghavan, and R. P. Sperline, *Langmuir* **16**, 3636 (2000).

156. (a) F. Kitamura, N. Nanbu, T. Ohsaka, and K. Tokuda, *J. Electroanal. Chem.* **452**, 241 (1998); (b) F. Kitamura, T. Ohsaka, and K. Tokuda, *Electrochim. Acta* **42**,

1235 (1997); (c) K. Arihara, F. Kitamura, T. Ohsaka, and K. Tokuda, *J. Electroanal. Chem.* **518**, 139 (2002); (d) F. Kitamura, T. Ohsaka, and K. Tokuda, *J. Electroanal. Chem.* **412**, 183 (1996).

157. W. R. Fawcett, A. A. Kloss, J. J. Calvente, and N. Marinkovic, *Electrochim. Acta* **44**, 881 (1998).

158. N. Marinkovic, J. J. Calvente, A. Kloss, Z. Kovaccova, and W. R. Fawcett, *J. Electroanal. Acta* **467**, 325 (1999).

159. N. Marinkovic, M. Hecht, J. S. Loring, and W. R. Fawcett, *Electrochim. Acta* **41**, 641 (1996).

160. P. W. Faguy and N. S. Marinkovic, *Anal. Chem.* **67**, 2791 (1995).

161. V. B. Paulissen and C. Korzeniewski, *J. Electroanal. Chem.* **290**, 181 (1990).

162. T. Bae, X. Xing, D. Scherson, and E. B. Yeager, *Anal. Chem.* **62**, 45 (1990).

163. F. Maroun, F. Ozanam, and J.-N. Chazalviel, *J. Electroanal. Chem.* **435**, 225 (1997).

164. K. C. Mandal, F. Ozanam, and J.-N. Chazalviel, *J. Electroanal. Chem.* **336**, 153 (1992).

165. K. Uosaki, Y. Shigematsu, H. Kita, and K. Kunimatsu, *J. Phys. Chem.* **94**, 4623 (1990).

166. P. A. Christensen, J. Eameaim, and A. Hamnett, *Phys. Chem. Chem. Phys.* **1**, 5315 (1999).

167. L. Kavan, P. Krtil, and M. Gratzel, *J. Electroanal. Chem.* **373**, 123 (1994).

168. K. E. Shaw, P. A. Christensen, and A. Hamnett, *Electrochim. Acta* **41**, 719 (1996).

169. P. Krtil, L. Kavan, I. Hoskovcova, and K. Kratochvilova, *J. Appl. Electrochem.* **26**, 523 (1996).

170. H.-Q. Li, S. G. Roscoe, and J. Lipkowski, *J. Electroanal. Chem.* **478**, 67 (1999).

171. K. Ataka, T. Yotsuyanagi, and M. Osawa, *J. Phys. Chem.* **100**, 10664 (1996).

172. N. S. Marinkovic, J. S. Marinkovic, and R. R. Adzik, *J. Electroanal. Chem.* **467**, 239 (1999).

173. Y. Shingaya, H. Kubo, and M. Ito, *Surf. Sci.* **427–428**, 173 (1999).

174. F. Maroun, F. Ozanam, and J.-N. Chazalviel, *J. Phys. Chem B* **103**, 5280 (1999).

175. F. Maroun, F. Ozanam, and J.-N. Chazalviel, *Surf. Sci.* **427–428**, 184 (1999).

176. Q. Fan and L. M. Ng, *J. Electrochem. Soc.* **141**, 3369 (1994).

177. D. Diesing, H. Winker, and A. Otto, *Phys. Stat. Solidi A* **159**, 243 (1997).

178. B. H. Erne, F. Ozanam, M. Shchakovsky, D. Vanmaekelbergh, and J.-N. Chazalviel, *J. Phys. Chem B* **104**, 5961 (2000).

179. I. Villegas and M. J. Weaver, *J. Am. Chem. Soc.* **118**, 458 (1996).

180. I. Villegas and M. J. Weaver, *J. Chem. Phys.* **103**, 2295 (1995).

181. (a) P. A. Thiel and T. E. Madey, *Surf. Sci. Rep.* **7**, 211 (1987); (b) P. A. Thiel, R. A. Depaola, and F. M. Hoffmann, *J. Chem. Phys.* **80**, 5326 (1984).

182. (a) J. O'M. Bockris and S. U. M. Khun, *Surface Electrochemistry*, Plenum Press, New York, 1993, Chapter 3; (b) K. Heinzinger, in J. Lipkowski and P. H. Ross (Eds.), *Structure of Electrified Interfaces*, VCH, New York, 1993, Chapter 7; (c) R. E. Verrall, in F. Franks (Ed.), *Water, A Comprehensive Treatise*, Vol. 3, Plenum, New York, 1976, Chapter 5; (d) O. A. El Seoud, *J. Mol. Liq.* **72**, 85 (1997).

183. E. A. Vogler, *Adv. Colloid Interface Sci.* **74**, 69 (1998).

184. A. M. Brodsky, M. Watanabe, and W. P. Reinhardt, *Electrochim. Acta* **36**, 1695 (1991).

185. K. Lum, D. Chandler, and J. D. Weeks, *J. Phys. Chem. B* **103**, 4570 (1999).

186. (a) G. L. Richmond, *Langmuir* **13**, 4804 (1997); (b) D. E. Gragson and G. L. Richmond, *J. Phys. Chem. B* **102**, 3848 (1998).

187. Q. Du, E. Freysz, and Y. R. Shen, *Phys. Rev. Lett.* **72**, 238 (1994).

188. (a) M. J. Shultz, *J. Phys. Chem. B* **103**, 2789 (1999); (b) Z. S. Nicholov, J. C. Earnshaw, and J. J. McGarvey, *Coll. Surf. A* **76**, 41 (1993).

189. Y. Marechal, in M.-C. Bellissent-Funel and J. C. Dore (Eds.), *Hydrogen Bond Networks*, Kluwer Academic, Netherlands, 1994, pp. 149–168.

190. K. Lum, D. Chander, and J. D. Weeks, *J. Phys. Chem. B* **103**, 4570 (1999).

191. M. Gerstein and R. M. Lynden-Bell, *J. Phys. Chem.* **97**, 2982 (1993).

192. M. F. Toney, J. N. Howard, J. Richer, G. L. Borges, J. G. Gordon, O. R. Meroy, D. G. Wiesler, D. Lee, and L. B. Sorensen, *Nature* **368**, 444 (1994).

193. (a) A. Bewick, K. Kunimatsu, and S. B. Pons, *Electrochim. Acta* **25**, 465 (1980); (b) A. Bewick and K. Kunimatsu, *Surf. Sci.* **101**, 131 (1980).

194. M. A. Habib and J. O'M. Bockris, *Langmuir* **2**, 388 (1986).

195. (a) R. G. Nuzzo, B. R. Zegarski, E. M. Korenic, and L. H. Dubois, *J. Phys. Chem.* **96**, 1355 (1992); (b) F. Bensebaa and T. H. Ellis, *Progr. Surf. Sci.* **50**, 73 (1995); (c) I. Engquist and B. Liedberg, *J. Phys. Chem.* **100**, 20089 (1996); (d) I. Engquist, M. Lestelius, and B. Liedberg, *Langmuir* **13**, 4003 (1997); (e) I. Engquist, I. Lundstrom, B. Liedberg, A. N. Parikh, D. L. Allara, *J. Chem. Phys.* **106**, 3038 (1997).

196. P. W. Faguy and W. N. Richmond, *J. Electroanal. Chem.* **410**, 109 (1996).

197. T. Iwasita and X. Xia, *J. Electroanal. Chem.* **411**, 95 (1996).

198. (a) I. V. Chernyshova and V. P. Tolstoy, *Appl. Spectrosc.* **49**, 665 (1995); (b) I. V. Chernyshova, *J. Phys. Chem. B* **105**, 8185 (2001).

199. T. Hasegawa, J. Nishijo, T. Imae, Q. Huo, and R. M. Leblanc, *J. Phys. Chem. B* **105**, 12056 (2001).

200. P.-A. Bergstrom, J. Lindgren, and O. Kristiansson, *J. Phys. Chem.* **95**, 8575 (1991).

201. C. J. Weinheimer and J. M. Lisy, *J. Chem. Phys.* **105**, 2938 (1996).

202. S. Pons, *J. Electron Spectrosc. Related Phenom.* **45**, 303 (1987).

203. N. Kizhakevariam, E. M. Stuve, and R. Döhl-Oelze, *J. Chem. Phys.* **94**, 670 (1991).

204. J. N. Chazalviel and F. Ozanam, *J. Appl. Phys.* **81**, 7684 (1997).

205. M. Niwano, Y. Kimura, and N. Miyamoto, *Appl. Phys. Lett.* **65**, 1692 (1994).

206. M. Niwano, T. Miura, Y. Kimura, R. Tajima, and N. Miyamoto, *J. Appl. Phys.* **79**, 3708 (1996).

207. A. Couto, A. Rincon, M. C. Perez, and C. Gutierrez, *Electrochim. Acta* **46**, 1285 (2001).

208. (a) S. A. Wasileski, M. J. Weaver, and M. T. M. Koper, *J. Electroanal. Chem.* **500**, 344 (2001); (b) M. T. M. Koper, R. A. van Santen, S. A. Wasileski, and M. J. Weaver, *J. Chem. Phys.* **113**, 4392 (2000).

209. J. K. Foley, C. Korzeniewski, and S. Pons, *Can. J. Chem.* **66**, 201 (1988).

210. M. J. Weaver, *J. Phys. Chem.* **100**, 13079 (1996).

211. K. Ashley, F. Weinert, and D. L. Feldheim, *Spectrochim. Acta* **36**, 1863 (1991).

212. T. Iwasita, A. Rodes, and E. Pastor, *J. Electroanal. Chem.* **377**, 215 (1994).

213. F. Huerta, E. Morallon, C. Quijada, J. L. Vazquez, and A. Aldaz, *Electrochim. Acta* **44**, 943 (1998).

214. C. Korzeniewski and S. Pons, *Langmuir* **2**, 468 (1986).

215. C. Korzeniewski and S. Pons, *J. Vac. Sci. Technol. B* **3**, 1421 (1985).

216. S. Pons, S. B. Khoo, A. Bewick, M. Datta, J. J. Smith, A. S. Hinman, and G. Zachmann, *J. Phys. Chem.* **88**, 3575 (1984).

217. F. C. Nart and T. Iwasita, *Electrochim. Acta* **41**, 631 (1996).

218. D. K. Lambert, *Electrochim. Acta* **41**, 623 (1996).

219. F. Ozanam and J.-N. Chazalviel, *J. Electron Spectrosc. Related Phenom.* **54–55**, 1219 (1990).

220. C. Korzenievski, S. Pons, P. P. Schmidt, and M. W. Severson, *J. Chem. Phys.* **85**, 4153 (1986).

221. Y. Sawatari, J. Inukai, and M. Ito, *J. Electron. Spectrosc. Related Phenom.* **64–65**, 515 (1993).

222. P. W. Faguy, N. S. Marinkovic, and R. R. Adzic, *J. Electroanal. Chem.* **407**, 41 (1996).

223. J. K. Sass, K. Bange, R. Dohl, E. Riltz, and R. Unwin, *Ber. Bunsenges. Phys. Chem.* **88**, 354 (1984).

224. J. K. Sass and K. Bange, *ACS Symp. Ser.* **378**, 54 (1988).

225. Y. Shingaya and M. Ito, *Surf. Sci.* **368**, 318 (1996).

226. Y. Shingaya, H. Kubo, and M. Ito, *Surf. Sci.* **427–428**, 173 (1999).

227. K. Yoshimi, M.-B. Song, and M. Ito, *Surf. Sci.* **368**, 389 (1996).

228. R. J. H. Clark, in R. J. H. Clark and R. E. Hester (Eds.), *Advances in Infrared and Raman Spectroscopy*, Vol. 4, Heyden, London, 1978, Chapter 4.

229. R. Gomez, J. M. Feliu, A. Aldaz, and M. J. Weaver, *Surf. Sci.* **410**, 48 (1998).

230. I. Villegas and M. J. Weaver, *Surf. Sci.* **367**, 162 (1996).

231. H. Kuzmany, in B. Schrader (Ed.), *Infrared and Raman Spectroscopy*, VCH, Weinheim, 1995, p. 272.

232. A. H. Reed and E. Yeager, *Electroanal. Acta* **15**, 1345 (1970).

233. N. J. Harrick, *Phys. Rev.* **125**, 1165 (1962).

234. J. I. Pankove, *Optical Processes in Semiconductors*, Prentice-Hall, Englewood Cliffs, NJ, 1971, pp. 67–76.

235. N. J. Harrick, *J. Appl. Phys.* **29**, 764 (1958).

236. A. H. Kahn, *Phys. Rev.* **97**, 1647 (1955).

237. N. J. Harrick, *Internal Reflection Spectroscopy*, Wiley, New York, 1967.

238. F. Ozanam, C. da Fonseca, A. Venkateswara Rao, and J.-N. Chazalviel, *Appl. Spectrosc.* **51**, 519 (1997).

239. G. Redmond and D. Fitzmaurice, *J. Phys. Chem.* **97**, 1426 (1993).

240. B. H. Erne, F. Ozanam, and J. N. Chalzaviel, *J. Phys. Chem. B* **104**, 11591 (2000).

241. Q. Dingrong, *Infrared Phys.* **33**, 127 (1992).

242. E. Lankinen, G. Sundholm, P. Talonen, T. Laitinen, and T. Saario, *J. Electroanal. Chem.* **447**, 135 (1998).

243. P. A. Christensen, A. Hamnett, A. R. Hillman, M. J. Swann, and J. S. Higgins, *J. Chem. Soc. Faraday Trans.* **88**, 595 (1992).

244. G. Samaggia and A. Nucciotti, *Phys. Rev.* **144**, 749 (1966).

245. A. V. Rao, J.-N. Chazalviel, and F. Ozanam, *J. Appl. Phys.* **60**, 696 (1986).

246. A. V. Rao and J.-N. Chazalviel, *J. Electrochem. Soc.* **134**, 2777 (1987).

247. T. Buffeteau, D. Blaudez, E. Pere, and B. Desbat, *J. Phys. Chem. B* **103**, 5020 (1999).

248. V. M. Dubin, F. Ozanam, and J.-N. Chazalviel, *Vib. Spectrosc.* **8**, 159 (1995).

249. A. Belaidi, M. Safi, F. Ozanam, J.-N. Chazalviel, and O. Gorochov, *J. Electrochem. Soc.* **146**, 2659 (1999).

250. J.-N. Chazalviel, A. Belaidi, M. Safi, F. Maroun, B. H. Erne, and F. Ozanam, *Electrochim. Acta* **45**, 3205 (2000).

251. D. M. Kolb, in R. J. Gale (Ed.), *Spectroelectrochemistry*, Plenum, New York, 1988, p. 87.

252. B. N. J. Persson, *Phys. Rev. B* **44**, 3277 (1991).

253. H. Noda, T. Minoha, L.-J. Wan, and M. Osawa, *J. Electroanal. Chem.* **481**, 62 (2000).

254. K. Ataka and M. Osawa, *J. Electroanal. Chem.* **460**, 188 (1999).

255. B. H. Erne, M. Shchakovsky, F. Ozanam, and J.-N. Chazalviel, *J. Electrochem. Soc.* **145**, 447 (1998).

256. I. V. Chernyshova, *J. Phys. Chem. B* **105**, 8178 (2001).

257. U. P. Fringeli, in F. M. Mirabella (Ed.), *Modern Techniques in Applied Molecular Spectroscopy*, Wiley, New York, 1998, p. 255.

258. J. Kritzenberger and A. Wokaum, *J. Molecular Catal. A* **118**, 235 (1997).

259. F. Ozanam and J.-N. Chazalviel, *Rev. Sci. Instrum.* **59**, 242 (1988).

260. J.-N. Chazalviel, V. M. Dubin, K. C. Mandal, and F. Ozanam, *Appl. Spectrosc.* **47**, 1411 (1993).

261. C. da Fonteca, A. Djebri, and J.-N. Chazalviel, *Electrochim. Acta* **41**, 687 (1996).

262. B. O. Budevska and P. R. Griffiths, *Anal. Chem.* **65**, 2963 (1993).

263. C. M. Pharr and P. R. Griffiths, *Anal. Chem.* **69**, 4665 (1997).

264. C. M. Pharr and P. R. Griffiths, *Anal. Chem.* **69**, 4673 (1997).

265. K. Ataka, Y. Hara, and M. Osawa, *J. Electroanal. Chem.* **473**, 34 (1999).

266. R. J. MacDonald, *Impedance Spectroscopy*, Wiley, New York, 1987.

267. J.-N. Chazalviel, K. C. Mandal, and F. Ozanam, *SPIE Conf. Proc.* **1575**, 40 (1992).

268. C. J. Manning and P. R. Griffiths, *Appl. Spectrosc.* **47**, 1345 (1993).

269. J. H. Duckworth, in J. Workman, Jr. and A. W. Springsteen (Eds.), *Applied Spectroscopy. A Compact Refence for Practitioners*, Academic, San Diego, 1998, pp. 93–165.

270. D. M. Haaland and E. V. Thomas, *Anal. Chem.* **60**, 1193 (1988).

271. R. S. Jackson and P. R. Griffiths, *Anal. Chem.* **63**, 2557 (1991).

272. J. K. Kauppinen, D. J. Moffatt, H. H. Mantsch, and D. G. Cameron, *Appl. Spectrosc.* **35**, 271 (1986).

273. D. G. Cameron and D. J. Moffatt, *J. Test. Eval. JTEVA* **12**, 78 (1984).

274. I. Noda, *Bull. Am. Phys. Soc.* **31**, 520 (1986).

275. I. Noda, *J. Am. Chem. Soc.* **111**, 8116 (1989).

276. I. Noda, *Appl. Spectrosc.* **44**, 550 (1990).

277. I. Noda, *Appl. Spectrosc.* **47**, 1329 (1993).

278. I. Noda, A. E. Dowrey, and C. Marcott, *Appl. Spectrosc.* **47**, 1317 (1993).

279. I. Noda, *Anal. Chem.* **21**, 1065A (1994).

280. I. Noda, *Appl. Spectrosc.* **54**, 994 (2000).

281. I. Noda, *Chemtracts-Macromol. Ed.* **1**, 89 (1990).

282. M. Osawa, K. Yoshii, Y. Hibino, T. Nakano, and I. Noda, *J. Electroanal. Chem.* **426**, 11 (1997).

283. A. Nabet and M. Pezolet, *Appl. Spectrosc.* **51**, 466 (1997).

284. D. L. Elmore and R. A. Dluhy, *Colloid Surf. A* **171**, 225 (2000).

285. R. A. Palmer, C. J. Manning, J. L. Chao, I. Noda, A. E. Dowrey, and C. Marcott, *Appl. Spectrosc.* **45**, 12 (1991).

286. Y. Ozaki, Y. Liu, and I. Noda, *Appl. Spectrosc.* **51**, 526 (1997).

287. M. A. Czarnecki, *Appl. Spectrosc.* **52**, 1583 (1998).

288. A. Gericke, J. G. Sergio, J. W. Brauner, and R. Mendelsohn, *Biospectroscopy* **2**, 341 (1996).

289. C. F. Bohren and D. R. Huffman, *Absorption and Scattering of Light by Small Particles*, Wiley, New York, 1983.

290. H. C. van de Hulst, *Absorption and Scattering of Light by Small Particles*, Wiley, New York, 1957.

291. B. Hapke, *Theory of Reflectance and Emittance Spectroscopy*, Cambridge University Press, Cambridge, 1993.

292. D. W. Shuerman (Ed.), *Light Scattering by Irregularly Shaped Particles*, Plenum, New York, 1980.

293. R. Ruppin and R. Englman, *Rep. Prog. Phys.* **33**, 149 (1970).

294. F. Abeles, Y. Borensztein, and T. Lopez-Rios, in P. Grosse (Ed.), *Advances in Solid State Physics*, Vol. 34, Vieweg, Braunschweig, 1984, pp. 93–117.

295. J. C. Garland and D. B. Tanner (Eds.), *Electrical Transport and Optical Properties of Inhomogeneous Media*, American Institute of Physics, New York, 1978.

296. J. Lafait, S. Berthier, M. Gadenne, and P. Gadenne, in G. C. Cody, T. H. Geballe, and P. Sheng (Eds.), *Physical Phenomena in Granular Materials*, MRS Symposia Proceedings No. 195, Materials Research Society, Pittsburgh, PA, 1990, p. 77.

297. L. Genzel and T. P. Martin, *Surf. Sci.* **34**, 33 (1973).

298. R. Ruppin, *Surf. Sci.* **34**, 20 (1973).

299. H. Metiu, *Prog. Surf. Sci.* **17**, 153 (1984).

300. C. F. Bohren, in M. Bass (Ed.), *Handbook of Optics*, McGraw-Hill, New York, 1995, pp. 6.1–6.21.

301. (a) J. E. Iglesias, M. Ocana, and C. J. Serna, *Appl. Spectrosc.* **44**, 418 (1990); (b) N. P. Morales, N. O. Nunez, R. Pozas, M. Ocana, and C. J. Serna, *Appl. Spectrosc.* **56**, 200 (2002).

302. P. Rennert, *Ann. Phys.* **47**, 27 (1990).

303. R. Fuchs, *Phys. Rev. B* **11**, 1732 (1975).

304. D. N. Napper, *Kolloidzeitschrift* **218**, 42 (1967).

305. A. L. Aden and M. Kerker, *J. Appl. Phys.* **22**, 1242 (1951).

306. A. Guttler, *Ann. Phys.* **11**, 5 (1952).

307. R. H. Morriss and L. F. Collins, *J. Chem. Phys.* **41**, 3357 (1964).

308. C. F. Eagen, *Appl. Opt.* **20**, 3035 (1981).

309. R. Bhandari, *Appl. Opt.* **24**, 1960 (1985).

310. J. Sinzig, U. Radtke, M. Quinten, and U. Kreibig, *Z. Physik D* **26**, 242 (1993).

311. J. Sinzig and M. Quinten, *Appl. Phys. A* **58**, 157 (1994).

312. M. Barabas, *J. Opt. Soc. Am.* **A2**, 2240 (1987).

313. R. Sharma and N. Balakrishnan, *Smart Materials Structures* **7**, 512 (1998).

314. P. Barnickel and A. Wokaun, *Mol. Phys.* **67**, 1355 (1989).

315. T. P. Martin, *Solid State Commun.* **17**, 139 (1975).

316. L. Chen, T. Goto, and T. Hirai, *J. Mater. Sci.* **25**, 4273 (1990).

317. M. Fujii, M. Wada, S. Hayashi, and K. Yamamoto, *Phys. Rev. B* **46**, 15930 (1992).

318. E. Wackelgard, *J. Phys. Condens. Matter* **8**, 4289 (1996).

319. T. Atsumi and M. Miyagi, *Electron. Commun. Jpn.* **81**(Part 2), 23 (1998).

320. A. Roos, G. Chinyama, and P. Hedenqvist, *Thin Solid Films* **236**, 40 (1993).

321. R. C. McPhedran and N. A. Nicorovici, *Physica A* **241**, 173 (1997).

322. F. J. Garcia-Vidal, J. M. Pitarke, and J. B. Pendry, *Phys. Rev. Lett.* **78**, 4289 (1997).

323. R. A. B. Devine, *Appl. Phys. Lett.* **68**, 3108 (1996).

324. (a) A. Brunet-Bruneau, S. Fisson, G. Vuye, and J. Rivory, *J. Appl. Phys.* **87**, 7303 (2000); (b) A. Brunet-Bruneau, S. Fisson, B. Gallas, G. Vuye, and J. Rivory, *Thin Solid Films* **377–378**, 57 (2000).

325. (a) R. A. B. Devine, *J. Vac. Sci. Technol. A* **6**, 3154 (1988); (b) G. Lucovsky, M. J. Manitini, J. K. Srivastava, and E. A. Irene, *J. Vac. Sci. Technol. B* **5**, 530 (1987).

326. T. Ohwaki, M. Takeda, and Y. Takai, *Jpn. J. Appl. Phys.* **36**(Part 1), 5507 (1997).

327. G. Srinivas and V. D. Vankar, *Mater.Lett.* **30**, 35 (1997).

328. S. A. Alterovitz, M. A. Drotos, and P. G. Young, *Mater. Res. Soc. Symp. Proc.* **284**, 45 (1993).

329. T. G. Fiske and L. B. Coleman, *Phys. Rev. B* **45**, 1414 (1992).

330. P. Pigeat, N. Pacia, and B. Weber, *Surf. Interface Anal.* **18**, 571 (1992).

331. Y. Okamoto, S. V. Ordin, and T. Miyakawa, *J. Appl. Phys.* **85**, 6728 (1999).

332. (a) J. M. Hammersley, in G. Deutscher, R. Zallen, and J. Adler (Eds.), *Percolation Structure and Processes, Annals of the Israel Physical Society*, Vol. 5, Hilger, Bristol, 1983, p. 47; (b) D. Stauffer and A. Aharony, *Introduction to Percolation Threshold*, 2nd ed., Taylor and Francis, London, 1992.

333. T. W. Noh, Y. Song, S.-I. Lee, J. R. Gaines, H. D. Park, and E. R. Kreidler, *Phys. Rev. B* **33**, 3793 (1986).

334. M. F. MacMillan, R. P. Devaty, and J. V. Mantese, *Phys. Rev. B* **54**, 4621 (1996).

335. H. S. Choi, J. S. Ahn, J. H. Jung, T. W. Noh, and D. H. Kim, *Phys. Rev. B* **43**, 13838 (1991).

336. (a) F. Brouters, *J. Phys. C* **19**, 7183 (1986); (b) S. Berthier, *Ann. Phys. (Paris)* **13**, 503 (1988).

337. S. Berthier, J. Lafait, C. Sella, and Thran-Khanh-Vien, *Thin Solid Films* **125**, 171 (1985).

338. P. Sheng, *Phys. Rev. Lett.* **45**, 60 (1980).

339. T. W. Noh, Y. Song, S.-I. Lee, J. R. Gaines, H. D. Park, and E. R. Kreidler, *Phys. Rev. B* **33**, 3793 (1986).

340. S. Berthier, J. Peiro, S. Fagnent, and P. Gadenne, *Physica A,* **241**, 1 (1997).

341. T. Robin and B. Souillard, *Europhys. Lett.* **8**, 753 (1989).

342. T. Robin and B. Souillard, *Physica A* **157**, 285 (1989).

343. Y. Yagil, M. Yosefin, D. J. Bergman, G. Deutscher, and P. Gadenne, *Phys. Rev. B* **43**, 11342 (1991).

344. R. A. Buhrman and H. G. Craighead, in L. E. Murr (Ed.), *Solar Materials Science,* Academic, New York, 1980, p. 277.

345. A. Heilmann, C. Hamann, G. Kampfrath, and V. Hopfe, *Vacuum* **41**, 1472 (1990).

346. L. Ward, *The Optical Constants of Bulk Materials and Films*, 2nd ed., Institute of Physics Publishing, Bristol, 1994.

347. W. T. McKenna and L. Ward, *Phys. Stat. Solidi (a)* **68**, K11 (1981).

348. E. J. Zeman and G. C. Schatz, *J. Phys. Chem.* **91**, 634 (1987).

349. T. R. Jensen, L. Kelly, A. Lazarides, and G. C. Schatz, *J. Cluster Sci.* **10**, 295 (1999).

350. T. R. Jensen, R. P. Van Duyne, S. A. Johnson, and V. A. Maroni, *Appl. Spectrosc.* **54**, 371 (2000).

351. E. Dobierzewska-Mozrzymas, P. Bieganski, and E. Pieciul, *Vacuum* **45**, 279 (1994).

352. G. W. Mbise, D. Le Bellac, G. A. Niklasson, and C. G. Granqvist, *J. Phys. D* **30**, 2103 (1997).

353. F. Yang, G. W. Bradberry, and J. R. Sambles, *Thin Solid Films* **196** (1991).

354. Y. Yagil, G. Deutscher, P. Gadenne, and C. Julien, *Phys. Rev. B* **46** (1993).

355. B. T. Sullivan and R. R. Parsons, *J. Vac. Sci. Technol. A* **5**, 3399 (1987).

356. A. Hartstein, J. R. Kirtley, and J. C. Tsang, *Phys. Rev. Lett.* **45**, 201 (1980).

357. R. Kellner, B. Mizaikoff, M. Jakusch, H. D. Wanzenböck, and N. Weissenbacher, *Appl. Spectrosc.* **51**, 495 (1997).

358. Ch. Kuhne, G. Steiner, W. B. Fischer, and R. Salzer, *Fresenius J. Anal. Chem.* **360**, 750 (1998).

359. C. W. Brown, Y. Li, J. A. Seelenbinder, P. Pilarnik, A. G. Rand, S. V. Letcher, O. J. Gregory, and M. J. Platek, *Anal. Chem.* **70**, 2991 (1998).

360. C. W. Brown, Y. Li, J. A. Seelenbinder, P. Pilarnik, and A. G. Rand, *Anal. Chem.* **71**, 1963 (1999).

361. M. Osawa and K. Yoshii, *Appl. Spectrosc.* **51**, 512 (1997).

362. H. Noda, K. Ataka, L. Wan, and M. Osawa, *Surf. Sci.* **427–428**, 190 (1999).

363. I. Taniguchi, S. Yoshimoto, Y. Sunatsuki, and K. Nishiyama, *Electrochemistry* **67**, 1197 (1999).

364. K. Ataka and M. Osawa, *J. Electroanal. Chem.* **460**, 188 (1999).

365. N. Matsuda, K. Yoshii, K. Ataka, M. Osawa, T. Matsue, and I. Uchida, *Chem. Lett.* **7**, 1385 (1992).

366. T. Kamata, A. Kato, J. Umemura, and T. Takenaka, *Langmuir* **3**, 1150 (1987).

367. J. Zhang, H. Z. Yu, J. Zhao, Z. F. Liu, and H. L. Li, *Appl. Spectrosc.* **53**, 1305 (1999).

368. J. Zhang, J. Zhao, H. X. He, H. L. Li, and Z. F. Liu, *Thin Solid Films* **327–329**, 287 (1998).

369. A. Badia, S. Singh, L. Demers, L. Cuccia, G. R. Brown, and R. B. Lennox, *Chemistry Eur. J.* **2**, 359 (1996).

370. S. J. Lee and K. Kim, *Vib. Spectrosc.* **18**, 187 (1998).

371. R. Aroca and B. Price, *J. Phys. Chem. B* **101**, 6537 (1997).

372. Y. Nakao and H. Yamada, *Surf. Sci.* **176**, 578 (1986).

373. Y. Nakao and H. Yamada, *J. Electron Spectrosc. Related Phenom.* **45**, 189 (1987).

374. W. Suetaka, *Surface Infrared and Raman Spectroscopy*, Plenum, New York, 1995, pp. 142–151.

375. A. E. Bjerke, P. R. Griffiths, and W. Theiss, *Anal. Chem.* **71**, 1967 (1999).

376. H. D. Wanzenbock, B. Mizaikoff, N. Weissenbacher, and R. Kellner, *J. Mol. Struct.* **410–411**, 535 (1997).

377. Y. Nishikawa, K. Fujiwara, and T. Shima, *Appl. Spectrosc.* **45**, 747 (1991).

378. Y. Suzuki, H. Seki, T. Inamura, T. Tanabe, T. Wadayama, and A. Hatta, *Surf. Sci.* **427–428**, 136 (1999).

379. M. Osawa and M. Ikeda, *J. Phys. Chem.* **95**, 9914 (1991).

380. S. Sato and T. Suzuki, *Appl. Spectrosc.* **51**, 1170 (1997).

381. M. Osawa, K. Ataka, M. Ikeda, H. Uchihara, and R. Nanba, *Anal. Sci.* **7**(Suppl.), 503 (1991).

382. M. Osawa, K. Ataka, K. Yoshii, and Y. Nishikawa, *Appl. Spectrosc.* **47**, 1497 (1993).

383. G. T. Merklin and P. R. Griffiths, *J. Phys. Chem. B* **101**, 5810 (1997).

384. Y. Nishikawa, K. Fujiwara, K. Ataka, and M. Osawa, *Anal. Chem.* **65**, 556 (1993).

385. Y. Nishikawa, K. Fujiwara, and T. Shima, *Appl. Spectrosc.* **45**, 752 (1991).

386. H. S. Han, C. H. Kim, and K. Kim, *Appl. Spectrosc.* **52**, 1047 (1998).

387. J. A. Seelenbinder, C. W. Brown, and D. W. Urish, *Appl. Spectrosc.* **54**, 366 (2000).

388. F. Maroun, F. Ozanam, J.-N. Chazalviel, and W. Treiss, *Vib. Spectrosc.* **19**, 193 (1999).

389. J. A. Seelenbinder, C. W. Brown, P. Pivarnik, and A. G. Rand, *Anal. Chem.* **71**, 1963 (1999).

390. S. Badilescu, P. V. Ashrit, V.-V. Truong, and I. I. Badilescu, *Appl. Spectrosc.* **43**, 549 (1989).

391. A. Hatta, Y. Suzuki, T. Wadayama, and W. Suetaka, *Appl. Surf. Sci.* **48–49**, 222 (1991).

392. Z. Y. Zhang and D. C. Langreth, *Phys. Rev. B* **39**, 10028 (1989).

393. H. D. Wanzenbock, B. Edl-Mizaikoff, G. Fdedbacher, M. Grasserbauer, R. Kellner, M. Arntzen, T. Luyven, W. Theiss, and P. Grosse, *Mikrochim. Acta* **14**(Suppl.), 665 (1997).

394. A. Otto, I. Mrozek, H. Grabhorn, and W. Akemann, *J. Phys. Condens. Matter* **4**, 1143 (1992).

395. R. G. Greenler, D. R. Snider, D. Witt, and R. S. Sorbello, *Surf. Sci.* **118**, 415 (1982).

396. G. T. Merklin and P. R. Griffiths, *Langmuir* **13**, 6159 (1997).

397. Y. Suzuki, M. Kobayashi, A. Hatta, and A. Otto, *Jpn. J. Appl. Phys.* **36**, 4446 (1997).

398. A. Hatta, N. Suzuki, Y. Suzuki, and W. Suetaka, *Appl. Surf. Sci.* **37**, 299 (1989).

399. R. Aroca and B. Proce, *J. Phys. Chem. B* **101**, 6537 (1997).

400. Y. Ishino and H. Ishida, *Appl. Spectrosc.* **42**, 1296 (1988).

401. B. W. Johnson, B. Pettinger, and K. Doblhofer, *Ber. Bunsenges.* **97**, 412 (1993).

402. (a) E. V. Alieva, L. A. Kuzik, G. Mattei, J. E. Petrov, and V. A. Yakovlev, *Physica Status Solidi A* **175**, 115 (1999); (b) G. Mattei, E. V. Alieva, J. E. Petrov, and V. A. Yakovlev, *Surf. Sci.* **428**, 235 (1999); (c) Y. Jiang, S. Sun, and N. Ding, *Chem. Phys. Lett.* **344**, 463 (2001).

403. M. Futamata and D. Diesing, *Vib. Spectrosc.* **19**, 187 (1999).

404. (a) I. Vanin, *Opt. Spectrosc.* **80**, 290 (1996); (b) S. Hayashi, R. Koga, M. Ohtuji, K. Yamamoto, and M. Fujii, *Solid State Commun.* **76**, 1067 (1990).

405. Y. Suzuki, M. Osawa, A. Hatta, and W. Suetaka, *Appl. Surf. Sci.* **33/34**, 875 (1988).

406. M. Kerker, D. S. Wang, H. Chew, and C. G. Blatchford, *Phys. Rev. B* **26**, 4052 (1982).

407. M. Osawa and K. Ataka, *Surf. Sci.* **262**, L118 (1992).

408. D. J. Bergman, *Phys. Rep. C* **43**, 377 (1978).

409. (a) M. Futamata, *Chem. Phys. Lett.* **317**, 304 (2000); (b) M. Futamata, *Chem. Phys. Lett.* **333**, 337 (2001).

410. (a) A. Roseler and E.-H. Korte, *Appl. Spectrosc.* **51**, 902 (1997); (b) A. Roseler and E.-H. Korte, *Thin Solid Films* **313–314**, 732 (1998); (c) E.-H. Korte, A. Roseler, and M. Buskuhl, *Talanta* **53**, 9 (2000); (d) A. Roseler and E.-H. Korte, *Fresenius J. Anal. Chem.* **362**, 51 (1998).

411. F. Brouers, A. K. Sarychev, S. Blacher, and O. Lothaire, *Physica A* **241**, 146 (1997).

412. Y. Zhu, H. Uchida, and M. Watanabe, *Langmuir* **15**, 8757 (1999).

413. (a) E. Zippel, M. W. Breiter, and R. Kellner, *J. Chem. Soc., Faraday Trans.* **87**, 637 (1991); (b) E. Zippel, R. Kellner, M. Krebs, and M. W. Breiter, *J. Electroanal. Chem.* **330**, 521 (1992).

414. A. M. Bradshaw and J. Pritchard, *Proc. R. Soc. London, Ser. A* **316**, 169 (1970).

415. K. P. Ishida and P. R. Griffiths, *Anal. Chem.* **66**, 522 (1994).

416. O. Krauth, G. Fahsold, N. Magg, and A. Pucci, *J. Chem. Phys.* **113**, 6330 (2000).

417. G. T. Merklin, L. T. He, and P. R. Griffiths, *Appl. Spectrosc.* **53**, 1448 (1999).

418. V. M. Agranovich and D. L. Mills (Eds.), *Surface Polaritons: Electromagnetic Waves at Surfaces and Interfaces*, North-Holland, Amsterdam, 1982.

419. A. S. Ilyinsky, G. Y. Slepyan, and A. Y. Slepyan, *Propagation, Scattering and Dissipation of Electromagnetic Waves*, Peter Peregrinus, London, 1993.

420. J. A. Sanchez-Gil, A. A. Maradudin, J. Q. Lu, V. D. Freilikher, M. Pustilnik, and I. Yurkevich, *J. Mod. Opt.* **43**, 435 (1996).

421. D. W. Berreman, *Phys. Rev.* **163**, 855 (1967).

422. S. K. Andersson and G. A. Niklasson, *J. Phys. Condens. Matter* **7**, 7173 (1995).

423. P. Pipoz, E. Bustarret, and T. Lopez-Rios, *Opt. Commun.* **149**, 33 (1998).

424. I. F. Salakhutdinov, V. A. Sychugov, A. V. Tishchenko, B. A. Usievich, O. Parriaux, and F. A. Pudonin, *IEEE J. Quantum Electron.* **34**, 1054 (1998).

425. M. J. Dignam and M. Moskovits, *J. Chem. Soc. Faraday Trans. II* **69**, 65 (1973).

426. (a) A. E. Bjerke, P. R. Griffiths, and W. Theiss, *Anal. Chem.* **71**, 1967 (1999); (b) H. Ishida, *Phys. Rev. B* **52**, 10819 (1995).

427. S. N. Voevodina and A. V. Tikhomorov, *Opt. Spektrosc.* **68**, 927 (1990).

428. J. Casset, *J. Opt. Am. Soc. Am.* **69**, 725 (1979).

429. P. Joensen, J. C. Irwin, J. F. Cochran, and A. E. Carzon, *J. Opt. Am. Soc. Am.* **63**, 1556 (1973).

430. J. M. Siqueros, L. F. Regalado, and R. Machorro, *Appl. Opt.* **27**, 4260 (1988).

431. R. Islem and D. R. Rao, *Opt. Mater.* **7**, 47 (1997).

432. R. Rusli and G. A. J. Amaratunga, *Appl. Opt.* **34**, 7914 (1995).

433. Q. H. Wu and I. Hodginson, *J. Opt. Am. Soc. Am. A* **10**, 2072 (1993).

434. R. M. A. Azzam and N. M. Bashara, *Ellipsomerty and Polarized Light*, North-Holland, Amsterdam, 1977.

435. R. M. A. Azzam (Ed.), *Selected Papers on Ellipsometry*, SPIE Milestone Series, Vol. MS27, SPIE Optical Engineering Press, Bellingham, 1991.

436. K. Vedam, *Thin Solid Films* **313–314**, 1 (1998).

437. J. M. Bennet, *J. Opt. Am. Soc. Am.* **54**, 612 (1964).

438. M. S. Shaalan, *J. Phys. D* **16**, 419 (1993).

439. C. Caliendo, E. Verona, and G. Saggio, *Thin Solid Films* **292**, 255 (1997).

440. N. A. Paraire, N. Moresman, S. Chen, P. Dansas, and F. Bertand. *Appl. Opt.* **36**, 2545 (1997).

441. E. Nitanai and S. Miyanaga, *Opt. Eng.* **35**, 900 (1996).

442. E. V. Alieva, G. Beitel, L. A. Kuzik, A. A. Sigarev, V. A. Yakovlev, G. N. Zhizhin, A. F. G. van der Meer, and M. J. van der Wiel, *J. Mol. Structure* **449**, 119 (1998).

443. G. N. Zhizhin, E. V. Alieva, L. Kuzik, V. A. Yakovlev, D. M. Shkrabo, A. F. G. van der Meer, and M. J. van der Wiel, *Appl. Phys. A* **67**, 667 (1998).

444. D. G. Walmsley, T. Bade, P. G. McCafferty, C. Rea, P. Dawson, R. J. Wallace, R. M. Bowman, and J. H. Clark, *Phys. C Superconduct.* **271**, 298 (1996).

445. L. Vriens and W. Rippens, *Appl. Opt.* **22**, 4105 (1983).

446. E. Elizalde and F. Rueda, *Thin Solid Films* **122**, 45 (1984).

447. M. Chang and U. J. Gibson, *Appl. Opt.* **24**, 504 (1985).

448. R. Swaenepoel, *J. Phys. E Sci. Instrum.* **16**, 1214 (1983).

449. (a) F. Abeles and M. L. Theye, *Surf. Sci.* **5**, 525 (1966); (b) J. M. Benett and M.-J. Booty, *Appl. Opt.* **5**, 41 (1966).

450. M. Kubinyi, N. Benko, A. Grofcsik, and W. J. Jones, *Thin Solid Films* **286**, 164 (1996).

451. Y. Wang and M. Miyagi, *Appl. Opt.* **36**, 877 (1997).

452. T. Buffeteau and B. Desbat, *Appl. Spectrosc.* **43**, 1027 (1989).

453. (a) J. P. Hawranek, P. Neelakantan, R. P. Young, and R. N. Jones, *Spectrochim. Acta* **32A**, 85 (1976); (b) D. Allara, A. Baca, and C. A. Pryde, *Macromolecules* **11**, 1215 (1978).

454. R. T. Graf, J. L. Koenig, and H. Ishida, *Appl. Spectrosc.* **39**, 405 (1985).

455. (a) R. Brendel, *Appl. Phys. A* **50**, 587 (1990); (b) G. K. Ribbegard and R. N. Jones, *Appl. Spectrosc.* **34**, 638 (1980); (c) R. Ferrini, G. Guizzetti, M. Patrini, A. Bosacchi, S. Franchi, and R. Magnanini, *Solid State Commun.* **104**, 747 (1997).

456. V. M. Zolotarev, *Optiko-mekhanicheskaya promyshlennost'* **8**, 46 (1984).

457. F. M. Mirabella, Jr., in F. M. Mirabella, Jr. (Ed.), *Internal Reflection Spectroscopy*, Marcel Dekker, New York, 1993, pp. 325–332.

458. V. P. Tolstoy, G. N. Kuznetsova, and I. I. Shaganov, *Zhurnal prikladnoi spektroskopii* **40**, 978 (1984).

459. J. R. Jasperse, A. Kahan, J. N. Plendle, and J. N. Mitra, *Phys. Rev.* **146**, 3, 526 (1966).

460. H. Hobert, H. H. Dunken, J. Meinschien, and H. Stafast, *Vib. Spectrosc.* **19**, 205 (1999).

461. S. S. Fouad and A. H. Ammar, *Physica B* **205**, 285 (1995).

462. M. Milosevic, *Appl. Spectrosc.* **47**, 566 (1993).

463. Z. M. Zhang, G. Lefever-Button, and F. R. Powell, *Int. J. Thermophys.* **19**, 905 (1998).

464. O. P. Konovalova, O. Y. Rusakova, and I. I. Shaganov, *Sov. J. Opt. Technol.* **55**, 402 (1988).

465. W. N. Hansen and W. A. Abdou, *J. Opt. Soc. Am.* **67**, 1537 (1977).

466. S. Selci, F. Ciccacci, G. Chiarotti, P. Chiaradia, and A. Cricenti, *J. Vac. Sci. Technol. A* **5**, 327 (1987).

467. K. N. Kuksenko and A. E. Chmel, *Zhurnal prikladnoi spektroskopii* **226**, 307 (1975).

468. C. R. Flach, A. Gericke, and R. Mendelsohn, *J. Phys. Chem. B* **101**, 58 (1997).

469. A. Gericke, A. V. Michailov, and H. Huhnerfuss, *Vib. Spectrosc.* **4**, 335 (1993).

470. D. Blaudez, T. Buffeteau, B. Desbat, P. Fournier, A.-M. Ritcey, and M. Pezolet, *J. Phys. Chem. B* **102**, 99 (1998).

471. C. W. Myers and S. L. Cooper, *Appl. Spectrosc.* **48**, 72 (1994).

472. A. W. Kleyn, *Progr. Sur. Sci.* **54**, 407 (1997).

473. C. W. Frank (Ed.), *Organic Thin Films: Structure and Applications*, American Chemical Society Washington, D.C., 1998.

474. (a) A. Ulman (Ed.), *Self-Assembled Monolayers of Thiols*, Academic, San Diego, 1998; (b) A. Ulman, *An Introduction to Ultrathin Organic Films: From Langmuir-Blodgett to Self-Assembly*, Academic, Boston, 1991.

475. J. C. Riviere and S. Mihra (Eds.), *Handbook of Surface and Interface Analysis*, Marcel Dekker, New York, 1998.

476. R. A. Dluhy, S. M. Stephens, S. Widayati, and A. D. Willams, *Spectrochim. Acta A* **51**, 1413 (1995).

477. R. H. Tredgold, *Order in Thin Organic Films*, Cambridge University Press, Cambridge, 1994.

478. B. Lecourt, D. Blaudez, and J. M. Turlet, *Thin Solid Films* **313–314**, 790 (1998).

479. D. Yang, Y. Sun, and D. Da, *Appl. Surf. Sci.* **144–145**, 451 (1999).

480. T. Schmitz-Hübsch, F. Sellam, R. Staub, M. Törker, T. Fritz, Ch. Kübel, K. Müllen, and K. Leo, *Surf. Sci.* **445**, 358 (2000).

481. I. Takanori Inoue, M. Masao, and O. Teiichiro, *Thin Solid Films* **350**, 238 (1999).

482. J. Stöhr, *NEXAFS Spectroscopy*, Vol. 25, Springer-Verlag, Berlin, 1992.

483. R. Giebler, B. Schulz, J. Reiche, L. Brehmer, M. Wuehn, Ch. Woell, A. P. Smith, S. G. Urquhart, H. W. Ade, and W. E. S. Unger, *Langmuir* **15**, 1291 (1999).

484. A. Kaito, T. Yatabe, S. Ohnishi, N. Tanigaki, and K. Yase, *Macromolecules* **32**, 5647 (1999).

485. K. K. Okudaira, E. Morikawa, D. A. Hite, S. Hasegawa, H. Ishii, M. Imamura, H. Shimada, Y. Azuma, K. Meguro, Y. Harada, V. Saile, K. Seki, and N. Ueno, *J. Electron Spectrosc. Related Phenom.* **101–103**, 389 (1999).

486. G. Weidemann, G. Brezesinski, D. Vollhardt, C. DeWolf, and H. Moehwald, *Langmuir* **15**, 2901 (1999).

487. Y. Wu, B. Zhao, W. Xu, G. Li, B. Li, and Y. Ozaki, *Langmuir* **15**, 1247 (1999).

488. C. C. Perry, B. G. Frederick, J. R. Power, R. J. Cole, S. Haq, Q. Chen, N. V. Richardson, and P. Weightman, *Surf. Sci.* **427–428**, 446 (1999).

489. I. M. Ward, *Adv. Polym. Sci.* **66**, 81 (1985).

490. M. Prakash, J. B. Peng, J. B. Ketterson, and P. Dutta, *Chem. Phys. Lett.* **128**, 354 (1986).

491. D. K. Schwartz, *Surf. Sci. Reports* **27**, 241 (1997).

492. C. M. Knobler and D. K. Schwartz, *Curr. Opinion Colloid Interface Sci.* **4**, 46 (1999).

493. C. Naselli, J. P. Rabe, J. F. Rabolt, and J. D. Swallen, *Thin Solid Films* **134**, 173 (1985).

494. C. Naselli, J. F. Rabolt, and J. D. Swallen, *J. Phys. Chem.* **82**, 2136 (1985).

495. A. Gericke and H. Huhnerfuss, *J. Phys. Chem.* **97**, 12899 (1993).

496. D. Blaudez, T. Buffereau, J. C. Cornut, B. Desbat, N. Escares, M. Pezolet, and J. M. Turlet, *Appl. Spectrosc.* **47**, 869 (1993).

497. Y. Oishi, Y. Takashima, K. Suehiro, and T. Kajiyama, *Langmuir* **13**, 2527 (1997).

498. V. A. Howarth, M. C. Petty, G. H. Davies, and J. Yarwood, *Langmuir* **5**, 330 (1989).

499. T. Takenaka, K. Harada, and M. Matsumoto, *J. Colloid Interface Sci.* **73**, 569 (1980).

500. X. Du, B. Shi, and Y. Liang, *Langmuir* **14**, 3631 (1998).

501. M. Subirade, M. Pézolet, and D. Marion, *Thin Solid Films* **284–285**, 326 (1996).

502. M. Methot, F. Boucher, C. Salesse, M. Subirade, and M. Pezolet, *Thin Solid Films* **284–285**, 627 (1996).

503. H. Hui-Litwin, L. Servant, M. J. Dignam, and M. Moskovits, *J. Phys. Chem. B* **102**, 5055 (1998).

504. S. Terashita, Y. Ozaki, and K. Iiyama, *J. Phys. Chem.* **97**, 10445 (1993).

505. S. Y. Park and E. Franses, *Langmuir* **11**, 2187 (1995).

506. Y. Tung, T. Gao, M. J. Rosen, J. E. Valentini, and L. J. Fina, *Appl. Spectrosc.* **47**, 1643 (1993).

507. A. N. Parikh and D. L. Allara, *J. Phys. Chem.* **96**, 927 (1992).

508. S. M. Stefens and R. A. Dluhy, *Thin Solid Films* **284–285**, 381 (1996).

509. R. G. Nuzzo, E. M. Korenic, and L. H. Dubois, *J. Phys. Chem.* **93**, 767 (1990).

510. L. H. Dubois, B. R. Zegarski, and R. G. Nuzzo, *J. Electron Spectrosc. Related Phenom.* **54–55**, 1143 (1990).

511. A. Gericke and R. Mendelsohn, *Langmuir* **12**, 758 (1996).

512. A. Gericke, J. W. Brauner, R. K. Erukulla, R. Bittman, and R. Mendelsohn, *Thin Solid Films* **284–285**, 428 (1996).

513. J. T. Buontempo and S. A. Rice, *J. Chem. Phys.* **98**, 5835 (1993).

514. J. T. Buontempo and S. A. Rice, *J. Chem. Phys.* **99**, 7030 (1993).

515. D. A. Myrzakozha, T. Hasegawa, J. Nishijo, T. Imae, and Y. Ozaki, *Langmuir* **15**, 6890 (1999).

516. S. Terashita, Y. Ozaki, and K. Iiyama, *J. Phys. Chem.* **97**, 10445 (1993).

517. (a) D. J. Ahn and E. I. Franses, *J. Phys. Chem.* **96**, 9952 (1992); (b) D. J. Ahn and E. I. Franses, *Thin Solid Films* **244**, 971 (1994).

518. H. Li, Z. Wang, B. Zhao, H. Xiong, X. Zhang, and J. Shen, *Langmuir* **14**, 423 (1998).

519. C. Jiang, W. He, J. Huang, Z. Tai, and Y. Liang, *Mater. Chem. Phys.* **62**, 236 (2000).

520. F. Bensebaa, T. H. Ellis, A. Badia, and R. B. Lennox, *Langmuir* **14**, 2361 (1998).

521. B. F. Sinnamon, R. A. Dluhy, and G. T. Barnes, *Colloids Surfaces A* **156**, 215 (1999).

522. B. F. Sinnamon, R. A. Dluhy, and G. T. Barnes, *Colloids Surfaces A* **156**, 49 (1999).

523. H. Sakai and J. Umemura, *Langmuir* **14**, 6249 (1998).

524. R. A. Dluhy, Z. Ping, K. Faucher, and J. M. Brockman, *Thin Solid Films* **327–329**, 308 (1998).

525. W. A. Hayes and D. K. Schwartz, *Langmuir* **14**, 5913 (1998).

526. A. Gericke, J. Simon-Kutscher, and H. Huhnerfuss, *Langmuir* **9**, 3115 (1993).

527. R. H. Terrill, T. A. Tanzer, and P. W. Bohn, *Langmuir* **14**, 845 (1998).

528. D. J. Neivandt, M. L. Gee, M. L. Hair, and C. P. Tripp, *J. Phys. Chem. B* **102**, 5107 (1998).

529. X. Du, B. Shi, and Y. Liang, *Langmuir* **14**, 3631 (1998).

530. L. K. Tamm and S. A. Tatulian, *Q. Rev. Biophys.* **30**, 365 (1997).

531. (a) H. L. Casal and H. H. Mantsch, *Biochim. Biophys. Acta* **779**, 381 (1984); (b) J. Umemura, D. G. Cameron, and H. H. Mantsch, *Biochim. Biophys. Acta* **602**, 32 (1980).

532. A. Horn, in R. J. H. Clark and R. E. Hester (Eds.), *Spectroscopy for Surface Science*, Vol. 26, Wiley, New York, 1998, pp. 273–339.

533. V. Kumar, S. Krishnan, C. Steiner, C. Maldarelli, and A. Couzis, *J. Phys. Chem. B* **102**, 3152 (1998).

534. J. Umemura, J. Takeda, T. Hasegawa, T. Kamata, and T. Takenaka, *Spectrochim. Acta A* **50**, 8 (1994).

535. D. Lin-Vien, N. B. Colthup, W. G. Fateley, and J. G. Grasselli, *Infrared and Raman Characteristic Frequencies of Organic Molecules*, Academic, San Diego, 1991, p. 140.

536. R. G. Snyder, *J. Phys. Chem.* **47**, 1316 (1967).

537. B. K. Vainshtein, *Modern Crystallography*, Vol. 1, Springer-Verlag, Berlin, 1981.

538. I. Kuzmenko, M. M. Kaganer, and L. Leiserowitz, *Langmuir* **14**, 3882 (1998).

539. A. I. Kitaigorodskii, *Organic Chemical Crystallography*, Consultants Bureau, New York, 1961.

540. L. H. Dubois and R. G. Nuzzo, *Annu. Rev. Phys. Chem.* **43**, 437 (1992).

541. D. Demus, J. Goodby, G. W. Gray, H.-W. Spiess, and V. Vill (Eds.), *Handbook of Liquid Crystals*, Vol. 1: *Fundamentals*, Wiley-VCH, Weinheim, 1998.

542. R. Mendelsohn, J. W. Brauner, and A. Gericke, *Annu. Rev. Phys. Chem.* **46**, 305 (1995).

543. G. G. Roberts, *Langmuir–Blodgett Films*, Plenum, New York, 1990.

544. E. B. Sirota, *Langmuir* **13**, 3849 (1997).

545. H. D. Sikes, J. T. Woodward, and D. K. Schwartz, *J. Phys. Chem.* **100**, 9093 (1996).

546. U. Pietsch, T. A. Barberka, U. Englisch, and R. Stömmer, *Thin Solid Films* **284–285**, 387 (1996).

547. M. Kramer and V. Hoffmann, *Opt. Mater.* **9**, 65 (1999).

548. R. G. Snyder, *J. Mol. Spectrosc.* **7**, 116 (1961).

549. R. G. Snyder, M. C. Goh, V. J. P. Srivatsavoy, H. L. Strauss, and D. L. Dorset, *J. Phys. Chem.* **96**, 10008 (1992).

550. F. Kimura, J. Umemura, and T. Takenaka, *Langmuir* **2**, 96 (1986).

551. J. Umemura, T. Kamata, T. Kawai, and T. Takenaka, *J. Phys. Chem.* **94**, 62 (1990).

552. Y. Liang, Y. Jiang, and Y. Tian, *Acta Phys. Chim. Sin.* **7**, 72 (1991).

553. D. G. Cameron, H. L. Casal, E. F. Gudgin, and H. H. Mantsch, *Biochim. Biophys. Acta* **596**, 463 (1980).

554. M. Shimomura, K. Song, and J. P. Rabolt, *Langmuir* **8**, 887 (1992).

555. R. F. Holland and J. R. Nielsen, *J. Mol. Spectrosc.* **9**, 436 (1962).

556. R. G. Snyder, G. L. Liang, H. L. Strauss, and R. Mendelsohn, *Biophys. J.* **71**, 3186 (1996).

557. A. Hjörtsberg, W. P. Chen, E. Butrstein, and M. Pomerantz, *Opt. Commun.* **25**, 65 (1978).

558. I. V. Chernyshova and K. Hanumantha Rao, *J. Phys. Chem. B* **105**, 810 (2001).

559. F. M. Mirabella, Jr., in F. M. Mirabella, Jr. (Ed.), *Internal Reflection Spectroscopy*, Marcel Dekker, New York, 1993, pp. 141–172.

560. K. Sakamoto, N. Ito, R. Arafune, and S. Ushioda, *Vib. Spectrosc.* **19**, 61 (1999).

561. R. Arafune, K. Sakamoto, and S. Ushioda, *Appl. Phys. Lett.* **71**, 2755 (1997).

562. M. P. Srinivasan and K. K. S. Lau, *Thin Solid Films* **307**, 226 (1997).

563. R. Maoz, J. Sagiv, D. Degenhardt, H. Möhwald, and P. Quint, *Supramol. Sci.* **2**, 9 (1995).

564. R. G. Greenler, *J. Phys. Chem.* **44**, 310 (1966).

565. (a) S. Enomoto, Y. Ozaki, and N. Kiramoto, *Langmuir* **9**, 3219 (1993); (b) P. A. Antunes, C. J. L. Constantino, R. Aroca, and J. Duff, *Appl. Spectrosc.* **55**, 1341 (2001).

566. I. Pelletier, H. Bourque, T. Buffeteau, D. Blaudez, B. Desbat, and M. Pezolet, *J. Phys. Chem. B* **106**, 1968 (2002).

567. T. Buffeteau, B. Desbat, E. Pere, and J. M. Turlet, *Mikrochim. Acta Suppl.* **14**, 631 (1997).

568. (a) D. Blaudez, F. Boucher, T. Buffeteau, B. Desbat, M. Grandbois, and C. Salesse, *Appl. Spectrosc.* **53**, 1299 (1999); (b) T. Buffeteau, E. Le Calvez, S. Castano, B. Desbat, D. Blaudez, and J. Dufourcq, *J. Phys. Chem. B* **104**, 4537 (2000).

569. (a) Y. P. Song, M. C. Petty, J. Yarwood, W. J. Feast, J. Tsibouklis, and S. Mukherjee, *Langmuir* **8**, 257 (1992); (b) Y. P. Song, J. Yarwood, J. Tsibouklis, W. J. Feast, J. Cresswell, and M. C. Petty, *Langmuir* **8**, 262 (1992).

570. Y. P. Song, M. C. Petty, and J. Yarwood, *Langmuir* **9**, 543 (1993).

571. Y. Ishino and H. Ishida, *Langmuir* **4**, 1341 (1988).

572. R. P. Sperline, S. Muralidhara, and H. Freiser, *Langmuir* **3**, 198 (1987).

573. J. Yarwood, R. Banga, A. M. Morgan, B. Evans, and J. Kells, *Thin Solid Films* **284–285**, 261 (1996).

574. C. Zannoni, in G. W. Gray and G. R. Luckhurst (Eds.), *Molecular Physics of Liquid Crystals*, D. Reidel, Dordrecht, Netherlands, 1983, p. 351.

575. D. A. Jarvis, I. M. Hutchinson, D. I. Bower, and I. M. Ward, *Polymer* **21**, 41 (1980).

576. G. R. Luckhurst, *Ber. Bunsenges. Phys. Chem.* **97**, 1169 (1993).

577. D. Dunmur and K. Toriyama, in D. Demus, J. Goodby, G. W. Gray, H.-W. Spiess, and V. Vill (Eds.), *Handbook of Liquid Crystals*, Vol. 1: *Fundamentals*, Wiley-VCH, Weinheim, 1998, pp. 189–203.

578. V. Tsvetkov, *Acta Physicochim. (USSR)* **16**, 132 (1942).

579. N. Everall, *Internet J. Vibrational Spectrosc.*, available on-line: http://www.ijvs.com/volume3/edition2/section1.html.

580. R. D. B. Fraser and T. P. MacRae, *Conformation in Fibrous Proteins and Related Synthetic Polypeptides*, Academic, New York, 1973.

581. T. Hasegawa, D. A. Myrzakozha, T. Imae, J. Nishijo, and Y. Ozaki, *J. Phys. Chem. B* **103**, 11124 (1999).

582. J. Huang and Y. Liang, *Thin Solid Films* **325**, 210 (1998).

583. E. H. Korte, *Mol. Cryst. Liq. Cryst.* **100**, 127 (1983).

584. E.-H. Korte, in B. Schrader (Ed.), *Infrared and Raman Spectroscopy*, VCH, Weinheim 1995, pp. 323–344.

585. H. Sakai and J. Umemura, *Langmuir* **14**, 6249 (1998).

586. J. B. Dence, *Mathematical Techniques in Chemistry*, Wiley-Interscience, New York, 1975, pp. 288–293.

587. M. Brogly, S. Bistac, and J. Schultz, *Macromol. Symp.* **141**, 129 (1999).

588. T. Hasegawa, J. Umemura, and T. Takenaka, *J. Phys. Chem.* **97**, 9009 (1993).

589. Y. Ren, C. W. Meuse, S. L. Hsu, and H. D. Stidham, *J. Phys. Chem.* **98**, 8424 (1994).

590. R. G. Nuzzo, F. A. Fusco, and D. L. Allara, *J. Am. Chem. Soc.* **109**, 2358 (1987).

591. R. G. Nuzzo, L. H. Dubois, and D. L. Allara, *J. Am. Chem. Soc.* **112**, 558 (1990).

592. A. Gericke, C. R. Flach, and R. Mendelsohn, *Biophys. J.* **73**, 492 (1997).

593. (a) T. Buffeteau, E. Le Calvez, B. Desbat, I. Pelletier, and M. Pezolet, *J. Phys. Chem. B* **105**, 1464 (2001); (b) F. Picard, T. Buffeteau, B. Desbat, M. Auger, and M. Pezolet, *Biophys. J.* **76**, 539 (1999).

594. D. D. Popenoe, S. M. Stole, and M. D. Porter, *Appl. Spectrosc.* **46**, 79 (1992).

595. J. L. Dote and R. L. Mowery, *J. Phys. Chem.* **92**, 1571 (1988).

596. T. Hasegawa, S. Takeda, A. Kawaguchi, and J. Umemura, *Langmuir* **11**, 1236 (1995).

597. L. J. Fina and Y. Tung, *Appl. Spectrosc.* **45**, 986 (1991).

598. H. Hoffmann, U. Mayer, and A. Krischanitz, *Langmuir* **11**, 1304 (1995).

599. H. Schopper, *Z. Phys.* **132**, 146 (1952).

600. M. S. Tomar and V. K. Srivastava, *Thin Solid Films* **15**, 207 (1973).

601. M. S. Tomar and V. K. Srivastava, *J. Appl. Phys.* **45**, 1849 (1974).

602. M. S. Tomar, *J. Phys. Chem.* **82**, 2736 (1976).

603. A. R. Elsharkawi and K. C. Kao, *J. Opt. Soc. Am.* **65**, 1269 (1975).

604. M. K. Dete, *J. Appl. Phys.* **55**, 3354 (1984).

605. H. L. Zhang, H. Zhang, J. Zhang, Z. Liu, and H. Li, *J. Colloid Interface Sci.* **214**, 46 (1999).

606. A. Ihs, K. Uvdal, and B. Liedberg, *Langmuir* **9**, 733 (1993).

607. L. Servant and M. J. Dignam, *Thin Solid Films* **242**, 21 (1994).

608. (a) V. Koppaka and P. H. Axelsen, *Langmuir* **17**, 6309 (2001); (b) M. J. Citra and P. H. Axelsen, *Biophys. J.* **71**, 1796 (1996).

609. M. Muller and F.-J. Schmitt, *Thin Solid Films* **310**, 138 (1997).

610. D. J. Neivandt, M. L. Gee, M. L. Hair, and C. P. Tripp, *J. Phys. Chem B* **102**, 5107 (1998).

611. Y. Cheng, N. Boden, R. J. Bushby, S. Clarkson, S. D. Evans, P. F. Knowles, A. Marsh, and R. E. Miles, *Langmuir* **14**, 839 (1998).

612. P. H. Axelsen, W. D. Braddock, H. L. Brockman, C. M. Jones, R. A. Dluhy, B. K. Kauman, and F. J. Puga II, *Appl. Spectrosc.* **49**, 526 (1995).

613. T. Takahashi, P. Miller, Y. M. Chen, L. Samuelson, D. Galotti, B. K. Mandal, J. Kumar, and S. K. Tripathy, *J. Polym. Sci. B* **31**, 165 (1993).

614. W.-H. Jang and J. D. Miller, *J. Phys. Chem.* **99**, 10272 (1995).

615. (a) D. Marsh, *Biophys. J.* **72**, 2710 (1997); (b) D. Marsh, *Biophys. J.* **77**, 2630 (1999).

616. S. Y. Park and E. I. Franses, *Langmuir* **11**, 2187 (1995).

617. J. M. F. Swart and P. C. M. Woerkom, *J. Mol. Struct.* **293**, 307 (1993).

618. I. V. Chernyshova, K. Hanumantha Rao, A. Vidyadhar, and A. V. Shchukarev, *Langmuir* **16**, 8071 (2000).

619. I. V. Chernyshova, K. Hanumantha Rao, A. Vidyadhar, and A. V. Shchukarev, *Langmuir* **17**, 775 (2001).

620. I. V. Chernyshova and K. Hanumantha Rao, *Langmuir* **17**, 2711 (2001).

621. D. Shin, M. Park, and S. Lim, *Thin Solid Films* **327–329**, 607 (1998).

622. Y. J. Chabal, *Surf. Sci. Rep.* **8**, 211 (1988).

623. X. Du and Y. Liang, *Chem Phys. Lett,* **313**, 565 (1999).

624. T. Miyashita and T. Suwa, *Langmuir* **10**, 3387 (1994).

625. H. Binder, T. Gutberlet, and A. Anikin, *J. Mol. Struct.* **510**, 113 (1999).

626. S. J. Ahn, D. H. Son, and K. Kim, *J. Mol. Struct.* **324**, 223 (1994).

627. V. Chechik, H. Schönherr, G. J. Vancso, and C. J. M. Stirling, *Langmuir* **14**, 3003 (1998).

628. M. Brogly, S. Bistac, and J. Schultz, *Macromol. Theory Simulations* **7**, 65 (1998).

629. D. Blaudez, T. Buffeteau, J. C. Cornut, B. Desbat, N. Escafre, M. Pizolet, and J. M. Turlet, *Appl. Spectrosc.* **47**, 869 (1993).

630. L. J. Fina and Y. Tung, *Appl. Spectrosc.* **45**, 986 (1991).

631. T. Hasegawa, Y. Kobayashi, J. Nishijo, and J. Umemura, *Vib. Spectrosc.* **19**, 199 (1999).

632. (a) B. A. Sexton, *Surf. Sci.* **94**, 435 (1980); (b) S. D. Ross, *Inorganic Infrared and Raman Spectra*, McGraw Hill, London, 1972.

633. (a) W. B. Fischer, A. Fedorowicz, and A. Koll, *Phys. Chem. Chem. Phys. (PCCP)* **3**, 4228 (2001); (b) W. B. Fischer and H.-H. Eysel, *J. Mol. Struct.* **415**, 249 (1997).

634. H. B. Librovich, V. P. Sakun, and N. D. Sokolov, in N. D. Sokolov (Ed.), *Hydrogen Bond*, Nauka, Moscow, 1981, pp. 174–211.

635. J. Kim, U. W. Schmitt, J. A. Gruetzmacher, G. A. Voth, and N. E. Scherer, *J. Chem. Phys.* **116**, 737 (2002).

636. K. Nakamoto, *Infrared and Raman Spectra of Inorganic and Coordination Compounds*, Part A, 5th edition, Wiley, New York.

637. J.-J. Max and C. Chapados, *J. Chem. Phys.* **115**, 2664 (2001).

638. J. R. Scherer, in R. J. H. Clark and R. E. Hester (Eds.), *Advances in Infrared and Raman Spectroscopy*, Heyden, London, Vol. 5, p. 149.

639. J. E. Bertie and E. Whalley, *J. Chem. Phys.* **40**, 1637 (1964).

640. A. Grodzicki and P. Piszczek, *Polish. J. Chem.* **68**, 2687 (1994).

641. B. Soptrajanov and V. M. Petrusevski, *J. Mol. Struct.* **408–409**, 283 (1997).

642. J. Cvetkovic, V. M. Petrusevski, and B. Soptrajanov, *J. Mol. Struct.* **408–409**, 463 (1997).

643. M. Trpkovska, B. Soptrajanov, and P. Malkov, *J. Mol. Struct.* **480–481**, 661 (1999).

644. S. Aleskovska, V. M. Petrusevski, and B. Soptrajanov, *J. Mol. Struct.* **408–409**, 413 (1997).

645. I. Gano, *Bull. Chem. Soc. Jpn.* **34**, 764 (1961).

646. B. Soptrajanov, G. Jovanovski, and L. Pejov, *J. Mol. Struct.* **613**, 47 (2002).

647. B. Soptrajanov, *J. Mol. Struct.* **555**, 21 (2000).

648. V. Ivanovski, V. M. Petrusevski, and B. Soptrajanov, *Vib. Spectrosc.* **19**, 425 (1999).

649. S. Baldelli, C. Schnitzer, D. J. Campbell, and M. J. Shultz, *J. Phys. Chem. B* **103**, 2789 (1999).

650. K. Kunimatsu and A. Bewick, *Ind. J. Technol.* **24**, 407 (1986).

651. M. Futamata, *Surf. Sci.* **427–428**, 179 (1999).

652. J. E. Crowell, J. G. Chen, D. M. Hercules, and J. T. Yates, Jr. *J. Chem. Phys.* **86**, 5804 (1987).

653. H. Ibach and S. Lehwald, *Surf. Sci.* **91**, 187 (1980).

654. Y. Shingaya, K. Hirota, H. Ogasawara, and M. Ito, *J. Electroanal Chem.* **409**, 103 (1996).

655. K. Shaw, P. Christensen, and A. Hamnett, *Electrochim. Acta* **41**, 719 (1996).

656. K. Hirota, M.-B. Song, and M. Ito, *Chem. Phys. Lett.* **250**, 335 (1996).

657. E.T. Wagner and T.E. Moylan, *Surf. Sci.* **182**, 125 (1987).

4

EQUIPMENT
AND TECHNIQUES

Fundamental concepts of FTIR spectroscopy and basic principles of spectrometer operation, spectrum acquisition, and spectral manipulation have been discussed in detail by Smith [1], Griffiths and de Haseth [2a], Chalmers and Griffiths [2b], and Mirabella [3]. At present, FTIR instrumentation, both commercial and home built, is developing rapidly, specifically with regards to better time and spectral resolution, SNR level, and scan speed. This is largely due to the availability of improved optical technologies and personal computers. Further information on modern commercially available FTIR spectrometers may be requested from individual manufacturers and is not included here. Improvements and developments in various components of FTIR spectrometers are routinely reported in journals such as *Applied Spectroscopy*, *Applied Optics*, *Spectroscopy Europe*, *Journal of Applied Spectroscopy*, and *Infrared Physics*. This chapter concentrates on spectroscopic techniques for measuring IR spectra, including the design and construction of accessories for studying ultrathin films, and original methods of sample preparation. Mainly the techniques compatible with a standard serial FTIR or dispersive spectrometer will be considered, with no variations in the optical arrangement. As was shown in Chapter 2, the choice of technique for recording IR spectra of ultrathin films in a given application is determined by several parameters. Among those are the optical parameters of the layer being studied and its thickness as well as the optical parameters of the substrate and the immersion medium. The choice also depends upon the shape and size of particles in the case of a dispersed sample or the degree of the surface roughness for bulk material. In this chapter, general recommendations will be proposed for choosing the most appropriate technique to solve a given problem in IR spectroscopy of ultrathin films.

4.1. TECHNIQUES FOR RECORDING IR SPECTRA OF ULTRATHIN FILMS ON BULK SAMPLES

4.1.1. Transmission and Multiple Transmission

Recording IR transmission spectra of ultrathin layers on the surface of transparent materials is perhaps the simplest IR spectroscopy technique. If the layer is located on the surface of a plane–parallel plate that is transparent in the IR region, it is sufficient to place it into an IR spectrometer at the focal point in the sample compartment and to measure the spectrum of the transmitted radiation in the usual manner.

The standard (*normal-incidence*) transmission technique was used in the early applications, for example, in microelectronics when dielectric layers ~50–200 nm in thickness were involved. However, to study thinner dielectric layers (<10 nm), the contrast achieved with this technique was insufficient. To enhance the sensitivity, a technique was adopted that records spectra of a stack of several samples (up to 15–20) [4]. In this way, a contrast spectrum was obtained of a monolayer of water adsorbed onto a quartz surface. However, when recording at an angle of incidence of $\varphi_1 = 0°$ the spectrum of a stack of semiconductor plates with a refractive index in the IR region larger than 3 (e.g., for Si), the proportion of the signal that is reflected from the front surface of the plate is fairly high (about 30–40%) (Fig. 1.11). Therefore, the intensity of the radiation passed through the plates furthest from the front plate is low, which is obviously a substantial limitation of the given technique. For substances with a high refractive index, this difficulty can be easily overcome by recording p-polarized spectra at an angle of incidence close to the Brewster angle (*transmission at the Brewster angle*) (Fig. 4.1a). Under these conditions, the energy losses upon reflection from the front plate are minimal. Thus a larger number of plates may be included in the stack (Fig. 4.1b) and the sensitivity will be enhanced. The other advantage of irradiating at the Brewster angle is p-polarizing of the resulting spectrum, because of the primary reflection of the s-component from the front face. This allows one to obtain p-polarized spectra of layers located on the plate surface without polarizers, increasing the level of useful signal by a few tens of percent and allowing the number of plates to be increased. To use this method, it is necessary that the Brewster angle be virtually independent of the layer thickness.

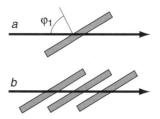

Figure 4.1. Scheme for recording IR spectra of layers by (*a*) single- and (*b*) multiple-transmission techniques.

It was shown in Refs. [5, 6] that in the case of ultrathin films this condition is met: The Brewster angle varies within the limits of the angular aperture for the real radiation beam.

The technique of transmission at the Brewster angle was suggested by Tolstoy [7] to measure spectra of SiO_2 layers located on both sides of a silicon plate 27 mm in diameter and 0.25 mm thick, which is used as a base for the manufacture of microelectronic chips. The plates were stacked in a holder that separates them by 0.5 mm, which then is placed into the sample compartment of the spectrometer at a specific angle with respect to the radiation beam. The spectra were measured with an ordinary dispersive spectrometer for both a single plate and stacks of 15 plates, treated by hydrofluoric acid and kept in air for 24 h. The spectra of the stack of 15 plates recorded at the Brewster angle are characterized by a ν_{LO} band at 1240 cm^{-1} (Fig. 4.2). The band position and intensity change upon annealing in air at temperatures from 200 to 500°C. The absorption band at 1100 cm^{-1} due to oxygen dissolved in silicon, and the ν_{TO} of the silicon oxide is also noteworthy. The position of the latter (ν_{TO}) proves to be less sensitive to changes in the lattice–chemical characteristics of the layer (Sections 5.1–5.3). The high sensitivity of these measurements is demonstrated by the fact that one can observe an absorption band in the 1170-cm^{-1} region in the spectrum of a Si sample immediately after etching in concentrated HF, at which point the Si surface is covered by a 0.2–0.4-nm layer of SiO_x under these conditions. Significant changes are observed in the spectra after subsequent thermal treatment of the sample in air at temperatures from 100 to 500°C. For example, annealing

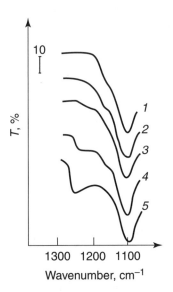

Figure 4.2. Multiple-transmission IR spectra of SiO_2 layers on silicon surface, $\varphi_1 = \varphi_B = 74°$: (1) after etching 1 min in HF; and after annealing in air at (2) 200, (3) 240, (4) 400, and (5) 500° C. Reprinted, by permission, from V. P. Tolstoy, L. P. Bogdanova, and V. B. Aleskovski, *Doklady AN USSR* **291**, 913 (1986), p. 914, Fig. 2. Copyright © 1986 Doklady AN USSR.

at 200°C causes the intensity of the wide band at 1170 cm^{-1} to increase slightly; at 240°C, a high-frequency shoulder at 1210 cm^{-1} becomes a separate band. The increase in the intensity of this band and the shift of its maximum to higher frequencies (up to 1245 cm^{-1}) are observed after annealing at 500°C.

It is also possible to measure spectra of layers on a real surface of silicon using the multiple-transmission technique at the Brewster angle for a single plate and even for a fraction of a plate [8]. This is done by placing it, as shown in Fig. 4.3, between two aluminum mirrors. In this case, the number of passages of the radiation through the sample is determined by the separation distance D_1 between the mirrors. One chooses the mirror length L in such a way as to minimize losses of beam intensity.

The intensity of p-polarized light transmitted through the system presented in Fig. 4.3 is described by the formula

$$I_p = I_p^0 (R^0)^N, \qquad N = \frac{L}{D_1 \tan \varphi_1},$$

where R^0 is the reflectance of the metal. The beam displacement upon passing through the plate, absorption by the layer, and the multiple reflections in the plate are all neglected.

The intensity of the s-polarized light transmitted through the same system proves to be considerably lower [8]. It is reflected from the front face of the semiconductor and multiply reflected in the narrow gap between the semiconductor and the metal. Therefore, when recording spectra on grating spectrometers (for which a significant polarization of the output radiation is observed), it is recommended to place a polarizer in front of the monochromator slit to maintain signal intensity.

In practice, recording transmission spectra at oblique angles of incidence requires special accessories, as determined by the geometry of the radiation beam in the sample compartment of the spectrometer: (i) the beam can be focused at the center of the sample compartment or (ii) on the entrance slit. One of the simplest accessories for both reflection and transmission measurements is shown in Fig. 4.4 [9]. It consists of a plateau (3) fixed in the sample compartment, a system of plane guiding mirrors (1 and 2), the sample (4), and directing mirrors (5).

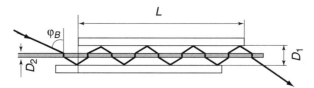

Figure 4.3. Scheme for recording IR spectra of layers by multiple-transmission technique: L = working length of the mirror, D_1 = separation between mirrors, and D_2 = thickness of plate under investigation. Reprinted, by permission, from V. P. Tolstoy, L. P. Bogdanova, and V. B. Aleskovski, *Doklady AN USSR* **291**, 913 (1986), p. 914, Fig. 1. Copyright © 1986 Doklady AN USSR.

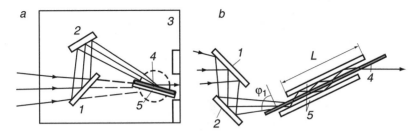

Figure 4.4. Schematic diagram of multipurpose accessory. Arrangement for recording (a) single-, double- and (b) multiple-transmission spectra: (1, 2) guiding mirrors; (3) plateau; (4) sample; (5) directing mirrors. Reprinted, by permission, from V. P. Tolstoy, *Methods of UV-Vis and IR Spectroscopy of Nanolayers*, St. Petersburg University Press, St. Petersburg, 1998, p. 128, Fig. 4.8. Copyright © St. Petersburg University Press.

The guiding mirrors may be mounted at the plateau in different positions. This accessory allows variation of transmission measurement conditions such as the number of beam passages through the sample, the angle of incidence, and the size of the sample. To record a single-transmission spectrum, it is sufficient to remove mirrors 1, 2, and 5 so that the radiation is directly incident upon the sample. The holder may serve as a goniometer, and angle of incidence can be varied from 15° to 80°. For a double-transmission spectrum, the optical scheme shown in Fig. 4.4a can be employed. In this case, the beam is incident upon the sample at an inclined angle, is reflected from the mirror located immediately behind the sample, and returns through the plate, arriving, finally, at the spectrometer. To record multiple-transmission spectra with this accessory, a Si sample plate and two mirrors (1 and 2) are mounted at $\varphi_B \approx 74°$ on the plateau together with two guiding mirrors (5), as shown in Figs. 4.4b. In this case, the plate under investigation may be larger than 10×25 mm; this means that the large silicon plates involved in ultra large-scale circuit (ULSC) technology may be probed.

A scheme based on the configuration of a double-beam spectrometer was proposed [10] for estimating the uniformity in thickness of a dielectric layer on the surface of a semiconductor plate (Fig. 4.5). In this technique, the ratio of the transmission spectra from two nearby points on the plate are obtained at oblique angles of incidence and in p-polarized radiation. Specifically, plane mirrors 2 and 5 focus both the sample and reference beams on plate 1. The holder of plate 1 is mounted kinematically and shifted by an electromechanical drive on the plateau in the focal plane. Passing through the sample at two nearby points, the beams are directed by mirrors 3 and 4 toward the spectrometer, the reference beam arriving into the sample channel and the sample beam into the reference one. Analyzing the differential transmission spectrum, one can estimate the degree of uniformity of the layer. Moreover, such estimation can be set up to be measured automatically if the electromechanical drive and the recorder are switched on. With dispersive and FTIR spectrometers, continuous and step-by-step drives should be used, respectively. In the former, the monochromator can be set to the frequency of the most intense absorption band.

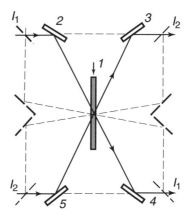

Figure 4.5. Schematic arrangement for studying uniformity in film: (1) plate, which can be moved with an electromechanical drive; (2–5) mirrors. Reprinted, by permission, from V. P. Tolstoy, *Methods of UV-Vis and IR Spectroscopy of Nanolayers*, St. Petersburg University Press, St. Petersburg, 1998, p. 130, Fig. 4.9. Copyright © St. Petersburg University Press.

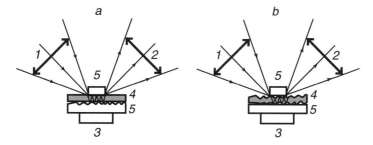

Figure 4.6. Schematic arrangement of conventional diffuse-reflectance accessory for measuring multiple-transmission spectra of microsamples. (1, 2) focusing optical system of accessory; (3) sample holder; (4) plate under study; (5) metallic mirrors. (*a*) Plate with polished surfaces placed between ground metallic surfaces. (*b*) Plate with ground surfaces placed between polished mirrors. Reprinted, by permission, from V. P. Tolstoy, *Methods of UV-Vis and IR Spectroscopy of Nanolayers*, St. Petersburg University Press, 1998, p. 130, Fig. 4.10. Copyright © 1998 St. Petersburg University Press.

When it is necessary to record the transmission spectrum of a microquantity of a material or a limited sample area, beam condensers are employed [11]. Alternatively, a diffuse-reflectance (DR) accessory may be used (Fig. 4.6). In this case, the radiation is multiply reflected between mirrors 5 and detected upon passing through the optical system 2. One of two optical schemes may be used: (i) the polished sample under study is placed between ground metallic mirrors 5 the surface roughness ensuring diffuse scattering of the transmitted radiation; (ii) the ground sample under investigation is placed between polished mirrors. In each configuration, the ground surface causes diffuse-scattered radiation to be multiply passed through the sample to obtain the transmission spectra, promoting enhancement of the spectral contrast.

Transmission measurements may be applied to ultrathin coatings on the polymer films that are self-supporting and those that have been deposited onto a transparent substrate [12]. A self-supporting film can be prepared by stretching the polymer. If the polymer is not already in the form of a film, the hot-pressing method is used, in which the material is molded and then pressed using a commercially available accessory in order to produce a film with a well-defined pathlength. If the polymer is soluble in a volatile solvent, it can be cast. To prepare a cast film, the polymer is first dissolved in an appropriate volatile solvent, and a few drops of the solution are then placed on the central part of a support. The solution spreads over the surface of the window (centrifuging can be used to give a uniform thickness), and the solvent is allowed to evaporate. The complete removal of solvent from the film is critical and must be verified by the absence of solvent bands in the spectrum [13]. The cast films can be studied on the support.

The interference fringes present in a transmission spectrum of a polymer film can be used to measure the film thickness. A discussion of this technique is given by Griffiths and de Haseth [2a]. The number of interference fringes, N, over a wavenumber range v_1 to v_2 can be related to the thickness of the sample, b, and its refractive index n by

$$b = \frac{1}{2\pi} \frac{N}{v_1 - v_2}.$$

To reduce spectral distortions caused by the interference fringes, the back surface of the film can be roughened, or the spectrum can be measured at the Brewster angle. Depending on the mechanical strength of the polymer, the minimum film thickness that can be achieved is usually 0.5–5 μm. A 10–20-μm film usually produces a good transmission spectrum.

4.1.2. IRRAS

To obtain the spectra of ultrathin films by IRRAS, the radiation from the spectrometer source must be directed at the film at a selected (optimum) angle of incidence (Sections 2.2 and 2.3), and the reflected radiation must be analyzed at the same angle. A single-reflection spectrum can be measured by IRRAS with the accessory shown in Fig. 4.4a by fixing the sample 4 to be investigated in the sample holder. The background spectrum is obtained from the bare substrate. To measure a multiple-reflection spectrum, two substrates are positioned parallel to one another on a plateau, as shown in Fig. 4.7a. For every metal, the number of reflections, N, between the parallel metallic plates under study is chosen so as to optimize the measurement (Section 2.2). For this purpose, varying the distance D between the plates varies N ($N = L/D \tan \varphi_1$). Commercial IRRAS accessories vary in the complexity of the directing optics and the angular range. For example, the FT-85 grazing-angle accessory (SpectraTech) and Nicolet Smart

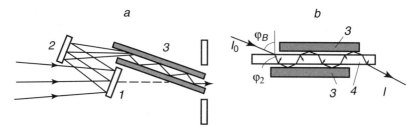

Figure 4.7. (*a*) Accessory and (*b*) modification of optical scheme for recording spectra by multiple metallic IRRAS: (1, 2) guiding mirrors; (3) samples; (4) silicon plate insert. (*a*) Reprinted, by permission, from V. P. Tolstoy, G. N. Kuznetsova, M. I. Ivanova, and A. I. Somsikov, *Pribory i tekhnika experimenta*, No. 3, 220–221 (1982); p. 220, Fig. 1. Copyright © 1982 Pribory i tekhnika experimenta; (*b*) Reprinted, by permission, from V. P. Tolstoy, in V. B. Aleskovski (Ed.), *Precision Synthesis of Solids*, Vol. 2, Leningrad State University Press, Leningrad, 1987, p. 76; p. 88, Fig. 13. Copyright © 1987 St. Petersburg University Press.

Refractor provide a fixed, high angle of incidence at the horizontally positioned sample by using silicon refracting optical elements instead of mirrors.

Polarization of the radiation is another condition to be specified when recording a spectrum by IRRAS. It is determined by a polarizer mounted on either the entrance window or the exit window of the sample compartment or in the IRRAS accessory. For a metal substrate, *p*-polarized light should be employed, as unpolarized radiation in IRRAS leads to spectral distortions [14]. This requirement is accounted for by introducing a built-in polarizer in the design of grazing-angle IRRAS accessories specialized for measurements on metals (e.g., in the FT-85 accessory of Spectra-Tech, the polarizing elements are silicon plates positioned in the beam at the Brewster angle of incidence). A number of modern accessories include a premounted, removable internal polarizer or polarizing plate for enhanced spectral contrast. To allow analysis of a small area of interest in a large sample, masks are often used.

The accessory attachments depicted in Figs. 4.4 and 4.8 may be used for a variety of different techniques. Thus, for example, the attachment shown in Fig. 4.8 [7] can be used to obtain single spectra by IRRAS, ATR spectra, multiple-reflection spectra, and single- and multiple-transmission spectra at variable angles of incidence. This unit contains a sample holder and two sets of parallel–plane mirrors mounted onto movable carriages (6 and 7). These mirrors are positioned perpendicular to the optical axis of the spectrometer. Mirrors 2 and 5 are fixed, and mirrors 3 and 4 can be moved along carriages 6 and 7 using special holders 8 and 9, respectively. Carriage 6 can also be moved as a unit along the optical axis. The required angle of incidence for samples of different sizes are set by moving carriage 6 over the plateau. The carriage in position 6 is used for the multiple-reflection technique, while the carriage in position 6^a may be used for the single-reflection method. In the former case, the angle of incidence can be varied from 45° to 80°, and in the latter from 30° to 80°. The position of the sample on the plateau for recording multiple-reflection spectra is

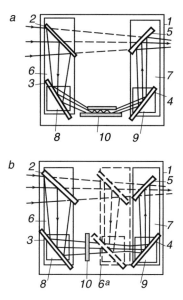

Figure 4.8. Optical scheme of multipurpose unit: (1) accessory plateau; (2–5) aluminum mirrors; (8, 9) movable holders which move with mirrors 3 and 4 perpendicular to optical axis of radiation beam; (10) sample. Unit allows transmission, multiple-transmission, multiple IRRAS, and MIR measurements to be made with carriage in position 6 and single IRRAS and ATR measurements when carriage is in position 6^a. Reprinted, by permission, from V. P. Tolstoy, in V. B. Aleskovski (Ed.), *Precision Synthesis of Solids*, Vol. 2, Leningrad State University Press, Leningrad, 1987, p. 90, Fig. 14. Copyright © 1987 St. Petersburg University Press.

shown in Fig. 4.8a. When single-reflection IRRAS or ATR spectra are recorded, the sample replaces mirror 4. The angle is set by turning mirrors 3 and 4 in a special holder around the imaginary axis defined by the intersection of the reflecting planes of these mirrors.

In general, to attain the required number of reflections in multiple-reflection spectra, it is necessary to use mirrors of considerable size (up to 100 mm), which is not always convenient. However, a special technique for recording multiple-reflection spectra has been developed to overcome this drawback [7]. In this technique, a solid plate that is transparent in the relevant spectral range and having a high refractive index, n_2, is placed between the metallic surfaces under investigation. Under these conditions, p-polarized light passes completely into the plate at the Brewster angle φ_B, with no reflection from its front surface. For high refractive indices, the Brewster angle ($\sim 67°$ for KRS-5, $\sim 74°$ for Si, and 76° for Ge) is close to the optimal angle of incidence $\varphi_1 = 75°$ used for recording spectra of metal surfaces by IRRAS (see Section 2.2). Therefore, placing a plate between the mirrors under investigation and directing the radiation onto the plate at the given angle φ_B (Fig. 4.7b) have two consequences. First, angles of incidence may be used that are close to the optimum value. Second, the required number of reflections may be obtained from smaller metal surfaces. [The light will propagate in the plate at an angle $\varphi_2 = \arcsin((\sin \varphi_B)/n_2)$, which is several times smaller than φ_B.] Calculations and experimental data show that this approach allows for sample sizes $2-3$ times smaller to be used without sacrificing sensitivity.

The multiple-reflection IRRAS technique holds a unique advantage in that ultrathin films on internal surfaces of metallic tubes can be probed nondestructively [15]. In the optical scheme of such "tubular" IRRAS, a metallic tube serves as the external multireflection element ("light pipe") guiding the IR beam as depicted in Fig. 4.9. The IR beam is coupled in and out of the multireflection element using spherical mirrors. The number of external reflections depends on the angle of incidence, the diameter and the length of the tube, which can be varied. Application of this technique benefits from coupling with SEIRA enabling an increase in the detection level of adsorbed p-nitrobenzoic acid compared to conventional IRRAS up to one order of magnitude [15].

In the case of transparent substrates, the sensitivity may be significantly increased with the BML-IRRAS technique (Section 2.3.3) or simply by placing a flat mirror under the substrate while the IR beam is incident on the other side [16]. The problem that can be met when measuring the IRRAS spectra of films on transparent substrates is interference fringes, which arise due to multiple reflections of the IR beam inside the substrate. To eliminate multiple reflections of the beam and to achieve the

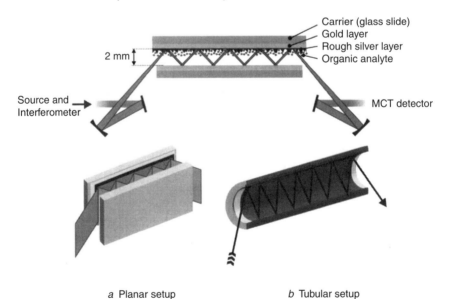

Optical Bench for Multiple External Reflectoin

a Planar setup   *b* Tubular setup

Figure 4.9. Setup for multiple IRRAS-SEIRA experiments showing (*a*) planar configuration and (*b*) tubular configuration. Analyte layer is deposited as thin film on top of silver film. Reprinted, by permission, from H. D. Wanzenböck, B. Mizaikoff, N. Weissenbacher, and R. Kellner, *Fresenius J. Anal. Chem.* **362**, 15 (1998), p. 17, Fig. 3. Copyright © 1998 Springer-Verlag.

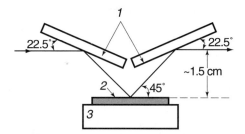

Figure 4.10. Scheme of single-reflection IRRAS accessory with horizontal sample: (1) directing mirrors; (2) sample; (3) sample holder. Reprinted, by permission, from J. Chatelet, H. H. Claasen, D. M. Gruen, I. Sheft, and R. B. Wright, *Appl. Spectrosc.* **29**, 185 (1975), p. 187, Fig. 1. Copyright © 1975 Society for Applied Spectroscopy.

properties of a semi-infinite substrate, the back side of the transparent substrate can be roughened.

A great variety of specialized accessories intended to solve specific scientific or technological problems have been reported in the literature, differing considerably from one another. Some examples include accessories that are designed to record spectra of samples in a vacuum or controlled-atmosphere chamber (Section 4.5), microreflecting IRRAS accessories that allow mapping or quality control of a sample surface [17], specialized FTIR IRRAS spectrometers providing effective background subtraction [18], and an apparatus for biophysical studies at a miniaturized air–water interface [19]. In practice, it is often convenient to place the sample horizontally, so as to avoid clamps or fasteners. A scheme of an IRRAS accessory with a horizontal sampling surface is shown in Fig. 4.10 [20]. A convenient IRRAS accessory with the horizontal sample position and ellipsoidal focusing optics is manufactured by Harrick. A grazing incidence IRRAS attachment, dubbed "Refractor," was described by Harrick and Milosevic [21]. A sample holder that facilitates reproducible substrate positioning was suggested by Stole and Porter [22].

A synchrotron radiation source, with its thousand-fold improvement in brightness over thermal sources, higher stability, and output IR spectroscopy, enables reflectivity changes to be measured from IRRAS with absolute baseline accuracy of $\pm0.2\%$ over a broad frequency range and with a spectral shape reproducibility of $\pm0.01\%$ in acquisition times of 100 s down to 50 cm^{-1} and with 1 cm^{-1} resolution. These advantages of synchrotron radiation were used in an IRRAS study of mode coupling in the CO−Cu(100) system [23].

The preparation of an optically smooth and clean substrate surface is considered in Section 4.10.2.

4.1.3. ATR

As in the case for IRRAS, ATR spectra require special accessories that are mounted in the sample compartment of a serial spectrometer. Central to all ATR accessories is the internal reflection element (IRE) with single or multiple reflections. This element is made from an IR transmitting material with a high and constant refractive index. Examples include Ge, thallium iodide–thallium bromide (KRS-5), ZnSe (Irtran-4), AgCl, ZnS (Cleartran), AgBr, diamond, GaAs, Si, CdTe (Irtran-6), Amorphous Material Transmitting IR radiation (AMTIR) (a Se–As–Ge glass), and ZrO_2 (Table A.2). To enhance the optical efficiency, the input–output sides of modern IREs are coated by antireflection material, while specialized coatings are used to increase hardness or chemical resistance. More information about the optical and physical properties of these materials can be found in Refs. [24–26]. For some systems IREs with low reflective indices such as BaF_2, quartz, or sapphire are more effective (Sections 2.4 and 2.5.4). IREs may be of various shapes and sizes [25, 27, 28]; the length of the working surface of the IRE crystal and the number of reflections will determine the sensitivity and SNR of the technique. The optical design of the ATR accessory is determined by the configuration of the IRE used. Each of these accessories (Figs. 4.11–4.14) contains two mirror systems: the first one to focus the IR source and direct it into the IRE and the second to collect the reflected radiation as it leaves the element and direct it into the instrument.

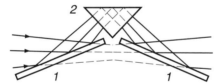

Figure 4.11. Optical scheme of ATR accessory with right-angle prism. (1) directing mirrors; (2) IRE. Reprinted, by permission, from N. J. Harrick, *Appl. Spectrosc.* **37**, 573–575 (1983), p. 574, Fig. 2. Copyright © 1983 Society for Applied Spectroscopy.

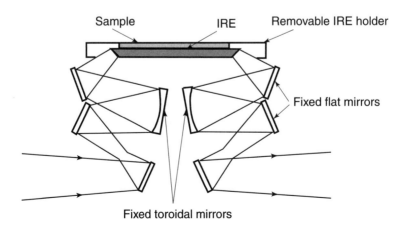

Figure 4.12. Optical scheme of horizontal ATR (HATR) accessory.

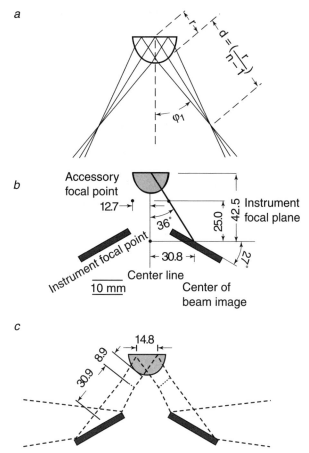

Figure 4.13. Optical scheme of ATR accessory with hemispherical IRE. (*a*) Collimation of diverging beam from point focus for hemisphere with refractive index of 1.3. (*b*) Physical layout and (*c*) geometric optics. All horizontal distances are with respect to accessory center and all vertical distances are with respect to instrument focal plane, (mm). Beam divergence of ±6° and focus diameter of 10 mm were used. Reprinted, by permission, from P. W. Faguy and N. S. Marinkovic, *Appl. Spectrosc.* **50**, 394 (1996), pp. 396 (Fig. 3) and 397 (Fig. 4). Copyright © 1996 Society for Applied Spectroscopy.

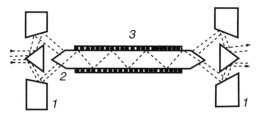

Figure 4.14. Experimental setup of CIRCLE ATR: (1) focusing mirrors; (2) IRE; (3) sample (coiled fiber). Reprinted, by permission, from A. M. Tiefenthaler and M. W. Urban, *Appl. Spectrosc.* **42**, 163 (1988), p. 163, Fig. 1. Copyright © 1988 Society for Applied Spectroscopy.

The principal requirements for all of the accessories is high accuracy in setting the angle of incidence and low level of radiative losses. The optical schemes of two ATR accessories, one with a right-angle prism and one with a trapezoidal prism, are shown in Figs. 4.11 and 4.12, respectively. Schemes for the hemicylindrical (or hemisphere) and cylindrical IREs are shown in Figs. 4.13 and 4.14, respectively. The advantage of hemicylindrical (or hemisphere) elements is that they collimate a diverging beam, allowing the angle of incidence to be precisely maintained (Fig. 4.13a). The position of the beam focus and the IRE (Fig. 4.13b) can be calculated using basic geometric optics. It has been shown [29] that the instrument focus can be reestablished by proper tuning of the reflection accessory. Thus, for a ZnSe hemisphere of radius 12.7 mm in contact with water, the optical scheme is that shown in Fig. 4.13c. In this geometry, the angle of incidence (36°) is slightly larger than the critical angle for the ZnSe–water interface (32°), which corresponds to the maximum spectral contrast (Section 2.5.4). Other characteristics of this optical arrangements are that (i) the beam diameter at the interface is ~1.5 times larger than the diameter of the beam at the focal point, (ii) the folding mirrors are at an angle of 27° from the instrument focal plane, and (iii) the beam entering and leaving the accessory fills the folding mirrors. The height of the sample surface is 42.5 mm above the instrument focal plane. This leaves enough room for the positioning of a polarizer oriented perpendicular to the beam direction at the focal point of the accessory [29].

To simplify routine analysis of bulk samples, the ATR accessories of the new generation have optimized optical configurations with predetermined angles of incidence. These accessories, however, are less effective for studying ultrathin films as compared to the older modifications, which allow one to optimize the angle of incidence for a given system. A more detailed information on different ATR accessories is presented, for example, in Refs. [25, 27, 30].

The CIRCLE unit (Fig. 4.14) was originally designed to study aqueous systems that are difficult to analyze because of the strong absorbance of water in transmission measurements [31]. This technique allows very low concentrations of solute (0.001 w/w) to be detected [32, 33] because of the large effective surface area of the cylindrical IRE. It also has high sensitivity toward thin films deposited onto the IRE directly [34]. For comparison, the surface area of a CIRCLE ATR element is greater than that of a rectangular ATR crystal by a factor of approximately 7.5, thereby giving it a superior SNR. As seen from Fig. 4.14, in the CIRCLE ATR accessory, a combination of mirrors focuses the spectrometer beam at an average angle of 45° on the surface of the ATR element. The light propagates through the crystal and is then redirected by another set of mirrors toward the detector. To apply CIRCLE ATR to a solid–liquid interface, the solution must be poured into a boat surrounding the element.

The major problem of ATR is to achieve optical contact between the IRE and the sample. Various approaches have been applied to this problem.

A. The simplest and most common approach involves an IRE made from the material of the film substrate (*reactive ATR*). This is possible for dielectrics such as sapphire [35], calcite [36], fluorite [37], ZnS, or weakly doped (intrinsic) semiconductors such as Ge, Si, and GaAs if the spectral range of interest is outside the region of their fundamental, lattice, or oxygen impurity (Si) bands. When the material is relatively transparent, the sensitivity level can be increased by using the MIR technique (Fig. 4.15). In addition, it is possible to study ultrathin films on a crystal surface of a definite crystallographic orientation with the reactive ATR.

In this technique there is the problem of cleaning and preparing a reproducible working surface before each experiment. For example, a quartz surface may be still highly contaminated with "grease", as revealed by the XPS method, even after polishing and washing with solvents. Cleaning procedures reported in the literature for Ge, Si, quartz (glass), GaAs, sapphire, fluorite, and KRS-5 are listed in Section 4.10.1.

B. When the IRE of a particular material cannot be made (e.g., because of high absorbance of the material, brittleness, or low availability), it has become common practice to coat a serial IRE from a film of this material (Fig. 4.15). This approach is also known as coated ATR spectroscopy. Since there exists a variety of methods of deposition and growth of polycrystalline or even crystalline films onto IREs [38, 39], the possibilities of the thin-film substrate ATR technique are very broad. Examples include Al_2O_3 [40] and TiO_2 [41] layers deposited by radio-frequency/direct current (rf/dc) magnetron reactive sputtering onto a ZnSe and Ge IRE, respectively, SiO_2 layers chemically produced on Ge and Si IREs [42], SiO_2 created by thermal growth on a Si IRE [43–45] and PbS deposited on Ge by the chemical bath [46] or vapor deposition [47] method. To simulate the surfaces of bones, teeth, and implants, apatite, titania, calcium phosphate [48–51] and polymer [52–58] layers were deposited onto Ge IREs.

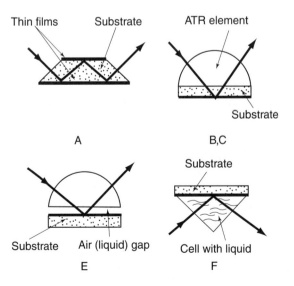

Figure 4.15. Schemes of ATR sampling techniques.

The use of a Si layer deposited on an IRE made from a transparent material (e.g., ZnSe [59] or Ge [60]) allows access to the $1550–1100$-cm^{-1} spectral region. If an accessory comprising a diamond IRE and a press (e.g., Golden Gate of Graseby Specac) is available, optical contact can be provided by pressing the substrate to the IRE (Fig. 4.16a). However, strong pressure can damage the film under study, as was observed for LB films [61]. In this case, a configuration depicted in Fig. 4.16b can be used [61, 62]. An IR-transparent plate is placed on the diamond IRE facing it with its clean surface while the opposite surface is carrying the film. To prevent the deformation of the substrate, a ring-shaped pressing tool is used. The IRE–islandlike metal film systems used for SEIRA measurements (Section 4.10.2) are another variant of technique B.

The optimum thickness of the substrate layer and the refractive index of the IRE as well as the angle of incidence and the number of reflections can be found only theoretically, as discussed in Sections 2.4 and 2.5.4. As a rule, in the case of opaque substrates (metals, doped semiconductors), the ATR technique with one reflection (the hemicylindrical ATR element in conjunction with a variable-angle ATR unit) is used (see, e.g., Refs. [63, 64]). However, one reflection is not a general requirement for such a system (see Sections 2.5.4 and 4.6.4 for more detail).

A serious problem in applying technique B is the interaction of an ATR element with the material of the substrate. Thus, it was found that Ge and GaAs with a Au layer form new intermediate phases with optical properties differing from those of pure Au; this is a possible explanation for the disagreement with spectral simulations [65]. Transient layers were observed for KRS-5, Ge, and ZnSe, which react with V_2O_5 gel to form a sol–gel bronze [66, 67]. Another problem with such multiphase configurations is mechanical stability. For example, BaF_2 is incompatible with Pt [65, 68]. This incompatibility can be overcome

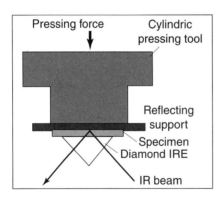
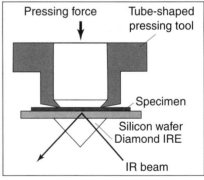

a b

Figure 4.16. ATR configurations with diamond IRE. Film substrate is pressed against IRE: (a) pressing deforming layer; (b) pressing not deforming layer. Adapted, by permission, from G. Muller and C. Riedel, *Fres. J. Anal. Chem.* **365**, 43 (1999), pp. 43 (Fig. 1) and 44 (Fig. 2). Copyright © 1999 Springer-Verlag.

by depositing a thin "wetting" layer. In the cases of steel films on a Ge IRE and Si films on a ZnSe IRE, intermediate layers of Cr_2O_3 [69] and Al_2O_3 [59], respectively, have been found to improve the substrate–film adhesion to the ATR element. Since direct deposition of Al_2O_3 on Ge produces a germanium oxide layer that is water soluble, it was suggested [70] to sputter a 30-nm passivation Si_3N_4 film directly onto a Ge surface followed by evaporation of a 60-nm Al_2O_3 film. Films of relatively hydrophobic polymers tend to peel off a hydrophilic IRE surface (e.g., in the case of Ge). This problem can be circumvented by pretreating the IRE with an organosilane coupling agent [71].

C. For surfaces of crystals, doped semiconductors, or natural objects (e.g., minerals) that cannot be synthetically reproduced by technique B or for which the deposition is technologically complex, it is possible to provide optical contact between a thin plate of the solid to be studied and an ATR element by lamination of a low-melting-point (80–90°C) commercial chalcogenide glass IKS-35 ($n = 2.37$) that serves as a glue [72]. After the glueing, the working surface of the plate can be polished and/or cleaned as described in Section 4.10.1. To obtain a contrast spectrum, the refractive index of the material for the ATR element must be close to that of the transition layer (IKS-24, KRS-5, or ZnSe would work well). It is convenient that the technique in question needs only one ATR element, to which any solid plate may be attached by slightly heating the chalcogenide glass.

The advantages of this technique are simplicity of sample preparation and possibility to study different crystallographic planes. Some applications of this technique are described in Section 7.5. However, as in technique B, such a multiphase sample may be mechanically or chemically unstable. The XPS spectra showed that after prolonged storage (about one year) of a 500-μm PbS plate fixed to an IKS-24 ATR element by IKS-35 glass, partial substitution of PbS lattice sulfur by Se (from the IKS-35 glass) on the outer surface of the PbS plate is observed.

D. The spectral contrast of ATR measurements using techniques A−C can be increased if a metal overlayer is deposited onto the film under study (MOATR) (Fig. 2.36) [73–75]. If overlayer deposition, which can damage the film, is not desirable, a metal-coated elastomer may be pressed against the IRE [76, 77]. Thus, Watanabe [77] reported enhancement of the ATR spectral contrast when a gold-coated poly(ethylene terephthalate) (PET) film was pressed against the Ge IRE covered by a 100-nm polyvinylacetate (PVA) layer. The air gap between the PVA layer and the Au layer was about 20 nm. The ATR spectra obtained at angles of incidence between 22.5° and 25.5° was 2.5 times more intense than those recorded by metallic IRRAS in p-polarized radiation at $\varphi_1 = 80°$ and 20 times more intense than transmission spectra, which is in agreement with the spectral simulations.

E. A thin film on a flat surface of a massive (metal [78], dielectric [79], or semiconductor [80]) plate can be studied using Otto's optical configuration (Fig. 2.36d). When this geometry is used for the in situ investigation of films at the solid–solution interfaces, it is referred to as the in situ IRRAS (external-reflection) configuration, by analogy with IRRAS at the air–metal interface (Section 2.5.4). To maximize surface sensitivity, the substrate should be pressed against the IRE with the film positioned between them, as illustrated in Fig. 4.16*a* (Section 2.5.4). Since most IR materials are brittle, such a pressing can break the IRE. The problem is overcome if a diamond IRE is used. A diamond ATR cell for IR measurements on metals is described in Ref. [81].

F. In liquid ATR [27, 82–90], the IRE is a liquid poured into a specially configured cell (e.g., prism), with the sample surface at the base of the cell (Fig. 4.15). This technique can be used for studying both the solid–solution and solution–solution interfaces. Spectroelectrochemical experiments at the interface of two immiscible liquids can be conducted by utilizing the cell shown in Fig. 4.17, which was designed for the UV/Vis optical region [88]. The IRE liquid should be transparent in the spectral region of interest and its refractive index should be higher than that of the substrate or the second liquid. This is a substantial limitation of the liquid ATR technique.

G. A description of the ATR sampling techniques would be incomplete without mentioning the waveguide technique [91], sometimes called the *evanescent wave absorption spectroscopy* (EWAS) (see the theory in Refs. [92–94]). This technique forms the basis of various chemical [95–97] and biological [98–101] sensors and is suitable for noninvasive and rapid (seconds) direct measurements of the spectra of tissues in vitro, ex vivo, and in vivo [102]. The waveguide technique is very promising for in situ studies of ultrathin coatings. The cell or reactor chamber may be positioned outside the spectrometer, which is an additional convenience for the measurements. Other advantages of EWAS include mechanical flexibility and relatively low cost with the new optical materials now available (e.g., silver halide [103, 104]).

EWAS uses either IR transparent optical fibers or planar waveguides (miniaturized IREs) as sensing elements. Such an element is coupled with either

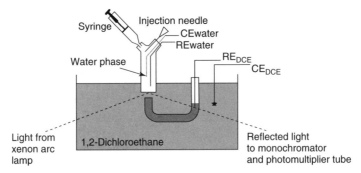

Figure 4.17. Experimental set-up employed in SEC measurements at liquid–liquid interface. CE refers to counter electrode and RE to reference electrode for two phases. Reprinted, by permission, from Z. Ding, R. G. Wellington, P. -F. Brevet, and H. H. Girault, *J. Electroanal. Chem.* **420**, 35 (1997). p. 37, Fig. 1. Copyright © 1997 Elsevier Science B.V.

a FTIR [91, 98] or tunable diode laser [105, 106] spectrometer. The coupling schemes can involve an IR microscope [98] or a mirror/lens [93] to focus the broad-band radiation onto the small aperture of the end of a planar waveguide or a fiber, respectively. A specially designed fiber sensor [104] consists of a fiber bent perpendicular to an almost full fiber turn or a helix with a different number of loops, in which particular fiber modes can be excited and used for detection. Besides other factors [91, 94], the length of the fiber controls the spectral sensitivity. Spectral distortions that arise in EWAS spectra obtained with multimode step-index optical fibers have been analyzed both theoretically and experimentally by Potyrailo et al. [107]. The surface of an optical fiber can be chemically modified, by analogy with technique B. For example, Taga and co-workers [108] described a fiber chemical sensor in which crystalline bacterial cell surface layers were used as an enzyme coupler. The reactive enzyme layer coating immobilized on the bacterial film served to catalyze chemical reactions. The chalcogenide fiber was coupled to a FTIR spectrometer, which yielded spectra in the range of $4000-800$ cm^{-1} at various stages of the chemical processes.

In the case of oversized samples (e.g., works of art), substrates with unusual shapes or "difficult to get at" samples such as board components or recessed sample areas, coatings can be remotely probed using a fiber-optic interface that directs the light from the source into a sampling probe head and thereafter collects the transmitted light into the detector of an IR spectrometer. The probe head (a miniturized ATR element or a reflectance probe with actual sample interface areas as

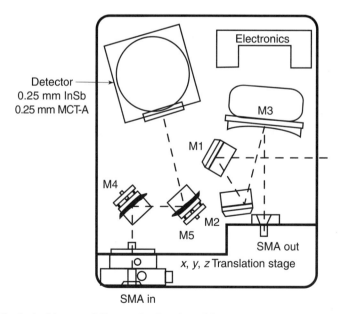

Figure 4.18. Optical layout of fiber-optics interface (Nicolet Instrument Inc.). Reprinted, by permission, from M. J. Smith and T. E. May, *Am. Lab.*, February 1992, p. RP-51. Copyright © 1992 Nicolet Instrument Corp.

small as 1–3 mm^2) is mounted into a tip looking like a pencil [109]. Figure 4.18 shows an optical diagram of a fiber-optics interface (Nicolet Instrument), which is positioned in the main sample compartment of a Nicolet spectrometer. Mirrors M1 and M2 are used to redirect the beam from the main bench into the beam-condensing mirror, M3. The beam-condensing mirror focuses the beam down onto the input of the fiber, which comes out of the front of the fiber-optics accessory. The return fiber connects to the accessory at an X, Y, Z translation stage at the front left of the accessory. Mirrors M4 and M5 are used to image the beam from the fiber onto the dedicated detector. The optics of the accessory was optimized for fibers that have a numerical aperture of approximately 0.21 and are about 200 μm in diameter. Connections between the fiber-optic accessory and the fibers are made through a standard fiber-optic communications Sub-Miniature-A (SMA-type) connector. FiberLink, another interface produced by Spectra-Tech, instead of traditional condensing optics that are difficult and tedious to align, uses nonimaging parabolic cones.

For probing the polymer surface, the sample can simply be pressed against the IRE, regardless of the film thickness (technique E). In this geometry, the sampling depth is controlled by the angle of incidence, the polarization, the refractive index of the IRE, and the thickness and optical parameters of the interlayer (Section 1.8.3). The optical contact can be improved by pressing the polymer when heated up to the softening temperature (the glass transition). To study adsorption on polymer films, an IRE can be readily coated with thin layers of the polymers by spin coating (technique B) [110]. In this case, the polymer film deposited on an IRE surface must be thinner than the penetration depth in order that the evanescent wave at the IRE–polymer interface may probe the adsorbed film. It has been shown [111] that if the polymer film thickness is much smaller than the penetration depth, the evanescent intensity is essentially the same as in the absence of the polymer. However, if the polymer absorption does not interfere with the spectrum of adsorbed molecules, the thickness of the polymer film can be varied up to the sampling depth, in order to limit the penetration depth into bulk solution and/or to minimize the signal from bulk solution, which is of primary importance in adsorption research.

Some recommendations of application of all the techniques mentioned above are generalized in Table 4.1.

Table 4.1. Possible combinations of ATR techniques and substrate material

Substrate	ATR technique						
	A	B	C	D	E	F	G
Dielectric	+	+	+	+	+	+	+
Semiconductor	+	+	+	+	+	−	+
Metal	−	+	−	−	+	−	+
Polymer	−	+	−	−	+	+	+

4.1.4. DRIFTS

Ultrathin films on rough (diffusely scattering) surface of bulk samples can be probed by DRIFTS using a "cup-on-the-saucer" accessory [112]. In such a scheme (Fig. 4.19), the light pipe serves both to direct the radiation from the interferometer normal onto the sample and to absorb the specularly reflected radiation. The diffusely reflected radiation is collected by the truncated ellipsoid of revolution, and the plane mirror deflects it onto the detector. The cup-on-the-saucer configuration has high optical throughput and is thus efficient for measuring the DRIFT spectra of scattering films with mirrorlike surfaces such as coatings by varnishes and enamels. The "openness" of the sample area permits the mapping of large surfaces. This principle has been employed in systems such as Spectropus [113] and SOC 400 Surface Inspection Machine/Infrared (SIMIR) [114], which are in addition specially fitted for analyses of remotely located samples of different sizes. The SIMIR is a small FTIR spectrometer coupled with a DRITFS accessory, designed to inspect large areas as well as highly curved surfaces. For metal surfaces this probe has a noise level of about 10^{-4} absorbance units, which is sufficient for detecting nanometer-thick organic film residuals on metals.

An interesting technique, referred to as Johnson's technique [115], was developed for detecting ultrathin films and surface contaminants on the substrates that cannot be placed directly into the instrument or when there is a strong overlap

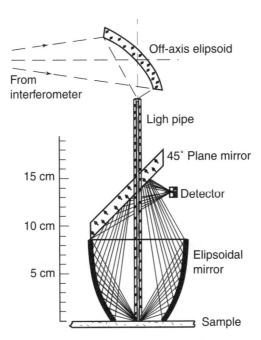

Figure 4.19. "Cup-on-the-saucer" optical geometry. Reprinted, by permission, from E. H. Korte and A. Otto, *Appl. Spectrosc.* **42**, 38 (1988), p. 39, Fig. 2. Copyright © 1988 Society for Applied Spectroscopy.

between the absorption of the substrate (e.g., a polymer) and that of the film. The surface (or the point) under study is gently ground using either a small amount of alkali halide powder [116] or the same powder covered by a nanolayer of silver (modified Johnson's method) [117]; these are analyzed by DRIFTS (though pressed-pellet transmission may also be applied). The surface extraction can be assisted by adding a few drops of a solvent. A nanometer-range sensitivity was demonstrated in the detection of organic contaminants on a polymer film and on a silicon wafer surface. An alternative procedure is to abrade the surface with silicon carbide paper and to measure its DRIFTS spectra.

4.2. TECHNIQUES FOR ULTRATHIN FILMS ON POWDERS AND FIBERS

A routine procedure for obtaining an IR spectrum of a film on a powder sample involves (i) measurement of a background spectrum of the uncovered powder, the mixture of the uncovered powder with KBr or pure KBr, and (ii) measurement of the spectrum of the analyte powder (or the mixture of the analyte powder with KBr in the same proportion as for the background spectrum). Normalizing the analyte spectrum with respect to the spectrum of a transparent matrix (KBr) decreases the spectral sensitivity to the adsorbate [118]. However, this approach can be useful in the following situations:

1. When absorption bands of the adsorbent and adsorbate are close to each other (or overlap) and the adsorbent is intrinsically heterogeneous (e.g., wood, fibers). In this case the spectrum measured against pure adsorbent is significantly distorted.

2. The KBr spectrum is employed in the internal-reference technique. In this case, a weak absorption band of the adsorbent (e.g., the overtones of the silica or zeolite lattice vibrations, $2100-1600$ cm^{-1} [119]) acts as the internal reference, and the ratio of adsorbate absorption to the reference absorption is proportional to the quantity of molecules adsorbed. The advantage of this is that it does not require to control precisely the sample weights [120].

The main problem in obtaining a quality spectrum of a thin film on a powder is eliminating the Fresnel components from the resulting spectrum (Section 1.10). As we will show below, existing practical solutions to this problem can be divided into two groups: (1) *invasive* methods, which require alterations to the sample (powder size, sample packing, dilution), and (2) *noninvasive* methods (optical and mechanical). The optimal parameters of the powder sample have been discussed in Section 2.7. Here, it needs only be stressed that (1) to achieve high reproducibility in the DR and DT spectra, the narrowest possible size distribution should be used, and (2) if the powder sample is diluted by KBr, the adsorbed layer can be significantly damaged (Fig. 2.52).

4.2.1. Transmission

The transmission method is the most simple from the technical viewpoint since, in contrast to the reflection techniques, it does not need special accessories. As distinct from the reflection spectra of powders, the transmission spectra provide a straightforward quantitative interpretation. However, the pellet technique requires more skill and time for sampling, as compared with sampling for the DRIFTS and ATR measurements. For studying metal-supported catalysts, the transmission method is less favorable than DRIFTS because of the high scattering and strong absorption of these powders (see Section 2.7.4 for further comparison with another IR spectroscopic techniques). As mentioned in Section 2.7.1, the light scattering can be reduced, especially by high-refractive-index powders, if the particle size is less than $\sim 1-2$ μm. In many cases, grinding in a vibrating mill for 1 min is sufficient to reduce the average particle size below 3 μm. Further grinding is ineffective: In many cases, once the particle size falls appreciably below about 1 μm, the particles start to stick together and behave optically as rather larger ones [121]. In addition, grinding changes the surface properties of the powder (e.g., the number of surface sites increases) and affects adsorption of reagents substantially [122]. The mechanical strength applied during grinding and pressing can induce polymorphical transitions in the powder material.

Several methods are used for preparing samples for the transmission spectrum measurement:

1. In pressed disks the particles are held together only by surface forces. Solids that can be easily pressed are silica derivatives (e.g., Cab-O-Sil and Aerosil), a number of oxides (e.g., Alumina C and Titania P25) and supported metals. Disregarding the technical problems concerning the disk preparation and their mechanical stability, which are discussed by Parkyns and Bradshaw [123], the essential disadvantages of this method for in situ studies are the formation of zones within the pressed matter, which are inaccessible for reaction with the surroundings, and changes in the support structure when the disk is pressed.

2. Squeezing between two IR-transparent windows has been described by Ahmed and Gallei [124]. Approximately 2 mg of sample powder is placed onto a horizontally positioned IR window and distributed uniformly over the window surface, then clamped by another window, and this whole construction is fixed in a holder so that the powder does not spill from the gap between the windows when in the vertical position. To avoid this, a simple mirror system can be used and the window with powder kept horizontal [125].

3. When the above-mentioned techniques cannot be used because, for example, of high absorption by the substance, a finely ground powder may be adhered to a window by depositing from a suspension and letting the solvent evaporate off. Once the deposit is dried, the particles adhere to the window by van der Waals forces, which are strong enough to prevent the substance from spilling while the plate is fixed in the spectrophotometer holder (if it is done carefully).

4. If the film on the powder is relatively thick or the powder has a large effective surface, the absorption can be "saturated" and, hence, distorted due to strong oscillators such as the phonon modes of oxides or the stretching vibration of the carboxylate group. If the powder is highly reflective toward IR radiation, the majority of the radiation is reflected from the front surface and the penetration depth is very shallow. As a result, the SNR of such spectra are very low. Both these problems can be solved by diluting the powder with a transparent matrix. This would be the alkali halide (KBr, KCl, CsI, AgCl) in sampling techniques 1 and 2 and hydrocarbon white oil (Nujol) or perfluorinated hydrocarbon oil (Fluorolube) in techniques 2 and 3. Otherwise, the dilution is not recommended (see analysis of artifacts in Sections 2.7.4 and 4.2.3).

One requirement of techniques 2 and 3 is that the powder layer must be continuous and without gaps within the cross section of the probing IR beam. Otherwise, a variable proportion of the radiation will not pass through the sample, which will lead to an apparent reduction of the band intensities in the measured spectrum. The error will be greater for samples with a greater absorption coefficient.

As seen in Fig. 4.20, for a given degree of scattering, an increase in the illumination of zone S will enhance the illumination of the entrance slit (aperture) and, as a consequence, the SNR of the measured spectrum. This effect was exploited by Tolstoy [126], who suggested using a KBr condenser of relative aperture 0.5 (Fig. 4.20b) to focus the incident beam onto the sample. This allowed the area of the source image on the sample to be reduced by a factor of >2. In place of the lens, any commercially available microfocus accessory may be employed.

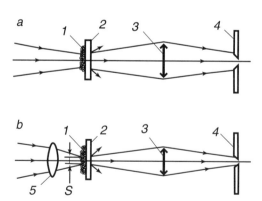

Figure 4.20. Passage of IR radiation through powder deposited on transparent plane–parallel plate with (a) standard technique and (b) additional focusing: (1) powder; (2) transparent plate; (3) focusing optics of monochromator; (4) monochromator entrance slit; (5) focusing lens. Reprinted, by permission, from V. P. Tolstoy, in V. B. Aleskovski (Ed.), *Precision Synthesis of Solids*, Vol. 2, Leningrad State University Press, Leningrad, 1987, p. 80, Fig. 5. Copyright © 1987 St. Petersburg University Press.

In the case of a dispersive spectrophometer, the sample can be additionally illuminated by focusing and redirecting the radiation reflected from the front surface of a beam attenuator onto the sample [127]. This attenuator is comprised of a simple comblike arrangement, which is driven in the reference beam automatically to equalize the intensities of the reference and sample beams. In the ordinary functioning regime of the spectrophometer, this radiation, whose energy reaches 95% of the basic beam energy, is scattered in the sample compartment. The simple modification of a dispersive instrument described above increases the useful signal by a factor of 2.0–3.5 and produces high-contrast spectra of a powdered sample in the region of maximum scattering even with a dispersive spectrophotometer. One example of this is the spectrum of silica gel in the absorption region of OH-containing functional groups and water [128].

However, this focusing has one drawback related to sample heating in the intense flux of IR radiation. This disadvantage can be overcome by using special light filters cutting off all radiation in the spectral region that is not used. Focusing lenses or entrance windows (in the case of thermovacuum cells) can also act as the required light filters.

4.2.2. Diffuse Transmission

Diffuse transmission (DT) can be measured with an integrating sphere, as is common in the UV/Vis/NIR spectral region [129–131]. An integrating sphere, whose concept dates to 1892 when Sumpner [132] wrote a paper describing such a device, is a hollow ball coated inside with a highly reflective and diffuse substance (Fig. 4.21). The sample and the detector are positioned on the wall of the sphere. In the mid-IR range, instead of barium sulfate and poly(tetrafluoroethylene) (PTFE), the sphere interior is covered by diffuse gold (InfraGold or InfraGold-LF). (One such product is the integrating sphere RT-060-SF by Labsphere). The sample is directionally illuminated while the scattered radiation is hemispherically (2π steradian) collected. If necessary, the regular transmission can be excluded by a light trap. Kuhn et al. [133] described quantitative measurements of the directional-hemispherical transmission of loose powder beds placed

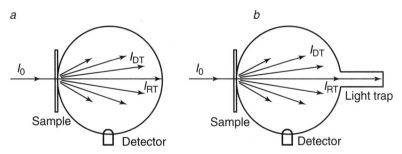

Figure 4.21. Measurement of (a) total and (b) diffuse transmittance with integrating sphere. I_{RT} and I_{DT} are regularly and diffusively transmitted radiation, respectively.

horizontally onto supporting layers of polyethylene (PE) and KBr using an integrating sphere 4 in. in diameter. However, such measurements are used mainly in metrology, due to the relatively low sensitivity of the IR-range detectors relative to photomultipliers and the relatively low power efficiency (2–6%). High-SNR total transmission spectra can be measured with a pyroelectric detector placed at the bottom of a cup containing the powder sample, as described by Boroumand et al. [134–136] or with an IR microscope [137].

A much simpler way of obtaining the DT spectra was suggested by Tolstoy et al. [138–140]. This technique involves measuring only the component I_{DT} passing through the powder layer (Fig. 1.22). To achieve this, the sample is illuminated by a beam focused at an angle 2α (α is the angular aperture of the spectrometer) using a special three-mirror system (Fig. 4.22). At this angle, the radiation components I_{RT} and I_0 do not reach the detector, being blocked by the entrance aperture, and only the component I_{DT} is detected. Component I_{DT} has a longer path through the sample than I_{RT} and has a greater contrast of absorption bands. The sample is prepared by one of techniques 2–4 (Section 4.2.1), but the layer need not be continuous (i.e., with no visible gaps, which is often not feasible) in techniques 2 or 3. However, since $I_{DT} < I_{DT} + I_{RT} + I_0$, the SNR of such DT spectra is relatively low and the technique can be used only with a FTIR spectrometer.

In another technique [139, 140], the components I_0 and I_{RT} are excluded from the detector because the powder support employed is not a plane–parallel plate but a wedge-shaped plate, with an angle

$$\beta \geq \arctan \frac{\sin 2\alpha}{n - \cos 2\alpha}, \tag{4.1}$$

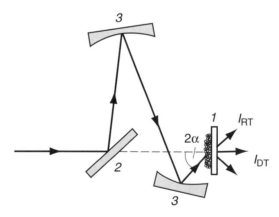

Figure 4.22. Optical scheme of diffuse-transmission accessory with side illumination of powder sample: (1) sample; (2, 3) directing and focusing mirrors, respectively. Reprinted, by permission, from V. P. Tolstoy, *Methods of UV-Vis and IR Spectroscopy of Nanolayers*, St. Petersburg University Press, St. Petersburg, 1998, p. 155, Fig. 4.25. Copyright © 1998 St. Petersburg University Press.

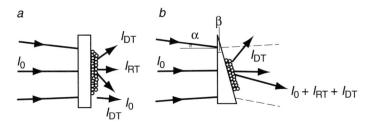

Figure 4.23. Passage of IR radiation through layer of powdered sample deposited on transparent (a) plane–parallel and (b) wedge-shaped plates: α = angular aperture of radiation beam; β = angle of inclination of wedge-shaped plate.

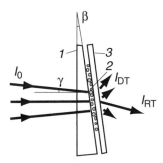

Figure 4.24. Cuvette for measuring DT spectra of powders: (1) ZnSe wedge-shaped window; (2) powder sample; (3) plane-parallel window. Reprinted, by permission, from S. P. Scherbakov, E. D. Kriveleva, V. P. Tolstoy, and A. I. Somsikov, *Pribory i tekhnika experimenta* No. 1, 206–208 (1992), p. 206, Fig. 1. Copyright © 1992 Pribory i tekhnika experimenta.

where n is the refractive index of the plate (Fig. 4.23). The powder is deposited on the wedge by sedimentation and can be held using a plane–parallel window made from a material with a low refractive index (e.g., LiF) (Fig. 4.24). As follows from simple geometric consideration, at angle β [Eq. (4.1)], a beam with an angular aperture α will not reach the entrance slit of the spectrometer, whereas the diffusely scattered component will. In a sense, the optical wedge fulfills the role of an optical filter, separating the noninformative component I_0 from the informative one, I_{DT}. In practical applications, the angle β must be chosen to obtain the highest quality spectra. In general, a real radiation beam has a finite width and angular aperture. This means that to totally exclude components I_0 and I_{RT} from the aperture, the wedge must deflect the beam by a somewhat larger angle than that calculated for the case of an ideal optical system. The angular aperture of the incident radiation beam should also be taken into account in Eq. (4.1). For a detector of size a, located at a distance ℓ from the sample, Eq. (4.1) may be represented by

$$\beta \geq \arctan \frac{\sin(\alpha + a/l) + n \sin(\alpha/n)}{n \cos(\alpha/n) - \cos(\alpha + a/l)}. \tag{4.2}$$

However, as the calculations show, the contribution to the value of β introduced by the additional terms involving a and l is less than $1°-2°$. Furthermore, the values of β must be increased by $1°-2°$ with respect to those calculated from Eq. (4.2) in any case, because of limited accuracy in the adjustment of the wedge in the sample compartment of the spectrometer. Therefore, to calculate the angle β, Eq. (4.1) is recommended, and in the small-angle approximation ($\sin\alpha \approx \alpha$, $\cos\alpha \approx 1$), it may be rewritten as

$$\beta \geq \frac{2\alpha}{n-1}. \tag{4.3}$$

Although the best angle β may be determined for each material by this formula, it should be kept in mind that in practice β should not be much larger than $2\alpha/(n-1)$, because the intensity of the total background of the scattered radiation will decrease for greater angles. In addition, there exists a limiting value of β at which total internal reflection occurs in the wedge. Therefore, values of $2\alpha/(n-1)$ increased by $1°-2°$ should be used for β.

To demonstrate the capabilities of this technique, $50-70$ μm powdered natural PbS (galena) was chosen, which, because of its high refractive index ($\sim$4), scatters IR radiation strongly. As was already noted in Section 2.7.4, such samples are the most difficult to investigate by transmission. Infrared spectra (100 scans, 4 cm^{-1} resolution) were recorded with a Perkin-Elmer 1760X FTIR spectrometer equipped with a MCT detector. The sample was prepared by sedimentation of a suspension in acetone on the wedge-shaped ZnSe plate. In addition, IR spectra were measured with a diffuse-reflection attachment and by pressing the powder in KBr. All the spectra are presented in Fig. 2.51. It can be seen that the absorption bands of the oxidized galena surface between 1600 and 800 cm^{-1} in the DT spectrum of galena have comparable intensity to those in the DRIFTS and transmission spectra (see Section 2.7.2 for more detail).

Advantages of the DT techniques shown in Figs. 4.22 and 4.23 are the absence of an expensive optical attachment and simple sampling. A disadvantage is a relatively low SNR, because only a part of the flux I_{DT} is used for the measurements (the rest of I_{DT} and I_{RT} are ignored); this can be partially overcome by using focusing optics or increasing the number of scans.

4.2.3. Diffuse Reflectance

As in the case of DT, diffuse reflectance (DR) can be measured in the mid-IR region with an integrating sphere [131, 141]. However, this technique is used mainly in reflectometry, where repeatable measurements with full hemispherical collection of radiation are needed [142]. Better performances, up to 12% efficiency, can be obtained using accessories with the so-called *biconical optical configuration*, which focuses the reflected radiation onto the IR detector. The SNR of such spectra is sufficient if such an accessory is used in conjunction with a FTIR spectrometer. The first DRIFTS accessory was described by Fuller and Griffiths in 1978 [143]. After that many forms of the DRIFTS accessories

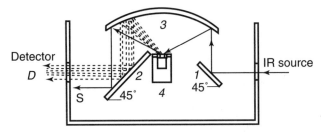

Figure 4.25. Optical diagram of DRIFTS accessory: (1) flat mirror (50 × 50 mm), (2) flat mirror (70 × 70 cm), (3) concave spherical reflector, and (4) sample cell; D = diffuse reflection; S = specular reflection. Dashed lines show optical path of diffuse reflection. Solid bold line shows optical path of specular reflection. Reprinted, by permission, from B. Li and R. D. Gonzalez, *Appl. Spectrosc.* **52**, 1488–1491 (1998), p. 1489, Fig. 1. Copyright © 1998 Society for Applied Spectroscopy.

were reported in the literature and developed by numerous optical instrument manufacturers. In a typical DR accessory (Fig. 4.25), the powder is placed in a cup. To avoid energy losses, the sample surface level should coincide perfectly with the optical plane of the accessory so that the sample is at the focal point of the accessory [144, 145]. A spherical, paraboloidal, or ellipsoidal reflector(s) focuses the IR radiation on the powder surface, collects the reflected radiation at the azimuthal angle of either 90° or 180° from the incident beam, and directs it to the detector.

The scheme with the azimuthal angle of 180° is called *on-axis* geometry: The optical elements are positioned so that the optical axis of the beam that exits from the interferometer and that of the collecting reflector coincide. In this scheme, the Fresnel component may be reduced by minimizing the solid angles of incidence and collecting radiation. This option has been used in several commercially produced DRIFT accessories, including those by Perkin-Elmer, Spectra-Tech (Collector), and Pike Instruments (EasiDiff). In the Perkin-Elmer diffuse-reflectance (PEDR) accessory shown in Fig. 4.26, the angle of incidence is less than 38° and the value of the Fresnel specular component is less than 5%. This accessory uses five flat reflectors, one of which (1) is double sided. Aspherical reflector 4 focuses the incident beam on the sample and collects the reflected beam with 8 × condensation power. The collecting angle of the PEDR accessory is a full π steradians, so that it is able to collect approximately 50% of the reflected energy. This accessory is mounted in the spectrometer sample compartment by placing its side slide support into the sample slide holder of the instrument. Access to the sample is accomplished from the front of the attachment by sliding the sample holder horizontally forward. The sample height is fully adjustable by means of a single focus adjustment knob. Apart from the simplicity and high energy throughput, the advantage of this configuration is improved reproducibility because the same ellipsoidal mirror is used for both illuminating and collecting the radiation. The DR accessory shown in Fig. 4.27 allows a collection efficiency up to 37%, with a concentric confocal ellipsoidal

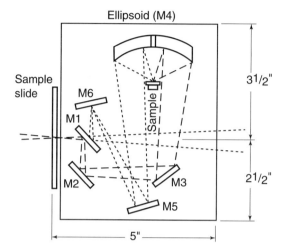

Figure 4.26. Schematic layout of Perkin-Elmer PEDR accessory: M_1 — double-sided mirror; M_2, M_3, M_5, M_6 — plane mirrors; M_4 — ellipsoidal mirror. Courtesy of Perkin-Elmer, Inc.

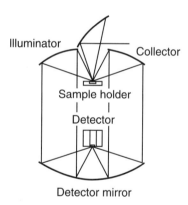

Figure 4.27. Optimized configuration for measuring diffuse reflectance. Reprinted, by permission, from T. Hirschfeld, *Appl. Spectrosc.* **40**, 1082–1085 (1986), p. 1083, Fig. 1. Copyright © 1986 Society for Applied Spectroscopy.

mirror arrangement, using a very large central opening in the collector mirror, and adapting the detector to the geometry of the collected beam [146].

If the azimuthal angle is 90°, the optical scheme is called *off-axis*, since the illuminating and collecting optics are in different planes. This geometry reduces the Fresnel specular component very efficiently [147]. The Praying Mantis (Harrick Scientific) (Fig. 4.28) incorporates two 6 : 1 90° off-axis ellipsoids which form a highly efficient DR collection system. The Praying Mantis is easily adaptable to accommodate large samples. It also can be configured with the appropriate reaction chamber to study materials and reactions in controlled environments.

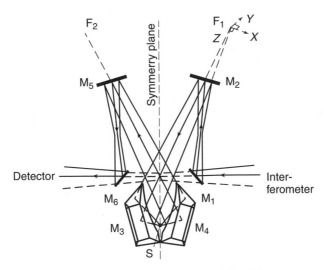

Figure 4.28. Optical diagram of Praying Mantis (Harrick Scientific Inc.) diffuse-reflectance accessory: M_1, M_2, M_5, M_6 — plane mirrors; M_3, M_4 — 6 : 1 90° off-axis focusing and collecting ellipsoids. Sample is placed at position S. Adapted, by permission, from K. Moradi, C. Depecker, and J. Corset, *Appl. Spectrosc.* **48**, 1491 (1994), p. 1492, Fig. 2. Copyright © 1994 Society for Applied Spectroscopy.

The effect of an increased specular component in the DR signal manifests itself as a deviation from linearity in the dependence of the KM response on concentration of an analyte dispersed in a transparent matrix [147]. The proportion of the Fresnel reflectance can be rather high for strongly absorbing powders in both types of accessories [148] and must be lowered by additional means. It has been found that both the Fresnel specular and Fresnel diffuse components can be reduced at the exit of an on- [149] and off-axis [150] accessory with crossed polarizers around the sample. This technique extends the linear region of the dependence of the absorbance (expressed in KM units) on the analyte concentration for the on-axis configuration [151]. However, this results in high energy losses: For a roughened PTFE sample these are 94%, 85% of which is due to the polarizers [135]. By contrast, according to the data of Brimmer and Griffiths [149] for neat quartz, the proportion of the Fresnel reflection exiting the off-axis accessory increases when the polarizers are crossed. The results of this work and the work of Yang et al. [153] have demonstrated that neither optical geometry nor polarization can totally eliminate the Fresnel reflectance from the front surface of powdered samples. However, the effect of specular reflectance on the KM intensity of a weakly absorbing analyte (e.g., for νCH_2 bands of organic compounds or overtones and combination bands of ionic compounds) is negligible when measured by DRIFTS employing off-axis as well as on-axis geometries [147]. Hence, for ultrathin films on weakly refracting substances (e.g., on oxides) the advantage of the Praying Mantis in reducing the overall specular intensity is redundant. If the throughputs of on- and off-axis configurations are compared, the on-axis configuration is found to be more efficient by a factor of 1.75 [153].

To reduce the specular reflectance collected by an on-axis accessory, different mechanical devices are used. One example is the patented Blocker, which represents a small gold-coated blade [152]. When placed across the sample surface so as to bisect it (Fig. 4.29), the Fresnel specular component is blocked from the collection mirror, and only the IR radiation penetrating into the powder reaches it. It should be taken into account that the Blocker can reduce radiation reflected from rough surfaces by 85% [152] or even more [153]. On the other hand, the Blocker is effective not only with powders but also with rough (textured) surfaces and was shown [153] to reduce the Fresnel diffuse component, along with the specular one.

Another mechanical solution to the problem of Fresnel specular reflectance is cutting off the specular component at the exit aperture of the accessory, as shown in Fig. 4.25 [154]. The waveguide that filters the KM component from the scattering radiation in the cup-on-the-saucer DRIFTS accessory (Fig. 4.19) may also be regarded as a mechanical device.

As for dilution with a transparent matrix, the recommendations are the same as for the transmission method. There is no point in diluting a powder transparent in the spectral region under study, since this decreases surface sensitivity (spectra *b* and *c* of Fig. 2.47) (see also Ref. [155]). Furthermore, this operation can remove/damage physically adsorbed species (Fig. 2.52) and change coordination of chemisorbed species (Section 7.4.2).

The DRIFTS spectrum of a powdered sample is greatly influenced by sample packing due to two different effects [156]. A surface particle size effect refers to the segregation of larger particles at the surface of the sample upon packing. This effect decreases the reflectance, as is also observed when the mean particle size increases. A more pronounced effect is the volume density effect, in which diffusely reflected light is strongly dependent on the powder density.

Many researchers use the "spatula" technique, which is sufficient for qualitative analysis. In this technique, the sample is simply poured into a diffuse-reflectance cup and the top is leveled with a razor blade or spatula. One thing to avoid is tapping the sample cup to make the particles settle, since this will cause bigger particles to rise to the top due to the surface particle effect [157].

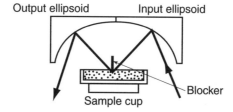

Figure 4.29. Diffuse-reflectance accessory with Blocker. Reprinted, by permission, from R. G. Messerschmidt, *Appl. Spectrosc.* **39**, 737 (1985). Copyright © 1985 Society for Applied Spectroscopy.

To increase reproducibility in DRIFTS measurements, all samples and references should be packed automatically. The powder is slightly pressed, although this sacrifices beam penetration depth and, hence, spectral contrast, which is greater for a loosely packed sample (Section 2.7.2). In this manner, DRIFTS spectra with coefficients of variation from 0.8 to 3.7% have been reported [158]. Murthy and Leyden [144] have demonstrated that reproducible packing of the sample improves the precision of band area ratio measurements. To achieve this, samples are prepared using an automatic, low-pressure packing unit with a mechanical packer in which packing time and pressure are controlled. TeVrucht and Griffiths [159] suggested a simple manual sample packing accessory that allows an experienced worker to prepare samples within 1–3 min that give approximately 3% scatter in the data. For quantitative analysis by DRIFTS of macroporous polymer particles, different pretreatments of spectra were compared by Liang et al. [160]. The best results were obtained by the following four-step procedure:

(1) Use the packing unit in order to reduce deviations due to surface differences for different packings.

(2) Define a zero-absorbance range (100% in reflectance spectrum) or some unchanged peak region to normalize the spectra in order to eliminate deviations between different packings.

(3) Correct baseline of the reflectance spectrum peak by peak.

(4) Finally, transform the baseline-corrected reflectance spectrum into KM format.

The relative deviations between the repeatedly measured spectra after this pretreatment were lower than 5%.

An interesting "cylinder" sampling technique for an on-axis DRIFTS accessory that can be extended to powders that pack easily was described by Hrebicik et al. [161]. It is based on the preparation of a cylinder (4 mm in diameter and 9 mm or more in length) from a mixture of the powder and KBr under a pressure of about 5.85 MPa. For pressing, a simple packing device is used. The measurements are performed with a routine on-axis baseline DR accessory (Spectra-Tech) with a stepper motor added to slide the sample holder. An Al mirror or a cylinder of pure KBr is used for the background. DRIFTS spectra are measured at several spots on the cylinder and averaged, the axis of the sample cylinder being perpendicular to the incident light for all spectra. The standard deviation of analyte spectra in absorbance units has been found to be better than 1.5%. However, the time needed for preparation and measurements for a cylinder is twice as long as for a cup.

Reeves [162] suggested that reproducibility in the DRIFTS spectra of samples with a high degree of inhomogeneity such as fibers, lignin, and protein could be improved by increasing the sample cup size up to 70 mm long by 9 mm wide by 8 mm deep and moving the sample linearly during the spectrum acquisition.

Moving the sample also reduces damage to the sample (discoloration and darkening) due to heating by IR irradiation.

When it is necessary to reduce the energy losses due to scattering from the front surface of a highly scattering powder sample or to profile a sample (e.g., a fiber), the KBr overlayer techniques can be used [163, 164]. In the classical set-up, the focal plane of the focusing hemispherical window coincides with the upper surface of the sample, and the KBr overlayer is above (Fig. 4.30a). As a result, a significant portion of the incoming light is scattered by the salt and the sample is illuminated by diffuse radiation. The scattered radiation is further defocused by the overlayer, collected, and sent to the detector. Because the difference in the refractive indices of the surroundings and the sample is reduced, a greater portion of radiation propagates through the interface and the spectral contrast increases.

In the modified version (Fig. 4.30b) the optical scheme of the DRIFTS accessory is changed slightly to minimize the incident and collecting angles. The entrance mirror is adjusted to direct radiation to the sample at smaller angles of incidence. The exit mirror is shortened to filter off the radiation scattered at large angles from the full cup. For DR measurements, the sample cup is filled so that the upper surface of the salt overlayer is in or below the focal plane. This sampling provides propagation of the incoming radiation from top to bottom and also enhances the collection of the radiation exiting the salt sample system at small angles. The radiation scattered at large angles from either a full cup or a cup filled 2 mm below the rim is blocked by the short mirror and the cup walls, respectively (Fig. 4.31). When the ratio of the underfilled cup signal to the filled cup signal is substituted in the KM formulas [Eq. (1.133)], the baseline approaches zero and all absorption peaks are visible. For depth profiling, the

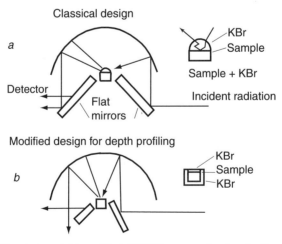

Figure 4.30. (a) Classical and (b) modified DRIFTS accessories for depth profiling of fibers. Reprinted, by permission, from F. Fondeur and B. S. Mitchell, *Spectrochim. Acta A* **56**, 467 (2000), p. 469, Fig. 2. Copyright © 2000 Elsevier Science B.V.

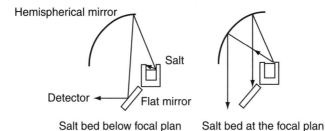

Salt bed below focal plan   Salt bed at the focal plan

Figure 4.31. Scheme of how energy is maintained nearly constant with different salt bed height in cup in technique shown in Fig. 4.30. Reprinted, by permission, from F. Fondeur and B. S. Mitchell, *Spectrochim. Acta A* **56**, 467 (2000), p. 469, Fig. 3. Copyright © 2000 Elsevier Science B.V.

thickness of the salt overlayer is varied. To quantify the results, the peak ratio of different components of the sample is used. This method has been applied to compare the effects of quenching alumosilicate fibers in mineral oil versus quenching in an air stream [164]. For this study, the fibers were chopped into small pieces with a knife until the pieces were $\sim$1 mm. The KBr particles were 200–300 μm. The broken fibers (10–15 pieces) were placed with their long axis parallel to the cup surface into a stainless steel cup already two-thirds filled with KBr and then an overlayer of KBr. This overlayer must be kept flat.

The scattering effects on the transformed spectra can be further eliminated mathematically. One of these methods is the multiplicative scatter correction, which normalizes the different spectra to an average scattering level [165–167].

The final point concerns the units in which a DR spectrum of adsorbed species should be represented for quantitative analysis. The application of DRIFTS for quantitative measurements is restricted not only by the variation in band intensities caused by irreproducible sample preparation but also by the nonlinear behavior of the reflectance due to scattering effects [168]. To increase the linear range, some authors [168–170] use KM units [Eq. (1.133)], while others [171–174] prefer absorbance, $-\log R_\infty$ (or $1/R_\infty$), where R_∞ is the reflectance spectrum of an "infinitely thick" sample relative to that of a nonadsorbing reference. For small values of K/S (where K and S are the KM absorption and scattering coefficients, respectively), such as is the case for DRIFTS of ultrathin films, Burger et al. [175] have shown that absorbance is proportional to the square root of KM units:

$$-\log R_\infty \approx 0.6\sqrt{\frac{K}{S}}.$$

It follows that KM units are more suitable for quantitative analysis. However, in some cases the transformation into absorbance leads to a more linear relation between $-\log R_\infty$ and analyte concentration [175]. Selecting units for representing spectra, it should be taken into account that the DRIFTS spectral pattern drawn in absorbance can be different from that drawn from the KM equation [172].

4.2.4. ATR

The ATR spectra of powders are measured routinely with HATR (Fig. 4.12), which can be improved as described by Messerschmidt [176]. In addition, a CIR-CLE (Fig. 4.14) or single-reflection diamond ATR [e.g., Golden Gate (Specac) or Atavar Single Bounce HATR (Nicolet)] accessory, where the IRE is positioned horizontally, can be used. A finely ground powder ($d < 5$ μm) can be conditioned externally or directly in the ATR cell [177–179]. To improve the optical coupling between the powder and the IR radiation, the powder can be pressed against the IRE using a special commercially available press (Fig. 4.32). Tight clamping is necessary to obtain spectra with a dispersion spectrophotometer. Except for diamond, IRE materials are usually brittle, and hence care must be taken when pressing. On the other hand, in the case of the diamond IRE, the IR radiation interacts with the sample only once, which lowers the spectral contrast relative to the multireflection HATR.

The ATR method is more efficient for in situ studies in the presence of solvent, which fulfills the function of the immersion medium (Section 2.7.3). For low solution concentrations ($<10^{-4}$ M), a typical experiment will involve (1) spreading a reference paste over the IRE surface (this paste contains pure powder wetted with pure solvent and is spread to a thickness comparable to the depth of penetration of IR radiation beyond the IRE–film interface), (2) scanning the background spectrum, (3) treating the paste directly in the accessory or changing

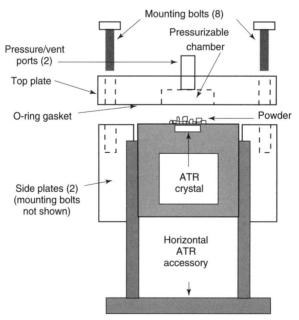

Figure 4.32. View of horizontal ATR accessory modified for measurements of IR spectra of powders. Reprinted, by permission, from C. M. Balik and W. H. Simendiger, *Polymer* **39**, 4723–4728 (1998), p. 4724, Fig. 1. Copyright © 1998 Elsevier Science Ltd.

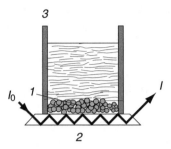

Figure 4.33. Cell for recording ATR spectra of the powder–liquid interface: (1) powder; (2) IRE, (3) cuvette.

the reference paste for one containing sample, and (4) scanning the sample spectrum. Before such an in situ experiment, a control spectrum of the IRE in contact with the reagent solution should be measured. For heavy particles the experiment can be simplified using a primitive in situ cell (Fig. 4.33) similar to an HATR accessory, where the IRE serves as the bottom. In this case, the pure powder is placed in a pure liquid and the background is measured when the particles have settled to the bottom. Then, a reagent is mixed into the cell, and after sedimentation of the powder, the sample spectrum is collected. To avoid shifting the powder on the IRE after mixing, the solid–liquid ratio of the suspension should be high. Additional considerations on sampling for in situ ATR measurements on powders can be found in Section 4.6.3.

In all cases, the ATR method gives poor SNR for hard, coarse particles. For these objects, DT or DR is preferable.

A CIRCLE cell can be used in studies of ultrathin films on fibers (Fig. 4.14) [180]. Comparing spectra recorded for the fibers that are (a) aligned along and (b) coiled around a cylindrical IRE provides information about orientation of surface species with respect to the fiber axis. This is because fibers are highly crystalline, and as IR radiation passes through them, it becomes polarized in parallel and perpendicular directions. As a result, the absorption bands due to the modes with different TDMs with respect to the fiber surface have different intensities for the two positions of the fibers with respect to the IRE.

4.3. HIGH-RESOLUTION FTIR MICROSPECTROSCOPY OF THIN FILMS

FTIR microspectroscopy (or FTIR microscopy or μ-FTIR) has been a conventional method for materials characterization since 1984, when Analect Instruments (now KVB) introduced a transmission microscope interfaced to its AQS FTIR [181]. Since then, FTIR microspectroscopy has developed into a greatly advanced tool for the analysis of thin films on a wide variety of substrates (including a single particle, cell, bacterium, or fiber) for scientific, industrial, and forensic applications [182–191]. Examples include oxide layers on technical Si wafers [192], organic films on Si (001) [193], organic [194–196] and inorganic [197] films, thin films on powders [198], the fiber–matrix [199] and adhesive–substrate [200–203] interfaces, lubricant films [204–206], and tribological surfaces [207, 208]. FTIR microspectroscopy has been employed for studying interactions in drug mixtures [209] and analysis of human tumor cell structures [210]. Modern instrumentation allows not only the identification of fibers but also dye types used to color them [211, 212]. This is potentially very beneficial in studying and in particular dating historical samples. One promising application is that of IR microscopes as detectors in TLC and HPLC [181, 213]. In situ FTIR microspectroscopy offers a number of advantages in spectroelectrochemical measurements (Section 4.6), studies of catalytic reactions in controlled gaseous atmospheres and temperatures [214], and characterization of defect formation during film deposition [215].

Spectral micrographs can be obtained in transmission, diffuse-transmission, IRRAS, ATR, and diffuse-reflection collection modes. The underlying theory of this technique, optical schemes of IR microscopes, sample-handling information, and possible sources of artifacts in the spectra are all discussed in Refs. [181, 183, 186, 216–219] and will not be repeated here. In general, optimal conditions for IR microscopic measurements are identical to those for ordinary FTIR measurements in the same geometry (Chapter 2). However, due to the fixed optical configuration and the larger angular apertures of IR microscopes, it is sometimes difficult to achieve optimal conditions. In this section, operation of IR microscopes in transmission, IRRAS, and ATR modes and the corresponding spatial resolution and detection limits will be presented and illustrated by experimental examples for thin films.

An IR microscope may be inserted into the sample compartment of an FTIR spectrometer or interfaced to the spectrometer as an external bench. All commercial IR microscopes utilize a reflecting lens with a Cassegrain-type configuration as an objective (with the lens prior to the detector). This configuration has high performance and flexibility. In most cases, the Cassegrain lens is used as a condenser (the lens prior to the sample), but an off-axis paraboloidal or ellipsoidal mirror may also be used. These mirrors generally have larger focal spots in the specimen plane than do the on-axis Cassegrains. Since the proportion of stray light increases as the focal size of the beam increases [181], the performance of the IR microscopes with a paraboloidal or ellipsoidal mirror condenser is generally poorer.

Modern IR microscopes are equipped with dedicated MCT detectors. Optimum focusing of the microscope is achieved when the image size in the detector plane is equal to the detector size. Since the sample size can vary over a wide range, the standard detector size is $250 \times 250 \ \mu m^2$, but other sizes are available. Without an aperture or mask, the signal seen by the detector is usually from the whole field of view. An aperture (either circular or rectangular in shape) is positioned at the primary image of the microscope in order to observe a specific sampling region that is smaller than and within the field of view. It has been shown [220] that when a single aperture is used, its optimal position to reduce stray radiation is in the condenser image plane, that is, prior to the sample. Since the image size is larger than the sampling area by a factor equal to the microscope lens magnification, the aperture size should be tuned to be equal to the size of the magnified image. Typical magnification of IR lenses is 15×, although higher magnifications up to 36× are available. [For example the grazing-angle objective (Spectra-Tech) has a magnification of 30×]. To view and select a specific sample area through the aperture, each IR microscope includes a visible-light microscope and can be equipped with a stereo monitor system. Additional capabilities such as a temperature-controlled stage, polarized radiation, and in situ cells are available.

4.3.1. Transmission

Figure 4.34 shows the typical geometry for transmission measurements [221]. A Cassegrain lens focuses radiation onto a sample and another Cassegrain lens collects the transmitted radiation. In this example, the Cassegrain lenses have 5× magnification, 0.6 numerical aperture, and a working distance of 24 mm, which gives an IR beam spot diameter on the sample of 0.1 mm and an angle of incidence of 9°–18°. The upper limit of the angular region is determined by the numerical aperture of the condenser, while the lower limit is due to blocking by the small secondary mirror in the Cassegrain lens. A variable knife-edge aperture, which is placed at the conjugate point of the sample, after the output Cassegrain mirror, is used for selecting a desired area of the sample for analysis. The sample should be positioned exactly at the focal plane of both the condenser and the objective; otherwise the spectrum will be distorted. In the IR microscope shown in Fig. 4.34, the sample is inserted in a diamond anvil cell (DAC), used

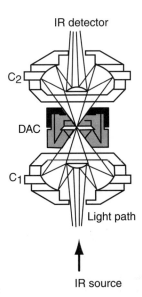

Figure 4.34. Schematic diagram of on-axis Cassegrain optics with diamond anvil cell (DAC) for IR transmission microspectroscopy; C_1 and C_2 are Cassegrain mirrors. Reprinted, by permission, from J. C. Chervin, B. Canny, J. M. Besson, and Ph. Pruzan, *Rev. Sci. Instrum.* **66**, 2595–2598 (1995), p. 2596, Fig. 1. Copyright © American Institute of Physics.

for reducing the thickness of an optically thick sample by pressing it between two diamond windows. The DAC flattens the substrate and prevents the beam from being refocused by a curved sample.

For illustration, Fig. 4.35 (curves 1 and 2) shows the transmission spectra of a 0.1-μm-thick film of an epoxy resin taken from different sampling areas.

4.3.2. IRRAS

For IRRAS at "near-normal" angles of incidence [e.g., the InspectIR (Spectra-Tech) accessory has an angular range of $16°-40°$] [222], IR microscopes employ an optical scheme with Cassegrain lenses and an additional aperture positioned between the source and the entrance aperture. The aperture splits the IR beam so that one-half of the Cassegrain lens delivers radiation to the sample and the other half collects the reflected radiation [181, 182, 217]. This approach utilizes ~50% of the incident energy available in transmission experiments, so that reflectance measurements require extended acquisition times in order to obtain the same SNR obtained in the transmission mode. Grazing angles of incidence ($70°-75°$ [183] or $65°-85°$ [222, 223]) are achieved with a special grazing-angle objective, which replaces the normal Cassegrain lens.

Figure 4.36 shows a scheme for an original IRRAS microscope, constructed on the basis of the DRIFTS accessory Collector (Spectra-Tech) [224]. Because there are no Cassegrain optics, the cost is low. This device consists of the Collector, a diaphragm, a monocular, and an x-y-z stage. The Collector is composed of four flat and two aspherical mirrors. The latter are off-axis ellipsoids for focusing and collecting the radiation. The ellipsoids condense the IR beam by a factor of 6, resulting in a spot size of IR radiation on the sample of about 3 mm. The sample is irradiated by the IR beam at an angle of about 60° with a wide angular dispersion. Access to the sample is provided from the top by sliding

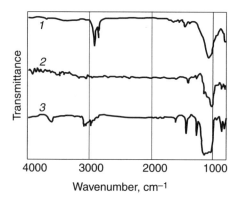

Figure 4.35. IR microscopic spectra of 100-nm layers of epoxy resin. Transmission spectra taken from circular area of (1) 100μm diameter and (2) 8×8 μm^2 area. (3) IRRAS from area of 40×40 μm^2. All spectra were obtained with Advanced Analytical Microscope (Spectra-Tech, Inc.) coupled with Bomem MB100 FTIR spectrometer. Each spectrum is average of 1000 scans. Reprinted, by permission, from P. Wilhelm, *Micron* **27**, 341 (1996), p. 343, Fig. 4. Copyright © 1996 Elsevier Science Ltd.

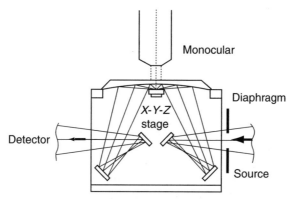

Figure 4.36. Optical configuration of reflectance FTIR microscope constructed on basis of Collector. Reprinted, by permission, from J. A. J. Jansen, J. H. Van Der Maas, and A. Posthuma De Boer, *Appl. Spectrosc.* **45**, 1149–1152 (1991), p. 1150, Fig. 1. Copyright © 1991 Society for Applied Spectroscopy.

the ellipsoids aside. The standard sample holder of the Collector is replaced by a manually controlled x-y-z stage (Microcontrol), adjustable in steps of 10 μm over a distance of 20 mm in three directions. The sample position is adjusted in such a way that the maximum energy of the detector signal is obtained. The single-beam spectra are normalized against a reflective aluminum mirror. Mounted at the top of the Collector is a monocular with a magnification of 30×. The monocular, with a reticule with scale divisions of 0.05 mm, is fixed with its focal point at the focal point of the IR radiation. An adjustable circular diaphragm is placed between the source of the spectrometer and the Collector. The diameter of the IR beam at the diaphragm is about 20 mm. The diameter of the diaphragm can be varied from 20 to 2 mm, giving an IR spot size in the Collector ranging from 3 to 0.3 mm. Further reduction of the spot size for this construction by employing a deuterated triglycine-sulfate (DTGS) detector was found to be not useful because of the low energy throughput of the IR radiation.

Curve 3 in Fig. 4.35 shows that the IRRAS spectra of an epoxy resin film were only distinguishable from a sampling area of 40×40 μm. Despite the strong dependence of the reflectance from metals on the angle of incidence in the 60–85° range (Fig. 1.11), spectra measured by μ-IRRAS can be subjected to quantitative analysis [223]. Applications of μ-IRRAS to investigating inhomogeneous ultrathin films can be found in Refs. [200, 201, 206, 207, 225].

4.3.3. DRIFTS and DTIFTS

A transmission or normal-reflection IR microscope can be used for DTIFTS or DRIFTS of films on powders or roughened surfaces, respectively. The spectrum of the clean powder is the reference spectrum. This approach can be quite sensitive to thin films. For example, Bouffard et al. [226] detected 50 ng of rhodamine B concentrated in a small area of zirconia. This capability would be particularly significant when considering an IR microscope as a detector in different types of chromatography [181].

4.3.4. ATR

The ATR mode is currently gaining popularity for studying microsamples [215, 227–236], owing to the minimal sample preparation, a reduced pathlength for highly absorbing materials, and a higher spatial resolution than the other modes (see below). Spectra are measured with a special ATR objective designed for a pointed hemisphere IRE. The conventional ATR objectives range from moderately priced systems employing the existing reflecting objective to higher priced devices employing specially designed ATR objectives [227]. Figure 4.37 illustrates a typical ATR geometry [234] in which a reflecting objective fulfils a dual role, by both introducing light into an IRE and collecting the internally reflected light. To measure an ATR spectrum, the specimen, placed on a support such as a microscope slide, is positioned on a stage and moved to the center of the viewing field. When the area to be analyzed has been centered in the field, the IRE crystal on the objective lens is slid into position, and the sample is raised to make contact with the IRE. In IR microscopes with a video imaging system (InspectIR Plus, Spectra-Tech), contact between the IRE and the sample is detected by an integral contact sensor system. Once contact is achieved, the ATR spectrum of the sample can be recorded. The reference is the ATR spectrum of the IRE in contact with air or with the clean surface without any film [225]. In some objectives, the incident radiation enters the IRE in approximately the $16°$–$40°$ angular range [225, 234], whereas others provide a range of $40°$–$51°$ [218, 234], which ensures that the ATR condition $\varphi_1 > \varphi_c$ is met for the majority of IRE–sample interfaces. In the case of the wide angular range ($16°$–$40°$) of the incident radiation, as the refractive index of the IRE decreases or that of the sample increases, a greater amount of light is transmitted rather than internally reflected [234]. However, the amount of transmitted light has no effect on the photometric accuracy if a background spectrum of the support is followed by the spectrum of the support covered by a film, although the SNR decreases.

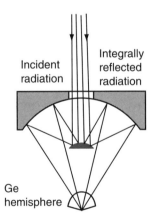

Figure 4.37. Configuration of typical ATR objective. Reprinted, by permission, from L. Lewis and A. J. Sommer, *Appl. Spectrosc.* **53**, 375–380 (1999), p. 376, Fig. 1. Copyright © 1999 Society for Applied Spectroscopy.

Micro-IREs made from diamond, Si, Ge, or ZnSe are available. Typical contact areas between the IRE and the sample range from 8×10^{-3} to 8×10^{-2} mm^2 (compared to several hundred square millimeters for conventional IREs). This allows small samples and also samples with rough surfaces to be studied. In addition, because of the reduced area, the actual applied pressure is greater and, therefore, more uniform, which yields a better optical contact and eliminates the baseline distortions associated with changing local pressures across much larger IREs.

While the study of small areas on large surfaces or large samples is relatively straightforward, the analysis of small isolated particles can be more problematic as the particle may move or roll away from the center of the crystal when pressure is applied [234]. In this case, the sample must be mounted or stabilized in some fashion, which may contaminate the sample. The sampling procedure for small isolated particles can be improved by employing cartridges with hemispherical internal reflection elements and an IR microscope equipped with a stereo monitor [234, 235]. The slide cartridge placed on the microscope stage in the appropriate orientation permits optimization of the energy through the crystals by fine adjustments to the x-y-z stage. An alternative approach to obtaining μ-FTIR ATR spectra is coupling an IR microscope with a planar waveguide [98].

An application of ATR IR microscopy in forensic science is provided in Fig. 4.38. It involves detecting cosmetic treatments found on hair fiber surfaces [237]. If this sample were to be analyzed by the transmission method, the spectrum would show predominantly protein from the bulk fiber. However, by μ-FTIR ATR spectroscopy, the contribution of the hair coating can be distinguished. For this example, a hair fiber was mounted on a glass slide and held in place at both ends with double-sided sticky tape. By placing the sample on a glass slide, the illumination from below the sample can be used to aid in viewing the sample with the survey mode of the ATR objective. A ZnSe IRE ($n = 2.4$)

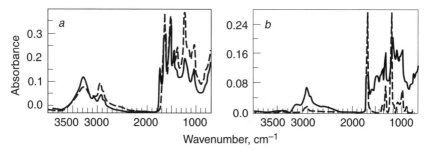

Figure 4.38. (*a*) IR ATR microscopic spectra of clean hair (solid line) and hair-sprayed hair (dashed line). (*b*) Difference spectrum (solid line) and reference spectrum of poly(vinylacetate) (dashed line). Spectra were measured with Nic-Plan microscope interfaced to Magna-IR (Nicolet Instrument Corp.) FTIR spectrometer. Sixty-four sample scans at resolution of 8 cm^{-1} were coadded and rationed against 64 background scans. Reprinted, by permission, from P. A. Martoglio, *Nicolet Application Note, AN-9694*, Nicolet Instrument, Madison, 1997. Copyright © Nicolet Instrument Corp.

yields a sampling diameter of 42 μm with a 2.5-mm-diameter upper aperture. The background spectrum is of the IRE−air interface. Figure 4.38*a* shows the μ-FTIR ATR spectra obtained from clean hair and hair coated with hair spray. The difference spectrum (Fig. 4.38*b*) allows one to conclude that the main resin in the hair spray is PVA.

4.3.5. Spatial Resolution and Smallest Sampling Area

Independently of the μ-FTIR method, the aperture and, therefore, the sampling area cannot be decreased below a certain threshold value that is controlled by diffraction from the aperture and, in the case of samples smaller than the wavelength, by diffraction from the sample. This effect does not reduce the energy throughput of the optical system but distorts the IR spectrum and limits the spatial (lateral) resolution of the microscope. Spectral distortions such as a baseline slope toward longer wavelengths that are due to diffraction from the aperture can be essentially reduced by recording the background spectrum with the identical aperture to that employed for the sample, but diffraction from the sample is more difficult to eliminate [181]. Diffraction effects in IR microscopes have been studied in detail in Refs. [181, 220, 234, 235]. Below, the problem is described briefly.

Diffraction arises when each point in the sample plane is imaged by the *Airy pattern* — a concentric series of rings, alternately bright and dark, around a central bright disk called the *Airy disk*. As the distance between two points in the sample plane decreases, their images begin to overlap. The *spatial resolution r* of the optical system is defined as the minimal distance between two points in the sample plane whose diffraction images (the Airy disks) can be resolved. For a point illumination source and a half-filled aperture of the objective, the value of *r* is estimated by the Rayleigh criterion [234, 238] as

$$r = \frac{0.61\lambda}{n_1 \sin\theta}, \tag{4.4}$$

where λ is the wavelength, n_1 is the refractive index of the measurement medium, and θ is the slope angle of the marginal ray exiting the lens. One can see that spatial resolution is lower for longer wavelengths. The quantity $n_1 \sin\theta$ is called the *numerical aperture* of the optical system. In a confocal arrangement, the theoretical diffraction-limited resolution is close to the wavelength (2.5–15 μm) [183, 218]. As the size of the sample approaches the "diffraction limit" (the wavelength of light), the portion of light falling outside the sampling area increases, and a larger physical area is probed by the beam than that defined by the aperture. Another characteristic of the IR microscope system is the smallest sample area from which a useful signal can be detected. This characteristic depends, apart from the spectrum intensity of the film in the system under study, upon the energy throughput of the microscope, which in turn is directly proportional to the numerical aperture: A larger numerical aperture gives a higher SNR. Typical numerical apertures are 0.5–0.7 for near-normal incidence and ∼1 for grazing angles.

4.3.6. Comparison of μ-FTIR Methods

In the transmission mode, spatial resolution was found to be $\sim$40 μm [183]. For an aperture corresponding to a sample size of 8 μm, the diameter of the sample actually viewed is about six times larger [239]. Curve 2 in Fig. 4.35 shows the transmission spectrum of a 0.1-μm-thick film of an epoxy resin taken from a sampling area 8 $\times$ 8 μm^2 [218]. For comparison, the transmission spectrum of a larger sample area is also shown (Fig. 4.35, curve 1). One can see that this film can be identified even when the sampling area approaches the theoretical diffraction limit. Sampling areas of 100–200 μm^2 have been reported in Refs. [239–241].

Sensitivity of standard (near-normal-incidence) transmission IR microscopy for nanometer-thick films is rather low (Section 2.1). However, this can be significantly improved for embedded ultrathin films using the optical arrangement described by Sassella et al. [242, 243]. The light beam impinges at near-normal incidence on the cross section (xz in the insert in Fig. 6.17) of the Si–SiO$_2$ nanolayer–Si structure, so that the beam lies in the plane of the film. Under these conditions, the maximum spectral contrast predicted by spectral simulations is achieved, since the radiation pathlength through the film formally approaches the sample length (2 mm). To improve spatial resolution, the spectra are taken using an IR microscope with a 36$\times$ objective and 13 $\times$ 100 μm^2 rectangular aperture. A spectrum of a bare Si is used as reference. As seem in Fig. 6.17, with this approach, a high-contrast spectrum of a 5-nm oxide layer can be obtained (see Section 6.4.3 for more details).

Another approach to increase surface sensitivity of transmission spectra is the use of mid-IR fiber optics as support material. Kellner et al. [244] have reported the IR microscopic transmission spectra of 3.6-nm LB bilayers of polyglutamate transferred onto chalcogenide glass fibers. The sensitivity was enhanced by geometric effects of the cylindrical substrate. This effect was confirmed by a theoretical approach based on a ray-tracing method. The interaction of the IR signal with the substrate lattice and the organic layer was simulated by approximating the dielectric functions of the materials by the harmonic oscillator formula [Eq. (1.46)]. Further experiments with different chemisorbed coatings on chalcogenide fibers in the same thickness range, including a coating of a methylacrylate–ethylacetate copolymer, have confirmed the potential of microtransmission for analysis of ultrathin organic layers on fibers [245].

It is evident that the divergence of the IR beam across the surface at grazing angles of incidence decreases spatial resolution of grazing-angle IRRAS microscopic spectra. Curve 3 in Fig. 4.35 confirms this; even the strongest bands in the IRRAS of an epoxy resin film were only distinguishable from a sampling area larger than 40 $\times$ 40 μm. Spectral resolutions of 40–60 μm have been reported [246], and with some types of microscopes, a resolution of 25 μm can be achieved [247]. However, compared with transmission spectra of the same film (Fig. 4.35, curve 2), the SNR in the higher wavenumber region in IRRAS is better. Micro-IRRAS has been used to measure [217] spectra of lubricant films

0.5 nm [206] and 0.36 nm [239] thick on magnetic disk media from areas of $100 \times 100 \ \mu m^2$.

In the case of the ATR mode, $n_1 > 1$. Therefore, as seen from Eq. (4.4), the resolution is higher by a factor of n_1 than that for standard transmission and IRRAS modes. This is because the IRE acts as a focusing lens. Apart from the magnification, the use of a hemispherical IRE offers the advantage of a higher numerical aperture (the radiation over a larger solid angle θ can be analyzed). For the ATR in a Ge IRE, the theoretical magnification factor of ~ 4 has been confirmed experimentally [234] for samples 60 μm in diameter and larger, while a spatial resolution of 22 μm has been found.

The minimal area required by ATR microscopes has been determined by Lewis and Sommer [234, 235] using a conventional IR microscope equipped with a specifically designed cartridge holding a Ge hemispherical IRE. For a thick polystyrene film, the minimal sampling area was $5 \times 5 \ \mu m^2$ (128 scans at 4 cm^{-1} resolution) [234], and for a 9-μm thick tie layer in a polymer multilayer laminate it was $6 \times 6 \ \mu m^2$ (512 scans at 4 cm^{-1} resolution) [235]. An aperture size corresponding to a 6×6-μm^2 sampling area was not sufficient to detect a usable signal in the normal-reflection geometry. Comparison of the ATR and transmission spectra from the same sampling area revealed [235] a higher SNR for the former.

Spatial resolution can be increased by introducing a dual aperturing (Redundant, Spectra-Tech). In this case, two apertures of identical size are placed at the source and sample images. This technique reduces the stray light by a factor of ~ 3 [183] by reducing the illuminated spot at the sample.

Better resolution but still controlled by diffraction can be obtained with a synchrotron IR source in a confocal arrangement. The intrinsically brighter synchrotron IR source allows areas as small as 3–4 μm to be probed [248–252, 266], which is very important for improving the quality of maps (vide infra). An additional advantage of synchrotron sources in orientational measurements is that the probing radiation is 100% polarized in the plane of the storage ring.

4.4. MAPPING, IMAGING, AND PHOTON SCANNING TUNNELING MICROSCOPY

Mapping a specimen allows comparison of the chemical identity and structure of a film at various points in the film. An early application of this technique was mapping the boundaries of gray matter, white matter, and the basal ganglia [253]. In these studies, all of the bands of white matter distinguishable were assigned, and the composition was in agreement with data obtained by other, biochemical methods. The technique of IR microspectroscopic mapping has found application in a variety of fields including coating technology [254], semiconductor manufacturing [255], and polymer science [256]. Maps of the distribution of different organic functional groups in the specimen [257–261] or chemical bonds in a mixture of inorganic compounds [262] have been constructed.

The mapping of a single cell from a wheat cross section (aleurine cell from a row of cells between the endosperm and seed coat) was achieved as early as 1993 [263]. With the use of a synchrotron source of IR radiation, maps of an aleurine cell and a cell wall with a spatial resolution of 5–6 μm [264, 265] and maps of a hair cross section with a spatial resolution of 3–4 μm [266] were obtained. The distribution of certain species such as proteins, lipids, and nucleic acids inside single living cells (some of which are mitotically active) [267, 268] was investigated.

To collect a map, an IR microscope with a small aperture is used, which allows a spectrum with reasonable SNR and spatial resolution to be measured. The aperture is moved in a regular manner, pausing after each step in the horizontal direction to collect a spectrum. At the end of each row the stage moves back to the start of the next row in the same direction. After all of the data have been collected, a characteristic absorption band is selected. The band absorbance plotted against the step distance is the map of the corresponding film feature [190]. When interpreting such a map, the spatial resolution of the measurement should be taken into account. In one case, a transmission map of a polymer boundary [269] revealed that the characteristic bands of these polymers advance from the boundary by 40-μm, and this was attributed to a 40 μm interdiffusion depth. However, since the spatial resolution is of the same value, the same band distribution would be observed if no interdiffusion took place.

Practical difficulties can arise when mapping soft samples if a pointed ATR probe deforms or damages the sample surface. To overcome this problem, Nakano and Kawata [232] suggested a photon tunneling IR microscope (see below for detail), in which the sample is mounted on the bottom of a Ge hemispherical IRE, and the IRE is then scanned with a piezoelectric controlled stage. This technique gives relatively high SNRs with remote apertures that are four times larger than the given sampling area, thereby reducing the diffraction effect. However, because of optical off-axis aberrations, the scan length is rather short (~100 μm). Moreover, the technique requires tracking both the source and sample apertures, which elongates and complicates the experiment. Esaki et al. [270] obtained longer scan lengths without the aberrations with a chevron-shaped IRE, but at the expense of the high signal throughput advantage associated with a hemispherical IRE. Lewis and Sommer [234] designed an IR-ATR microscopic experiment for mapping an area of 600×600 μm^2 with a conventional IR mapping microscope. Instead of moving the source aperture with a hemisphere, a single aperture after the sample and global illumination of the sample are used. However, this decreases spatial resolution.

Creating a map as described above is a time-consuming task if both high spatial resolution and a high SNR are required. Conventional mapping is also inefficient for samples that are smaller than the standard MCT detector element size, due to excess noise in the spectrum. This will place a limit on the quality of spectra collected from small sample areas. *Imaging* with array detectors that cover suitable spectral regions can improve the spectra [271, 272]. The IR image of a sample can be built with image systems, which are now commercially available, either in the transmission or reflection mode. For example, the Bio-Rad FTS Stingray 6000 IR imaging system is a Bio-Rad FTS 6000 IR spectrometer coupled with a UMA 500 IR microscope [273]. A focal plane array (FPA) detector is mounted on the microscope. The MCT FPA detector with its 64×64-element array can view a sample approximately 7×7 mm per pixel or 400×400 mm overall using the standard Bio-Rad $15\times$ microscope objective. The FTS 6000 spectrometer is operated in step-scan (S^2) mode for the imaging application. Although the integration time for the FPA is less than a millisecond, due to the time required for readout of the entire array, the use of the step-scan mode is necessary. The data from the array detector are transferred directly to memory via a digital frame grabber card that is installed in a single Windows NT based computer that controls both the spectrometer and the array detector.

A FTIR imaging technique using a rapid-scanning spectrometer instead of a S^2 one has been recently described by Snively et al. [274]. This technique can collect a data set consisting of 64×64 spectra with a 4-cm^{-1} spectral resolution over a 1360-cm^{-1} spectral range in 34 s. This was demonstrated by imaging adsorbates on different supported catalyst materials. Bellamy and co-workers [275] suggested to combine a moveable two-dimensional Hadamard encoding mask and an FTIR spectrometer for obtaining chemical maps and spectra of individual pixels of the maps. One can achieve resolution approaching the diffraction limit at rapid data collection by coupling an IR diode laser to a conventional IR microscope [276]. This technique was demonstrated by mapping the distribution of an additive in a layered polymer sample that had been contaminated by my migration of this additive.

Photon scanning tunneling microscopy (PSTM) [277–280] circumvents the diffraction limit and allows mapping with micrometer-scale spatial resolution. This method is a development of scanning near-field optical microscopy (SNOM) [281]. The near field is the evanescent field arising from total internal reflection (1.4.10°). Figure 4.39 shows a home-built set-up for near-field optical microscopy in the IR range. In the PSTM configuration, the angle of incidence φ is larger than the critical angle φ_c and the light beam is totally reflected inside a substrate that is transparent in the IR range. A silicon wafer ($50 \times 15 \times 2$ mm^3) with smooth surfaces and one side cut at an angle φ of $20°$ was used as a substrate in one case [277, 278]. The evanescent field created at the substrate–air interface is probed by the tip of a fluoride glass optical fiber [282] which is transparent in the IR range up to $\lambda = 6$ μm. The near field is then locally converted into a propagating one by the fiber and guided to a MCT detector. Micrometric screws and a piezoelectric transducer provide the coarse and fine adjustments of the fiber position relative to the surface, respectively. A dc motor was preferred for in-plane displacements and two-dimensional scanning. In this set-up, the fiber tip has a diameter of 10 μm, which determines the possible size of area analyzed to be approximately 10×10 μm^2.

The sensitivity of the PSTM experiment can be significantly improved in two ways. In one, the absorption from the most intense band in the spectrum can be

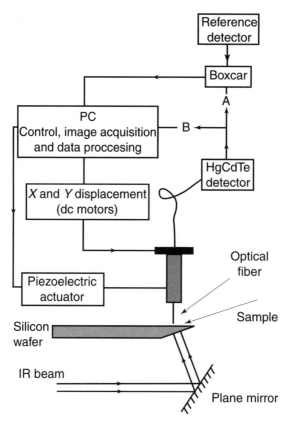

Figure 4.39. Experimental set-up for scanning near-field optical microscopy (SNOM) in IR range. Reprinted, by permission, from A. Piednoir, C. Licoppe, and F. Creuzet, *Opt. Commun.* **129**, 414–422 (1996), p. 415, Fig. 1. Copyright © 1996 Elsevier Science B.V.

used for mapping. In the other, when the IR band of interest is clearly defined, the use of a more appropriate detector (e.g., InSb or bolometer at helium temperature) may provide an enhancement of about 10, and therefore, an increase in the SNR by a factor of 300 might be expected, which means that the IR spectrum could be acquired with a 1-μm fiber tip and thus 1 μm resolution. This method has been applied to studies of polymer diffusion in hydrogel laminates [283, 284].

Hammiche and co-workers [285] described a technique in which a miniaturized Wollaston wire resistive thermometer is used as a probe to record IR absorption spectra by detecting photothermally induced temperature fluctuations at the sample surface. These authors claimed that such an approach opens the way to spatial resolution extended beyond the diffraction limit by a few hundred nanometers. As an alternative, Palanker et al. [280] suggested to use tipless probing.

4.5. TEMPERATURE-AND-ENVIRONMENT PROGRAMMED CHAMBERS FOR IN SITU STUDIES OF ULTRATHIN FILMS ON BULK AND POWDERED SUPPORTS

Infrared spectroscopy is one of a very few methods that permit real-time monitoring of deposition and/or evolution of ultrathin films in situ with full control over the composition and structure at the atomic level. This has been widely exploited for synthesizing high-quality films in a controlled way and for obtaining an understanding of catalyst reactivity (Chapters 5 and 7). One requirement for in situ measurements is a temperature-and-environment programmed chamber or cell. Various chambers for transmission, IRRAS, ATR, and DRIFTS measurements, including their technical advantages and disadvantages, are considered in detail in several monographs [123, 286–289]. Below, several recently reported chambers will be described.

Various designs for IR reactor cells for transmission measurements are described in many references [290–313]. Figure 4.40 shows a cell for studying a working catalyst surface [314]. The cell consists of a central stainless steel hollow cylinder with 3-mm-thick walls and two flanges welded to both ends. The catalyst sample is pressed as a thin disk and placed in the center of the cylinder. The distance between the catalyst sample and the flanges is 42.5 mm, which is sufficient to keep the temperature at the central portion at 773 K and to maintain the temperature at the flange below 523 K with a cooling water jacket. The heating is provided with a 4-ft-by-$\frac{1}{2}$-in. Barnstead Termolyne standard insulated samox heating tape wrapped around the cylinder and insulated. One effective approach for direct heating of the sample is the use of internal heating elements

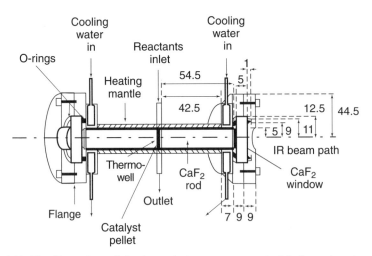

Figure 4.40. The IR reactor cell for transmission measurements (all dimensions in millimeters). Reprinted, by permission, from S. S. C. Chuang, M. A. Brundage, M. W. Balakos, and C. Srinivas, *Appl. Spectrosc.* **49**, 1151–1163 (1995), p. 1152, Fig. 1. Copyright © 1995 Society for Applied Spectroscopy.

such as a U-shaped tunnel [315] and tungsten wire grid, which are in direct contact with the catalyst sample. The use of internal heating elements increases the dead volume of the reactor, and it becomes difficult to eliminate a possible role of the heating element in the reaction. A J-type thermocouple is placed within a thermowell so that its tip touches the catalyst disk.

The catalyst disk sample is placed in the hollow cylinder between two CaF_2 rods 54.5 mm in length and 10 mm in diameter, polished on both ends. The volume between the two CaF_2 rods is 125 mm^3. The catalyst disk fills 75 mm^3, leaving 50 mm^3 as the void volume of the reactor. CaF_2 was chosen for the rod and window because its transmission range of $4000-1200$ cm^{-1} is ideal for CO hydrogenation and NO–CO reaction studies. CaF_2 was also chosen because of its low solubility in water (<2 mg/100 g H_2O), its operability at elevated temperatures (up to 873 K), and its ability to form a seal that can withstand high pressures. The reactor is rated at 6.0 MPa with a safety factor of 4. The CaF_2 rods decrease the reactor volume, thus reducing the average residence time of gaseous species, allowing for rapid response of gaseous reactants and products during transient studies. The small reactor volume minimizes the amount of gaseous reactants and products present within the reactor, decreasing the IR intensity of gaseous species, allowing observation of unambiguous IR absorbance signals for the adsorbate on the catalyst surface. The CaF_2 rods also act as a brace to hold the catalyst disk. CaF_2 windows 25 mm in diameter and 5 mm thick are placed in a machined channel within each flange. Both faces of the CaF_2 window are polished and are pressed by flanges with Viton O-rings, which fit within a machined groove on the flanges.

Measurements under controlled temperature, environment, and pressure in the DRIFTS geometry are complicated by energy losses and the increase in the specular Fresnel component in the reflected radiation, because of the windows isolating the reactor volume. Minimizing this component severely limits the cell design. One of the first DRIFTS reactor systems was described by Hamadeh et al. [316]. There are currently commercial cells available that are compatible with the DRIFTS accessories. One example is the high-temperature/high-pressure chamber produced by Spectra-Tech for the Collector, where the temperature can be varied up to 900°C and the pressure up to 1500 psi. The temperature chambers used in conjunction with the Praying Mantis (Harrick) allow studies at pressures from 10^{-6} torr to 2 or 3 atm and at temperatures from $-150°C$ to 600°C (under vacuum). A simple cell enclosed in the low-cost DRIFTS accessory (Fig. 4.25) is depicted in Fig. 4.41 [154]. The advantage of this accessory–cell system is that the incident and reflected beams are normal to the accessory windows; this eliminates energy losses due to reflection from the windows. The reactor cell is composed of three parts: a window, a stainless steel body with a sample well, and a stainless steel cover. It has a gas inlet and outlet. The sample well is 12.7 mm in diameter and 9 mm in depth. The volume of the cell is approximately 1.15 cm^3. The average residence time in the reactor is 1.15 s with a flow rate of 60 mL $\cdot$ min^{-1}. The small dead volume is essential for transient response studies with single-pass differential flow capabilities. For transient studies, the

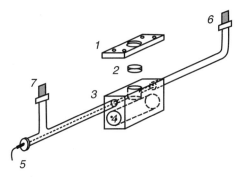

Figure 4.41. Schematic diagram of cell for temperature- and atmosphere-controlled measurements using DRIFTS accessory shown in Fig. 4.25: (1) cell cover ($38.1 \times 21 \times 3.2$ mm); (2) window (16×2 mm); (3) cell assembly ($38.1 \times 21 \times 29.2$ mm); (4) cartridge heater (12.7×38.1 mm); (5) thermocouple; (6) gas inlet ($\phi = 6.35$ mm); (7) gas outlet ($\phi = 6.35$ mm). Reprinted, by permission, from B. Li and R. D. Gonzalez, *Appl. Spectrosc.* **52**, 1488–1491 (1998), p. 1489, Fig. 2. Copyright © 1998 Society for Applied Spectroscopy.

average residence time should be equal to or less than the time required to record a single spectrum. A stainless steel screen (100 mesh), located inside the well next to the wall along the axial direction, was used to confine the catalyst within the sample cell. An insulated thermocouple was inserted through the gas outlet and positioned close to the catalyst bed. The window was a $16 \times$ 2-mm NaCl optical lens. Depending on the wavelength requirement, the optical lenses could be constructed from NaCl, KBr, or CaF_2. The lens was compression sealed to the sample well with two high-temperature gaskets ($600°C$). A cartridge heater 12.7 mm in diameter and 38.1 mm in length was placed underneath the well. The sample cell was designed with a rectangular shape because it is more easily positioned, and it enables the use of a smaller volume, saving space for the sample holder and mirror movement. The sample cell was positioned on a sample holder that was also constructed from aluminum plates. It had a set of grooves positioned in such a way that the sample holder could easily be inserted into the correct position and could be conveniently removed. The system was thoroughly insulated to avoid damage to the IR spectrometer. Precise adjustment was essential in order to obtain reproducible spectra.

Murthy et al [317] described artifacts in the variable-temperature DRIFTS spectra, which are caused by changing sample height due to the thermal expansion as temperature is changed. To overcome this problem, Venter and Vannice [318] suggested a modification of DRIFTS accessories. The reader can find other temperature-controlled DRIFTS cells in Refs. [316, 319–326].

Reactor chambers for IRRAS at low pressures are optically simple [327–332]. For complex diagnostics such a chamber is coupled with another instrument(s) (e.g., mass spectrometer for collecting temperature-programmed reaction spectra and Auger electron spectrometer [333] and a cell for in situ IRRAS and quartz microbalance studies of atmospheric corrosion of metals [334]). Figure 4.42 shows one design of a chamber for measuring the IRRAS during physical vapor

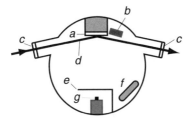

Figure 4.42. Physical vacuum deposition apparatus for real-time in situ IRRAS measurements: (a) substrate; (b) thickness monitor; (c) ZnSe windows; (d) IR beam; (e) shutter; (f) mercury lamp; (g) crucible. Reprinted, by permission, from M. Tamada, H. Koshikawa, and H. Omichi, *Thin Solid Films* **292**, 164–168 (1997), p. 165, Fig. 1. Copyright © 1997 Elsevier Science S.A.

deposition (PVD) [335–337]. The distance between the substrate and the crucible is 250 mm. A glass slide ($72 \times 26 \times 1$ mm) coated with 100 nm Ag was used as substrate. A low-pressure mercury lamp of 20 W was set in the vacuum chamber to induce the polymerization of a monomer film of N-vinylcarbazol (NVCz) (see Section 7.7 for more detail). The deposition rate was monitored by a thickness monitor maintained at 261 K. The NVCz in the crucible was maintained at 320 K for a deposition rate of 22 nm $\cdot$ min^{-1}. The IRRAS spectra of the film as it was deposited onto the substrate at various temperatures were continually measured, with a resolution of 4 cm^{-1}. For this, the IR beam was introduced onto the substrate through one ZnSe window, and the reflective IR beam was directed to an MCT external detector from an FTIR instrument through the other ZnSe window. The angle of incidence of the IR beam was 80°. The PVD was carried out under a vacuum of 5×10^{-4} Pa.

In contrast to the above example, in situ formation of ultrathin films at high pressures is difficult to follow with IRRAS because of interfering absorption by the ambient atmosphere [338–341]; to remove this obstacle, as in the in situ spectroscopy of the solid–liquid interface, the PM method can be applied [342–344] (Section 4.7). As an alternative, s- and p-polarized spectra are measured successively or by taking alternatively a few scans with each polarization until the desired SNR is obtained and the resulting spectrum is represented in A_{sp} units [Eq. (7.1)].

For in situ studies under controlled external conditions, the ATR method is generally used in the MIR geometry because of its high surface sensitivity and energy throughput [345]. A typical ATR chamber (see also Refs. [346, 347]) constructed for real-time monitoring of surface chemical reactions at pressures as low as 3×10^{-9} torr during plasma-enhanced chemical vapor deposition (PECVD) is shown in Fig. 4.43 [348]. An IR beam from a FTIR spectrometer impinges onto a Si(100) multiple IRE through a polarizer and a BaF$_2$ window. The size of the IRE was $50 \times 20 \times 2$ mm^3. The ATR cell adapted to the commercial Specac Golden-Gate unit with a diamond IRE was devised by Muller and Riedel for studying photochemical reactions in thin films on Si [62]. This cell coupled with a Bio-Rad FTS 6000 FTIR spectrometer was used in the real-time ATR-FTIR studies of fast photopolymerization reactions induced by UV radiation with time resolution on the order of 1 s at temperatures up to 200°C [349].

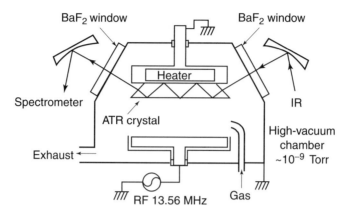

Figure 4.43. Scheme of ATR system for monitoring surface reactions in PECVD reactor. Reprinted, by permission, from Y. Miyoshi, Y. Yoshida, S. Miaziki and M. Hirose, *J. Non-Crystalline Solids* 198–200, 1029 (1996), p. 1030, Fig. 1. Copyright © 1996 Elsevier Science B.V.

Due to better compatibility with gaseous environment, the ATR geometry is technically more advanced as compared to IRRAS for experiments at high pressure. For example, an ATR technique for measuring simultaneously the CO_2 sorption and the consequent swelling of a polymer film was described by Flichy et al. [350]. MacLaurin et al. [351] described a fiber-optics probe that is heatable to 230°C. The probe has chalcogenide fibers and a ZnSe IRE, allowing access to a wide range of standard laboratory reaction vessels and fume cupboard arrangements. The performance was demonstrated via the in situ analysis of an acid-catalyzed esterification reaction in toluene at 110°C. Particular emphasis was given to the quantitative interpretation of the spectroscopic data using gas chromatographic reference data.

The main technical problem for measurements at nonambient temperatures is the large difference in thermal expansion coefficients of common IR window materials and the chamber body. Use of materials that substitute cements, which are commonly used to glue a window, is discussed by Parkins and Bradshaw [123]. An additional problem that arises when performing high-temperature spectroscopic experiments is that the sample itself becomes an IR emitter [316] that saturates the signal reaching the detector. MCT detectors were found [352] to be more sensitive to this effect than deuterated triglycine sulfate (DTGS) ones. To reduce this interfering factor, the sample spectrum should be collected at the same temperature as the background spectrum.

4.6. TECHNICAL ASPECTS OF IN SITU IR SPECTROSCOPY OF ULTRATHIN FILMS AT SOLID–LIQUID AND SOLID–SOLID INTERFACES

The solid–liquid interface can be probed by transmission, IRRAS, and ATR, though the former method has not been used frequently so far. General technical requirements of the cells for in situ measurements are (1) optical layout allowing optimal detection of interfacial species, (2) compatibility with a host reflection accessory, (3) possibility to exchange window materials, (4) easy dismantling and cleaning, (5) high resistance to solution components, and (6) absence of leakage. In addition, for the spectroelectrochemical (SEC) studies, (7) adequate electro-chemical performance is necessary. Provided a suitable cell is available, the main experimental problem remains that, with the exception of ATR-SEIRA, in situ IR spectra of ultrathin films can be obscured by the absorption of the solution and, in the case of the "thin-layer" optics, have low reproducibility due to diffusion constraint. Both these characteristics depend on the design and implementation of the in situ experiment. Specific measurement protocols are followed to improve this. In this section, these technical aspects will be outlined. The advanced mod-ulation approaches to reducing the limitations described above and extending capabilities of IR spectroscopy are discussed in Sections 4.7 and 4.9. Another problem for SEC studies — the preparation and maintenance of the electrode — is considered in Section 4.10. Complementary information can be found in several reviews [353–368].

4.6.1. Transmission

The simplest and earliest cells are those for transmission measurements of adsorbates on microporous glass and powders in the mutual transparency (low-absorbancy) spectral region of the substrate and solvent [369, 370]. For example, for the silica(quartz)/alumina$-CCl_4$ system, this would be at $\nu >$ 2800 cm^{-1} [369–372]. In this case, the surface (interfacial) sensitivity can be increased significantly by using cells with long pathlengths in which radiation is directed along (not perpendicular to) the substrate plate [373]. A cell for measurements on self-supporting pressed disks was described first by Hasegawa and Low [374]. The contribution of the liquid to in situ transmission spectra of pressed disks (which are porous) can be reduced simply by wetting. For this, the disk is isolated between two IR transparent windows, and its lower end is dipped into the liquid [375]. Alternatively, wetting capillaries may be used [372]. It should be kept in mind that when a disk is immersed in solution, the transmittance increases because the scattering of radiation by the disk is decreased. For discriminating between the absorption due to the dissolved and absorbed reagent, the following procedure was described [376]. After treatment with the solution, the solution is replaced in the cell with pure solvent and the transmission spectrum is measured.

SEC transmission measurements are complicated by the concurrent require-ments of semitransparency and electrochemical performance of the whole sys-tem (windows–electrolyte–electrode). To minimize the beam path, the so-called thin-layer arrangement is used. However, this can introduce unfavorable electro-chemical effects such as large ohmic potential (iR) drops, nonuniform potential (current) distribution over the electrode surface, and slow response rate of the electrode following a change in the applied potential (i.e., large time constant of

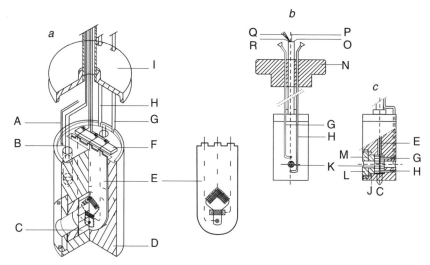

Figure 4.44. (*a*) Low-temperature optically transparent thin-layer electrochemical (LT-OTTLE) cell for variable-temperature UV-Vis/IR spectroelectrochemical studies. (A) Pt-100 thermocouple (Overcheck B. V.) for temperature control. Another Pt-100 positioned close to thin-layer compartment for measuring temperature of copper block. (B) One of two heaters (40W). (C) CaF_2 window (5-mm thickness) separated by indium gasket from distance copper ring. (D) Copper block. (E) Polyethylene spacer (0.18 mm thick) with melt-sealed electrodes (right-angle-shaped Pt minigrid auxiliary, Pt minigrid working, Ag wire pseudoreference) and contact Ag wires. (F) Insulating Teflon plate. (G, H) Quarter-inch outlet and inlet tubes of cell body, respectively. (I) Upper cover plate. (*b*) Front and (*c*) side views of LT-OTTLE cell with details of filling ports in cross section. (J) Distance copper ring. (K) Optical pathway across masked working electrode. (L) Copper ring for fixation of windows. (M) Indium gaskets. (N) Fixed stopper with O-ring protecting inner chamber of cryostate from water condensation. (O, P, R) Wire contacts to temperature control unit (heaters, Pt-100 thermocouples). (Q) Insulated leads connecting sealed electrodes with potentiostat. (C, E, G, H) see (*a*). Reprinted, by permission, from F. Hartl, H. Luyten, H. A. Nieuwenhuis, and G. C. Schoemaker, *Appl. Spectrosc.* **48**, 1522 (1994), p. 1524, Figs. 2 and 3. Copyright © 1994 Society for Applied Spectroscopy.

the cell) due to slow diffusion of reactants and products and the accumulation of undesirable products in the thin layer [354, 356, 372].

A thin-layer arrangement that includes an optically transparent electrode (OTE) sandwiched between two IR windows is called an optically transparent thin-layer electrochemical (OTTLE) cell. The thin-layer cell is in contact with a large container holding the reference and auxiliary electrodes. Since its introduction by Heineman et al. in 1967 [377], a wide variety of OTTLE cells have been reported in the literature (see Refs. [378–380] and references therein). Improvements in cell design include the use of fiber-optic guides for illumination and collection of radiation [381] and miniaturizing the cell for IR microscopic detection [382]. To overcome poor surface sensitivity of the normal-incidence transmission OTE technique, stemming from the short pathlength, the so-called parallel absorption SEC method [383] can be used, in which the radiation beam is directed parallel to the electrode surface.

There are two types of OTEs [384]. A metal microgrid with small (10–30-μm) holes, which allows ≥50% of the radiation to be transmitted, or an interdigitized array metal electrode [385] may be used for identification of products or intermediates in redox systems [386, 387]. The other type is comprised of a metal film deposited on a transparent support. The thickness of this film is a compromise between its electrical conductivity and optical transparency. This type of OTE can be used for surface analysis, particularly when the metal of the working electrode can provide the surface enhancement (Section 3.9.4).

Decreasing the temperature of an electrochemical reaction with short-lived redox products can slow the reaction down enough to follow the products by changes in the spectrum. One example is a low-temperature (LT) OTTLE cell devised by Hartl et al. [378] for chemical and SEC studies at temperatures down to 183 K, which maintains a preselected temperature constant to within ±0.5 K. The complete experimental setup consists of two components, the outer nitrogen bath cell (cryostat) and the inner (LT-OTTLE) sample cell containing electrodes housed in a copper block. The scheme of the inner cell is shown in Fig. 4.44. The working electrode is a Pt minigrid (32 wires/cm, 80% transmittance). Platinum is also used for the auxiliary and pseudoreference electrodes.

The diffuse-transmission technique, which uses a wedge-shaped entrance window (Fig. 4.24), allows for the investigation of layers on a surface of powders placed in a liquid. In this case, the greatest effect is expected for substances with a high refractive index, such as chalcogenides of heavy metals and Ge. For such materials, in spite of the fact that they are immersed in a liquid, the difference between their refractive index and that of the liquid will be fairly large (about 1–2), and the scattered light intensity will also be high.

4.6.2. In Situ IRRAS

The optical geometry consisting of the transparent window–solution layer–film–substrate system, which is used for studying ultrathin films on bulk metals or doped semiconductors in situ, is usually referred to as IRRAS [354–356, 362, 365], which implies that the angle of incidence at the window–electrolyte interface is lower than the critical angle φ_c. The spectra measured using the thin-layer cell with the CaF$_2$ window at the angle of incidence of 60° (see, e.g., Refs. [388–391]) fall in this category, since for the CaF$_2$–water interface $\varphi_c \approx 70°$. However, the maximum contrast of the film reflection spectrum is observed at an angle of incidence larger than the critical angle for the window–solution interface (Section 2.5.4). In this case, the film is probed by the evanescent wave established at the window–solution interface, and ATR geometry in the Otto configuration is employed, rather than IRRAS. This optical condition is met when the cell is equipped with a ZnSe window (the critical angle is ~30°) [392–395] or with the CaF$_2$ window but the angle of incidence is 76° [396]. Therefore, to avoid confusing the reader, the whole optical geometry, including the Otto-ATR and the IRRAS set-up, will be referred to as in situ IRRAS.

Typical in situ reflection cells containing disk electrodes are shown in Figs. 4.45 and 4.46. The edges and back of the electrode are sealed to ensure that

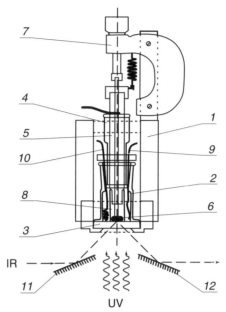

Figure 4.45. Schematic diagram of spectro-photoelectrochemical reflection cell with slab window: (1) metallic supporting frame; (2) glass body of cell; (3) CaF$_2$ window; (4) outer glass tube; (5) thin internal tube holding working electrode; (6) disc-shaped working electrode; (7) micrometer gauge; (8) Teflon holder; (9) reference electrode; (10) Pt counter electrode; (11, 12) plane mirrors. Reprinted, by permission, from J. Klima, K. Kratochvilova, and J. Ludvik, *J. Electroanal. Chem.* **427**, 57 (1997), p. 58, Fig. 1. Copyright © 1997 Elsevier Science S.A.

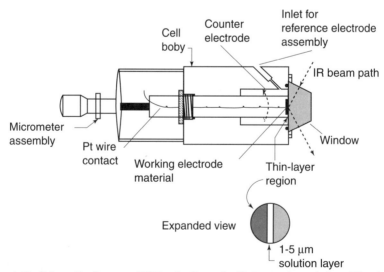

Figure 4.46. Schematic diagram of SEC reflection cell with Dove prism window. Slice through center of cell viewed from top is shown. Expanded view of thin-layer cavity region also appears. Reprinted, by permission, from C. Korzeniewski and M. W. Severson, *Spectrochim. Acta A* **51**, 499 (1995), p. 501, Fig. 1. Copyright © 1995 Elsevier Science Ltd.

the electrochemical reactions occur only at the front surface. Electrical contact to the disk is made through a wire that is spot welded to the back of the electrode. The electrode is positioned against a window. As shown in Section 2.5.4, the film signal decreases as the thickness of the solution layer thickness increases above $1-2$ μm, even for a transparent solution. In practice, an optimal thickness of the electrolyte interlayer is $1-5$ μm [392, 397, 398], which yields a time constant for the cell of ~5 ms for a 1 M electrolyte [356].

The window (IRE) material and the optimum angle of incidence are selected using spectral simulations (Section 2.5.4) while not neglecting to consider the chemical reactivity of the window material and the angular range of the ATR accessory. One question to address is which shape of window (IRE) to use. Historically, a slab-shaped window has been popular (see, e.g., Refs. [399–404]). Figure 4.45 shows a three-electrode cell with a slab-shaped CaF_2 window devised for in situ spectro-photo-electrochemical studies on photoexcited semiconductors [405]. The working electrode is a 10-mm-diameter titanium disk covered by a polycrystalline TiO_2 (anatase) film. In this arrangement, the slab-shaped window provides additional irradiation of the electrode by UV light. Slab-shaped windows are used in SEC microreflection cells designed for IR microscopy [406–408].

Although inexpensive, flat windows have a serious disadvantage in that a high proportion of radiation is reflected from the air–IRE interface. These energy losses can be significantly reduced if the radiation at the air–IRE interface is at normal incidence and the IRE is a Dove prism (Fig. 4.46) [365, 397, 409–411] or a hemisphere (hemicylinder) (Fig. 4.13) [29, 393, 412–414]. The latter geometry is arguably the most flexible and effective since it allows the angle of incidence to be varied without varying the angle of refraction. An additional advantage of hemispherical/hemicylindrical IREs is that they can improve the collimation of the incidence beam if it is focused at the proper distance from the solution–IRE interface (Fig. 4.13) [27, 415]. A beam collimated at the electrolyte–window interface can provide better sensitivity than when the beam is focused at the electrode surface [29, 393, 413]. The sensitivity and reproducibility of a SEC cell with a hemisphere window (the ZnSe IRE–water–Pt system) were studied by Faguy and Marinkovic [413]. The conditions were 4096 scans at a resolution of 16 cm^{-1} and laser modulation rate of 100 kHz of an ATI-Mattson Galaxy 8020 rapid-scanning FTIR spectrometer equipped with a 45° Michelson interferometer, a water-cooled globar source, and a narrow-bandpass MCT detector. It was found that at $\varphi_1 = 36°$ and the maximum signal throughput the smallest detectable signal, which was statistically determined, is 0.0075% and 3.3×10^{-5} in reflectivity and absorbance units, respectively. However, this limit does not depend on the optical throughput and should not be used as a detection limit.

A crucial aspect is alignment of the cell with the host reflection accessory. One alignment procedure [29, 413] is based on the changes that arise in the interferogram as the thin-layer cell is established. In the first step, the whole system (cell plus reflection accessory) is aligned without an electrode to maximize the throughput of p-polarized radiation. Then the electrode is introduced and the electrode itself is aligned again to increase the throughput. When aligning a cell with the hemisphere (hemicylinder) or Dove prism window, it must be taken into account that such a window acts as a lens and changes the focal point (the larger the refractive index of the window material, the larger the change) [412]. To illustrate the effect of the cell alignment on the experimental reproducibility, Fig. 4.47 shows in situ IRRAS spectra of a Pt(111) surface in 0.05 M solution of H_2SO_4 [413] measured in s- and p-polarization at two optical throughputs. The spectra are represented in $-\Delta R/R$ units [Eq. (4.5)]. The major feature in the p-polarized spectra is a 1200–1280-cm^{-1} band assigned to the ν_1 stretching mode of adsorbed bisulfate, which shifts to higher wavenumbers upon increasing potential due to the Stark effect and/or donation tuning (see the interpretation of Fig. 3.43). At the same time, a negative-going band at 1040 cm^{-1}, which is associated with the depletion of bisulfate from the thin layer due to adsorption, is only seen in the case of the higher optical throughput (Fig. 4.47b). This is due to increased sensitivity arising from better optical throughput. Additionally, positive-going bands associated with solution-phase species, $\sim$1200–1000 cm^{-1}, are detected to different degrees in the two different experiments. These bands are observed only at potentials below the range of the bisulfate adsorption and are due to ionic migration and local pH change [416–419] (Section 3.7). Namely, when

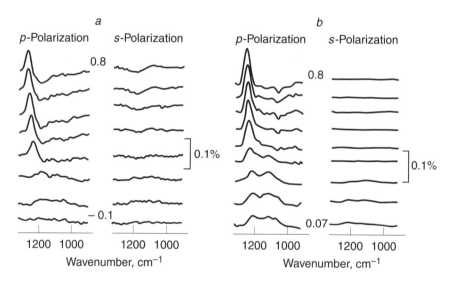

Figure 4.47. The s- and p-polarized SNIFTIRS for Pt(111) in 0.05 M H_2SO_4: (a) partial and (b) thorough optical alignment. Reference potential was -0.25 V (Ag|AgCl). Sample potential range is shown. Spectra were acquired using ATI-Mattson Galaxy 8020 rapid-scanning FTIR spectrometer equipped with 45° Michelson interferometer, water-cooled globar source, and narrow-band-pass MCT detector every (a) 129 mV and (b) 67 mV. Each spectrum is average of 4096 scans generated from 32 cycles of 128 scans each alternately collected in sample and background spectra at resolution of 16 cm^{-1}. Total acquisition time for one SNIFTIR spectrum was 15 min and each measurement run included from 7 to 12 spectra. Reprinted, by permission, from P. W. Faguy and N. S. Marinkovic, *Anal. Chem.* **67**, 2791–2799 (1995), p. 2796, Figs. 3 and 4. Copyright © 1995 American Chemical Society.

the electrode is polarized from the reference potential of -0.25 V (Ag$-$AgCl) to the sample potential, adsorbed hydrogen discharges and anions are attracted toward the electrode surface. Based on these and other data, it was concluded that both reproducibility and sensitivity depend strongly on optical throughput. Other sources of error are electrode fouling, transient loss of potential control, changes in the SEC cell configuration, and contributions from water vapor.

The foregoing example also shows that, as in transmission measurements, thin-layer cells lead to the thin-layer problem (Section 4.6.1). The diffusion decoupling [420, 363] of the very thin layer between the working electrode and the IR window (retardation of the free exchange of ions with the rest of the electrolyte) can result in accumulation/depletion of the reaction products/reactants whose absorption is superimposed on the spectrum of the adsorbed species. Furthermore, the accumulated products of any electrode reaction can distort the spectra measured by IRRAS [412, 421].

The following technical ways to circumvent this problem have been suggested.

The *barrel-plunger* cell design (Figs. 4.45 and 4.46) allows the working electrode to be withdrawn into the bulk electrolyte during the potential step and then returned to the window to record the spectrum. The electrode can be moved manually [422], as shown in Figs. 4.45 and 4.46, or by computer control [423]. Since its development by Pons et al. [424], this technique has been widely applied [356, 425–428].

In the conventional (stagnant) thin-layer configuration, changing the electrode potential yields, among other features (Section 3.7), bipolar bands that reflect potential-dependent changes in the interfacial and solution-phase adsorbate composition that arise from adsorption–desorption equilibria. To interpret these bands can be difficult, especially if the position shift upon adsorption is insufficient to avoid overlapping of the opposite-polarity solution and interfacial bands. However, in the presence of sufficient hydrodynamic flow, only the unipolar band component for the adsorbed species remains, as the solution composition is invariant. The bipolar bands from the solution phase can be extracted by appropriate subtraction of corresponding spectra obtained in the presence and absence of solution flow. These tactics are realized employing flow-through cells [429–431]. To allow the electrolyte to flow, a hole is drilled in the center of the electrode [431] or in the center of the IR window [432]. The capability of these flow cells has been demonstrated in a study of the adsorption of azide and cyanate ion onto polycrystalline silver [432]. The dependence of the spectra on the flow speed can provide information on the reaction kinetics [433]. The flow cell tactics can be utilized to monitor irreversible Faradaic processes either by maintaining a continuous supply of reactant solution or by removing products from the thin layer [430, 431, 432]. One drawback to this approach is that, in general, the flow in such cells is ill-defined, which makes the SEC results mostly qualitative in nature [434].

An invasive method such as to use a highly concentrated buffer of an appropriate composition also allows one to avoid significant potential-induced migration [435] (Section 3.7).

The time constant of a SEC cell can be reduced to microseconds using microelectrodes (25–200 μm size) [436, 437]; this is beneficial in time-resolved studies (Section 4.9) [406–408]. Since the iR drop in a microcell is smaller than in a SEC cell of conventional size, the thickness of the thin layer can be decreased, which in turn improves the SNR. However, this approach utilizes an IR microscope to record the spectra; this requires additional experimental skills. Alternatively, instead of a conventional interferometer, a continuously tunable IR laser with a small focusing spot can be used in conjunction with a flow-through cell [438]. The ohmic drop is reduced and the ion exchange between the thin layer and the bulk solution is facilitated if the working disk electrode is replaced by a microarray electrode. The microarray electrode described in Ref. [439] is composed of (nine) Pt microdisks that are themselves cross sections of a 1-mm-diameter Pt wire. These microdisks are separated from each other by several grooves that facilitate mass transport to each microdisk. As in the conventional cell, Pt foil is the counter electrode and an Ag–AgCl electrode separated from the bulk solution by a glass frit is the reference electrode.

To improve the mass transport in reactions involving high electric currents and the evolution of gas, a working electrode with small perforations can be employed [440]. Temperature-controlled SEC cells that operate between ambient and 57–70°C are described in Refs. [361, 441, 442]. The sensitivity of this method in the far-IR region can be substantially increased by the use of synchrotron radiation [443].

An assembly for substractively normalized interfacial FTIR spectroscopy (SNIFTIRS) (Section 4.6.4), based on the commercial spectrophotometer Unicam Research Series 1, was presented in Ref. [444]. The sample chamber has been modified to achieve a vertical orientation of the cell and a special support was constructed for the same purpose. The original path of the IR radiation has been modified by using two black anodized Al 30° wedges, both supporting a parabolic mirror. This allows the radiation to reach the working electrode forming 60° with the vertical. The cell permits an easy manipulation of either the solution or the electrodes in the sample chamber of the spectrometer, making it unnecessary to open this chamber.

Finally, IRRAS of the electrode–electrolyte interface can be measured using the emersed electrode technique. This technique is based on the fact [445] that the compact DL remains intact upon removal of the electrode from electrolyte. To overcome the loss of the potential control upon emersion it was suggested [446] that an electrode that is in partial contact with electrolyte be rotated. In a more advanced double-cell technique [447], a cell similar to that shown in Fig. 4.45 is employed. The difference is that the thin internal tube holding the working electrode (Pt disk) serves also as a container for the second cell. Both the main outer and internal auxiliary cells are filled with the same solution and have reference and counter electrodes inside. Since the Pt electrode is a good conductor its surfaces are equipotential. Therefore, any potential applied on the electrode back side is established at the outer surface that is pressed against the window for IR measurements.

4.6.3. ATR

The concept of ATR at the interface of two media is described in 1.4.10° and Section 1.8.3. In situ ATR measurements of ultrathin films started in the mid-1960s with studies of the adsorption of a stearic acid monolayer from D_2O onto Ge [448], and chemical [449] and electrochemical [450] oxidation of Ge, where a Ge multiple internal reflection element (MIRE) acts as both the substrate and the electrode. Later, "coated" ATR [60, 451–454] and MOATR with the SEIRA effect [455] were introduced in in situ experiments. The principal advantage of the ATR geometry is that the corresponding in situ cells are free from diffusion effects (the volume of solution phase in contact with the IRE is arbitrary), which is useful when studying time-dependent phenomena (Section 4.9.1).

Semiconductors. The standard ATR measurement on moderately doped Ge, Si, or GaAs involves multiple internal reflection (MIR) geometry (Fig. 4.48*a*) and the "reactive" ATR sampling technique (Section 4.1.3). Cell configurations with MIREs have been described in Refs. [415, 456–461]. To study processes at polar semiconductors containing light elements (e.g., oxides) in the low-wavenumber region and doped semiconductors, the coated ATR sampling technique and single-reflection geometry are most suitable. An important practical detail is that metallic electrical contacts on the semiconductor electrode have to be located outside the path of the IR light through the semiconductor; otherwise most IR light would be absorbed by the metal [367]. Current homogeneity can be reached by evaporating a gold grid on a back side of a semiconductor electrode [462]. An increase in sensitivity of in situ ATR spectra can be obtained by incorporating a reference beam measurement [463].

In general, in situ ATR at nonmetal substrates suffers from poor surface selectivity as a result of the relatively large sampling depth (Table 1.3). Nevertheless, there is no special problem to distinguish the spectral contribution

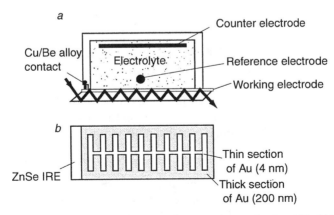

Figure 4.48. (*a*) Schematic diagram of SEC cell with multiple-reflection IRE. (*b*) Top view of patterned Au–ZnSe IRE electrode by Bae et al. [417]. Numbers in parentheses are thicknesses of Au deposits in two sections.

from a surface species when adsorption produces large band shifts. Otherwise, diluted solutions of the adsorbing species can be used. For example, for the Si$-$SiO$_2$$-$adsorbed CTAB$-$solution interface at $\varphi_1 = 45°$ and a CTAB concentration of 5.5×10^{-5} M it was found [464] that less than 5% absorption is due to the solvated surfactant, which is tolerable for the spectrum analysis in terms of the adsorbed film. Also, the sampling depth can be varied by adjusting the angle of incidence and, in the case of coated ATR, matching the refractive indices of the IRE and the substrate layer as well as selecting an appropriate thickness of the substrate coating.

Polymers. In situ investigations of the degradation of a polymer$-$substrate interface by water and measurements of the water accumulated at the interface were undertaken with a special technique developed by Nguyen et al. based on the MIR method [466a, b, c]. A water chamber is attached to the coated IRE, and water is added to it. MIR spectra are taken automatically at specified time intervals, without disturbing either the specimen or the instrument, and quantitative analysis of the data made use of the two-layer Fresnel formulas. A cell with the same optical geometry was constructed by Balik and Simendinger [467] for analysis of gas diffusion through polymer films. In this cell, optical contact between the polymer film and an IRE is maintained by pressurized gas. Another technical solution [468a] is to sandwich a thin layer of penetrant (e.g., a lubricant) between the IRE and the polymer sample. This allows accurate control and measurement of the thickness of the lubricant layer, which, in turn, facilitates subsequent data analysis. The diffusion is studied by monitoring the time-resolved change in absorbance of either a unique polymer or penetrant band. A feature of this technique is that it can provide an estimate of solubility, as well as an estimate of the diffusivity of the penetrant in the polymer. To study intermolecular diffusion across the polymer$-$polymer interface [468b], a $\sim$3-μm film of one polymer was cast directly onto a hemispherical ZnSe IRE. A thick film of the second polymer was placed on an Au foil and then pressed against the coated IRE. This assembly was then clamped in a heating cell and placed in the ATR attachment. A technique for quantifying transport along the polymer$-$substrate interface and interfacial hydrolytic stability of polymeric composites and systems exposed to water and high relative humidities is described in Ref. [466d]. This technique can distinguish water transport through the film from that along the interface. Spectroscopic analyses of fractured surfaces of poor- and well-bonded polymer$-$substrate systems after water exposure indicate that the technique is capable of discerning a hydrolytically-stable interface from a water-susceptible interface.

Metals. A thin metal electrode deposited on the IRE, as for transmission measurements, is called an OTE, while the whole IRE$-$OTE$-$sample layer system is referred to as an extended internal reflection element (EIRE). Metal coatings provide the shortest penetration depths (up to a few nanometers). Another important advantage of the OTE is the enhancement of IR absorption of species in the first few monolayers (SEIRA) (Section 3.9.4). This effect has been exploited in a series of SEC single-reflection geometry studies [469–480].

A sketch of a single-reflection cell from the Osawa group [471] used in SEC measurements is shown in Fig. 4.49. The hemicylindrical IRE coated with a metal film electrode is attached to the Kel-F cell body by sandwiching a silicone rubber sheet and a copper foil contact. The cell body was configured to be mounted in the ATR accessory, with inlet and outlet tubes for gas bubbling, reference and counter electrodes, and a Luggin capillary. To detect a species at the metal–solution interface, p-polarized radiation was focused at the IRE base plane. The angle of incidence was either 70° [478] or 60° [469, 471].

As discussed in Sections 2.4 and 2.5.4, the spectral contrast can be increased using MIR-OTEs. This approach has been used, for example, in the ATR studies of adsorption of lipid bilayers to Au [481], p-nitrobenzoic acid on Ag and Au [482] and electrochemical reactions on Au [434], Cu [483], steel [484], Pt [485], and iron [486] electrodes. The ATR technique with a Si MIRE covered by Ag and Au films was applied to the study of thiocyanate adsorption on silver and gold [487]. Enhanced SNR has been reported for the MIR in situ spectra of proteins adsorbed onto a Cu-coated cylindrical internal reflection element [488]. Since the optical path of the beam through a MIR OTE strongly depends on the wavelength, optical schemes with a fixed incident beam are not applicable in this case [362].

The main drawback of an OTE is instability and poor conductivity of the islandlike metal film. To circumvent this problem, a Pt grid evaporated on the surface of a ZnSe IRE was used in ATR studies of the electropolymerization of

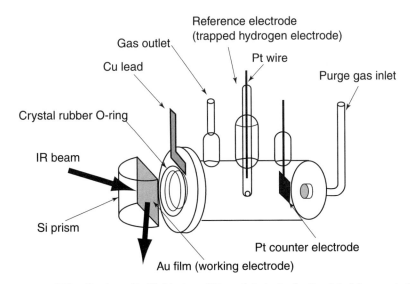

Figure 4.49. SEC reflection cell with thin-layer (20-nm Au) electrode. Reprinted, by permission, from K. Ataka, T. Yotsuyanagi, and M. Osawa, *J. Phys. Chem.* **100**, 10664 (1996), p. 10665, Fig. 1. Copyright © 1996 American Chemical Society.

aniline at a gold surface [489, 490]. Bae et al. [434] reported a MIR technique using a patterned metal film electrode (Fig. 4.48b). This is obtained as follows. A mask in the form of a zipper is placed onto one of the large faces of the MIRE (ZnSe, $50 \times 20 \times 3$ mm, $45°$, in this case), and a thick ($\sim$200-nm) layer of the metal (Au) is sputtered. After removing the mask, a very thin ($\sim$4-nm) film is sputtered onto the whole patterned surface to create, over certain areas, a electrically conducting thin film connected to the thick film. The ATR on such a MIR OTE was compared with in situ IRRAS on a bulk gold electrode (a CaF_2 Dove prism as a window, $\varphi_1 = 60°$) using 2,5-dihydroxybenzyl mercaptan (DHBM) adsorbed on gold as a model system [434]. This species undergoes a reversible change in oxidation state in a potential region in which gold behaves as an ideal polarizable electrode. The layer of DHBM was adsorbed on the gold substrate, then the cell was washed with water to remove nonadsorbed DHBM and filled with 0.1 M $HClO_4$. The in situ spectra (Fig. 4.50), were measured while sweeping the potential linearly with a rate of 1 mV $\cdot$s^{-1} and using +0.40 V (SHE) as a reference and represented in $-\Delta R/R$ units [Eq. (4.5)]. At +0.81 V, where the DHBM is expected to be fully oxidized, the spectra measured by both ATR and IRRAS exhibit positive- and negative-going bands assigned to the oxidation of DHBM. In addition, IRRAS detects a positive-going band at $\sim$1112 cm^{-1} associated with migration of perchlorate into the thin layer during

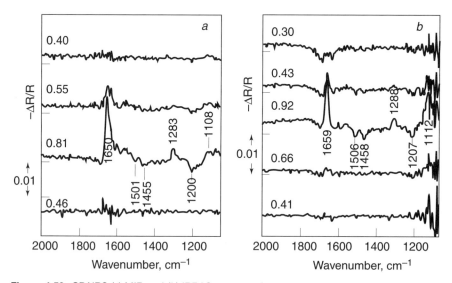

Figure 4.50. SPAIRS (a) MIR and (b) IRRAS spectra of oxidation of 2,5-dihydroxybenzyl mercaptan on (a) patterned Au–ZnSe electrode and (b) solid Au. Spectra were obtained using IBM IR-98 FTIR spectrometer (Bruker 113v) equipped with MCT detector at 4-cm^{-1} resolution with unpolarized and p-polarized radiation for MIR and IRRAS, respectively, during single voltammetric scan at 1 mV $\cdot$ s^{-1}. Labels in each curve correspond to average potential values during each coaddition. Reprinted, by permission, from I. T. Bae, M. Sandifer, Y. W. Lee, D. A. Tryk, C. N. Sukenik, and D. A. Scherson, *Anal. Chem.* **67**, 4508 (1995), pp. 4512 (Fig. 6) and 4513 (Fig. 7). Copyright © 1995 American Chemical Society.

the oxidation. The migration is reversible, as evidenced by the disappearance of the 1112-cm^{-1} band upon fully reducing the adsorbate layer. There is no spectral feature at +0.40 and +0.46 V, which provides evidence that the redox process is fully reversible. Thus, it is important from the technical viewpoint that the SNRs and composition of the spectra measured by these two different methods are very similar, except for the extra band due to the bulk solution measured by IRRAS.

Powders. Commercial availability of HATR accessories [491] (Fig. 4.12) has made in situ studies of the powder surfaces a routine procedure. For these measurements, the suspension is spread onto the ATR element (Section 4.2.4). For heavy powders, a cell coupled with a horizontal ATR unit, as shown in Fig. 4.33, can be used. To measure in situ HATR spectra of coatings on powders without changing the sample surface area under stirring of the solution in contact, a special protocol for adhering colloidal metal oxide particles to the ZnSe IRE was developed [179, 492]. In the case of smectite (hydrous aluminosilicate) fine particles, a stable coating 25–50 nm thick adheres spontaneously on a ZnSe IRE upon exposure to a dilute aqueous suspension (solids concentration of 10 g · L^{-1}) [493]. The adhered smectite particles are highly oriented with the (001) face parallel to the surface of the IRE. A more universal sampling technique for powders is described by Ninness et al. [494]. The technique involves first formulating a coating comprised of high-surface-area silica particles and a polymeric binder in a suitable solvent. The resultant coating is applied to the surface of an IRE and mounted in an ATR apparatus. The technique is demonstrated with the ZnSe IRE coated with fumed silica particles in a polyethylene (PE) matrix. Access of the silica surface in the matrix to adsorbates was evaluated by comparing the gas-phase reaction of silanes on silica–PE-coated CsI windows in transmission with silica–PE-coated ZnSe in an ATR evacuable cell. It is shown that the PE weakly perturbs about 25% of the surface hydroxyl groups, and that all surface groups are available for reaction with silanes. The silica/PE is indefinitely stable in an aqueous environment and has advantages of at least two orders higher sensitivity and a wider spectral range over studies using oxidized silicon wafers. To measure spectral pH envelopes and adsorption isotherms on powdered oxides, a flow cell [177] can be used. A special cell for monitoring in situ metal oxide surface charge with a single-reflection prism ZnSe IRE coated with a sol–gel TiO$_2$ film was described by Dobson and co-workers [495]. A CIRCLE cell with a ZnSe IRE has been employed for in situ surface studies of colloidal particles suspended in aqueous solutions [496–499]. It should be mentioned that the PTFE O-rings that assemble CIRCLE cells are characterized by two strong absorption bands at 1150 and 1200 cm^{-1}, which can contribute in the integral ATR spectrum if the rings are deformed in the course of the in situ experiment [500].

For SEC studies, the surface of a powder semiconductor electrode can be pressed against an IRE. Such a cell was devised by Leppinen et al. [501] for studying redox reactions on semiconducting sulfides (Fig. 4.51). The cell consists of a Teflon cell body and a movable working electrode holder, at the end of which the mineral bed electrode is attached. The Pt counter electrode is placed

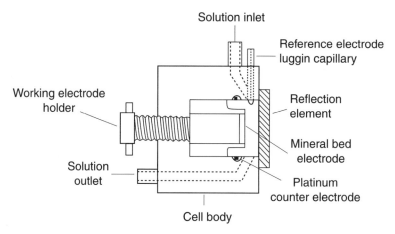

Figure 4.51. SEC–MIR cell constructed for in situ FTIR studies of reactions on mineral-particle bed electrodes. Reprinted, by permission, from J. O. Leppinen, C. I. Basilio, and R. H. Yoon, *Int. J. Min. Process.* **26**, 259 (1989), p. 260, Fig. 1. Copyright © 1989 Elsevier Science Publishers B.V.

along the inner wall of the cell near the working electrode. The capillary tip for the reference electrode is placed close to the working electrode. The cell body is attached to a stainless steel cell holder (not shown) that is on the other side of the IRE. The working electrode is made of finely ground mineral powder embedded in carbon paste. The electrolyte solution is circulated through the cell during the selected polarization period. Afterward, the solution flow is stopped and the working electrode is pressed against the Ge IRE to obtain the ATR spectra while applying a potential ($< \sim 100$ mV, SHE) (see Section 7.5.2 for more detail). In another technique, the powder electrode is prepared by pressing the conducting powder onto a metal mesh attached to an IRE [502].

4.6.4. Measurement Protocols for SEC Experiments

The signal that is measured in SEC experiments is a difference signal generated from reflectance spectra taken at two different (sample and reference) potentials. Depending on the measurement protocol, this may be a *static* or *dynamic difference*. In this section, we will consider protocols for measuring the static potential-difference spectra, while the dynamic ones are the subject of Section 4.7. A typical static measurement using a conventional continuous-scan FTIR spectrometer involves first acquiring a reference spectrum at the initial potential (also called the "base") and then changing the electrode potential by a preselected staircase program, measuring a "sample" spectrum at each potential. The spectra are represented on an absorbance scale of $-\log(R_{\mathrm{sam}}/R_{\mathrm{ref}})$ or in $-\Delta R/R$ units,

$$-\frac{\Delta R}{R} = -\frac{R_{\mathrm{sam}} - R_{\mathrm{ref}}}{R_{\mathrm{ref}}}, \tag{4.5}$$

where R_{sam} and R_{ref} are single-beam spectra at the reference and sample potentials, E_{ref} and E_{sam}, respectively. The reference potential is usually one of the limits of the potential range under study, which is selected in a nonelectroactive region. Positive bands in the resulting spectrum indicate the formation of an intermediate or product, and negative bands indicate loss of a species at the sample potential relative to the reference potential (see the discussion of Fig. 3.43). This approach is sometimes called potential-difference IR (PDIR) spectroscopy [503] or single-potential alternation IR spectroscopy (SPAIRS) [504, 505]. SPAIRS can be applied to follow any electrochemical reaction, regardless of its reversibility. The spectral variations associated with a specific potential change can be extracted numerically. In particular, when the difference between SPAIRS spectra is analyzed (Figs. 7.34, 7.35, 7.41, and 7.43), the low-frequency drift of the baseline and uncompensated absorption of the atmosphere may be partially eliminated. Modern spectrometers are able to provide SPAIRS spectra with a good SNR at a spectral resolution of $8-16$ cm^{-1} for $1-15$ s [410, 471, 476, 477], simultaneously with slow-scanning ($1-5$-mV $\cdot$ s^{-1}) cyclic voltammograms (Figs. 3.41 and 4.50), provided that the experimental geometry is optimal. The potential is linearly changed, while spectra are collected at regular intervals over a much shorter time than the time of the potential scan. Therefore, such spectra are averages over a certain potential range.

A better SNR for in situ spectra of interfacial species formed in reversible reactions can be achieved by a modification of the PDIR protocol, originally named subtractively normalized interfacial FTIR spectroscopy (SNIFTIRS) [424, 506]. SNIFTIRS involves multiple alternation of the electrode potential between the reference potential E_{ref} and the sample potential E_{sam} and collecting N interferograms (about $5-100$) at each potential. This cycle is repeated M times so that in total $N \times M$ interferograms are averaged for each potential. The number M determines the SNR. Experimentally, the modulation of the electrode potential is synchronized with the acquisition of the interferograms by connecting the external trigger port of the potentiostat to the communication port of the FTIR instrument computer. The electrode potential can be changed in various ways so as to affect the reaction. For example (Fig. 4.47), E_{ref} and E_{sam} can be multiply alternatively switched, and $5-20$ scans are acquired at each switch. The reference and sample spectra are separately coadded and then recalculated in one SNIFTIRS spectrum represented in $-\Delta R/R$ units. Afterward, this procedure is repeated for another sample potential. However, recent trends are toward recording spectra successively during a potential sweep, which can have either a stepwise or linear waveform (Figs. 3.43a and 3.44). The sweep is repeated as many times as necessary for achieving an acceptable SNR. At each sweep, the spectra are measured at the same potentials for a few seconds and coadded.

SNIFTIRS is able to provide detection limits for in situ IRRAS of $10^{-5}-10^{-4} \Delta R/R$ with a spectral resolution of $8-16$ cm^{-1} and several hours of data collection [361]. In general, application of this technique is restricted to reversible electrochemical systems. However, flow cell tactics enable one to utilize this method even when examining irreversible Faradaic processes if the net accumulation/consumption of the product/reactant in the thin layer can be avoided [431].

Faguy and Marinkovic [413] studied how different potential-change protocols affect spectra measured in situ by IRRAS in the case of Faradaic reactions. These authors found that SPAIRS is more sensitive to the components in the diffuse layer. Therefore, a combination of these two protocols can provide additional information about the DL [416].

In general, a routine quantitative analysis of adsorbed species with in situ FTIR spectroscopy is problematic, primarily due to the difficulties in determining the baseline. Nevertheless, several approaches to quantifying the SPAIRS [503, 504] and SNIFTIRS [393–396, 507] have been recently reported. In the latter case, quantitative analysis based on the BLB law in conjunction with Gouy–Chapman theory is applied to estimate the ionic charge density in the diffuse layer, near the point of zero charge. For quantitative analysis of the SPAIRS data, the polarization rotation at the window–solution–electrode interface and the relative linear polarization throughput for the spectrometer should be taken into account [508].

As an alternative, the electrode potential can be modulated quickly and a step-scan (S^2) FTIR spectrometer or a dispersive spectrophotometer is used to measure spectra (Section 4.9).

4.7. POLARIZATION MODULATION SPECTROSCOPY

The major problem encountered in in situ spectroscopy of ultrathin films is relatively low surface sensitivity and SNR because of interfering absorption by the electrolyte and/or the reactor (ambient) atmosphere and low-frequency noise (fluctuations of the source and the $1/f$ noise of the detector). To improve the SNR, increase the surface sensitivity, and filter out low-frequency noise, modulation methods were incorporated into FTIR spectroscopy [509–514]. The modulation methods employ the *ac advantage* [510], which states that a small difference signal can be measured more accurately as the amplitude of a periodically varying ac signal than as the difference between two time-dependent dc signals. The ac signal is measured by electronics tuned to the ac modulation frequency. The signal processing removes all the modulation-independent signals (such as the strong absorption by water vapor) and interfering absorption of solution species and noise whose frequencies lie either outside the bandpass filter or else at the modulation frequency but out of phase with the measured signal.

In the continuous-scan mode, a Michelson interferometer generates modulation of the radiation at each wavenumber v with a frequency $F = 2Vv$, where V is the mechanical velocity of the scanning mirror in centimeters per second. The frequency F is called the Fourier frequency. A typical speed of the scanning mirror is $0.1–10$ cm $\cdot$ s^{-1}, so that signals within the IR spectral region fall into the $10^2–10^4$-Hz range. Since in most measurement schemes a low-pass filter is used to separate the Fourier and modulation frequencies, the modulation frequency should satisfy the sampling theorem. Specifically, the modulation rate of the sample excitation must be no less than twice the highest Fourier frequency in the spectral range. There are two ways of achieving high-frequency modulation of an FTIR signal: *polarization modulation* (PM) and *absorption modulation* [510], though in the case of chiral molecules polarization modulates the sample absorption [511]. In this section, a technical background for polarization modulation FTIR spectroscopy is provided and the advantages in the studies of interfacial films are discussed. Technical aspects of application of this method to the air–water interface are the subject of Section 4.8. Potential modulation is considered in Section 4.9.2.

The PM method, developed independently by Pritchard in 1972 [512, 513] and Blanke in 1975 [514], relies on the fact that, unlike an isotropic immersion medium, a film selectively absorbs *s*- and *p*-polarized radiation. Already the first experimentation [515–519] demonstrated that PM is capable of increasing surface sensitivity on a metallic substrate by several orders of magnitude and discriminate between the film spectrum and the absorptions due to the randomly oriented gas-phase or liquid molecules anywhere in the optical path. The PM method was first applied using a commercial FTIR spectrometer by Dowrey and Marcott [520] in 1982 to measure a spectrum by IRRAS of a 10-Å film of cellulose acetate adsorbed onto polished copper. The combination with FTIR spectroscopy allowed measurements of good-quality PM spectra of ultrathin films in real time. This advantage has been exploited in the studies of gaseous-phase reactions on metals [521–523] and transparent solid substrates [524–526], the structure of SAMs on gold [527–531] and silicon [532–535], and proteins at the liquid–liquid interface [536]. To date, several research groups have employed PM-IRRAS to study L monolayers of amphiphile molecules on water [537–546], the protein–lipid interactions [547–556], and SEC experiments [388, 557–563].

To provide polarization modulation of the probing beam, in 1969 Kemp [564] suggested a device called a photoelastic modulator (PEM). The PEM consists of two components [565], the modulator head and the control unit. The modulator head, made from an IR-transparent crystal with an isotropic refractive index when unstressed (ZnSe or CaF_2), is placed in the beam path. The control unit generates high-frequency oscillations that are used to drive stress transducers attached to the polarization modulator crystal. The stress applied along one axis of the PEM crystal induces an anisotropy in the refractive index of the crystal, which results in a rotation of the polarization of the transmitted beam and causes the polarization state of the IR beam to alternate. A fixed polarizer in the optical path before the PEM provides either *s*- or *p*-polarized radiation (Fig. 4.52). The sample is placed directly after the PEM to ensure that the polarization of the beam is not altered before absorption by the sample. After the sample, the signal is directed to a MCT detector, followed by a bandpass filter, a lock-in amplifier (LIA), and electronic processing (demodulation).

The PEM is not achromatic, so modulation of the polarization can only be achieved at one wavelength, determined by the stress voltage applied to the PEM. The operating PEM frequency depends on the crystal (either 37 or 50 kHz for ZnSe), and the polarization modulation is at twice the operating frequency of

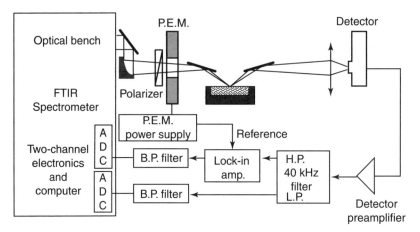

Figure 4.52. Schematic of optical potential-modulation (PM)-IRRAS setup and of two-channel electronic processing. PEM = photoelastic modulator, BP = bandpass, LP = low-pass, HP = high-pass. Reprinted, by permission, from D. Blaudez, T. Buffereau, J. C. Cornut, B. Desbat, N. Escares, M. Pezolet, and J. M. Turlet, *Appl. Spectrosc.* **47**, 869 (1993), p. 871, Fig. 2. Copyright © 1993 Society for Applied Spectroscopy.

the PEM. At nearby wavelengths, the PEM introduces a large background signal in the spectra that varies slowly in a sinusoidal fashion. To reduce this limitation, special procedures for the baseline correction can be used [565]. However, a more cardinal means consists of adjusting the modulation frequency with the modulator control box to the spectral range of interest and performing the PM experiments separately for several narrower ranges. As an alternative, Polavarapu and Deng [566] suggested chromatic potential modulation using an interferometer based on the principle of polarization division (introduced by Martin and Puplett [567] for far-IR), rather than a conventional Michelson interferometer operating on the principle of amplitude division of the incoming light coupled with the PEM. It was reported that this approach provided higher sensitivity to ultrathin films than the photoelastic PM.

Electronic processing of the output of the IR detector yields a PM-IRRAS signal. For an ultrathin film on a metallic substrate the PM-IRRAS signal is [568, 569]

$$S = C \frac{R_p - R_s}{R_p + R_s} J_2(\phi_0), \qquad (4.6)$$

where C is a constant, J_2 is the second-order Bessel function of the maximum dephasing ϕ_0 introduced by the PEM, and R_s and R_p are the s- and p-polarized reflectances of the sample, respectively. The PM-IRRAS signal contains spectral information about the metal surface because species in the bulk solution/atmosphere, which are randomly oriented, absorb IR radiation isotropically ($R_s = R_p$) and do not contribute to S. As a consequence, the absorption spectrum of an ultrathin film can be measured without any further reference to the spectrum of the bare substrate. This overcomes the problem of preparing the perfect clean reference substrate, which is problematic due to the presence of various contaminant hydrocarbons in the real laboratory atmosphere conditions. Furthermore, it is possible to obtain an in situ IRRAS spectrum of an interfacial species at a single potential. Nevertheless, two PM spectra measured at two different potentials are typically used to eliminate the slowly varying background due to the PEM.

In the case of a nonmetal substrate, both R_s and R_p contain contributions from absorption by the film, and interpretation of the PM-IRRAS signal in the form of Eq. (4.6) is less straightforward. Moreover, a dielectric substrate has its own large, specific, and spectrally dependent PM-IRRAS signal [532]. To eliminate this contribution, the PM-IRRAS of an ultrathin film at a dielectric substrate is normalized with the PM-IRRAS signal of the clean dielectric [532, 568, 569, 585]:

$$\frac{\Delta S}{S} = \frac{S(d) - S(0)}{S(0)}, \tag{4.7}$$

where $S(d)$ and $S(0)$ are the PM-IRRAS signals of the substrate covered by a film and the clean substrate, respectively, and d is the film thickness. The values of these quantities are extracted from the measured interferograms by using, instead of the approximation in Eq. (4.6), the explicit expression [568, 569]

$$S = C \frac{J_2(\phi_0)(R_p - R_s)}{(R_p + R_s) + J_0(\phi_0)(R_p - R_s)}, \tag{4.8}$$

where J_0 and J_2 are, respectively, the zero- and second-order Bessel functions of the maximum dephasing ϕ_0 introduced by the PEM and C is a constant accounting for the different amplification of the two parts during the two-channel electronic processing.

As for conventional metallic IRRAS, PM-IRRAS on metallic substrates is measured at the grazing angles of incidence. In the case of nonmetallic substrates, the optimum angle of incidence φ_{1opt} corresponds to the maximum of the difference PM-IRRAS signal $\Delta S = S(d) - S(0)$ [Eq. (4.7)]. It was found theoretically [569] that the value of φ_{1opt} for PM-IRRAS of a monolayer on a substrate with the refractive index of 1.5 (glass) at 1700 cm^{-1} is ~76°, while for substrates having a higher refractive index (e.g., ZnSe or Si) its value shifts toward higher (>85°) angles, which makes the studies on these substrates more technically difficult. Nevertheless, the PM-IRRAS of a monolayer on Si measured using the FTIR spectrometer IFS88 (Bruker) equipped with a MCT detector at 80° by coadding 512 scans with a resolution of 4 cm^{-1} is characterized by the peak-to-peak noise level of 8.5×10^{-6} absorbance units at 1500 cm^{-1} [535].

An important advantage of PM-IRRAS is that information about molecular orientations in ultrathin films on nonmetallic substrates can be obtained by analyzing the direction and intensity of the absorption bands (Section 3.11.4). However, in the general case, the interpretation of the PM spectrum can be rather ambiguous. To overcome this limitation, it was suggested [532] to measure PM-IRRAS only with s-polarization. For these measurements, the set-up shown in Fig. 4.52 is modified by replacing the flat directing mirrors with two ZnSe plates on which the beam is incident at the Brewster angle. These plates split the incident radiation into two parts: s-polarized radiation is reflected onto the sample surface, while p-polarized radiation is transmitted through the plates and directly to the detector.

Instead of a LIA, the signal reaching the detector can be demodulated with special electronics, such as those developed by Corn et al. [557, 558]. This approach, which is referred to as real-time PM (RTPM), provides better common signal rejection capabilities, higher SNR (for the same acquisition time) and long-term stability, relative to PM systems employing a LIA. Richmond and co-workers [388] have compared efficiency of conventional (static) and RTPM FTIR SPAIRS measurements for probing in situ adsorbed species at a metal electrode. The spectra of adsorbed thiocyanate ion, imidazole, and glucose on Cu were obtained with both techniques under identical electrochemical conditions. The spectra from RTPM SPAIRS were measured at a PEM frequency of 1975 cm^{-1}. The results for the SCN$^-$–Cu electrode system are shown in Fig. 4.53. The signals from atmospheric gases are completely eliminated by the PM method, but the absorption from the atmosphere is rather strong in the static spectra. Moreover, unlike PM, the background of the static spectra is distorted. Though the spectra provided by both techniques exhibit the band at 2175 cm^{-1} characteristic of adsorbed thiocyanate, the intensity of the band in RTPM SPAIRS is higher, indicating a higher sensitivity of this technique toward interfacial species relative to that of the static method. A productive approach is to combine both the

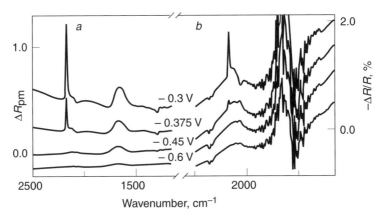

Figure 4.53. In situ IRRAS spectra of SCN$^-$ adsorbed on Cu electrode: (a) PM and (b) ordinary SPAIRS spectra. Reference is -1.2 V (Ag–AgCl) and sample potentials are marked. Experiments were performed on Mattson RS-1 spectrometer configured with external bench analogous to that shown in Fig. 4.51. Photoelastic modulator was Hinds International ZnSe Series II modulator, operating at 37 kHz. The PM wavefront was sampled in real time with ATI Instruments real-time sampling accessory. The MCT detector with D^* of 5×10^{10} cm · W^{-1} · s$^{-1/2}$ was used. Spectra are represented in absorption depth of PM signal (ΔR_{pm}). Reprinted, by permission, from W. N. Richmond, P. W. Faguy, R. S. Jackson, and S. C. Weibel, *Anal. Chem* **68**, 621 (1996), p. 625. Copyright © 1996 American Chemical Society.

conventional (static) and RTPM FTIR SPAIRS techniques, which allows better discrimination between surface, bulk solution, and DL effects [570]. At the same time, PM is ineffective for rejecting the electrolyte absorption in the ATR spectra of the semiconductor–solution interface due to similar spatial selectivity of s- and p-polarized radiation under these conditions.

4.8. IRRAS OF AIR–WATER INTERFACE

IRRAS is among very few spectroscopic methods such as sum frequency generation (SFG) and second harmonic generation (SHG) that are suited for analysis of chemical functional groups and conformations within ultrathin films at the air–water (AW) interface. This method was first applied by Dluhy and Cornell in 1985 [571]. The sampling aspects of this type of IR measurement have been thoroughly discussed in earlier reviews [572–574]. In the spectral region from 800 to 4000 cm^{-1}, the maximum value of the absorption index of water is 0.212 for the νOH mode at $\sim$3400 cm^{-1}, while the normal-dispersion refractive index varies within the range of $\sim$1.25–1.35 [575–578]. Therefore, according to its optical properties, water is a weakly refracting transparent/weakly absorbing substrate. The optimization of the IRRAS measurements on such substrates is discussed in Section 2.3.1, while the general features of the spectrum interpretation can be found in Sections 3.3.2 and 3.4.

The optical layout of the standard experiment is relatively simple: The IR beam is directed at a selected angle of incidence onto the horizontal water surface, collected at the same angle after the reflection, and afterward detected. The spectrum measured in the presence of a film at the interface is normalized with respect to the reflectance spectrum of the bare water (Fig. 3.27). To perform such measurements, commercial variable-angle IRRAS accessories with a horizontal container for a liquid are currently available (Harrick, Graseby Specac). The optics for IRRAS can be coupled with a Langmuir film balance to study L monolayers [572a, 579]. Since reflectance of water is low, the spectra are measured using a MCT detector by coadding 1000–2000 scans with resolution of 4–8 cm^{-1}. The IR bandpass filters can be placed just before the water to reduce any localized heating effect [580].

Typically, the angle of incidence is 30°–35°. A technique was reported [581] to enhance the spectrum contrast by measuring at the angle of incidence of 52° ($\sim\varphi_B$) and focusing and collecting the reflected radiation with off-axis parabolic mirrors. However, due to the vanishing SNR (Section 2.3.1), the p-polarized IRRAS in the vicinity of the Brewster angle puts in enhanced demands to the optical accessory, including more than 99% efficiency of the polarizer (to make the p-polarized contribution dominant in the radiation passing through the system) and the angular divergence of the incident radiation of <1° (which can be set with Iris diaphragms) [573b]. An alternative is the differential measurements (Section 2.3.1). In any case, obtaining accurate data is a time-consuming procedure: It takes at best a day to measure IRRAS at a single angle of incidence [573b]. In the multiangular experiments that are required for the spectrum fitting method of determining the molecular orientation (Section 3.11.3), a great deal of time is consumed for reassembling the setup before collecting spectra at each angle of incidence. A timesaver in such measurements can be a computer-controlled variable-angle IRRAS accessory (Bruker). Selecting the angle of incidence it should be kept in mind that according to spectrum simulations the maximum SNR in both s- and p-polarized IRRAS of an isotropic film at the AW interface is achieved at the angle of incidence of $75°$–$80°$.

Since the SNR becomes significantly lower when the spectra are measured with polarized radiation [573, 582, 583], the number of averaged spectra, N, should be increased (SNR $\propto \sqrt{N}$). As a rule, an appropriate number of scans is $\sim$2000 and 4000 for s- and p-polarization, respectively, increasing up to 8000–9000 for p-polarization in the vicinity of the Brewster angle [583]. However, when the number of scans increases, data collection time increases proportionally, becoming up to several hours for one pair of polarized spectra. For prolonged acquisition times, the interference fringes can appear in the IRRAS spectra that result from water evaporation (Fig. 4.54). To reduce another technical problem — interference from water vapor — low resolution and the correct apodization are recommended. The use of a triangular apodization was found to lead to worse results than the strong Blackman–Harris apodization, although the latter shows the disadvantage of slightly broader bands [584]. An additional improvement in water vapor compensation can be obtained using fine regulation of the humidity in the sample compartment with a low flow of dry nitrogen in the chamber [582]. As an alternative, a "shuttle" accessory described in Ref. [19a] can be employed. The inherent limitation of IRRAS of the AW interface is the presence of strong negative absorption bands of bulk water, which interferes with the spectral analysis in these regions (see the discussion in Section 3.7). These bands arise from normalizing the sample spectrum with respect to the reflectance spectrum of the bare water subphase (Fig. 3.27).

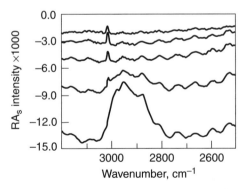

Figure 4.54. Ratios of single-beam s-polarized IRRAS spectrum of water surface covered by film against the same, but at different heights of surface. Heights are controlled by time t (in hours) of evaporation. From top to bottom: $-\log(R_{t=1}/R_{t=0})$; $-\log(R_{t=2}/R_{t=0})$; $-\log(R_{t=3}/R_{t=0})$; $-\log(R_{t=4}/R_{t=0})$; $-\log(R_{t=5}/R_{t=0})$. Reprinted, by permission, from Y.-S. Tung, T. Gao, M. J. Rosen, J. E. Valentini, and L. J. Fina, *Appl. Spectrosc.* **47** (1993), p. 1648, Fig. 10. Copyright © 1993 Society for Applied Spectroscopy.

To increase the SNR of the method, remove absorption from the atmosphere, and reduce the baseline distortions, in 1993 Desbat and co-workers [569, 585] introduced PM (Section 4.7) in IRRAS of the AW interface. PM-IRRAS requires the set-up depicted schematically in Fig. 4.52, in which the IR beam is directed from the PEM onto the water surface, and the reflected beam is redirected to the MCT detector with two flat mirrors. The most appropriate from the viewpoint of both surface and orientational sensitivity (Section 3.11.5) is the angle of incidence of $\sim 75°$ [569]. PM provides a good SNR in the spectra acquired within 5–10 min with a 4-cm^{-1} resolution.

4.9. DYNAMIC IR SPECTROSCOPY

Infrared spectroscopy can yield information not only about the chemical composition of and structural changes in ultrathin films and complex interactions taking place on surfaces but also about the kinetics and dynamics of these phenomena. Below we will briefly consider time-resolved spectroscopic methods that are amenable to ultrathin films.

4.9.1. Time Domain

In a time-domain experiment, a reaction is triggered by a perturbation of the sample such as irradiation, heating, or stepwise changing the potential, and the response (relaxation) is followed spectroscopically as a function of time. Depending on the apparatus and the measurement technique, a transient process can be characterized with a time resolution (TR) of up to 1 ps [586]. A dispersive spectrometer with a high-temperature ceramic IR light source, a photovoltaic MCT detector, and a specifically developed low-noise wide-band preamplifier can provide nanosecond TR and sensitivity in $\Delta R/R$ of 10^{-6} [587]. TRs up to 1 ps are attainable with a pump-probe-type of experiment, in which absorption changes at one or several wavenumbers are monitored using an IR picosecond-pulsed tunable laser, instead of dispersive optics with a conventional continuous IR source [587–589]. In this type of time-resolved experiment, excitation pulses create vibrational excess population resulting in absorption changes at specific wavenumbers. Recent progress in nonlinear pump-probe spectroscopy for studying the dynamics of adsorbed molecules has been reviewed by Ueba [590].

However, in the last decade, efforts have increasingly focused on adapting FTIR interferometry for IR spectroscopic TR measurements, largely due to the throughput (Jaquinot) and multiplex advantages of these systems, together with high spectral resolution (<0.1 cm^{-1}) [2a] and ease and speed of measurement. Time resolution FTIR spectroscopic measurements can be performed in either the continuous (rapid) scan or S^2 mode. Depending on the TR required and the FTIR instrument available, several strategies can be used for time-resolved spectroscopic measurements.

Continuous-scan interferometry is the conventional mode in which the mirror is moved with a constant velocity, modulating IR radiation at the Fourier frequencies [2a]. The major advantage of this type of interferometry is its capability to follow irreversible reactions, while the major disadvantage is relatively low TR. The scanning time t is connected with the mirror velocity v and the spectral resolution $\Delta \nu$ as [2a]

$$t = \frac{1}{2v\Delta\nu}. \tag{4.9}$$

In conventional rapid-scan FTIR spectrometers, the mirror velocities range from 0.16 to 3.16 cm $\cdot$ s^{-1} [591], providing TR on the order of 10^{-1} s at a spectral resolution on the order of 16 cm^{-1} or greater. These specifications are satisfactory, for example, for collecting IR spectra of an electrode–electrolyte interface simultaneously with cyclic voltammograms with rates of 1–10 mV $\cdot$ s^{-1}. In particular, with one sample scan at a resolution of 8 cm^{-1} for 0.6 s, Ataka et al. [471] achieved an adequate SNR in the ATR-SEIRA spectra of a chromophore (heptylviologen) on a silver electrode measured in situ using a Bio-Rad FTS-60A/896 instrument operated with the continuous mode. By IRRAS with a TR of 500 ms and 2 cm^{-1} resolution, Sushchikh et al. [592] obtained the isotopic-jump spectra of CO on Ir(111). The measurements were performed using a Perkin-Elmer 1725X FTIR spectrometer and the software permitting to obtain data at rate of two spectra per second. SNIFTIRS has been used for kinetic studies with a time resolution of several seconds, since the potential can be modulated with frequencies typically on the order of 10^{-3}–10^{-1} Hz [404, 593, 594]. TR up to 10^{-3} s at 4 cm^{-1} spectral resolution can be realized using a purpose-built ultra-rapid-scanning interferometer [595].

The stroboscopic method [596], developed by Mantz [597, 598] in 1976, coupled with rapid-scan interferometry has a TR of 10^{-3}–10^{-6} s. The N-times improvement in TR is achieved as follows. If it takes a certain time t to obtain a complete interferogram of length X, then it only requires a time t/N to record a segment X/N of the interferogram. Thus, the pump probe is synchronized with the analog-to-digital converter (A/D) of the FTIR spectrometer (the synchronous sampling), and N interferograms are collected for the process N times, triggered with different fixed time delays relative to the starting point of the interferometer mirror. This means that the ith interferogram has a delay of it/N. The TR spectra are obtained by slicing the interferograms into segments that are then numerically resorted to produce a set of isochronical interferograms, each representing a fixed time delay (see Refs. [599–601] for more detail). The stroboscopic method requires no special equipment other than a means of connecting the interferometer controller to the reaction initiator. Typical measuring times for the stroboscopic method are 10–15 h. Steeman [602] and Ekgasit et al. [603] suggested another technique to monitor dynamic changes of reversible phenomena without synchronization between the externally applied perturbation and spectral acquisition provided the sampling rate is carefully chosen with regards to the perturbation period. The experimentation time of this technique is on the order of 1–2 h, which is significantly shorter than that of the S^2 technique (see below). More recently, Masutani et al. [604, 605] described an asynchronous (uncorrelated) sampling system for studying repetitive phenomena with microsecond TR, each time-resolved spectrum requiring $\sim$6 min. The key of this technique is the use of a low-pass filter. The dynamic interferograms of a sample perturbed with a succession of pulses are sent to a gate circuit with a trigger pulse delayed from the excitation pulse by a predetermined period of time. The low-pass filter filters out high-frequency components of the signal to convert the output from the gate circuit to an analog signal, which becomes equivalent to the isochronical interferogram at the selected delay time. Further improvement in TR (up to 1 ps) can be achieved by using synchrotron radiation, which intrinsically contains a rapid succession of pulses [606, 607].

Since the advent of S^2 interferometry by Sakai and Murphy in 1978 [608] and its commercialization, first by Bruker (model IFS 88) in 1987 and then by other manufacturers [609], TR S^2-FTIR spectroscopic measurements have become relatively routine. The principal difference in operation between S^2 and continuous-scan FTIR interferometry is the manner in which the interferometer mirror is moved [610–612]. The S^2 interferometer mirror is moved successively to a set of equidistant data collection positions. A reproducible and reversible event is triggered at each stop of the moving mirror. The data are collected as a function of time relative to the start of the event. The procedure is repeated for each mirror position. After the moving mirror has completed a full series of steps (until a suitable interferometer path difference generates the desired spectral resolution), the interferograms collected corresponding to different time slices are sorted and Fourier transformed into time-resolved spectra. Ideally, with this measurement procedure, S^2 interferometry imposes no limit on TR. However, in practice, the TR depends on the signal strength, the detector sensitivity and response speed, the speed of the digitizing electronics of the spectrometer, and the performance of the system used to initiate the transient process [613]. The actual limit, determined by the detector response, is 10 ns. Thus, Hun and Spiro [614] were able to resolve photochemical reactions on a 40-ns time scale using the 1064-nm radiation from a 9-ns pulsed Nd:YAG laser as the pump in a S^2 FTIR difference spectrometer. Chen and Palmer [615] reported good SNR S^2 FTIR spectra collected at 30-ns intervals with a setup that included a modified Bruker IF88 S^2 FTIR spectrometer and a 355-nm 10-ns pulsed Nd:YAG laser.

To improve the SNR, several signals measured at the same mirror position are averaged. As a result, several thousandfold triggering of the reaction is required to achieve appropriate spectral resolution and SNR. For example, to study the generation of the monocation radical of heptylviologen ($HV^{\bullet+}$) upon a potential step from -0.2 to -0.55 V (Ag−AgCl) on a gold electrode with 100-μs-resolved ATR-SEIRA spectroscopy (Fig. 4.55), potential modulation cycles were repeated about 2000 times with spectral resolution of 8 cm^{-1} [473].

To eliminate the technical difficulties connected with the dead stop of the moving mirror and the transient mechanical vibrations that are associated with the above-mentioned kind of movement, an alternative mode can be used in which the fixed mirror oscillates with a rectangular wavefront, while the other mirror is moved at a constant velocity [616].

S^2 measurements have several advantages when compared to rapid-scan stroboscopic TR measurements [599]. First, because the detector samples

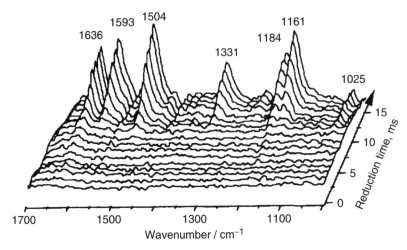

Figure 4.55. Series of time-resolved ATR-SEIRA spectra of the electrochemically generated HV$^{\bullet+}$ at gold electrode collected after potential step from -0.2 to -0.55 V (Ag–AgCl). Each spectrum was obtained with resolution of 8 cm^{-1} and acquisition time of 100 μs, but only spectra of every 1-ms interval are shown. Spectra were measured with Bio-Rad FTS-60A/896 operated with S^2 mode and equipped with linearized dc-coupled MCT detector (Bio-Rad). About 2000 times potential modulation cycles were repeated to obtain set of data. Reprinted, by permission, from M. Osawa, K. Yoshii, Y.-I. Hibino, T. Nakano, and I. Noda, *J. Electroanal. Chem.* **426**, 11–16 (1997), p. 13, Fig. 2. Copyright © 1997 Elsevier Science S.A.

according to an external clock, it is possible to take signals at irregular intervals. Second, the TR interferograms are collected in a linear fashion as opposed to the complicated data sorting of the stroboscopic method. Third, the experiment need not be controlled by the interferometer; instead, once the mirror has stopped, the experiment may be triggered independently. Finally, in the S^2 experiments the lack of reversibility can be compensated for by a perfect regeneration of the event by some scheme [617]; practically this is rarely feasible.

However, for the same data collection time, S^2 interferometry is more sensitive to multiplicative noise (i.e., noise proportional to the signal) than continuous-scan interferometry [591]. To eliminate the multiplicative and $1/f$ noise, phase modulation (at ∼400 Hz) of IR radiation in conjunction with LIA demodulation is used [591]. Since the LIA and some IR detectors need the IR signal to be modulated at a single carrier frequency, a mechanical chopper, phase modulation (when at each position the fixed mirror is dithered at a fixed frequency), or modulation of absorption of the sample is used to produce a carrier frequency. In this case, the TR measurement is referred to as a synchronous multiple-modulation experiment. Multiple modulation is unnecessary if the so-called dc coupled detector which does not require a varying signal is used.

Possible sources of errors and artifacts leading to misinterpretation of TR S^2 FTIR spectra are discussed by Rödig and Siebert [618]. These include failure of the moving mirror to hold its position for the predetermined period with an accuracy better than a few nanometers, as well as interference of the power line frequency. Moreover, if the reaction is triggered by laser pulses, heating of the sample by the laser excitation contributes to IR intensity changes at a fixed sampling position and can cause considerable fluctuations of the time-dependent IR signals. Chin and Lin [613] suggested that improved data acquisition in S^2 interferometry may be achieved with a controller with a transient digitizer. In the case of a slow process, the measuring time can become rather long, which is impractical because of instrumental instabilities such as drifting of dc-coupled electronic devices, misalignment of the Michelson interferometer, and fluctuations in the position of the movable interferometer mirror, which all introduce additional noise and error. Rödig and Siebert [619] circumvented this problem with the help of a wheel that allows 10 samples to be placed successively into the IR beam. Thus the average of the signals from 10 samples, instead of signals from only one sample, is taken, with a repetition rate larger than the relaxation period of the reaction.

4.9.2. Frequency Domain: Potential-Modulation Spectroscopy

In a frequency-domain experiment (modulation/demodulation), a periodic perturbation is applied, and changes in the absorption are recorded using LIA detection. Historically, the first potential-modulation technique, electrochemically modulated IR spectroscopy (EMIRS), was developed for increasing surface sensitivity in the SEC measurements with dispersive spectrometers, and the first in situ IR spectra of electrochemical interfaces were obtained with this technique [420, 424, 620]. The EMIRS experiment involves rapid modulation of the electrode potential during slow measurement of the spectrum, detection by a LIA of the associated change in intensity of the reflected radiation, demodulation of the signal obtained, and representation of it as a function of a wavelength. Discrimination against the background signal in a potential-modulation spectrum is based on the fact that band position for species either at or near the surface (roughly within the Debye length) shifts with changes in the applied potential because of the Stark and donation/backdonation effects (Section 3.7), while the background of the spectrum remains the same. The resultant spectrum, represented as the normalized change in reflectivity [Eq. (4.5)], contains bipolar bands that reflect the difference in the surface IR spectra at the two potentials [621]. Typically, the modulation frequency is ~ 10 Hz or less. The sensitivity of the technique is estimated as $\sim 10^{-6} \Delta R/R$ [361]. An approach to an additional reduction of noise and increase of accuracy in the determination of EMIRS band centers was described by Beden [622].

Applicability of the modulation methods for TR measurements is limited only by the modulation frequency. This can be understood using the analogy of a movie camera [599], where the modulation frequency separates between adjacent temporal events. Assume that a movie is being made of a ball being dropped. If the rate at which the film is passing the lens is slow enough, then the ball will appear as a blur on each frame. By increasing the rate at which the film passes the lens, the image of the falling ball on each frame becomes less and less blurred.

When the film is moving fast enough, the image of the ball becomes resolved to an image of the ball itself; the ball moves a small amount on each frame.

TR in potential modulation experiments is limited to 10^{-1}–10^{-2} s by the response of the thin-layer cell and, in the case of ATR, to 10^{-6} s by the response of the electrochemical equipment, especially the potentiostat (0.5–10 μs) [478] and the A/D converter [473] used. The first millisecond time-resolved IR spectra of an ultrathin film at the electrode–electrolyte interface were also measured with a dispersive spectrometer by repetitive measurements of IR absorption at fixed wavenumbers and conversion of the data into TR spectra [623]. The disadvantage of this technique was long collection times (on the order of a few tens of hours).

As mentioned in Section 4.7, to separate the Fourier and modulation frequencies, the modulation frequency should satisfy the sampling theorem. Thus, it is possible to detect high-frequency changes induced by ultrasonic perturbation of the sample with a continuous-scan FTIR spectrometer. However, the response of many interesting systems is in the audio-frequency range, which overlaps with the Fourier frequencies. For example, the double-layer recharging takes place on the 10^{-3}-s time scale. To overcome this, a special laboratory-built interferometer with a custom-built low scanning speed (~6 μm · s^{-1}) [624–627] and a S^2 interferometer [628–630] have both been used.[†] In the former case, at the modulation frequency of 1 kHz and the single-reflection sensitivity, $\Delta R/R$, of 10^{-6} an acquisition time of ~2 h for a typical spectrum with 7 cm^{-1} resolution (Fig. 3.47) was required [627].

The principal advantage of potential-modulation S^2 spectroscopy is that since the interferometer is moved stepwise and both ac and dc signals are collected at every optical retardation, the potential can be modulated with a single frequency over the entire spectral range, with no problem separating the modulation frequency from the Fourier frequencies (Section 4.9.1). A schematic representation of the instrumentation used by Ataka and co-workers [478] for the S^{2}FT-EMIRS measurements is depicted in Fig. 4.56. The signal is demodulated by an EG&G model 5210E LIA. The potential modulation is achieved by applying a sinusoidal wave from a waveform generator in the LIA. The LIA measures the in-phase and quadrature components of the IR signal, which represent the absorption changes that occur synchronously and 90° out of phase, respectively, with the applied electrochemical modulation (vide infra). These components and the dc component of the signal passing through a low-pass filter are digitized with an A/D converter in the spectrometer. A set of in-phase and quadrature spectra was constructed with a single S^2 spectrometer scan in about 1 h. The measurements were repeated, changing the potential-modulation frequency. The use of a dc-amplified MCT detector makes the measurements easier than those of Budevska and Griffiths [628], which required phase modulation. The quick response of the ATR SEC cell shown in Fig. 4.49 to the applied potential meant

---

[†] A recent development successively applied to a conventional continuous-scan type FTIR is based on the interferogram amplitude modulation [see T. Imajo, S. Inui, K. Tanaka, and T. Tanaka, *Chem. Phys. Lett.* **274**, 99 (1997).]. This technique uses a narrow-bandpass filter to select the modulation frequency in the range 1–100 kHz.

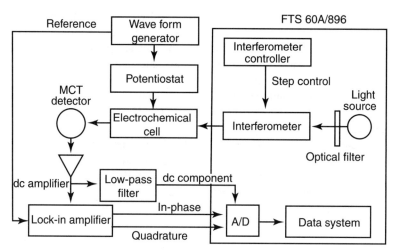

Figure 4.56. Block diagram for experimental setup for potential modulated FTIR measurements. Reprinted, by permission, from K. Ataka, Y. Hara, and M. Osawa, *J. Electroanal. Chem.* **473**, 34 (1999), p. 36, Fig. 1. Copyright © 1999 Elsevier Science S.A.

that the electrode potential could be modulated at a frequency between 40 and 100 kHz. An additional advantage is that the high-frequency modulation shortens the time constant of the LIA. As a result, sufficient spectra can be acquired over a shorter period. Potential-modulation spectroscopic data can be represented either as a collection of impedance dependences, each one corresponding to a given wavenumber [478], or as a collection of in-phase and an out-of-phase spectra (Figs. 7.45 and 7.46, respectively).

Thus, for electrochemical systems with time constants higher than 1 s, the kinetic and dynamic data can be obtained using a conventional continuous-scan FTIR. To accelerate the time scales over which measurements can be made, either EMIRS or S²-FTEMIRS can be used, the time constant of the SEC cell being the only upper limit for TR.

4.10. PREPARATION OF SUBSTRATES

Since surface preparation protocols are of critical importance in dictating the ultrathin deposition, in this section we will consider protocols for cleaning IREs and preparing metal substrates.

4.10.1. Cleaning of IREs

The cleaning protocol may include polishing, ultrasonic treatment in different solvents and solutions, plasma and UV/ozone etching, or sputtering a fresh surface from the IRE material, depending on the system under study. The surface cleanliness may be checked by comparing the ATR spectrum of the "cleaned" sample IRE with the spectrum of a "new" reference IRE.

A great deal of literature is devoted to different cleaning procedures for Si and their effect on the surface structure (see, e.g., Refs. [631–633] and Section 6.6.2). Germanium and Si IREs can be cleaned [634] by sonicating them at 58°C in ethanol, trichloroethylene, ethanol, and deionized water respectively for 15 min each. This treatment is repeated once and then followed by drying in an Ar jet. For hydrophilic Si IREs, two additional 15-min sonications, first in $NH_4OH-H_2O_2-H_2O$ (1:1:5 v/v) and then $HCl-H_2O_2-H_2O$ (1:1:5) are done between the two cleaning sequences, and the second sequence is shortened (only 5 min for each step). For a hydrophobic Si surface the IRE surface is prepared by oxidation ($H_2O_2 + H_2SO_4$ at 80°C for 10 min) and oxide dissolution in either HF or nitrogen-bubbled electronic grade NH_4F [368]. Rinsing in HF leads to a hydrogenated, atomically rough surface (characterised by SiH, SiH_2 and SiH_3 groups, corresponding to a roughness on a scale of a few angstroms), whereas a 10-min rinse in concentrated NH_4F (40%, pH 8) is known to lead to an atomically flat Si(111) surface, terminated by an ordered array of SiH bonds oriented perpendicular to the surface.

Another cleaning protocol for Ge consists of the following steps [635]: (1) degreasing by sonication in ethyl acetate bath (5 min); (2) rinsing with glass-distilled, deionized water; (3) treatment with 5% aqueous HF solution ($\sim$1 min); (4) rinsing with deionized water and blow drying with N_2 gas; (5) oxygen plasma treatment (5 min); and (6) cooling to room temperature under low oxygen pressure.

Glass and Si can also be cleaned as follows [636]. First the substrates are sonicated in freshly prepared Nichromix solution for 30 min. The Nichromix solution is prepared by dissolving the Nichromix crystals in sulfiric acid and stirring the solution until it becomes clear. Following the Nichromix treatment, the glass plate is rinsed and sonicated in water for 30 min. Cleaning of the Si IREs involves three steps to remove organic contaminants without stripping the native oxide layer. The first step involves sonication in acetone for 30 min. The second step is the same as for cleaning glass. In the third step each side of the substrate is treated in a plasma cleaner for 10 min. The plasma treatment is performed using argon at 100 W.

Brunner et al. [637] cleaned silica glass and Si by sonication in a $H_2SO_4-H_2O_2$ (4:1 v/v) solution. (*Caution*: $H_2SO_4-H_2O_2$ mixtures react violently with organic materials and should be handled with great care!) To obtain a hydrophilic, contaminant-free Si surface with a native oxide layer of 12–14 Å thick, the following has been suggested [638]: (1) sonication of the substrates in toluene, (2) rinsing with acetone and ethanol, and (3) blow drying in high-purity N_2. After this the substrate is exposed for 15 min to a UV-ozone atmosphere in a commercial cleaning chamber equipped with a low-pressure mercury quartz lamp (λ_{max} of 185 and 254 nm).

GaAs may be cleaned by successive ultrasonication in ethanol, acetone, and dichloromethane for 10 min each at 20°C [639]. The effect of different aqueous solutions used for etching and cleaning GaAs(100) surfaces was investigated by Gutjahr et al. [640]. Etching in HF solutions has been shown to result in clean and smooth surfaces. These are of a mixed hydrophilic/hydrophobic character due to the presence of Ga—H and OH bonds on the surface. Etching in H_2SO_4–H_2O_2 and NaOH–H_2O_2 mixtures gives smooth surfaces as well. It has been found [641] that Na_2S and $(NH_4)_2S$ remove the GaAs oxide and deposit a film that includes various AsS, GaS, SO, AsO, and GaO species.

Miller et al. [35, 37] cleaned the sapphire and fluorite IREs by immersion in 0.1 M KOH solution and heating (40°C) and sonicating for 30 min. Next, the IREs were placed in a 50% (volume) mixture of chloroform and ethanol and simultaneously heated and sonicated for 10 min. Finally, the IREs were subjected to a low-temperature oxygen plasma for 30 min on each long face of the IRE and immediately placed in the ATR accessory.

Since KRS-5 is slightly soluble in hot water, its outer layer can be uniformly dissolved off together with any adsorbed species. Therefore, polishing plus treatment with hot water in ultrasonic cleaner is superior to the usual grinding for this material.

4.10.2. Metal Electrode and SEIRA Surfaces

The use of well-defined metal surfaces allows correlation between the geometry of the surface and the electrochemical reaction occurring on it to be established and group frequencies of the adsorbed species to be determined [642]. Low-index crystallographic faces of metal electrodes such as the (100), (110), and (111) planes are widely used in SEC studies because of their low surface free energies, high symmetries, and relative stabilities [643]. Three procedures have been suggested to prepare the well-ordered surfaces. In one method [644], single crystals are grown by zone refining, oriented by the Laue back-reflection technique, and then cut along the desired crystal plane. A second procedure [645, 646] involves annealing a Pt electrode surface in a H_2–O_2 flame at about 1200°C, quenching by millipore water [404, 647, 648] (or annealing in a propane–air flame and cooling in a H_2–Ar stream [390]), and transferring into the SEC cell under a droplet of a pure water. To simplify this treatment before each measurement, a half-sphere single-crystal electrode has been used [404]. Atomically smooth (111) faces of Ag and Au can be grown by epitaxial vapor deposition onto a hot (400°C) mica substrate [649–651]. In all cases the samples are subjected to either thermal or electrochemical annealing. Thermal annealing in UHV serves a dual purpose: first, to bring contaminants from the bulk onto the surface (where they can be oxidatively desorbed or sputtered away) and, second, to achieve atomic smoothness. Electrochemical annealing is based on the fact that, at appropriate potentials, disordered interfacial atoms can be induced to diffuse to stable (ordered) sites [652–657]. The damaged surface layers of reactive metals (such as Ag in NaCN–H_2O_2 solution [658] or Cu in acidic media [659]) can be dissolved to expose the ordered underlayers by electropolishing. When a clean surface is ready, it should be recognized that prolonged exposure to an electrolyte or extensive anodic oxidation will result in an accumulation of impurities or a deterioration of its order, respectively. Recleaning and reordering can be accomplished in UHV by high-temperature oxygenation or Ag^+ ion sputtering followed by a thermal treatment to restore atomic smoothness or in situ by electropolishing [Cu(111)] or electrochemical annealing [Au(111) and Pd(111)]. A Hg electrode for in situ IRRAS can be prepared [660] by dipping a gold disk embedded in one end of a Teflon rod in a clean mercury pool. The excess amount of mercury on the surface is shaken off, and the surface is thoroughly washed with ultra-pure water. The remaining mercury is spread over the surface to show a metallic luster like a mirror.

Manufacturing of metal OTEs has been described in detail by Bauhofer [362]. Apart from the demands of the optical and electrical performance and the purity of the materials, the OTE–substrate system must meet the mechanical and chemical stability requirements (including the adhesion of the film to the substrate). To obtain a well-ordered, clean single-crystal metal surface (Pt, Ir Au,) the sample can be sputtered with Ar^+ ions, oxidized in 10^{-7} mbar of oxygen at 1000 K to remove surface carbon, and annealed at 1400 K [661, 662].

The following method of preparation of a SEIRA 20-nm-thick Au film with a preferentially oriented (111) surface was described by Osawa and co-workers [471]. The film was prepared on top of a Si hemicylinder by evaporation in a vacuum of 5×10^{-5} Pa from a tungsten basket by thermal heating. The deposition rate was kept at $0.01 \ nm \cdot s^{-1}$. After the deposition, the film was shortly flame annealed (about 10 s) to yield the (111) surface. The STM analysis revealed [663] that these films consist of metal particles with dimension of 50–80 nm. The surface of these films are cleaned in situ in the test solution by cycling the potential repeatedly between 0 V (SHE) (which is slightly positive of the hydrogen evolution) and 1.5 V (oxide formation region). The film thickness was measured with a quartz crystal microbalance. Kellner et al. [482] deposited Ag films by magnetron sputtering under ambient conditions and a target current of 120 mA and an argon pressure of 4×10^{-3} mbar.

However, PVD and sputtering are expensive high vacuum methods requiring precise and extensive control of all deposition parameters to ensure the reproducibility of SEIRA-active surfaces. To overcome this disadvantage, the research group of Mizaikoff [15] suggested to deposit SEIRA films electrochemically. To produce low-cost IRRAS-SEIRA surfaces (Fig. 4.9), a glass slide was first covered with a 0.1-μm Au layer using a sputtering process. This substrate provided the reference surface for the SEIRA measurements. Galvanization of the Au surface with silver was performed at a potential of 3.0 V and a current of 30 mA in a cyanidic silver bath with the composition 2.5 g $AgNO_3$, 3.5 g KCN and 4.0 g Na_2CO_3 in 100 mL H_2O. After 5 s of metal deposition the substrate was removed from the galvanic bath and the metal surface was rinsed with water and methanol. These process parameters stand for the deposition of a silver layer of roughly 20 nm thickness. Repeated deposition of metal or etching with a freshly-prepared mixture of $NH_3 : H_2O_2 = 1 : 1$ was used for further increasing the roughness of the metal surface. The SEIRA effect from Ag films deposited electrochemically from the different electrolytes under different conditions was studied by Maroun et al. [664] (Figs. 3.63 and 3.64 in Section 3.9.4). The highest SEIRA activity was achieved with Ag films deposited from a 10^{-2} $AgNO_3$ solution in a 0.1 M

EDTA buffer with pH 4. The nucleation potential was taken at -0.7 V (SCE) (duration 1 or 2 s) and the deposition potential was in the $-0.20-0$-V region. For silver deposition, chloride ions and an AgCl electrode should be excluded to avoid precipitation of AgCl.

4.10.3. BML Substrate

Preparation of a good BML structure is the main problem of the BML-IRRAS technique. The buffer layer can be grown by thermal oxidation [665] or deposited by CVD [666–668], ion beam sputtering [666], vacuum evaporation [669], and molecular beam epitaxy (MBE) [670, 671]. Vacuum evaporation can produce a $[c(2 \times 2)Si]$ structure, while MBE was found [670] to produce the Si surface with clear 2×1 patterned and ordered steps with regular terrace width, which can be used in the structural investigation of surface species on Si. To prepare a $SiO_2(50 \text{ nm})-W(70 \text{ nm})-n$-$Si(100)$ sample [672], W was deposited by CVD on a $Si(100)$ wafer and then SiO_2 was deposited on top of the W layer by electron cyclotron resonance (ECR)–plasma CVD. An alternative method is the generation of metal layers within a dielectric substrate by ion implantation [673].

REFERENCES

1. B. C. Smith, *Fundamentals of Fourier Transform Infrared Spectroscopy*, CRC Press, Boca Raton, FL, 1996.
2. (a) P. R. Griffiths and J. A. de Haseth, *Fourier Transform Infrared Spectrometry*, Wiley, New York, 1986; (b) J. Chalmers and P. Griffiths (Eds.), *Handbook of Vibrational Spectroscopy*, Vols. 1 and 2, Wiley, New York, 2002.
3. F. M. Mirabella (Ed.), *Modern Techniques in Applied Molecular Spectroscopy*, Wiley, New York, 1998.
4. L. N. Kuraeva, V. M. Zolotarev, and Y. V. Lisizin, *Kolloidnyj zhurn. (Russ.)* **1**, 138 (1979).
5. V. A. Kisel, *Zhurnal eksperimentalnoi i teoreticheskoi phyziki* **26**, 228 (1954).
6. V. A. Kisel, *Zhurnal eksperimentalnoi i teoreticheskoi phyziki* **29**, 658 (1955).
7. V. P. Tolstoy, *Precision Synthesis of Solids*, Vol. 2, Leningrad State University Press, Leningrad, 1987, p. 76 (in Russian).
8. V. P. Tolstoy, L. P. Bogdanova, and V. B. Aleskovski, *Doklady AN USSR* **291**, 913 (1986).
9. V. A. Skryshevsky and V. P. Tolstoy, *IR Spectroscopy of Semiconductor Structures*, Lybid', Kiev, 1991 (in Russian).
10. V. A. Skryshevsky, V. I. Strikha, V. P. Tolstoy, Iu. A. Averkin, and N. K. Kardamonov, *Russ. Inventor Certificate, No. 1539608, Bull. Izobr.* **4**, 100 (1990).
11. V. P. Tolstoy, *Methods of UV-Vis and IR Spectroscopy of Nanolayers*, St. Petersburg University Press, St. Petersburg, 1998.
12. J. Coates, in J. Workman, Jr. and A. W. Springsteen (Eds.), *Applied Spectroscopy. A Compact Reference for Practitioners*, Academic, San Diego, 1998, pp. 50–92.

13. W. D. Perkins, in P. B. Coleman (Ed.), *Practical Sampling Techniques for Infrared Analysis*, CRC Press, Boca Raton, FL, 1993, pp. 11–54.

14. Y. P. Song, M. C. Petty, and J. Yarwood, *Vib. Spectrosc.* **1**, 305 (1991).

15. H. D. Wanzenböck, B. Mizaikoff, N. Weissenbacher, and R. Kellner, *Fresenius J. Anal. Chem.* **362**, 15 (1998).

16. T. Urisu, Y. Yamada, and T. Hattori, *Interface Control of Electrical, Chemical, and Mechanical Properties, Symposium*, Material Research Society, Boston, MA, 1993, pp. 413–418.

17. (a) N. E. Lindsay, R. O. Carter III, P. J. Schmitz, L. P. Haack, R. E. Chase, J. E. deVries, and P. A. Willermet, *Spectrochim. Acta A* **49**, 2057 (1993); (b) F. Fondeur and J. L. Koenig, *Appl. Spectrosc.* **47**, 1 (1993).

18. Z. Xu and J. T. Yates, Jr., *J. Vac. Sci. Technol. A* **8**, 3666 (1990).

19. (a) C. R. Flach, J. W. Brauner, and R. Mendelsohn, *Appl. Spectrosc.* **47**, 982 (1993); (b) C. R. Flach, Z. Xu, X. H. Bi, J. W. Brauner, and R. Mendelsohn, *Appl. Spectrosc.* **55**, 1060 (2001).

20. J. Chatelet, H. H. Claasen, D. M. Gruen, I. Sheft, and R. B. Wright, *Appl. Spectrosc.* **29**, 185 (1975).

21. N. J. Harrick and M. Milosevic, *Appl. Spectrosc.* **44**, 519 (1990).

22. S. M. Stole and M. D. Porter, *Appl. Spectrosc.* **44**, 1418 (1990).

23. G. P. Williams, *Surf. Sci.* **368**, 1 (1996).

24. M. Thomas and T. J. Harris, in M. Bass (Ed.), *Handbook of Optics*, Vol. 2, McGraw-Hill, New York, 1995, pp. 33.3–33.101.

25. J. P. Coates, in F. M. Mirabella (Ed.), *Internal Reflection Spectroscopy*, Marcel Dekker, New York, 1993, pp. 53–96.

26. F. M. Mirabella, in F. M. Mirabella (Ed.), *Modern Techniques in Applied Molecular Spectroscopy*, Wiley, New York, 1998, Chapter 4.

27. N. J. Harrick, *Internal Reflection Spectroscopy*, Wiley, New York, 1967.

28. Y. Q. Lu, M. R. Yalamanchili, and J. D. Miller, *Appl. Spectrosc.* **52**, 851 (1998).

29. P. W. Faguy and N. S. Marinkovic, *Appl. Spectrosc.* **50**, 394 (1996).

30. S. V. Compton and D. A. C. Compton, in P. B. Coleman (Ed.), *Practical Sampling Techniques for Infrared Analysis*, CRC Press, Boca Raton, FL, 1993, pp. 55–92.

31. W. C. Cambell and J. M. Ottaway, *Talanta* **21**, 837 (1974).

32. J. P. Byrne, *Austr. J. Chem.* **32**, 249 (1979).

33. R. E. Sturgeon, C. L. Chakrabarti, and C. H. Langford, *Anal. Chem.* **48**, 1792 (1976).

34. A. M. Tifentaller and M. W. Urban, *Appl. Spectrosc.* **42**, 163 (1988).

35. W. M. Cross, J. J. Kellar, and J. D. Miller, in H. Schubert (Ed.), *Five Particles Processing in Flotation*, Preprints of XVII International Mineral Processing Congress, Dresden, Sept. 23–28, Bergakademi, Freiberg, Dresden, 1991, Part II, pp. 319–338.

36. C. A. Young and J. D. Miller, *Int. J. Mineral Process.* **58**, 331 (2000).

37. J. D. Miller and J. J. Kellar, in K. V. Sastry and M. C. Fuerstenau (Eds.), *Challenges in Mineral Processing*, Society of Mining Engineers of American Institute of Mining, Metallurgical and Petroleum Engineers, Littleton, CO, 1989, pp. 109–120.

38. D. A. Glockner and S. Ismat (Eds.), *Handbook of Thin Film Process Technology*, Institute of Physics Publishing, Bristol, 1995.

39. F. C. Matacotta and G. Ottaviani (Eds.), *Science and Technology of Thin Films*, World Scientific, Singapore, 1995.

40. R. P. Sperline, Y. Song, and H. Freiser, *Colloids Surf. A* **93**, 111 (1994).

41. J. Drelich, Y. Q. Lu, L. Y. Chen, J. D. Miller, and S. Guruswamy, *Appl. Surf. Sci.* **125**, 236 (1998).

42. M. J. Azzopardi and H. Arribart, *J. Adhesion* **46**, 103 (1994).

43. P. Eaton, P. Holmes, and J. Yarwood, *J. Appl. Polymer Sci.* **82**, 2016 (2001).

44. T. Bayer, K. J. Eichhorn, K. Grundke, and H. J. Jacobasch, *Macromolecular Chem. Phys.* **200**, 852 (1999).

45. S. A. Sukhishvili, A. Dhinojwala, and S. Granick, *Langmuir* **15**, 8474 (1999).

46. K. Laajalehto, P. Nowak, A. Pomianowski, and E. Suoninen, *Colloids Surf.* **57**, 319 (1991).

47. H. B. Mark and S. B. Pons, *Anal. Chem.* **38**, 119 (1966).

48. J. L. Ong, K. K. Chittur, and L. C. Lucas, *J. Biomed. Mater. Res.* **28**, 1337 (1994).

49. E. Ruckenstein, S. Gourisankar, and R. E. Baier, *J. Colloid Interface Sci.* **96**, 245 (1983).

50. H. Zeng, K. K. Chittur, and W. Lacefield, *Biomaterials* **20**, 377 (1999).

51. J. L. Ong, K. K. Chittur, and L. C. Lucas, *J. Biomed. Mater. Res.* **28**, 1337 (1994).

52. J. S. Jeon, R. P. Sperline, and S. Raghavan, *Appl. Spectrosc.* **46**, 1644 (1992).

53. T. A. Giroux and S. L. Cooper, *J. Colloid Interface Sci.* **146**, 179 (1991).

54. M. Muller, C. Werner, K. Grundke, K. J. Eichhorn, and H. J. Jacobasch, *Mikrochim. Acta Suppl.* **14**, 671 (1997).

55. R. Kellner, G. Gidaly, and F. Unger, *Adv. Biomater.* **3**, 423 (1982).

56. J. S. Jeon, R. P. Sperline, and S. Raghavan, *Appl. Spectrosc.* **46**, 1644 (1992).

57. A. M. Baty, P. A. Suci, B. J. Tyler, and G. G. Greesey, *J. Colloid Interface Sci.* **177**, 307 (1996).

58. P. A. Suci and G. G. Geesey, *Colloid Surf. B* **22**, 159 (2001).

59. R. P. Sperline, J. S. Jeon, and S. Raghavan, *Appl. Spectrosc.* **49**, 1178 (1995).

60. E. P. Boonekamp, J. J. Kelly, J. van der Ven, and A. H. M. Sondag, *J. Electroanal. Chem.* **344**, 187 (1993).

61. G. Muller and C. Riedel, *Fresenius J. Anal. Chem.* **365**, 43 (1999).

62. G. Muller and C. Riedel, *Appl. Spectrosc.* **53**, 1551 (1999).

63. Y. L. Cheng, N. Boden, R. J. Bushby, S. Clarkson, S. D. Evans, P. F. Knowles, A. Marsh, and R. E. Miles, *Langmuir* **14**, 839 (1998).

64. A. Hatta, Y. Suzuki, T. Wadayama, and W. Suetaka, *Appl. Surf. Sci.* **48/49**, 222 (1991).

65. B. W. Johnson, J. Bauhofer, K. Doblhofer, and B. Pettinger, *Electrochim. Acta* **37**, 2321 (1992).

66. N. Bay, P. Tien, and V. Truong, *Appl. Spectrosc.* **49**, 1279 (1995).

67. S. Badilescu, Y. Djaoued, P. V. Ashrit, F. E. Girouard, G. Bader, and V. V. Truong, *Solid State Ionics* **74**, 189 (1994).

68. B. W. Johnson, B. Pettinger, and K. Doblhofer, *Ber. Bunsen. Phys. Chem.* **97**, 412 (1993).

69. A. J. Pedraza, M. J. Godbale, and P. J. Bremer, *Appl. Spectrosc.* **47**, 161 (1993).

70. A. Couzis and E. Gulari, in P. M. Holland and D. N. Rubingh (Eds.), *Mixed Surfactant Systems*, ACS Symposium Series 501, American Chemical Society, Washington DC, 1992, pp. 354–365.

71. A. J. Gellman, B. M. Naacz, R. G. Schmidt, M. K. Chaudhuri, and T. M. Gentle, *J. Adhes. Sci. Technol.* **4**, 597 (1990).

72. I. V. Chernyshova and V. P. Tolstoy, *Appl. Spectrosc.* **49**, 665 (1995).

73. Y. Ishino and H. Ishida, *Appl. Spectrosc.* **42**, 1296 (1988).

74. T. Hasegawa, J. Umemura, and T. Takenaka, *Thin Solid Films* **210**, 583 (1992).

75. T. Hasegawa, J. Umemura, and T. Takenaka, *Appl. Spectrosc.* **47**, 379 (1993).

76. C. G. L. Khoo and H. Ishida, *Appl. Spectrosc.* **44**, 512 (1990).

77. A. Watanabe, *Appl. Spectrosc.* **47**, 156 (1993).

78. Y. Suzuki, S. Shimada, A. Hatta, and W. Suetaka, *Surf. Sci.* **219**, L595 (1989).

79. Y. Lu, J. Drelich, and J. D. Miller, *J. Colloid Interface Sci.* **202**, 462 (1998).

80. A. Mielczarski, E. Mielczarski, J. Zachwieja, and J. M. Cases, *Langmuir* **11**, 2787 (1995).

81. C. E. Gigola and G. L. Haller, *Appl. Spectrosc.* **44**, 159 (1990).

82. A. Cibly, J. Barr, and B. Crowford, *J. Phys. Chem.* **70**, 1520 (1966).

83. G. V. Saidov and M. E. Iudovich, *Opt. Spectrosc.* **36**, 1216 (1974).

84. T. Ohnishi and H. Tsubimura, *Chem. Phys. Lett.* **41**, L77 (1976).

85. R. P. Sperline and H. Freiser, *Langmuir* **6**, 344 (1989).

86. J. M. Perera, J. K. McCulloch, B. S. Murray, F. Grieser, and G. W. Stevens, *Langmuir* **8**, 366 (1992).

87. J. M. Perera, G. W. Stevens, and F. Grieser, *Colloids Surf. A* **95**, 185 (1995).

88. Z. Ding, R. G. Wellington, P.-F. Brevet, and H. H. Girault, *J. Electroanal. Chem.* **420**, 35 (1997).

89. D. J. Fermin, Z. Ding, and H. H. Girault, *J. Electroanal. Chem.* **447**, 125 (1998).

90. S. Tsukahara and H. Watarai, *Chem. Lett.* No. 1, 89 (1999).

91. C. W. Brown, in F. M. Mirabella (Ed.), *Modern Techniques in Applied Molecular Spectroscopy*, Wiley, New York, 1998, Chapter 10.

92. D. Marcuse, *Theory of Dielectric Optical Waveguides*, Academic, New York, 1991.

93. A. Messica, A. Greenstein, and A. Katzir, *Appl. Opt.* **35**, 2274 (1996).

94. Y. Xu, N. B. Jones, J. C. Fothergill, and C. D. Hanning, *J. Modern Opt.* **46**, 2007 (1999).

95. C. Malins, A. Doyle, B. D. MacCraith, F. Kvasnik, M. Landl, P. Simon, L. Kalvoda, R. Lukas, K. Pufler, and I. Babusik, *J. Environ. Monitor.* **1**, 417 (1999).

96. F. Regan, B. D. MacCraith, J. E. Walsh, K. ODwyer, J. G. Vos, and M. Meaney, *Vib. Spectrosc.* **14**, 239 (1997).

97. J. S. Namkung, M. Hoke, R. S. Rogowski, and S. Albin, *Appl. Spectrosc.* **49**, 1305 (1995).

98. M. S. Braiman and S. E. Plunkett, *Appl. Spectrosc.* **51**, 592 (1997).

99. F. Regan, M. Meaney, J. G. Vos, B. D. MacCraith, and J. E. Walsh, *Anal. Chim. Acta* **334**, 85 (1996).

100. J. D. Andrade, R. A. Van Wagenen, D. E. Gregonis, K. Newby, and N. J. Lin, *IEEE Trans. Electron. Devices* **32**, 1175 (1985).

101. T. Vo-Dinh, B. J. Tromberg, G. D. Griffin, K. R. Ambrose, M. J. Sepaniak, and E. M. Gardenhire, *Appl. Spectrosc.* **41**, 735 (1987).

102. N. I. Afanasveva, *Macromol. Symp.* **141**, 117 (1999).

103. M. Akusch, B. Mizaikoff, and R. Kellner, *Sens. Actuat. B* **38**, 83 (1997).

104. H. M. Heise, A. Bittner, L. Kupper, and L. N. Butvina, *J. Mol. Structure* **410–411**, 521 (1997).

105. I. Schnitzer, A. Katzir, U. Schiessl, W. J. Riedel, and M. Tacke, *Mater. Sci. Eng. B* **5**, 333 (1989).

106. P. Hahn, M. Tacke, M. Jakusch, B. Mizaikoff, and A. Katzir, *Appl. Spectrosc.* **55**, 39 (2001).

107. R. A. Potyrailo, V. P. Ruddy, and G. M. Hieftje, *Appl. Opt.* **35**, 4102 (1996).

108. K. Taga, R. Kellner, U. Kainz, and U. B. Sleytr, *Anal. Chem.* **66**, 35 (1994).

109. *FT-IR Fiber Optics for Remote Sampling in Mid-Infrared*, Nicolet Specifications, Nicolet Instrument, Madison, 1999.

110. T. J. Lenk, B. D. Ratner, K. K. Chittur, and R. M. Gendreau, *Langmuir* **7**, 1755 (1991).

111. W. M. Reichert, *CRC Rev. Biocompatibility* **5**, 173 (1989).

112. E. H. Korte and A. Otto, *Appl. Spectrosc.* **42**, 38 (1988).

113. G. L. Powell, M. Milosevic, J. Lucania, and N. J. Harrick, *Appl. Spectrosc.* **46**, 111 (1992).

114. G. L. Powell, R. L. Cox, T. E. Barber, and J. T. Neu, in J. T. Hoggatt (Ed.), *Technology Transfer in a Global Community. 28th International SAMPE Technical Conference*, Vol. 28, SAMPE, Covina, CA, 1996, pp. 1171–1182.

115. N. T. M. Johnson, *Off. Dig. Oil Colour Chem. Assoc.* **32**, 1067 (1960).

116. K. Mukai and N. Makino, *Jpn. J. Appl. Phys.* **31**, 4579 (1992).

117. N. Makino, K. Mukai, and Y. Kataoka, *Appl. Spectrosc.* **51**, 1460 (1997).

118. K. W. van Every and P. R. Griffiths, *Appl. Spectrosc.* **45**, 347 (1991).

119. R. N. Cochran and M. Deeba, U.S. Patent 4, 398, 041 (1983).

120. S. R. Culler, in P. B. Coleman (Ed.), *Practical Sampling Techniques for Infrared Analysis*, CRC Press, Boca Raton, FL, 1993, pp. 93–105.

121. W. Maddams, *Int. J. Vib. Spec.* **1**, Sect. 1 (1999).

122. Y. Lu, J. Drelich, and J. D. Miller, *J. Colloid Interface Sci.* **202**, 462 (1998).

123. N. D. Parkyns and D. I. Bradshaw, in H. A. Willis, J. H. van der Maas, and R. G. J. Miller (Eds.), *Laboratory Methods in Vibrational Spectroscopy*, 3rd ed., Wiley, Chichester, 1991, pp. 363–410.

124. A. Ahmed and E. Gallei, *Appl. Spectrosc.* **28**, 430 (1974).

125. D. Treibmann and G. Sadowski, *Jena Rev.* 226 (1980).

126. V. P. Tolstoy, *Pribory i tekhnika experimenta* **3**, 214 (1989).

127. V. P. Tolstoy and A. I. Somsikov, *USSR Inventor Certificate No. 1245898, Bull. Izobr.* **27**, 131 (1986).

128. S. V. Khabibova, V. P. Tolstoy, A. A. Malkov, and A. A. Malygin, *Zhurnal prikladnoi khimii* **62**, 302 (1989).

129. J. M. Palmer, in M. Bass (Ed.), *Handbook of Optics*, Vol. 2, McGraw-Hill, New York, 1995, pp. 25.1–25.25.

130. A. Springsteen, in J. Workman and A. Springsteen (Eds.), *Applied Spectroscopy. A Compact Reference for Practitioners*, Academic, San Diego, 1998, pp. 249–268.

131. W. Ho and C. Ma, *Infrared Phys. Technol.* **38**, 123 (1997).

132. W. Sumpner, *Proc. Phys. Soc. London* **12**, 10 (1892).

133. J. Kuhn, S. Korder, M. C. Arduinischuster, R. Caps, and J. Fricke, *Rev. Sci. Instrum.* **64**, 2523 (1993).

134. F. Boroumand, H. Vandenbergh, and J. E. Moser, *Anal. Chem.* **66**, 2260 (1994).

135. F. Boroumand, J. E. Moser, and H. Vandenbergh, *Appl. Spectrosc.* **46**, 1874 (1992).

136. A. Mandelis, F. Boroumand, and H. Vandenbergh, *Spectrochim. Acta A* **47**, 943 (1991).

137. D. J. J. Fraser and P. R. Griffiths, *Appl. Spectrosc.* **44**, 193 (1990).

138. V. P. Tolstoy, *Zhurnal prikladnoi spektroskopii* **49**, 162 (1988).

139. V. P. Tolstoy and S. P. Shcherbakov, *J. Appl. Spectrosc.* **57**, 577 (1992).

140. S. P. Shcherbakov, E. D. Kriveleva, V. P. Tolstoy, and A. I. Somsikov, *Pribory i tekhnika experimenta.* No. 1, 159 (1990).

141. W. Richter, S. M. Sarge, and F. Kammer, *Appl. Opt.* **33**, 1270 (1994).

142. L. M. Hanssen and S. Kaplan, *Anal. Chim. Acta* **380**, 289 (1999).

143. M. P. Fuller and P. R. Griffiths, *Anal. Chem.* **50**, 1906 (1978).

144. R. S. S. Murthy and D. E. Leyden, *Anal. Chem.* **58**, 1228 (1986).

145. K. Moradi, C. Depecker, and J. Corset, *Appl. Spectrosc.* **48**, 1491 (1994).

146. T. Hirschfeld, *Appl. Spectrosc.* **40**, 1082 (1986).

147. D. M. Hambree and H. R. Smyrl, *Appl. Spectrosc.* **43**, 267 (1989).

148. P. J. Brimmer, P. R. Griffiths, and N. J. Harrick, *Appl. Spectrosc.* **40**, 258 (1986).

149. P. J. Brimmer and P. R. Griffiths, *Appl. Spectrosc.* **41**, 791 (1987).

150. F. Boroumand, J. E. Moser, and H. Vandenbergh, *Appl. Spectrosc.* **46**, 1874 (1992).

151. B. J. Brimmer and P. R. Griffiths, *Appl. Spectrosc.* **42**, 242 (1988).

152. R. S. Messerschmidt, *Appl. Spectrosc.* **39**, 737 (1985).

153. P. W. Yang, H. H. Mantsch, and F. Baudais, *Appl. Spectrosc.* **40**, 974 (1986).

154. B. Li and R. D. Gonzalez, *Appl. Spectrosc.* **52**, 1488 (1998).

155. H. S. Shah, *Proc. SPIE–Int. Soc. Opt. Eng.* **1983**, 338 (1993).

156. S. A. Yeboah, S.-H. Wang, and P. R. Griffths, *Appl. Spectrosc.* **38**, 259 (1984)

157. B. C. Smith, *Fundamentals of Fourier Transform Infrared Spectroscopy*, CRC press, Boca Raton, FL, 1996, p. 111.

158. M. L. E. TeVrucht and P. R. Griffiths, *Appl. Spectrosc.* **43**, 1293 (1989).

159. M. L. E. TeVrucht and P. R. Griffiths, *Appl. Spectrosc.* **43**, 1492 (1989).

160. Y. Z. Liang, A. A. Christy, A. K. Nyhus, S. Hagen, J. S. Schanche, and O. M. Kvalheim, *Vib. Spectrosc.* **20**, 47 (1999).

161. M. Hrebicík, G. Budínová, T. Godarská, D. Vláil, S. B. Vogenseh, and K. Volka, *J. Mol. Structure* **410–411**, 527 (1997).

162. J. B. Reeves III, *Appl. Spectrosc.* **50**, 965 (1996).

163. F. Fondeur and B. S. Mitchell, *J. Am. Ceram. Soc.* **79**, 2469 (1996).

164. F. Fondeur and B. S. Mitchell, *Spectrochim. Acta A* **56**, 467 (2000).

165. P. Geladi, D. MacDougall, and H. Martens, *Appl. Spectrosc.* **39**, 491 (1985).

166. T. Isaksson and T. Nos, *Appl. Spectrosc.* **42**, 1273 (1988).

167. A. H. Aastveit and P. Marum, *Appl. Spectrosc.* **47**, 463 (1993).

168. K. W. van Every and P. R. Griffiths, *Appl. Spectrosc.* **45**, 347 (1991).

169. E. Pere, H. Cardy, O. Cairon, M. Simon, and S. Lacombe, *Vib. Spectrosc.* **25**, 163 (2001).

170. P. Persson, N. Nilsson, and S. Sjoberg, *J. Colloid Interface Sci.* **177**, 263 (1996).

171. T. J. Porro and Pattacini, *Appl. Spectrosc.* **44**, 1170 (1990).

172. S. J. Lee, S. W. Han, M. Yoon, and K. Kim, *Vib. Spectrosc.* **24**, 265 (2000).

173. I. V. Chernyshova, K. Hanumantha Rao, A. Vidyadhar, and A. V. Shchukarev, *Langmuir* **16**, 8071 (2000).

174. I. V. Chernyshova, K. Hanumantha Rao, A. Vidyadhar, and A. V. Shchukarev, *Langmuir* **17**, 775 (2001).

175. T. Burger, H. J. Ploss, J. Kuhn, S. Ebel, and J. Fricke, *Appl. Spectrosc.* **51**, 1323 (1997).

176. R. G. Messerschmidt, *Appl. Spectrosc.* **40**, 632 (1986).

177. D. Peak, R. G. Ford, and D. L. Sparks, *J. Colloid Interface Sci.* **218**, 289 (1999).

178. P. E. Poston, D. Rivera, R. Uibel, and J. M. Harris, *Appl. Spectrosc.* **52**, 1391 (1998).

179. S. J. Hug, *J. Colloid Interface Sci.* **188**, 415 (1997).

180. A. M. Tiefenthaler and M. W. Urban, *Appl. Spectrosc.* **42**, 163 (1988).

181. J. E. Katon, in F. M. Mirabella (Ed.), *Modern Techniques in Applied Molecular Spectroscopy*, Wiley, New York, 1998, pp. 267–290.

182. J. E. Katon and A. J. Sommer, *Appl. Spectrosc. Rev.* **25**, 173 (1989–90).

183. J. E. Katon, *Micron* **27**, 303 (1996).

184. D. A. Clark, in A. Townshend (Ed.), *Encyclopedia of Analytical Science*, Vol. 5, Academic, London, 1995, pp. 3174–3182.

185. P. J. Treado and M. D. Morris, *Appl. Spectrosc. Rev.* **29**, 1 (1994).

186. J. Chalmers and P. Griffiths (Eds.), *Handbook of Vibrational Spectroscopy*, Vol. 2, Wiley, New York, 2002, Chapter 8.

187. J. E. Katon and A. J. Sommer, *Anal. Chem.* **64**, 931 (1992).

188. F. J. Bergin, *Appl. Spectrosc.* **43**, 511 (1989).

189. J. S. Matthew and T. C. Richard, *Appl. Spectrosc.* **43**, 865 (1989).

190. B. Cook, *Internet J. Vib. Spectroscopy*, available on-line: ijvs.com/volume2/edition4/section1.htm.

191. N. Ferrer, *Mikrochim. Acta Suppl.* **14**, 329 (1997).

192. P. Grosse, B. Harbecke, H. Meyer, and R. Meyer, *Appl. Phys. A* **39**, 257 (1986).

193. Y. Wang, J. Shan, and R. J. Hamers, *J. Vac. Sci. Technol. B* **14**, 1038 (1996).

194. M. C. Grieve, *Sci. Justice* **35**, 179 (1995).

195. M. W. Tundol, E. G. Batrick, and A. Montaser, *Spectrochim. Acta. Part B* **46**, 1535E (1991).

196. M. W. Tundol, E. Batrick, and A. Montaser, *J. Forestic Sci.* **36**, 1027 (1991).

197. W. Grahlert, B. Leuport, and V. Hopfe, *Vib. Spectrosc.* **19**, 353 (1999).

198. A. Fong and G. M. Hieftje, *Appl. Spectrosc.* **48**, 394 (1994).

199. H. D. Mavrich, A. Fondeur, H. Ishida, J. L. Koenig, and H. D. Wagner, *J. Adhesion* **46**, 91 (1994).

200. F. Fondeur and J. L. Koenig, *Appl. Spectrosc.* **47**, 1 (1993).

201. F. Fondeur and J. L. Koenig, *J. Adhesion* **43**, 263 (1993).

202. A. Taboudoucht, *Diss. Abstr. Int.* **50**(9), 105 (March 1990).

203. B. George, F. Touyeras, Y. Grohens, and J. Vebrel, *Int. J. Adhesion Adhesives* **17**, 121 (1997).

204. J. Molenda, M. Gradkowski, and C. Kajdas, *Tribologia* **28**, 773 (1997).

205. N. E. Lindsay, R. O. Carter, P. J. Schmitz, L. P. Haack, R. E. Chase, J. E. Devries, and P. A. Willermet, *Spectrochim. Acta A* **49**, 2057 (1993).

206. C. J. Mastrangelo, L. Von Shell, and Y. Lee, *Appl. Spectrosc.* **44**, 1415 (1990).

207. L. S. Helmick, L. J. Gschwender, S. K. Sharma, C. E. Snyder, J. C. Liang, and G. W. Fultz, *Tribology Trans.* **40**, 393 (1997).

208. D. L. Wooten and D. W. Hughes, *Lubrication Eng.* **43**, 736 (1987).

209. E. M. Suzuki, *J. Forensic Sci.* **37**, 467 (1992).

210. E. Benedetti, L. Teodori, M. L. Trinca, P. Vergamini, F. Salvati, F. Mauro, and G. Spremola, *Appl. Spectrosc.* **44**, 1276 (1990).

211. M. C. Grieve, R. M. E. Griffin, and R. Malone, *Sci. Justice* **38**, 1988, 27.

212. J. Zieba-Palus, *Mikrochim. Acta Suppl.* **14**, 361 (1997).

213. P. L. Lang and L. J. Richwine, in P. B. Coleman (Ed.), *Practical Sampling Techniques for Infrared Analysis*, CRC Press, Boca Raton, FL, 1993, pp. 145–163.

214. V. A. Self and P. A. Sermon, *Rev. Sci. Instrum.* **67**, 2096 (1996).

215. F. L. Williams, G. A. Petersen, C. K. Carmiglia, and B. J. Pond, *J. Vac. Sci. Technol. A* **10**, 1472 (1992).

216. R. G. Messerschmidt and M. A. Harthcock (Eds.), *Infrared Microspectroscopy: Theory and Applications*, Marcel Dekker, New York, 1988.

217. H. J. Humecki (Ed.), *Practical Guide to Infrared Microspectroscopy*, 2nd ed., Marcel Dekker, New York, 1995.

218. P. Wilhelm, *Micron* **27**, 341 (1996).

219. Don Clark, *Internet J. Vib. Spectrosc.*, available on-line: ijvs.com/volume2/edition4/section1.htm.

220. A. J. Sommer and J. E. Katon, *Appl. Spectrosc.* **45**, 527 (1991).

221. J. C. Chervin, B. Canny, J. M. Besson, and Ph. Pruzan, *Rev. Sci. Instrum*, **66**, 2595 (1995),

222. J. A. Reffner and P. A. Martoglio, in H. J. Humecki (Ed.), *Practical Guide to Infrared Microspectroscopy*, 2nd ed., Marcel Dekker, New York, 1995, Chapter 2.

223. S. V. Pepper, *Appl. Spectrosc.* **49**, 354 (1995).

224. J. A. J. Jansen, J. H. Van Der Maas, and A. Posthuma De Boer, *Appl. Spectrosc.* **45**, 1149 (1991).

225. J. A. Reffner, G. Ressler, D. W. Schiering, and W. T. Wihlbord, *Mikrochim. Acta Suppl.* **14**, 333 (1997).

226. S. P. Bouffard, A. J. Sommer, J. E. Katon, and E. Godber, *Appl. Spectrosc.*, **48**, 1387 (1994).

227. J. A. Reffner, C. C. Alexay, and R. W. Hornlein, *Proc. SPIE Int. Soc. Opt. Eng.* **1575**, 301 (1992)

228. J. A. Reffner, W. T. Wihlbord, and S. W. Strang, *Am. Lab.* **April**, 46 (1991).

229. N. J. Harric, M. Milosevic, and S. L. Berets, *Appl. Spectrosc.* **45**, 944 (1991).

230. P. A. Martoglio, in J. L. Friel (Ed.), *Proceeding of 28th Annual Meeting of the Microbeam Society*, VCH, New York, 1994, p. 95.

231. Y. Esaki, K. Nakai, and T. Araga, *Toyota Chuo Kenkyusho R&D Rebyu* **30**, 57 (1995).

232. T. Nakano and S. Kawata, *Scanning* **16**, 368 (1994).

233. B. Yan, J. Fell, and G. Kumaravel, *J. Org. Chem.* **61**, 7467 (1996).

234. L. Lewis and A. J. Sommer, *Appl. Spectrosc.* **53**, 375 (1999).

235. L. Lewis and A. J. Sommer, *Appl. Spectrosc.* **54**, 324 (2000).

236. T. Gal and P. Toth, *Can. J. Appl. Spectrosc.* **37**, 55 (1992).

237. P. A. Martoglio, *Infrared Microspectroscopy in Forensic Science*, Nicolet Application Note AN-9694, Nicolet Instrument, Madison, 1996.

238. S. Inoue and R. Oldenbourg, in M. Bass (Ed.), *Handbook of Optics*, Vol. 1, McGraw-Hill, New York, 1995, pp. 17.1–17.52.

239. A. J. Sommer and J. E. Katon, *Appl. Spectrosc.* **45**, 527 (1991).

240. D. W. Schiering, T. J. Tague, J. A. Reffner, and S. H. Vogel, *Analysis* **28**, 46 (2000).

241. R. Bhargava, S.-Q. Wang, and J. L. Koenig, *Appl. Spectrosc.* **52**, 323 (1998).

242. A. Sassella, A. Borghesi, and B. Pivac, *Mikrochim. Acta Suppl.* **14**, 343 (1997).

243. A. Borghesi and A. Sassella, *Phys. Rev. B* **50**, 17756 (1994).

244. R. Kellner, B. Mizaikoff, K. Taga, W. Theiss, and P. Grosse, *Fresenius J. Anal. Chem.* **346**, 612 (1993).

245. B. Mizaikoff, K. Taga, and R. Kellner, *Appl. Spectrosc.* **47**, 1476 (1993).

246. O. Ruau, P. Landais, and J. L. Gardette, *Fuel* **76**, 645 (1997).

247. J. A. Reffner, *Appl. Spectrosc. Mater. Sci.* **89**, 1437 (1991).

248. G. L. Carr and G. P. Williams, *SPIE Conf. Proc.* **3153**, 51 (1997).

249. J. A. Reffner, P. A. Martoglio, and G. P. Williams, *Rev. Sci. Instrum.* **66**, 1298 (1995).

250. J. A. Reffner, G. L. Carr, and G. P. Williams, *Rev. Sci. Instrum.* **66**, 1490 (1995).

251. N. Guilhaumou, P. Dumas, G. L. Carr, and G. P. Williams, *Appl. Spectrosc.* **52**, 1029 (1998).

252. J. A. Reffner, G. L. Carr, and G. P. Williams, *Mikrochim. Acta Suppl.* **14**, 339 (1997).

253. D. L. Wizel and S. M. LeVine, *Spectroscopy* **8**, 40 (1993).

254. R. T. Carl, in P. E. Russel (Ed.), *Microbeam Analysis*, San Fransisco Press, San Fransisco, 1989, pp. 163–165.

255. R. T. Carl, *Application of Infrared Microimaging to Measurement of Photoresist Thickness*, Nicolet Application Note AN9034, Nicolet Instrument, Madison, 1990.

256. M. A. Harthcock and S. C. Atkin, *Appl. Spectrosc.* **42**, 449 (1988).

257. D. L. Witzel and S. M. LeVine, *Science* **285**, 1224 (1999).

258. J. Kressler, R. Schafer, and R. Thomann, *Appl. Spectrosc.* **52**, 1269 (1998).

259. C. M. Snively and J. L. Koenig, *Appl. Spectrosc.* **53**, 170 (1999).

260. W. J. Haap, T. B. Walk, and G. Jung, *Angewandte chemie* **37**, 3311 (1998).

261. J. J. Sahlin and N. A. Peppas, *J. Appl. Polym. Sci.* **63**, 103 (1997).

262. J. R. Schoonover, F. Weesner, G. J. Havrrilla, M. Sparrow, and P. Treado, *Appl. Spectrosc.* **52**, 1505 (1998).

263. D. L. Wetzel and J. A. Refner, *Cereal Foods World* **38**, 9 (1993).

264. D. L. Wetzel, in J. A. De Haseth (Ed.), *FT-IR Spectroscopy*, American Institute of Physics, Woodbury, NY, 1998, p. 354.

265. D. L. Wetzel, J. A. Reffner, and G. P. Williams, *Mikrochim. Acta Suppl.* **14**, 353 (1997).

266. G. L. Carr, *Vib. Spectrosc.* **19**, 53 (1999).

267. Special issue on FT-IR microspectroscopy, *Cell Mol. Biol.* **44** (February 1998).

268. N. Jamin, P. Dumas, J. Moncuit, W. H. Fridman, J. L. Teillaud, G. L. Carr, and G. P. Williams, *Proc. Natl. Acad. Sci. USA.* **95**, 4837 (1998).

269. T. Nishioka, T. Nakano, and N. Teramae, *Appl. Spectrosc.* **46**, 1904 (1992).

270. Y. Esaki, K. Nakai, and T. Araga, *Toyota Chuo Kenkyusho R & D Rebyu* **30**, 57 (1995).

271. E. N. Lewis, P. J. Treado, R. C. Reeder, G. M. Story, A. E. Dowrey, C. Marcott, and I. W. Levin, *Anal. Chem.* **67**, 3377 (1995).

272. J. M. Chalmers, N. J. Everall, K. Hewitson, M. A. Chesters, M. Pearson, A. Grady, and B. Ruzicka, *Analyst* **123**, 579 (1998).

273. N. A. Wright. *Internet J. Vib. Spectroscopy*, available on-line: www.ijvs.com/volume2/edition4/section1.htm.

274. C. M. Snively, S. Katzenberger, G. Oskarsdottir, and J. Lauterbach, *Opt. Lett.* **24**, 1841 (1999).

275. M. K. Bellamy, A. N. Mortensen, R. M. Hammaker, and W. G. Fateley, *Appl. Spectrosc.* **51**, 477 (1997).

276. J. A. Bailey, R. B. Dyer, D. K. Graff, and J. R. Schoonover, *Appl. Spectrosc.* **54**, 159 (2000).

277. A. Piednoir, C. Licoppe, and F. Creuzet, *Opt. Commun.* **129**, 414 (1996).

278. A. Piednoir and F. Creuzet, *Microne* **27**, 335 (1996).

279. D. V. Palanker, G. M. H. Knippels, T. I. Smith, and H. A. Schwettman, *Nucl. Instrum. Methods Phys. Res. B* **144**, 240 (1998).

280. D. V. Palanker, G. M. H. Knippels, T. I. Smith, and H. A. Schwettman, *Opt. Commun.* **148**, 215 (1998).

281. E. L. Buckland, P. J. Moyer, and M. A. Paesler, *J. Appl. Phys.* **73**, 1018 (1993).

282. D. Courjon, K. Sarrayeddine, and M. Spajer, *Opt. Commun.* **71**, 23 (1989).

283. N. A. Peppas and J. J. Sahlin, *Macromolecules* **29**, 7124 (1996).

284. N. A. Peppas and J. J. Sahlin, *J. Biomater. Sci. Polym. Ed.* **8**, 421 (1997).

285. A. Hammiche, H. M. Pollock, M. Reading, M. Claybourn, P. H. Turner, and K. Jewkes, *Appl. Spectrosc.* **53**, 810 (1999).

286. L. H. Little, *Infra-red Spectra of Adsorbed Species*, Academic, London, 1966.

287. A. V. Kiselev and V. I. Lygin, *Infra-red Spectra of Surface Compounds*, Wiley, New York, 1975.

288. J. Chalmers and P. Griffiths (Eds.), *Handbook of Vibrational Spectroscopy*, Vol. 2, Wiley, New York, 2002, Chapter 7.

289. (a) J. Ryczkowski, *Catalysis Today* **68**, 263 (2001); (b) J. Ryczkowski, *Internet J. Vib. Spectrosc.* [www.ijvs.com] **6**, 2, 1 (2002).

290. A. A. Tsyiganenko, *Pribory i Tehnika. Experimenta.* **1**, 255 (1990).

291. H. Miura and R. Gonzalez, *J. Phys. E Sci. Instrum.* **15**, 373 (1982).

292. W. Herzog and O. Hess, *Chem. Ing. Tech.* **55**, 213 (1983).

293. R. M. Friedman and H. C. Dannhardt, *Rev. Sci. Instrum.* **56**, 1589 (1985).

294. O. C. T. Connor, J. C. Q. Fletcher, and M. W. Rantenbach, *J. Phys. E Sci. Instrum.* **19**, 367 (1986).

295. M. Nagai, L. L. Lucietto, Y.-E. Li, and R. D. Gonzalez, *J. Catal.* **101**, 522 (1986).

296. H. Arakawa, T. Fukushima, and M. Ichikawa, *Appl. Spectrosc.* **40**, 884 (1986).

297. A. J. Dabrowski and J. Biechonsky, *React. Kinet. Catal. Lett.* **34**, 345 (1987).

298. R. A. Prokopowich, P. E. Silveston, F. L. Baudais, D. E. Jrish, and R. R. Hudgins, *Appl. Spectrosc.* **42**, 385 (1988).

299. F. P. Larkins and M. R. Nordin, *Appl. Spectrosc.* **42**, 906 (1988).

300. G. J. Suppes and M. A. McHugh, *Rev. Sci. Instrum.* **60**, 666 (1989).

301. T. Szilagyi, T. I. Koranyi, Z. Paul, and J. Tilgner, *Catal. Lett.* **2**, 287 (1989).

302. W. J. Phillips, J. H. Welch, and J. Brashear, *Rev. Sci. Instrum.* **63**, 2174 (1992).

303. T. J. Bruno, *Rev. Sci. Instrum.* **63**, 4459 (1992).

304. O. Bache and M. Ystenes, *Appl. Spectrosc.* **48**, 985 (1994).

305. V. A. Self and P. A. Sermon, *Rev. Sci. Instrum.* **67**, 2096 (1996).

306. M. Komiyama and Y. Obi, *Rev. Sci. Instrum.* **67**, 1590 (1996).

307. P. E. Field, R. J. Comb, and R. B. Knapp, *Appl. Spectrosc.* **50**, 1307 (1996).

308. K. Iinuma, N. Sasaki, and Y. Satoh, *Rev. Sci. Instrum.* **68**, 2305 (1997).

309. R. Mariscal, H. R. Reinhoudt, A. D. Van langeveld, and J. A. Moulijn, *Vib. Spectrosc.* **16**, 119 (1998).

310. S. M. Massick, and P. C. Ford, *Organometallics* **18**, 4362 (1999).

311. G. M. Underwood, T. M. Miller, and V. H. Grassian, *J. Phys. Chem. A* **103**, 6184 (1999).

312. J. A. Lercher, V. Veefkind, and K. Fajerwerg, *Vib. Spectrosc.* **19**, 107 (1999).

313. M. Komiyama and Y. Obi, *Rev. Sci. Instrum.* **67**, 1590 (1996).

314. S. S. C. Chang, M. A. Brundage, M. W. Balakos, and G. Srinivas, *Appl. Spectrosc.* **49**, 1151 (1995).

315. S. D. Worely, J. P. Wey, and W. C. Neely, in D. J. Dwyer and F. M. Hoffman (Eds.), *Surface Science of Catalysis*, American Chemical Society, Washington, DC, 1992, p. 251.

316. I. M. Hamadeh, D. King, and P. R. Griffiths, *J. Catal.* **88**, 264 (1984).

317. R. S. S. Murthy, J. P. Blitz, and D. E. Leyden. *Anal. Chem.* **58**, 3167 (1986)

318. J. J. Venter and M. A. Vannice, *Appl. Spectrosc.* **42**, 1096 (1988).

319. J. P. Blitz and S. M. Augustine, *Appl. Spectrosc.* **45**, 1746 (1991).

320. M. Grathwol, *Naturwiss. Rdsch.* **26**, 147 (1973).

321. V. Yu. Borovkov and V. K. Kasanski, *Doklady AN USSR* **261**, 1374 (1981).

322. T. Hirschfeld, *Appl. Spectrosc.* **40**, 1082 (1986).

323. J. J. Venter, and M. A. Vannice, *Appl. Spectrosc.* **42**, 1096 (1988).

324. S. Dai, J. P. Young, and G. Mamantov, *Appl. Spectrosc.* **45**, 1056 (1991).

325. Q. Fan, C. Pu, and E. S. Smotkin, *J. Electrochem. Soc.* **143**, L21 (1996).

326. W.-F. Lin, J.-T. Wang, and R. F. Savinell, *J. Electrochem. Soc.* **144**, 1917 (1997).

327. C. M. Aubuchon, B. S. Davison, A. M. Nishimura, and N. J. Tro, *J. Phys. Chem.* **98**, 240 (1994).

328. D. Petersson and C. Leygraf, *J. Electrochem. Soc.* **140**, 1256 (1993).

329. M. Niwano, *Surf. Sci.* **428**, 199 (1999).

330. N. P. Magtoto and H. H. Richardson, *Surf. Interface Anal.* **25**, 81 (1997).

331. K. Nishikawa, K. Ono, M. Tuda, T. Oomori, and K. Namba, *Jpn. J. Appl. Phys. Part 1* **34**, 3731 (1995).

332. C. H. F. Peden, *ACS Symp. Ser.* **482**, 143 (1992).

333. B. C. Wiegand, S. P. Lohokare, and R. G. Nizzo, *J. Phys. Chem.* **97**, 11553 (1993).

334. T. Aastrup and C. Leygraf, *J. Electrochem. Soc.* **144**, 2986 (1997).

335. M. Tamada, H. Omichi, and N. Okui, *Thin Solid Films*, **260**, 168 (1995).

336. M. Tamada, H. Koshikawa, and H. Omichi, *Thin Solid Films*, **292**, 113 (1997).

337. M. Tamada, H. Koshikawa, and H. Omichi, *Thin Solid Films*, **292**, 164 (1997).

338. G. A. Beitel, A. Laskov, H. Oosterbeek, and E. W. Kuipers, *J. Phys. Chem.* **100**, 12494 (1996).

339. M. D. Weisel, F. M. Hoffmann, and C. A. Mims, *J. Electron Spectrosc. Related Phenom.* **64-5**, 435 (1993).

340. F. M. Hoffmann and M. D. Weisel, *Surf. Sci.* **270**, 495 (1992).

341. F. M. Hoffmann and M. D. Weisel, *ACS Symp. Ser.* **482**, 202 (1992).

342. V. M. Bermudez, *Thin Solid Films* **347**, 195 (1999).

343. V. M. Bermudez, *J. Vac. Sci. A* **16**, 2572 (1998).

344. T. Ohtani, J. Kubota, J. N. Kondo, Ch. Hirose, and K. Domon, *Shinku* **40**, 717 (1997).

345. M. Niwano, *Surf. Sci.* **427–428**, 199 (1999).

346. D. W. Sting, Filed patent application GB 2228 083 A, Great Britain.

347. U. Wolf, R. Leiberich, and J. Seeba, *Catal. Today* **49**, 411 (1999).

348. Y. Miyoshi, Y. Yoshida, S. Miaziki, and M. Hirose, *J. Non-Cryst. Solids* **198–200**, 1029 (1996),

349. T. Scherzer and U. Decker, *Vib. Spectrosc.* **19**, 385 (1999).

350. N. M. B. Flichy, S. G. Kazarian, C. J. Lawrence, and B. J. Briscoe, *J. Phys. Chem. B* **106**, 754 (2002).

351. P. MacLaurin, N. C. Crabb, I. Wells, P. J. Worsfold, and D. Coombs, *Anal. Chem.* **68**, 1116 (1996).

352. R. Lin and R. L. White, *Anal. Chem.* **66**, 2976 (1994).

353. J. Chalmers and P. Griffiths (Eds.), *Handbook of Vibrational Spectroscopy*, Vol. 4, Wiley, New York, 2002, Chapter 4.

354. A. J. Bard and L. R. Faulkner, *Electrochemical Methods*, Wiley, New York, 1980.

355. A. Bewick and S. Pons, in R. J. H. Hester and R. E. Clark (Eds.), *Advances in Infrared and Raman Spectroscopy*, Wiley-Heyden, London, 1985, p. 1.

356. J. Foley, C. Korzeniewski, J. L. Daschbach, and S. Pons, in A. J. Bard (Ed.), *Electroanalytical Chemistry*, Vol. 14, Marcel Dekker, New York, 1986, p. 309.

357. S. Pons, J. K. Foley, J. Russell, and M. Severson, in J. O'M. Bockris, B. E. Conway, and R. E. White (Eds.), *Modern Aspects of Electrochemistry*, Vol. 17, Marcel Dekker, New York, 1986, p. 223.

358. C. Korzeniewski and S. Pons, *Progr. Anal. At. Spectrosc.* **10**, 1 (1987).

359. S. M. Stole, D. D. Popenoe, and M. D. Porter, in H. D. Abruna, *Electrochemical Interfaces: Modern Techniques for In-Situ Interface Characterization*, VCH, New York, 1991, pp. 339–410.

360. K. Ashley, *Talanta* **38**, 1209 (1991).

361. R. J. Nichols, in J. Lipkowski and P. N. Ross (Eds.), *Adsorption of Molecules at Metal Electrodes*, VCH, New York, 1992, pp. 347–389.

362. J. Bauhofer, in F. M. Mirabella (Ed.), *Internal Reflection Spectroscopy*, Marcel Dekker, New York, 1993, pp. 233–254.

363. T. Iwasita and F. C. Nart, in H. Gerischer and C. W. Tobias (Eds.), *Advances in Electrochemical Science and Engineering*, Vol. 4, VCH, Weinheim, 1995, pp. 123–216.

364. T. Iwasita and F. C. Nart, *Progr. Surf. Sci.* **55**, 271 (1997).

365. C. Korzeniewski, *Crit. Rev. Anal. Chem.* **27**, 81 (1997).

366. P. Christensen and A. Hamnett, *Electrochim. Acta* **45**, 2443 (2000).

367. (a) J. N. Chazalviel, B. H. Erne, F. Maroun, and F. Ozanam, *J. Electroanal. Chem.* **502**, 180 (2001); J. N. Chazalviel, B. H. Erne, F. Maroun, and F. Ozanam, *J. Electroanal. Chem.* **509**, 108 (2001).

368. J.-N. Chazalviel, S. Fellah, and F. Ozanam, *J. Electroanal. Chem.* **524–525**, 137 (2002).

369. V. N. Filimonov and A. N. Terenin, *Doklady AN USSR* **109**, 982 (1956).

370. V. N. Filimonov, *Optika i Spektroskopiya USSR* **1**, 490 (1956).

371. C. H. Rochester, *Progr. Colloid Polym. Sci.* **67**, 7 (1979).

372. A. D. Buckland, J. Graham, R. Rudham, and C. H. Rochester, *J. Chem. Soc. Faraday Trans. I* **77**, 2845 (1981).

373. N. J. Simmons and M. D. Porter, *Anal. Chem.* **69**, 2866 (1997).

374. M. Hasegawa and M. J. D. Low, *J. Colloid Interface Sci.* **26**, 95 (1968).

375. A. K. Mills and J. A. Hockey, *J. Chem. Soc. Faraday Trans. I* **77**, 1945 (1981).

376. M. Hasegawa and M. J. D. Low, *J. Colloid Interface Sci.* **30**, 378 (1969).

377. R. W. Murray, W. R. Heineman, and G. W. O'Dom, *Anal. Chem.* **39**, 1666 (1967).

378. F. Hart, H. Luyten, H. A. Nieuwenhuis, and G. C. Schoemaker, *Appl. Spectrosc.* **48**, 1522 (1994).

379. Y. J. Ma, J. M. Zheng, and S. M. Zhu, *Chinese J. Inorg. Chem.* **15**, 61 (1999).

380. P. B. Graham and D. J. Curran, *Anal. Chem.* **64**, 2688 (1992).

381. A. Chimura, J. Naka, and T. Kitagawa, *Denki Kagaku* **62**, 489 (1994).

382. Z. L. Li and X. Q. Lin, *J. Electroanal. Chem.* **386**, 83 (1995).

383. R. L. McCreely, in B. Rossiter and J. F. Hamilton (Eds.), *Physical Methods of Chemistry*, Vol. 2: *Electrochemical Methods*, Wiley, New York, 1986, pp. 591–661.

384. J. Wang, *Analytical Electrochemistry*, VCH, New York, 1994.

385. R. M. Blanchard, A. R. Noble-Luginbuhl, and R. G. Nuzzo, *Anal. Chem.* **72**, 1365 (2000).

386. J. Niu and S. Dong, *Electrochim. Acta* **40**, 823 (1995).

387. P. B. Graham and D. J. Curran, *Anal. Chem.* **64**, 2688 (1992).

388. W. N. Richmond, P. W. Faguy, R. S. Jackson, and S. C. Weibel, *Anal. Chem.* **68**, 621 (1996).

389. C. S. Kim and C. Korzeniewski, *Anal. Chem.* **69**, 2349 (1997).

390. F. Huerta, E. Morallon, C. Quijada, J. L. Vazquez, and A. Aldaz, *Electrochim. Acta* **44**, 943 (1998).

391. H.-Q. Li, S. G. Roscoe, and J. Lipkowski, *J. Electroanal. Chem.* **478**, 67 (1999).

392. J. A. Mielczarski, E. Mielczarski, J. Zachwieja, and J. M. Cases, *Langmuir* **11**, 2787 (1995).

393. N. S. Marinkovic, M. Hecht, J. S. Loring, and W. R. Fawcett, *Electrochim. Acta* **41**, 641 (1996).

394. W. R. Fawcett, A. A. Kloss, J. J. Calvente, and N. Marinkovic, *Electrochim. Acta* **44**, 881 (1998).

395. J. J. Calvente, N. S. Marinkovic, Z. Kovacova, and W. R. Fawcett, *J. Electroanal. Chem.* **421**, 49 (1997).

396. N. S. Marinkovic, J. S. Marinkovic, and R. R. Adzic, *J. Electroanal. Chem.* **467**, 291 (1999).

397. C. Korzeniewski and M. W. Severson, *Spectrochim. Acta* **51A**, 499 (1995).

398. C. M. Pharr and P. R. Griffiths, *Anal. Chem.* **69**, 4673 (1997).

399. B. Beden and C. Lamy, in R. J. Gale (Ed.), *Spectroelectrochemistry: Theory and Practice*, Plenum, New York, 1988, p. 189.

400. G. L. J. Trettenhahn, G. E. Nauer, and A. Neckel, *Ber. Bunsenges. Phys. Chem.* **97**, 422 (1993).

401. A. E. Russell, L. Rubasinghama, T. H. Ballinger, and P. L. Hagans, *J. Electroanal. Chem.* **422**, 197 (1997).

402. A. E. Russell, L. Rubasingham, P. L. Hagans, and T. H. Ballinger, *Electrochim. Acta* **41**, 637 (1996).

403. A. Zimmermann and L. Dunsch, *J. Mol. Structure* **410–411**, 165 (1997).

404. S.-G. Sun and Y. Lin, *Electrochim. Acta* **44**, 1153 (1998).

405. J. Klima, K. Kratochvilova, and J. Ludvik, *J. Electroanal. Chem.* **427**, 57 (1997).

406. S. G. Sun, S. J. Hong, S. P. Chen, G. Q. Lu, H. P. Dai, and X. Y. Xiao, *Sci. China B* **42**, 261 (1999).

407. M. E. Rosa-Montanez, H. De Jesus-Cardona, and C. R. Cabrera-Martinez, *Anal. Chem.* **70**, 1007 (1998).

408. Z. L. Li and X. Q. Lin, *J. Electroanal. Chem.* **386**, 83 (1995).

409. P. A. Brooksby, N. W. Duffy, A. J. McQuillan, B. H. Robinson, and J. Simpson, *J. Organomet. Chem.* **582**, 183 (1999).

410. Y. Honda, M.-B. Song, and M. Ito, *Chem. Phys. Lett.* **273**, 141 (1997).

411. H.-Q. Li, S. G. Roscoe, and J. Lipkowski, *J. Electroanal. Chem.* **478**, 67 (1999).

412. D. D. Popenoe, S. M. Stole, and M. D. Porter, *Appl. Spectrosc.* **46**, 79 (1992).

413. P. W. Faguy and N. S. Marinkovic, *Anal. Chem.* **67**, 2791 (1995).

414. P. W. Faguy and W. R. Fawcett, *Appl. Spectrosc.* **44**, 1309 (1990).

415. J. Fahrenfort, *Spectrochim. Acta* **17**, 698 (1961).

416. J. D. Roth and M. J. Weaver, *Anal. Chem.* **63**, 1603 (1991).

417. I. T. Bae, X. Xing, E. B. Yeager, and D. A. Scherson, *Anal. Chem.* **61**, 1164 (1989).

418. I. T. Bae, D. A. Scherson, and E. B. Yeager, *Anal. Chem.* **62**, 45 (1990).

419. T. Iwasita and F. C. Nart, *J. Electroanal. Chem.* **295**, 215 (1990).

420. A. Bewick, K. Kunimatsu, and B. S. Pons, *Electrochim. Acta* **25**, 465 (1980).

421. H. Seki, K. Kunimatsu, and W. G. Golden, *Appl. Spectrosc.* **39**, 437 (1985).

422. C. Kvarnstrom and A. Ivaska, *Synth. Met.* **62**, 125 (1994).

423. T. M. Vess and D. W. Wertz, *J. Electroanal. Chem.* **313**, 81 (1994).

424. S. Pons, T. Davidson, and A. Bewick, *J. Electroanal. Chem. Interfacial Electrochem.* **140**, 211 (1982).

425. S. Pons, T. Davidson, and A. Bewick, *J. Am. Chem. Soc.* **105**, 1802 (1982).

426. S. Pons, M. Datta, J. F. McAleer, and A. S. Hinman, *J. Electroanal. Chem. Interfacial Electrochem.* **160**, 369 (1984).

427. (a) J. K. Foley and S. Pons, *Anal. Chem.* **57**, 945A (1985); (b) J. K. Foley, S. Pons, and J. J. Smith, *Langmuir* **1**, 697 (1985).

428. (a) C. Korzeniewski and S. Pons, *J. Vac. Sci. Technol. B* **3**, 1421 (1985); (b) S. Bkhoo, J. K. Foley, C. Korzeniewski, and S. Pons, *J. Electroanal. Chem. Interfacial Electrochem.* **233**, 223 (1987).

429. B. R. Clark and D. H. Evans, *J. Electroanal. Chem. Interfacial Electrochem.* **69**, 181 (1976).

430. J. O'M. Bockris and B. Yang, *J. Electroanal. Chem.* **252**, 209 (1988).

431. R. J. Nichols and A. Bewick, *Electrochim. Acta* **33**, 1691 (1988).

432. J. D. Roth and M. J. Weaver, *Anal. Chem.* **63**, 1603 (1991).

433. V. Theile, C. H. Hamann, and R. Holze, *Z. Phys. Chem. Int. J. Res. Phys. Chem. Chem. Phys.* **190**, 241 (1995).

434. I. T. Bae, M. Sandifer, Y. W. Lee, D. A. Tryk, C. N. Sukenik, and D. A. Scherson, *Anal. Chem.* **67**, 4508 (1995).

435. D. S. Corrigan and M. J. Weaver, *J. Electroanal. Chem.* **239**, 55 (1988).

436. A. Szulborska and A. Baranski, *J. Electroanal. Chem.* **377**, 269 (1994).

437. A. S. Baranski and A. Moyana, *Langmuir* **12**, 3295 (1996).

438. D. K. Roe, J. K. Sass, D. S. Bethune, and A. C. Luntz, *J. Electroanal. Chem.* **216**, 293 (1987).

439. H.-Q. Zhang and X.-Q. Lin, *J. Electroanal. Chem.* **434**, 55 (1997).

440. J. Zhang, J. Lu, C. Cha, and Z. Feng, *J. Electroanal. Chem.* **265**, 329 (1989).

441. (a) J. Huang and C. Korzeniewski, *J. Electroanal. Chem.* **471**, 146 (1999); (b) D. Kardash, J. Huang, and C. Korzeniewski, *J. Electroanal. Chem.* **476**, 95 (1999).

442. W. F. Lin, P. A. Christensen, and A. Hammet, *J. Phys. Chem. B* **104**, 120002 (2000).

443. (a) A. Russell, L. Rubasingham, P. L. Hagans, and T. Ballinger, *Electrochim. Acta* **41**, 637 (1996); (b) A. Russell, L. Rubasingham, P. L. Hagans, and T. Ballinger, *J. Electroanal. Chem.* **422**, 197 (1997).

444. J. D. Mozo, M. Dominguez, E. Roldan, and J. M. R. Mellado, *Electroanalysis* **12**, 767 (2000).

445. (a) J. E. Pemberton and S. D. Garvey, in P. Vanysek (Ed.), *Modern Techniques in Electroanalysis*, Wiley, New York, 1996, Chapter 2; (b) W. N. Hansen, C. L. Wang, and T. W. Humphreys, *J. Electroanal. Chem.* **93**, 87 (1978); (c) D. M. Kolb and W. N. Hansen, *Surf. Sci.* **79**, 205 (1979).

446. (a) G. J. Hansen and W. N. Hansen, *Ber. Bunsenges.* **91**, 317 (1987); (b) A. Shen, J. E. Pemberton, *J. Electroanal. Chem.* **479**, 12 (1999).

447. C. A. Melendres, F. Hahn, and C. Lamy, *Electrochim. Acta* **46**, 3493 (2001).

448. A. H. Reed and E. G. Yeager, *Appl. Opt.* **7**, 451 (1965).

449. K. H. Beckmann, *Ber. Bunsenges.* **70**, 842 (1966).

450. A. H. Reed and E. Yeager, *Electrochim. Acta* **15**, 1345 (1970).

451. D. E. Tallant and D. H. Evans, *Anal. Chem.* **41**, 835 (1969).

452. B. D. Cahan, J. Horkans, and E. Yeager, *Symp. Faraday Soc.* **4**, 36 (1970).

453. T. Takamura, K. Takamura, W. Nippe, and E. Yeager, *J. Electrochem. Soc.* **117**, 626 (1970)

454. I. A. Vainshenker and E. D. Kriveleva, *Obogashchenie Rud [J. Mineral Process.]* **3**, 24 (1972).

455. A. Hatta, Y. Suzuki, T. Wadayama, and W. Suetaka, *Appl. Surf. Sci.* **48**, 222 (1991).

456. C. da Fonseca, F. Ozanam, and J. N. Chazalviel, *Surf. Sci.* **365**, 1 (1996).

457. U. Wolf, R. Leiberich, and J. Seeba, *Catal. Today* **49**, 411 (1999).

458. P. H. Axelsen, W. D. Braddock, H. L. Brockman, C. M. Jones, R. A. Dluhy, B. K. Kaufman, and F. J. Puga, *Appl. Spectrosc.* **49**, 526 (1995).

459. J. Schmitt and H.-C. Flemming, *Int. Biodeterioration Biodegradation* **41**, 1 (1998).

460. J. S. Jeon, R. P. Sperline, and S. Raghavan, *Appl. Spectrosc.* **46**, 1644 (1992).

461. A. Venkateswara Rao, J. N. Chazalviel, and F. Ozanam, *J. Appl. Phys.* **60**, 696 (1986).

462. A. V. Rao, F. Ozanam, and J. N. Chazalviel, *J. Electron Spectrosc. Related Phenom.* **54**, 1215 (1990).

463. D. R. Mattson, *Am. Lab.*, March 1989; reprint from Colora Messtechnik GmbH, D-7073 Lorch, Postfach 1240.

464. D. J. Neivandt, M. L. Gee, M. L. Hair, and C. P. Tripp, *J. Phys. Chem. B* **102**, 5107 (1998).

465. P. Bruesch, T. Stockmeier, F. Stucki, P. A. Buffat, J. K. N. Linder, *J. Appl. Phys.* **73**, 7701 (1993).

466. (a) T. Nguyen, E. Byrd, and C. Lin, *J. Adhes. Sci. Technol.* **5**, 697 (1991); (b) T. Nguyen, E. Byrd, and D. Bentz, *J. Coat. Technol.* **66**, 39 (1994); (c) T. Nguyen, E. Byrd, D. Bentz, and C. J. Lin, *Progr. Organic Coatings* **27**, 181 (1996); (d) I. Linossier, F. Gaillard, M. Romand, and T. Nguyen, *J. Adhesion* **70**, 221 (1999).

467. C. M. Balik and W. H. Simendinger, *Polymer* **39**, 4723 (1998).

468. (a) Y. Yi, K. Nerbonne, and J. Pellegrino, *Appl. Spectrosc.* **56**, 509 (2002); (b) C. M. Laot, E. Marand, and H. T. Oyama, *Polymer* **40**, 1095 (1999).

469. M. Osawa, K. Yoshii, K. Ataka, and T. Yotsuyanagi, *Langmuir* **10**, 640 (1994).

470. Y. Suzuki, K. Sagisaka, and A. Hatta, *Appl. Surf. Sci.* **84**, 1 (1995).

471. K. Ataka, T. Yootsuyanagi, and M. Osawa, *J. Phys. Chem.* **100**, 10664 (1996).

472. M. Osawa and K. Yoshii, *Appl. Spectrosc.* **51**, 512 (1997).

473. M. Osawa, K. Yoshii, Y. Hibino, T. Nakano, and I. Noda, *J. Electroanal. Chem.* **426**, 11 (1997).

474. M. Futamata, *Surf. Sci.* **427–428**. 179 (1999).

475. M. Futamata and D. Diesing, *Vib. Spectrosc.* **19**, 187 (1999).

476. F. Maroun, F. Ozanam, and J. N. Chazalviel, *Surf. Sci.* **427–428**, 184 (1999).

477. H. Noda, K. Ataka, L. Wan, and M. Osawa, *Surf. Sci.* **427–428**, 190 (1999).

478. K. Ataka, Y. Hara, and M. Osawa, *J. Electroanal. Chem.* **473**, 34 (1999).

479. L. Wan, M. Terashima, H. Noda, and M. Osawa, *J. Phys. Chem. B* **104**, 3563 (2000).

480. H. Noda, T. Minoha, L.-J. Wan, and M. Osawa, *J. Electroanal. Chem.* **481**, 62 (2000).

481. Y. Cheng, N. Boden, R. J. Bushby, S. Clarkson, S. D. Evans, P. F. Knowles, A. Marsh, and R. E. Miles, *Langmuir* **14**, 839 (1998).

482. H. D. Wanzenbock, B. Edl-Mizaikoff, G. Fdedbacher, M. Grasserbauer, R. Kellner, M. Arntzen, T. Luyven, W. Theiss, and P. Grosse, *Mikrochim. Acta Suppl.* **14**, 665 (1997).

483. T. Iwaoka, P. R. Griffiths, J. T. Kitasaka, and G. G. Geesey, *Appl. Spectrosc.* **40**, 1062 (1986).

484. A. J. Pedraza, M. J. Godbole, and P. J. Bremer, *Appl. Spectrosc.* **47**, 161 (1993).

485. M. C. Pham, F. Adami, P. C. Lacaze, J. P. Doucet, and J. E. Dubois, *J. Electroanal. Chem.* **201**, 413 (1986).

486. H. Neugebauer, G. Nauer, and N. Brinda-Konopik, *J. Electroanal. Chem.* **122**, 381 (1981).

487. D. B. Parry, J. M. Harris, and K. Ashley, *Langmuir* **6**, 209 (1990).

488. T. Iwaoka, P. R. Griffith, J. T. Kitasaka, and G. G. Geesey, *Appl. Spectrosc.* **40**, 1062 (1986).

489. A. Zimmermann and L. Dunsch, *J. Mol. Structure* **410–411**, 165 (1997).

490. Z. Ping, H. Neugebauer, and A. Neckel, *Electrochim. Acta* **41**, 767 (1996).

491. R. E. Baier, *J. Soc. Cosmet. Chem.* **29**, 283 (1978).

492. S. J. Hug and B. Sulzberger, *Langmuir* **10**, 3587 (1994).

493. C. T. Johnston and G. S. Premachandra, *Langmuir* **17**, 3712 (2001).

494. B. J. Ninness, D. W. Bousfield, and C. P. Tripp, *Appl. Spectrosc.* **55**, 655 (2001).

495. K. D. Dobson, P. A. Connor, and A. J. McQuillan, *Langmuir* **13**, 2614 (1997).

496. M. I. Tejedor-Tejedor and M. A. Anderson, *Langmuir* **2**, 203 (1986).

497. M. I. Tejedor-Tejedor, L. D. Tickanen, and M. A. Anderson, *Electrochim. Acta* **36**, 1891 (1991).

498. B. C. Barja, M. I. Tejedor-Tejedor, and M. A. Anderson, *Langmuir* **15**, 2316 (1999).

499. M. I. Tejedor-Tejedor, E. C. Yost, and M. A. Anderson, *Langmuir* **8**, 525 (1992).

500. P. A. Suci, J. D. Vrany, and M. W. Mittelman, *Biomaterials* **19**, 327 (1998).

501. J. O. Leppinen, C. I. Basilio, and R. H. Yoon, *Int. J. Min. Process.* **26**, 259 (1989).

502. A. Fikus, H. Dietz, and W. Plieth, *Z. Phys. Chem. Int. J. Res. Phys. Chem. Chem. Phys.* **193**, 185 (1996).

503. D. S. Corrigan and M. J. Weaver, *Langmuir* **4**, 599 (1988).

504. L.-W. H. Leung and M. J. Weaver, *J. Phys. Chem.* **92**, 4019 (1988).

505. D. S. Corrigan, L. W. H. Leung, and M. J. Weaver, *Anal. Chem.* **68**, 621 (1996).

506. K. Ashley and S. Pons, *Chem. Rev.* **88**, 673 (1988).

507. N. S. Marinkovic, J. J. Calvente, Z. Kovacova, and W. R. Fawcett, *J. Electrochem. Soc.* **143**, L171 (1996).

508. P. W. Faguy and N. S. Marinkovic, *Surf. Sci.* **339**, 329 (1995).

509. L. A. Nafie and M. Diem, *Appl. Spectrosc.* **33**, 130 (1979).

510. L. A. Nafie and D. W. Vidrine, in J. R. Ferraro and L. J. Basile (Eds.), *Fourier Transform Infrared Spectroscopy*, Vol. 3, Academic, New York, 1982, pp. 83–123.

511. M. Niemeyer, G. G. Hoffmann, and B. Schrader, *J. Mol. Struct.* **349**, 451 (1995).

512. J. Pritchard, *Surface and Defect Properties of Solids*, Vol. 1, Specialist Periodical Reports, Chemical Society, London, 1972.

513. M. A. Chesters, J. Pritchard, and M. L. Sims, in F. Ricca (Ed.), *Adsorption-Desorption Phenomena*, 1971 Conference at Florence, Academic, New York, 1972.

514. J. F. Blanke, Ph.D. Thesis, University of Minnesota, 1975.

515. R. W. Stobie, B. Rao, and M. I. Dignam, *Surf. Sci.* **56**, 334 (1976).

516. A. M. Bradshaw and F. M. Hoffmann, *Surf. Sci.* **72**, 513 (1978).

517. W. G. Golden, Ph.D. Thesis, University of Minnesota, 1978.

518. K. W. Hipps and G. A. Crosby, *J. Phys. Chem.* **83**, 555 (1979).

519. W. G. Golden, D. S. Dunn, and J. Overend, *J. Catal.* **71**, 395 (1981).

520. A. E. Dowrey and C. Marcott, *Appl. Spectrosc.* **36**, 414 (1982).

521. G. A. Beitel, A. Laskov, and E. W. Kuipers, *J. Phys. Chem.* **100**, 12494 (1996).

522. T. Ohtani, J. Kubota, J. N. Kondo, Ch. Hirose, and K. Domen, *Shinku* **40**, 717 (1997).

523. P. W. Faguy, W. N. Richmond, and J. H. Payer, *Appl. Spectrosc.* **52**, 567 (1998).

524. T. Wadayama, Y. Maiwa, H. Shibata, and A. Hatta, *Appl. Surf. Sci.* **100–101**, 575 (1996).

525. V. M. Bermudez, *J. Vac. Sci. Technol. A* **16**, 2572 (1998).

526. V. M. Bermudez, *Thin Solid Films* **347**, 195 (1999).

527. R. V. Duevel and R. M. Corn, *Anal. Chem.* **164**, 337 (1992).

528. R. V. Duevel, R. M. Corn, and M. D. Liu, *J. Phys. Chem.* **96**, 468 (1992).

529. D. G. Hanken and R. M. Corn, *Anal. Chem.* **67**, 3767 (1995).

530. M. R. Anderson, M. N. Evaniak, and M. Zhang, *Langmuir* **12**, 2327 (1996).

531. B. L. Frey and R. M. Corn, *Anal. Chem.* **68**, 3187 (1996).

532. M. Gatin and M. R. Anderson, *Vib. Spectrosc.* **5**, 255 (1993).

533. T. Buffeteau, B. Desbat, E. Pere, and J. M. Turlet, *Microchim. Acta* **14**(Suppl.), 627 (1997).

534. G. Steiner, O. Savchuk, H. Moller, D. Ferse, H.-J. Adler, and R. Salzer, *Macromol. Symp.* **164**, 158 (2001).

535. G. Steiner, H. Moller, O. Savchuk, D. Ferse, H.-J. Adler, and R. Salzer, *J. Mol. Structure* **563–564**, 273 (2001).

536. Y. Cheng, L. Murtomaki, and R. M. Corn, *J. Electroanal. Chem.* **483**, 88 (2000).

537. Y. Ren, Y. K.-i. Iimura, and T. Kato, *Chem. Phys. Lett.* **325**, 503 (2000).

538. Y. Ren, Md. M. Hossain, and T. Kato, *Chem. Phys. Lett.* **332**, 339 (2000).

539. Y. Ren, Y. K.-i. Iimura, and T. Kato, *J. Chem. Phys.* **114**, 1949 (2001).

540. Y. Ren, K.-i. Iimura, and T. Kato, *Langmuir* **17**, 2688 (2001).

541. Y. Ren, Y. K.-i. Iimura, and T. Kato, *J. Chem. Phys.* **114**, 6502 (2001).

542. C. Alonso, D. Blaudez, B. Desbat, F. Artzner, B. Berge, and A. Renault, *Chem. Phys. Lett.* **284**, 446 (1998).

543. A. Dicko, H. Bourque, and M. Pezolet, *Chem. Phys. Lipids* **96**, 125 (1998).

544. H. Bourque, I. Laurin, M. Pezolet, J. M. Klass, R. B. Lennox, and G. R. Brown, *Langmuir* **17**, 5842 (2001).

545. R. Johann, D. Vollhardt, and H. Mohwald, *Colloid Surf. A* **182**, 311 (2001).

546. E. Le Calvez, D. Blaudez, T. Buffeteau, and B. Desbat, *Langmuir* **17**, 670 (2001).

547. W.-P. Ulrich and H. Vogel, *Biophys. J.* **76**, 1639 (1999).

548. S. E. Taylor, B. Desbat, and G. Schwarz, *Biophys. Chem.* **87**, 63 (2000).

549. M. Grandbois, B. Desbat, and C. Salesse, *Langmuir* **15**, 6594 (1999).

550. M. Grandbois, B. Desbat, and C. Salesse, *Biophys. Chem.* **88**, 127 (2000).

551. M. Laguerre, J. Dufourcq, S. Castano, and B. Desbat, *Biochim. Biophys. Acta* **1416**, 176 (1999).

552. S. Castano, B. Desbat, and J. Dufourcq, *Biochim. Biophys. Acta* **1463**, 65 (2000).

553. I. Vergne and B. Desbat, *Biochim. Biophys. Acta* **1467**, 113 (2000).

554. E. Bellet-Amalric, D. Blaudez, B. Desbat, F. Graner, F. Gauthier, and A. Renault *Biochim. Biophys. Acta* **1467**, 131 (2000).

555. L. Dziri, B. Desbat, and R. M. Leblanc, *J. Am. Chem. Soc.* **121**, 9618 (1999).

556. I. Estrela-Lopis, G. Brezesinski, and H. Mohwald, *Biophys. J.* **80**, 749 (2001).

557. B. J. Barner, M. J. Green, E. I. Saez, and R. M. Corn, *Anal. Chem.* **63**, 55 (1991).

558. M. J. Green, B. J. Barner, and R. M. Corn, *Rev. Sci. Instrum.* **62**, 1426 (1991).

559. B. L. Frey, D. G. Hanken, and R. M. Corn, *Langmuir* **9**, 1815 (1993).

560. E. I. Saez and R. M. Corn, *Electrochim. Acta* **38**, 1619 (1993).

561. D. G. Hanken, R. R. Naujok, J. M. Gray, and P. M. Corn, *Anal. Chem.* **69**, 240 (1997).

562. W. N. Richmond, P. W. Faguy, and S. C. Weibel, *J. Electroanal. Chem.* **448**, 237 (1998).

563. K. Uosaki, *Electrochemistry* **67**, 1105 (1999).

564. J. C. Kemp, *J. Opt. Soc. Am.* **59**, 950 (1969).

565. B. Beccard, *Polarization Modulation FT-IR Spectroscopy*, Nicolet Application Note AN-9592 4/99, Nicolet Instrument, Madison, 1995.

566. P. L. Polavarapu and Z. Deng, *Appl. Spectrosc.* **50**, 91 (1996).

567. D. H. Martin and E. Puplett, *Infrared Phys.* **10**, 105 (1969).

568. T. Buffeteau, B. Desbat, and J. M. Turlet, *Appl. Spectrosc.* **45**, 380 (1991).

569. (a) D. Blaudez, T. Buffeteau, J. C. Cornut, B. Desbat, N. Escafre, M. Pezolet, and J. M. Turlet, *Appl. Spectrosc.* **47**, 869 (1993); (b) D. Blaudez, J. M. Turlet, J. Dufourcq, D. Bard, T. Buffeteau, and B. Desbat, *J. Chem. Soc. Faraday Trans.* **92**, 525 (1996).

570. P. W. Faguy and W. N. Richmond, *J. Electroanal. Chem.* **410**, 109 (1996).

571. R. A. Dluhy and D. G. Cornell, *J. Phys. Chem.* **89**, 3195 (1985).

572. (a) R. A. Dluhy, S. M. Stephens, S. Widayati, and A. D. Williams, *Spectrochim. Acta A* **51**, 1413 (1995); (b) R. Mendelsohn and C. R. Flach, in J. Chalmers and P. Griffiths (Eds.), *Handbook of Vibrational Spectroscopy*, Vol. 2, Wiley, New York, 2002, p. 1028.

573. R. Mendelsohn, J. W. Brauner, and A. Gericke, *Annu. Rev. Phys. Chem.* **46**, 305 (1995).

574. D. Blaudez, T. Buffeteau, B. Desbat, and J. M. Turlet, *Curr. Opinion Colloid Interface Sci.* **4**, 265 (1999).

575. J. E. Bertie, M. K. Akmed, and H. H. Eysel, *J. Phys. Chem.* **93**, 2210 (1989).

576. J. E. Bertie and Z. Lan, *Appl. Spectrosc.* **50**, 8 (1996).

577. T. Iwata, J. Koshoubu, C. Jin, and Y. Okubo, *Appl. Spectrosc.* **51**, 9 (1997).

578. E. D. Palik, *Handbook of Optical Constants of Solids* Academic, New York, 1998.

579. B. F. Sinnamon, R. A. Dluhy, and G. T. Barbes, *Colloids Surfaces A* **146**, 49 (1999).

580. H. Sakai and J. Umemura, *Langmuir* **13**, 512 (1997).

581. J. T. Buontempo and S. A. Rice, *Appl. Spectrosc.* **46**, 725 (1992).

582. A. Gericke, A. V. Michailov, and H. Huhnerfuss, *Vib. Spectrosc.* **4**, 335 (1993).

583. C. R. Flach, A. Gericke, and R. Mendelsohn, *J. Phys. Chem. B* **101**, 58 (1997).

584. A. Gericke and H. Huhnerfuss, *J. Phys. Chem.* **97**, 12899 (1993).

585. L. Dziri, B. Desbat, and R. M. Leblanc, *J. Am. Chem. Soc.* **121**, 9618 (1999).

586. R. A. Palmer, G. D. Smith, and P. Chen, *Vib. Spectrosc.* **19**, 131 (1999)

587. T. Yuzawa, C. Kato, M. W. George, and H. O. Hamaguchi, *Appl. Spectrosc.* **48**, 684 (1994).

588. P. O. Stoutland, W. H. Dyer, and W. H. Woodruff, *Science* **257**, 1913 (1992).

589. R. J. H. Clark and R. E. Hester (Eds.), *Time-Resolved Spectroscopy*, Wiley, New York, 1989.

590. H. Ueba, *Progr. Surf. Sci.* **55**, 115 (1997).

591. C. J. Manning and P. R. Griffiths, *Appl. Spectrosc.* **51**, 1092 (1997).

592. M. Sushchikh, J. Lauterbach, and W. H. Weinberg, *Surf. Sci.* **393**, 135 (1997).

593. S. I. Yanger and D. W. Vidrine, *Appl. Spectrosc.* **40**, 174 (1986).

594. J. Daschbach, D. Heisler, and B. S. Pons, *Appl. Spectrosc.* **40**, 489 (1986).

595. P. R. Griffiths, B. L. Hirsche, and C. J. Manning, *Vib. Spectrosc.* **19**, 165 (1999).

596. H. Toriumi, H. Sugisana, and H. Watanabe, *Jpn. J. Appl. Phys.* **27**, L935, 1988.

597. A. W. Mantz, *Appl. Spectrosc.* **30**, 459 (1976).

598. A. W. Mantz, *Appl. Opt.* **17**, 1347 (1978).

599. B. Lerner and M. Daun, *Time Resolved Spectroscopy: Stroboscopic and Step Scan Methods*, Nicolet FTIR Technical Note TN-9252, Nicolet Instrument, Madison, 1992.

600. J. A. Graham, W. M. Grim III, and W. G. Fateley, *J. Mol. Struct.* **113**, 311 (1984).

601. J. J. Sloan, *SPIE Proc.* **1145**, 10 (1989). ·

602. P. A. M. Steeman, *Appl. Spectrosc.* **51**, 1688 (1997).

603. S. Ekgasit, H. W. Siesler, and P. A. M. Steeman, *Appl. Spectrosc.* **53**, 1535 (1999).

604. K. Masutani, H. Sugisawa, A. Yokota, Y. Furukawa, and M. Tasumi, *Appl. Spectrosc.* **46**, 560 (1992).

605. K. Masutani, K. Numahata, K. Nishimura, S. Ochiai, Y. Nagasaki, N. Katayama, and Y. Ozaki, *Appl. Spectrosc.* **53**, 588 (1999).

606. R. A. Palmer, G. D. Smith, and P. Chen, *Vib. Spectrosc.* **19**, 131 (1999).

607. G. L. Carr, *Vib. Spectrosc.* **19**, 53 (1999).

608. H. Sakai and R. E. Murphy, *Appl. Opt.* **17**, 1347 (1978).

609. Bio-Rad Laboratories, Inc., Cambridge, MA; Bruker Instruments, Inc. Billerica, MA; Nicolet Instrument Corp., Madison, WI; Manning Applied Technology, Moscow, ID.

610. R. A. Palmer, J. L. Chao, R. M. Dittmar, V. G. Gregoriou, and S. E. Plunkett, *Appl. Spectrosc.* **47**, 1297 (1993).

611. G. V. Hartland, W. Xie, and H. L. Dai, *Rev. Sci. Instrum.* **63**, 3261 (1992).

612. T. J. Johnson, A. Simon, J. M. Weil, and G. W. Harris, *Appl. Spectrosc.* **47**, 1376 (1993).

613. T.-L. Chin and K.-C. Lin, *Appl. Spectrosc.* **53**, 22 (1999).

614. X. H. Hun and T. G. Spiro, *Laser Chem.* **19**, 141 (1999).

615. P. Chen and R. A. Palmer, *Appl. Spectrosc.* **51**, 589 (1997).

616. T. Nakano, T. Yokoyama, and H. Totiumi, *Appl. Spectrosc.* **47**, 1374 (1993).

617. S. Boulas, R. Rammelsberg, K. Gerwert, and H. Chorongiewski, *Vib. Spectrosc.* **19**, 143 (1999).

618. C. Rödig and F. Siebert, *Appl. Spectrosc.* **53**, 893 (1999).

619. C. Rödig and F. Siebert, *Vib. Spectrosc.* **19**, 271 (1999).

620. S. Pons, *Electroanal. Chem.* **150**, 495 (1983).

621. A. Bewick, K. Kunimatsu, B. S. Pons, and J. W. Russell, *J. Electroanal. Chem.* **160**, 47 (1984).

622. B. Beden, *J. Electroanal. Chem.* **345**, 1 (1993).

623. M. Nakamura, H. Ogasawara, J. Inukai, and M. Ito, *Surf. Sci.* **283**, 248 (1993).

624. F. Ozanam and J.-N. Chazalviel, *Rev. Sci. Instrum.* **59**, 242 (1988).

625. J.-N. Chazalviel, V. M. Dubin, K. C. Mandal, and F. Ozanam, *Appl. Spectrosc.* **47**, 1411 (1993).

626. C. da Fonteca, A. Djebri, and J.-N. Chazalviel, *Electrochim. Acta* **41**, 687 (1996).

627. F. Ozanam, C. da Fonseca, A. Venkateswara Rao, and J.-N. Chazalviel, *Appl. Spectrosc.* **51**, 519 (1997).

628. B. O. Budevska and P. R. Griffiths, *Anal. Chem.* **65**, 2963 (1993).

629. C. M. Pharr and P. R. Griffiths, *Anal. Chem.* **69**, 4665 (1997).

630. C. M. Pharr and P. R. Griffiths, *Anal. Chem.* **69**, 4673 (1997).

631. Y. J. Chabal, in F. M. Mirabella (Ed.), *Internal Reflection Spectroscopy*, Marcel Dekker, New York, 1993, pp. 191–232.

632. L. Zazzara and J. F. Evans, *J. Vac. Sci. Technol. A* **11**, 934 (1993).

633. K. Arima, K. Endo, T. Kataoka, Y. Oshikane, H. Inoue, and Y. Mori, *Surf. Sci.* **446**, 128 (2000).

634. D. J. Ahn and E. I. Franses, *J. Phys. Chem.* **96**, 9952 (1992).

635. P. Frantz and S. Granick, *Macromolecules* **27**, 2553 (1994).

636. V. Kumar, S. Krishnan, C. Steiner, C. Maldarelli, and A. Couzis, *J. Phys. Chem. B* **102**, 3152 (1998).

637. H. Brunner, U. Mayer, and H. Hoffmann, *Appl. Spectrosc.* **51**, 209 (1997).

638. H. Brunner, T. Vallant, U. Mayer, and H. Hoffmann, *Surf. Sci.* **368**, 279 (1996).

639. T. Hasegawa, J. Umemura, and T. Takenaka, *J. Phys. Chem.* **97**, 9009 (1993).

640. K. Gutjahr, M. Reiche, and U. Gosele, *Materials Science Forum* **196–201**, 1967 (1996).

641. J. Yota and V. A. Burrows, *J. Vac. Sci. Technol. A* **11**, 1083 (1993).

642. T. Iwasita, F. C. Nart, A. Rodes, E. Pastor, and M. Weber, *Electrochim. Acta* **40**, 53 (1995).

643. M. P. Soriaga and J. L. Stickney, in P. Vanysek (Ed.), *Modern Techniques in Electroanalysis*, Wiley, New York, 1996, pp. 1–58.

644. E. A. Wood, *Crystal Orientation Manual*, Colombia University Press, New York, 1963.

645. J. Clavilier, *J. Electroanal. Chem.* **107**, 205 (1980).

646. B. Kaischew and B. Mutaftschiiew, *Z. Phys. Chem.* **204**, 334 (1955).

647. S.-G. Sun, A.-C. Chen, T.-S. Huang, J.-B. Li, and Z.-W. Tian, *J. Electroanal. Chem.* **340**, 213 (1992).

648. N.-H. Li and S.-G. Sun, *J. Electroanal. Chem.* **448**, 5 (1998).

649. P. O. Nilsson and D. E. Eastman, *Phys. Scr.* **8**, 113 (1973).

650. M. S. Zei, Y. Nakai, G. Lehmpfuhl, and D. M. Kolb, *J. Electroanal. Chem.* **150**, 201 (1983).

651. C. E. Chidsey, D. N. Loiacono, T. Sleator, and S. Nakahara, *Surf. Sci.* **200**, 45 (1988).

652. D. M. Kolb, G. Lempfuhl, and M. S. Zei, *J. Electroanal. Chem.* **179**, 289 (1984).

653. D. J. Trevor, C. E. D. Chidsey, and D. N. Loicono, *Phys. Rev. Lett.* **62**, 929 (1989).

654. E. Holland-Moritz, J. Gordon II, K. Kanazawa, and R. Sonnnenfeld, *Langmuir* **7**, 1981 (1991).

655. J. F. Rodriguez, M. E. Bothwell, G. L. Cali, and M. P. Soriaga, *J. Am. Chem. Soc.* **112**, 7392 (1990).

656. G. L. Cali, G. M. Berry, M. E. Bothwell, and M. P. Soriaga, *J. Electroanal. Chem.* **197**, 523 (1991).

657. M. E. Bothwell, G. L. Cali, G. M. Berry, and M. P. Soriaga, *Surf. Sci.* **249**, L322 (1991).

658. A. Bewick and B. Thomas, *J. Electroanal. Chem.* **65**, 911 (1975).

659. J. L. Stickney, I. Villegas, and C. B. Ehlers, *J. Am. Chem. Soc.* **111**, 6473 (1989).

660. K. Arihara, F. Kitamura, T. Ohsaka, and K. Tokuda, *J. Electroanal. Chem.* **488**, 117 (2000).

661. J. Lauterbach, R. W. Boyle, M. Schick, W. J. Mitchell, B. Meng, and W. H. Weinberg, *Surf. Sci.* **350**, 32 (1996).

662. I. Villegas and M. J. Weaver, *J. Phys. Chem. B* **101**, 5842 (1997).

663. M. Terashima, H. Noda, and M. Osawa, *J. Phys. Chem. B* **104**, 3563 (2000).

664. F. Maroun, F. Ozanam, J.-N. Chazalviel, and W. Theiss, *Vib. Spectrosc.* **19**, 193 (1999).

665. O. Seiferth, K. Wolter, B. Dillmann, G. Klivenyi, H.-J. Freund, D. Scarano, and A. Zecchina, *Surf. Sci.* **421**, 176 (1999).

666. Y. Imaizumi, Y. Zhang, Y. Tsusaka, T. Urisu, and S. Sato, *J. Mol. Structure* **352/353**, 447 (1995).

667. V. M. Bermudez, *Thin Solid Films* **347**, 195 (1999).

668. M. Firon, C. Bonnelle, and A. Mayeux, *J. Vac. Sci. Technol. A* **14**, 2488 (1996).

669. M. McGonigal and V. M. Bermudez, *Surf. Sci.* **241**, 357 (1991).

670. Y. Kobayashi, K. Sumitomo, K. Prabhakaran, and T. Ogino, *J. Vac. Sci. Technol. A* **14**, 2263 (1996).

671. Y. Kobayashi, K. Sumitomo, and T. Ogino, *Surf. Sci.* **427–428**, 229 (1999).

672. K. Fukui, H. Miyauchi, and Y. Iwasawa, *Chem. Phys. Lett.* **274**, 133 (1997).

673. L. M. Stuck, J. Eng, B. E. Bent, Y. J. Chabal, G. P. Williams, A. E. White, S. Christman, E. E. Chaban, K. Raghavachari, G. W. Flynn, K. Radermacher, and S. Mantel, *Mater. Res. Soc. Symp. Proc.* **386**, 395 (1995).

5

INFRARED SPECTROSCOPY OF THIN LAYERS IN SILICON MICROELECTRONICS

5.1. THERMAL SiO$_2$ LAYERS

In the processing of integrated circuits, silicon dioxide (SiO$_2$) can be used as a mask during ion implantation or diffusion of impurity into silicon, for passivation, for protection of the device surface, as interlayers for multilevel metallization, or as the active insulating material — the gate oxide film in metal–oxide–semi-conductor (MOS) devices [1, 2]. At the present time, several methods have been developed for the formation of SiO$_2$ layers, including thermal and chemical oxidation, anodization in electrolyte solutions, and various chemical vapor deposition (CVD) techniques [2, 3].

When a low density of surface states at the Si–SiO$_2$ interface is desired, the thermal oxidation method is employed. Silicon dioxide layers to be used as gate insulators in Si devices are generally grown at temperatures of at least 700°C. The chemical reactions that occur during the thermal oxidation of silicon in oxygen or in water vapor are shown below:

$$\text{Si}_{\text{solid}} + \text{O}_2 \longrightarrow \text{SiO}_2, \qquad \text{Si}_{\text{solid}} + 2\text{H}_2\text{O} \longrightarrow \text{SiO}_2 + 2\text{H}_2. \qquad (5.1)$$

These reactions involve a redistribution of the valence electrons between Si and O. Because of the bulk expansion accompanying the Si oxidation, the external surface of the SiO$_2$ layer does not coincide with the initial silicon surface, and the Si–SiO$_2$ interface moves into the Si substrate. (The SiO$_2$ layer will increase in thickness by d at the expense of a silicon layer $0.44d$ thick [2].) The Si oxidation takes place at the Si–SiO$_2$ interface, with oxidant diffusing through the SiO$_2$ layer. An overview of the oxidation models is presented in Refs. [2, 4–6]. The kinetics of the oxidation process is described by the linear-parabolic law, according to which the oxide thickness d obeys the equation

$$d^2 + Ad = B(t + t_1),   (5.2)$$

where t_1 is the shift along the time axis, which is necessary to account for the initial oxide layer, and B and B/A are, respectively, the parabolic and linear constants of the oxidation rate.

A number of questions concerning the thermal oxidation still remain unanswered, including the problem of transport of charged and neutral oxidants, the formation of structural defects at the Si–SiO₂ interface, and the role of surface states [2, 5, 7, 8]. The Si oxidation process depends on several factors, such as the crystallographic structure at the surface, the type and level of doping in the substrate, the nature of defects, and the surface roughness [2].

Investigation of the oxide layers using He⁺ backscattering revealed that the bulk of the oxide layers obtained with different methods usually have a chemical composition close to SiO₂, but at the Si–SiO₂ interface, there is a layer depleted of oxygen. The X-ray structural analysis and the electron diffraction spectra of this interface show that the nearest-neighbor structure in the SiO₂ layer consists of SiO₄ tetrahedra, whereas the bulk of the SiO₂ layer has an amorphous structure, although the appearance of a microcrystalline structure with a microcrystallite size of ∼1 μm is possible depending on the SiO₂ formation process. The structural configurations of SiO₂ were considered, for example, in Ref. [9].

At normal incident radiation, the IR spectra of the layers of thermally formed SiO₂ have three characteristic absorption bands arising from the Si–O–Si groups [10–12]. The low-frequency band (around $v_3 = 450$ cm⁻¹) is caused by the rocking mode corresponding to the out-of-plane motion of the oxygen atom. The weakest absorption, of intermediate frequency ($v_2 = 800$ cm⁻¹), is connected with bending vibrations, in which the oxygen atom motion occurs in the Si–O–Si plane and along the Si–O–Si angle bisector. The antisymmetric stretching vibrations, in which the oxygen atom motion occurs in the Si–O–Si plane and parallel to a line joining the two silicon atoms, are responsible for the strongest absorption, near $v_1 = 1080$ cm⁻¹ (Fig. 5.1, [10]). The stretching mode band is asymmetric with a broadening toward low frequencies [13]; for thick SiO₂ a long high-frequency tail is observed [14], and the frequency corresponding to the maximum of this peak increases with increasing oxidation temperature (Fig. 5.1) [10].

The absorption spectra of the thermally formed amorphous SiO₂ layers are similar to crystalline quartz spectra but can be distinguished by larger full-width-at-half-maximum (FWHM) values [9, 13, 15]. This confirms that the nearest-neighbor structure, which affects IR absorption spectra, is the same for amorphous SiO₂ and crystalline quartz.

At oblique incident IR radiation the stretching mode absorption band is split into two bands with peaks around 1090 (v_{TO}) and 1250 cm⁻¹ (v_{LO}) (Sections 3.1 and 3.2 and Refs. [14, 16–24]). Figure 5.2 shows the IR absorption spectrum of a thermal SiO₂ layer at normal and 75° oblique incidence in p-polarized light.

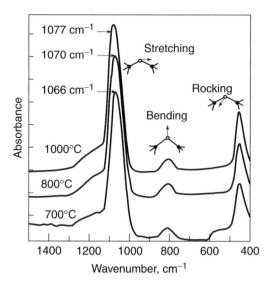

Figure 5.1. IR absorption spectra of SiO$_2$ layer grown by annealing Si in dry oxygen at temperatures 700, 800, and 1000°C. Reprinted, by permission, from G. Lucovsky, M. J. Manitini, J. K. Srivastava, and E. A. Irene, *J. Vac. Sci. Technol. B* **5**, 530 (1987). Copyright © 1987 American Vacuum Society.

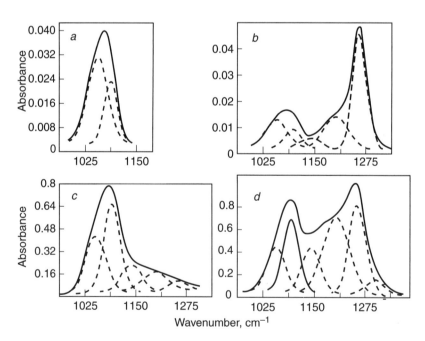

Figure 5.2. Result of deconvolution into Gaussian profiles of stretching absorption band of SiO$_2$ films 27 nm (a, b) and 600 nm (c, d) thick. Band was recorded at $\varphi = 0°$ (a, c) and in p-polarized light at $\varphi = 75°$ (b, d).

The frequencies of LO and TO modes of the asymmetric O stretch are given by the central and noncentral force approximations [25]:

$$\nu_{1TO} = \left(\frac{2(\alpha_c \sin^2 \theta/2 + \beta \cos^2 \theta/2)}{m} \right)^{1/2} \tag{5.3}$$

$$\nu_{1LO} = \left(\frac{2(\alpha_c \sin^2 \theta/2 + \beta \cos^2 \theta/2 + \gamma^{ss})}{m} \right)^{1/2} \tag{5.4}$$

where $\gamma^{ss} = Z^2 \rho / [\varepsilon_\infty \varepsilon_0 (2m + M)]$ is an electrostatic term, M and m are the atomic masses of silicon and oxygen, respectively, θ is the average angle of the Si−O−Si bridging bond, α_c and β are the central and noncentral force constants, ρ is the density of amorphous SiO_2, ε_∞ is the relative optical permittivity, ε_0 is the absolute permittivity of free space (1.1.1°), and Z is the transverse dynamic effective charge related to the stretching motion of the oxygen. The force constants do not depend on the bond angle but are connected with the frequencies ν_3 and ν_2 of rocking and bending modes, respectively [25]:

$$\nu_3 = \left(\frac{2\beta}{m} \right)^{1/2}, \tag{5.5}$$

$$\nu_2 = \left(\frac{4(\alpha_c + 2\beta)}{3m} \right)^{1/2}. \tag{5.6}$$

Taking into account that $\nu_3 = 450$ cm^{-1} and $\nu_2 = 800$ cm^{-1}, the ratio β/α_c can be calculated and is equal to 0.16. As can be seen from Eqs. (5.3) and (5.4), the shift of the bands at ν_1 can be ascribed to the change in the bond angle and the force constants. It was supposed in Ref. [26] that the red shift of the absorption band of thermal SiO_2 is connected with the decrease of the force constant α_c, which is due to the increase in the length of the Si−O bond. Further, it was supposed that the Si−O−Si bond angle remains constant and equal to $\theta = 144°$, as for stoichiometric SiO_2. However, it is currently assumed that the shift of the band at ν_1 is caused by the change in the bond angle, and the force constant values remain unchanged. According to Eqs. (5.3) and (5.4), the decrease in θ must lead to a shift of ν_1 toward lower frequencies; this has been observed experimentally [10, 14, 26–30]. The calculated shift of the LO and TO modes as a function of the Si−O−Si bond angle is presented in Fig. 5.3 [14]. The Si−O−Si bond is flexible, and the bond angle varies over a wide range ($120° < \theta < 180°$) depending on the microstructure within the film, whereas the Si−O bond length is practically constant (1.59–1.62 Å) [31]. Hence, the discrete set of Si−O−Si angles is responsible for the variety of peak positions observed.

Application of such ultrasensitive methods of FTIR spectroscopy as multiple transmission and IRRAS allows one to obtain the absorption spectra of ultrathin (several monolayers) layers of silicon dioxide [20, 22, 23, 32–37] which reveal a number of new absorption bands of the Si−O bond [9, 13, 37]. Questions

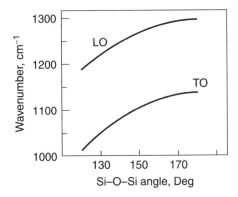

Figure 5.3. Frequencies of TO and LO modes calculated from Eqs. (5.3) and (5.4) versus Si—O—Si bond angle. Reprinted, by permission, from I. P. Lisovskii, V. G. Litovchenko, V. G. Lozinskii, and G. I. Steblovskii, *Thin Solid Films* **213**, 164 (1992). Copyright © 1992 Elsevier Science.

Table 5.1. Parameters of SiO_2 stretching bands

Peak Number	Position of Maximum (cm^{-1})	FWHM (cm^{-1})	Mode Type	Si—O—Si Angle (deg)
1	1056	58	TO	132
2	1091	46	TO	144
3	1147	52	TO	180
4	1200	68	LO	126
5	1252	44	LO	142
6	1300	57	LO	180

Source: Reprinted, with permission, from I. P. Lisovskii, V. G. Litovchenko, V. G. Lozinskii, and G. I. Steblovskii, *Thin Solid Films* **213**, 164 (1992). Copyright © 1992 Elsevier Science.

concerning the deconvolution of the asymmetric stretch band into Gaussian components are discussed in Refs. [13, 14, 33, 34]. The degree of asymmetry was found to be independent of the layer thickness d for relatively thick ($d > 10$ nm) thermal layers of SiO_2 on Si. However, for thinner layers, the location of the maximum, the FWHM, and the degree of asymmetry were all found to depend strongly on d.

It was established in Refs. [13, 14] that at the normal propagation of light, the absorption bands of thin oxide layers ($d < 40$ nm) may be described with a high degree of accuracy by the sum of two Gaussian profiles having maxima at 1056 and 1091 cm^{-1}. The presence of two elementary bands implies that at least two types of amorphous SiO_2 structures exist simultaneously. In the case of thicker oxide films, a high-frequency broadening of the absorption band appears, and six Gaussian components are necessary for its description. In addition to the two components mentioned above, four more must be included with maxima at 1147, 1200, 1252, and 1300 cm^{-1}. Table 5.1 lists averaged parameters for all six elementary peaks [14]. Figure 5.2 gives examples of the deconvolution into Gaussian profiles of absorption bands in the transmission spectra for thick and thin SiO_2 films measured at normal and oblique incidence of 75°.

Although the total area of the peaks increases linearly with increasing film thickness d, a change in the shape of the spectrum is caused by a change in the relative contributions of each Gaussian component to the absorption band. The behavior of the bands at 1056 and 1091 cm^{-1} is not trivial; their relative contributions to the absorption depends on d. It is seen that the band at 1056 cm^{-1} makes the main contribution to the absorption in the case of thin films, whereas the band at 1091 cm^{-1} dominates for thick films. This may cause the observed shift in the absorption maximum toward high frequencies as the oxide thickness increases. Furthermore, this indicates that the two types of oscillators associated with the bands at 1056 and 1091 cm^{-1} are distributed nonuniformly over the thickness. The first type is determined by the structure of the oxide layer adjacent to the $Si-SiO_2$ interface while the second one is determined mainly by the structure of the oxide bulk [38].

According to Eqs. (5.3) and (5.4), the frequencies of the TO and LO modes lie in the ranges of 1010–1140 and 1190–1300 cm^{-1}, respectively. Therefore, the IR absorption spectra in the stretching mode range can display at least three types of pairs of the LO–TO absorption bands corresponding to three types of oxide structures. These structures differ from each other by the bridge oxygen bond angles and yield six vibration bands as a result of LO–TO splitting (Table 5.1). Authors of Ref. [14] consider the sixfold ring configuration (quartzlike structure) that is characterized by the 1091- and 1252-cm^{-1} absorption bands and the $Si-O-Si$ bond angle $\theta = 143°$; planar threefold rings or packed fourfold rings (coesitelike structure) that reveal the 1056- and 1200-cm^{-1} absorption bands and $\theta = 129°$; and β-crystobalite and fragments of the $Si-O-Si$ chains, which probably exist together with the rings and manifest the 1147- and 1300-cm^{-1} absorption bands and $\theta = 180°$ [39]. The increasing thickness of the SiO_2 film causes the relative contributions made by each component of the ν_{LO} bands to change, which reflects the corresponding change in the relationships between the different types of the above-mentioned oxide structures.

5.2. LOW-TEMPERATURE SiO₂ LAYERS

Low-temperature dielectric film deposition has been actively studied in recent years in an attempt to reduce amount of energy consumed in the manufacture of circuits. Advances in very large scale and ultra large scale integrated (VLSI and ULSI) technology in the submicrometer range generate an increasing need for an alternative to high-temperature thermal SiO_2 production to reduce the dopant impurity diffusion [40]. Among low-temperature deposition methods the most notable are the various CVD techniques. Typical reactions used to form CVD oxides are [2, 3] the reaction of silane and oxygen ($SiH_4 + O_2 \rightarrow SiO_2 + 2H_2$) at $T = 400–450°C$, the decomposition of tetraethoxysilane (TEOS) [$Si(OC_2H_5)_4 \rightarrow SiO_2 + secondary products$] at $T = 650–750°C$, the reaction of dichlorosilane with nitrous oxide ($SiCl_2H_2 + 2N_2O \rightarrow SiO_2 + 2N_2 + 2HCl$) at $T = 850\text{--}900°C$, the reaction of silane and nitrous oxide ($SiH_4 + 4N_2O \rightarrow SiO_2 + 4H_2 + 2H_2O$) at $T = 200\text{--}350°C$, and so on.

The last reaction is particularly convenient for the passivation of device surfaces because of the low temperature required. Low-temperature layers can also be formed by the vacuum evaporation of SiO_2 powder or by ion-plasma sputtering of Si in an argon–oxygen mixture. The anodic oxidation of silicon in an ethylene glycol solution of KNO_3 and various chemical oxidation reactions of Si (e.g., in H_2O_2) can also be used.

At normally incident radiation, the IR absorption spectra of the low-temperature oxide layers exhibit the same three bands of stretching, bending, and rocking vibrations as the thermal layers, at 1080, 800, and 450 cm^{-1}, respectively. However, features of these bands are strongly dependent on the preparation conditions of the layer [26–29, 41–48].

It has been revealed by radioactivation analysis and XPS that the low-temperature layers have a chemical composition SiO_x in which $0 < x < 2$ [26–28, 48]. The density of the oxide (and x) can be increased by thermal annealing of the layers. Increasing the temperature from 450 to 1000–1100°C will increase x up to 2 and is accompanied by a decrease in the refractive index n of the layer from 4 ($x = 0.4$) to 2.5 ($x = 1$) and 1.43 ($x = 2$), by an increase in the band gap E_g from 1.1 eV ($x = 0$) to 8.2 eV ($x = 2$), and by a shift of the absorption band maximum from 980 to 1082 cm^{-1} [27, 28, 48]. The spectra of annealed layers for which $x = 2$ are similar to those of the oxide that has been thermally grown at 1000°C. The frequency corresponding to the TO absorption peak obeys the linear relationship [27, 49–51]

$$\nu(SiO_x) = 900 + 90x \quad (cm^{-1}). \tag{5.7}$$

Figure 5.4 shows the Si–O–Si stretching frequency as a function of the oxygen content x in the SiO_x film [49]. Note that by using the calibration curve shown in Fig. 5.4, one can determine the oxygen content in a SiO_x layer of a given thickness from the frequency of the band maximum, assuming the layer is homogeneous [26–28, 45].

To explain the composition of amorphous SiO_x layers, two models were proposed [27, 52]. The model of unordered bonds of tetrahedra assumes that the nonstoichiometric SiO_x layers are a monophase system consisting of a mixture of highly dispersed, randomly bonded tetrahedra of $Si-(Si_yO_{4-y})$, where $y = 0, \ldots, 4$. Alternatively, in the "mixture" model, the layer consists of a simple mixture of Si with SiO_2, but this model is not corroborated experimentally. In particular, the edge of the absorption band monotonically shifts with increasing x. In the mixture model, this absorption band must coincide with that of silicon and be independent of the composition [26, 41]. By contrast, the model of irregularly bound tetrahedra describes the experimental dependence of the SiO_x/SiO_2 concentration ratio on x rather well [27, 41].

The frequency of the Si–O–Si stretching band of a thick CVD oxide is close to that of a thermal oxide of the same thickness [28, 53–55]. Thus, the

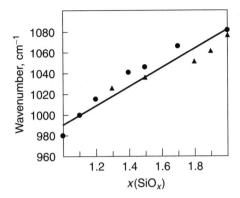

Figure 5.4. Band maximum of TO stretching mode of Si–O–Si bonds: symbols — experiment; solid line — calculation from Eq. (5.7). Reprinted, by permission, from H. Ono, T. Ikarashi, K. Ando, and T. Kitano, *J. Appl. Phys* **84**, 6064 (1998). Copyright © 1998 American Institute of Physics.

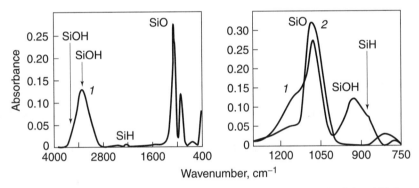

Figure 5.5. IR absorption of (1) CVD oxide (SiH₄ with 60% concentrated H₂O₂) and (2) thermal oxide produced using O₂ at 1050°C. Reprinted, by permission, from F. Gaillard, P. Brault, and P. Brouquet, *J. Vac. Sci. Technol. A* **15**, 2478 (1997). Copyright © 1997 American Vacuum Society.

IR absorption spectra of the CVD oxide 150 nm thick, formed from SiH₄ and H₂O₂, and of the thermal oxide demonstrate the same 1080-cm⁻¹ maximum [53]. However, the FWHM of this peak for the CVD oxide (118 cm⁻¹) is considerably broader than that of the thermal oxide (76 cm⁻¹) due to the presence of the silanol (SiOH) group (Fig. 5.5). Indeed, after plasma or thermal treatment, which transforms the SiOH groups into SiO₂, the FWHM of the Si–O–Si stretching band approaches that of a thermal oxide, and the oxide layer density increases. This latter effect, which is related to the increased disorder in oxide films, is attributed to the incorporation of OH and H₂O into the oxide films during their growth and is observed for anodic oxides as well [56].

The spectra of low-temperature oxides show a band at 940 cm⁻¹ due to the SiOH group [44, 53]. A band at 3350 cm⁻¹ may be assigned to either adsorbed water molecules or SiOH groups [29, 40, 41, 44, 57, 58]. The higher frequency band at 3650 cm^{-1} is also attributed to silanol [40, 53, 58]. A peak at 1640 cm^{-1} can appear when absorbed water molecules are present in a standard CVD oxide [20, 40]. The quantitative assessment of the hydrogen content in films is based on the measurement of the absorbance at 3650 cm^{-1} (SiOH) and 3350 cm^{-1} (associated with H$_2$O) according to the following expressions [58]:

$$S = (179A_{3650} - 41A_{3350}),$$

$$W = (-14A_{3650} + 89A_{3350}), \tag{5.8}$$

where S and W are the weight percents of OH (from silanol) and water, respectively, and A_ν is the optical density per micrometer of the film at the frequency ν. These values can readily be converted to atomic percents (at. %) [59]:

$$\text{at. \% H(SiOH)} = \frac{5.9S}{5 + 0.07S + 0.12W}$$

$$\text{at. \% H(H}_2\text{O)} = \frac{11.1W}{5 + 0.07S + 0.12W} \tag{5.9}$$

The amounts of water and silanol depend strongly on the conditions of the oxide formation, its thickness, and the annealing procedure [40, 59, 60]. For example, for a CVD film formed in the SiH$_4$ + O$_2$ mixture at 100°C, the water and silanol contents are reduced from 6.6 and 6.3%, respectively, to about 1.6% for both after annealing at 600°C for 10 s [60]. The amount of silanol groups decreases continuously with increased annealing temperature over the 250–500°C range [59].

The appearance of absorption bands from other impurities depends on the method of oxide deposition. Thus, the spectra of low-temperature SiO$_2$ layers formed by CVD of organic compounds such as tetramethoxysilane (TMOS) or tetraethylorthosilicate can result in absorption bands at 2870 and 2960 cm^{-1}, corresponding to C–H stretching vibrations [29, 61], and bands at 1195 and 840 cm^{-1} attributed to C–O and Si–OCH$_3$, respectively [61, 62]. The bands arising from N–H (3300 cm^{-1}) and Si–N bonds (850 cm^{-1}) appear in the spectra of CVD oxides produced by the reaction of silane with nitrous oxide [28]. Incomplete oxidation of silane leads to the presence in the spectra of SiH peaks at 880 and 2150 cm^{-1} (Fig. 5.5), which disappear after plasma or thermal annealing at temperatures higher than 300°C [53, 59].

The presence of impurities, especially bonded hydrogen, can produce undesirable side effects. Incorporation of hydrogen in the SiO$_2$ network can significantly alter the chemical and electric properties of the oxide, to the detriment of microelectronic device performance. The aging of SiO$_2$ films formed by the CVD method was compared with aging of thermal oxides [63]. No changes in the features of the 1080-cm^{-1} absorption band were observed for thermal oxides within the period of observation (2 years). By contrast, the exposition of a CVD film to humid atmosphere over a few months resulted in the shift of the peak position of the ν_{TO} band toward higher frequencies and a decrease of the FWHM. Apparently, the aging leads to a structural relaxation of the SiO₂ continuous random network.

The effect of the experimental parameters of plasma-enhanced CVD (PECVD) — substrate temperature and SiH₄–O₂ ratio — on the ν_{LO} band (Fig. 5.6) was studied in Ref. [64]. With a change in the flow rate of oxygen, V_{O_2}, into the reactor from 7 to 30 L · h^{-1} (V_{SiH_4} = 12L · h^{-1} = const), the maximum of the ν_{LO} band of the SiO₂ layer shifts from 1247 to 1250 cm^{-1}, the FWHM

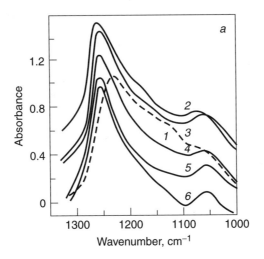

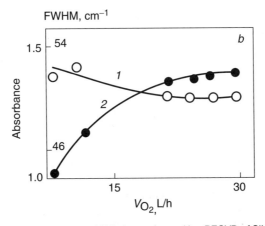

Figure 5.6. (a) IR absorption spectra of SiO₂ layers formed by: PECVD of SiH₄ + O₂ at (1) 20°C and (2–5) 300°C at various flow rates of reactants into reactor — V_{O_2} and V_{SiH_4}, respectively, L/h): (1) 18.0 and 9.6; (2) 26.4 and 12.0; (3) 10.8 and 10.8; (4) 12.0 and 6.9; (5) 18.6 and 1.8; (6) thermal oxidation in O₂ at 1000°C with oxide thickness 200 nm. (b) Dependence of (1) FWHM and (2) absorbance at LO stretching mode frequency on V_{O_2}. Reprinted, by permission, from Yu. A. Averkin, N. K. Karmadonov, V. A. Skryshevsky, V. I. Strikha, and A. V. Kharlamov, *Ukrainian J. Phys.* **34**, 1817 (1989). Copyright © 1989 Institute of Theoretical Physics, National Academy of Science of Ukraine.

decreases from 53 to 50 cm^{-1}, and the absorbance A_m increases by 1.4 times. These changes result in an absorption band similar to that of the thermally grown SiO$_2$ layer, for which $\nu = 1250$ cm^{-1} and FWHM $= 45$ cm^{-1}. Thus, the properties of CVD layers produced in the reaction of SiH$_4$ decomposition in O$_2$ at 300°C are close to those of the thermally grown SiO$_2$ layers. The linear dependence of the absorbance A_m on the layer thickness (measured ellipsometrically) testifies to the sufficiently high homogeneity of these layers for different proportions of reactants in the mixture.

Analyzing the features of the absorption bands for layers formed under constant V_{SiH_4} and variable V_{O_2} a limiting value of V_{O_2} is found, above which the absorption band features remain unchanged; excess O$_2$ influences neither the parameters of the bands nor the properties of the resulting oxides. Deposition of oxide layers on a cold silicon substrate produces broader absorption bands (FWHM $= 65$ cm^{-1}) of SiO$_2$ at the relatively low frequency of $\nu = 1230$ cm^{-1}. The influence of the oxide deposition parameters on the absorption caused by the TO stretching bonds were considered for the case of CVD and PECVD for various reactants: SiH$_4$ + N$_2$O [59], SiH$_4$ + O$_2$ [60], TMOS [61], TEOS [65, 66], and tetraethylorthosilicate [67].

In recent years, the IR absorption of fluorinated silicon oxide (F$_x$SiO$_y$) has been actively studied [68–74]. These films are very easily deposited by several PECVD or liquid-phase deposition (LPD) methods and are characterized by a low dielectric constant, which decreases with increased concentration of fluorine in the film. Decreasing the dielectric constant of the intermetal dielectric film is the most efficient way to reduce the adjacent wiring capacitance, which will improve the performance of submicrometer integrated circuits. However, the F$_x$SiO$_y$ films become reactive to water as the fluorine concentration increases. The film desorbs H$_2$O and HF under thermal annealing after humidification, which causes reliability problems in the VLSI fabrication [68].

The transmittance spectra of the LPD F$_x$SiO$_y$ film are shown in Fig. 5.7 [70]. Incorporation of F to about 10–12% results in changes in the IR absorption spectra with respect to the spectra of the SiO$_2$ layer, including, first, the appearance of a weak band of the Si−F bond stretching mode at 935 cm^{-1}; second, changes in the shape of the Si−O−Si asymmetric stretching mode at 1080 cm^{-1} (a shift of the peak position to the higher frequency with increasing F content, which is accompanied by a decrease in the FWHM [72–74]); and third, the reduction in the intensities of the IR absorption bands at 800 and 450 cm^{-1} with increasing F content [71]. These changes are accompanied by a drop in the dielectric constant from 3.8 to 3.3 for a F concentration of 7–10 at. %. Note that at the largest concentration of F, the absorption peak of the Si−F stretching mode is split [68, 69, 71]. A model in which the fluorine atoms are bonded in an Si−O−F sequence has been invoked to explain the substantial changes that occur in the IR spectra of F$_x$SiO$_y$ films [71].

It is assumed [70] that LPT F$_x$SiO$_y$ contains strongly H bonded SiO−H··· F−Si structures that are restructured during annealing. Annealing at temperatures between 300 and 500°C will break SiO−H bonds to form SiO, because the

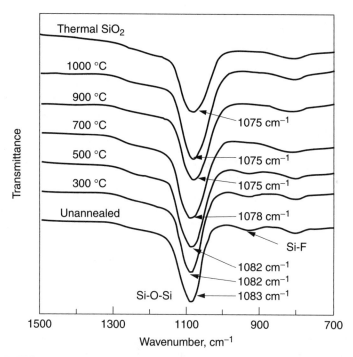

Figure 5.7. FTIR transmission spectra of LPD F$_x$SiO$_y$ oxide at different annealing temperatures. Reprinted, by permission, from C.-F. Yeh and C.-L. Chen, *J. Electrochem. Soc.* **142**, 3579 (1995). Copyright © 1995 Electrochemical Society, Inc.

SiO−H bond energy is only 111 kcal · mol^{-1}, but by the same token, few Si−F bonds are broken as the Si−F bond energy is 143 kcal · mol^{-1}. At annealing temperatures higher than 700°C, most of the Si−F bonds are broken, and the Si−F peak disappears (Fig. 5.7). This suggests that the LPD oxide has become more dense, similar to a thermal oxide. Increasing the annealing temperature causes the Si−O−Si bond absorption to increase gradually and shift toward lower frequencies from 1083 cm^{-1} (unannealed LPD oxide) to 1075 cm^{-1} (at the 1000°C annealing), while the FWHM increases from 63 to 84 cm^{-1}, which is similar to the FWHM (85 cm^{-1}) of the corresponding band of the thermal SiO₂ (Fig. 5.7) [70].

According to Ref. [72], the addition of Ar gas during the deposition can improve the thermal stability of the incorporated fluorine, because sputtering in the Ar atmosphere enhances the removal of the weakly bonded silicon fluoride from the as-deposited film surface. The IR absorption of F$_x$SiO$_y$ after interaction with water is discussed in Ref. [68].

5.3. ULTRATHIN SiO₂ LAYERS

Infrared spectroscopy of ultrathin oxide films on Si exploits the Berreman effect, arising when *p*-polarized radiation is incident at an inclined angle (Sections 3.1

and 3.3). According to thin-film approximation (Table 1.2), absorption at ν_{LO} and ν_{TO} is proportional to the thickness of the oxide layer [20, 36, 75]. This law is obeyed for the ν_{TO} band at 1080 cm^{-1} (for thicknesses of 2–35 nm) [76, 77], the ρ_{TO} band at 450 cm^{-1} (for thicknesses of 10–120 nm) [78], and the ν_{LO} band at 1250 cm^{-1} (for thickness of 2–12 nm) [16].

In the case of ultrathin oxide layers, it is more convenient to study IR absorption in the region of LO vibrations (1250 cm^{-1}) for two reasons. First, the intensity of the ν_{LO} band is higher than that of the ν_{TO} band. Second, the TO stretching mode of SiO$_2$ (1070 cm^{-1}) lies close to the absorption band of O$_i$ in silicon (1105 cm^{-1}), which renders any studies of weakly absorbing ultrathin films of silicon oxide more difficult.

Absorption measurements in the region of the LO vibrations reach sensitivities <1 nm, allowing one to follow the natural oxidation of monocrystalline silicon in air [79]. Under exposure of silicon to air over 1–2 months, the native oxide attains thicknesses on the order of 1.0 nm. An absorption band arises at 1100 cm^{-1} after 8 h exposure to air followed by the water rinse. Over a three-day period following this treatment, the band shifts to 1180 cm^{-1}, and another band appears at 1050 cm^{-1}. During the subsequent 30 days, the 1180-cm^{-1} peak shifts to 1210 cm^{-1}, while only very small frequency shifts are observed for the 1050-cm^{-1} peak.

The analyses of ultrathin SiO$_2$ layers (1–5 nm) obtained by several methods have been performed by IR multiple-transmission spectroscopy using p-polarized radiation incident at the Brewster angle [22, 23, 32, 80]. The absorption spectra for different thicknesses of silicon oxide grown thermally at 600°C with the addition of HCl exhibit peaks at 1240 and 1160 cm^{-1} (Fig. 5.8a) [22]. As shown in Section 5.1, these peaks correspond to either two different structures of silicon dioxide (six- and fourfold ring configurations) or SiO$_x$ with $x = 1$ (the peak at 1160 cm^{-1}) and $x = 2$ (the peak at 1240 cm^{-1}). One can see from Fig. 5.8 that the absorbance at the maximum of these bands increases linearly with increasing oxide thickness. This allows one to normalize the absorbance to the oxide thickness. (The normalization factor depends upon the number of passages of IR radiation through the Si wafers.)

When d increases from 1.3 to 5.0 nm, the maximum of the high-frequency band is shifted from 1235 to 1248 cm^{-1} (Fig. 5.8b) with a corresponding reduction in the band FWHM. At $d = 5.0$ nm, the oxide layer properties are similar to those of bulk SiO$_2$. This frequency shift of the ν_{LO} band can be attributed to distorting surface effects in which the angle of the Si−O−Si bonds change and also with the existence of a nonstoichiometric oxide at the SiO$_2$−Si interface [81], since the optical effects for SiO$_2$ films in this thickness range account for only 2 cm^{-1} of the shift (Section 3.4). According to Eq. (5.4), the increase in ν_{LO} may result from an increase in the bridging bond angle θ, the network density ρ, and the electrostatic term Z.

As seen from Fig. 5.8, the absorption at 1160 cm^{-1} is weaker than that at 1240 cm^{-1}. The unchanged position of the maximum and considerably less pronounced increase in the peak intensity with increasing total layer thickness (see

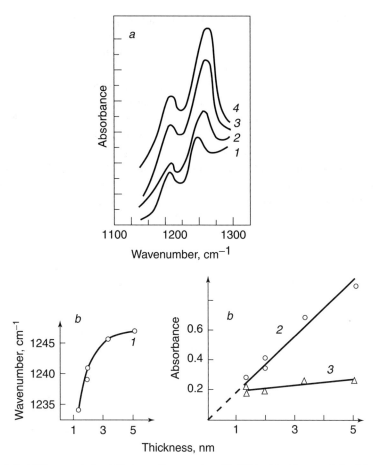

Figure 5.8. (a) The *p*-polarized IR absorption spectra with 75° incident beam of ultrathin silicon oxide layers grown at 600°C on Si in presence of HCl. Oxide thicknesses: (1) 1.3, (2) 1.9, (3) 3.2, and (4) 5.0 nm. (b) Experimental (1) 1240 cm⁻¹ peak position and (2, 3) absorbance intensity A_m of absorption bands at (2) 1240 and (3) 1160 cm⁻¹ as functions of layer thickness. Reprinted, by permission, from S. Kilchitskaya, S. V. Litvinenko, V. A. Skryshevsky, V. I. Strikha, and V. P. Tolstoy, *Poverhnost (Surface)* **4**, 99 (1987). Copyright © 1987 Institute of Solid State Physics, Russian Academy of Science.

Fig. 5.8*b*) are characteristic for this band. The observed difference in the behavior of the bands at 1240 and 1160 cm⁻¹ can be related to an increase in the homogeneity of the oxide layer with increasing thickness (the SiO mole fraction decreases throughout the layer) as well as to the fact that there is a transition SiO layer between silicon and SiO₂ [82, 83] whose thickness also increases with increasing thickness of the oxide layer, but to a lesser extent.

Note that a thermal oxide layer formed at higher temperatures (800–950°C in dry O₂) exhibits a higher frequency for SiO₂ — 1253 cm⁻¹ — and an improved ratio $A_{m1250}/A_{m1160} = 3$ for $d = 2.0$ nm [16]. However, for thicknesses greater than 5 nm, this peak ceases to shift toward higher frequencies as for layers formed at lower temperatures [18].

The general form of the spectra of ultrathin anodic oxides (formed at the potentiostatic mode in a KNO_3–ethylene glycol solution) testifies to the fact that an anodic oxide is less perfect than a thermal oxide. Specifically, the peak position in the spectrum of an anodic oxide can vary from 1224 cm^{-1} for applied voltage $E = 4$ V to 1231 cm^{-1} (at $E = 30$ V). For oxides of equal thickness, the absorption band maximum of the anodic oxide is shifted to lower frequencies relative to the thermal oxide. Figures 5.9a,b show that the variation of peak intensity of the 1240-cm^{-1} band with anodization time t depends on the anodization potential [22]. The anodization time at which the peak intensity and by extension the thickness become independent of t is longer for higher E. For $t = 20$ min, the peak intensity varies linearly with the anodization potential for potentials less than 12 V; for longer anodization times, the region of linear dependence is shifted toward higher potentials.

The dependences of the 1240-cm^{-1} peak position and the band FWHM on the anodization potential E are presented in Fig. 5.9c. As seen, the peak frequency increases and the FWHM decreases with increasing potential. For the absorption band at 1160 cm^{-1}, the frequency of the peak and the FWHM are practically independent of E and t. The peak intensity of this band increases more weakly with increasing E and t compared to the peak intensity of the band at 1240 cm^{-1} (see Fig. 5.9a).

Analogous results were obtained in the analysis of the TO mode absorption [77]. The anodic oxide film thickness increased with applied voltage, the peak position of the ν_{TO} band shifted from 1054 cm^{-1} (for the 1-nm oxide film) to 1061 cm^{-1} (for the 10-nm oxide film), and the FWHM decreased with the oxide thickness.

The thickness of the anodic oxide layer obtained from a two-step applied voltage ($E = 1$ V for 10 min and then $E = 8$ V for another 10 min) is that of the anodic oxide grown at a constant potential $E = 7$ V over 20 min. But in the IR absorption spectra of the oxide being grown with the two-step mode, the 1240-cm^{-1} peak is at a higher frequency, the FWHM is less, and the ratio A_{m1240}/A_{m1160} is larger, indicating that this film is of higher quality than that grown at a single voltage [22].

Ultrathin anodic SiO_2 films grown under potentiostatic and galvanostatic conditions were compared in Ref. [40]. For potentiostatic oxides relative to galvanostatic oxides of the same thickness, the maximum of the ν_{TO} band absorption is observed at slightly higher frequencies ($\Delta\nu = 2$ cm^{-1}) and is somewhat narrower. For both types of films, the frequency of the maximum increases with increasing film thickness, while the corresponding FWHM decreases. As-grown 11-nm-thick films of both types contain 3.8% OH from isolated silanol with an absorption peak at 3650 cm^{-1} and 5.0% OH from H_2O and/or associated silanol with an absorption peak at 3300 cm^{-1}. The empirical formula of the as-grown anodic oxides is $SiO_{1.93}(OH)_{0.14} \times 0.18H_2O$. Thermal annealing at a temperature

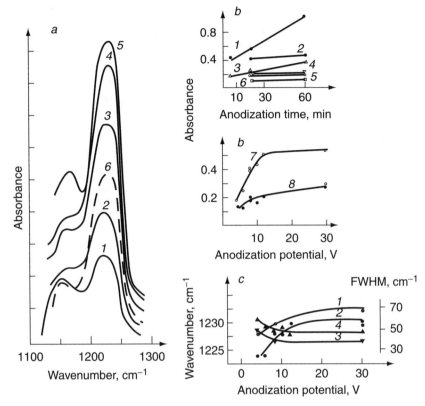

Figure 5.9. (a) IR absorption spectra of anodic silicon oxide grown at $t = 20$ min and different applied voltages: (1) 4 V; (2) 6 V; (3) 8 V, (4) 12 V; (5) 30 V; (6) 1 V, $t = 10$ min; 8 V, $t = 10$ min. (b) Absorbance of 1240-cm^{-1} band (1, 2, 4, 7) and 1160-cm^{-1} band (3, 5, 6, 8) versus anodization time t and anodization potential: (1, 3) 30 V; (2, 5) 8 V; (4, 6) 4 V at $t = 20$ min. (c) (1, 2) Peak position and (3, 4) FWHM of 1240-cm^{-1} band versus anodization potential at anodization times (1, 3) 60 and (2, 4) 20 min. Reprinted, by permission, from S. Kilchitskaya, S. V. Litvinenko, V. A. Skryshevsky, V. I. Strikha, and V. P. Tolstoy, *Poverhnost (Surface)* **4**, 99 (1987). Copyright © 1987 Institute of Solid State Physics, Russian Academy of Science.

below 500°C eliminates water and the associated silanol. At annealing temperatures above 700°C, the anodic oxide becomes stoichiometric.

Table 5.2 lists features of the absorption bands for oxides produced by various methods [22]. Assuming that high-quality oxide is indicated by the high frequencies of the 1240-cm^{-1} peak (approaching bulk SiO₂), low FWHM values, and high ratios A_{m1240}/A_{m1160} (implying homogeneity), one can conclude that the most perfect oxide is the thermal oxide and the most imperfect oxide is the chemical one. Furthermore, the homogeneity of all the oxides improves as their thickness increases. These conclusions are supported by data concerning the effect of thickness on the composition for chemical and thermal silicon oxides [84, 85] obtained by He$^+$ backscattering and XPS.

Table 5.2. Characteristics of IR absorption spectra of SiO_2 layer obtained by various methods

N	Method of Oxidation	Average Thickness (nm)	LO Peak Position	FWHM (cm^{-1})	A_{m1240}/A_{m1160} Ratio
1	Chemical in H_2O_2	2.4	1217	50	1.1
2	Thermal at 600°C in ambient air	2.4	1220	66	1.3
3	Anodic in KNO_3	2.4	1224	64	1.3
4	Thermal at 600°C with HCl	2.4	1243	40	1.9
5	Thermal at 600°C in ambient air	3.5	1242	50	1.7
6	Anodic KNO_3	3.5	1231	56	1.9
7	Thermal at 600°C with HCl	3.5	1247	32	2.6

Reprinted, by permission, from S. Kilchitskaya, S. V. Litvinenko, V. A. Skryshevsky, V. I. Strikha, and V. P. Tolstoy, *Poverhnost (Surface)* **4**, 99 (1987). Copyright © 1987 Institute of Solid State Physics, Russian Academy of Science.

Native silicon oxide films (1.5 nm) were formed by dipping Si in solutions of (a) ultrapure water with ozone O_3, (b) $NH_4OH-H_2O_2-H_2O$, and (c) $H_2O_2-H_2O$ and analyzed by ATR [86]. The chemical oxides all demonstrate some degree of imperfection, as indicated by the frequency of the ν_{LO} band (1214, 1213, and 1173 cm^{-1} for films from solutions a, b, and c, respectively) (Fig. 5.10). The

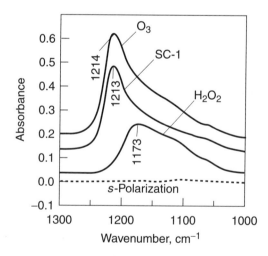

Figure 5.10. S-(dashed line) and p-(solid lines) polarized absorption spectra of native silicon oxide formed by several methods. Reprinted, by permission, from T. Ohwaki, M. Takeda, and Y. Takai, *Jpn. J. Appl. Phys.* **36**, 5507 (1997). Copyright © 1997 Publication Board of Japanese Journal of Applied Physics.

structure of the native oxide formed in O_3 solution is closest to that of bulk SiO_2 compared to the other oxides. A model in which the native oxides consist of SiO_2 and voids is presented.

Analysis of the ν_{TO} band indicated that a relatively low quality of native oxide 0.5–0.7 nm thick was formed by dipping in water. The asymmetric Si–O–Si stretching peak is centered at 1014 cm^{-1} for samples immersed in $H_2{}^{18}O$ and at 1046 cm^{-1} for samples immersed in $H_2{}^{16}O$ [87].

A comparison has been made using IRRAS of chemical oxides formed by the following solutions: H_2SO_4–H_2O_2 (SPM), HNO_3, NH_4OH–H_2O_2–H_2O (SC1), and HCl–H_2O_2–H_2O (SC2) [88]. The frequency of the ν_{LO} band for chemical oxides formed in SPM (1220 cm^{-1}) is the highest of these oxides; the others have maxima at 1200 cm^{-1}. The FWHM of the ν_{LO} band is smallest for the oxide formed in SPM and largest for the oxide formed in SC2. It is concluded that the chemical oxide formed in SPM resembles thin thermally oxidized SiO_2 layers, whereas the oxide formed in SC2 have spectra that differ significantly from the spectra of thermally oxidized SiO_2 in terms of the parameters of the ν_{LO} band.

The region of oxide restructuring near the interface with silicon was assessed by IR transmission spectra of ultrathin layers of various thicknesses [18, 22, 49, 80, 89, 90]. It is evident that this interface region consists of a Si-rich oxide, because of its resistance to HF etching [49]. The reconstruction of the silicon crystal surface during the hydrogen treatment shifts the LO absorption band of the native oxide 2.0–3.0 nm thick from 1230 to 1220 cm^{-1} [91]. The thermally grown oxide–Si interface appears to extend up to 5 nm, as determined by the shift in the 1250-cm^{-1} peak (Fig. 5.8) [22]. However, the actual spatial distribution of frequencies generated by this region may be much narrower, because the IR absorption occurs in a heterogeneous medium. The absorbance A, which is measured experimentally, can be written as [18]

$$A = \log\left[\exp\left(\sum_i \frac{k\Delta x}{1 + [2(\nu - \nu_i)/\Delta\nu]^2}\right)\right] \qquad (5.10)$$

Here, the oxide layer is subdivided into i slices of thickness Δx. A Lorentzian peak of absorption centered at ν_i with a FWHM of $\Delta\nu$ is associated with each of the i slices. The fitting and deconvolution of the measured peak shifts as a function of oxide thickness [Eq. (5.10)] indicated that the region in which the stoichiometry varied from SiO_2 was less than 1.6 nm thick for the thermal oxide–Si interface [18, 49]. Since both the ν_{TO} and ν_{LO} bands have the same thickness dependence, the peak shifts are attributed to changes in stoichiometry and not to any stress or increased density [18].

These results are in good agreement with the data obtained by Auger electron spectroscopy (AES), secondary ion mass spectrometry (SIMS), Rutherford backscattering (RBS), and XPS [82, 83, 92, 93], which indicate the presence of a nonstoichiometric SiO_2 layer 0.2–4 nm thick between the Si and SiO_2. The model suggested in Ref. [93] describes the Si–SiO_2 interface as consisting of the

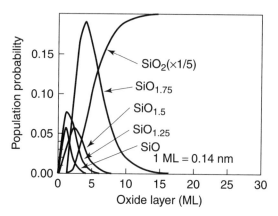

Figure 5.11. Population probability of suboxide structure SiO_x. Reprinted, by permission, from H. Ono, T. Ikarashi, K. Ando, and T. Kitano, *J. Appl. Phys.* **84** 6064 (1998). Copyright © 1998 American Institute of Physics.

Si matrix, the Si disordered surface region, and the regions of structurally rearranged SiO_x (x varies from $x = 0$ for Si to $x = 2$ for SiO_2), highly disordered SiO_2, and regular amorphous SiO_2. Population probabilities of suboxide structure SiO_x calculated from the multilayer interface model are shown in Fig. 5.11 [49].

5.4. SILICON NITRIDE, OXYNITRIDE, AND CARBON NITRIDE LAYERS

Stoichiometric and hydrogenated amorphous silicon nitrides (Si_3N_4, and a-SiN_x:H, respectively) have many uses in the microelectronics industry, such as to passivate various capless devices and to act as a diffusion barrier against species such as H_2O or Na^+. Silicon nitride is also employed as a diffusion barrier in selective silicon oxidation (because of the low oxidation rate) and as an interlayer insulator material [2].

The CVD of silicon nitride can be obtained in several ways, including from the reaction between silane and ammonia ($3SiH_4 + 4NH_3 \rightarrow Si_3N_4 + 12H_2$) at atmospheric pressure and temperatures between 300 and 700°C and the reaction of dichlorsilane with ammonia ($3SiCl_2H_2 + 4NH_3 \rightarrow Si_3N_4 + 6HCl + 6H_2$) at reduced pressures and temperatures between 700 and 800°C. PECVD enables one to manufacture the layers at very low substrate temperatures (100–400°C). For this purpose, one uses the reaction of silane with ammonia in an Ar^+ plasma ($SiH_4 + NH_3 \rightarrow SiNH + 3H_2$) or silane injection into a nitrogen discharge ($2SiH_4 + N_2 \rightarrow 2SiNH + 3H_2$). The CVD layers were found to contain 10–35 at. % of hydrogen [2].

Features of the resulting layers such as the density, the specific resistance, the breakdown voltage, the band gap, the etching rate, and the permittivity can be correlated to their chemical composition, which is dictated by the deposition conditions. For example, silicon nitride deposited at low ratios of ammonia to dichlorsilane contains a high concentration of silicon atoms, which undermines the dielectrical properties of the layer. The presence of oxygen in the layer gives rise to an increase in the etching rate. An increase in the Si/H ratio lowers the resistance appreciably. Deposition at high temperatures can lead to the formation of crystalline phases [2]. One can control the deposition process by varying the temperature and pressure in the reactor, the precursor concentrations, and, in the case of CVD deposition, the frequency and power of the discharge, the geometry of the reactor, the partial pressures of the reactants, the substrate temperature, and the pumping rate.

IR spectroscopy is convenient for the analysis of the chemical composition of silicon nitride layers. A typical IR transmission spectrum of a low-temperature CVD film is presented in Fig. 5.12 [94]; the peak positions and their assignment are shown in Table 5.3. Thin films deposited at low temperatures are rich in both nitrogen and hydrogen and exhibit only Si−N stretching (850 cm^{-1}), N−H stretching (3330 cm^{-1}), Si−H stretching (2180 cm^{-1}), and N−H bending (1150 cm^{-1}) absorption peaks (Fig. 5.12). The observation of the additional peaks listed in Table 5.3 depends strongly on the deposition conditions.

Silicon oxynitride (SiO$_x$N$_y$) films exhibit properties that fall somewhere between those of SiO$_2$ and those of Si$_3$N$_4$ films and have diverse applications in microelectronics. The oxynitride layers can be obtained if nitrogen oxide (N$_2$O) is involved in the reaction of silane with ammonia [108, 109, 117−119], when silicon nitride is deposited onto an oxidized silicon substrate (silicon dioxide nitrification is incomplete at 800°C [100, 107]), or upon addition of gaseous oxygen during the CVD of silicon nitride [64, 100, 104, 120].

In the IRRAS spectra of the SiNO layers one can observe a broad peak between 840 and 1100 cm^{-1} (Fig. 5.13) [110], whose maximum shifts from 1080

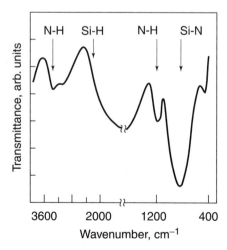

Figure 5.12. IR transmission spectrum of thin silicon nitride film produced by CVD of SiH$_4$ + N$_2$ at 65°C. Reprinted, by permission, from T. Inukai and K. Ono, *Jpn. J. Appl. Phys.* **33**, 2593 (1994). Copyright © 1994 Publication Board of Japanese Journal of Applied Physics.

Table 5.3. Absorption peaks of silicon nitride

Peak Position (cm^{-1})	Peak Assignment	References
450–462	Si–N wagging	96–98
800–900	TO Si–N stretching (asymmetric)	64, 94–114
1100	LO Si–N stretching	95, 98
1150–1200	N–H bending	94, 111, 113, 114
1550	N–H$_2$ bending	114, 115
2000–2260	Si–H stretching	94, 96, 103, 110, 112–114, 116
3300–3400	N–H stretching	94, 96, 98, 103, 109–114, 116
3445	N–H$_2$ stretching	115

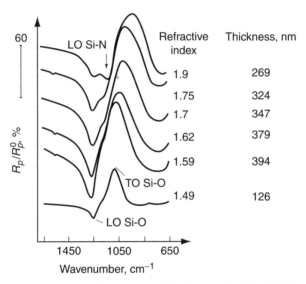

Figure 5.13. The p-polarized IRRAS spectra at 70° of various SiO$_x$N$_y$ thin films on Si wafer. Refractive indices are shown on right of spectra; R_p of bare silicon wafer in p-polarization is used as references. Reprinted, by permission, from M. Firon, C. Bonnelle, and A. Mayeux, *J. Vac. Sci. Technol. A* **14**, 2488 (1996). Copyright © 1996 American Vacuum Society.

to 840 cm^{-1} as the oxygen content is reduced [98, 110, 121]. When the layer is a composite of SiO$_2$ and Si$_3$N$_4$, two well-resolved absorption peaks corresponding to the vibrations of the Si–O (1080 cm^{-1}) and Si–N (840 cm^{-1}) bonds appear in the spectral region of 400–1500 cm^{-1} [99, 108, 109]. In addition, with increased oxygen content, the peaks corresponding to the vibrations of the N–H, and Si–H bonds shift from 3340 cm^{-1} (Si$_3$N$_4$) to 3375 cm^{-1} (SiNO), and from 2170 to 2205 cm^{-1}, respectively [98, 118]. The peaks of the Si–O bending (810 cm^{-1}) and Si–O rocking (450 cm^{-1}), are also observed in the spectra of SiNO layers [97, 98, 116].

The peak intensity of the 840 cm^{-1} band was found to vary linearly with the layer thickness [100]. This effect can be utilized to monitor the thickness of a silicon nitride layer being used as a mask. The oxidation of CVD silicon nitride during its annealing in oxygen is studied in Ref. [64] for layers deposited under different conditions. The IR absorption spectra measured in p-polarized radiation at $\varphi = 75°$ of the Si$_3$N$_4$ (140 nm)/SiO$_2$ (30 nm)/Si structure, formed under different deposition conditions ($M = 1.6$–4.6, where $M = (4\%\ \mathrm{SiH_4} + 96\%\ \mathrm{Ar^+})/$ NH$_3$) are presented in Fig. 5.14. The general shape of the spectra, and the pronounced absorption bands at 1250 cm^{-1} and 1070 cm^{-1} (SiO$_2$), and the broad band at 870 cm^{-1} (Si$_3$N$_4$) confirm that the layers contain SiO$_2$, and Si$_3$N$_4$. High-temperature annealing in wet oxygen leads to an increase in the intensities of the absorption bands of SiO$_2$, and a reduction in the intensity of the Si$_3$N$_4$ absorption band, as well as the appearance of structural peculiarities in the band at 870 cm^{-1} due to the formation of the crystalline phase of Si$_3$N$_4$.

Variation of M can affect the absorption spectra substantially. Although the thickness of the SiO$_2$ layer as determined by the 1250 cm^{-1} peak absorption is virtually independent of M prior to the annealing, the annealing process causes a considerable increase in thickness for all values of M, excepting the narrow range of $M = 3.1$–3.3 (Fig. 5.15). For these values of M, the intensity of the band corresponding to the Si$-$O stretches in the 1200–1250 cm^{-1} range does not vary significantly (Fig. 5.14). The intensity of the Si$_3$N$_4$ absorption band

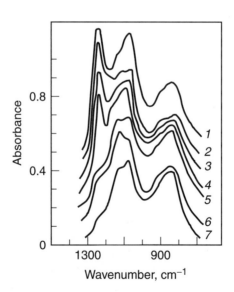

Figure 5.14. IR absorption spectra of Si$_3$N$_4$ (140 nm)–SiO$_2$ (30 nm)–Si structure after annealing at 1050°C for 5 h in wet oxygen at $M = (4\%\ \mathrm{SiH_4} + 96\%\ \mathrm{Ar})/\mathrm{NH_3}$: (1) 4.6, (2) 2.5, (3) 1.6, (4) 4.5, (5) 3.0, (6) 3.3, and (7) without annealing at $M = 3.0$. Reprinted, by permission, from Yu. A. Averkin, N. K. Karmadonov, V. A. Skryshevsky, V. I. Strikha, and A. V. Kharlamov, *Ukrainian J. Phys.* **34**, 1817, (1989). Copyright © 1989 Institute of Theoretical Physics, National Academy of Science of Ukraine.

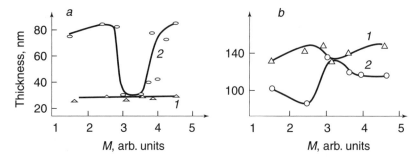

Figure 5.15. Thicknesses of (a) SiO_2 and (b) Si_3N_4 layers obtained from IR absorption spectra (1) before and (2) after annealing in wet oxygen at 1050°C during 5 h as functions of $M = $ (4% SiH_4 + 96% Ar)/NH_3. Reprinted, by permission, from Yu. A. Averkin, N. K. Karmadonov, V. A. Skryshevsky, V. I. Strikha, and A. V. Kharlamov, *Ukrainian J. Phys.* **34**, 1817, (1989). Copyright © 1989 Institute of Theoretical Physics, National Academy of Science of Ukraine.

after annealing remains practically unchanged for values of $M = 3.1$–3.3 and decreases appreciably for $M > 3.3$ and $M < 3.1$. This may arise because at $M \approx 3.1$ the composition of the layers is closest to the stoichiometric composition (with the maximum refractive index $n = 2.1$ [99]). Deviation from this value of M ($M > 3.3$ or $M < 3.1$) leads to an increase in the rate of growth and a decrease in its refractive index n, implying that the NH_3 content is no longer optimal for SiH_4 decomposition. These layers are easily oxidized during thermal annealing in oxygen (Fig. 5.15) [64]. Applying IR absorption spectroscopy enables one to resolve the relative contributions of SiO_2 and Si_3N_4 to the total layer thickness over a range of M values.

The integral of the extinction coefficient at 3330 cm^{-1} was shown to vary approximately linearly with hydrogen content [94], and the absorption at 840 cm^{-1} depends linearly on the value of x in SiN_x [95, 96]:

$$N_N = x N_{Si} = A I_N \qquad (5.11)$$

where the integration $I_N = \int \alpha(v)\, dv/v$ is carried out over the whole absorption band. Here, $\alpha(v)$ is the extinction coefficient at 840 cm^{-1}, $A = (6.3$–$9.2) \times 10^{18}$ cm^{-1}, and N_N and N_{Si} are the concentrations of nitrogen and silicon atoms (it is assumed that $N_{Si} = 5 \times 10^{22}$ cm^{-3} = const).

The band shapes in the spectra of amorphous nitride layers have been investigated [95, 100, 102, 105]. The band ascribed to Si−N stretching is the superposition of three Gaussian components with frequencies 750, 840, and 960 cm^{-1} [95] and that of the Si−H bond is the superposition of two Gaussian components with frequencies 2000 and 2100 cm^{-1} [102]. The shape of the absorption bands of the SiN_x:H films has been found to depend on x. The band at 2000 cm^{-1} appears in spectra of layers with $x < 0.4$ and corresponds to Si−H stretching, as in the spectra of a-Si:H. Upon the formation of SiN_x:H, the SiH band in the spectrum shifts from 2000 to 2080–2090 cm^{-1}. In the case of a-Si:H layers, the multihydrides SiH_2, and SiH_3 are responsible for absorption at these frequencies. The replacement of Si in a-SiN_x:H films by the more electronegative atom of N leads to a larger band shift than in the case of a-Si:H. Only one Gaussian component with a maximum at 2100 cm^{-1} is observed for $x > 0.4$.

For $x \leq 0.5$, absorption at 750 and 840 cm^{-1} is observed, corresponding to the vibrations of Si$-$N bonds. The maxima are independent of x. As x increases, the intensity of the band at 750 cm^{-1} decreases rapidly, whereas the intensity of the bands at 840 and 2100 cm^{-1} increases. Therefore, the band at 750 cm^{-1} is attributed to isolated Si$-$N bonds surrounded by Si atoms. The bands at 840 and 2100 cm^{-1} can be attributed to N$-$Si$-$H groups in which the Si is surrounded by one H atom and one N atom. The behavior of these bands indicates an increase in the number of N$-$Si$-$H bonds with increasing x up to 0.5.

For $x > 0.5$, bands at 840, 960, and 2100 cm^{-1} are observed. The band at 960 cm^{-1} arises when $x = 0.5$; peaks at 960 and 2100 cm^{-1} shift toward higher frequencies with increasing x (the band at 2100 cm^{-1} shifts to 2200 cm^{-1} at $x > 1.0$). As x increases, the intensities of the bands at 840 and 960 cm^{-1} increase, whereas the intensity of the band at 2100 cm^{-1} decreases. The behavior of these bands points to further restructuring due to the replacement of the H and Si atoms embedded at the Si site of the N$-$Si$-$H bond by N atoms. As the result, the band at 840 cm^{-1} can be assigned to the Si$-$N$-$Si$-$N$_m-$, and N$-$Si$-$Si$-$N$_m-$bonds. In this case, the bands at 960 and 2100 cm^{-1} are ascribed to the N$_m-$Si$-$H$-$bonds, with higher values of m leading to higher frequencies.

Interest in carbon nitride films was first stimulated by theoretical studies of a solid analogous with β-S_3N_4. However, C_3N_4 has also attracted attention due to its application in microelectronics as a barrier against corrosion and general wear. In the FTIR analysis containing different quantities of nitrogen [122$-$124], the IR spectra of CN_x films show absorption bands at 1350, 1500, and 2200 cm^{-1}, corresponding to C$-$N, C$=$N, and C$\equiv$N stretching modes, respectively [124].

5.5. AMORPHOUS HYDROGENATED FILMS

5.5.1. a-Si:H Films

Using advanced thin a-Si:H film processing technology, modern microelectronics industry has developed a number of electronic devices such as solar cells, thin-film field-effect transistors (FETs), visible light-emitting diodes (LEDs), and color detectors [125$-$128]. Methods of a-Si:H film production include the decomposition of silane in a glow discharge plasma in the presence of doping gases, reactive silicon sputtering, and CVD followed by hydrogenization [125].

The structure of amorphous semiconductors such a-Si with tetrahedral geometry is characterized by high rigidity accompanied by topological disorder. This leads to such defects as dangling bonds, which form defect states in the band gap. Hydrogen atoms incorporated into a-Si saturate the silicon dangling bonds, decreasing the density of defect states in the gap and reducing the structural disorder. Consequently, the presence of hydrogen is crucial in improving the electrical and optical properties of a-Si:H [125,127,129]. Irrespective of the synthesis method, materials with acceptable electronic properties contain by necessity between 2 and 16 at. % hydrogen, and in films its content can exceed 50% [125]. It is generally accepted that the total or partial loss of hydrogen leads to the appearance of defects and the deterioration of the electronic properties.

The chemical activity of hydrogen depends not only on the hydrogen concentration but also on such parameters as the substrate temperature, the discharge input power, the proportion of the precursors, and the rate of their flow. Information about hydrogen inclusion is important, because even the type of hydrogen bonding can determine the electrophysical parameters of the films [125, 130–135]. It is known that hydrogen can form mono-, di-, and trihydride bonds as well as different polymeric configurations [SiH, SiH_2, SiH_3, $(SiH_2)_n$]. The presence of Si—H bonds indicates saturation of dangling bonds in the material, whereas the polymeric complexes correspond to recombination or association centers that increase the degree of disorder in the lattice structure, leading to a decrease in the photocarrier mobility and general electrical quality [125, 130, 131].

IR absorption spectroscopy of a-Si:H and its alloys has been applied to optimize the deposition parameters and produce a-Si:H films with low defect density ($<10^{16}$ cm^{-3}) [125, 126, 129, 132–145]. This can be achieved at deposition temperatures ranging from 150 to 300°C under standard optimized plasma conditions — pure silane, low deposition rates (0.1 $nm \cdot s^{-1}$), low gas pressures (3–10 Pa), and low power (10 $mW \cdot cm^{-2}$) [129].

The IR absorption spectra of several a-Si:H films are shown in Fig. 5.16 [129]. The standard a-Si:H film displays three characteristic IR absorption regions: (1) the stretching absorption band that can be resolved into two Gaussian components centered at 2000 cm^{-1} (SiH) and 2080 cm^{-1} (SiH_2), (2) bending mode bands at 840 and 880 cm^{-1}, and (3) a wagging mode at 630 cm^{-1} (Fig. 5.16).

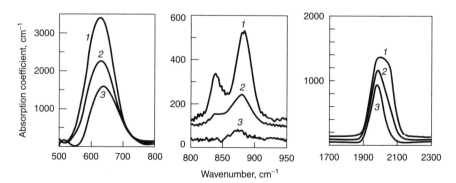

Figure 5.16. Typical IR absorption spectra of PECVD a-Si:H film obtained in pure silane at several substrate temperatures: (1) 50, (2) 100, and (3) 250°C. Reprinted, by permission, from K. Zellama, L. Chahed, P. Sladek, M. L. Theye, J. H. von Bardeleben, and P. Roca i Cabarrocas, *Phys. Rev. B* **53**, 3804 (1996). Copyright © 1996 American Physical Society.

The injection of additional hydrogen atoms leads to changes in the spectra; specifically the stretching vibrations are shifted from 2000 cm^{-1} (SiH) to 2080 cm^{-1} (SiH$_2$) and 2140 cm^{-1} [SiH$_3$ or (SiH$_2$)$_n$]. This effect is thought to be connected with an increase in the electronegativity of the SiH groups, since the stretching vibration frequencies depend linearly on the sum of the electronegativities of the molecular groups [125]. When a second H atom is joined to the SiH group, an absorption caused by the bending vibration (880 cm^{-1}, SiH$_2$) arises. In polysilane chains, the SiH$_2$ vibrations give rise to a doublet at 880 and 840 cm^{-1}. The doublet position can differ for different configurations of the polysilane chains. Films characterized by an absorption spectrum of the polysilane type have a columnar structure and consist of two phases: one of a-Si:H with 5–10 at. % H$_2$ and the other of polysilane.

The hydrogen concentration in hydrogenated amorphous films is often determined from the IR transmission spectra [145–148]. The spectra are fitted with a single Gaussian for each mode $\alpha_i(\nu)$. Then, the integral absorption coefficient I_i is defined by

$$I_i = \int \frac{\alpha_i(\nu)\,d\nu}{\nu}, \tag{5.12}$$

where the integral is taken over the spectral region centered at the band frequency. The concentration of Si–H bonds N_i in the mode i is proportional to I_i:

$$N_i = A_i I_i, \tag{5.13}$$

where A_i is a proportionality constant for mode i. If the Si–H unit is considered as a harmonic oscillating dipole embedded in a solid, one can write the proportionality constant A_i as [144, 149]

$$A_i = \frac{cn\nu_i\mu}{(4\pi^2\,e_i^{*2})} \quad (\text{cm}^{-2}), \tag{5.14}$$

where c is the velocity of light, n is the refractive index, μ is the reduced mass, ν_i is the frequency of mode i, and e_i^* is the effective charge or the oscillator strength of the dipole.

The following proportionality constants were reported [148]: $A_{640} = 2.1 \times 10^{19}$ cm^{-2}, $A_{2000} = 9 \times 10^{19}$ cm^{-2}, and $A_{2080} = 2.2 \times 10^{20}$ cm^{-2}. This allows one to estimate the total bound hydrogen content C_H from the wagging band at 640 cm^{-1}, the concentration C_{H1} of hydrogen bound as isolated monohydride groups SiH contributing to the 2000-cm^{-1} band, and the concentration C_{H2} of SiH$_2$ that contributes to the 2080-cm^{-1} band.

For PECVD a-Si:H films, the total hydrogen content decreases with increasing substrate temperature. Increasing the annealing temperature from 50 to 500°C results in a decrease of the total bound hydrogen content C_H and the concentration C_{H1} of hydrogen bound as isolated monohydride groups [129]. The optical gap of a-Si:H and the optical defect density in the gap increase monotonically as functions of the SiH$_2$ content [145].

The accuracy of determining the hydrogen content from IR absorption spectra can be undermined in several ways. For example, light-induced degradation of photoconductivity of weakly doped films and alloys of amorphous hydrogenated silicon, known as the Staebler–Wronski effect [135], is clearly observed upon illumination by radiation with energy $h\nu > 1.17$ eV over several hundreds of hours [150]. Usually, the formation of dangling bonds when weak Si–Si bonds are broken as a result of the nonradiative recombination of an electron–hole pair in the tails of the conductivity and valence bands is responsible for the effect [133, 151].

The mechanism has not been completely elucidated, and it has been established that this effect is not necessarily connected with hydrogen effusion from the film. The change observed in the IR spectra in the Si–H absorption region after extended illumination is considerable. For example, illumination by a 16 AM1† solar spectrum over 400 h decreases the absorption coefficient in the 2000–2100-cm^{-1} region by a factor of 1.5, although subsequent thermal annealing restores it to its initial value [133].

The dependence of IR transmission of the a-Si:H films on electron beam irradiation was studied [38]. The results obtained show that for films produced by SiH_4 decomposition in a high-frequency plasma, an extended irradiation period (90–120 min) leads to an increase in the absorption of 10–15% in the region corresponding to SiH and SiH_2 stretching modes for a substrate temperature $T = 380°C$ and 30% for a substrate temperature of 280°C. On the basis of the correlation between the absorption coefficient in this spectral region, the spin density caused by the dangling bonds, and the photoconductivity [133, 152], one can conclude that "soft" electron irradiation restores the "weak" Si–Si bonds (as does the thermal annealing), and the hydrogen concentration in the films remains unchanged. Hydrogen effusion takes place at annealing temperatures >350°C [130]. The transformation of IRRAS spectra of a-Si:H films after thermal CO_2 laser treatment is presented in Fig. 5.17 [153].

It has been suggested that, because IR absorption spectra depend on the local electric fields created by defects in the material, the hydrogen content determination from the IR spectra can be in error by about 50% [154]. However, IR spectroscopy is irreplaceable in determining the coordination of hydrogen in films, which, as noted above, determines the electrophysical parameters and characteristics of devices made on the basis of these films.

The problem of a-Si:H oxidation was considered in a number of works [155–159]. Production of MIS structures based on a-Si:H films precludes the application of thermal oxidation, as hydrogen effusion occurs at the high temperatures necessary for the thermal growth of an oxide. Therefore, low-temperature methods of SiO_2 deposition, namely anodic and chemical oxidation, are employed [125, 154, 155, 160, 161].

The spectrum of superthin anodic silicon oxide grown on an a-Si:H surface in an ethylene glycol solution of KNO_3 (0.04 mol·L^{-1}) with additional illumination

---

† AM1 = 1 atmosphere mass.

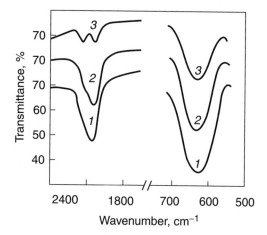

Figure 5.17. IR transmission spectra of a-Si:H film (1) before and (2) after 3 s and (3) 10 s annealing by CO_2 laser ($\lambda = 10.6 \, \mu m$, $p = 100 \, W \cdot cm^{-2}$). Reprinted, by permission, from H. Gleskova, V. V. Ilchenko, V. A. Skryshevsky, and V. I. Strikha, *Czech J. Phys.* **43**, 169 (1993). Copyright © 1993 Czech Journal of Physics.

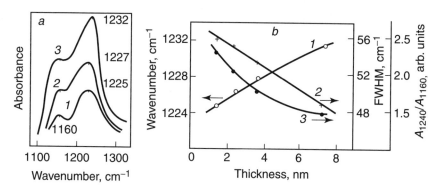

Figure 5.18. (a) IRRAS of anodic oxide SiO_2 grown on a-Si:H layer for various SiO_2 thicknesses d: (1) 1.5, (2) 3.6, and (3) 7.2 nm. (b) Various spectral parameters as function of oxide layer thickness d: (1) ν_m, (2) FWHM(1240), and (3) A_{1240}/A_{1160}. Reprinted, by permission, from V. A. Skryshevsky, V. I. Strikha, and H. Gleskova, *Czech J. Phys.* **42**, 331 (1992). Copyright © 1992 Czech Journal of Physics.

from an incandescent lamp is presented in Fig. 5.18 [161]. The analysis of the absorption band arising from the stretching vibrations of the Si—O bonds showed that with an increase in the oxide thickness from 1.5 to 7.2 nm, the position of the band maximum shifts from 1225 to 1232 cm^{-1}, and the FWHM decreases from 56 to 48 cm^{-1}. But, unlike spectra of the anodic oxides on the c-Si wafer, the intensity of the band at 1160 cm^{-1} increases with increasing SiO_2 thickness to a greater extent than the intensity of the 1230-cm^{-1} band. This may be explained by some peculiarities associated with the a-Si:H oxidation process. First, the dangling bonds in a-Si:H easily capture oxygen, and so the oxidation rate of a-Si:H is enhanced [157]; second, diffusion of oxygen into an amorphous material is rapid. Therefore, in contrast to c-Si, a-Si:H oxidation can also occur inside the film [158, 159].

5.5.2. a-SiGe:H

Alloying of a-Si:H with germanium is a convenient method to modify the energy band gap (from 1.7 to 1.0 eV) for specific applications [141]. The hydrogenated amorphous silicon–germanium alloy (a-SiGe:H) is especially suitable for application in terrestrial cascade solar cells, because the low band gap of a-Si$_{1-x}$Ge$_x$:H adjacent to an a-Si:H layer can collect both low- and high-energy photons.

Figure 5.19 shows IR absorption spectra of a-SiGe:H films deposited on a Si substrate by a glow discharge in a mixture of GeH$_4$ + SiH$_4$ + H$_2$ with various values of X_g = GeH$_4$/(GeH$_4$ + SiH$_4$) [126]. As X_g changes from 0 to 1, the peaks of the Si—H stretching mode at 2090 and 2000 cm^{-1} are replaced by peaks at 1980 and 1890 cm^{-1}, corresponding to the stretching modes of Ge—H [147]. If X_g increases from 0 to 0.2, the peaks of the (SiH$_2$)$_n$ bending modes at 890 and 845 cm^{-1} increase. However, as X_g increases further, these two peaks are gradually replaced by the peaks of the Ge—H bending vibrations at 830 and 760 cm^{-1} [126, 147], and the Si—H wagging mode at 640 cm^{-1} is gradually replaced by the Ge—H wagging mode at 570 cm^{-1} (Fig. 5.19). The quantity of bound H in a-SiGe:H films can be calculated by integrating the absorption peak of

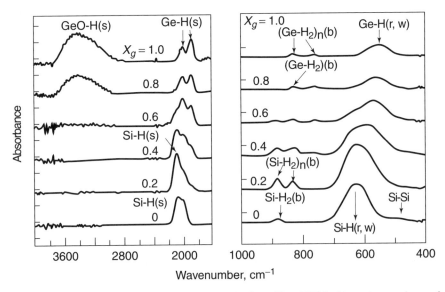

Figure 5.19. IR absorption of a-SiGe:H deposited at T_s = 200°C for various values of X_g = GeH$_4$/(GeH$_4$ + SiH$_4$). Reprinted, by permission, from Y.-P. Chou and S.-C. Lee, *J. Appl. Phys.* **83**, 4111 (1998). Copyright © 1998 American Institute of Physics.

Si—H and Ge—H wagging modes (630 and 570 cm^{-1}) on the basis of Eq. (5.13) using calibration factors given in Refs. [141, 147, 162]. For the Ge—H bond, the proportionality constant $A_i = 1.1 \times 10^{19}$ cm^{-2}.

The IR absorption spectra of oxidized a-SiGe:H films depend strongly on both the Ge content in the film and the oxidation conditions. With high GeH$_4$ concentrations during deposition, the resulting films have a tendency to rapid oxidation under atmospheric conditions, after which they will exhibit the absorption of GeO—H and SiO—H stretching modes (the broad peak at 2800–3700 cm^{-1}). With higher X_g values, the O$_2$Si—H and O$_3$Si—H stretching bands (at 2145 and 2245 cm^{-1}, respectively) are replaced by the O$_2$Ge—H and O$_3$Ge—H stretching bands (at 2025 and 2055 cm^{-1}, respectively). The bands of O$_2$Si—H$_2$ bending (930 and 975 cm^{-1}), O$_3$Si—H bending (830 cm^{-1}), Si—O—Si bending and stretching (820 and 1010 cm^{-1}, respectively), Ge—O—Ge stretching (870 cm^{-1}), and Ge—O stretching (760 cm^{-1}) are also observed [126].

5.5.3. a-SiC:H Films

Devices based on a-SiC:H have certain advantages over other semiconductor materials in a number of applications in optoelectronics, such as thin-film light-emitting diodes, coatings for laser facets, and a broadband window material for amorphous solar cells [163–165]. These applications exploit the fact that the optical energy gap and the refractive index of the films can be varied by changing their chemical composition.

IR spectroscopy has been applied to study the structural properties of a-Si$_{1-x}$C$_x$:H films prepared by magnetron or RF sputtering and the CVD technique [163-170]. Table 5.4 lists peak assignments for spectra of films with various carbon contents.

The IR absorption spectra of CVD films obtained at various proportions of silane and methane, as defined by $X = 100 \times$ SiH$_4$/(SiH$_4$ + CH$_4$), are shown in

Table 5.4. Peak assignments of a-Si$_{1-x}$C$_x$:H

Peak Position (cm^{-1})	Peak Assignment	References
640–650	SiH$_n$ wagging	166–169
760–800	SiC stretching	163, 166–169
845	SiH$_2$ rocking	166
880–900	SiH$_2$ scissors	166, 167, 169
950–960	Si–CH$_n$ wagging and rocking	163, 166, 169
980–1000	CH$_n$ wagging and/or rocking	166, 167
1200–1500	CH$_n$, Si(CH$_3$), C(CH$_3$) bending and scissors	163, 170, 171
1500–1600	C—C stretching	163, 171
2000	SiH stretching	163, 164, 167–170
2080	SiH$_2$ stretching	163, 164, 167–170
2090–2150	C—SiH stretching	169
2800–2900	CH$_n$ stretching	163–167, 170

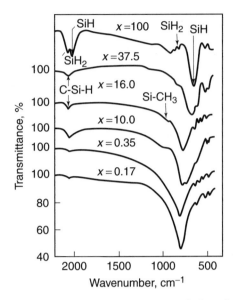

Figure 5.20. IR absorption spectra of a-SiC:H as function of silane fraction $X = SiH_4/(SiH_4 + CH_4) \times 100\%$ in gas mixture. Reprinted, by permission, from A. S. Kumbhar, D. M. Bhusari, and S. T. Kshirsagar, *Appl. Phys. Lett.* **66**, 1741 (1995). Copyright © 1995 American Institute of Physics.

Fig. 5.20 [169]. One can see that with a decrease in X, the spectrum evolves from that of a-Si:H to that of a-Si$_{1-x}$C$_x$:H as the SiC stretching band shifts from 760 to 800 cm^{-1} and becomes more asymmetric in the higher frequency region and the Si—H stretching absorption peak shifts from 2000 cm^{-1} (as in pure a-Si:H) to 2090 cm^{-1} (as in an a-SiC:H alloy). The gradual shift of the Si—H stretching absorption band may be a result of the induction effect as the number of (more electronegative) carbon atoms attached to Si—H increases [149]. Variations in the IR absorption spectra are accompanied by the growth of the optical band gap from 1.7 eV (pure a-Si:H) to 2.6 eV (a-Si$_{1-x}$C$_x$:H alloy with $X = 0.17$) [169]. The carbon content in the film can be calculated from the SiC stretching absorption band at 760 cm^{-1} by using Eqs. (5.13) and (5.14) and the coefficient $A_i = 2.1 \times 10^{19}$ cm^{-2} [171].

5.6. FILMS OF AMORPHOUS CARBON, BORON NITRIDE, AND BORON CARBIDE

5.6.1. Diamondlike Carbon

The two crystalline forms of carbon are diamond, which has a tetrahedrally bonded sp^3 configuration, and graphite, which has a trigonally bonded sp^2 configuration. The various forms of noncrystalline carbon, such as hard and soft forms of hydrogenated amorphous carbon a-C:H, can be considered as hybrids of graphite, diamond, and polyethylene $(CH_2)_n$. Diamondlike carbon (DLC) films are metastable amorphous materials containing a mixture of sp^3 and sp^2 *and* sometimes sp^1 coordinated carbon atoms in a disordered network [172]. DLC films are often deposited by CVD using a dilute mixture of hydrocarbons (often including methane) [173]. These films have excellent potential in industrial applications such as hard protective coatings, electric insulators, passivation layers, and antireflection coatings in silicon solar cells [174, 175].

Depending on the precursors and the deposition technique, DLC films may contain hydrogen in amounts ranging from 10 to 60%. The hydrogen content determines the film structure and is responsible for the wide optical gap, the high electrical resistance, and the passivation of dangling bonds in amorphous material [172, 176]. IR spectroscopy has been applied to study the hydrogen and carbon contents and their optimal configuration in DLC films [172–174, 177–183].

The IR absorption spectrum of a typical DLC film is shown in Fig. 5.21 [173], and the relevant absorption peaks are assigned in Table 5.5. All DLC films show a characteristic absorption in the two-phonon range of 1300–2700 cm^{-1}; DLC films containing defects display an additional absorption band in the symmetry-forbidden one-phonon region below 1350 cm^{-1} and another band between 2750 and 3300 cm^{-1} (C$-$H stretching region) [173]. The C$-$H absorption region is typical for CVD films, and the C$-$N absorption region is commonly observed in spectra of DLC films grown by CVD using a gaseous mixture of hydrocarbons and nitrogen as precursors [174, 179].

The CH stretching region (2750–3300 cm^{-1}) gives information about the hydrogen bound to carbon and impurity atoms. The concentration of bound hydrogen in films can be calculated from the CH_x stretching band in the region of 2800–3000 cm^{-1} using the method of integral CH_x peak absorption [180]:

$$N = \frac{A}{2900d} \int \alpha(\nu)\, d\nu, \tag{5.15}$$

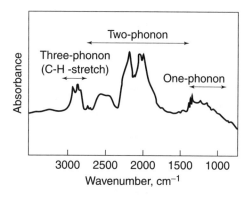

Figure 5.21. IR absorption spectrum of DLC films obtained by CVD. Reprinted, by permission, from K. M. McNamara, B. E. Williams, K. K. Gleason, and B. E. Scruggs, *J. Appl. Phys.* **76**, 2466 (1994). Copyright © 1994 American Institute of Physics.

Table 5.5. Assignments of peaks of DLC films

Peak Position (cm^{-1})	Peak Assignment	References
1030	O—CH$_3$ deformation	173
1130–1150	O—CH$_3$ rocking	173
1220	C—N stretching	173
1250	C—N nitrogen-vacancy pair	173
1370	N—CH$_3$ scissor	173
	sp^3 C—CH$_3$ deformation	172
	sp^3 CH	182
1420	CH$_3$ bending	174
1450	sp^3 C—CH$_3$ deformation	172
	N—CH$_3$, O—CH$_3$ deformation	173
	sp^3 CH$_2$ scissor	173, 182
1500–1580	C=C aromatic stretching	173, 182
1600–1620	C=C olefinic stretching	174, 182
	C=C isolated stretching	173
	C=N stretching	174
1740	C=O stretching	172, 173
2100	C≡C isonitrile group	172, 179
2200	C≡N stretching	174
2850	sp^3 CH$_2$	172, 173, 180, 182
2880	sp^3 CH$_3$	172, 173, 179, 180, 182
2920	sp^3 CH$_2$, sp^3 CH	172, 173, 180, 182
2960	sp^3 CH$_3$	172, 173, 180, 182
2980	sp^2 CH$_2$	172, 173
3025	sp^2 CH	172, 173
3085	sp^2 CH$_2$	173
3350	N—H stretching	174, 179

where N is the hydrogen density (cm^{-3}), α is the wavelength-dependent absorbance, A is a constant determined from a paraffin standard to be 7×10^{20} cm^{-2}, d is the film thickness, and the integral is taken over the 2800–3000-cm^{-1} range [184].

The frequencies of the C—H bond modes are shifted slightly, which allows one to distinguish between sp^3 and sp^2 configurations [173]. Some diamond films contain hydrogen bound to sp^2 carbon, while in other films the sp^2 carbon is below the detection limit [180]. The $sp^3 \rightarrow sp^2 \rightarrow sp^1 \rightarrow sp^2$ (aromatic) transformation of carbon was studied by FTIR [181]. It was shown that this transformation correlates with the release of hydrogen from the films during bombardment by various ionic species.

5.6.2. Boron Nitride and Carbide Films

Like carbon, boron nitride (BN) exists in two main crystalline structures, a hexagonal structure (h-BN) and a cubic zinc blend structure (c-BN). The c-BN is the second hardest known material and hence offers itself for numerous possible applications in the microelectronics industry, due to its extremely large bandgap, low conductivity, high dielectric constant, and high chemical inertness even at a high temperatures; in addition, it represents an ideal cutting-tool material. Hexagonal BN has also been used for insulating layers in optical devices and high-temperature electronics. Thin films of BN have been produced by different techniques, including sputtering, evaporation, and plasma-assisted CVD [185–187]. CVD at low substrate temperatures ($T < 400°C$) results in the growth of h-BN exclusively, whereas a thin film of c-BN can be deposited at $T > 500°C$ [188]. The IR absorption spectra of various forms of BN have been studied [185–191].

The IR transmission spectra of c-BN and h-BN films deposited by CVD on a Si substrate [188] are shown in Fig. 5.22. The TO stretching peak at 1383 cm^{-1} and the bond-bending peak at 770 cm^{-1} dominate the h-BN spectrum. The TO stretching band at 1055 cm^{-1} arises from the c-BN phase (Table 5.6). For films deposited on a Si wafer, the presence in the IR spectra of SiO$_2$ absorption (1080 cm^{-1}) near the c-BN peak must be taken into account when estimating the c-BN content in the film [188, 190]. The amorphous hydrogenated boron nitride grown by CVD at low temperatures ($T < 250°C$) exhibits three absorption peaks: 1371 cm^{-1} (a-B:H) and 1263 and 1505 cm^{-1} (a-B:N) [187]. The TO–LO splitting of BN absorption spectra was studied by polarized IRRAS [189]. It

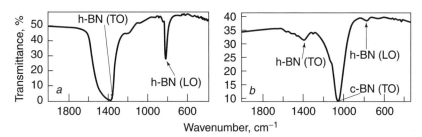

Figure 5.22. IR transmission spectra of h-BN and c-BN CVD films on Si substrates. Reprinted, by permission, from D. H. Berns and M. A. Cappelli, *Appl. Phys. Lett.* **68**, 2711 (1996). Copyright © 1996 American Institute of Physics.

Table 5.6. Assignments of peaks of BN

Peak Position (cm^{-1})	Peak Assignment	References
760–800	h-BN bending	185, 186, 188, 189
1364–1400	h-BN stretching (TO)	185, 186, 188, 189
1600	h-BN stretching (LO)	189
1055–1110	c-BN stretching (TO)	186, 188–191
1300	c-BN stretching (LO)	189
1263–1350	a-B:N	187, 189
1505–1550	a-B:N	187, 189

should be noted that the position of BN absorption peaks depends strongly on the deposition parameters (Table 5.6).

Amorphous hydrogenated boron carbide (a-B:C:H) was first grown as a p-type doped amorphous hydrogenated carbon (a-C:H). This material has become an important industrial wear-resistant coating that is widely used to protect computer hard disks from damage due to crashing magnetic heads. The IR absorption spectra of a-B:C:H deposited by plasma onto a Si substrate show a broad peak near 1300 cm^{-1} due to boron icosahedra [192]. Lower carbon concentrations cause a slight red shift from 1333 to 1295 cm^{-1}. Weaker peaks at 2930 and 2560 cm^{-1} are due to the stretching modes of C$-$H and B$-$H bonds, respectively.

5.7. POROUS SILICON LAYERS

Light-emitting devices are currently manufactured using expensive A^3B^5 direct gap semiconductors. However, use of monocrystalline silicon, especially after its conversion to porous silicon (PS), could lower the cost of optoelectronic devices substantially. This material was obtained more than 40 years ago by electrochemical etching of silicon in HF-based electrolytes [193] and has been used for thick (>1-μm) insulator layers in silicon-on-insulator (SOI) technology for integrated circuit (IC) processing [194]. Then in 1990, Canham [195] reported intense visible photoluminescence (PL) of PS at room temperature. Since then, LED devices [196, 197], metal$-$PS$-$silicon photodetectors [198], and solar cells [199–202] have all been fabricated using PS. PS is also of potential interest as optical links in VLSI circuits are explored and optical data processing systems based on silicon IC technology are developed.

PS is a spongelike material with randomly positioned pores and a sparse silicon network of nanocrystallites. Quantum-size effects in this materials are responsible for the transformation of silicon from indirect ($E_g = 1.1$ eV) to direct ($E_g = 1.9$–2.2 eV) band gap material [203] and for emission of light in the visible region [195, 204, 205]. The quantum nature of these changes is supported by the theoretical calculation of the energy levels in nanometer wires and by several different experiments confirming the presence of quantum-sized crystallites in PS [205–207]. Other possible causes of the observed luminescence have been proposed, including a layer of hydrogenated amorphous silicon on the sample surface [207, 208] and the formation of chemical compounds such as siloxene ($Si_6O_3H_4$) during etching [209].

IR absorption spectra of PS layers show a great deal of variation, depending on composition and morphology of the layers as well as on the method of formation used. Figure 5.23 presents absorption spectra of PS layers formed in various solutions on a Si wafer [210]. The as-prepared PS layer on a silicon wafer exhibits the characteristic absorption bands of the Si substrate, namely the bulk Si$-$Si (610–616-cm^{-1}) stretching mode [211, 212], and asymmetric (1105-cm^{-1}) stretching bands of the interstitial oxygen in Si [213–217].

Anodized PS is almost completely passivated with hydrogen [210, 218] and is characterized by strong absorption of SiH$_2$ and SiH wag modes at 622 and

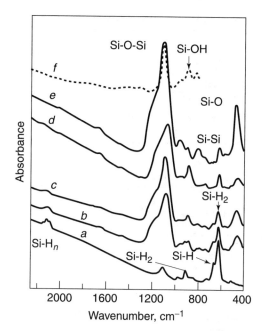

Figure 5.23. IR absorption spectra: (*a*) as-prepared PS; (*b–d*) PS anodized for 5 min in (*b*) 1HCl–7H$_2$O, (*c*) 0.04 *M* KNO$_3$–ethylene glycol, (*d*) 0.02 *M* KNO$_3$–ethylene glycol, (*e*) PS anodized for 30 min in 0.02 *M* KNO$_3$–ethylene glycol, and (*f*) flat Si anodized for 5 min in 0.02 *M* KNO$_3$–ethylene glycol. Reprinted, by permission, from M. Shimura, M. Katsuma, and T. Okumura, *Jpn. J. Appl. Phys.* **35**, 5730 (1996). Copyright © 1996 Publication Board of Japanese Journal of Applied Physics.

664 cm^{-1} [210, 215, 217, 219–221]; the twisting, bending, and scissor bands of SiH$_2$ at 804, 848, and 906 cm^{-1}, respectively [213, 215, 217, 219, 221, 222]; and strong stretching mode absorption of SiH (2087 cm^{-1}), SiH$_2$ (2106 cm^{-1}), and SiH$_n$(2140 cm^{-1}) [213, 215, 217, 219, 220, 223–225]. Oxygen species are not usually observed in as-prepared films. Weak bands of SiF such as the SiF$_x$ ($x = 1, 2, 3$) stretching band may also be detected [215, 226–228], but most SiF bonds are rapidly hydrolyzed to Si—OH by exposure to water [229, 230]. The chemical composition of films after various treatments is presented in Table 5.7 [205] and the peak assignments in Table 5.8.

Thermal annealing in vacuum leads to a loss of hydrogen from the PS layer; the peaks in the 2080–2140-cm^{-1} range corresponding to the monohydride, dihydride, and trihydride disappear after annealing at 450°C [207, 215, 218, 239, 240] as do the Si–H bands in the 620–910-cm^{-1} range. Elimination of hydrogen from the PS layer corresponds to a decrease in the PL intensity, suggesting that the Si–H species plays an important role in PS luminescence [218, 241]. It is known that hydrogen passivation reduces the number of dangling bonds that act as nonradiative recombination centers [218]. The correlation between silicon hydride species and the PS luminescence has also been observed in other

Table 5.7. Surface chemical composition depending on PS treatment

PS Formation and Treatment	Chemical Composition
In situ in HF during and after forming	SiF_xH_y
Freshly etched in inert ambient conditions	SiH_x
Chemically or anodically oxidized	SiO_xH_y
Rapidly thermally oxidized at high temperatures	SiO_xH_y
Aged in ambient air for months to year	$SiO_xH_yC_z$

Table 5.8. Absorption bands observed in spectra of PS layers

Peak Position (cm^{-1})	Peak Assignment	References
484	Si−O−Si rocking	210, 215, 231
515	Si−O−Si symmetric stretching of interstitial oxygen in Si	214
616	Bulk Si−Si stretching	211, 212
622	SiH_2 wag	210, 215, 220, 221, 231, 232
664	SiH wag	210, 215, 217, 219−221, 231−233
804	SiH_2 twist or SiH (Si_2O) bending	214 234
832	SiF, SiF_2 stretching	228
843	SiH (SiO_2) bending	232, 234
880	O_3−SiH bending	210, 215, 219, 232
906	SiH_2 scissors	210, 213, 215, 217, 219, 221, 222, 231−233, 235
946	SiF_3 stretching	215, 228
1060	Si−O−Si stretching (TO)	210, 215, 231, 236
1105	Si−O−Si asymmetric stretching of interstitial oxygen in Si	213, 216, 217, 235
1170	Si−O−Si stretching (LO)	236
2087	SiH stretching	210, 213, 215, 217, 219, 220, 223−225, 228, 233, 235
2106	SiH_2 stretching	210, 213, 215, 217, 219, 220, 223−225, 228, 233, 235
2140	SiH_n stretching	213, 217, 223−225, 231, 233
2190	SiH−SiO_2 structural group	215, 217, 231
2250	O_3SiH stretching	213, 215, 219, 223, 231, 232
2853	CH_2 symmetric stretching	214
2921	CH_2 asymmetric stretching	214, 237
2955	CH_3 symmetric stretching	214, 237
3400−3500	O−H stretching, water vapor	206, 222, 224, 238
3660	O−H stretching, Si(OH)	231

experiments [235, 238, 242, 243]. For example, the adsorption of organoamone molecules significantly affects the concentration of both SiH and SiH_2 stretch modes of surface species and leads to a decrease in the quantum yield of photoluminescence [243].

Oxidized PS was investigated in detail by IR spectroscopy; results of thermal oxidation [213, 234, 239, 244–248], anodic oxidation [210, 219, 238], and chemical oxidation [231, 249, 250] were obtained. During oxidation the PS samples gradually lose the absorption peaks corresponding to silicon hydrides (Fig. 5.23). However, the oxidation leads to the formation of a SiO_x layer, characterized by the presence of the rocking (484-cm^{-1}) and stretching ν_{TO} (1060-cm^{-1}) and ν_{LO} (1170-cm^{-1}) bands [210, 215, 236, 251]. Bands can appear at 700–900, 2190, and 2250 cm^{-1} that are ascribed to more complex oxygen structural groups [213, 215, 219, 223, 231, 236].

Silicon oxide species are also detected in IR spectra after aging PS films in ambient air [233, 240, 252, 253]. However, the native oxide growth on PS can be suppressed by treating the surface in a HF solution [253]. The aging of PS layers can also lead to the appearance of absorption bands (2850–3000 cm^{-1}) associated with hydrocarbon contamination [237]. Aged samples exhibit a broad IR absorption band in the 3000–3600-cm^{-1} region due to adsorption of water molecules, giving SiOH complexes [240].

UV irradiation leads to an increase in the intensity of the Si—O stretching bands and the H—Si—O_3 deformation and stretching bands [236, 251]. This is typical of photooxidation of PS [217]. The decrease in intensity of these four bands after etching in HF indicates the formation of an oxide layer under UV illumination. Its appearance is also accompanied by a reduction in the intensity of all Si—H bands (664, 906, 2087, 2106, and 2140 cm^{-1}). Figures 5.24 and 5.25 show the increase of oxide bands and the decrease of SiH_n bonds upon UV illumination of the PS layer [236, 251].

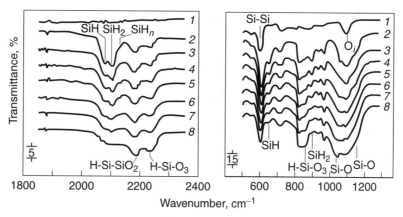

Figure 5.24. IR normal-incidence transmission spectra of (1) Si-substrate and (2–8) PS samples upon UV illumination for various times: (2) 0, (3) 10, (4) 20, (5) 30, (6) 40, (7) 60, and (8) 100 min.

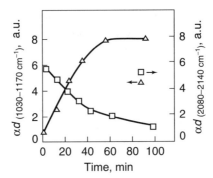

Figure 5.25. Absorption of PS layer in 1030–1170- and 2080–2140-cm^{-1} spectral regions as function of duration of UV illumination.

Recently, the application of PS to chemical sensors has been considered [254–256]. The main advantages of this material for gas sensor and actuator applications are a unique combination of (i) a crystalline structure, (ii) a huge internal surface (200–500 m^2·cm^{-2}) [218, 257] that enables one to enhance the adsorbate effects, and (iii) a highly reactive surface, allowing efficient modification of the PS surface by various treatments such as contact with organic solvent, thermal annealing, or an illumination.

The decomposition of H_2O and D_2O on a PS surface was studied by FTIR [258]. The spectra reveal that H_2O (D_2O) dissociates upon adsorption at 300 K to form SiH (SiD) and SiOH (SiOD) groups. The decomposition of the surface species formed from H_2O (D_2O) adsorption can be monitored using the SiH (SiD) stretching mode at 2090 (1513) cm^{-1}, the SiOH (SiOD) stretching mode at 3680 (2707) cm^{-1}, and the Si–O stretching mode at 900–1100 cm^{-1}. Changes in the chemical composition of the surface and its morphology during etching and the formation of a hydrophobic silicon surface upon immersion in an aqueous solution of HF were analyzed in [226, 259].

IR spectroscopy was used for the analysis of adsorption and decomposition of trichlorosilane and trichlorogermane on PS [260]. Trichlorosilane dissociates upon adsorption, forming SiH, $SiCl_x$, ClSiH, and Cl_2SiH. Exposure of a PS surface to organic solvents results in both the modification of IR absorption of PS in the 900–1250-cm^{-1} region due to C–Si–O groups and the appearance of weak absorption peaks at 2850 and 2950 cm^{-1} due to OCH_3 groups [238]. The adsorption of alkyl groups [261], benzene vapor [262], propanol [235], methanol [225, 238], and other gases or solutions [215, 238, 263] also leads to the modification of the IR absorption spectra.

5.8. OTHER DIELECTRIC LAYERS USED IN MICROELECTRONICS

5.8.1. CaF$_2$, BaF$_2$, and SrF$_2$ Layers

The dielectric layers CaF_2, BaF_2, and SrF_2 have crystalline lattice parameters similar to those of Si and do not exhibit an excessive concentration of defects at the interface with silicon, which is of importance in silicon microelectronics [264, 265]. The TO phonon absorption bands of these cubic crystals are in the far-IR region (see Table 5.9).

The frequencies of the ν_{LO} absorption bands of these materials, determined from the LST law [Eq. (1.41)], lie at higher frequencies (e.g., 465 cm^{-1} for CaF$_2$). Therefore, the structural and chemical properties of this material can be easily analyzed using standard IR spectrophotometers operating in this spectral region.

Examples of such analyses are presented in Fig. 5.26. CaF$_2$ layers with thicknesses ranging from 30 to 460 nm were obtained by molecular beam epitaxy on n-Si. The ν_{LO} band measured in p-polarized radiation is a symmetric Gaussian with a maximum at 450 cm^{-1}. The absorbance increases linearly with an increase in the thickness, while the maximum shifts from 452 cm^{-1} for a 30-nm layer up to 472 cm^{-1} for a 460-nm layer, and the FWHM decreases from 62 to 28 cm^{-1} (Fig. 5.27).

These changes in the absorption bands cannot be attributed to a chemical interaction between the layer and the silicon substrate because under molecular

Table 5.9. Optical parameters of CaF$_2$, BaF$_2$, and SrF$_2$

	CaF$_2$	SrF$_2$	BaF$_2$
ν_1(TO), cm^{-1}	257	217	184
α_1, cm^{-1}	3.9×10^4	3.3×10^4	2.8×10^4
ν_2(TO), cm^{-1}	328	316	278
α_2, cm^{-1}	1.7×10^3	0.5×10^3	0.4×10^3
ε_{st}	6.7	6.6	7.2
ε_∞	2.05	2.07	2.16
ν_{LO}, cm^{-1}(calculated)	463	374	326

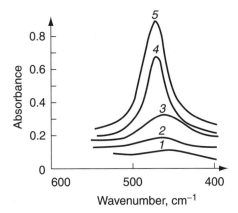

Figure 5.26. The p-polarized absorption spectra of CaF$_2$ layers on Si substrates at 75° incidence for different thicknesses: (1) 30, (2) 45, (3) 100, (4) 330, and (5) 460 nm.

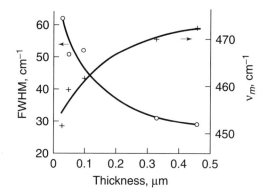

Figure 5.27. Frequency and the FWHM of ν_{LO} band as functions of thickness of CaF$_2$ layer.

beam epitaxy this interaction is weak [266]. The lattice constant of the CaF$_2$ layer ($a = 5.46$ Å) is close to that of Si ($a = 5.43$ Å), and therefore the influence of the substrate on the length and the angle of the Ca−F bond may be neglected. By the methods of AES and electron diffraction, a nonstoichiometric layer was detected at the CaF$_2$−Si interface [267, 268], and it seems likely that the observed change in the absorption band is caused by this change in the structure of the CaF$_2$ film as its thickness increases, as for silicon oxide layers.

5.8.2. GeO$_2$ Film

Amorphous GeO$_2$ is a promising material for optical waveguides and various optical elements in integrated optical systems [269]. The usual methods of the film fabrication are based on reactive sputtering of a GeO$_2$ target and laser ablation deposition techniques [270]. Most sputtered–deposited GeO$_x$ films require thermal annealing to produce defect-free stoichiometric films of GeO$_2$. IR transmission spectra of GeO$_x$ thin films are shown in Fig. 5.28 [270]. The structure of a-GeO$_2$ can be described as Ge−Ge$_y$O$_{4-y}$ tetrahedra bonded by bridging oxygen atoms. As the partial pressure of oxygen in the sputtering mixture is increased, the Ge atoms surrounding the central Ge atom are replaced by O atoms [271], and the ν_{TO} band of the asymmetric Ge−O−Ge stretch mode in GeO$_2$ that appears at 885 cm^{-1} shifts linearly from 885 to 740 cm^{-1} for $y = 0, \ldots, 4$ according to the empirical equation [271]

$$\nu(x) = 72.4x + 743 \text{ cm}^{-1}, \tag{5.16}$$

where x is the oxygen content ranging from 2 to zero. Spectra in Fig. 5.28 show the transformation from GeO$_x$ to GeO$_2$ via thermal annealing. According to Eq. (5.16), the Ge−O−Ge asymmetric mode absorption band shifts toward higher energies as the substrate temperature increases to 300°C, implying an increase of oxygen in the film matrix ($x \approx 2$ at 300°C). Furthermore, the Ge−O−Ge bond angle θ changes from 93° to 118° [271]. The broadening of the absorption band at high temperatures may be a result of the interfacial strain due

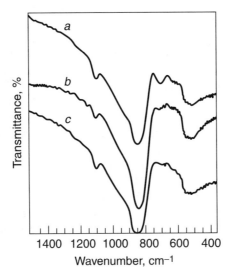

Figure 5.28. IR transmittance spectra of laser deposited GeO$_x$ thin films on Si substrate at three substrate temperatures: (a) ambient, (b) 100°C, and (c) 300°C. Reprinted, by permission, from S. Witanachchi and P. J. Wolf, *J. Appl. Phys.* **76**, 2185 (1994). Copyright © 1994 American Institute of Physics.

to a thermal expansion mismatch between Si and GeO$_2$ [270]. IR transmission spectra of a-Si$_x$Ge$_y$O$_z$ alloy films and a-SiO$_x$:H/a-GeO$_x$:H multilayers were studied [272, 273].

5.8.3. Metal Silicides

Metal silicides are widely used in microelectronic devices for the increase of the potential barrier height in metal–silicon Schottky contact. Titanium silicide is favored for its low resistivity and thermal stability. IR spectroscopy is not typically used for investigation of metal silicides, because they do not have a characteristic and selective absorption in the IR range. Nevertheless, FTIR spectroscopy has been used to measure the thickness of Ti and titanium silicide films [274] as well as to monitor titanium silicide formation during the reaction of Ti films on a Si wafer [275].

Only the as-deposited and/or annealed ($T = 300°C$) Ti films show wide absorption bands near 3000 cm^{-1} [275]. At different stages of the silicidation reaction, the films reveal different nonselective absorption curves in the 400–4000-cm^{-1} range. Thus, it was shown that the absorption at 4000 cm^{-1} increases monotonically as the annealing temperature of the sample increases and the absorption at 400 cm^{-1} increases monotonically with increasing sample conductivity. Thus the composition of a silicide may be deduced; however, in calculations of absorbance it is necessary to take into account the strong dependence of the reflectance on both the frequency ($R = 95\%$ at 400 cm^{-1}, $R = 60\%$ at 4000 cm^{-1}) and the silicide composition [275].

5.8.4. Amorphous Ta₂O₅ Films

Ta_2O_5 films with a high dielectric constant $\varepsilon \cong 25$ are suitable for use as memory cell dielectrics in dynamic random-access memories (DRAMs) [276, 277]. They may also successfully replace SiO_2 layers in gate dielectrics in metal–oxide–semiconductor field-effect transistor (MOSFET) devices. Ta_2O_5 films are produced using either low-pressure CVD or PECVD [278, 279]. Upon deposition of a Ta_2O_5 film on a silicon substrate, an interstitial silicon oxide layer may be formed as a result of the diffusion of oxygen atoms through Ta_2O_5 during the deposition or annealing processes.

The IR absorbance spectra of as-deposited 84-nm-thick Ta_2O_5 film on a Si substrate are shown in Fig. 5.29 [277]. The amorphous film is characterized by ν_{TO} bands at 519 and 642 cm^{-1}. The ν_{LO} bands at 795 and 940 cm^{-1} arise as supplementary peaks at oblique incident radiation (65° to the normal). Recrystallization at 900°C for 30 min in nitrogen atmosphere causes peaks to appear at 514, 576, 700, and 820 cm^{-1} (normal incidence) and 511, 576, 635, 802, and 940 cm^{-1} (oblique incidence). The appearance of SiO_2 absorption peaks at 1072 and 1250 cm^{-1} indicates the formation of the interstitial layers of silicon oxide during the Ta_2O_5 deposition, and thermal annealing increases the thickness of the SiO_2 layer. Therefore, the resulting structure can be considered as Ta_2O_5–SiO_2–Si. In the CVD process using a $Ta(OC_2H_5)_5$ precursor, the peaks at 2300 and 1500 cm^{-1} related to carbon species are identified in the IR absorbance spectra [276].

5.8.5. SrTiO₃ Film

The $SrTiO_3$ films have even larger dielectric constants than Ta_2O_5 (up to 270 [280]) and are also of interest for use as storage capacitors in DRAM and for silicon VLSI applications. The $SrTiO_3$ directly deposited on a Si substrate reveals a much lower effective dielectric constant due to the formation of a SiO_2

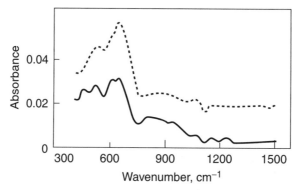

Figure 5.29. IR absorbance spectra of as-deposited 84-nm Ta_2O_5 film on Si wafer measured at normal incidence (dashed line) and 65° angle of incidence (solid line). Reprinted, by permission, from R. A. B. Devine, *Appl. Phys. Lett.* **68**, 1924 (1996). Copyright © 1996 American Institute of Physics.

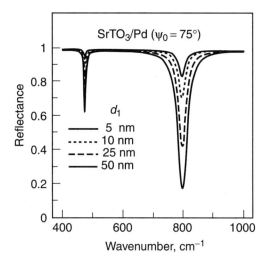

Figure 5.30. The p-polarized IRRAS spectra of $SrTiO_3$-Pd-Si recorded at 75° angle of incidence for different $SrTiO_3$ film thicknesses. Reprinted, by permission, from H. Myoren, T. Matsumoto, and Y. Osaka, *Jpn. J. Appl. Phys.* **31**, L1425 (1992). Copyright © 1992 Publication Board Japanese Journal of Applied Physics.

interfacial layer [281]. To avoid this and produce a thin-film capacitor with a high-capacitance density, a metal barrier layer on the Si wafer has been proposed [280]. Polarized IRRAS was used to analyze the dielectric properties of the $SrTiO_3$ film deposited by RF magnetron sputtering on a Pd−Si substrate.

Figure 5.30 shows the p-polarized reflection spectra of $SrTiO_3$ recorded at an angle of incidence of 75° for various film thicknesses [280]. In these spectra, the absorption bands at 480 and 800 cm^{-1} (ν_{LO} band) and the very weak absorption at 544 cm^{-1} (ν_{TO} band) are observed. The absorption band of LO modes increases linearly with increasing angle of incidence. The minimum thickness of the $SrTiO_3$ film, detectable by absorption at the ν_{LO} band frequency, was less than 5 nm.

5.8.6. Metal Nitrides

Metal nitride thin films on a Si substrate display interesting optoelectronic, thermal, and acoustic properties [282]. For example, aluminum nitride AlN with a band gap of 6.2 eV and a high surface acoustic wave (SAW) velocity (6×10^5 cm/s) has potential for UV-light-emitting and SAW devices [283]. AlN also exhibits high thermal conductivity, stability, and resistivity (10^{13} Ω cm), and its hardness and thermal coefficient of expansion (2.6×10^{-6}/K) are comparable to that of silicon. Metal nitride films have been prepared by many methods, including CVD, plasma-assisted CVD, metallorganic CVD, sublimation of bulk material, plasma-assisted molecular beam epitaxy (MBE), and laser vapor deposition [283–287].

Figure 5.31 shows the absorbance of AlN films deposited on Si by electron cyclotron resonance plasma CVD at two reaction pressures [286]. The strong

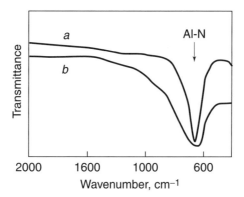

Figure 5.31. IR transmission spectra of AlN films deposited by CVD on Si substrate at (a) 67 Pa and 390°C and (b) 2.7 Pa and 340°C. Reprinted, by permission, from W. Zhang, Y. Someno, M. Sasaki, and T. Hirai, *Jpn. J. Appl. Phys.* **32**, L116 (1993). Copyright © 1993 Publication Board of Japanese Journal of Applied Physics.

absorption peak observed at 665 cm^{-1} is due to the TO phonon modes of Al$-$N [283, 286, 288]. This value is in good agreement with that obtained for single-crystal AlN (667 cm^{-1}) [289]. The film deposited at the higher reaction pressure has a narrower absorbance peak owing to higher degree of crystallinity. The LO phonon peak of Al$-$N is positioned at 888 cm^{-1} [287, 290]. For PECVD deposition, the AlN films display the N$-$H (3205 cm^{-1}) and Al$-$H (2108 cm^{-1}) absorption bands [291].

The IR reflectance spectra of thin GaN and AlGaN heterostructures with a buffer AlN layer were measured [287]. Besides the TO/LO phonon modes of Al$-$N, the LO mode of Ga$-$N (744 cm^{-1}) was observed. This correlates with the calculated phonon modes of GaN (LO: 734 and 743 cm^{-1}; TO: 532 and 561 cm^{-1}) [292]. The vibrational features of α-GaN and β-GaN epilayers have been discussed [285], and infrared studies of TiN thin film are reported [293].

5.9. MULTI- AND INHOMOGENEOUS DIELECTRIC LAYERS: LAYER-BY-LAYER ETCHING

Various multilayer dielectric films are commonly used in microelectronics, including, for example, "combination oxides" $Si_3N_4-SiO_2$. Combination oxides, which consist of a thin thermal SiO_2 layer (to achieve interface properties equivalent to thermal SiO_2) followed by sputtered bulk oxide (produced at low temperatures), were studied by IR spectroscopy [294]. The peak position and the FWHM of the stretching mode of combination oxides only differ by 6$-$8 cm^{-1} from those of dry thermal oxides. This minor deviation indicates that the combination oxides and thermal ones have a comparable stoichiometry.

Films of silicon nitride deposited on a-Si:H employed in thin-film transistors and superlattice structures have been studied by FTIR-ATR [295]. Hydrogen effusion from a-Si:H is induced by the SiN deposition and varies with substrate temperature. The incorporation and release of hydrogen during a-Si:H growth on a SiO_2 substrate were investigated [296, 297]. The IR grazing-angle internal reflection (GIR) spectroscopy was applied to study ultrathin SiN films (0.4–4.0 nm) on ultrathin SiO_2 layers (used as an antireflection coating in MIS solar cells) [298]. The $Si-O-Si$ stretching absorption of the 1.3-nm SiO_2 layer is significantly altered by SiN deposition using direct PECVD, while the vibrational properties of the ultrathin oxide are unchanged after silicon nitride deposition by remote PECVD. The absorption band broadening following the direct PECVD is attributed to the formation of a silicon oxynitride film due to ion bombardment. The evolution of IR transmission spectra of $a-SiO_x:H-a-GeO_x:H$ multilayers as a function of the thermal annealing regimes were studied [273].

In general, ex situ IR spectroscopy with layer-by-layer etching is not widely used in the analysis of diffusion profiles in semiconductor structures, because it is exceeded in sensitivity by AES with the layer-by-layer ionic etching. However, IR spectroscopy can be informative in studies of structural and/or chemical inhomogeneity of multilayer or heterogeneous films. Furthermore, selective control over the etching of SiO_2 with respect to Si is required in certain processing steps in VLSI manufacturing [299].

The distribution of inhomogeneities in anodic SiO_2 layers on Si substrate has been studied by p-polarized multiple transmission and IRRAS at $\varphi = 75°$ [23, 64, 80]. A decrease in the thickness of the silicon oxide upon etching (in highly dilute HF) is accompanied by a reduction in the intensities of the 1240- and 1160-cm^{-1} peaks, a broadening of the band at 1240 cm^{-1}, and a shift of its maximum from 1240 cm^{-1} for $d = 13$ nm to 1210 cm^{-1} for $d = 1$ nm (Fig. 5.32) [80]. The changes in the intensity, shape, and position of the 1160-cm^{-1} band are less pronounced, so that at an oxide thickness of 1.5 nm, the intensities of the 1240- and 1160-cm^{-1} bands become comparable. This allows one to conclude that the SiO_2 layer is inhomogeneous, containing silicon oxides of different stoichiometry to a depth of at least several monolayers from the Si substrate, while growth of the layer involves mainly the formation of stoichiometric SiO_2. The qualitative changes in the SiO_x absorption spectra occurring during the layer-by-layer etching coincide with those observed in the spectra of silicon oxide layers of various thicknesses on silicon [22, 80].

Evidence for inhomogeneity in thin plasma-grown oxides was also obtained from the deconvolution of the ν_{LO} band combined with chemical etchback in a dilute solution of HF [24, 90]. The shift of the ν_{LO} band from 1251 cm^{-1} ($d_{SiO_2} = 70$ nm) to 1245 cm^{-1} ($d_{SiO_2} = 5$ nm) is assumed to be the result of structural modifications in the SiO_2 network. These modifications include an increase in density (+3.8%) and a change in the average bridging bond angle (−1.6°) in the slice furthest from the SiO_2-Si interface. The oxide close to the SiO_2-Si interface is changed most significantly (−2.3° in the bridging bond angle and +2% in the density). The probability of the appearance of suboxide SiO_x structures ($x = 1, 1.25, 1.5, 1.75, 2$) in a thermal oxide film 4.0 nm thick was calculated according to the multilayer interface model and with allowance made

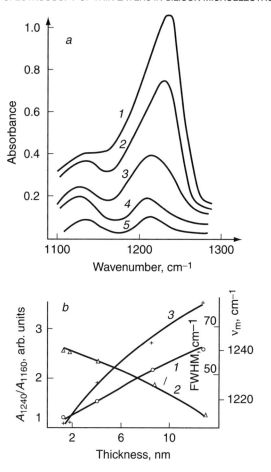

Figure 5.32. (a) Absorption spectra of silicon oxide after layer-by-layer etching. The SiO_2 layer thicknesses d are (1) 13.0, (2) 8.6, (3) 3.8, (4) 1.6, and (5) 1.0 nm. (b) Parameters of absorption spectra: (1) ν_m, (2) FWHM(1240), (3) A_{1240} A_{1160}. Reprinted, by permission, from S. S. Kilchitskaya, T. S. Kilchitskaya, V. A. Skryshevsky, V. I. Strikha, and V. P. Tolstoy, *Zhurnal prikladnoi spektrokopii* **48**, 445 (1988). Copyright © 1988 Publishing House "Nauka i technika".

for the influence of the layer-by-layer etching on the position of the ν_{TO}/ν_{LO} peaks [49]. The quality of the ultrathin Si_3N_4-SiO_2 films was analyzed using the dependence of the ν_{TO} band parameters on the etched-back oxide thickness [300]. The etching of the thermal SiO_2 layer from 12 to 1.5 nm decreases the frequency of the TO Si-O-Si stretching mode from 1075 to 1057 cm^{-1} and increases the FWHM from 75 to 110 cm^{-1}.

The IRRAS spectra of thermal SiO_2-$(n^+)Si$ structures under layer-by-layer chemical etching are shown in Fig. 5.33 [23]. In the spectra of thick layers bands are observed at 1250 and 1100 cm^{-1}, corresponding to the ν_{LO} and ν_{TO} bands, respectively. As the etching is continued, the band at 1100 cm^{-1} decreases

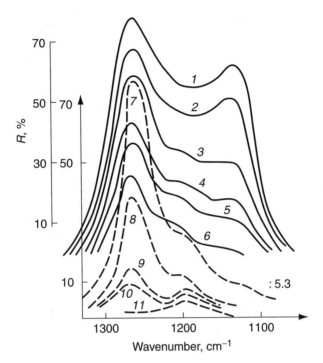

Figure 5.33. IRRAS spectra of SiO$_2$–(n^+)Si structures at oxide thicknesses d of (1) 400, (2) 348, (3) 307, (4) 223, (5) 192, (6) 138, (7) 76, (8) 38, (9) 9, (10) 6, and (11) 2.5 nm.

rather rapidly (quenching of ν_{TO} bands by a degenerated semiconductor or metal layers; Section 3.2.2), and a band at 1160 cm^{-1} appears. The ratio of the 1250- and 1100-cm^{-1} band intensities is determined by experimental conditions and is closely related to the angle of incidence φ and the layer thickness d, as follows from Fig. 3.20. Indeed, it can be shown that $R_{LO}/R_{TO} \sim (1/d)^2 \sin \varphi$ [301]. The profile of homogeneity is defined by the ratio R_{1160}/R_{1250}.

At $d > 20$ nm, the SiO$_2$ layers are practically homogeneous, and the ratio R_{1160}/R_{1250} is independent of the thickness. For a thinner SiO$_2$ layer, the band at 1250 cm^{-1} is broader, its maximum is shifted toward lower frequencies, and the ratio R_{1160}/R_{1250} is increased. These effects are all a result of an oxygen deficiency at the initial stages of thermal silicon oxide formation, as in the case of anodic oxides, which leads to the appearance of a nonstoichiometric SiO$_x$ layer, as observed in the IRRAS spectra [23].

The profile of the multilayer Si$_3$N$_4$–SiO$_x$–Si structure annealed in wet oxygen was studied with layer-by-layer chemical etching and concurrent measurement of IR absorption [23, 64]. As seen from Fig. 5.34 [64], during the layer-by-layer etching, the intensities of the bands at 1250 and 820 cm^{-1} (proportional to the SiO$_2$ thickness) decrease sharply, and the intensity of the band at 870 cm^{-1} (proportional to the Si$_3$N$_4$ thickness) remains virtually unchanged. This is because Si$_3$N$_4$ represents an efficient diffusion barrier [2]. On further etching, the intensities of the bands at 1250, 1070, and 870 cm^{-1} decrease uniformly. This suggests that the layer being studied is the heterogeneous SiNO system consisting of the SiO$_x$ and Si$_3$N$_4$ phases, the SiO$_x$ oxide apparently containing some Si$_3$N$_4$, causing the silicon oxide band to be shifted toward lower frequencies (Fig. 5.34).

The distribution of oxygen impurities in the a-Si:H films was analyzed by IR transmission spectroscopy in combination with ionic etching [302]. It was established that the aging of these films in air at room temperature results in the appearance of a strong absorption from the Si−O−Si bonds, but the oxygen is not localized at the surface or in the near-surface region. The existence of pores in the a-Si structure leads to the migration of the oxygen impurities to depths on the order of micrometers from the film surface.

The chemical content of the a-Si:H surface during CF$_4$−H$_2$ plasma etching was analyzed by MIR [299]. The C−F$_x$ and C=C stretching vibrations at 1230 and 1700 cm^{-1}, respectively, appear following plasma etching due to fluoro-carbon film formation. MIR was also applied to study the sublayer structure resulting from the etching in CF$_4$−O$_2$ plasma of a 70-nm-thick a-Si:H layer on a Si substrate [303]. A hydrogen-rich surface sublayer at the film-free surface, a hydrogen-rich interface layer with Si dangling bonds at the a-Si:H−c-Si interface, and a "transition sublayer" with higher SiH$_2$ concentration at depths over 30 nm from the a-Si:H−c-Si interface were all identified.

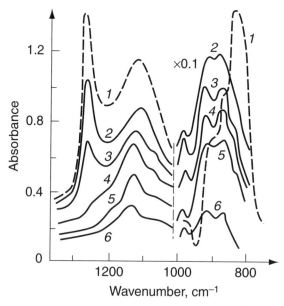

Figure 5.34. Variations in IR absorption spectra of SiO$_2$–Si$_3$N$_4$–SiO$_x$–Si structure after layer-by-layer chemical etching (curve number increases with decreasing film thickness). Reprinted, by permission, from Yu. A. Averkin, N. K. Karmadonov, V. A. Skryshevsky, V. I. Strikha, and A. V. Kharlamov, *Ukrainian J. Phys.* **34**, 1817, (1989). Copyright © 1989 Institute of Theoretical Physics, National Academy of Science of Ukraine.

REFERENCES

1. S. M. Sze, *Physics of Semiconductor Devices*, Vol. 1, Wiley, NewYork, 1981.

2. S. M. Sze (Ed.), *VLSI Technology*, Vol. 1, McGraw-Hill, NewYork, 1983.

3. W. Kern and V. S. Ban, in J. L. Vossen and W. Kern (Eds.), *Chemical Vapor Deposition of Inorganic Thin Films, Thin Film Process*, Academic, New York, 1978, p. 257.

4. E. A. Irene, in M. Froment (Ed.), *Passivity Metal, and Semicond.*, Proc. 5th Int. Symp., Elsevier, Amsterdam, 1983, p. 11.

5. E. A. Irene and R. Ghez, *Appl. Surf. Sci.* **30**, 1 (1987).

6. E. A. Lewis and E. A. Irene, *J. Vac. Sci. Technol.* **A4**(3, Pt. 1), 916 (1986).

7. S. E. Blum, K. H. Brown, and R. Srinivasan, *Appl. Phys. Lett.* **43**, 1026 (1983).

8. W. A. Tiller, *J. Electrochem. Soc.* **128**, 689 (1981).

9. V. G. Litovchenko and A. P. Gorban, *Bases of Physics of Microelectronics Systems of Metal-Insulator-Semiconductor*, Naukova Dumka, Kiev, 1978.

10. G. Lucovsky, M. J. Manitini, J. K. Srivastava, and E. A. Irene, *J. Vac. Sci. Technol.* **B5**, 530 (1987).

11. F. L. Galeener, *Phys. Rev. B* **19**, 4292 (1979).

12. P. N. Sen and M. F. Thorpe, *Phys. Rev. B* **15**, 4030 (1977).

13. I. W. Boyd, *Appl. Phys. Lett.* **51**, 418 (1987).

14. I. P. Lisovskii, V. G. Litovchenko, V. G. Lozinskii, and G. I. Steblovskii, *Thin Solid Films* **213**, 164 (1992).

15. W. A. Pliskin and H. S. Lehman, *J. Electrochem. Soc.* **112**, 1013 (1965).

16. H. Shirai and R. Takeda, *Jpn. J. Appl. Phys.* **35**, 3876 (1996).

17. A. Sassella, B. Pivac, T. Abe, and A. Borghesi, *Mater. Sci. Eng.* **B36**, 221 (1995).

18. R. A. B. Devine, *Appl. Phys. Lett.* **68**, 3108 (1996).

19. C. H. Bjorkman, T. Yamazaki, S. Miyazaki, and M. Hirose, *J. Appl. Phys.* **77**, 313 (1995).

20. P. Grosse, B. Harbecke, B. Heinz, and R. Meyer, *Appl. Phys.* **A39**, 257 (1986).

21. U. Teschner and K. Hubner, *Phys. Stat. Solidi* **B159**, 917 (1990).

22. S. S. Kilchitskaya, S. V. Litvinenko, V. A. Skryshevsky, V. I. Strikha, and V. P. Tolstoy, *Poverhnost (Surface)* **4**, 99 (1987).

23. V. A. Skryshevsky and P. Tolstoy, in B. Muravsky (Ed.), *Physics, and Application of Metal-Semiconductor Contact*, Kubansky University, Krasnodar, 1989, p. 51.

24. C. Martinet, R. A. B. Devine, and M. Brunel, *J. Appl. Phys.* **81**, 6996 (1997).

25. A. Lehmann, L. Schumann, and K. Hubner, *Phys. Stat. Solidi* **B117**, 689 (1983).

26. A. L. Shabalov and M. S. Feldman, *Thin Solid Films* **151**, 317 (1987).

27. M. Nakamura, Y. Mochizuki, K. Usami, Y. Itoh, and T. Nozaki, *Solid State Commun.* **50**, 1079 (1984).

28. P. G. Pai, S. S. Chao, Y. Takagi, and G. Lucovsky, *J. Vac. Sci. Technol.* **A4**, 689 (1986).

29. Yu. Bekeris, R. Bendere, R. Kalnynia, and I. Feltyn, *Izvestia Akademii Nauk Latv. SSR, Phys Techniques* **6**, 97 (1984).

30. L. Schumann, A. Lehmann, and H. Sobotta, *Phys. Stat. Solidi* **B110**, K69 (1982).

31. S. T. Pantelides and W. A. Harrison, *Phys. Rev. B* **13**, 2667 (1976).

32. V. P. Tolstoy, L. P. Bogdanova, and V. B. Aleskovsky, *Dokladi Akademii Nauk USSR, Phys. Khimia* **291**, 913 (1986).

33. I. W. Boyd and J. Wilson, *J. Appl. Phys.* **62**, 3195 (1987).

34. I. W. Boyd, *Appl. Phys. Lett.* **50**, 320 (1987).

35. K. Ishikawa, H. Ogawa, and S. Fujimura, *J. Appl. Phys.* **86**, 3472 (1999).

36. B. Harbecke, B. Heinz, and P. Grosse, *Appl. Phys.* **A38**, 263 (1985).

37. K. Ishikawa, H. Ogawa, and S. Fujimura, *J. Appl. Phys.* **85**, 4076 (1999).

38. V. A. Skryshevsky and V. P. Tolstoy, *Infrared Spectroscopy of Semiconductor Structures*, Lybid, Kiev, 1991.

39. V. V. Tarasov, *Problems of Glass Physics*, Stroiizdat, Moscow, 1979.

40. K. B. Clark, J. A. Bardwell, and J.-M. Baribeau, *J. Appl. Phys.* **76**, 3114 (1994).

41. S. K. J. Al-Ani, M. A. R. Sarkar, J. Beynon, and C. A. Hogarth, *J. Mater. Sci.* **20**, 1637 (1985).

42. C. Dominguez, J. A. Rodriguez, F. J. Munoz, and N. Zine, *Thin Solid Films* **346**, 202 (1999).

43. L. Zajickova, J. Janca, and V. Perina, *Thin Solid Films* **348**, 49 (1999).

44. M. Nakamura, R. Kanzawa, and K. Sakai, *J. Electrochem. Soc.* **133**, 1167 (1986).

45. M. J. Rack, D. Vasileska, D. K. Ferry, and M. Sidorov, *J. Vac. Sci. Technol.* **B16**, 2165 (1998).

46. R. Nonogaki, S. Nakai, S. Yamado, and T. Wade, *J. Vac. Sci. Technol.* **A16**, 2827 (1998).

47. K. Usami, S. Hyashi, Y. Uchida, and M. Matsumura, *Jpn. J. Appl. Phys.* **37**, L97 (1998).

48. A. L. Shabalov and M. S. Feldman, *Thin Solid Films* **110**, 215 (1983).

49. H. Ono, T. Ikarashi, K. Ando, and T. Kitano, *J. Appl. Phys.* **84**, 6064 (1998).

50. G. Lucovsky, S. Y. Lin, P. D. Richard, S. S. Chao, Y. Takagi, P. Pai, J. E. Keem, and J. E. Tyler, *J. Non-Cryst. Solids* **75**, 429 (1985).

51. K. A. Buckle, K. Pastor, C. Constanine, and D. Johnson, *J. Vac. Sci. Technol.* **B10**, 1133 (1992).

52. H. R. Philipp, *J. Appl. Phys.* **42**, 787 (1971).

53. F. Gaillard, P. Brault, and P. Brouquet, *J. Vac. Sci. Technol.* **A15**, 2478 (1997).

54. G. Giroult-Matlakowski, C. Charles, A. Durandet, R. W. Boswell, S. Armand, H. M. Persing, A. J. Perry, P. D. Lloyd, and S. R. Hyde, *J. Vac. Sci. Technol.* **A12**, 2754 (1994).

55. S. Rojas, L. Zanotti, A. Borghesi, A. Sasselle, and G. U. Pignatel, *J. Vac. Sci. Technol.* **B11**, 2081, (1993).

56. I. Montero, L. Galan, O. Najmi, and J. M. Albella, *Phys. Rev. B* **50**, 4881 (1994).

57. B. Subramaniam, L. E. Halliburton, and J. J. Martin, *J. Phys. Chem. Solids* **45**, 575 (1984).

58. W. A. Pliskin, in H. R. Huff and R. R. Burgess (Eds.), *Semiconductor Silicon*, Electrochemical Society, Princeton, NJ, 1973, p. 506.

59. J. D. Chapple-Sokol, W. A. Pliskin, and R. A. Conti, *J. Electrochem. Soc.* **138**, 3723 (1991).

60. H. M. Dauplaise, K. Vaccaro, B. R. Bennett, and J. P. Lorenzo, *J. Electrochem. Soc.* **139**, 1684 (1992).

61. K. Awazu and H. Onuki, *Appl. Phys. Lett.* **69**, 482 (1996).

62. A. Nara and H. Itoh, *Jpn. J. Appl. Phys.* **36**, 1477 (1997).

63. E. G. Parada, P. Gonzalez, J. Pou, J. Serra, D. Fernandez, B. Leon, and M. Perez-Amor, *J. Vac. Sci. Technol.* **A14**, 436 (1996).

64. Yu. A. Averkin, N. K. Karmadonov, V. A. Skryshevsky, V. I. Strikha, and A. V. Kharlamov, *Ukrainskii Fizicheskii Zhurnal [Ukrainian Phys. J.]* **34**, 1817 (1989).

65. K. H. A. Bogart, S. K. Ramiraz, L. A. Gonzales, and G. R. Bogart, *J. Vac. Sci. Technol.* **A16**, 3175 (1998).

66. K. Okimura, and N. Maeda *J. Vac. Sci. Technol.* **A16**, 3157 (1998).

67. D. A. DeCrosta, J. J. Hackenberg, and J. H. Linn, *J. Electrochem. Soc.* **143**, 1079 (1996).

68. M. Yoshimaru, S. Koizumi, and K. Shimokawa, *J. Vac. Sci. Technol.* **A15**, 2915 (1997).

69. T. Tamura, J. Sakai, M. Satoh, Y. Inoue, and H. Yoshitaka, *Jpn. J. Appl. Phys.* **36**, 1627 (1997).

70. C.-F. Yeh and C.-L. Chen, *J. Electrochem. Soc.* **142**, 3579 (1995).

71. G. Lucovsky and H. Yang, *Jpn. J. Appl. Phys.* **36**, 1368 (1997).

72. K. M. Chang, S. W. Wang, C. H. Li, T. H. Yeh, and J. Y. Yang, *Appl. Phys. Lett.* **70**, 2556 (1997).

73. J. Lubguban, Jr., A. Saitoh, Y. Kurata, T. Inokuma, and S. Hasegawa, *Thin Solid Films* **337**, 67 (1999).

74. J.-H. Kim, S.-H. Seo, S.-M. Yun, H.-Y. Chang, K.-M. Lee, and C.-K. Choi, *Appl. Phys. Lett.* **68**, 1506 (1996).

75. D. W. Berreman, *Phys. Rev.* **130**, 2193 (1963).

76. A. Slaoui, J. P. Ponpon, and P. Siffert, *Appl. Phys.* **A43**, 301 (1987).

77. E. M. Allegretto and J. A. Bardwell, *J. Vac. Sci. Technol.* **A14**, 2437 (1996).

78. A. Singh, *Rev. Sci. Instrum.* **56**, 1481 (1985).

79. H. Shirai, *Jpn. J. Appl. Phys.* **33**, L94 (1994).

80. S. S. Kilchitskaya, T. S. Kilchitskaya, V. A. Skryshevsky, V. I. Strikha, and V. P. Tolstoy, *Zhurnal prikladnoj spectroscopii [J. Appl. Spectrosc.]* **48**, 445 (1988).

81. G. N. Kuznetsova, *IR-Spectroscopy of Multiple Reflection*, Leningrad Technological Institute, Leningrad, 1982.

82. C. R. Helms, *J. Vac. Sc. Technnol.* **16**, 608 (1979).

83. J. Zolomy, *Period. Polytechnic Elec. Eng.* **28**, 281 (1984).

84. M. Sobolewski and C. R. Helms, *J. Vac. Sci. Technol.* **A3**, 1300 (1985).

85. E. Henzel, K. Wollschlager, D. Schulze, U. Kreissing, W. Skorupa, and J. Finster, *J. Surf. Interace Anal.* **7**, 205 (1985).

86. T. Ohwaki, M. Takeda, and Y. Takai, *Jpn. J. Appl. Phys.* **36**, 5507 (1997).

87. M. L. W. van der Zwan, J. A. Bardwell, G. I. Sproule, and M. J. Graham, *Appl. Phys. Lett.* **64**, 446 (1995).

88. C. R. Inomata, H. Ogawa, K. Ishikawa, and S. Fujimura, *J. Electrochem. Soc.* **143**, 2995 (1996).

89. Y. Okuno and K. Park, *Appl. Phys. Lett.* **69**, 541 (1996).
90. C. Martinet and R. A. B. Devine, *Appl. Phys. Lett.* **67**, 2696 (1995).
91. L. Zhong, L. Ling, and F. Shimura, *Appl. Phys. Lett.* **63**, 99 (1993).
92. F. Yano, A. Hiraoka, T. Itoga, H. Kojima, and K. Kanehori, *J. Vac. Sci. Technol.* **A13**, 2671 (1995).
93. A. Lehmann, L. Schumann, and K. Hubner, *Phys. Stat. Sol*. **B121**, 505 (1984) 99.
94. T. Inukai and K. Ono, *Jpn. J. Appl. Phys.* **33**, 2593 (1994).
95. T. Li and J. Kanicki, *Appl. Phys. Lett.* **73**, 3866 (1998).
96. S.-L. Wang, R. G. Cheng, and M.-W. Qi, *J. Non-Cryst. Solids* **97–98**, 1039 (1987).
97. J. A. Diniz, P. J. Tatsch, and M. A. A. Pudenzi, *Appl. Phys. Lett.* **69**, 2214 (1996).
98. L.-N. He, T. Inokuma, and S. Hasegawa, *Jpn. J. Appl. Phys.* **35**, 1503 (1996).
99. V. I. Bely, L. A. Vasilieva, and V. A. Gritcenko, *Silicon Nitride in Electronics*, Nauka, Novosibirsk, 1982.
100. Yu. P. Sitonite, V. V. Ostrikova, and Yu. Yu. Vaitkus, *Litovskii Phys. Sbornik* **26**, 207 (1986).
101. B. F. Hanyaloglu and E. S. Aydil, *J. Vac. Sci. Technol.* **A16**, 2794 (1998).
102. S. Hasegawa, M. Matuura, H. Anbutsu, and Y. Kurata, *Philos. Mag.* **B56**, 633 (1987).
103. W. R. Knolle and J. W. Osenbach, *J. Appl. Phys.* **58**, 1248 (1985).
104. M. Modreanu, N. Tomozeiu, P. Cosmin, and M. Gartner, *Thin Solid Films* **337**, 82 (1999).
105. M. Maeda and H. Nacamura, *J. Appl. Phys.* **58**, 484 (1985).
106. M. Maeda and H. Nacamura, *J. Appl. Phys.* **55**, 3068 (1984).
107. M. L. Naiman, C. T. Kirk, and R. J. Aucoin, *J. Electrochem. Soc.* **131**, 637 (1984).
108. P. Pan, J. Abernathey, and C. Schaefer, *J. Electron. Mater.* **4**, 617 (1985).
109. B. Claflin and G. Lucovsky, *J. Vac. Sci. Technol.* **B16**, 2154 (1998).
110. M. Firon, C. Bonnelle, and A. Mayeux, *J. Vac. Sci. Technol.* **A14**, 2488 (1996).
111. N. Watanabe, M. Yoshida, Y.-C. Jiang, T. Nomoto, and I. Abiko, *Jpn. J. Appl. Phys.* **30**, L619 (1991).
112. A. Feifar, J. Zemek, and M. Trchova, *Appl. Phys. Lett.* **67**, 3269 (1995).
113. C. Juang, J. H. Chang, and R. Y. Hwang, *J. Vac. Sci. Technol.* **B10**, 1221 (1992).
114. W.-S. Liao, C.-H. Lin, and S. C. Lee, *Appl. Phys. Lett.* **65**, 2229 (1994).
115. A. D. Bailey III and R. A. Gottscho, *Jpn. J. Appl. Phys.* **34**, 2172 (1995).
116. A. Sassella, A. Borghesi, F. Corni, A. Monelli, G. Ottaviani, R. Tonini, B. Pivac, M. Bacchetta, and L. Zanotti, *J. Vac. Sci. Technol.* **A15**, 377 (1997).
117. C. M. M. Denisse, K. Z. Troost, and J. B. O. Elferink, *J. Appl. Phys.* **60**, 2536 (1986).
118. C. M. M. Denisse, K. Z. Troost, and F. H. P. M. Habraken, *J. Appl. Phys.* **60**, 2543 (1986).
119. W. M. Arnoldbik, C. H. M. Maree, and F. H. P. M. Habraken, *Appl. Surf. Sci.* **74**, 103 (1994).
120. J. Olivares-Rosa, O. Sanchez, J. M. Albella, *J. Vac. Sci. Technol.* **A16**, 2757 (1998).
121. E. Fogarassy, C. Fuchs, A. Slaoui, S. de Unamuno, J. P. Stoquert, W. Marine, and B. Lang, *J. Appl. Phys.* **76**, 2612 (1994).

122. S. Trusso, C. Vasi, and F. Neri, *Thin Solid Films* **355–356**, 219 (1999).

123. M. Jelinek, J. Zemek, M. Trchova, V. Vorlicek, J. Lancok, R. Tomov, and M. Simeckova, *Thin Solid Films* **366**, 69 (2000).

124. Z. J. Zhang, S. Fan, J. Huang, and C. M. Lieber, *J. Electron. Mater.* **25**, 57 (1996).

125. J. D. Joannopoulos and G. Lucovsky (Eds.), *The Physics of Hydrogenated Amorphous Silicon*, Springer, Berlin, 1984.

126. Y.-P. Chou and S.-C. Lee, *J. Appl. Phys.* **83**, 4111 (1998).

127. R. B. Wehrspohn, S. C. Deane, I. D. French, I. Gale. J. Hewett, M. J. Powell, and J. Robertson, *J. Appl. Phys.* **87**, 144 (2000).

128. R. A. Street, *Hydrogenated Amorphous Silicon*, Cambridge University Press, Cambridge, 1991.

129. K. Zellama, L. Chahed, P. Sladek, M. L. Theye, J. H. von Bardeleben, and P. Roca i Cabarrocas, *Phys. Rev. B* **53**, 3804 (1996).

130. M. H. Brodsky (Ed.), *Amorphous Semiconductors*, Springer-Verlag, Berlin, 1979.

131. O. A. Golikova, M. M. Mezdrogina, V. X. Kudoyarova, and L. P. Seregin, *Phys. Techn. Semicond.* **21**, 1464 (1987).

132. B. Garrido, A. Perez-Rodriguez, J. Morante, A. Achiqa, F. Gourbilleau, and R. Madelon, *J. Vac. Sci. Technol.* **B16**, 16 (1998).

133. C. S. Hong and H. L. Hwang, *J. Appl. Phys.* **61**, 4593 (1987).

134. D. C. Marra, E. A. Edelberg, R. L. Naone, and E. S. Aydil, *J. Vac. Sci. Technol.* **A16**, 3199 (1998).

135. G. N. Parsons, D. V. Tsu, and G. Lucovsky, *J. Non-Cryst. Solids* **97–98**, 1375 (1987).

136. G. Lucovsky and W. B. Pollard, *Physica* **BC117**, 865 (1983).

137. T. Satoh and A. Hiraki, *Jpn. J. Appl. Phys.* **24**, L491 (1985).

138. S. C. Shen and Q. L. Jue, *Physica* **BC117**, 868 (1983).

139. H. Wagner and W. Beyer, *Solid. State Commun.* **48**, 585 (1983).

140. Z.-Q. Wu, C.-Y. Xu, and W.-P. Zhang, *J. Non-Cryst. Solids* **59–60**, 217 (1983).

141. A. R. Middya, S. Ray, S. J. Jones, and D. L. Williamson, *J. Appl. Phys.* **78**, 4966 (1995).

142. H. Rinnert, M. Vergnat, G. Marchal, and A. Burneau, *Appl. Phys. Lett.* **69**, 1582 (1996).

143. S. A. McQuaid, S. Holgado, J. Garrido, J. Martinez, J. Piqueras, R. C. Newman, and J. H. Tucker, *J. Appl. Phys.* **81**, 7612 (1997).

144. J. D. Ouwens, R. E. I. Schropp, and W. F. van der Weg, *Appl. Phys. Lett.* **65**, 204 (1994).

145. C. Manfredotti, F. Fizzotti, M. Boero, P. Pastorino, P. Polesello, and E. Vittone, *Phys. Rev. B* **50**, 18046 (1994).

146. E. C. Freeman and W. Paul, *Phys. Rev. B* **18**, 4288 (1978).

147. M. Cardona, *Phys. Stat. Solidi* **B118**, 463 (1983).

148. A. H. Mahan, P. Raboisson, and R. Tsu, *Appl. Phys. Lett.* **50**, 335 (1987).

149. H. Wieder, M. Cardona, and C. R. Guarieri, *Phys. Stat. Solidi* **B92**, 99 (1979).

150. J. F. Tian, D. S. Jiang, and B. R. Zeng, *Solid. State Commun.* **57**, 543 (1986).

151. E. S. Sabisky, *J. Non-Cryst. Solids* **87**, 43 (1986).

152. X. Xu, A. Morimoto, M. Kumeda, and T. Shimizu, *Jpn. J. Appl. Phys.* **26**, 661 (1986).

153. H. Gleskova, V. V. Ilchenko, V. A. Skryshevsky, and V. I. Strikha, *Czech. J. Phys.* **43**, 169 (1992).

154. S. Oguz, D. A. Anderson, and W. Paul, *Phys. Rev. B* **22**, 880 (1980).

155. S. Arimoto, H. Ohno, and H. Hasegawa, *Technical Digest — 1st International Photovoltaic Science and Engineering Conference*, Japan Society of Applied Physics, Tokyo, 1984, p. 175.

156. S. Arimoto, H. Yamamoto, H. Ohno, and H. Hasegawa, *J. Appl. Phys.* **57**, 4778 (1985).

157. K. Yokota, T. Kageyamaand, and S. Katoyama, *Solid State Electron.* **28**, 893 (1985).

158. R. R. Koropecki and R. Arce, *J. Appl. Phys.* **60**, 1802 (1986).

159. H. Ohsaki, K. Miura, and Y. Tatsumi, *J. Non-Cryst. Solids* **93**, 395 (1987).

160. H. Gleskova, V. A. Skryshevsky, J. N. Bullock, S. Wagner, and J. Stuchlik, *Mater. Lett.* **16**, 305 (1993).

161. V. A. Skryshevsky, V. I. Strikha, and H. Gleskova, *Czech. J. Phys.* **42**, 331 (1992).

162. D. Bermejo and M. Cardona, *J. Non-Cryst. Solids* **32**, 421 (1979).

163. T. Friessnegg, M. Boudreau, P. Mascher, A. Knights, P. J. Simpson, and W. Puff, *J. Appl. Phys.* **84**, 786 (1998).

164. L. Jiang, X. Chen, X. Wang, L. Xu, F. Stubhan, and K. H. Merkel, *Thin Solid Films* **352**, 97 (1999).

165. X. Redondas, P. Gonzales, B. Leon, M. Perez-Amor, J. C. Soares, and M. F. da Silva, *J. Vac. Sci. Technol.* **A16**, 660 (1998).

166. N. V. Tzenov, M. B. Tzolov, and D. I. Dimova-Malinovska, *Semicond. Sci. Technol.* **9**, 91 (1994).

167. P. I. Rovira and F. Alvarez, *Phys. Rev. B* **55**, 4426 (1997).

168. W. K. Choi, F. L. Loo, C. H. Ling, F. C. Loh, and K. L. Tan, *J. Appl. Phys.* **78**, 7289 (1995).

169. A. S. Kumbhar, D. M. Bhusari, and S. T. Kshirsagar, *Appl. Phys. Lett.* **66**, 1741 (1995).

170. Y. L. Chen, C. Wang, G. Lucovsky, D. M. Maher, and R. J. Nemanich, *J. Vac. Sci. Technol.* **A10**, 874 (1992).

171. D. K. Basa and F. W. Smith, *Thin Solid Films* **192**, 121 (1990).

172. S.-C. Seo, D. C. Ingram, and H. H. Richardson, *J. Vac. Sci. Technol.* **A13**, 2856 (1995).

173. K. M. McNamara, B. E. Williams, K. K. Gleason, and B. E. Scruggs, *J. Appl. Phys.* **76**, 2466 (1994).

174. M. Nakayama, Y. Matsuba, J. Shimamura, Y. Yamamoto, H. Chihara, H. Kato, K. Maruyama, and K. Kamata, *J. Vac. Sci. Technol.* **A14**, 2418 (1996).

175. V. G. Litovchenko, N. I. Klyui, A. B. Romanyuk, and V. A. Semenovich, in J. Schmid, H. A. Ossenbrink, P. Helm, E. Ehman, and E. D. Duneop (Eds.), *Proc. 2nd World Conference and Exhibition on Photovoltaic Solar Energy Conversion*, Vienna, Joint Research Centre, European Commission, 1998, p. 3715.

176. J. Hong, A. Goullet, and G. Turban, *Thin Solid Films* **364**, 144 (2000).

177. M. McGonigal, J. N. Russell, Jr., P. E. Pehrsson, H. G. Maguire, and J. E. Butler, *J. Appl. Phys.* **77**, 4049 (1995).

178. F. Fuchs, C. Wild, K. Schwarz, W. Muller-Sebert, and P. Koidl, *Appl. Phys. Lett.* **66**, 177 (1995).

179. G. Sreenivas, S. S. Ang, and W. D. Brown, *J. Electron. Mater.* **23**, 569 (1994).

180. M. S. Haque, H. A. Naseem, J. L. Shultz, W. D. Brown, S. Lal, and S. Gangopadhyay, *J. Appl. Phys.* **83**, 4421 (1998).

181. M. J. Paterson, K. G. Orrman-Rossiter, S. Bhargava, and A. Hoffman, *J. Appl. Phys.* **75**, 792 (1994).

182. A. J. M. Buuron, M. C. M. van de Sanden, W. J. van Ooij, R. M. A. Driessens, and D. C. Schram, *J. Appl. Phys.* **78**, 528 (1995).

183. F. Mauri and A. DalCorso, *Appl. Phys. Lett.* **75**, 644 (1999).

184. R. Erz, W. Dotter, K. Jung, and H. Ehrhardt, *Diamond Relat. Mater.* **4**, 469 (1995).

185. H. Jensen, U. M. Jensen, and G. Sorensen, *Appl. Phys. Lett.* **66**, 1489 (1995).

186. P. Scheible and A. Lunk, *Thin Solid Films* **364**, 40 (2000).

187. S.-H. Lin and B. J. Feldman, *Solid State Commun.* **96**, 29 (1995).

188. D. H. Berns and M. A. Cappelli, *Appl. Phys. Lett.* **68**, 2711 (1996).

189. M. F. Plass, W. Fukarek, S. Mandl, and W. Moller, *Appl. Phys. Lett.* **69**, 46 (1996).

190. M. Ben el Mekki, M. A. Djonadi, V. Mortet, E. Guiot, G. Nouet, and N. Mestres, *Thin Solid Films* **355**, 89 (1999).

191. H. Hofsass, C. Ronning, U. Griesmeier, M. Gross, S. Reinke, and M. Kuhr, *Appl. Phys. Lett.* **67**, 46 (1995).

192. S.-H. Lin, B. J. Feldman, and D. Li, *Appl. Phys. Lett.* **69**, 2373 (1996).

193. D. R. Turner, *J. Electrochem. Soc.* **105**, 402(1958).

194. C. Oules, A. Halimaoui, J. L. Regolini, A. Perio, and G. Bomchil, *J. Electrochem. Soc.* **139**, 3595 (1992).

195. L. T. Canham, *Appl. Phys. Lett.* **57**, 1046 (1990).

196. A. Richter, W. Lang, S. Steiner, F. Kozlowski, and H. Sandmaier, *Mat. Res. Soc. Symp. Proc.* **256**, 209 (1992).

197. F. Namavar, H. P. Maruska, and N. M. Kalkhoran, *Appl. Phys. Lett.* **60**, 2514 (1992).

198. J. P. Zheng, K. L. Jiao, W. P. Shen, W. A. Anderson, and H. S. Kwok, *Appl. Phys. Lett.* **61**, 459 (1992).

199. Y. S. Tsuo, Y. Xiao, M. J. Heben, X. Wu, F. J. Pern, and S. K. Deb, *23rd Photovoltaic Specialists Conference*, Institute of Electrical and Electronic Engineers (IEEE), Electron Devices Society, Louisville, 1993, p. 287.

200. S. Bastide, M. Cuniot, P. Williams, Q. N. Le, D. Sarti, and C. Levy-Clement, in R. Hill, W. Palz, and P. Helm (Eds.), *Proc. 12th European Photovoltaic Solar Energy Conf., Amsterdam*, H. S. Stephens & Associates, Bedford, 1994, p. 780.

201. V. A. Skryshevsky and A. Laugier, *Thin Solid Films* **346**, 254 (1999).

202. V. A. Skryshevsky, A. Laugier, S. V. Litvinenko, and V. I. Strikha, in J. Schmid, H. A. Ossenbrink, P. Helm, E. Ehman, and E. D. Duneop (Eds.), *Proc. 2nd World Conference, and Exhibition on Photovoltaic Solar Energy Conversion*, Joint Research Centre, European Commission, Vienna, 1998, p. 1611.

203. Z. Chen, T. Y. Lee, and G. Bosman, *Appl. Phys. Lett.* **64**, 3446 (1994).

204. L. E. Friedersdorf, P. C. Searson, S. M. Prokes, O. J. Glembocki, and J. M. Macaulay, *Appl. Phys. Lett.* **60**, 2285 (1992).

205. A. G. Gullis, L. T. Canham, and P. D. J. Calcott, *J. Appl. Phys.* **82**, 909 (1997).

206. R. T. Collins, M. A. Tischler, and J. H. Stathis, *Appl. Phys. Lett.* **61**, 1649 (1992).

207. S. M. Prokes, O. J. Glembocki, V. M. Bermudez, R. Kaplan, L. E. Friedersdorf, and P. S. Searson, *Phys. Rev. B* **45**, 13788 (1992).

208. T. Matsumoto, M. Daimon, T. Futagi, and H. Mimura, *Jpn. J. Appl. Phys.* **31**, L619 (1992).

209. M. S. Brandt, H. D. Fuchs, M. Stutzmann, J. Weber, and M. Cardona, *Solid State Commun.* **81**, 307 (1992).

210. M. Shimura, M. Katsuma, and T. Okumura, *Jpn. J. Appl. Phys.* **35**, 5730 (1996).

211. M. H. Brodsky, M. Cardona, and J. J. Cuomo, *Phys. Rev. B* **16**, 3356 (1977).

212. R. J. Collins and H. Y. Fan, *Phys. Rev.* **93**, 674 (1954).

213. Y. H. Seo, H.-J. Lee, H. I. Jeon, D. H. Oh, K. S. Nahm, Y. H. Lee, E.-K. Suh, H. J. Lee, and Y. G. Kwang, *Appl. Phys. Lett.* **62**, 1812 (1993).

214. Z. C. Feng, A. T. S. Wee, in Z. C. Feng and R. Tsu (Eds.), *Porous Silicon*, World Scientific, Singapore, 1994, p. 175.

215. Y. Xiao, M. J. Heben, J. M. McCullough, Y. S. Tsuo, J. I. Pankove, and S. K. Deb, *Appl. Phys. Lett.* **62**, 1152 (1993).

216. W. Kaiser, P. H. Keck, and C. F. Lange, *Phys. Rev.* **101**, 1264 (1956).

217. J. M. Lavine, S. P. Sawan, Y. T. Shieh, and A. J. Bellezza, *Appl. Phys. Lett.* **62**, 1099 (1993).

218. K. Li, D. C. Diaz, J. C. Campbell, and C. Tsai, in Z. C. Feng, and R. Tsu (Eds.), *Porous Silicon*, World Scientific, Singapore, 1994, p. 261.

219. S. Shih, K. H. Jung, D. L. Kwong, M. Kovar, and J. M. White, *Appl. Phys. Lett.* **62**, 1780 (1993).

220. T. Tamura, A. Takazawa, and M. Yamada, *Jpn. J. Appl. Phys.* **32**, L322 (1993).

221. C. Tsai, K. H. Li, J. Sarathy, S. Shih, J. C. Campbell, B. K. Hance, and J. M. White, *Appl. Phys. Lett.* **59**, 2814 (1991).

222. M. A. Tischler and R. T. Collins, *Solid State Commun.* **84**, 819 (1992).

223. A. Nakajima, T. Itakura, S. Watanabe, and N. Nakayama, *Appl. Phys. Lett.* **61**, 46 (1992).

224. H. Nikj, Y. Ogata, T. Sakka, and M. Iwasaki, *Bull. Inst. Atom. Energy. Kyoto Univ.* **84**, 34 (1993).

225. J. M. Rehm, G. L. McLendon, L. Tsybeskov, and P. M. Fauchet, *Appl. Phys. Lett.* **66**, 3669 (1995).

226. J. N. Chazalviel and F. Ozanam, *J. Appl. Phys.* **81**, 7684 (1997).

227. C. J. Fang, L. Ley, H. R. Shanks, K. J. Gruntz, and M. Cardona, *Phys. Rev. B* **22**, 6140 (1980).

228. Q.-S. Li, R.-C. Fang, in Z. C. Feng and R. Tsu (Eds.), *Porous Silicon*, World Scientific, Singapore, 1994, p. 235.

229. T. Takahagi, A. Ishitani, H. Kuroda, and Y. Nagasawa, *J. Appl. Phys.* **69**, 803 (1991).

230. M. Stutzmann, M. S. Brandt, E. Bustarret, H. D. Fuchs, M. Rosenbauer, and J. Weber, *J. Non-Cryst. Solids* **186**, 931 (1993).

231. Y. H. Ogata, T. Tsuboi, T. Sakka, and S. Naito, in L. T. Canham and V. Parkhutik (Eds.), *Proc. Int. Conf. Porous Semicond, Science & Technology, Mallorca*, Technical University of Valencia, Valencia, 1998, p. 28.

232. L. Tsybeskov, S. P. Duttagupta, and P. M. Fauchet, *Solid State Commun.* **95**, 429 (1995).

233. T. Maruyama and S. Ohtani, *Appl. Phys. Lett.* **65**, 1346 (1994).

234. H. Chen, X. Hou, G. Li, F. Zhang, M. Yu, and X. Wang, *J. Appl. Phys.* **79**, 3282 (1996).

235. K.-H. Li, C. Tsai, J. C. Campbell, M. Kovar, and J. M. White, *J. Electron. Mater.* **23**, 409 (1994).

236. V. A. Skryshevsky, V. I. Strikha, V. A. Vikulov, A. V. Kozinetc, A. V. Mamikin, and A. Laugier, in R. Ciach and S. V. Svechnikov (Eds.), *New Photovoltaic Materials for Solar Cells*, Proc. of First Polish–Ukrainian Symposium, European Materials Research Society (E-MRS), Cracow, 1996, p. 202.

237. A. Loni, A. J. Simons, P. D. J. Calcott, J. P. Newey, T. I. Cox, and L. T. Canham, *Appl. Phys. Lett.* **71**, 107 (1997).

238. M. A. Hory, R. Herino, M. Ligeon, F. Muller, F. Gaspard, I. Mihalcescu, and J. C. Vial, *Thin Solid Films* **255**, 200 (1995).

239. F. Moller, M. Ben Chorin, and F. Koch, *Thin Solid Films* **255**, 16 (1995).

240. N. H. Zoubir, M. Vergnat, T. Delatour, A. Burneau, Ph. de Donato, and O. Barres, *Thin Solid Films* **255**, 228 (1995).

241. M. B. Robinson, A. C. Dilon, D. R. Haynes, and S. M. George, *Appl. Phys. Lett.* **61**, 1414 (1992).

242. C. Tsai, K.-H. Li, D. S. Kinosky, R.-Z. Qian, T.-C. Hsu, J. T. Irby, S. K. Banerjee, A. F. Tasch, J. C. Campbell, B. K. Hance, and J. M. White, *Appl. Phys. Lett.* **60**, 1700 (1992).

243. J. L. Coffer, S. C. Lilley, R. A. Martin, and L. A. Files-Sesler, *J. Appl. Phys.* **74**, 2094 (1993).

244. J. Yan, S. Shih, K. H. Jung, D. L. Kwong, M. Kovar, J. M. White, B. E. Gnade, and L. Magel, *Appl. Phys. Lett.* **64**, 1374 (1994).

245. Y. Kanemitsu, T. Futagi, T. Matsumoto, and H. Mimura, *Phys. Rev. B* **49**, 1473 (1994).

246. B. Gellor, *Appl. Surface Sci.* **108**, 449 (1997).

247. L. Tsybeskov and P. M. Fauchet, *Appl. Phys. Lett.* **64**, 1983 (1994).

248. J. Salonen, V.-P. Lehto, and E. Laine, *Appl. Phys. Lett.* **70**, 637 (1997).

249. R. Czaputa, R. Fritzl, and A. Popitsch, *Thin Solid Films* **255**, 212 (1995).

250. C. Canaria, A. Wun, and M. J. Sailor, in L. T. Canham and V. Parkhutik (Eds.), *Proc. Int. Conf. Porous Semicond, Science & Technology, Mallorca*, Technical University of Valencia, Valencia, 1998, p. 203.

251. V. A. Skryshevsky, V. A. Vikulov, V. I. Strikha, and L. Tristani, in M. Balkanski and I. Yanchev (Eds.), *Fabrication, Properties, and Application of Low Dimensional Semiconductor*, NATO ASI Series, Kluwer, Dordrecht, 1995, p. 173.

252. W. Theiss, M. Arntzen, S. Hilbrich, M. Wernke, R. Arens-Fischer, and M. G. Berger, *Phys. Stat. Solidi* **190**, 15 (1995).

253. C.-K. Kim, C.-H. Chung, and S. H. Moon, *J. Appl. Phys.* **78**, 7392 (1995).

254. I. Schechter, M. Ben-Chorin, and A. Kux, *Anal. Chem.* **67**, 3727 (1995).

255. J. M. Laueraas and M. J. Sailor, *Science* **261**, 1567 (1993).

256. V. Polishchuk, E. Souteyrand, J. R. Martin, D. Nicolas, V. Strikha, and V. Skry-shevsky, *Anal. Chim. Acta* **375**, 205 (1998).

257. B. Hamilton, *Semicond. Sci. Technol.* **10**, 1187 (1995).

258. P. Gupta, A. C. Dillon, A. S. Bracker, and S. M. George, *Surf. Sci.* **245**, 360 (1991).

259. T. Ya. Gorbach, G. Yu. Rudko, P. S. Smertenko, S. V. Svechnikov, M. Ya. Valakh, V. P. Bondarenko, and A. M. Dorofeev, *Semicond. Sci. Technol.* **11**, 601 (1996).

260. A. C. Dillon, M. L. Wise, M. B. Robinson, and S. M. George, *J. Vac. Sci. Technol.* **A13**, 1 (1995).

261. J. M. Buriak and M. J. Allen, *J. Am. Chem. Soc.* **120**, 1339 (1998).

262. T. F. Harper and M. J. Sailor, *J. Am. Chem. Soc.* **119**, 6943 (1997).

263. T. F. Young, J. F. Liou, C. P. Chen, Y. L. Yang, and T. C. Chang, in L. T. Canham and V. Parkhutik (Eds.), *Proc. Int. Conf. Porous Semicond, Science & Technology, Mallorca*, Technical University of Valencia, Valencia, 1998, p. 212.

264. E. M. Voronkova, *Optical Materials for Infrared Techniques*, Nauka, Moscow, 1965.

265. W. Kaiser, W. G. Spitzer, R. H. Kaeser, and L. E. Howarth, *Phys. Rev.* **127**, 1950 (1962).

266. A. A. Velichko and S. K. Noak, *Rev. Electron Technique* **1397**, 47 (1988).

267. J. Zegenhagen and J. R. Patel, *Phys. Rev. B* **141**, 5315 (1990).

268. J. Wollschlager and A. Meier, *Appl. Surface Sci.* **104–105**, 392 (1996).

269. Z. Yi Lin and B. K. Garside, *Appl. Opt.* **21**, 4324 (1982).

270. S. Witanachchi and P. J. Wolf, *J. Appl. Phys.* **76**, 2185 (1994).

271. D. A. Jishiashvili and E. R. Kutelia, *Phys. Stat. Solidi* **B143**, K147 (1987).

272. M. Zacharias, F. Stolze, T. Drusedau, and W. Bock, *Phys. Stat. Solidi* **B189**, 409 (1995).

273. H. Freistedt, F. Stolze, M. Zacharias, J. Blasing, and T. Drusedau, *Phys. Stat. Solidi* **B193**, 375 (1996).

274. J.-J. Lee, C.-O. Lee, and S. Ernst, *J. Vac. Sci. Technol.* **B6**, 1533 (1988).

275. K. L. Saenger, C. Cabral, Jr., L. A. Clevenger, and R. A. Roy, *J. Appl. Phys.* **77**, 5156 (1995).

276. D. Laviale, J. C. Oberlin, and R. A. B. Devine, *Appl. Phys. Lett.* **65**, 2021 (1994).

277. R. A. B. Devine, *Appl. Phys. Lett.* **68**, 1924 (1996).

278. G. Q. Lo, D. L. Kwong, and S. Lee, *Appl. Phys. Lett.* **60**, 3286 (1992).

279. Z.-W. Fu, L.-Y. Chen, and Q.-Z. Qin, *Thin Solid Films* **340**, 164 (1999).

280. H. Myoren, T. Matsumoto, and Y. Osaka, *Jpn. J. Appl. Phys.* **31**, L1425 (1992).

281. H. Ishiwara and K. Azuma, *Mat. Res. Soc. Symp. Proc.* **116**, 369 (1988).

282. C. L. Aardahl, J. W. Rogers, Jr., H. K. Yun, Y. Ono, D. J. Tweet, and S. T. Hsu, *Thin Solid Films* **346**, 174 (1999).

283. R. D. Vispute, J. Narayan, H. Wu, and K. Jagannadham, *J. Appl. Phys.* **77**, 4724 (1995).

284. C. Wetzel, D. Volm, B. K. Meyer, K. Pressel, S. Nilsson, E. N. Mokhov, and P. G. Baranov, *Appl. Phys. Lett.* **65**, 1033 (1994).

285. G. Mirjalili, T. J. Parker, S. F. Shayesteh, M. M. Bulbul, S. R. P. Smith, T. S. Cheng, and C. T. Foxon, *Phys. Rev. B* **57**, 4656 (1998).

286. W. Zhang, Y. Someno, M. Sasaki, and T. Hirai, *Jpn. J. Appl. Phys.* **32**, L116 (1993).

287. C. Wetzel, E. E. Haller, H. Amano, and I. Akasaki, *Appl. Phys. Lett.* **68**, 2547 (1996).

288. C. Carlone, K. M. Lakin, and H. R. Shanks, *J. Appl. Phys.* **55**, 4010 (1984).

289. A. T. Collins, E. C. Lightowlers, and P. J. Dean, *Phys. Rev.* **158**, 833 (1967).

290. J. A. Sanjurjo, E. Lopez-Cruz, P. Vogl, and M. Cardona, *Phys. Rev. B* **28**, 4579 (1983).

291. N. Azema, J. Durand, R. Berjoan, J. L. Balladore, and L. Cot, *J. Phys. IV* **1**, 405 (1991).

292. T. Kozawa, T. Kachi, H. Kano, Y. Taga, M. Hashimoto, N. Koide, and K. Manabe, *J. Appl. Phys.* **75**, 1098 (1994).

293. K.-I. Hanaoka, H. Ohnishi, and K. Tachibana, *Jpn. J. Appl. Phys.* **34**, 2430 (1995).

294. R. K. Bhan, R. Ashokan, and P. C. Mathur, *Jpn. J. Appl. Phys.* **33**, 2708 (1994).

295. T. Matsumoto, J. Watanabe, T. Tanaka, and Y. Mishima, *Appl. Phys. Lett.* **58**, 39 (1991).

296. M. Katiyar, Y. H. Yang, and J. R. Abelson, *J. Appl. Phys.* **77**, 6247 (1995).

297. S. S. Lee, M. J. Kong, S. F. Bent, C.-M. Chiang, and S. M. Gates, *J. Phys. Chem.* **100**, 20015 (1996).

298. T. Balz, R. Brendel, and R. Hezel, *J. Appl. Phys.* **76**, 4811 (1994).

299. D. C. Marra and E. S. Aydil, *J. Vac. Sci. Technol.* **A15**, 2508 (1997).

300. D. Landheer, Y. Tao, J. E. Hulse, T. Quance, and D.-X. Xu, *J. Electrochem. Soc.* **143**, 1681 (1996).

301. V. M. Agranovich and D. L. Mills (Eds.), *Surface Polaritons: Electromagnetic Waves at Surfaces, and Interfaces*, North-Holland, Amsterdam, 1982.

302. R. R. Koropecky, R. Arce, and J. Ferron, *Appl. Surf. Sci.* **25**, 321 (1986).

303. G. Fameli, D. della Sala, F. Roca, and F. Pascarella, *J. Appl. Phys.* **78**, 7269 (1995).

6

APPLICATION OF INFRARED SPECTROSCOPY TO ANALYSIS OF INTERFACES AND THIN DIELECTRIC LAYERS IN SEMICONDUCTOR TECHNOLOGY

6.1. ULTRATHIN OXIDE LAYERS IN SILICON SCHOTTKY-TYPE SOLAR CELLS

Production of metal–semiconductor or Schottky-type solar cells requires no high-temperature technological processes, and they use the high-energy end of the solar spectrum more effectively than $p-n$ junctions. However, Schottky-type solar cells have smaller energy conversion efficiencies η and open-circuit voltages U_{oc}. The latter can be increased by introducing a dielectric layer of tunneling thickness (2–3 nm) between the metal and semiconductor. This increases the current of minority charge carriers relative to the majority charge carriers and can increase both the absolute value of the photocurrent and the potential barrier and hence U_{oc} and η [1–4]. It should be noted that the solar cell characteristics are highly dependent on the properties of the ultrathin dielectric layer between the metal and the semiconductor, including its thickness, chemical composition, and structure. The interface in the metal–semiconductor contact is essentially the bottleneck in the optimization and control of the solar cell parameters.

Besides the standard methods of SIMS, AES, and XPS for investigation of these ultrathin dielectric layers in Schottky-type solar cells, IRRAS has also been applied [5–8]. Ultrathin (2–10-nm) SiO_2 films in metal–oxide–silicon structures have also been investigated by IRRAS [9–14].

The parameters of $Ti-SiO_x-p$-Si solar cells and the composition of the tunneling SiO_x layer were found to be influenced by the manufacturing process [5, 6]. Initially, 3–4-nm oxide layers are produced by anodization of p-Si(100). The

Schottky contact is formed by electron beam deposition of Ti at a deposition rate of between 0.1 and 5.0 nm·s^{-1}. The latter may be adjusted by varying the current density or the energy of electrons bombarding the target. The silicon substrate temperature T_s was also varied between 300 and 650°C during the metal deposition.

The results of the IRRAS study show that the absorption bands of the interfacial oxide layer differ substantially from those of the initial oxide on a free silicon surface, depending on the substrate temperature during the titanium deposition. Specifically, the intensity of the 1240-cm^{-1} (SiO$_2$) and 1160-cm^{-1}(SiO) bands decreases, and the 1240-cm^{-1} peak shifts to lower frequencies. The dependence of the intensities of the absorption peaks of SiO$_2$, SiO, and TiO$_x$ on the temperature of the silicon substrate during the Ti deposition is shown in Fig. 6.1. Extrapolations to absorbances of the initial oxide layer on the free silicon surface are indicated by the dashed line.

The decrease in the intensities of the 1240- and 1160-cm^{-1} absorption peaks, indicating a decrease in the oxide film thickness during the Ti deposition, is connected with the formation of titanium oxide, as evidenced by the appearance of the broad absorption band in the region of 1000–400 cm^{-1}. The possibility of titanium oxide formation in these structures has been proven by theoretical and experimental investigations of the Ti−SiO$_2$−Si system [15, 16]. The data correlate well with the results of chemical analysis of the Ti−Si and Ti−SiO$_2$−Si interfaces by Raman spectroscopy [17], XPS, and AES [8, 16, 18]. It was shown that the formation and structural reconstruction of the TiO$_x$ (0.14 < x < 1.7), TiSi$_x$, and SiO$_x$ phases is governed by the substrate temperature.

Upon deposition of Ti, the absorption band of SiO shows less change than that of SiO$_2$, apparently because of the difference in bond energies of SiO and

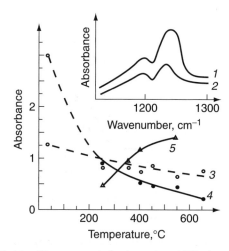

Figure 6.1. Effect of Ti deposition on p-polarized spectra of SiO$_2$ interface layer recorded by IRRAS at 75° incidence: (1) before and (2) after Ti deposition. Intensities of (3) 1160-, (4) 1240-, and (5) 700-cm^{-1} absorption bands as functions of silicon substrate temperature.

SiO_2. The energy of the SiO bond is 190 kcal·mol^{-1}, whereas that in the reaction $SiO_2 \rightarrow SiO + O$ is 112 kcal·mol^{-1} [5].

A comparison of the distribution profiles for the Ti$-SiO_2-$Si and Ti$-$Si structures by SIMS shows that higher temperatures accelerate the diffusion of silicon into titanium (until Si atoms appear on the titanium surface of the Ti$-$Si structures at $T > 500°C$) and titanium into SiO_x and Si. The fraction of TiO_2 in the oxide layer increases with increasing temperature, particularly in the region close to the titanium$-$oxide layer interface, due to the decrease in the SiO_2 fraction. The profile of the SiO distribution is virtually unaffected by changes in substrate temperature.

The metal$-$oxide interaction also depends upon the rate of the metal deposition. Experimental results show that an increase in the deposition rate affects the structure and composition of the interfacial layer in much the same way as an increase in substrate temperature [6]. The effects of the rate of metal deposition, the substrate temperature, and the layer thickness on the electrical and physical properties of solar cells may be understood in terms of the physical and chemical interactions in Ti$-SiO_x-$Si cells [5, 6, 8].

The effect of UV irradiation on Ti$-SiO_2-$Si solar cells with insulator layers of various thicknesses has been investigated by dark and light current$-$voltage ($I-V$) and capacitance$-$voltage ($C-V$) measurements as well as by IRRAS [7]. The observed change in the diode parameters under UV irradiation is shown to arise from reconstruction of the interface, specifically changes in the insulator layer thickness, in its composition, and in the density of surface states. In particular, the initial insulator thickness is shown to play an important role in determining the device parameters.

The absorption at the frequencies of the LO modes of SiO (1170 cm^{-1}) and SiO_2 (1250 cm^{-1}) is presented in Fig. 6.2 for structures with oxide layer thickness $d = 2.5$ nm (curves 1 and 2) and $d = 5.0$ nm (curves 3 and 4) before

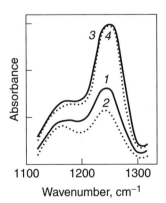

Figure 6.2. IR absorption spectra for Ti–SiO_x–p-Si solar cells with oxide layer thicknesses of (1, 2) 2.5 nm and (3, 4) 5.0 nm (1, 3) before and (2, 4) after UV irradiation for 200 h. Reprinted, by permission, from S. S.Kilchitskaya, T. S.Kilchitskaya, G. D.Popova, and V. A.Skryshevsky, *Thin Solid Films* **346**, 226 (1999). Copyright ©1999 Elsevier Science.

(solid line) and after (dashed line) UV irradiation for 200 h. For the structure with a 5.0-nm oxide layer, UV irradiation causes a change in the oxide layer composition, as revealed by the change in the SiO/SiO_2 ratio (Fig. 6.2, curves 3 and 4). On the other hand, for the 2.5-nm oxide layer, the absorption of the SiO and SiO_2 modes decreases equally. The 2.5-nm oxide layer shrinks by 0.3–0.5 nm after 100 h of the UV irradiation; further UV treatment has no effect. These data are in agreement with the results obtained from $C-V$ characteristics.

Structures with oxide layers of 2.5 and 5.0 nm are characterized by different behavior under UV irradiation, indicating that they have different interfacial state densities, as calculated from $C-V$ characteristics. Structures with $d = 2.5$ nm exhibit a greater density of surface states than structures with $d = 5.0$ nm, which agrees very closely with the reported dependence of the interfacial state density on the oxide layer thickness [19]. After UV irradiation, the interfacial state density decreases only negligibly for structures with $d = 2.5$ nm but increases drastically for structures with $d = 5.0$ nm. This behavior may be caused by various interfacial reconstructions occurring in the structures, which is confirmed by IRRAS.

Such variation in the density of surface states can be explained by the existence of defect centers associated with dangling bonds arising from the discontinuity of Si—O or Si—H bonds and oxygen dangling bonds (—O) [20]. For structures with $d = 5.0$ nm, the increase in the density of surface states may be due to the simultaneous formation of two types of defects, trivalent Si atoms ($\equiv$Si·) and oxygen dangling bonds (—O) [21]. The defects arise when the Si—O bonds cleave under UV irradiation, as proven by the decrease in the ν_{LO} absorption band of Si—O modes (see Fig. 6.2).

Using GIR to investigate the thickness of the ultrathin oxide layer in Al—SiO_x—Si devices, the number of Al—O bonds (band of 849 cm^{-1}) was found to increase with thermal annealing [13] (Fig. 6.3). The decrease in the SiO_x thickness is a temperature-activated process with an activation energy of 0.98 eV and can be well described within the framework of an Al—AlO_y—SiO_x—Si model. In addition, contamination by aluminum can considerably influence the characteristics of the SiO_x—Si interface, causing an increase in the density of electronic states, the effect being more pronounced for thinner silicon oxide layers [22, 23].

For the development of cheap solar cells, layers of a heavily doped semiconductor like ITO, which may act as a barrier-forming semiconductor, a wide band gap window, and an antireflection coating in the heterostructure [1, 24], attract particular interest. It was shown [25] that the application of vacuumless technologies such as pyrolysis in air for the production of the above-mentioned layers can substantially influence the properties of interfacial layers. The IR spectra of In_2S_3—SiO_x—Si heterostructures in solar cells were studied. The In_2S_3 films were produced by thermal decomposition of a complex of the general form L_3In (where L is a ligand). Thermal decomposition of this compound on a heated Si substrate ($T = 230–270°C$) produces a film of indium sulfide with good adhesion

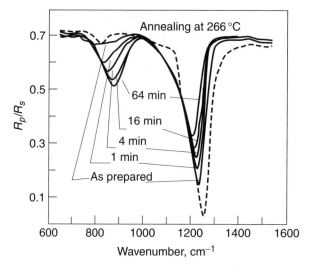

Figure 6.3. GIR spectra of Al–SiO$_x$–Si structure at 80° incidence after different annealing times at 266°C; R_p/R_s denotes ratio of reflectance of p-polarized to s-polarized light. Thickness of SiO$_x$ is 1.3 nm. Broken line shows GIR spectrum calculated for SiO$_2$ film 1.3 nm thick with dielectric function determined for thick SiO$_2$ film oxidized at 1000°C in dry O$_2$. Reprinted, by permission, from R. Brendel and R. Hezel, *J. Appl. Phys.* **71**, 4377 (1992). Copyright © 1992 American Institute of Physics.

and resistivity of $1 \times 10^3 \, \Omega \cdot$cm. The thickness of the In$_2$S$_3$ layer was found to be 60–80 nm.

Figure 6.4 shows the p-polarized spectra from IRRAS of SiO$_x$–Si structures in the 1100–1300-cm^{-1} region before and after deposition of a layer of In$_2$S$_3$ for different thicknesses of silicon oxide [25]. The deposition of In$_2$S$_3$ on the heated substrate of SiO$_x$–Si leads to a qualitative change in the shape of the absorption bands in the spectra of anodic silicon oxide. An increase in the 1240-cm^{-1} absorption peak intensity is observed, which is evidence of an increase in the oxide layer thickness (e.g., from 4.0 nm on a free surface of Si to 6.0 nm in the presence of an In$_2$S$_3$ layer). At the same time, the frequency of the absorption peak of SiO$_2$ shifts to higher energies, from 1240 cm^{-1} (for $d_{\text{SiO}_2} = 4.0$ nm) to 1250 cm^{-1} ($d_{\text{SiO}_2} = 6.0$ nm). With the deposition of In$_2$S$_3$, the intensity of the 1160-cm^{-1} absorption peak increases to a greater extent (Fig. 6.5) [25].

As was noted in Chapter 5, thermal annealing improves the stoichiometry of the oxide SiO$_x$, as nonstoichiometric SiO$_x$ ($0 < x < 2$) decomposes into Si and SiO$_2$. IR spectra exhibit an increase in the absorbance of the ν(SiO$_2$) band and in the ratio of the peak intensities, $A_{\text{SiO}_2}/A_{\text{SiO}}$. A comparison with the experimental data shows that deposition of In$_2$S$_3$ on a heated SiO$_x$–Si substrate leads to an increase in the total thickness of silicon oxide (as indicated by an increase in A_{SiO_2} and ν_{SiO_2}) and to a reconstruction of the oxide structure due to an increase in the molar fraction of SiO. The latter is considerably higher for thinner films of silicon oxide in the In$_2$S$_3$–SiO$_x$–Si structure (Fig. 6.5).

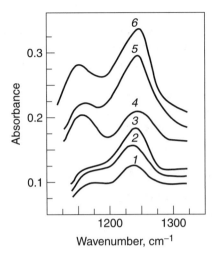

Figure 6.4. The *p*-polarized spectra from IRRAS of silicon oxide on Si at 75° incidence (1–3) before and (4–6) after In_2S_3 deposition. Thickness of silicon oxide, *d*: (1) 1.0, (4) 2.5, (2) 2.0, (5) 4.5 (3), 4.0, and (6) 6.0 nm. Reprinted, by permission, from S. S. Kilchitskaya, T. S. Kilchitskaya, V. A. Skryshevsky, V. I. Strikha, and V. P. Tolstoy. *Zhurnal prikladnoi spektroskopii* **48**, 445 (1988). Copyright © 1988 Publishing House "Nauka i technika."

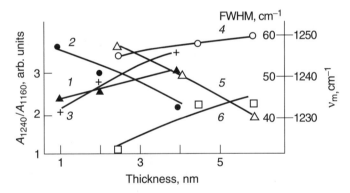

Figure 6.5. Parameters of 1240- and 1160-cm^{-1} absorption peaks of silicon oxide as functions of thickness (1–3) before and (4–6) after deposition of In_2S_3 layer: (1, 4) ν_{1240}; (2, 5) FWHW (1240 cm^{-1}); (3, 6) A_{1240}/A_{1160}. Reprinted, by permission, from S. S. Kilchitskaya, T. S. Kilchitskaya, V. A. Skryshevsky, V. I. Strikha, and V. P. Tolstoy. *Zhurnal prikladnoi spektroskopii* **48**, 445 (1988). Copyright © 1988 Publishing House "Nauka i technika."

6.2. CONTROL OF THIN OXIDE LAYERS IN SILICON MOS DEVICES

MIS structures form the basis of the majority of modern planar microelectronics and very large scale integrated (VLSI) devices. Because the stability and reliability of these devices as well as their output characteristics are determined to a large extent by the surface effects at metal–oxide and oxide–semiconductor interfaces, the structure and chemical properties of the oxide layers must be well understood in order to control the manufacturing processes in MIS technology [26, 27]. Physical and chemical interactions between the metal, dielectric, and semiconductor during device processing can significantly alter the resulting oxide layers relative to those on a free surface of the semiconductor. The minute dimensions (<10 μm) of the active elements in integrated circuits (IC) render the recording of IR absorption spectra difficult, and either microscopy methods or larger test structures must often be used.

6.2.1. CVD Oxide Layers in Al–SiO$_x$–Si Devices

Low-temperature deposition of CVD oxides is an attractive option in IC production because of its low cost and it minimizes degradation of the conducting layers of aluminum or other metals. The influence of precursor concentrations on the composition and density of the SiO$_x$ layers in Al−SiO$_x$−Si devices produced by TEOS decomposition in oxygen atmosphere at $T = 300°C$ has been studied [28]. The Al contacts are deposited by electron beam evaporation. The IRRAS results in the spectral range of 1000–1300 cm^{-1} are presented in Fig. 6.6. As seen, the shape of the absorption bands depends substantially on the ratio of gaseous reactants, $M = C_{O_2}/C_{TEOS}$.

As M increases from 0 to 0.8, a linear increase in the intensity of the SiO$_2$ absorption (1250 cm^{-1}) is observed, signaling a linear increase in the thickness of the SiO$_2$ film from 0.006 to 0.2 μm. At $M > 0.8$, the increasing O$_2$ concentration does not influence the thickness of the SiO$_2$ layer: The dependence of d_{SiO_2} on M appears to saturate (Fig. 6.7) [28].

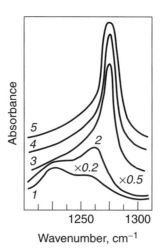

Figure 6.6. The *p*-polarized IRRAS at 75° incidence of Al–SiO$_x$–Si devices containing CVD oxide grown with reactant concentrations in ratio $M = N_{O_2}/N_{TEOS}$ of (1) ~0, (2) 0.3, (3) 0.8, (4) 1.6, and (5) 2.0.

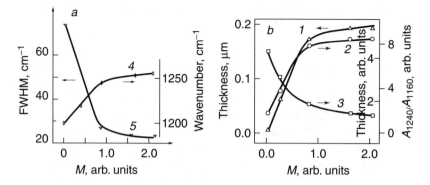

Figure 6.7. Dependence of (1) d_{SiO_2}, (2) A_{1240}/A_{1160}, (3) $d_{Al_2O_3}$, (4) ν_{1240} and (5) FWHM on M.

An increase in the percentage of oxygen in the reactant mixture leads to an increased density of the SiO_2 layer. This is indicated by the shift of the absorption peak from 1200 cm^{-1} ($M \approx 0$) to 1250 cm^{-1} ($M \geq 0.8$) and by the corresponding decrease in FWHM and increase in d_{SiO_2}/d_{SiO}. This is attributed to incomplete decomposition of TEOS when $M < 0.8$ and the oxygen is insufficient, leading to a large quantity of SiO in the deposited layers. An increased oxygen content gives more complete TEOS decomposition and thus a higher molar proportion of SiO_2 in the layer; the reaction is saturated at $M = 0.8$. Excess oxygen has virtually no effect on the decomposition reaction or the density of the resulting layers. A comparison of FWHM and ν_m of the SiO_x absorption bands for thermal and CVD oxides shows that their densities are similar when $M \geq 0.8$.

Upon deposition of Al on a SiO_x layer, the solid-phase interaction at the Al–SiO_x interface and the formation of Al_2O_3 can be described by the following reaction: $3SiO_2 + 4Al \rightarrow 2A_2O_3 + 3Si$ [13, 29]. The exact position of the Al_2O_3 absorption peak between 850 and 970 cm^{-1} is determined by the substrate temperature [9, 10, 13]. The thickness of the Al_2O_3 layer also depends upon the SiO_x film formation procedure, as indicated by the dependence of the Al_2O_3 band intensity on the oxygen percentage M in the reactant mixture. When M is increased from 0 to 0.8, the intensity of the A_2O_3 band decreases by a factor of approximately 3, signaling a corresponding decrease in the thickness of the Al_2O_3 layer (Fig. 6.7) [28]. It can be concluded that the more dense SiO_x layers are less reactive toward Al.

Low-temperature silicon dioxide films prepared by evaporating silicon in an oxygen atmosphere onto Mo were studied [30]. It was found that the stretching Si–O bond appears as a broad asymmetric peak centered at 1178 cm^{-1} in IRRAS of the film prepared at 50°C. Upon heating, this peak gradually shifts to higher frequencies up to 1252 cm^{-1} for a film annealed at 1200°C. The correlation of the electric parameters of MOS capacitors with the dielectric and IR properties of the thermal and low-temperature oxides, as well as with their stoichiometry, were investigated [31, 32].

6.2.2. Monitoring of Aluminum Corrosion Processes in Al–PSG Interface

Phosphosilicate glass (PSG) is widely used in IC manufacturing as a solid-state source of diffusion, as effective protection against alkali ions, and as an intermediate dielectric between metal interconnections and underlying structures. Incorporation of phosphorus pentoxide (P_2O_5) into silicon oxide in the form of PSG decreases the temperature necessary for the glass reflow appreciably, which ensures metallic lay-outs of higher quality [26, 33, 34]. An increase in the phosphorus content leads to the lowering of the reflow temperature but simultaneously promotes the corrosion of aluminum owing to the interaction with phosphoric acid produced from the reaction of phosphorus pentoxide with water [35].

IRRAS was used to monitor the interaction between an aluminum film and PSG under different processing treatments [36]. The PSG films ($d = 400$ nm) were produced on n-Si substrates by CVD from gas mixtures of SiH_4, PH_3, and O_2 in an Ar atmosphere at $T = 300°C$ in the reaction $SiH_4 + 4PH_3 + 7O_2 \rightarrow SiO_2 + 2P_2O_5 + 2H_2O + 6H_2$. The concentration of phosphorus in the films was changed by varying the PH_3 content in the reactant mixture.

The p-polarized IR transmission spectra ($\varphi = 75°$) for PSG layers with different phosphorus content are shown in Fig. 6.8 [36]. In addition to the SiO_2 bands, an intense peak at 1325 cm^{-1} from P=O bonds was observed [33–38]. The ratio of the absorbance at 1325 cm^{-1} to that at 1080 or 1250 cm^{-1} depends linearly on the phosphorus concentration determined by methods such as chemical or neutron activation analysis. The same dependence is observed in spectra of borosilicate glass for the ratio of the absorbance at 1350 cm^{-1} to that at 1080 cm^{-1}; this is used for determination of the boron concentration in borosilicate glass films [38–40]. This determination of the absolute concentration of phosphorus in PSG films by IR spectra was developed in [41].

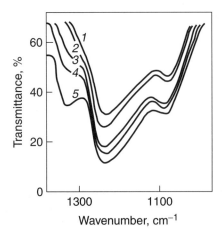

Figure 6.8. p-polarized IR transmission spectra at $\varphi = 75°$ of PSG for different phosphorus concentrations in the film (in weight percent): (1) 5.5; (2) 7.0; (3) 8.3; (4) 9.3; (5) 10.6.

Evaporation of a 1-μm Al layer causes the absorbance at 1080 cm^{-1} to decrease (the "metallic quenching" effect, Section 3.2.2) and the ν_{LO} band of SiO$_2$ to shift toward higher frequencies. The most pronounced changes in the spectra of these structures were observed after annealing at $T \geq 450°C$ for 1 h. The splitting of the complex band at 1100–1350 cm^{-1} into two peaks at 1325 and 1250 cm^{-1} and a substantial increase in the intensity of the first of them (Fig. 6.9) [36] are characteristic for one group of samples. This appears to indicate an increase in the concentration of P$_2$O$_5$ in PSG, owing to the further oxidation of atomic phosphorus through interaction with SiO$_2$. This is corroborated by the fact that the 1170-cm^{-1} band from SiO is clearly observed by IRRAS when the Al layer has been etched.

For the second group of structures, the bands caused by the Si–O and P=O bonds virtually disappear after thermal annealing, but a weak peak in the region of 920 cm^{-1} appears (see Fig. 6.9). This peak is ascribed to the absorption of Al$_2$O$_3$ [13] and becomes more intense after elimination of the Al layer by chemical etching (Fig. 6.10) [36]. Hence it follows that for samples in the second group thermal annealing leads to the reaction $3SiO_2 + 4Al \rightarrow 2Al_2O_3 + 3Si$, accompanied by the sublimation of P$_2$O$_5$ at 360°C [42]. Another possible reaction is that between orthophosphoric acid and Al film: $P_2O_5 + 3H_2O + Al \rightarrow 2AlPO_4 + 3H_2$ [33, 43].

The Auger electron spectra of the samples of both groups (after annealing and elimination of Al by etching) showed the Auger peak at 86 eV of silicon (SiLVV) bonded to the oxygen in SiO$_x$ [44]. The very weak Auger peak of Al (AlLVV) at 63 eV is characteristic for the first group of structures. At the same time, for the second group of samples, the Auger peak of Al was two orders of magnitude stronger and was shifted toward lower energies (58 eV). The peak at 58 eV has been assigned to the aluminum in Al$_2$O$_3$ [45], and the peak at 63 eV may be either from metallic Al or from the Al in AlO$_x$. Thus, the Auger

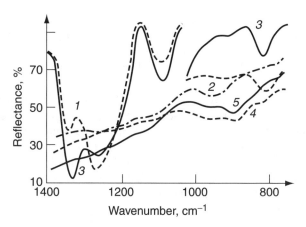

Figure 6.9. IRRAS ($\varphi = 75°$) of Al–PSG–Si structures after thermal annealing at $T \geq 450°C$ for 1 h (curves labeled as in Fig. 6.8).

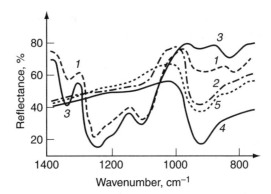

Figure 6.10. The p-polarized IR transmission spectra ($\varphi = 75°$) of PSG–Si structures after annealing and removal of Al film by chemical etching (curves labeled as in Fig. 6.8).

electron spectroscopic data confirm the formation of Al_2O_3 in the second group of structures.

It can be concluded that annealing of Al–PSG structures can lead to an increase in the P_2O_5 concentration in PSG owing to the further oxidation of phosphorus reacting with SiO_2. In addition, Al and PSG may react, resulting in the transformation of the PSG film into various aluminum compounds, in particular Al_2O_3. It should also be noted that under the high-temperature treatment used for the glass reflow, the P and B content in the PSG and BSG films may change. IR absorption spectroscopy showed lower P and B concentrations during annealing in water vapor, in dry oxygen, in nitrogen, and in air. Furthermore, unannealed films lose B and P during prolonged storage in air [33, 38].

6.2.3. Determination of Metal Film and Oxide Layer Thicknesses in MOS Devices

A method to determine the thicknesses of the semitransparent metallic film and the thin dielectric layer in a MOS structure simultaneously has been developed and approved by the authors. This method is based on the observed decrease in IR absorption by the TO peak of the oxide layer when a metallic film is deposited onto its surface.

The reflections are measured in p-polarized light at the Brewster angle from the side of the semiconductor substrate. The measurements are carried out at three frequencies. The first is in the region where there is no selective absorption of the oxide (I_1, I_4). The next two are at the peaks of the TO (I_2, I_5) and LO (I_3, I_6) phonon absorptions of the oxide layer, first prior to the deposition of a metallic coating (with the help of an external mirror) ($I_1 - I_3$) and then after the deposition of the metal ($I_4 - I_5$). The relevant intensities are given by

$$I_1 = I_0(1 - R_1)^2(1 - R_2)^2(1 - R_4)^2 R_5, \tag{6.1}$$

$$I_2 = I_0(1 - R_1)^2(1 - R_2)^2(1 - R_4)^2 R_5 \exp(-2\alpha_{01}d_1), \tag{6.2}$$

$$I_3 = I_0(1 - R_1)^2(1 - R_2)^2(1 - R_4)^2 R_5 \exp(-2\alpha_2 d_1), \qquad (6.3)$$

$$I_4 = I_0(1 - R_1)^2(1 - R_2)^2 R_3, \qquad (6.4)$$

$$I_5 = I_0(1 - R_1)^2(1 - R_2)^2 R_3 \exp(-2\alpha_1 d_2), \qquad (6.5)$$

$$I_6 = I_0(1 - R_1)^2(1 - R_2)^2 R_3 \exp(-2\alpha_2 d_2). \qquad (6.6)$$

Here, R_1–R_5 are the reflectances at the air–semiconductor, semiconductor–dielectric, dielectric–metal, and dielectric–air interfaces and metallic mirror, respectively; $2d_1$ and $2d_2$ are the effective thicknesses of the dielectric layer corresponding to the optical pathlength of the light beam before and after the metal deposition; and α_{01}, α_1, and α_2 are the absorption coefficients of the dielectric layer at the TO (before and after the metal deposition) and LO frequencies, respectively. The intensity of the incident IR radiation I_0 is assumed to be independent of the frequency.

Because $R_1 = 0$ for the measurements in p-polarized radiation at $\varphi = \varphi_{Br}$ and assuming that the chemical composition of the dielectric layer does not change after the deposition of a metal, it follows from Eqs. (6.1)–(6.6) that

$$d_2 = d_1 \frac{\ln(I_4/I_6)}{\ln(I_1/I_3)}, \qquad (6.7)$$

$$2d_1(\alpha_1 - \alpha_{01}) = \left(\frac{\ln(I_4/I_5)}{\ln(I_4/I_6)}\right) \ln\left(\frac{I_1}{I_3}\right) - \ln\left(\frac{I_1}{I_2}\right). \qquad (6.8)$$

The absorption in the region of the LO mode of the oxide is practically independent of the presence of a metal film and is governed only by the dielectric thickness. Rather good correlation between the absorption at ν_{LO} and the thickness of a thin oxide film was obtained from measurements in p-polarized light at $\varphi = \varphi_{Br}$ [46]. Finally, the thickness of the oxide layer in the MOS structure [Eq. (6.7)] and the contribution of changes in its thickness to changes in the TO band intensity may be determined using an independent measurement of the LO band intensity in the spectrum of the oxide. Equation (6.8) characterizes the change in absorbance at the TO band resulting from a change in the thickness (conductivity) of the evaporated metal film.

Figure 6.11 shows the change in the spectra from IRRAS of an 80-nm SiO_2 layer after electron beam evaporation of a 100-nm Al film. Total quenching of the TO absorption band of SiO_2 (1070 cm^{-1}) is observed, while the intensities of the ν_{LO} bands of SiO_2 (1250 cm^{-1}) and SiO (1170 cm^{-1}) increase. This can be attributed to the conditions of light propagation through the structure [because the absorbance at LO frequencies depends considerably on the angle of incidence (Section 1.5 and Ref. [47]), unlike the absorbance at TO frequencies], to the partial oxidation of the Si wafer during the deposition and subsequent annealing of Al, or to the solid-state interaction between Al and SiO_2 in the annealing process [48]. The ellipsometric and IR spectroscopic study of these structures after the removal of Al by chemical etching confirms that the thickness of the oxide under the Al layer exceeds its initial value, and thus the observed increases

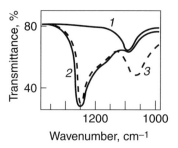

Figure 6.11. The p-polarized spectra from IRRAS (at $\varphi = 75°$) of (1) Si wafer, (2) Al–SiO$_2$–Si, and (3) SiO$_2$–Si.

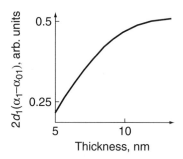

Figure 6.12. Experimental dependence of $2d_1(\alpha_1 - \alpha_{01})$ on d_{Au} for Au–SiO$_2$–Si structure.

in the 1250- and 1170-cm^{-1} bands are caused by the increases in the SiO$_2$ and SiO thicknesses, respectively.

The correlation between the ν_{TO} band intensity and the metallic layer thickness measured by independent methods is shown in Fig. 6.12. Clearly, IRRAS allows effective determination of the metallic layer thickness down to ≤ 10 nm.

6.3. MODIFICATION OF OXIDES IN METAL–SAME-METAL OXIDE–InP DEVICES

Same-metal oxides of A^{3}B^5 materials are not normally used in MOS devices, due to their heterogeneity and nonstoichiometry (chemical composition depends significantly on the conditions of both production and subsequent thermal treatment [49–51]). Therefore, at present, preference is given to MOS structures containing silicon oxides. Examples of such systems include SiO$_2$ on substrates made of GaAs, InP, and InSb, which were studied by IR absorption [52–54]. Nevertheless, MOS structures with same-metal oxides A^{3}B^5 are still considered to be promising for a number of applications in optoelectronics, such as solar cells [1].

The influence of the deposition of metallic coatings and the thermal treatment on the structural and chemical properties of same-metal oxides formed on

InP(111) substrates by anodic, chemical, and thermal oxidation was studied [55]. The anodization was carried out in an aqueous solution of ammonium citrate and propylene glycol at an applied potential of 120 V. Concentrated nitric acid was used for the chemical oxidation at 80°C for 15 min, and the thermal oxidation occurred in a flow of wet oxygen, $T_{O_2} = 450$°C, $T_{H_2O} = 95$°C. The oxide thickness was 80–150 nm. The effect of a thermally deposited metallic coating (Au or Al, $d = 0.2$–0.3 μm) on the oxide chemical composition was studied by IRRAS. Conditions of optimum and degraded electrophysical characteristics of the MOS structures were achieved by annealing at 300 and 530°C, respectively, for 1 h in a mixture of N_2 and O_2.

Spectra of oxides formed under different conditions show a wide band lying between 1400 and 800 cm^{-1} with characteristic absorption peaks at 940, 1010, 1090, 1150, and 1220 cm^{-1} (Fig. 6.13) [55]. These spectra were compared with standard IR absorption spectra of the bulk compounds In_2O_3, P_2O_5, and $InPO_4$ [56, 57], which, as is known from XPS studies and thermodynamic analyses [49, 51, 58], can contain InP oxide and will have characteristic absorption bands at 940, 1140, and 1090 cm^{-1} ($InPO_4$); 1226, 1150, and 1015 cm^{-1} (P_2O_5); and 1090 and 1050 cm^{-1} (In_2O_3). This comparison reveals that although all the above-mentioned IR absorption peaks in the experimental spectra of InP oxides are observed, the experimental IR spectra of InP oxide and the individual standard spectra are not identical. Thus, the InP oxide must be a multiphase structure representing a mixture of In and P oxides ($In_2O_3 + P_2O_5 + InPO_4$). The proportion of each component in InP oxide is not constant and depends on the oxide formation procedure, as is seen from the intensities of the individual peaks.

In the case of thermal oxides grown in a wet oxygen flow, the absorption spectra show two peaks of approximately equal intensity at 940 and 1140 cm^{-1}, which are observed in bulk IR spectra of $InPO_4$ [56], testifying that $InPO_4$ is the dominant phase in these films. This correlates with data from XPS [49, 58, 59] and Raman spectroscopy [60], according to which the thermal oxides grown

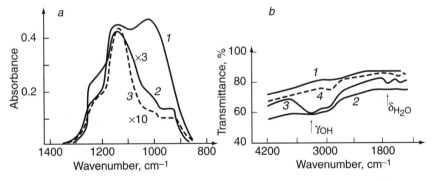

Figure 6.13. (a) IRRAS spectra in p-polarized radiation at 75° of InP oxides: (1) thermal, (2) chemical, and (3) anodic oxide. (b) IR transmission spectra of InP oxides: (1) thermal, (2) chemical, (3) anodic oxide, and (4) anodic oxide after thermal annealing at 300°C for 1 h.

at $T > 400°C$ consist primarily of $InPO_4$, with a small proportion of In_2O_3. IR spectroscopy data also indicate the presence of In_2O_3 by the absorption bands at 1090 and 1050 cm^{-1}. The appearance of a weak peak at 1220 cm^{-1} in the spectrum of thermal oxides suggests the presence of P_2O_5.

In the spectra of anodic oxides, the absorption peaks at 1150, 1090, and 1220 cm^{-1} dominate, whereas the peaks at 940 and 1010 cm^{-1} that dominate the spectra of thermal oxides are much less pronounced. The peaks at 1150, 1090, and 1220 cm^{-1} are connected with the absorption of P_2O_5 and In_2O_3, and thus it can be concluded that anodic oxides consist of P_2O_5 and In_2O_3, the latter being more abundant. The weak peak at 940 cm^{-1} suggests that $InPO_4$ can also appear in anodic oxides, even before annealing, although it was earlier assumed that $InPO_4$ is formed only after a thermal treatment of low-temperature oxides [49, 51, 61].

The spectra of chemical oxides are similar to those of anodic oxides, indicating that these oxides are close in chemical composition. The absorption bands at 1220, 940, and 1010 cm^{-1} are more intense in the spectra of chemical oxides, implying a greater proportion of $InPO_4$ and P_2O_5 in the chemical oxides.

In spectra of InP oxides between 2900 and 3500 cm^{-1} and at 1650 cm^{-1}, one can distinguish absorption peaks of the stretching and bending vibrations of the O—H groups; these are most pronounced for the anodic oxide, considerably weaker for the chemical oxide, and practically absent for the thermal oxide and disappear upon slight thermal heating ($T \approx 200–300°C$) (Fig. 6.13 [55]). From this, it can be concluded that these peaks result from O—H groups of adsorbed water. Same-metal oxides do not contain detectable amounts of hydroxyl ion salts in the form $In(OH)_3$ (the probability of their formation upon oxidation of InP in water solutions has been discussed [49]).

Low-temperature annealing ($T = 300°C$) causes the absorption spectra of anodic and chemical oxides to undergo substantial changes, namely an increase in the 940- and 1220-cm^{-1} absorption peak intensities by a factor of 1.3–2.0. Thus, annealing results not only in the dehydration of same-metal oxides but also in an increase in the concentrations of P_2O_5 and $InPO_4$, which occurs in the oxide simultaneously with a decrease in the In_2O_3 concentration [55]. It is these latter effects that allow for the optimization of the electrophysical characteristics of MOS devices.

P_2O_5 and $InPO_4$ have wider energy gaps E_g and higher resistivities than In_2O_3, with potential barriers at the contact with InP; furthermore they do not contain the traps that cause long-time relaxation of MOS structure characteristics [62]. Analyzing the variations in intensity of each IR absorption peak and taking into account all the possible processes that could lead to the observed effects, it was concluded that, in addition to the reactions of the oxide itself, namely $In_2O_3 + P_2O_5 \rightarrow 2InPO_4$ or at the oxide–InP interface $4P_2O_5 + 5InP \rightarrow 5InPO_4 + 8P$ and $4In_2O_3 + 3InP \rightarrow 3InPO_4 + 8In$, the diffusion of phosphorus atoms from indium phosphide into the oxide takes place. In particular, this occurs in the transition region where, from the Auger spectroscopy data, accumulation

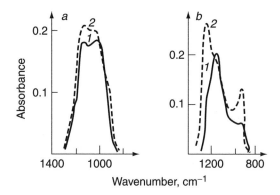

Figure 6.14. IRRAS spectra in p-polarized radiation at 75° of (a) thermal and (b) chemical oxides: (1) unannealed; (2) after thermal annealing at 300°C for 1 h.

of excess phosphorus atoms occurs. This allows the increase in the P_2O_5 concentration occurring simultaneously with an increase in $InPO_4$ concentration and a decrease in In_2O_3 concentration to be attributed to oxidation of the phosphorus atoms that penetrate into the oxide (Fig. 6.14 [55]).

High-temperature annealing ($T = 530°C$), which leads to the degradation of the electrophysical characteristics of MOS structures [63], results in the increase in intensity of the absorption peak at 940 cm^{-1} ($InPO_4$) and the decrease in intensity of the absorption peak of In_2O_3, indicating that $InPO_4$ continues to be formed in the oxide. However, as for $T = 300°C$, the IR absorption spectrum of InP oxide is not transformed into the $InPO_4$ spectrum; thus, even at such temperatures, anodic and chemical oxides do not become a single-phase material consisting of $InPO_4$, as was previously assumed [51].

Another peculiarity in the IR absorption spectra of InP oxide annealed at $T = 530°C$ is the absorption peak at 1220 cm^{-1}, which is considerably less intense than in the spectra of the oxide annealed at $T = 300°C$, suggesting that P_2O_5 is lost. As a result, the band at 1090 cm^{-1}, attributed to In_2O_3, again becomes dominant in the IR absorption spectra. The intensity of the peak at 1220 cm^{-1} does remain greater than in IR absorption spectra of unannealed InP oxides. Since the temperature of P_2O_5 sublimation, T_{sub}, is 360°C [42], P_2O_5 sublimation will occur at $T > T_{sub}$, and the lowering of the resistivity of the oxide observed by us on annealing at 450°C is not surprising.

A metal deposited onto an oxide has an appreciable effect on oxides annealed at 300°C and 530°C (see Fig. 6.15); this is most pronounced for Al, which clearly shows an absorption band at 1220 cm^{-1}, caused by the presence of P_2O_5 in the oxide films, even at $T = 530°C$. After deposition of a Au coating and annealing at $T = 530°C$ (when the exfoliation of the Au film is observed), the intensity of the absorption band at 1220 cm^{-1} is similar to that observed for oxides annealed with no metallic coating.

From the above analysis of IR spectra, it follows that the deposition of a metal film onto a same-metal InP oxide can effectively stabilize the properties of such

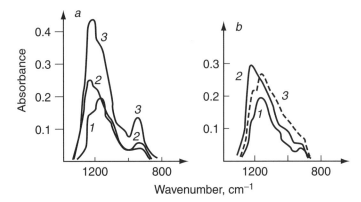

Figure 6.15. IRRAS spectra in p-polarized radiation at 75° of (a) Al–chemical oxide–InP and (b) Au–chemical oxide–InP: (1) after metal deposition; (2) after thermal annealing at 300°C for 1 h; (3) after thermal annealing at 530°C for 1 h.

oxides during thermal treatment by hindering the sublimation of P_2O_5, but this sealing effect depends upon the physicochemical and mechanical properties of the metals employed.

Degradation of the surface upon formation of a metallic contact with InP, as occurs in the production of solar cells, was also observed [64]. IR absorption showed a change in the chemical bonding; in particular, the lowering of the P concentration, the appearance of an imperfect nonstoichiometric surface structure, and the formation of P—H bonds were recorded. IR spectra of same-metal GaAs oxides were studied [65, 66], and it was established that they represent a more complicated chemical structure than a simple mixture of Ga—O and As—O bonds. The differences between the chemical structure of anodic and chemical oxides were also considered [65].

6.4. DIELECTRIC LAYERS IN SANDWICHED SEMICONDUCTOR STRUCTURES

6.4.1. Silicon-on-Insulator

The relatively small geometry that is possible for integrated circuits made on silicon-on-insulator (SOI) gives them a significant advantage over similar circuits fabricated on Si [67]. Recently, SOI technology has demonstrated promising results in the fabrication of gigabyte DRAM devices, primarily due to the fact that both sides of the SOI structure may be used, which improves packing density considerably [68]. A SOI structure can also be used as a support substrate in the fabrication of thin-film silicon solar cells [69] and in hybrid bipolar and complementary metal–oxide–semiconductor devices (biCMOS) [70].

To manufacture an oxide layer for a SOI structure, either the local oxidation of silicon (LOCOS) process or the separation by implanted oxygen (SIMOX) technique is used [68]. A SIMOX structure is fabricated by a high dosage oxygen implantation ($>10^{18}$ cm^{-2}) followed by thermal annealing. By changing the type of implanted ions, buried silicon nitride and carbide layers may be formed [70].

Three-dimensional SOI structures represent successive stacks of silicon and dielectric layers, and the analysis of their structural and chemical and geometric parameters is still a problem. However, it can be solved using the reflectance spectra and interpreting them by either the matrix method or recursion relationships (Section 1.7). In one example [68], the thicknesses of all the layers in a six-layer SOI structure in which half of the layers are absorbing were determined. However, the technical problem occurring in reflectance spectra, specifically the interference of the beam reflected from the front surface and from the buried SiO$_2$ layer, must also be addressed. A detailed theoretical analysis of interference fringes has been done that yields the refractive index profiles and oxygen concentration profile [70].

One of the fundamental problems associated with materials fabricated by SIMOX is the high density of dislocations remaining in the top silicon layer after annealing [67]. A significant reduction in the dislocation density may be achieved by techniques of sequential implantation and annealing (SIA) [71]. The resulting substrates have dislocation densities several orders of magnitude lower than those obtained with a standard, single implantation at about 550°C followed by annealing [71, 72].

The analysis of FTIR absorbance spectra of SOI structures produced by the SIMOX and SIA technologies has been performed for different implantation and annealing protocols [72]. In the spectra of a 300-nm buried oxide layer, one can observe three main absorption ν_{TO} peaks from the vibration modes of Si$-$O$-$Si at 465, 810, and 1083 cm^{-1}. The positions of these peaks are quite close to the corresponding absorption bands of thermal SiO$_2$. For structures produced by SIA technology, the stretching absorption peak is shifted to higher frequency (1094 cm^{-1}, Fig. 6.16 [72]). The ν_{TO} band of the stretching mode of the SIMOX layer is shifted toward lower frequencies. This may be the result of a compression-induced stress in the oxide, which could decrease the bond angle [73], or the presence of oxygen precipitates in SIA structures, which have an absorption band close to 1105 cm^{-1} [74].

6.4.2. Polycrystalline Silicon–c-Si Interface

Semi-insulating polycrystalline silicon (SIPOS) is found in many applications, including the electrical passivation of high-voltage planar devices, emitters in heterojunction transistors, and thin-film silicon solar cells on cheap foreign substrates. SIPOS is usually deposited by CVD onto a Si substrate. The composition of the SIPOS SiO$_x$ film varies from $x = 0$ (polycrystalline or amorphous Si) to $x = 2$ (SiO$_2$). The untreated film consists of a nonrandom mixture of Si and SiO$_2$ and relatively small amounts of an intermediate oxide of SiO$_{1-\Delta}$ ($\Delta = 0.14$) [75, 76].

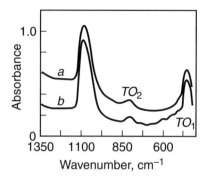

Figure 6.16. IR absorbance spectra of SOI structures made by (a) SIMOX and (b) SIA processes. Reprinted, by permission, from A. Perez, J. Samitier, A. Cornet, J. R. Morante, P. L. F. Hemment, and K. P. Homewood, *Appl. Phys. Lett.* **57**, 2443 (1990). Copyright © 1990 American Institute of Physics.

The interface layer formed in the CVD process between the silicon substrate and the SIPOS film can lead to unstable $I-V$ characteristics or to an electric breakdown in high-voltage planar devices. The modified ATR method was applied to study this interfacial layer between a Si substrate and polycrystalline Si [77]. IR reflectance spectra were recorded in p- and s-polarized radiation at 80°. The normalized reflectance spectra $R = R_p/R_s$ were compared to the spectra of a test (reference) structure with a known thickness of SiO_2 and were calculated by the three-phase model, with an absorbing layer characterized by a complex dielectric constant and two transparent silicon layers with real dielectric constants. The analysis of the ν_{LO} band at 1210 cm^{-1} revealed that the interface layer is comprised of silicon oxides, mainly SiO_2, with a thickness of 0.7 nm [77].

6.4.3. SiO$_2$ Films in Bonded Si Wafers

Infrared transmission methods have been developed to characterize thin oxide layers sandwiched between two silicon substrates. In particular, the sensitivity of this method has been improved for films that are thinner than 10 nm [78–80]. This work was motivated in part by the many applications of such silicon-sandwiched devices, including in the base structure for gas sensors, SOI, and micromechanical structures. The Si wafer bonding occurs if two free wafers are brought into contact and subsequently annealed. Before bonding, wafers are cleaned in a standard RCA solution with a deionized water rinse. Initial joining is performed under vacuum at room temperature by applying gentle pressure. Permanent bonding is then achieved via interfacial chemical reactions [81]. The thin oxide layer on the bonded Si surface arises from preliminary oxidation [78] or as the result of thermal annealing of the bonded wafers [79].

IR transmission of the Si-bonded wafers at a grazing angle of ≈90° is a highly sensitive technique and does not require multiple internal reflections (which would necessitate a beveled Si substrate). The resulting spectra are the best for

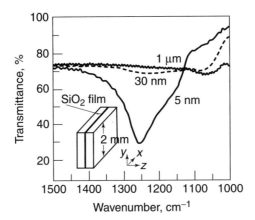

Figure 6.17. Transmission spectra of SiO$_2$ films 5 nm (solid line), 30 nm (dashed line), and 1 μm (dotted line) inserted into bonded Si wafers. Spectra were measured with IR microscope directing unpolarized radiation onto cross section of Si–SiO$_2$–Si stratified structure (see Section 4.3.6 for more detail). Reprinted, by permission, from A. Borghesi, A. Sassella, and T. Abe, *Jpn. J. Appl. Phys.* **34**, L1409 (1995). Copyright © 1995 Publication Board of Japanese Journal of Applied Physics.

ultrathin SiO$_2$ films [78]. The spectral contrast is strongly enhanced at oblique angles of incidence, because the absorption of the silicon substrate can be neglected if the relationship between the spectra recorded in *p*- and *s*-polarized radiation is considered [80]. The 5-nm film has a very strong absorption at 1256 cm^{-1}, the 30-nm film only a weak absorption, and the 1-μm film no absorption in the range of the LO mode (Fig. 6.17) [78]. These results can be easily explained by the fact that the Berreman thickness for such a configuration is about 20 nm (Section 3.2). For films thicker than 20 nm, the absorbance at the LO mode frequency vanishes because the surface modes no longer interact and the thin-film approximation is not valid [78, 82].

Hydrophobic and hydrophilic bonded silicon wafers were studied by the methods of multiple internal reflection spectroscopy (MIR) and multiple internal transmission spectroscopy (MIT) [79, 81, 83, 84]. The sensitivity of the MIT configuration is 20–40 times higher than that of the traditional MIR configuration [81]. However, both of these techniques require beveled Si wafers.

The modes of all three hydrides SiH, SiH$_2$, and SiH$_3$ are observed simultaneously in the 2100-cm^{-1} range; however, the absorption intensities differ for hydrophobic and hydrophilic wafers. For a hydrophobic surface, the ratio $\Delta R/R$ is approximately 100 times higher than for a bonded hydrophilic sample [79]. In addition to Si−H, O$_2$−Si−H (2200 cm^{-1}) and O$_3$−Si−H (2250 cm^{-1}) vibration modes and O−H stretch modes (3725 cm^{-1}) are also observed (Fig. 6.18 [84]).

6.4.4. Quantum Wells

Multiquantum wells (QWs) are multilayered structures consisting of sequences of interchanging thin layers (3−10 nm) of two different semiconductors, each

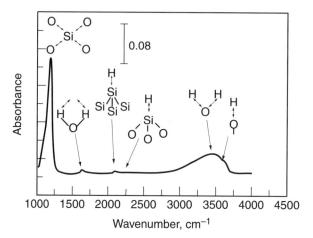

Figure 6.18. Typical MIT IR spectrum of two hydrophilic Si(111) wafers bonded at room temperature. Reprinted, by permission, from M. K. Weldon, V. E. Marsico, Y. J. Chabal, D. R. Hamann, S. B. Christman, and E. E. Chaban, *Surf. Sci.* **368**, 163 (1996). Copyright © 1996 Elsevier Science.

with different band gaps. They are usually grown by the method of vapor-phase epitaxy (VPE) or molecular-beam epitaxy (MBE) on a semiconductor substrate. Long-wavelength IR detectors that are highly sensitive in the 8–12-μm atmospheric spectral window have been developed based on intersubband absorption of GaAs–AlGaAs, GaAs–GaInP, and SiGe–Si QWs [85–88] and for the 3–5-μm spectral range InGaAs–InAlAs and InGaAs–InP QWs [88, 89]. Usually, A^3B^5 infrared photodetectors are based on the photoionization of carriers from the ground state of *n*-doped QWs and the subsequent carrier transport in the presence of an applied electric field. The *n*-type QWs have excellent IR sensitivity due to the low electron effective mass and high electron mobility. However, the quantum-mechanical selection rule for the intersubband transition requires that the electric field of the incoming radiation have a component perpendicular to the QW plane, and special optical coupling methods such as that using planar metal grating structures are needed [89]. Recently, *p*-type QW detectors, which use intervalence subband transitions, have been considered [90]. Due to mixing of the light and heavy-hole states, optical transitions between the valence subbands are allowed for radiation of normal incidence [89].

QW structures are ideal for IR detectors, since their absorption wavelength can be systematically varied by changing their well width and barrier height. Figure 6.19 shows a typical absorption spectrum of a QW consisting of 10 periods of 4-nm *n*-GaAs, $n = 2 \times 10^{18}$ cm^{-3}, and 30-nm undoped $Ga_{0.5}In_{0.5}P$ barriers, formed by metal–organic VPE on a semi-insulating undoped GaAs [85]. The absorption was measured in a multipass waveguide configuration at 45°. Absorption is observed at 8 μm (155 meV) with FWHM = 82 meV. Other QWs have analogous absorption spectra [88, 91] but may become more complicated in the presence of subbands. This is characteristic, for example, for SiGe–Si QWs [86].

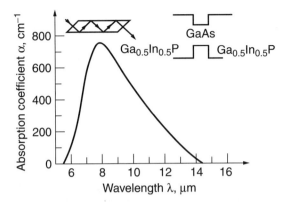

Figure 6.19. Intersubband absorption spectrum recorded at room temperature. Insets show 45° multipass waveguide configuration and band of tuning of lattice-matched GaAs–$Ga_{0.5}In_{0.5}P$ heterostructure. Reprinted, by permission, from S. D. Gunapala, B. F. Levine, R. A. Logan, T. Tanbun-Ek, and D. A. Humphrey, *Appl. Phys. Lett.* **57**, 1802 (1990). Copyright © 1990 American Institute of Physics.

6.5. IR SPECTROSCOPY OF SURFACE STATES AT SiO₂–Si INTERFACE

The SiO_2–Si interface is a crucial feature of MOS structures because it determines their electric characteristics. For VLSI and ULSI devices, defects and the generation of defects in the region close to the SiO_2–Si interface can affect the reliability of the devices more than defects in the bulk of the oxide layer [92]. Therefore, the most important parameters to monitor during the fabrication of MOS structures are the chemical composition and the structural disordering at the SiO_2–Si interface. One of the main types of defects occurring at the interface that can affect the output characteristics of MOS structures is silicon suboxide species on the SiO_2 side of the interface (E' defects or oxygen-deficient silicon centers), which are primarily responsible for trapped positive charges in silicon substrate oxides. Another is unsaturated silicon atoms (trivalent silicon) at the interface, which can induce interfacial electron states near the middle of the Si band gap (named P_b centers) [93–96].

Energetically speaking, these interfacial electron states are traps within the Si gap, localized at the interface, which can interact with free charge carriers in the semiconductor substrate. During accumulation, the traps are filled by the majority charge carriers, but upon inversion or depletion, their charge changes. The density of interfacial electron states reaches a maximum at the interface and extends by tunnelling several angstroms into the SiO_2 layer (Fig. 6.20). Interfacial defects can also create traps in the bulk, N_t, in a space charge region (SCR) extending to a depth on the order of micrometers (Fig. 6.20). Usually the concentration of the bulk traps is smaller than the concentration of the doping impurity, N_d, but the total concentration N_t in the SCR can be compared with the density of the interfacial electron states.

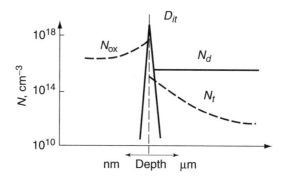

Figure 6.20. Diagram of spatial distribution of electron states in interfacial region. Here, D_{it} is density of surface state, N_t, N_{ox}, and N_d are concentrations of bulk traps, oxide traps, and dopant, respectively.

In recent years improvements in the technological processing has lowered the density of surface electron states in the best MOS structures from 10^{12} to 10^8 cm^{-2}·eV^{-1}. The methods commonly used to investigate surface electron states are the field effect, the frequency dependence of the surface conductivity and $C - V$, deep level transient spectroscopy (DLTS), and electron paramagnetic resonance (EPR) methods [27, 95–99].

Harrick [100] proposed to use internal reflection spectroscopy to investigate the surface electron states in the dielectric–semiconductor systems. In contrast to the indirect methods of the field-effect and surface conductivity measurements, measurement of the optical absorption allows one to determine the spectral distribution of the surface states directly. However, assuming that the surface state density is 10^{10} cm^{-2} and the absorption cross section is 10^{-16} cm^2, the absorption by surface states will cause changes in the reflectance of only 10^{-6}. Therefore, modulation spectroscopy (Section 4.7) is used, in which only the change in absorption under an alternating current is amplified and the filling of a surface state is changed by application of a constant potential difference.

The absorption on surface states can cause the electron transitions from filled surface states to the c-band (λ_1) and from filled states of the v-band to free states (λ_2); for n-type surfaces $\lambda_1 > \lambda_2$, and for p-type surfaces $\lambda_1 < \lambda_2$ (Fig. 6.21*b*). Since carriers can undergo transitions from any filled state of the v-band to any free surface states and from any filled surface states to a free state in the c-band, narrow absorption bands are not usually observed, and only the boundaries of absorption are fixed. When the value of the constant voltage bias is changed, a change in the filling of states close to the Fermi level is possible, causing a shift of the absorption boundaries. Inversion of the surface conductivity from n- to p-type results in a change of the sign of the signal. In addition, the broadband signal of absorption, $\sim\lambda^2$, caused by free carriers is always superimposed on the surface state absorption. The surface state absorption may be distinguished from the absorption of free carriers by observing the dependence of the spectral shape on the applied voltage bias (for the latter, absorption is independent of the

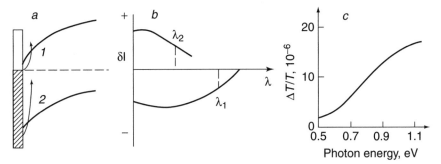

Figure 6.21. (a) Scheme of electronic transitions; (b) general shapes of IR absorption for surface with n-type conduction; (c) absorption of defects of trivalent silicon in SiO_2–Si.

voltage). The absorption of surface states caused by the defects of trivalent silicon in a thermal SiO_2–Si system was obtained with this method [101] (Fig. 6.21c).

At energies higher and lower than the band gap of a bulk material, the method of differential surface reflection [102–105] is also applied to study electronic transitions in surface states. This method involves measuring the reflection from a clean sample surface and then from the sample surface after exposure to the atmosphere of any gas (usually oxygen). The results are presented in $\Delta R/R_d$ units, representing the relative change of the reflectance: $\Delta R/R_d = [R(0) - R(d)]/R(d)$. [Here, $R(d)$ is the value of the reflectance for a "saturated" surface, after long exposure to an oxygen atmosphere.] This is related to the surface dielectric function, representing the transition between filled and free surface electron states. Qualitatively, the reflectance of a clean surface is higher than that of a surface after oxidation when the surface states are released. Since transitions of the oxide layer do not appear in the spectral region where the surface state transitions are observed (0.3–4 eV), the spectrum $\Delta R/R_d$ represents the distribution of states on the clean surface. Because of a small contribution from surface effects ($\sim 10^{-2}$) to the total reflectance from a semiconductor surface, R must be measured very accurately, preferably in a high vacuum. Here, $\Delta R/R_d$ can be quantitatively analyzed on the basis of the macroscopic theory developed for a three-layer model, in which the surface is considered to be an absorbing layer between the substrate and the surroundings of thickness $d \ll \lambda$ [106] (Eq. (3.41) in Section 3.10).

The spectrum $\Delta R/R_d$ for Si(111) 2 × 1 obtained in nonpolarized radiation is presented in Fig. 6.22. The results show the presence of an energy gap in the zone structure arising from dangling bonds on the reconstructed surface of silicon. The surface is strongly anisotropic with the principal axis of the permittivity tensor lying along the direction 2 × 1 of the reconstruction. The dependence of the Si(111) 2 × 1 surface reflectance on the radiation polarization has been analyzed in detail in Ref. [107]. In the theoretical analysis of the spectra $\Delta R/R_d$ [Eq. (3.41)], it was assumed that the effect of the oxidation is only to release surface states. However, this is not necessarily the case even for nonabsorbing oxides. In reality, the adsorption of oxygen on a semiconductor results

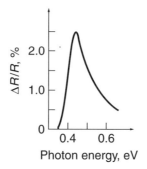

Figure 6.22. Differential spectra of Si(111) 2 × 1 surface at $T = 295$ K.

in a shift of the surface Fermi level, implying a change in the surface electric field. Hence, the change in $\Delta R/R_d$ can also be influenced by the Franz–Keldysh effect. However, a typical contribution from this effect to the change of $\Delta R/R_d$ does not exceed 0.2%, which is smaller than the contribution caused by the transitions connected with surface states. Nevertheless, the Franz–Keldysh effect induced by oxidation has been observed in GaAs and GaP, which are materials with a high electroreflection coefficient [102].

The properties of surface states at the SiO_2–Si interface could also be inferred by analysis of IR absorption spectra of silicon oxide [94, 108–110]. Although the IR absorption of the Si–O–Si stretching mode does not allow for direct determination of the density of surface states at the interface or the traps in the oxide, it does supplement techniques such as the C–V method, RBS analysis, and AES.

The correlation between the midgap interfacial state density D_{it} and the thickness strain in thermally grown SiO_2 films has been studied [110]. The D_{it} decreases with the oxidation temperature and with the oxide film thickness. The frequency of the Si–O stretching vibrations increases with increasing oxidation temperature and film thickness. The peak position v is expressed as a function of the average Si–O–Si bond angle as

$$v \approx v_0 \sin\theta, \qquad (6.9)$$

where $v_0 = 1134$ cm^{-1}. The interatomic distance d_{Si-Si}, which can be considered as an internal strain parameter [73], is also proportional to the Si–O–Si bond angle:

$$d_{Si-Si} = 2r_0 \sin\theta, \qquad (6.10)$$

where r_0 is the Si–O bond length. Since v and d_{Si-Si} are proportional to $\sin\theta$, the change in v gives the thickness strain [111]:

$$\xi(SiO_2) = 1 - \frac{v}{v_1}. \qquad (6.11)$$

Here, $\nu_1 = 1079$ cm^{-1} is the bond-stretching frequency of a fully relaxed film and ν is the corresponding bond-stretching frequency in the stressed oxide. The calculation of the microscopic strain from the Si–O–Si stretching frequency shows the existence of a linear relationship between D_{it} and the thickness strain in SiO$_2$ [110].

The structural and electrical properties of SiO$_2$ structures irradiated by high-energy ions was studied [94]. When the flux of Xe or Ni is higher than 10^{11} ions/cm^2, the absorption peak at 1079 cm^{-1} shifts toward lower wavenumbers (Fig. 6.23 [94]). In addition, the optical density at the frequency of the Si–O stretch mode decreases while the FWHM increases. From the IR absorption peak, the variation in the Si–O–Si bond angle was estimated. The Si–O–Si angle drops from 144.5° to 140.9° and 134.3° after irradiation by $5 \cdot 10^{12}$ ions/cm^2 of Ni and Xe, respectively. Simultaneously, the interatomic d_{Si-Si} distance decreases from 3.04 to 3.01 and 2.95 Å, respectively [94]. Therefore, the high-energy ions not only induce oxygen vacancies and other defects in SiO$_2$ but also distort the bulk network [112]. The midgap interface state density D_{it} and the oxide charge density N_{ox} as determined by electrical analysis increase linearly with increasing ion flux. The same results were obtained for neutron-irradiated silica, namely the Si–O bond vibration peak shifts by 15 cm^{-1} toward lower wavenumbers, and the intertetrahedral Si–O–Si bond angle decreases by 4° [113].

As known, saturation of defects connected with dangling Si–Si bonds by hydrogen atoms decreases the defect concentration [114]. The number of silicon dangling bonds responsible for electron traps may also be reduced by oxygen introduced during the deposition process [115]. The electron traps related to the presence of hydrogen-bond silanol group (SiOH) and adsorbed water are also undesirable in MOS devices [116]. Thus, thermal annealing may decrease the interfacial trap density D_{it} by decreasing the concentration of H$_2$O and OH in SiO$_2$ [109].

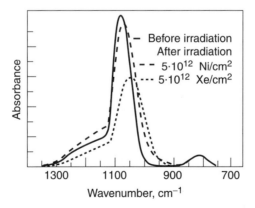

Figure 6.23. IR absorption of thermal SiO$_2$ before and after irradiation by Xe and Ni at 5×10^{12} ions/cm^2. Reprinted, by permission, from M. C. Busch, A. Slaoui, P. Siffert, E. Dooryhee, and M. Toulemonde, *J. Appl. Phys.* **71**, 2596 (1992). Copyright © 1992 American Institute of Physics.

6.6. IN SITU INFRARED CHARACTERIZATION OF Si AND SiO₂ SURFACES

Infrared spectroscopy is particularly useful in the study of adsorption and decomposition of molecules on a Si surface, an important application in the semiconductor industry, which has a variety of processes that involve the reaction of gas-phase molecules with a Si surface. There is a plethora of measurement techniques in IR spectroscopy that have good surface sensitivity. The chemical information available, including identification and determination of concentrations of species plus structural information, can be obtained with very high resolution and yet with no undesirable physical or chemical changes in the material of interest caused by the measurement [117].

There are several reviews and books devoted to the theory and techniques of surface IR spectroscopy [100, 102, 117, 118]. The classical scheme proposed by Harrick uses an internal reflection element [100]. Later, the ATR method was developed to improve the surface sensitivity. The optimum conditions for recording ATR spectra and experimental techniques are described here in Chapters 2 and 4.

6.6.1. Monitoring of CVD of SiO₂

The deposition of SiO₂ by several CVD processes and surface reactions of gaseous reactants with a SiO₂ surface has been monitored in situ by ATR and the standard FTIR transmission methods [53, 119–121]. Using these methods, information about the IR absorption of the surface species, the film, and the ambient gas in the reactor during the film growth can be obtained. It is well known that CVD parameters such as the quality of the film, the rate of deposition, and the profile of chemical composition are sensitive to the transport and reactions of species in the plasma as well as to the surface reactions. The IR absorption of CVD silicon dioxide after deposition (ex situ) was considered in Section 5.2.

Table 6.1 presents the assignment of IR absorption peaks observed after various CVD processes of SiO₂. The table includes absorption peaks of the organic and organometallic gas phases, adsorbed species, and chemical bonds of films, with the exception of peaks attributed to Si—O bonds, which have already been discussed in the relevant contexts.

The type of surface species and their relative surface concentration depend on the deposition conditions such as the identity and concentration of the gaseous reactants in the plasma reactor, the temperature of the substrate, the duration of the deposition process, and the quality of the Si surface before deposition. The ATR-FTIR studies of the SiO₂ surface exposed to TEOS at various temperatures show that below $100°C$ TEOS is physisorbed and above $100°C$ TEOS is irreversibly chemisorbed onto the SiO₂ surface [53]. Therefore, in low-temperature CVD, physically adsorbed TEOS can be trapped in the growing oxide, resulting in more ethoxy and OH species. This leads to a poor-quality film, with increased porosity and instability. High-quality OH-free films can be produced by limiting the TEOS flux to the surface during the deposition [53]. Absorption peaks from TEOS

Table 6.1. Peaks assignment after CVD from TEOS and SiH$_4$

Deposition Process	Peak Position (cm^{-1})	Peak Assignment	References
1. TEOS in gas phase	810	CH$_2$ rocking	53, 122
	1300	CH$_2$ twisting	53
	1395	CH$_2$ wagging	53
	966, 1175	CH$_3$ rocking	53, 123
	1087, 1117	C−O stretching	53
	2901, 2943	CH$_2$ stretching	53, 120
	2936, 2981	CH$_3$ stretching	53, 120
2. TEOS absorbed on SiO$_2$ at 200°C	1107 1080	C−O stretching in−OCH$_2$CH$_3$ species	53
	960, 1160	CH$_3$ rocking	53
	2896, 2939, 2981	C−H stretching of O−C$_2$H$_5$	53
	1299, 1370, 1396, 1450, 1460, 1490	C−H of −CH$_3$ and −CH$_2$	53, 124
	3746	Isolated OH stretching	53
	3360	Hydrogen-bonded OH	53
3. PECVD from SiH$_4$ + O$_2$ + Ar	2250	Si−H stretching in HSiO$_3$	119, 125, 126
	2206	Si−H stretching in H$_2$SiO$_2$	119, 125, 126
	2200	Si−H stretching in H$_3$SiO or HSi(SiO$_2$)	119, 125, 126
	2160	Si−H stretching in H$_2$Si(SiO)	119, 125, 126
	2140	Si−H stretching in H$_3$SiSi	119, 125, 126
	2127	Si−H stretching in HSi(Si$_2$O)	119, 125, 126
	2110	Si−H stretching in H$_2$SiSi$_2$	119, 125, 126
	2090	Si−H stretching in HSiSi$_3$	119, 125, 126
	3740	OH stretching of isolated SiOH	119, 125, 126
	3600–3000	OH stretching of associated SiOH	119, 125, 126

molecules (indicating physical adsorption) and peaks from chemically adsorbed TEOS are presented in Table 6.1. The problems associated with monitoring the delivery of condensable gases and the SiO$_2$ composition during the CVD process were also studied for triethylsilane (TES) + H$_2$ [121], TEOS + ozone O$_3$ [120], and trimethylphoshine (TMP) [120]. Adsorbed TMP absorbs IR radiation at 1041 and 753 cm^{-1}. The FTIR spectra of TES in the gas phase and adsorbed onto the Si surface reveal absorption peaks from C−H and Si−H (2100 cm^{-1}) and Si−C$_2$H$_5$ (900–1250 cm^{-1}).

In situ ATR-FTIR spectra of films produced by CVD in an O$_2$-rich mixture (SiH$_4$ + O$_2$) show that the oxide surface is largely covered with isolated and associated hydroxyls. Thermal annealing at 250°C removes weakly bound OH from the surface without additional Si−O−Si bond formation. At that time, the Ar^+ bombardment during Ar^+-assisted deposition releases H_2O from SiOH, forming Si$-$O$-$Si bonds due to the reaction $Ar^+ + 2SiOH \rightarrow Ar^+ + SiOSi + H_2O$. The reactions between surface species (hydroxyl groups) and gas-phase molecular fragments of SiH_4 (SiH_x, $x = 1, 2, 3$) produce surface hydrides $HSiO_3$, H_2SiO_2, and H_3SiO. The silane fragments also react with Si on the surface, producing silicon hydrides $HSi(SiO_2)$, $H_2Si(SiO)$, $HSi(Si_2O)$, and H_3SiSi [119]. Table 6.1 gives the absorption peaks of silicon hydrides observed after CVD from a mixture of SiH_4 and O_2.

6.6.2. Cleaning and Etching of Si Surfaces

It is well known that hydrocarbons and other contaminants of Si surfaces can prevent defect-free epitaxial growth and can cause significant degradation of the MOS device parameters [127]. To successfully effect the oxidation or epitaxial growth of Si for the fabrication of ICs, it is very important to start with a clean silicon surface that is chemically stable and free from impurities. The preferred cleaning technique usually includes chemical oxidation and H treatment in water followed by oxide removal in a hydrofluoric acid (HF) solution or fluorine plasma. The resulting surfaces are hydrophobic and quite resistant to chemical attacks, as most of the dangling bonds are terminated by hydrogen atoms, with a small fraction terminated by either fluorine atoms or hydroxyl groups [128–131]. The chemical state of Si surfaces treated in water and a HF-based solution or fluorine plasma was examined in situ by MIR spectroscopy [128–130, 132–137], IRRAS [138, 139], and ATR spectroscopy [140–143].

The IR absorption spectrum of a Si surface recorded during immersion in deionized water (DI water) exhibits a broad peak at 2100 cm^{-1} and a weak peak at 2080 cm^{-1} that can be attributed to dihydride Si (SiH_2) and monohydride Si (SiH), respectively [128, 130, 132, 134, 136]. In addition, the IR spectra of a Si surface treated with ultrapure water or an SC1 solution ($NH_4OH-H_2O_2-H_2O$, $0.25 : 1 : 5$ at $70°C$) display the 2140-cm^{-1} absorption peak of trihydride Si (SiH_3) [136, 142]. Hence it can be concluded that rinsing in water leads to the hydrogen termination of Si surface. The ATR spectra of a deuterium-terminated Si surface formed by a similar wet method has also been considered [143].

The immersion of a hydrogen-terminated Si wafer into a flow of HF solution decreases the SiH peak intensity, indicating that after immersion in HF solution the surface is not completely terminated by hydrogen. Moreover, a broad peak around 2230 cm^{-1} is observed. This peak is assigned to hydrogen-associated Si fluorides $SiH(SiF_2)$ [128, 130], based on a semiempirical formula for calculating the Si$-$H stretching frequencies for various substituted silanes and experimental values of the Si$-$H stretching vibration frequencies for the fluorine-substituted silanes SiH_3F, SiH_2F_2, and $SiHF_3$ (2206, 2245, 2314 cm^{-1}, respectively) [144]. Thus, the Si surface may be partially covered by hydrogen-associated Si fluorides. When the surface is rinsed in water after immersion in the HF solution, the hydrogen-associated Si fluorides are removed and the surface Si$-$F bonds are converted to Si$-$H bonds, leading to the complete hydrogen termination of the surface [128]. There is doubt about this conclusion because the absorption band at 2230 cm^{-1} is present in the spectra of both Ge and Si wafers and probably arises from electrolyte adsorption [133] (see also Section 3.7.2).

It is possible to directly observe the presence and behavior of Si−F bonds by IRRAS. Two absorption peaks may be observed between 905 and 925 cm^{-1} for a Si surface treated in the HF solution [139], attributed to the SiF$_2$ symmetric stretching mode (918 cm^{-1}) and the SiH$_2$ bending mode (910 cm^{-1}). The absorption peak of SiF$_2$ decreases after a DI water rinse, but no change in SiH$_x$ bonds is detected [139]. It was found that the concentration of fluorine on a Si surface increases with increasing HF concentration, reaching 2.6×10^{14} atoms/cm^2 upon etching in 50% HF [131]. Most of the Si−F bonds are rapidly hydrolyzed to Si−OH by rinsing the wafer in water and are also gradually hydrolyzed by exposure to the moisture in air.

Other agents may be used in IC technology to clean the silicon surface prior to the gate oxidation. In situ FTIR spectroscopy was applied to characterize the H-terminated Si surface during immersion in NH$_4$F [129, 134] and NaF [135] solutions. While HF dipping is clearly effective in removing native oxide, hydrocarbon contamination (CH$_2$ and CH$_3$ peaks of stretching modes in the range 2850–2950 cm^{-1} and bending modes in the range 1350–1460 cm^{-1}) can be worse after HF treatment [132]. To remove hydrocarbons at low temperatures, plasma etching is effective. FTIR spectroscopy was applied to study several plasma treatments, including F$_2$ [138, 142, 145], XeF$_2$ [140], NH$_3$ [127], and Cl$_2$ [146].

During treatment of Si in F$_2$ gas, the Si−F absorption bands attributed to SiF$_2$, SiF$_3$, and SiF$_4$ arise between 850 and 880 cm^{-1}. The F$_2$ treatment at room temperature removes hydrogen from the Si surface (the coverage of Si−H bonds is found to decrease) and forms Si−F bonds [138, 145]. The following reaction is thought to occur: S−H(surface) + F$_2$(gas) → Si−F(surface) + HF(gas). The reaction is terminated at one-monolayer Si−F coverage [140, 141]. The monolayer of Si−F is converted to 70% Si−H and 30% Si−OH by immersing in water for 1 min [141].

Possible contaminants of the Si surface during H$_2$ and NH$_3$ plasma cleaning may be monitored in situ by ATR-FTIR [127]. Peaks assigned to these contaminants are presented in Table 6.2.

H$_2$ plasma cleaning minimizes hydrocarbon contamination by using only atomic hydrogen at room temperature. However, atomic hydrogen at $T < 350°C$ does not remove SiO$_2$ from Si, and ion bombardment is required to break Si−O bonds. However, excessive ion bombardment may result in a highly damaged Si surface, as evidenced by a strong Si−H absorption. Although the H$_2$ plasma etching reduces the hydrocarbon content, removes SiO$_2$, and forms Si−H, the H atoms also react with reactor walls, producing H$_2$O [127].

The problem of distinguishing between species in the gas phase and species on the Si surface during Cl$_2$ plasma etching is considered for in situ IRRAS measurements [146]. Unsaturated silicon chlorides SiCl$_x$ ($x = 1, 2, 3$) exhibit a series of peaks in the 500–600-cm^{-1} spectral region.

Table 6.2. Absorption peaks of possible contaminants

Peak Position (cm^{-1})	Peak Assignment	References
1670	H$_2$O scissor	147, 148
2100	Si$-$H stretching	127
2200	O$_2$SiH$_2$	127, 149
2256	HSiO$_2$	127, 150
2850	C$-$H stretching	127
2917	C$-$H stretching	127
3000–3700	O$-$H stretching in physisorbed water	127
3660	O$-$H stretching in SiOH	127

Chemomechanical polishing (CPM) of Si is studied using the IR absorption of surface chemical species as measured in the MIR geometry [151, 152]. CPM uses an alkaline suspension of colloidal silica and combines a mechanical grinding action with chemical etching to produce smooth, defect-free starting surfaces for subsequent device patterning. The IR absorption spectrum of a Si surface after CPM displays the results of H termination, namely the presence of the SiH (2070–2090-cm^{-1}), SiH$_2$ (2090–2120-cm^{-1}) and SiH$_3$ (2120–2150-cm^{-1}) absorption peaks. Among the IR absorption peaks of contaminants are IR absorption peaks of hydrocarbons (CH$_2$ stretching modes at 2930 and 2855 cm^{-1}, CH$_3$ stretching mode at 2970 cm^{-1}), hydroxyl termination at 3100 cm^{-1}, and a variety of oxidation states of H$-$Si$-$O that absorb close to 2250 cm^{-1}.

6.6.3. Initial Stages of Oxidation of H-Terminated Si Surface

In order to produce reliable ULSI devices, the general chemical stability of a HF-treated Si surface exposed to gases and in particular the mechanism of its oxidation must be characterized. The kinetics of the initial stages of oxidation of a H-terminated Si surface exposed to several gases was studied by MIR and ATR spectroscopies and IRRAS [153–157]. The growth of native oxide on a H-terminated surface has also been studied in situ [158, 159].

A H-terminated Si surface exhibits peaks corresponding to SiH (2080 cm^{-1}) and SiH$_2$ (2110 cm^{-1}) immediately after HF treatment, with the latter predominant (Fig. 6.24) [153]. As the exposure time in air is increased, the SiH and SiH$_2$ modes decrease in intensity, suggesting that the surface SiH bonds are attacked by oxidant present in air and converted to Si$-$O$-$Si bonds. Additional peaks appear at 2200 and 2250 cm^{-1} that are attributed to SiH$_2$(O$_2$) and SiH(O$_3$), respectively. The number of these intermediate oxidation species initially increases and then drops [154]. In addition, generation of SiH(O$_3$) follows that of SiH$_2$(O$_2$). Based on these trends, a two-stage model of the oxidation kinetics was proposed [153–155]. At the initial stages of oxidation, oxygen attacks both the Si$-$H and back bonds of a Si atom in the outermost layer, and intermediate oxidation species are produced. IR absorption spectra for (111) and (100) Si surfaces demonstrate a strong dependence on the crystallographic orientation of the concentration of intermediate oxidation species [156].

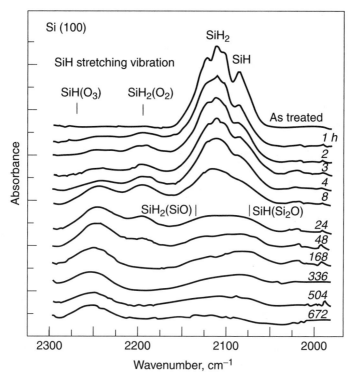

Figure 6.24. Absorption spectra of Si–H stretching modes of H-terminated Si(100) surfaces after exposure to air for various times. Reprinted, by permission, from M. Niwano, J. Kageyama, K. Kinashi, J. Sawahata, and N. Miyamoto, *Surf. Sci. Lett.* **301**, L245 (1994). Copyright © 1994 Elsevier Science.

The water present in air is the major oxidizing agent of SiH and SiH$_2$ bonds. Oxidation induced by water probably generates SiOH and Si$-$O$-$Si bridging bonds on the surface via the following reactions: SiH + H$_2$O → SiOH + H$_2$ and SiH + SiOH → Si$-$O$-$Si + H$_2$. As a consequence, the surface, which was initially hydrophobic, becomes hydrophilic. The hydrophilic surface reacts much strongly with O$_2$ and H$_2$O than the H-terminated hydrophobic surface, and the rate of oxidation increases, as was observed in IR spectra [154, 155]. Rinsing the Si surface treated in HF solution with water causes the fluorine bonds to be rapidly replaced by oxygen-containing species via the reaction SiF + H$_2$O → SiOH + HF. The kinetics of the oxidation of a H-terminated Si surface in water are investigated by monitoring the IR absorbance at the frequencies of the Si$-$H and Si$-$O$-$Si modes [159].

The thermal nitration of a H-terminated Si surface and the CVD of silicon nitride were studied in situ by FTIR-IRRAS [160, 161]. The adsorption and thermal decomposition of phosphine (PH$_3$) on a Si surface was also studied by IR absorption; depending on the coverage and the exposure to the flux, PH$_3$ adsorbs both nondissociatively and dissociatively, and the IR absorption peaks of phosphorus-containing species in both molecular and dissociated forms lie between 2060 and 2330 cm^{-1} [162]. The adsorption and decomposition of disilane (Si_2H_6) on a Si surface was analyzed by MIR-FTIR [163]. The adsorption of methanol and its subsequent decomposition upon heating were studied by IRRAS [164].

REFERENCES

1. V. I. Strikha and S. S. Kilchitskaya, *Solar Cells Based on Metal-Semiconductor Contact*, Energoizdat, St. Petersburg, 1992.

2. J. Shewchun, D. Burk, and M. Spitzer, *IEEE Trans. Electron Devices* **ED27**, 705 (1980).

3. M. A. Green and P. B. Godfrey, *Appl. Phys. Lett.* **29**, 610 (1976).

4. P. Victorovich, G. Kamarinos, and P. Even, *Phys. Stat. Solidi* **A48**, 137 (1978).

5. S. S. Kilchitskaya, S. V. Litvinenko, V. A. Skryshevsky, V. I. Strikha, and V. P. Tolstoy, *Izvestiya vysshikh uchebnykh zavedennij* **9**, 86 (1988).

6. V. V. Eremenko, G. V. Kuznecov, S. S. Kilchitskaya, S. V. Litvinenko, V. A. Skryshevsky, and V. I. Strikha, *Geliotechnika [Solar Eng.]* **4**, 20 (1989).

7. S. S. Kilchitskaya, T. S. Kilchitskaya, G. D. Popova, and V. A. Skryshevsky, *Thin Solid Film* **346**, 226 (1999).

8. T. A. Vdovenkova, A. P. Vetrov, S. S. Kilchitskaya, T. S. Kilchitskaya, G. D. Popova, and V. I. Strikha, *Solid State Electron.* **38**, 929 (1995).

9. V. P. Tolstoy and V. N. Krilov, *Opt. Spectrosc.* **55**, 1066 (1983).

10. V. P. Tolstoy and S. N. Grusinov, *Opt. Spectrosc.* **63**, 823 (1987).

11. P. Grosse, in U. Rossler (Ed.), *Advanced in Solid State Physics*, Vol. 31, Vieweg, Braunschweig, 1991, p. 77.

12. R. Brendel, *Appl. Phys.* **A50**, 587 (1990).

13. R. Brendel and R. Hezel, *J. Appl. Phys.* **71**, 4377 (1992).

14. Y. Matsui, Y. Miyagawa, J. Izumitani, M. Okuyama, and Y. Hamakawa, *Jpn. J. Appl. Phys.* **31**, 369(1992).

15. R. Butz and H. Wagner, *Phys. Stat. Solidi* **A94**, 71 (1986).

16. M. A. Taubenblatt and C. R. Helms, *J. Appl. Phys.* **53**, 6308 (1982).

17. D. B. Aldrich, C. L. Jahncke, R. J. Nemanich, and D. E. Sayers, *Mater. Res. Soc. Symp. Proc.* **22**, 343 (1992).

18. E. Buzaneva, T. Vdovenkova, S. Litvinenko, V. Makhnjuk, V. Strikha, V. Skryshevsky, P. Shevchuk, V. Nemoshkalenko, A. Senkevich, and A. Shpak. *J. Electron. Spectrosc. Related Phenom.* **68**, 707 (1994).

19. L. S. Olsen, *Solid-State Electron.* **20**, 741 (1977).

20. D. A. Buchanan, *Appl. Phys. Lett.* **65**, 1257 (1994).

21. G. Greaves, *J. Non-Cryst. Solids* **32**, 119 (1979).

22. T. Itoga, H. Kojima, and A. Hiraiwa, *Solid State Devices and Materials*, Extended Abstract of 1992 Int. Conf., Tsukuba, Japan Society of Applied Physics, Tokyo, 1992, p. 434.

23. E. T. P. Benny and J. Majhi, *Semicond. Sci. Technol.* **7**, 154 (1992).

24. S. P. Singh, A. K. Saxena, I. M. Tiwari, and O. P. Agnihotry, *Thin Solid Films* **127**, 77 (1985).

25. S. S. Kilchitskaya, T. S. Kilchitskaya, V. A. Skryshevsky, V. I. Strikha, and V. P. Tolstoy, *Zurnal Prikladnoj Spectroscopii [J. Appl. Spectrosc.]* **48**, 445 (1988).

26. S. M. Sze (Ed.), *VLSI Technology*, Vol. 1, McGraw-Hill, New York, 1983.

27. S. M. Sze, *Physics of Semiconductor Devices*, Wiley, New York, 1981.

28. Y. A. Averkin, N. K. Karmadonov, A. V. Kharlamov, and V. A. Skryshevsky, *Dielectr. Semicond.* **40**, 40 (1991).

29. S. Iwata, K. Nakata, and A. Kikushi, *J. Jpn. Inst. Metals* **50**, 287 (1986).

30. X. Xu and D. W. Goodman, *Appl. Phys. Lett.* **61**, 774 (1992).

31. A. Toriumi, H. Satake, N. Yasuda, and T. Tanamoto, *Appl. Surface Sci.* **117–118**, 230 (1997).

32. T. Tamura, J. Sakai, M. Satoh, Y. Inoue, and H. Yoshitaka, *Jpn. J. Appl. Phys.* **36**, 1627 (1997).

33. R. A. Bowling and G. B. Larrabee, *J. Electrochem. Soc.* **132**, 141 (1985).

34. S. Rojas, R. Gomarasca, L. Zanotti, A. Borghesi, A. Sassella, G. Ottaviani, L. Moro, and P. Lazzeri, *J. Vac. Sci. Technol.* **B10**, 633 (1992).

35. D. P. Poenar, N. van der Puil, P. J. French, and R. F. Wolffenbuttel, *J. Electrochem. Soc.* **143**, 968 (1996).

36. V. A. Skryshevsky, *Microelectron. Ser. 6* **5**, 81 (1990).

37. A. S. Tenney and M. Ghezzo, *J. Electrochem. Soc.* **120**, 1276 (1973).

38. K. H. Hurley, *Solid State Technol.* **30**, 103 (1987).

39. W. Kern, *RCA Rev.* **32**, 429 (1971).

40. W. Kern and G. L. Schnable, *RCA Rev.* **43**, 423 (1982).

41. B. Wangmaneerat, J. A. McGuire, T. M. Niemczyk, D. M. Haaland, and J. H. Linn, *Appl. Spectrosc.* **46**, 340 (1992).

42. G. V. Samsonov (Ed.), *Physico-Chemical Properties of Oxides*, Metallurgia, Moscow, 1978.

43. H. Itoh, Y. Ohmori, and M. Horiguchi, *J. Non-Cryst. Solids* **88**, 83 (1986).

44. L. E. Davis, N. C. MacDonald, P. W. Palmberg, G. E. Riach, and R. E. Weber, *Handbook of Auger Electron Spectroscopy*, Physical Electranics Industries, Eden Prairie MN, 1976.

45. W. J. Gignac, R. S. Williams, and S. P. Kowalczyk, *Phys. Rev.* **B32**, 1237 (1985).

46. H. Shirai and R. Takeda, *Jpn. J. Appl. Phys.* **35**, 3876 (1996).

47. V. M. Agranovich and D. L. Mills (Eds), *Surface Polaritons: Electromagnetic Waves at Surfaces and Interfaces*, North-Holland, Amsterdam, 1982.

48. S. Iwata, K. Nakata, and A. Kikushi, *J. Jpn. Inst. Metals* **50**, 287 (1986).

49. G. Hollinger, E. Bergignat, J. Joseph, and Y. Robach, *J. Vac. Sci. Technol.* **A3**, 2082 (1985).

50. G. S. Korotchenkov, V. J. Tsvitsinsky, and V. A. Mikhailov, *J. Vac. Sci. Technol.* **B3**, 981 (1985).

51. M. Faithpour, P. K. Boyer, G. J. Collins, and C. W. Wilmsen, *J. Appl. Phys.* **57**, 637 (1985).

52. S. C. Shei, Y.-K. Su, C.-J. Hwang, and M. Yokoyama, *Jpn. J. Appl. Phys.* **34**, 476 (1995).

53. S. C. Deshmukh and E. S. Aydil, *J. Vac. Sci. Technol.* **A13**, 2355 (1995).

54. Y. K. Su and U. H. Liaw, *J. Appl. Phys.* **76**, 4719 (1994).

55. G. S. Korotchenkov and V. A. Skryshevsky, *Mat. Res. Soc. Symp. Proc.* **355**, 497 (1995).

56. J. E. Griffits, G. P. Schwartz, W. A. Sunder, and H. Schonhorn, *J. Appl. Phys.* **53**, 1832 (1982).

57. M. Yamaguchi, *J. Appl. Phys.* **53**, 1834 (1982).

58. A. Nelson, K. Geib, and C. W. Wilmsen, *J. Appl. Phys.* **54**, 4134 (1983).

59. N. Shibata and H. Ikoma, *Jpn. J. Appl. Phys.* **31**, 3976 (1992).

60. G. P. Schwartz, W. A. Sunder, and J. E. Griffits, *Appl. Phys. Lett.* **37**, 925 (1980).

61. M. Fathipour, W. H. Makky, and J. McLaren, *J. Vac. Sci. Technol.* **A1**, 662 (1983).

62. I. F. Wager, K. M. Geib, and C. W. Wilmsen, *J. Vac. Sci. Technol.* **B1**, 778 (1983).

63. V. F. Skryshevsky and V. P. Tolstoy, *Infrared Spectroscopy of Semiconductor Structures*, Lybid, Kiev, 1991.

64. J. Henry and J. Livingstone, *Infrared Phys. Technol.* **36**, 655 (1995).

65. C. T. Lenczycki and V. A. Burrows, *Thin Solid Films* **193–194**, 610 (1990).

66. C. Huang, A. Ludviksson, and R. M. Martin, *Surface Sci.* **265**, 314 (1992).

67. *Silicon on Insulator and Buried Metals in Semiconductors*, J. C. Sturm, C. K. Chen, L. Pfeiffer, P. L. F. Hemment (Eds.), *Mater. Res. Soc. Symp. Proc.* **107** (1988).

68. J. M. Leng, J. J. Sidorowich, Y. D. Yoon, J. Opsal, B. H. Lee, G. Cha, J. Moon, and S. I. Lee, *J. Appl. Phys.* **81**, 3570 (1997).

69. A. Takami, A. Arimoto, H. Morikawa, S. Hamamoto, T. Ishihara, H. Kumabe, and T. Murotani, in R. Hill, W. Palz, P. Helm, and H. S. Stephens (Eds.), *Photovoltaic Solar Energy*, Proc. 12th European Conf., Bedford, Amsterdam, 1994, p. 59.

70. N. Hatzopoulos, D. I. Siapkas, and P. L. F. Hemment, *J. Appl. Phys.* **77**, 577 (1995).

71. J. Margail, J. Stoemenos, C. Jaussaud, and M. Bruel, *Appl. Phys. Lett.* **54**, 526 (1989).

72. A. Perez, J. Samitier, A. Cornet, J. R. Morante, P. L. F. Hemment, and K. P. Homewood, *Appl. Phys. Lett.* **57**, 2443 (1990).

73. G. Lucovsky, M. J. Manitini, J. K. Srivastava, and E. A. Irene, *J. Vac. Sci. Technol.* **B5**, 530 (1987).

74. P. Gaworzewski, E. Hild, F. G. Kirscht, and L. Vecsernyes, *Phys. Stat. Solidi* **A85**, 133 (1984).

75. P. Bruesch, Th. Stockmeier, F. Stucki, and P. A. Buffat, *J. Appl. Phys.* **73**, 7677 (1993).

76. P. Bruesch, Th. Stockmeier, F. Stucki, P. A. Buffat, and J. K. N. Lindner, *J. Appl. Phys.* **73**, 7690 (1993).

77. P. Bruesch, Th. Stockmeier, F. Stucki, P. A. Buffat, and J. K. N. Lindner, *J. Appl. Phys.* **73**, 7701 (1993).

78. A. Borghesi, A. Sassella, and T. Abe, *Jpn. J. Appl. Phys.* **34**, L1409 (1995).

79. M. Reiche, S. Hopfe, U. Gosele, H. Strutzberg, and Q.-Y. Tong, *Jpn. J. Appl. Phys.* **35**, 2102 (1996).

80. R. Brendel, *J. Appl. Phys.* **72**, 794 (1992).

81. M. K. Weldon, Y. J. Chabal, D. R. Hamman, S. B. Christman, E. E. Chaban, and L. C. Feldman, *J. Vac. Sci. Technol.* **B14**, 3095 (1996).

82. A. Borghesi and A. Sassella, *Phys. Rev. B* **50**, 17756 (1994).

83. D. Feijoo, Y. J. Chabal, and S. B. Christman, *Appl. Phys. Lett.* **65**, 2548 (1994).

84. M. K. Weldon, V. E. Marsico, Y. J. Chabal, D. R. Hamann, S. B. Christman, and E. E. Chaban, *Surf. Sci.* **368**, 163 (1996).

85. S. D. Gunapala, B. F. Levine, R. A. Logan, T. Tanbun-Ek, and D. A. Humphrey, *Appl. Phys. Lett.* **57**, 1802 (1990).

86. T.-S. Liou, T. Wang, and C.-Y. Chang, *J. Appl. Phys.* **77**, 6646 (1995).

87. M. O. Manasreh, F. Szmulowicz, D. W. Fischer, K. R. Evans, and C. E. Stutz, *Appl. Phys. Lett.* **57**, 1790 (1990).

88. G. Hasnain, B. F. Levine, D. L. Sivco, and A. Y. Cho, *Appl. Phys. Lett.* **56**, 770 (1990).

89. S. A. Stoklitsky, Q. X. Zhao, P. O. Holtz, B. Monemar, and T. Lundstrom, *J. Appl. Phys.* **77**, 5256 (1995).

90. B. F. Levine, S. D. Gunapala, J. M. Kuo, S. S. Pei, and S. Hui, *Appl. Phys. Lett.* **59**, 1864 (1991).

91. S. Sauvage, P. Boucaud, F. H. Julien, J.-M. Gerard, and J. Y. Marzin, *J. Appl. Phys.* **82**, 3396 (1997).

92. D. J. DiMaria, *Microelectron. Eng.* **28**, 63 (1995).

93. Y. Y. Kim and P. M. Lenahan, *J. Appl. Phys.* **64**, 3551 (1988).

94. M. C. Busch, A. Slaoui, P. Siffert, E. Dooryhee, and M. Toulemonde, *J. Appl. Phys.* **71**, 2596 (1992).

95. R. C. Helms and B. E. Deal, *The Physics and Chemistry of SiO2 and the Si−SiO2 Interface*, Plenum, New York, 1988.

96. W. Daum, H. J. Krause, U. Reichel, and H. Ibach, *Phys. Rev. Lett.* **71**, 1234 (1993).

97. A. Stesmans, *Solid State Commun.* **96**, 397 (1995).

98. M. Schulz, *Surf. Sci.* **132**, 422 (1983).

99. V. I. Strikha, E. V. Buzaneva, and I. A. Radzievsky, *The Semiconductor Devices with Schottky Barrier*, Soviet Radio, Moscow, 1974.

100. H. J. Harrick, *Internal Reflection Spectroscopy*, Harrick Scientific, Ossining, NY, 1979.

101. N. M. Johnson, W. B. Jackson, and M. D. Moyer, 13th Int. Conf. Defects Semiconduct., Coronado, 1985, p. 499.

102. Y. J. Chabal, *Surf. Sci. Rep.* **8**, 211 (1988).

103. Y. J. Chabal, M. K. Weldon, and V. E. Marsico (Applications of Infrared Absorption Spectroscopy to the Microelectronic Industry), *J. Phys.* **IV7**, C6−3 (1997).

104. S. Mannarone, P. Chiaradia, F. Ciccacci, *Solid State Commun.* **33**, 593 (1980).

105. S. Selci, F. Ciccacci, and G. Chiarotti, *J. Vac. Sci. Technol.* **A5**, 327 (1987).

106. J. D. E. McIntyre and D. E. Aspnes, *Surf. Sci.* **24**, 417 (1971).

107. S. Selci, P. Chiaradia, and F. Ciccacci, *Phys. Rev. B* **31**, 4096 (1985).

108. S. S. Kim, D. J. Stephens, G. Lucovsky, G. G. Fountain, and R. J. Markunas, *J. Vac. Sci. Technol.* **A8**, 2039 (1990).

109. N. Sano, M. Sekiya, M. Hara, A. Kohno, and T. Sameshima, *Appl. Phys. Lett.* **66**, 2107 (1995).

110. C. H. Bjorkman, J. T. Fitch, and G. Lucovsky, *Appl. Phys. Lett.* **56**, 1983 (1990).

111. C. H. Bjorkman, J. T. Fitch, and G. Lucovsky, *MRS Symp. Proc.* **146**, 197 (1989).

112. J. P. Durand, F. Jollet, Y. Langevin, and E. Dooryhee, *Nucl. Instrum. Methods* **B32**, 248 (1988).

113. I. Simon, in J. D. Mackenzie (Ed.), *Modern Aspects of the Vitreous State*, Butterworths, London, 1960.

114. Z. Jing, G. Lucovsky, and J. L. Whitten, *J. Vac. Sci. Technol.* **B13**, 1613 (1995).

115. D. Xu and V. J. Kapoor, *J. Appl. Phys.* **70**, 1570 (1991).

116. E. H. Nicollians, C. N. Berglund, P. F. Schmidt, and J. M. Andrews, *J. Appl. Phys.* **42**, 5654 (1971).

117. V. A. Burrows, *Solid State Electron.* **35**, 231 (1992).

118. A. L. Smith, *Applied Infrared Spectroscopy*, Wiley, New York, 1976.

119. S. M. Han and E. S. Aydil, *J. Vac. Sci. Technol.* **A14**, 2062 (1996).

120. J. A. Oneill, M. L. Passow, and T. J. Cotler, *J. Vac. Sci. Technol.* **A12**, 839 (1994).

121. K. Kawamura, S. Ishizuka, H. Sakaue, and Y. Horiike, *Jpn. J. Appl. Phys.* **30**, 3215 (1991).

122. C. A. M. Mulder and A. A. J. M. Damen, *J. Non-Cryst. Solids* **93**, 169 (1987).

123. M. C. Matos and L. M. Ilharco, *J. Non-Cryst. Solids* **147–148**, 232 (1992).

124. L. L. Teder, G. Lu, and J. E. Crowell, *J. Appl. Phys.* **69**, 7037 (1991).

125. Y. Ogata, H. Niki, T. Sakka, and M. Iwasaki, *J. Electrochem. Soc.* **142**, 195 (1995).

126. Y. Kato, T. Ito, and A. Hiraki, *Jpn. J. Appl. Phys.* **27**, L1406 (1988).

127. Z.-H. Zhou, E. S. Aydil, R. A. Gottscho, Y. J. Chabal, and R. Reif, *J. Electrochem. Soc.* **140**, 3316 (1993).

128. M. Niwano, Y. Kimura, and N. Miyamoto, *Appl. Phys. Lett.* **65**, 1692 (1994).

129. M. Niwano, Y. Takeda, Y. Ishibashi, K. Kurita, and N. Miyamoto, *J. Appl. Phys.* **71**, 5646 (1992).

130. M. Niwano, T. Miura, Y. Kimura, and N. Miyamoto, *J. Appl. Phys.* **79**, 3708 (1996).

131. T. Takahagi, A. Ishitani, H. Kuroda, and Y. Nagasawa, *J. Appl. Phys.* **69**, 803 (1991).

132. L. Ling, S. Kuwabara, T. Abe, and F. Shimura, *J. Appl. Phys.* **73**, 3018 (1993).

133. J.-N. Chazalviel and F. Ozanam, *J. Appl. Phys.* **81**, 7684 (1997).

134. P. Jakob, Y. J. Chabal, K. Raghavachari, P. Dumas, and S. B. Christman, *Surf. Sci.* **285**, 251 (1993).

135. J. Rappich and H. J. Lewerenz, *J. Electrochem. Soc.* **142**, 1233 (1995).

136. O. Vatel, S. Verhaverbeke, H. Bender, M. Caymax, F. Chollet, B. Vermeire, P. Mertens, E. Andre, and M. Heyns, *Jpn. J. Appl. Phys.* **32**, L1489 (1993).

137. P. Jakob, P. Dumas, and Y. J. Chabal, *Appl. Phys. Lett.* **59**, 2968 (1991).

138. M. Okuyama, M. Nishida, and Y. Hamakawa, *Jpn. J. Appl. Phys.* **34**, 737 (1995).

139. Y. Yamada, T. Hattori, T. Urisu, and H. Ohshima, *Appl. Phys. Lett.* **66**, 496 (1995).

140. Y. Morikawa, K. Kubota, H. Ogawa, T. Ichiki, A. Tachibana, S. Fujimura, and Y. Horiike, *J. Vac. Sci. Technol.* **A16**, 345 (1998).

141. M. Nakamura, T. Takahagi, and A. Ishitani, *Jpn. J. Appl. Phys.* **32**, 3125 (1993).

142. H. Kanaya, K. Usuda, and K. Yamada, *Appl. Phys. Lett.* **67**, 682 (1995).

143. H. Luo and C. E. D. Chidsey, *Appl. Phys. Lett.* **72**, 477 (1998).

144. G. Lucovsky, *Solid State Commun.* **29**, 571 (1979).

145. M. Nishida, M. Okuyama, and Y. Hamakawa, *Appl. Surf. Sci.* **79–80**, 409 (1994).

146. K. Nishikawa, K. Ono, M. Tuda, T. Oomori, and K. Namba, *Jpn. J. Appl. Phys.* **34**, 3731 (1995).

147. Y. J. Chabal, in F. M. Mirabela, Jr. (Ed.), *Internal Reflection Spectroscopy: Theory and Applications*, Marcel Dekker, New York, 1992.

148. Y. J. Chabal and S. B. Christman, *Phys. Rev. B* **29**, 6974 (1984).

149. Y. Nagasawa, I. Yoshii, K. Naruke, K. Yamamato, H. Ishida, and A. Ishitani, *J. Appl. Phys.* **68**, 1429 (1990).

150. H. Ogawa and T. Hattori, *Appl. Phys. Lett.* **61**, 577 (1992).

151. G. J. Pietsch, Y. J. Chabal, and G. S. Higashi, *J. Appl. Phys.* **78**, 1650 (1995).

152. G. J. Pietsch, Y. J. Chabal, and G. S. Higashi, *Surf. Sci.* **331–333**, 395 (1995).

153. M. Niwano, J. Kageyama, K. Kinashi, J. Sawahata, and N. Miyamoto, *Surf. Sci. Lett.* **301**, L245 (1994).

154. M. Niwano, J. Kageyama, K. Kurita, K. Kinashi, I. Takahashi, and N. Miyamoto, *J. Appl. Phys.* **76**, 2157 (1994).

155. T. Miura, M. Niwano, D. Shoji, and N. Miyamoto, *J. Appl. Phys.* **79**, 4373 (1996).

156. M. Niwano, J. Kageyama, K. Kinashi, N. Miyamoto, and K. Honma, *J. Vac. Sci. Technol.* **A12**, 465 (1994).

157. M. Nishida, Y. Matsui, M. Okuyama, and Y. Hamakawa, *Jpn. J. Appl. Phys.* **32**, 286 (1993).

158. H. Ogawa, K. Ishikawa, C. Inomata, and S. Fujimura, *J. Appl. Phys.* **79**, 472 (1996).

159. E. P. Boonekamp, J. J. Kelly, J. van de Ven, and A. H. M. Sondag, *J. Appl. Phys.* **75**, 8121 (1994).

160. T. Watanabe, A. Ichikawa, M. Sakuraba, T. Matsuura, and J. Murota, *J. Electrochem. Soc.* **145**, 4252 (1998).

161. A. D. Bailey III and R. A. Gottscho, *Jpn. J. Appl. Phys.* **34**, 2172 (1995).

162. J. Shan, Y. Wang, and R. J. Hamers, *J. Phys. Chem.* **100**, 4961 (1996).

163. K. J. Uram and U. Jansson, *Surf. Sci.* **249**, 105 (1991).

164. W. Ehrley, R. Butz, and S. Mantl, *Surf. Sci.* **248**, 193 (1991).

7

ULTRATHIN FILMS AT GAS–SOLID, GAS–LIQUID, AND SOLID–LIQUID INTERFACES

Characterization of ultrathin films on substrates with different optical properties (semiconductors, metals, and dielectrics) and different forms (powders, substrates with mirrorlike and roughened surfaces) is of tremendous importance for research in areas such as catalysis, corrosion protection, flotation, detergency, depollution, oil recovery, flocculation–dispersion, lubrication, adhesion, biotechnology, and biocompatibility. There already exist numerous monographs and reviews devoted to the application of IR spectroscopy in many ultrathin film research activities. In this chapter, we will demonstrate through examples the application of different IR spectroscopic techniques to particular scientific or technological problems. The emphasis will be on the advanced experimental and interpretation approaches such as in situ and time-resolved measurements, measurements under optimal experimental geometry, and interpretation of measured spectra by using spectral simulations and 2DIR. The examples chosen are representative of the current state of the field, including some from our own work. No overlap with recent reviews is intended, and references, particularly reviews and monographs, are provided to introduce specialized literature on specific applications of IR spectroscopy. Because this is an increasingly active field, such a list of references cannot be expected to be complete but can act as a guideline for the reader interested in a particular area from which more detailed information can be found.

7.1. IR SPECTROSCOPIC STUDY OF ADSORPTION FROM GASEOUS PHASE: CATALYSIS

Adsorption — selective attachment of an adsorbate species to the surface of an adsorbent from the bulk — results from energetically favorable interactions between the adsorbate and the adsorbent, which are determined by all the components in the system [1]. When a molecule becomes attached to a surface functional group (*adsorbed*), the degrees of its vibrational and rotational motion and their symmetry change and will depend upon the new surface structures, which will impose new vibrational selection rules. To determine these rules, normal coordinate analysis is applied, as for conventional compounds. This rather complex problem, which is discussed extensively in Refs. [2–6], is beyond the scope of the present handbook. However, in many cases, interpretation of the experimental spectra may be simplified if two types of vibrational modes of the adsorbed molecule are distinguished. The frequencies and the relative intensities of the *intramolecular* modes can change significantly and new, *intermolecular* modes (due to the bonds between the adsorbate and the atom(s) on the surface of the adsorbent) can appear when *chemisorption* — covalent bonding between the adsorbate and the adsorbent — occurs. Such a process is usually irreversible and is limited to one monolayer. These intermolecular modes provide direct information about the nature and strength of the covalent bonding, which will influence many properties of the newly formed films, including their stability. However, the intermolecular bonds may be relatively weak, and the corresponding bands are often in the far-IR region of the spectrum, so generally it is not trivial to detect them [7]. For *physisorption*, in which no covalent bond is formed and the adsorption is weak, the intramolecular modes are usually only slightly perturbed, and the IR band intensities of the adsorbed layers are controlled only by the surface selection rules and the dipole–dipole coupling effect (Section 3.6). Physisorption is usually a reversible process. A distinction between physisorption and chemisorption can usually be made from the temperature dependence of the adsorption process [1, 8]. Physisorption generally decreases with temperature, and chemisorption increases with temperature, since the latter process involves an activation stage. However, the distinction between physisorption and chemisorption is somewhat arbitrary, and in many cases intermediates or more complex processes (initially physisorption and then chemisorption) are observed.

Although the first IR spectra of adsorbed molecules were reported in the late 1930s [9], IR spectroscopy became more widely used for studying adsorbed molecules on the surface of disperse and porous materials after the publications of Terenin and co-workers in the late 1940s [10, 11]. They described adsorption of organic molecules on porous silicate glasses, studied in the near-IR (NIR) spectral range by transmission. However, the position and intensity of absorption bands in the NIR depend strongly on the anharmonicity of the mode; this makes the interpretation of NIR spectra less straightforward than that for spectra in the mid-IR. Indeed, since the publication by Sidorov in 1954 [12], the vast majority of IR studies of catalysts have been performed in the mid-IR. These have been discussed in a series of monographs [2–6, 13] and reviews [14–33].

7.1.1. Adsorption on Powders

Until the mid-1980s, the studies of adsorption on powders have been performed by transmission of pressed self-supporting disks [3–5]. Since then, in situ detection of adsorbed submonolayers on powders in gaseous environment with DRIFTS has become commonplace (Table 7.1). Sensitivity of DRIFTS to species adsorbed on supported metal or metal oxide is much higher than that for unsupported oxides. With silver particles, the SNR is improved by the SEIRA effect [34].

In order to unravel adsorption mechanisms, a detailed knowledge of the composition and reactivity of the adsorption centers on the initial adsorbent is imperative [35–37]. It is well known [37–39] that fractured surfaces can be covered by cations and anions with unoccupied orbitals that act as Lewis acid and Lewis base centers, respectively. After adsorption of water molecules, with the evolution of hydroxyl and hydrogen species, the Lewis centers are transformed into Bronsted centers, which will influence surface properties. In many cases, characterizing the adsorption sites is complicated, and controversy still exists in the interpretation of IR spectra of functional groups, even for extensively studied oxides such as silica and alumina [6, 40, 41].

As an example, we will consider the hydroxyl cover on γ-Al_2O_3. This oxide is commonly used as a catalyst and as a support for catalysts. It contains tetrahedrally and octahedrally coordinated Al^{3+} ions arranged in a slightly distorted face-centered-cubic (fcc) lattice. The hydroxyl groups, whose density is about $2–12$ nm^2 [42], induce many chemical reactions at the oxide surface, either by establishing active sites or by H-bonding molecules. IR absorption of hydroxyls on dehydrated γ-Al_2O_3 was first interpreted by Peri in 1965 [43, 44]. However, it turned out to be inconsistent with a number of the properties of the alumina surface, and since that time a series of models have been proposed in the literature. Tsyganenko and Filimonov [45] recognized three types of surface hydroxyls coordinated by one, two, or three Al^{3+} ions. Knozinger and Ratnasamy [46] developed this idea, suggesting that the number of possible types of OH groups on alumina should be six, since each Al^{3+} ion can be coordinated to oxygen either tetrahedrally or octahedrally. Recently, Tsyganenko and Mardilovich [47] proposed that the number of possible surface hydroxyls upon dehydroxylation is even larger, because a central Al^{3+} ion can be either three- or five-coordinated. The following interpretation of the spectrum of calcined γ-Al_2O_3 under N_2 in the DRIFTS chamber (Fig. 7.1, solid line) was proposed by Liu and Truitt [48]. The highest frequency absorption band at 3791 cm^{-1} is attributed to the OH group bonded to one tetrahedrally coordinated Al^{3+} ion (labeled according to Knozinger's notation [49] as I_4)[†], and that at 3768 cm^{-1} corresponds to the OH group bonded to one octahedrally coordinated Al^{3+} ion (I_6). The 3731- and 3690-cm^{-1} bands are due to OH groups coordinated to two Al^{3+} ions, the former being with two Al^{3+} ions with octahedral coordination (II_{66}), and the latter being with one Al^{3+} ion with octahedral coordination and the other with tetrahedral (II_{64}) coordination.

[†] Different OH configurations are usually designated types I, II, and III, where the Roman number indicates the coordination number.

Table 7.1. IR spectroscopic studies of adsorption sites and catalytic processes

Surface Species	Adsorbent	Method[†]	Comments	References
Surface sites	γ-Al$_2$O$_3$	DRIFTS	Observation of three types of Lewis acid sites	919
	Cu exchanged Y zeolites	T, power XRD, XANES, EXAFS	CO as probe molecule	920
	Dealuminated ZSM-12 and β zeolites	NH$_3$-STPD, T at 150–550°C, MAS NMR	Bronsted/Lewis acidity as function of Si/Al ratio	921
	H-dealuminated ZSM-5 zeolites	T	H/D exchange, acidity	922
	MCM-41	T, ^{29}Si MAS NMR, XRD	Identification, acidity, CO as probe molecule	923
	Metal substituted AlPO$_4$ sieves	In situ μ-FTIR	Benzene and strong bases as probe molecules, spatial resolution of 20 μm × 20 μm	924
	Mono- and bi-pillared smectites	T	i-Propanol, n-butane, NH$_3$, and nitrile as probe molecules	925
	Montmorillonite	ATR	Bronsted/Lewis acidity	926
	Microcristalline TiO$_2$ (anatase)	T, high resolution TEM	Relationships between morphology and acid/base character of surface sites	927
	Mixed alumina-silicas	T	Enhanced surface acidity, low-temperature studies	928
	Mordenites	T, NMR	CO and C$_6$H$_6$ as probe molecules	929
	Nanocrystalline TiO$_2$	OS, T, XPS		930

Table 7.1. (*Continued*)

Surface Species	Adsorbent	Method[†]	Comments	References
	Protonated H-ZSM-5, H-β, H-Y, and dealuminated H-Y zeolites	T	H$_2$, N$_2$, and CO as probe molecules, demonstrated that catalytic and spectroscopic characterization of acidity is consistent only within same class of zeolites	931
	Sulfated sol-gel ZrO$_2$	XRD, TG, in situ DRIFTS	Isomerization of n-butane and adsorpion of pyridine as probe reactions	932
	Ultrastable Y zeolites	T	CO, CD$_3$CN, and C$_5$H$_5$N as probe molecules, problems of selection of suitable probe molecule	933
	VAlON	TPD-mass spectrometry, DRIFTS, XPS	NH$_3$ and CDCl$_3$ as probe molecules, basic center described as VO$-^+$H$_4$N, correlation between catalytic activity, NH$_4^+$ band intensity, and $\Delta\nu$CD of CDCl$_3$ measured by DRIFTS	934
OH groups	MFI-type zeolites	In situ DRIFTS	Band assignment in 2000–6200-cm^{-1} spectral range	935
	Natural and synthetic quartz, γ-Al$_2$O$_3$, feldspars	DRIFTS	Band assignment, CO$_2$ as probe molecule	380
	Re$_2$O$_7$, CrO$_3$, MoO$_3$, V$_2$O$_5$, TiO$_2$, Nb$_2$O$_5$ alumina supported	T	Effect of oxide deposition on alumina OH bands, applicability of CO$_2$ chemisorption technique for determining monolayer coverage of alumina supported metal oxides	936
	Y$_2$O$_3$, Er$_2$O$_3$, Ho$_2$O$_3$	DRIFTS	Specification	937
	ZrO$_2$	T	After thermal treatment	938

OH groups, H_2O	$Cs_xH_{3-x}PW_{12}O_{40}$, $H_3PW_{12}O_{40}$	T	Dehydration mechanism up to 627K	939
H_2O	Alkali-metal cation-exchanged X zeolites	TP DRIFTS, TPD	Band assignment; desorption profiles calculated from DRIFTS were identical to those obtained by TPD	940
	Bentonite	DT μ-FTIR, HATR	Dehydration	941
	Cr_2O_3	T, quasielastic neutron scattering, dielectric relaxation measurements	Dynamic properties of adsorbed water molecules at around critical temperature of 303 K	942
	Epitaxially grown La_2O_3(001)	BML-IRRAS	Dissociative adsorption with accompanying conversion of surface oxide ions to hydroxyls of one type	943
	Supported vanadium oxide	DRIFTS, UV/Vis, Raman	Effect of adsorbed water on surface structure	944
	TiO_2 sol-gel film	MOATR-SEIRA	In situ observation of photoenhanced adsorption	945
$H_2^{16}O$, $H_2^{18}O$	H-ferrierite	T	Observation of strong H bonding	946
H_2O, alcohols and alkenes	Porous Si or mechanically abraded Si wafers	T	Preparation of fully D-terminated silicon wafers, kinetics of reaction of O-deuterated alcohols and water on porous silicon	947
D_2O	Mordenite zeolite DM-20	TR T, pump-probe	Dynamic characteristics of vibrationally excited water	948

Table 7.1. (*Continued*)

Surface Species	Adsorbent	Method[†]	Comments	References
H_2O_2	Ti silicalite sieve	T	Thermally stable TiOOH species oxidizes small olefins at room temperature in dark or under photoexcitation	949
CO	$Au-TiO_2$, $Au-ZrO_2$	DRIFTS, pulse thermal analysis, TEM, XPS, XRD	CO oxidation, probable mechanism	950
	Cu supported on ZnO, Al_2O_3, SiO_2	T	Dependence on reducing conditions	951
	Cu exchanged zeolites	DRIFTS	Analysis of νCO, $2\nu CO$, and combination modes	952
	Cu-ZSM-5	T at 85 K	Formation of $Cu^{2+}(CO)_2$ and small amount of $Cu^+(CO)_3$	953
	Ion-exchanged ZSM-5	T	Dependence of CO adsorption on partner gas (CH_4, NO, CO_2)	954
	NaZSM-5	T	Study of role balancing cations in entrapment of CO in zeolite, results suggest two different kinds of pores	955
	NaZSM-5 zeolite	T at 300–470 K and 5–500 Torr	Effect of exchanging charge balancing cation in NaZSM-5 with H^+ or Ca^{2+} was evaluated	956
^{12}CO, ^{13}CO	TiO_2-SiO_2 mixed oxide	T at 85 K	Adsorption	957
CO, $^{15}N_2$	Alkali cation exchanged EMT zeolites	T at 85 K	Adsorption and coadsorption	958
CO, CO_2	Tetragonal and monoclinic ZrO_2	T, TPD	Comparison of adsorption capacity and adsorbed species	959

Species	Material	Technique	Description	Ref.
CO, CO_2	Nonstoichiometric Ni–Mn spinel oxides	In situ DRIFTS under steady-state and transient conditions	Detailed mechanism is proposed, effects of pretreatment are explained on basis of observed kinetics and proposed mechanism	960
CO, N_2	NaY zeolite	T at 85 K	Adsorption and co-adsorption, evidence of coordination of two molecules to one Na^+ site	961
CO_2	$Cr_2O_3(0001)/Cr(110)$	BML-IRRAS	Comparison with adsorption on polycrystalline α-Cr_2O_3	962
	NaX zeolite	DRIFTS	Observation of two types of adsorption sites	963
H_2	Faujasites	DRIFTS	H_2 as molecular probe for alkaline metal ions in faujasites	964
H_2, D_2	Na form of faujasites	DRIFTS	Local environment and adsorpotion sites	965
N_2	Diamond	DRIFTS	CH/CD exchange	966
	NaY zeolites	T at 85 K	Formaiton of geminal dinitrogen species	967
N_2, O_2	Cation-exchanged zeolites	DRIFTS, Raman	Direct interaction with cations at room temperature, KM intensity of absorption mirrors adsorption isotherm	968
N_2, O_2, and D_2	LiX, NaX, and NaLiX zeolites	T, ab initio calculations	Low-temperature adsorption	969
NO with NH_3	V_2O_5–TiO_2 catalyst	TP DRIFTS	Reduction	970
NO, NO + O_2	TiO_2	T	Species formed after adsorption of NO and coadsorption	971
NO, NO + O_2	ZrO_2, sulfated ZrO_2	T	Detection of N_2O and small amounts of nitro species, nitrates, and nitric acid; addition of oxygen results in formation of nitrites and nitrates	972

Table 7.1. (Continued)

Surface Species	Adsorbent	Method[†]	Comments	References
N_2O	CuZSM-5 and NaZSM-5	DRIFTS up to 773 K	Characterization prior to and during decomposition	973
NO_2, HNO_3	Al_2O_3	DRIFTS	Formation of nitrate, intermediate nitrite (AlOOH$\cdots$ NO_2) adduct, and acidic OH groups.	974
O_2	CeO_2	T	Band assignment	975
O_3	CeO_2	T at 77–300 K	Acidity of surface sites probed using CO_2, pyridine, acetonitrile, or methanol at 293 K or CO at 77 K	976
H_2S and CH_3SH	SiO_2	T	Sorption from liquid O_2	977
	SiO_2, γ-Al_2O_3, TiO_2, and ZrO_2	T	Adsorption and resulting changes of surface acidity, CO and 2,6-dimethylpyridine as probe molecules	978
Acetone	TiO_2	In situ T	Adsorption and photocatalytic oxidation, Aldol condensation reaction followed by dehydration	979
Acetyl-acetonate of Cu	SiO_2	DRIFTS	Integrated Kubelka-Munk (KM) intensities vary linearly with concentration of adsorbed complex	980
Acrolein and methacrolein	Mo–V oxide catalysts	DRIFTS	Mechanism of selective partial oxidation of unsaturated aldehydes to corresponding carboxylic acids	981
Alcohols	SiO_2	DRIFTS	Adsorption and oxidation of alcohols	982

Adsorbate	Adsorbent/System	Method	Comment	Ref.
Ethylene	CuNa-Y	T, DRIFTS, cluster model calculations	Adsorption-induced activation of νCH, νCC, and δCH2 modes was interpreted on basis of two chemically nonequivalent adsorption sites interacting with ethylene	983
Fluoroaceto-phenone	MFI structure type zeolites	T, ^{19}F MAS NMR	Specification of interaction	984
Formate	TiO$_2$(110)	p-Polarized IRRAS	Distinguishing of two types of adsorbed species at surface coverage of 0.59 ML	985
Hydrocarbons	H-ZSM-5	T at 100–773 K	Study of reactivity hydrocarbons on acid sites of zeolite	986
	ZSM-5 zeolites	FTIR, TG	Dependence on SiO$_2$/Al$_2$O$_3$ ratios and moisture, effect of La addition	987
Isocyanic acid	Oxides	T, micro calorimetry	Acid generated in situ by thermal decomposition of nitromethane	988
Methanol	Aerosil 50 and Aerosil 300	Gas-volumetric and in situ T	Checking of validity of BLB equations	989
	CrAPO-5 zeolite	Interference microscopy and μ-FTIR T	Two-dimensional concentration profiles, nonhomogeneous distribution of adsorbate in zeolite crystal under equilibrium with adsorbate vapor	990
	H-ZSM-5	T, TG	Isotopic studies on coke formation	991
	Supported V$_2$O$_5$	In situ IR, Raman, and UV/Vis DRS	Correlation between molecular structures of surface adsorbent species and bonding of adsorbate	992
3-Aminopropyl-triethoxysilane	Silica gel	DRIFTS	Polymerization	993

Table 7.1. (Continued)

Surface Species	Adsorbent	Method†	Comments	References
Benzene, toluene, 2-and 4-picoline	Hydrated and dehydrated silica	T, Raman	Mechanisms of sorption, models of volatile aromatic pollutant interactions	994
1-Butene	H-Ferrierite zeolite	T, DRIFTS, UV/Vis 300–670 K	Mechanisms of interaction at increasing temperature	995
CH-acids	MgO	T	Observation of νCH shift upon adsorption	996
C_5H_5N, CO	Cu_2O/ZnO junction, Pt/Al_2O_3 pellets	μ-FTIR T	Cell for surface studies evaluated using adsorption/desorption of pyridine on a Cu_2O/ZnO junction and adsorption of CO on Pt/alumina pellets	997
CF_3CFCl_2	γ-Al_2O_3	T from 298 K up to 573 K	Surface trifluoroacetate species are considered as intermediates for complete oxidation of CF_3CFCl_2	998
1-Naphthyl acetate	NaY zeolite	TR S^2 FTIR T	Photodissociation, observation of acetyl radical, microsecond TR	999
p-Xytlene	Silicate crystal particles	Polarized μ-FTIR T	Orientation of molecules relative to host particle	1000
Organosilanes	Fumed silica	T	In supercritical liquid CO_2	1001

† OS — optical spectroscopy, T — IR transmission, TP — temperature-programmed, TR — time resolved, other abbreviations are spelled out in Acronyms.

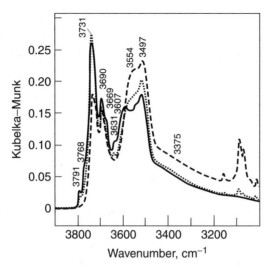

Figure 7.1. DRIFTS spectra of γ-alumina calcined at 700°C for 1 h under flowing N_2 before pyridine adsorption (dark solid line), after pyridine adsorption at room temperature (dash line), and after pyridine adsorption followed by heating at 400°C for 1 h (dotted line). All spectra recorded at room temperature. Adapted, by permission, from X. Liu and R. E. Truitt, *J. Am. Chem. Soc.* **119**, 9856 (1997), p. 9860, Fig. 4a. Copyright © 1997 American Chemical Society.

The bands at 3669, 3631, and 3607 cm^{-1} are assigned to the OH group bonded with three Al^{3+} ions (III).

On a given surface, it is possible to differentiate between Lewis acid sites (coordinatively unsaturated metal cations acting as electron acceptors) and Bronsted acid sites (proton donors) and to determine the site density and strength by analyzing the spectra of the so called *probe molecules*, that is, simple molecules such as NH_3, pyridine, CO, or NO that provide detectable, interpretable, and quantifiable spectral response upon interaction with different sites [2, 5, 6, 33, 37, 50]. One criterion when adsorbing pyridine is the formation of a pyridinium ion (pyN$^+$−H) at a Bronsted acid site, characterized by the 1540-cm^{-1} band, or weak coordination to a Lewis acid site, characterized by the 1450-cm^{-1} band. The presence of several types of sites results in a multiplicity of the bands mentioned. The ratio of the 1540- and 1450-cm^{-1} band intensities is proportional to the ratio of Bronsted and Lewis sites on the surface [51, 52]. The probe-molecule characterization methods for basic sites (typically constituted by surface OH$^-$ or O^{2-} anions on oxides) are much less developed [33]. Additional information about the adsorption sites can be extracted from the analysis of the spectral changes for the neighboring functional groups that are not involved directly in the bonding. Figure 7.1 [48] shows how the spectrum of the hydroxyl groups on γ-Al_2O_3 is disturbed upon adsorption of pyridine. The bands at 3791 and 3768 cm^{-1} disappear and the band at 3731 cm^{-1} significantly decreases. Bands at 3690 and 3669 cm^{-1} are practically unchanged, while those around 3550 cm^{-1} increase and shift to 3561 and 3504 cm^{-1}. At the same time, three hydrogen-bonded bands at 3554, 3493, and 3375 cm^{-1} are present, indicating that there are three types of Lewis acidic sites, in agreement with the hypothesis of Tsyganenko and Mardilovich [47]. The pyridine desorption revealed that the most easily removed pyridine molecules are those interacting with the Lewis sites in the proximity of the II_{66} type and II_{64} type OH groups.

In many cases, however, the steady-state in situ spectral measurements do not allow one to differentiate reactivities of surface species. This problem can be solved by comparing the transient responses of adsorbed species to an external perturbation (temperature, pressure, probe reactants). However, a change in the chemical equilibrium will change the reactivity of the system under study (an example is adsorption-assisted desorption). To circumvent this problem, the temporal spectral changes resulting from pulsed injections of an isotope of the reactant can be analyzed; this is called isotopic jump or steady-state isotopic transient kinetic analysis [53–55]. For example, this technique was used by Chuang et al. [56] to study the NO–CO reaction on Rh–SiO_2. The time-resolved IR spectra (Fig. 7.2) measured in situ (experimental set-up shown in Fig. 4.40) indicate that linear Rh^+–CO at 2110 cm^{-1}, geminal (gem)-dicarbonyl CO at 2093

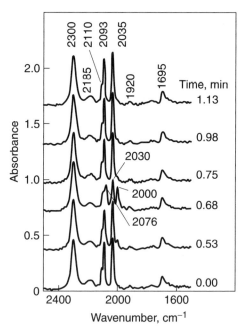

Figure 7.2. Transient time-resolved transmission IR spectra of pulsed injection of 1 cm^3 of ^{13}CO into 10 $cm^3 \cdot min^{-1}$ ^{12}CO, 10 $cm^3 \cdot min^{-1}$ NO, and 15 $cm^3 \cdot min^{-1}$ He flow with 2% Ar during NO–CO reaction over 4 wt % Rh–SiO_2 at 473 K and 0.1 MPa. Spectra measured with resolution of 4 cm^{-1} using Nicolet 5SXC equipped with DTGS (deuterated triglycine sulfate) detector and in situ cell shown in Fig. 4.40, by averaging two or four scans. Reprinted, by permission, from S. S. C. Chuang, M. A. Brundage, M. W. Balakos, and G. Srinivas, *Appl. Spectrosc.* **49**, 1151 (1995), p. 1161, Fig. 6b. Copyright © 1995 Society for Applied Spectroscopy.

and 2035 cm^{-1}, Rh$-$NO$^+$ at 1920 cm^{-1}, Rh$-$NO$^-$ at 1695 cm^{-1}, Rh$-$NCO at 2185 cm^{-1}, and Si$-$NCO at 2300 cm^{-1} are present on the surface of the catalyst before the pulse injection of ^{13}CO into a steady-state flow of ^{12}CO. The formation of Si$-$NCO, which is a result of spillover of Rh$-$NCO, is a slow process at 473 K. Pulse injection of ^{13}CO into ^{12}CO led to the partial replacement of Rh$^+-$CO and gem-dicarbonyl. The intensity of ^{13}C gem-dicarbonyl did not increase to the level of ^{12}C gem-dicarbonyl when gaseous ^{12}CO was replaced by gaseous ^{13}CO. The results indicate that the exchange between CO and gem-dicarbonyl is not rapid at 473 K. The constant IR absorption due to Rh$-$NCO and Si$-$NCO during the pulse injection of ^{13}CO suggests that these species do not participate in the catalytic cycles of formation of ^{13}CO$_2$ from ^{13}CO.

Water molecules are usually H bonded to surface hydroxyls because these bonds are stronger than those between water molecules themselves. This water monolayer on ZnO, SnO$_2$, Cr$_2$O$_3$, NaF, and SrF$_2$ can be considered as a homogeneous two-dimensional phase [57]. On the basis of FTIR transmission and quasi-elastic neutron scattering measurements, it was established [57] that the transition from solid to liquid for the two-dimensional adsorbed state of water on hydroxylated Cr$_2$O$_3$ occurs at 303 K. A different situation arises with the MgO (100) surface that is not wetted [58]: The IR spectra suggest that water is adsorbed physically via three-dimensional island formation. Other applications of IR spectroscopy in studying catalytic processes are reviewed in Refs. [58–84]. The vibrational frequencies and structures resulting from adsorption of CO, CO$_2$, O$_2$, and NO on various oxides have been collected by Urban [2]. References to some of the original work carried out between 1988 and 2002 are collected in Table 7.1.

7.1.2. Adsorption on Bulk Metals

IRRAS is the preferred method for probing ultrathin films on bulk metals. It can be applied in situ and is experimentally simple (Section 4.1.2), while its optical theory is relatively well understood (Section 2.2). Interpretation of metallic IRRAS data and extracting molecular orientation information is straightforward (Section 1.8.2). Other approaches to these systems include ATR in Otto's configuration (Fig. 2.36) and SEW spectroscopy. However, ATR is virtually restricted to the surfaces of islandlike films of metals, while the theoretical background of this technique remains to be developed (Section 3.9.4). SEW spectroscopy is highly sensitive toward films on metal surfaces due to the macroscopic propagation of radiation along the surface [85–88], as predicted theoretically by Bell et al. for CO on Pt in 1975 [89]. However, it is experimentally rather sophisticated, requiring intense, highly collimated sources.

As discussed in Section 2.2, the maximum sensitivity in IRRAS is achieved with p-polarized radiation at grazing angles of incidence. With a single reflection, the method can detect CO adsorbed on Ir at 0.002-monolayer (ML) coverage [90]. A SNR on the order of 1000 can be reached, even with a conventional double-beam dispersive spectrophotometer [91]. This is crucial for studies of catalysis and flotation phenomena. However, if the oscillator strength of the adsorbed molecule is weak to medium, it may be difficult to detect less than 0.05 ML. For highly reflecting metals, the spectral contrast can be increased with the multireflection technique (Section 4.1.2), but in practice, a single reflection is most frequently used, particularly when other techniques are employed in conjunction with IRRAS. The sensitivity can be further enhanced by the immersion technique (Section 2.5.2) and polarization modulation (Section 4.7). The latter technique is also effective in distinguishing adsorbed from nonadsorbed species for in situ studies since the s- and p-components interact identically with the nonadsorbed molecules but only the p-component interacts with the adsorbed film. This is the major advantage of IRRAS over other methods such as high-resolution electron energy loss spectroscopy (HREELS). Moreover, resolution of HREELS is usually no better than 50 cm^{-1}, while for IRRAS it is around 10 cm^{-1} ($\sim$1 meV). As compared to the sum frequency generation (SFG) method, IRRAS has a higher SNR, which allows obtaining spectra at lower coverages and from weaker oscillators [92] for a time of 0.5–1 s [93]. Operation at grazing angles of incidence requires supports several centimeters large, limiting the number of systems that can be studied with a dispersive spectrometer. FTIR microscopy (Section 4.3) extends IRRAS to surfaces as small as 100 μm.

IRRAS studies of adsorption on metal surfaces began by Greenler [94, 95] and Low and McManus [96] in the 1960s. Since that time, IRRAS has become a relatively routine method that is used in numerous laboratories for investigating such phenomena as catalysis, corrosion inhibitors, self-assembly, and lubrication. Application of IRRAS to the study of adsorbed gases has been discussed in Refs. [25, 26, 97–121]. A summary can be found in Refs. [122–126].

One of the most thoroughly studied molecules in the history of IRRAS is CO, because it possesses only one internal vibrational mode, which is a strong oscillator, and can be isotopically labeled at either end. Thus far, adsorption of CO on Cu(100) [127–136], Cu (111) [137, 138], Cu (110) [139], and polycrystalline Cu [140] has been studied. Due to a strong interaction between adsorbed CO and Pt, Ni, Rh, Ir, and Pd, fine details of this phenomenon have been extensively investigated. In particular, adsorption of CO on Pt (111) [141–155], Pt (110) [92, 156–158], and Pt (335) [159, 160]; coadsorption of CO and Xe on Pt (335) [161, 162], CO and acetonitrile on Pt (111) [163], and CO and ethene on Pt (111) [164]; adsorption of CO on polycrystalline Ni [165, 166], Ni(111) [167], Ni (100) [168–170], and NiO(111)/Ni(111) [171]; coadsorption of CO and H$_2$ on Ni(110) [172], CO and NO on Ni(100) [173], CO and NO on Ni(111) [174], CO and H$_2$O on Ni (110) [175], CO and NO on NiO–Ni (1100) [176], CO on Pd(110) [177, 178], CO on Pd(100) and Pd(111) [179], CO on Pd(111) [180, 181], and CO on polycrystalline Pd [182]; and coadsorption of CO and NO on Pd(100) and Pd(111) [183], CO on Ir(110) [184], CO on Ir(111) [90, 93], CO and H$_2$ on Ir (111) [185], CO on polycrystalline Ir [186], CO on Rh(100) [187], CO on Co(1010) [188], CO on Co(110) [189–191], CO on Mo(100) [192], CO on polycrystalline W [193], CO on Ru (001) [194–197], and CO and O$_2$ on Ru(001) [198] have been studied.

As a rule, an increase in the coordination number of the CO site decreases the νCO frequency: The band between 1700 and 1900 cm^{-1} is typical for the threefold bridge, 1900–2000 cm^{-1} for the twofold bridge, and 2000–2100 cm^{-1} for the atop position. For example, CO molecules occupy exclusively atop sites on Ir(111) (Fig. 3.36) [90] and Pt(110) [92]. The observed band shift between the zero-coverage limit and the saturation coverage [62 and 50 cm^{-1} for Ir(111) and Pt(110), respectively] is attributed solely to dipole coupling, based on the Persson–Ryberg modeling and the isotopic dilution (Hammaker's method) data (Section 3.6). Bands at 1840–1857 and 1810 cm^{-1} due to the bridge configurations are observed, for example, for Pt(111) at 150 K [199]. Only the former band is present at 95 K, and both bands overlap at 300 K. In contrast, the position of the single νCO band due to the atop configuration is practically independent of surface coverage, when CO is adsorbed onto both atomically flat single-crystal and polycrystalline Cu surfaces [108]. For a Cu(100) substrate, the spectrum from IRRAS shows a narrow band at 2085 cm^{-1}, which shifts only slightly on saturation to 2094 cm^{-1} without broadening. This effect could be attributed to [108] (1) a weak dipole–dipole coupling of the adsorbed molecules, (2) the islandlike mechanism of the adsorbed film growth (to explain how, even at low coverages, the coupling is strong), or (3) mutual compensation of a blue shift from mode coupling and a red shift due to chemical effects (backdonation). Using Hammaker's method, Hollins and Pritchard [108] observed for Cu(111) that at constant isotope composition, the ^{12}CO and ^{13}CO bands shift to the lower frequencies as the total surface coverage increases, which is consistent with the third hypothesis. This interpretation has been confirmed by the IRRAS measurements of Persson and Ryberg [200], who varied the isotopic composition at constant coverage. Recently, Cook, McCash and co-workers [132, 133] studied the temperature dependence of the CO band width for CO adsorbed perpendicular to Cu(100) as a $(7\sqrt{2} \times \sqrt{2})R$ 45° overlayer. A band shift from 2086.7 to 2084.7 cm^{-1} and an increase in the FWHM from 6.9 to 9 cm^{-1} as temperature increased from 23 to 120 K was observed. This was interpreted according to existing theories [201, 202] in terms of anharmonic coupling between the νCO mode and the frustrated translation of the CO molecule within the potential trap of the adsorption site.

The position of the νCO band depends on the Cu crystallographic plane [108]. The low-index Cu(100) and Cu(111) surfaces give bands at 2080 and 2076 cm^{-1}, respectively, while the Cu(110), Cu(311), Cu(211), and Cu(755) surfaces give bands between 2096 and 2110 cm^{-1}, close to the band positions for evaporated polycrystalline Cu and supported Cu (2100–2103 cm^{-1}). This effect has been attributed [108] to a predominance of stepped or higher index planes at the surface of the evaporated films and supported Cu.

The spectra of CO adsorbed at terminal (2010–2030-cm^{-1}) and bridge (1880–1965-cm^{-1}) sites on Ni(100) at 300 K are shown in Fig. 7.3 [203]. Population of these sites depends on the CO coverage. Bridge sites appear to be preferred at low and high coverages. Correlating these data with the formation of the well-ordered c(2 × 2)-CO structure slightly below and at 0.5 ML, observed by LEED,

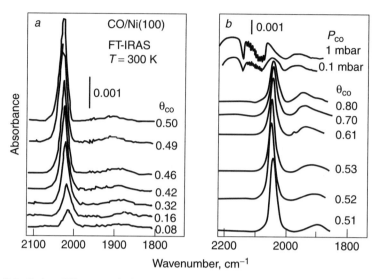

Figure 7.3. Series of IR spectra for increasing CO coverages above 0.5 ML on Ni(100) at 300 K. Upper two spectra display rotational fine structure of gas phase CO. Reprinted, by permission, from A. Grossman, W. Erley, and H. Ibach, *Surface Science*, **330**, L646–L650 (1995), p. L647, Fig. 1. Copyright © 1995 Elsevier Science B.V.

the minimum of the bridge fraction was assigned to the preferential occupation of atop adsorption sites at ∼0.5 ML. Since the coverage-dependent blue shift is insignificant, it is probable that two types of ordered clusters coexist on the surface with CO molecules occupying only one site in each phase. As surface coverage further increases, bridge sites are predominantly occupied in order to achieve a denser CO overlayer.

An interesting feature for CO adsorbed on Pt(110)-(1 × 2) at 30 K and low surface coverage is two bands at 2056 and 2042 cm^{-1} (Fig. 7.4) [158]. The band at 2056 cm^{-1} was assigned to the singleton frequency for CO molecules bonded in the most stable sites — the atop sites on the Pt ridges. To assign the second component, the spectra as a function of surface coverage and temperature were measured. As coverage increases, the relative intensities of these bands change, with the low-frequency component disappearing. Simultaneously, the high-frequency component shifts to 2091 cm^{-1} due to dipole coupling, while a new band at 2144 cm^{-1} assigned to physisorbed CO arises (Fig. 7.4). The spectrum changes in a complex manner upon heating, and one band at 2061 cm^{-1} is observed eventually at 75 K (Fig. 7.5). These spectral changes were interpreted to arise from rearrangement of the molecules from the initial adsorbed state in the form of one-dimensional gas and a small fraction of chains (or one-dimensional islands) at 30 K into the chains at 40 K. On further heating, CO molecules evaporate from the chains, and the original isolated CO singletons are re-formed.

The second most popular molecule is NO, which together with CO is the major constituent of automotive exhaust effluent. Interactions of NO with Pt(100) [204],

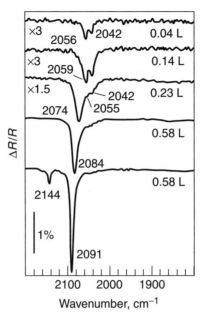

Figure 7.4. Spectra showing adsorption of CO on Pt(110) at 30 K. Exposures are given. Reprinted, by permission, from R. K. Sharma, W. A. Brown, and D. A. King, *Chem. Phys. Let.*, **291**, 1–6 (1998), p. 2, Fig. 1. Copyright © 1998 Elsevier Science B.V.

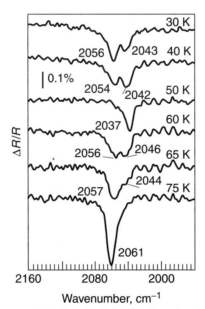

Figure 7.5. Spectra showing results of heating Pt(110) surface that has been exposed to 0.12 L of CO at 30 K. Heating to 50 K leads to disappearance of low-frequency band, and further heating to 65 K results in reappearance of higher frequency band. Reprinted, by permission, from R. K. Sharma, W. A. Brown, and D. A. King, *Chem. Phys. Let.*, **291**, 1–6 (1998), p. 3, Fig. 3. Copyright © 1998 Elsevier Science B.V.

Ni(111) [205], Pd(111) [206], and Ag(111) [207] have been studied as well as with oxygen coadsorbed onto Ni(111) [208] and Ru(001) [209]. The general rule for interpretation of IRRAS seems not to apply for NO adsorbed on Ni(111). At 85 K, spectra from IRRAS have been interpreted to show that three distinct bonding states exist: bridge bonding, both tilted and perpendicular, and terminal or atop bonding [210]. However, Aminpirooz et al. [211], using a combination of surface-sensitive extended X-ray absorption fine structure (SEXAFS) and LEED data, together with Asensio et al. [212], using photoelectron diffraction (PED), showed that only one structure is present at the surface and it is neither a bridge nor an atop or terminal site. According to Asensio et al., all of the NO molecules occupy identical fcc threefold hollow sites at all coverages. In contrast, Aminpirooz et al. have suggested that both threefold fcc and hexagonal close-packed (hcp) hollow sites are occupied. The latter model has been confirmed by Mapledoram et al. [213].

Of particular interest to chemists is adsorption on clean well-defined metal surfaces that model the surface processes occurring with catalysts. Thus, adsorption of methane [214–216], ethane [217–219], alkyl halides [220], vinyl iodide [221], CH_3Br [222], n-azopropane [223], ICH_3 [224], PF_3 and CO, NH_3, and Xe [225], and perfluoroalkanes [226], on Pt(111); acetaldehyde and O_2 on Pt (111) and Pt(100) [227]; organosilanes [228] and n-alkanes [229] on Pt (111); H_2O on Ni(110) [230]; N_2 on Ni(111) [231] and Ni(110) [232]; CH_3OH on NiO–Ni(110) [233]; C_2H_5OH on Ni(100) [234]; toluene on Ni(111) [235]; methyl formate, ethyl formate, and methyl acetate on Ni(111) [236]; ethyl formate on Ni(111) [237]; ethylene on Ni(111) [238]; esters of formic acid on Ni(111) [239]; cyclohexane on Ni(111) [240]; methylamine and trimethylamine on Ni(111) [241]; C_2D_6 on Cu [242]; acetonitrile on Cu (100) [243]; H_2 on Cu (111) [244]; H_2 and ethene on Cu (111) [245]; ethene on Cu (110) [246]; methyl acetate and preadsorbed oxygen on Cu(110) [247]; fluorinated ethoxides on Cu (111) [248]; methanol on Cu (110) [249]; O_2 on Cu (110) [250, 251] and ethene on Cu (110) [252]; ethylene on Pd(111) [253] and Pd(001) [254]; H_2O [255, 256], ethene [257], ethylidene [258], and 1-hexene [259] on Ru(001); NCO on Rh(111) [260]; H_2 on W(100) [261], W(110), and W(111) [262]; CH_2I_2 [263], NO_2 [264], NH_3 and ND_3 [265], 1,3-butadiene [266], 1,1,1,2,2,3,3,-heptafluoro-7,7-dimethyl-4,6-octanedione [267], and quaterthiophene on Ag(111) [268]; Cl_2 and CH_4 on Ag(100) [269]; Cl_2 on Ag(100) [270]; acrylonitrile on Au(111) [271]; acroleine on Au(111) [272]; and NO_2, N_2O_3, and N_2O_2 on Au(111) [273] have all been investigated. IRRAS measurements of polycrystalline Ag after adsorption of Cl_2 [274], acrolein and acrylic acid [275], and O_2 [276] and polycrystalline Au after adsorption of H_2O [277], HNO_3 [278], tricyclohexylphosphine [279], and silane [280] have been reported.

7.2. NATIVE OXIDES: ATMOSPHERIC CORROSION AND CORROSION INHIBITION

As discussed extensively in Section 3.1, the key feature in interpretating spectra of ultrathin dielectric and oxide layers on the surface of metals is the strong

Berreman effect, which is the essential difference between spectra of ultrathin films and spectra of the absorption index of the film material. Several studies of dielectric films on metals [281–287] have illustrated this concept. In spite of this, a number of attempts have been made to assign the high-frequency bands to various structures in the dielectric [288–291] or, which is also inconsistent with optical theory (Section 3.2), to the LO film mode [292].

Furthermore, experimental spectra of layers of the same composition, depend on the density and homogeneity of the layer substance and on the degree of crystallinity. Such correlations are discussed in Sections 5.1 and 5.3 for SiO_2 films. At the present time, a wide range of metal oxides have already been studied by IRRAS. The ν_{LO} frequencies and composition of the corresponding oxides arising at the surface as a result of oxidation in air at different temperatures, corrosion in different liquid media, and anodization have been determined. Table 7.2 includes relevant references. Below, IRRAS of different oxide layers on metals and a corrosion-protective film on Cu will be considered.

It is known that amorphous Al_2O_3 layers can be obtained by polishing the Al surface with absolute alcohol and by oxidizing Al. Annealing the amorphous Al_2O_3 layer at $500°C$ in argon results in its partial crystallization into the γ-Al_2O_3 form, whereas boiling in water leads to a transformation into the boehmite form. The spectra from IRRAS of layers prepared according to these recipes are shown in Fig. 7.6 [282]. In fact, the band positions of 950 and 960 cm^{-1} in the IRRAS of the layers (curves 1 and 4, respectively) are within the frequency range 960–920 cm^{-1} reported for the ν_{LO} band of amorphous alumina films [281]. For the film partially crystallized into γ-Al_2O_3 (curve 2), the ν_{LO} band shifts to 990 cm^{-1}. The additional band at 1150 cm^{-1} was ascribed to the ν_{LO} band of the boehmite microphase in the film partially transformed into boehmite (curve 3). Based on these correlations, it can be concluded that polishing of the Al surface in the presence of water (curve 5) produces an amorphous aluminum oxide film.

The high sensitivity of IRRAS allowed the composition of natural protective films caused by polishing surfaces of Al, Mg, and Cu to be investigated [293]. Prior to recording the spectra, the ground metallic plates were polished with diamond or Cr_2O_3 paste. In the latter case, alcohol or water was used as a solvent. The flatness and the quality of the final surface were never worse than ± 3 μm and $\nabla 12$, respectively. The spectra were recorded by IRRAS in p-polarized radiation on a UR-20 spectrophotometer with a specially developed accessory shown in Fig. 4.7. A polarizer was mounted in front of the entrance slit. The size of the metallic samples was 20×92 mm, and the number of reflections $N = 26$.

Comparison of the band positions in the experimental spectra presented in Figs. 7.7 and 7.8 with the ν_{LO} frequencies of the possible oxides (Table A.1) suggests that polishing with diamond and Cr_2O_3 paste produced oxide layers on the surfaces of all the metals studied. The Al_2O_3 layer (curves 3–5 in Fig. 7.7) is characterized by the 940-cm^{-1} band, MgO (curves 1 and 2 in Figs. 7.7 and 7.8) by the 680-cm^{-1} band, and Cu_2O (curves 3 and 4 in Fig. 7.8) by the 640-cm^{-1} band. In addition, in the spectra of all the samples polished with the Cr_2O_3 paste, an absorption band at 600 cm^{-1} was observed and assigned to chromium

Table 7.2. IR spectroscopic studies of oxide films on metals

Metal	Film	Method of Deposition	References
Al	Amorphous Al_2O_3	Anodization	1002, 1003
		Evaporation	1004
		Oxidation	1005
	$Cr–PO_4$, $Al–PO_4$	Conditioning with acidic solution of chromates and phosphates	1006
	MgO	Vacuum evaporation	1007
	WO_3	PECVD	1008
Al alloy	$Al–PO_4$	Conditioning with solutions of chromates and phosphates	1009
Mg	MgO	Oxidation	1010
Ti	TiO_2	Oxidation at $700°C$	1011
Ta	Ta_2O_5	Anodization	1012, 1013
Nb	Nb_2O_5	Anodization	1014
Pb	$Pb–SO_4$	Conditioning with a 0.1 M sulfate solution	1015
Steel	Cr_2O_3, Al_2O_3	Vacuum evaporation	1016
	FeOOH and other products	Atmospheric corrosion	1017–1019
Stainless steel	α-Fe_2O_3, Fe_3O_4	Atmospheric corrosion at high temperature	
	$FeCr_2O_4$, $(Fe,Cr)_2O_3$	Oxidation	1020–1022
Steel	Phosphates of Zn and Fe	Corrosion inhibition study in neutral chloride solutions	1023
Zr	ZrO_2	Oxidation	1024
Ag	Ag-O	Adsorption of O_2	1025
Cu	$CuSO_3Cu_2SO_3 \cdot 2H_2O$	Air containing SO_2	297
	Cu_2O	Atmospheric corrosion	1026, 1027
	Cu_2O, CuO	Oxidation	1028–1030
Pt	Ni oxide/hydroxide electrochromic films (α-$Ni(OH)_2$, β-$Ni(OH)_2$, and β-NiOOH)	Sol-gel	1031
Ni	$NiSO_4$ of two types, carbonates and incorporated water	Rain-induced atmospheric corrosion	1032
Ni	NiO	Oxidation	1033
Ni	$NiSO_4 \cdot nH_2O$	Atmospheric corrosion	1034
Zn	$Zn_4CO_3(OH)_6 \cdot H_2O$, $ZnSO_3 \cdot nH_2O$, $Zn_4SO_4(OH)_6 \cdot nH_2O$	Atmospheric corrosion	1034
Fe	Carbonate green rust	Electrochemical oxidation	1035
Fe	$CaCO_3$	Electrodeposition	1036
Au	$Zr(HPO_4)_2$	Adsorption from solution	1037
Cr	Cr_2O_3	Oxidation	1038–1040
Sn	Corrosion films	Dipping in 0.10 M NaOH, 0.15 M NaCl, and pH 4.3 phosphate buffer solutions and exposing in air	1041
Ru	Al_2O_3	Vacuum evaporation	1042

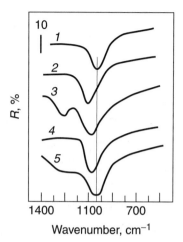

Figure 7.6. Spectra measured by IRRAS of (1) an amorphous Al_2O_3 layer obtained by polishing Al with absolute alcohol, (2) same as 1 and then partially crystallized into γ-Al_2O_3 form by annealing at 500°C in argon, (3) same as 1 and transformed into boehmite form by boiling in water, (4) amorphous Al_2O_3 layer obtained by sputtering Al in vacuum on quartz and then oxidized in air, and (5) Al surface after polishing with water. Reprinted, by permission, from V. P. Tolstoy, *Methods of UV-Vis and IR Spectroscopy of Nanolayers*, St. Petersburg University Press, St. Petersburg, 1988, p 196, Fig. 6.1. Copyright © 1998 V. P. Tolstoy.

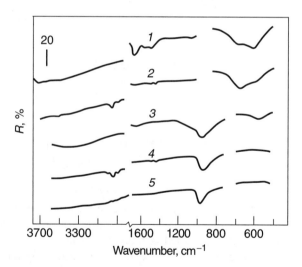

Figure 7.7. Spectra measured by IRRAS of metal surfaces polished by various means, obtained in p-polarized radiation at $\varphi \approx 75°$ with number of reflections $N = 26$: (1) Mg polished with Cr_2O_3 paste under water; (2) Mg polished with Cr_2O_3 paste under alcohol; (3) Al polished with Cr_2O_3 paste under water; (4) Al polished with diamond paste; (5) Al polished with diamond paste and then cleaned by technique described in text. Reprinted, by permission, from V. P. Tolstoy, G. N. Kuznetsova, S. I. Koltsov, and V. B. Aleskovskii, *Zhurnal prikladnoi khimii* **53**, 2353 (1980), p. 2354, Fig. 1. Copyright © 1980 V. P. Tolstoy.

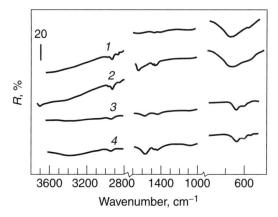

Figure 7.8. Spectra measured by IRRAS of Mg (1, 2) and Cu (3, 4) surfaces polished with Cr$_2$O$_3$ paste under alcohol, obtained in p-polarized radiation at $\varphi \approx 75°$ with number of reflections $N = 26$: (1, 3) surfaces as prepared; (2, 4) after subsequent exposure in air for 5 days. Reprinted, by permission, from V. P. Tolstoy, G. N. Kuznetsova, S. I. Koltsov, and V. B. Aleskovskii, *Zhurnal prikladnoi khimii* **53**, 2353 (1980), p. 2355, Fig. 2. Copyright © 1980 V. P. Tolstoy.

oxide that had penetrated into the surface layer. At the same time, in the spectra from IRRAS of all the samples polished with alcohol as a solvent, bands typical for νCH (2800–3000 cm^{-1}) and δCH (1460–1470 cm^{-1}) modes are present, and when water was used, the bands in the 3200–3700- and 1640-cm^{-1} regions caused by adsorbed water are visible. In the spectra of Al and Mg, there are the bands at 1150 and 1480 cm^{-1} attributed to the corresponding hydroxides. Polishing with a diamond paste leads to the formation of the oxide layers with organic contaminants incorporated (the bands at 1480, 1580, and 2800–3000 cm^{-1}). The composition of the natural oxide layer was found to change during exposure of the samples to ambient atmosphere. For example, after storage for five days, CuOH (the 1580-cm^{-1} band) and MgOH (1480 cm^{-1}) were detected on Cu and Mg, respectively, and adsorbed water (1640 and 3200–3700 cm^{-1}) appeared on both surfaces (Fig. 7.8).

Curve 5 in Figure 7.7 demonstrates the use of IRRAS to monitor the purity of metallic surfaces before chemical modification. The samples were polished with diamond paste, washed in CCl$_4$ and hot water, etched in HNO$_3$ in order to remove the damaged layer, swilled in bidistilled water, and then annealed for 1 h in an oil-free vacuum at $t = 200°$C. The spectra show only the bands of the natural oxides and no bands from impurities.

IRRAS can be extremely useful for studying in situ the corrosion and anticorrosion mechanisms [295]. For example, in order to understand high-temperature corrosion processes on AISI type 304 stainless steel, Guillamet et al. [284] measured the spectra by IRRAS of a steel plate exposed for 1 min to air at high temperatures. Comparison with the ν_{LO} bands of a series of oxides indicated that the main product is α-Fe$_2$O$_3$ (not Fe$_3$O$_4$, as suggested earlier for corrosion of iron [294]) and the minor products are Cr_2O_3 and a spinel phase of Me(Me = Mn, Fe)Cr_2O_4. Spectra measured in situ by IRRAS of Cu exposed to air at 80% relative humidity (RH) and 25°C revealed that the corroded film consists of Cu_2O (the ν_{LO} band at 645 cm^{-1}) and physisorbed water [296].

An excellent methodology involving application of 2DIR, spectral simulations, and XPS for interpreting IR spectra was demonstrated by Itoh et al. [297] in the study by IRRAS of atmospheric corrosion and of the growth mechanism of a corrosion layer on Cu in air containing water vapor and SO_2. Spectra were obtained in situ in the atmosphere with 80% RH and 8.7 ppm SO_2, at the angle of incidence of 80°, by averaging 320 interferograms, with a resolution of 8 cm^{-1}. To eliminate interfering signals caused by fluctuations in partial pressures of water vapor and SO_2, p- and s-polarized spectra were alternatively measured every 16 scans, and the resulting spectra were presented as a difference spectrum of the two measurements:

$$A_{sp} = -\log\left(\frac{S_p}{S_s}\right) - \left[-\log\left(\frac{R_p}{R_s}\right)\right] \qquad (7.1)$$

where R and S indicate reference and sample spectra, respectively, and subscripts p and s relate to p- and s-polarization, respectively. It took 15 min to acquire a pair of spectra using p- and s-polarization. Reference spectra were measured before the introduction of the corrosive gas to the test cell, and after introduction of the corrosive gas sample, spectra were acquired at the prescribed times. A series of spectra measured by IRRAS with increasing exposure time is shown in Fig. 7.9. The band at around 1060 cm^{-1} and the shoulder at about 990 cm^{-1} are attributed to sulfite, and the band at around 1130 cm^{-1} is assigned to sulfate. XPS data indicated that the main components of the corrosion film are sulfite in the form of Chevreul's salt ($CuSO_3Cu_2SO_3 \cdot 2H_2O$) and sulfate $CuSO_4 \cdot 5H_2O$. The small band at around 1450 cm^{-1} is signed to carbonate. The 1650- and 1580-cm^{-1} bands are assigned to δOH of physically and chemically adsorbed water, respectively. The wavenumber of the 1580-cm^{-1} band suggests a partial charge transfer from oxygen lone-pair electrons [298]. To make temporal changes of the individual components in the manifolds shown in Figure 7.9 more tractable, the 2DIR technique (Section 3.8) was used. As a result, an additional band at $\sim$930 cm^{-1} due to both sulfite and sulfate was revealed, and the positions of the sulfite and sulfite bands were adjusted. On this basis, the absorption spectrum of sulfite alone was reconstructed from three components at 1054, 990, and 930 cm^{-1} and compared with the simulated IRRAS spectra of Chevreul's salt. The optical coefficents employed in the simulations were determined using the KK relations from transmission spectra of Chevreul's salt. As can be seen in Fig. 7.10, both spectra are similar in shape and position, but the reconstructed spectrum is wider. This difference was attributed to poor crystallization of Chevreul's salt and/or coexistence of Chevreul's salt and $Cu_2SO_3 \cdot 0.5H_2O$ as a precursor in the corrosion layer. In addition, 2DIR allowed assignment of the 1580-cm^{-1} band to both sulfate and sulfite, while the 1650-cm^{-1} band was found to be asynchronously correlated with both these species. Thus it was concluded

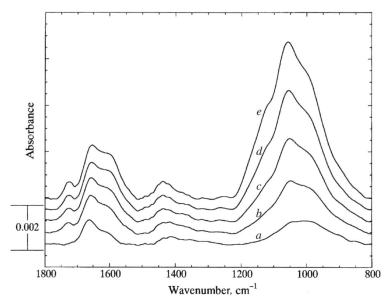

Figure 7.9. Series of IRRAS spectra on copper in air containing 80% relative humidity (RH) and 8.7 ppm SO_2. Exposure time (*a*) 1, (*b*) 4, (*c*) 8, (*d*) 12, and (*e*) 16 h. Reprinted, by permission, from J. Itoh, T. Sasaki, T. Ohtsuka, and M. Osawa, *J. Electroanal. Chem.* **473**, 256–264 (1999), p. 259, Fig. 3. Copyright © 1999 Elsevier Science S.A.

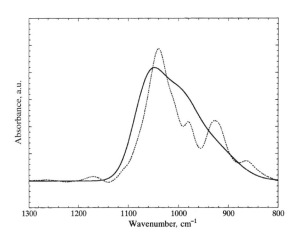

Figure 7.10. Spectra reconstructed from three component bands of sulfite (———) and calculated from transmission spectrum of Chevreul's salt (. . . . . .). Reprinted, by permission, from J. Itoh, T. Sasaki, T. Ohtsuka, and M. Osawa, *J. Electroanal. Chem.* **473**, 256–264 (1999), p. 262, Fig. 8. Copyright © 1999 Elsevier Science S.A.

Table 7.3. IR studies of corrosion inhibitors

Inhibitor	Metal	Comments	References
$(C_nH_{2n+1})SO$ aliphatic sulfoxides	Fe	Protective properties studied in $1-10$ N H_2SO_4; sorption process model is suggested	1043
Two-dimensional polymer film prepared by modification of alkanethiols	Fe	Film characterized by XPS and FTIR refection spectroscopy and contact angle measurement; protective efficiency of film found to be high both against anodic and atmospheric corrosion	1044
γ-Aminopropyltriethoxysilane and bis-triethoxysilyl ethane	Fe	Effect of the amine functional group on corrosion rate in 3% NaCl solutions	1045
Plasma polymer deposited from a mixture of hexamethyldisilane and argon	Steel	Characterization of Fe oxide and plasma polymer by FTIR, XPS, and QCM, testing of corrosion performance of coated samples according to kinetics of cathodic delamination which was measured in-situ by scanning Kelvin probe (SKP).	1046
Different silane coupling agents	Epoxy–steel interface	Effect of heat and humid treatment on polymer degradation and steel corrosion	1047
2-Carboxyethyl phosphonic acid in absence and presence of Zn^{2+}	Carbon steel	Protective film analyzed by X-ray diffraction, Fourier transform infrared, and luminescence spectra	1048
Zinc salt/phosphonic acid mixture	Steel	Electrochemical measurements (steady-state current–voltage curves and ac impedance) coupled with IRRAS and XPS to investigate inhibition of corrosion of carbon steel;	1049

Table 7.3. *(Continued)*

Inhibitor	Metal	Comments	References
		FTIR spectrum indicates reaction of phosphonic acid with zinc hydroxide and iron oxide to produce metal salts	
New organic molecule with chelating groups	Zn	XPS, SEM-EDS, and FTIR	1049
Sodium benzoate, sodium *N*-dodecanoylsarcosinate, sodium *S*-octyl-3-thiopropionate, 8-quinolinol, and 1,2,3-benzotriazole	Zn	XPS and IRRAS	1050
Benzotriazole, 2-mercaptobenzotriazole	Cu	Corrosion protection studied in 0.001 N HCl	1051, 1052
Benzimidazole	Cu	Composition and thickness of protective film determined	1053
Benzotriazole	Cu	Oxidation of Cu in presence of inhibitor studied	1054
Azole	Cu	Orientation of inhibitor determined	1055
Benzotriazole	Cu	IRRAS and XPS studies of adsorbed layer	1056
Undecylimidazole	Cu	Thermal stability	1057
Azole	Cu	Thermal stability	1058
2-Mercaptobenzoxazole	Cu	Potentiodynamic polarization, STM, QCM, XPS, IRRAS. and AES	1059
3-Mercaptopropyltrimethoxysilane (MPS)		Inhibition studied by IRRAS as function of MPS pretreatment concentration in ethanol	1060
8-Hydroxyquinoline	Cu	XPS, FTIR, and SEM stated that protective film is formed on surface by polymerization of Cu(II)–hydroxyquinoline complexes, films that play essential role in inhibition of Cu corrosion	1061

Table 7.3. (*Continued*)

Inhibitor	Metal	Comments	References
Copolymer of vinyl imidazole (VI) and vinyl trimethoxy silane (VTS)	Cu	Effect of copolymer composition on copper corrosion protection at elevated temperatures investigated by IRRAS and SEM; in addition, an adhesion test performed to characterize interface between copper and polymer film after heat treatment	1062
Imidazole, 2-methylimidazole, 2-ethylimidazole, 2-propylimidazole, 2-undecylimidazole, 2-heptadecylimidazole	Cu	Form of reagent adsorption	1063–1065
Undecylimidazole	Cu	Thermal stability of 150-nm-thick undecylimidazole layer on copper surface studied by IRRAS; cleavage of imidazole ring and structure of progressive degradation of decomposition products proposed	1066
Imidazole	Cu	On Cu electrodes	1067
Oxime group: salycilaldoxime and benzoinoxime	Cu	SEM, IRRAS, and XPS studies of protection in neutral aqueous NaCl solutions; inhibitory mechanism on Cu dissolution process related to both chelating effect of Cu(II) ions close to copper surface and blocking action of organic complex surface	1068

that adsorption of water precedes the formation of sulfite and sulfate, and the formation rate of sulfite is higher initially than that of sulfate, but the rate declines to become nearly equal to that of sulfate within the experimental period of 16 h.

To protect the metal against atmospheric corrosion, one can block the active adsorption centers by chelating agents and/or prevent diffusion of oxidative

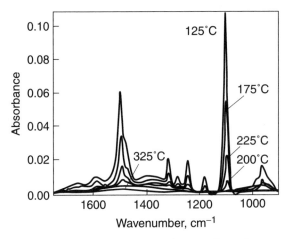

Figure 7.11. Single-reflection IRRAS of imidazole film on Cu under vacuum at series of temperatures as film is heated from room temperature to 325°C in 50°C increments. Film was held at each temperature for 15 min before increasing temperature. Reprinted, by permission, from R. L. Opila, H. W. Krautter, B. R. Zegarski, and L. H. Dubois, *J. Electrochem. Soc.* **142**, 4074–4077 (1995), p. 4075, Fig. 3. Copyright © 1999 Electrochemical Society.

reagents to the surface by coating the metal with a protecting organic film [299–301]. In order to understand the underlying mechanisms, in situ IRRAS is useful, as demonstrated in early works of Little [302], Thibalt and Talbot [303], and Friberg and Muller [304]. Some systems studied in the last decade are listed in Table 7.3. For example, a 5-nm imidazole layer can limit the oxidation of the Cu surface. To illustrate this, Fig. 7.11 shows a series of spectra measured by IRRAS of a Cu plate covered by an imidazole film heated from 25 to 325°C in 50°C increments in vacuum. The sample was held at each temperature for 15 min. The most intense features in the IRRAS spectra were assigned as follows: 1470 cm^{-1}, ring stretching; 1320 cm^{-1}, CH bending; 1170 cm^{-1}, ring breathing; 1090 cm^{-1}, CH in-plane bending; and 950 cm^{-1}, ring bending of imidazole. In vacuum, the intensity of all the bands decreases proportionately, indicating desorption of the film. The kinetics of desorption were monitored using the most intense band at 1094 cm^{-1} as a measure of imidazole coverage. In doing so, the kinetics under vacuum and under N$_2$ were found to be nearly identical. Even after 120 min at 200°C the imidazole concentration had decreased by only half. However, this process is greatly accelerated when the heating is performed in air. Thus, it can be concluded that oxygen plays a direct role in the decomposition of the imidazole films.

7.3. ADSORPTION ON FLAT SURFACES OF DIELECTRICS AND SEMICONDUCTORS

Adsorbed layers on flat, transparent substrates can be studied by transmission, IRRAS, and ATR methods. Historically the first measurements were by the

ATR method in the MIR configuration, which, with an Si IRE and ≥ 100 more reflections, gave good-quality spectra of OH and CH groups at submonolayer coverage, even with a conventional dispersive spectrometer [305]. This is why the majority of ATR studies of adsorption have been performed by the "reactive" ATR technique (Section 4.1.3) with IREs of intrinsic semiconductors (Si, Ge, and GaAs) [26, 306]. The first MIR measurements involved Ge and Si substrates, on which the adsorption of stearic acid and its salts was studied [305, 307–310]. Dielectrics whose surface is accessible by this technique are halogen salts of alkali and alkali-earth metals, AgCl, and KRS-5(6). In one example, MIR spectroscopy with polarized radiation was used to determine the orientation of a dye (1,1-diethyl-2,2-carbocyanide bromide) adsorbed on AgCl [311]. Adsorbed aggregates (micelles) were revealed in which the molecular planes are perpendicular to the substrate surface.

The transparency range of oxides and diamond in the IR region is narrower than that of salts and intrinsic semiconductors (Table A.2). For example, centimeter-thick quartz is transparent only above 2700 cm^{-1}, which still allows detection of the νCH and νOH bands of the surface species in the ATR geometry. This approach was used [312, 313] in studies of the hydroxyl covers formed on a quartz surface during such processes as dehydration (the loss of physisorbed water), dehydroxylation (the loss of hydroxyl groups), heating in UHV, conditioning by water in the reactor camera for hydrothermal synthesis, exposure to HF vapors and solutions, and treatment by a glow discharge. The MIR geometry was used to differentiate between different types of hydroxyl groups on the (4150), (0001), and (1123) surfaces of α-Al$_2$O$_3$ [314]. The bonding geometry of hydrogen on (100) [315, 316] and (110) single-crystal diamond [317] was studied using the phenomenon of electron emission by the hydrogen-terminated diamond from the conduction band edge directly into vacuum. It was shown [316] that with the E sampling technique (Section 4.1.3) 0.1 ML of the C−D surface groups can be detected at a resolution of <1 cm^{-1}, which is much higher than that of HREELS applied to the same system. Based on the dichroic ratio (DR) values for the νCD and δCD bands, it was inferred that the C−D bonds deposited in a microwave plasma are perpendicular to the surface. A sensitivity of 0.03 ML was achieved by Yang et al. [318] for stearic acid adsorbed from CCl$_4$ on a α-Al$_2$O$_3$(0001) surface. The kinetics of the adsorption/desorption process were measured as well as the adsorption isotherms and the molecular orientation (MO). Heats of adsorption of 0.4 ML of stearic acid and n-amyl alcohol on α-Al$_2$O$_3$(4150) were measured by Haller and Rice [314].

As discussed in Section 4.1.3, the transparency range of oxides can be expanded by using the "coated" ATR technique, in which the bulk oxides of interest are modeled by thin polycrystalline films deposited onto an IRE or by native oxides on Si or Ge IRE. This approach was suggested by Zolotarev et al. [319] and applied to the adsorption of water on SiO$_2$, MgF$_2$, ZnS, ZrO$_2$, GeO$_2$, Al$_2$O$_3$, and CaF$_2$ films [320, 321]. The intensity of the absorption bands of water was used for estimating the porosity of these films and elucidating the effect of deposition rate, pressure of the residue gases, and substrate temperature on the porosity.

The adsorption kinetics of a mixture of surfactants (sodium dodecyl sulfate and sodium laurate) onto a vacuum-deposited Al_2O_3 film and diphenylchlorosilane [322] and phthalocyanines [323] on chemically deposited silicon oxide on silicon was also studied [324]. Coated ATR was also effective in the IR spectroscopic studies of adsorption on polymers (Section 7.7).

IRRAS is used less often to study adsorption on transparent substrates because of the lower SNR relative to that of MIR (Section 2.6). Nevertheless, the sensitivity of this method was shown to be sufficient for ∼0.3 ML of a long-chain (C_{12}) adsorbate on silicates in the νCH spectral region [325] and 0.2 ML of a surfactant on sulfides in the head group spectral region (Table 7.4). Applicability of IRRAS to organic monolayers on transparent substrates was first analyzed theoretically by Udagawa et al. [326], who confirmed the results predicted by the experimental spectra of stearic acid on quartz and ITO. IRRAS studies of xanthate adsorption on metal sulfides were performed by Mielczarski et al. [327–330] (Section 7.4.4), with optimum conditions determined by spectral simulations. Mielczarski and Mielczarski [331] first demonstrated the possibility of including in the spectral analysis the absorption bands of a (rather thick) film in the spectral region of the phonon bands of the substrate (Section 2.3.2).

IRRAS is the leading spectroscopy for characterizing the structure and properties of monolayers adsorbed at the air–water interface (Section 4.8). Research activity in this area [332–334] is largely motivated by the potential applications of transferred Langmuir (L) monolayers in molecular electronics and nonlinear optics and by fundamental interest in the organization and dynamics of quasi

Table 7.4. Sensitivity of in situ and ex situ IR spectroscopic techniques in studies of xanthate adsorption

Technique and Conditions	Xanthate and Sulfide	Minimum Detectable Surface Coverage	References
DRIFTS, 85% KBr, powder size $d_{50\%} = 4$ μm	AX, galena	∼0.5 ML	490
In situ ATR on powders, Ge multiple IRE, $\varphi_1 = 45°$[a]	EX, pyrite	∼0.1 ML	483
In situ ATR, unpolarized, $\varphi_1 = 36°$	n-Butyl-X, galena	0.1 ML	518
Ex situ ATR measured with sphalerite MIRE, $\varphi_1 = 45°$	Iso-propyl-X, sphalerite	∼1 ML	499
Ex situ IRRAS, p-polarization, $\varphi_1 = 70°$	EX, chalcopyrite	0.5 ML	327
	EX, chalcocite	0.2 ML	611
In situ IRRAS, p-polarization, ZnSe window, $\varphi_1 = 45°$, 13 reflections	EX, chalcocite	0.3 ML	611

[a] Angle of incidence. X = xanthate, E = ethyl, A = amyl.

two-dimensional systems which model biological membranes (Section 7.8). To increase the SNR and surface selectivity and to reduce instrumental drift, this method is coupled with polarization modulation (Section 4.7)

The simplest method of obtaining IR spectra of molecules adsorbed on transparent bulk solids is *transmission*, which is practical only with FTIR spectrometry. In particular, adsorption of CO_2 [335–340], CO [341–346], H_2 [347], H_2O [348], and SO_2 [349] on the (100) surface of NaCl; N_2O on NaCl(001) [350]; CO [351] and NH_3 [352] on MgO(100); CO_2 on MgO(100) [353]; CO on ZnO [354]; H_2O on the quartz surface [355]; and *n*-pentane on α-Al_2O_3(0001) [356] was studied. Transmission studies of the adsorption of gases on bulk dielectrics have been reviewed by Ewing [357].

Transmission of polarized radiation at suitable angles of incidence permits the determination of MO. Figure 7.12 [345] shows the *s*- and *p*-polarized spectra of CO adsorbed at a coverage of 0.6 ML on NaCl(100) measured at 45°. The band position is 2149 and 2156 cm^{-1} in the *s*- and *p*-polarized spectra, respectively. From the value of the DR, Heidberg et al. [345] concluded that the molecules are adsorbed with the TDMs parallel to the surface.

Even at normal incidence, the transmission spectra may be relatively sensitive. To demonstrate this, Fig. 7.13 shows the νOH band of water adsorbed on NaCl(100) at $-27°$C, which is lower than the freezing point of a saturated NaCl solution, and a pressure of 0.2 mbar (35% RH) [358]. The spectrum was recorded by the multiple-transmission technique. For this purpose, 11 crystals cleaved along the (100) faces of NaCl were collected in a pile with spacers of 0.1-mm Ta wire, so that the NaCl faces did not touch each other. To reduce reflection from the external air–NaCl boundary, wedged silicon windows were attached to both ends of the pile. Comparison of the band position of adsorbed water with that of water in various phases (the top of Fig. 7.13) suggests that at this temperature the adlayer is mostly in the liquid state. On this basis, the surface coverage can be evaluated by introducing the absorption index of bulk water in the modified BLB relationship:

$$\tilde{A} = \frac{j\sigma S_{H_2O}}{2.303}$$

Here, $\tilde{A}$ is the integrated absorbance of the adsorbed water in reciprocal centimeters, S_{H_2O} is the surface density of the adsorbed water, j is the number of exposed NaCl(100) faces, and σ is the integrated cross section in centimeters and is related to the absorption index k as

$$\sigma = \frac{4\pi}{\rho} \int_{band} k\nu \, d\nu,$$

where ρ is the molecular density of water at the given temperature. For water at room temperature, it was found that $\sigma = 1.4 \times 10^{-16}$ cm/molecule. With the known quantity S_{H_2O}, the coverage $\theta = S_{H_2O}/S_{NaCl}$ was calculated using the value

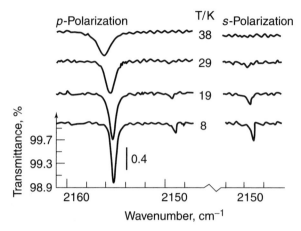

Figure 7.12. Transmission spectra of 0.6 ML of CO adsorbed at NaCl(100). Angle of incidence is 45°. Reprinted, by permission, from J. Heidberg, E. Kampshoff, R. Kuhnemuth, M. Suhren, and H. Weiss, *Surf. Sci.* **269/270**, 128 (1992), p. 131, Fig. 2. Copyright © 1999 Elsevier Science B.V.

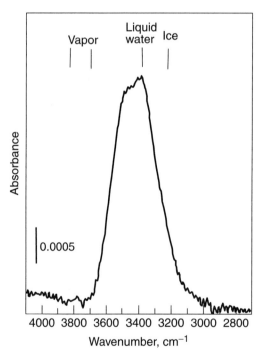

Figure 7.13. Infrared absorption of water adsorbed to NaCl(100) at −27°C and 0.2 mbar of water. Coverage is calculated to be 1.5 ML. Reprinted, by permission, from M. Foster and G. E. Ewing, *Surf. Sci.* **427/428**, 102 (1999), p. 102, Fig. 1. Copyright © 1999 Elsevier Science B.V.

of 6.4×10^{14} ion pairs/cm^2 for the density of adsorption sites at the NaCl(100) face, S_{NaCl}. For the adlayer, whose spectrum is shown in Fig. 7.13, the coverage was found to be 1.5 ML.

The information that IR spectroscopy provides about surface species on transparent and nontransparent substrates is discussed in more detail in Sections 7.4 and 7.5 using the example of flotation systems.

7.4. ADSORPTION ON MINERALS: COMPARISON OF DATA OBTAINED IN SITU AND EX SITU

There is significant interest in understanding the mineral–aqueous interaction in view of its crucial role in environmental and geochemical processes and extractive technologies [359]. Surface reactions occurring in natural and technological aqueous systems are complex, variable, and competitive [42]. IR spectroscopy has been very useful in tackling this problem [4, 22, 360–369], primarily due to its high surface sensitivity ($<1\%$ ML, Section 7.1), the large amount of structural information it provides, the simplicity of in situ applications, flexibility, and its general availability. It is interesting to note that the original motivation for the development of IR spectroscopy of adsorbed molecules was in the acquisition of IR spectra of water adsorbed on clay minerals in 1937 [9]. However, the drawback of IR spectroscopy is its lack of sensitivity toward elemental composition. Therefore, an optimal choice for studying mineral–aqueous interactions would seem to be a combination of IR and XPS spectroscopies [325, 370–378]. Furthermore, by combining these two methods with techniques such as STM (or AFM), complementary nanoscale-resolution data on surface topology could be obtained.

This section is divided into four parts. In the first one, application of IR spectroscopy to speciation and coordination of mineral surfaces after grinding and treatment with salt solutions is demonstrated. In the next parts, IR spectroscopic data on mineral–surfactant interactions that are of interest for flotation are reviewed. To date, all the IR spectroscopic methods have been applied to mineral surfaces. It is interesting that in some cases the data obtained in situ and ex situ on powders and plane surfaces are different. Thus it is also possible to compare the effectiveness of different IR techniques.

7.4.1. Characterization of Mineral Surface after Grinding: Adsorption of Inorganic Species

To characterize the mineral surface that is created by grinding and conditioning in aqueous solutions, IR spectroscopic techniques for powders (Section 4.2) are more suitable than techniques used for probing slab surfaces. In addition to a higher surface sensitivity, the former deal with real sample surfaces that differ from fractured, polished, and artificial (e.g., sputtered or chemically deposited) surfaces of minerals in terms of composition/configuration, distribution, and surface density of adsorption sites. As an illustration of this, it was found [379] that abraded galena is negligibly oxidized in water, whereas powdered samples conditioned in the same way are easily oxidized. This discrepancy has been ascribed to organic contaminants (carboxylic acids and their salts) inhibiting oxidation of the surface. Stirring the suspension can remove the contaminants and enable the oxidation to occur.

DRIFTS studies have been used to determine the hydroxyl coating on quartz, γ-alumina, and feldspar [380], the surface composition of silicates after grinding [381–383]. and the composition of complex oxidized films on galena [375–377, 384–386]. Figure 7.14 (curve 1) shows the DRIFTS spectrum of wet, ground quartz (5 μm) measured against KBr. The narrow band at 3745 cm^{-1} is assigned to the stretching νOH vibrations of surface-isolated silanol groups [40, 380]. The complex structured absorption band in the 3000–3700-cm^{-1} spectral region due to adsorbed hydroxyls H bonded to water is typical for some types of natural fine quartz powders [380]. A significant decrease in the number of hydroxyls on the surface of quartz conditioned in HCl solution (pH 2) is demonstrated by curve 2. These observations are in agreement with the fact that pH $\approx$ 2 is PZC for the quartz surface [359].

Since air drying of the mineral powder may affect the coordination and speciation of adsorbed species [387–389], more realistic data on adsorption from solution should be collected in situ. This can be accomplished with the HATR [387] or CIRCLE [390] accessory (Section 4.2.4). Wijnja and Schulthess [391, 392] used this approach for studying in situ the aging and adsorption of carbonate at the γ-Al$_2$O$_3$–water interface. Despite the fact that this system is extensively studied [359, 393], the speciation and coordination of carbonate species adsorbed on metal oxides and their correlation with the macroscopic characteristics, such as proton stoichiometry and surface charge, are still a matter of debate. Both inner sphere complex formation through direct coordination to a surface group and outer sphere complex formation through weaker electrostatic interactions have

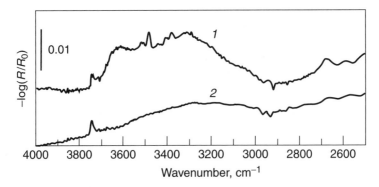

Figure 7.14. DRIFT spectra of (1) untreated quartz and (2) quartz after 10 min treatment in HCl solution, pH 2. Both spectra were measured against KBr. Reprinted, by permission from I. V. Chernyshova, K. Hamunantha Rao, A. Vidyadhar, and A. V. Shchukarev, *Langmuir* **16**, 8071–8084 (2000) p. 8075, Fig. 3. Copyright © 2000 American Chemical Society.

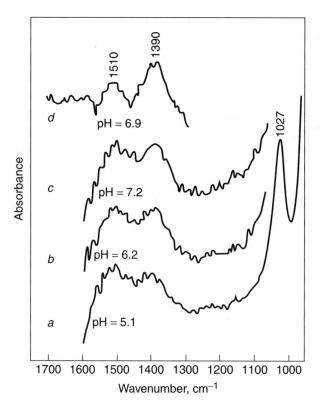

Figure 7.15. ATR-FTIR spectra of Al oxide with carbonate species adsorbed. Initial $[NaHCO_3] = 0.001\ M$ for spectra *a*, *b*, and *c*. Initial $[NaHCO_3] = 0.010\ M$ in D_2O for spectrum *d*. Band at 1027 cm^{-1} in spectrum A is 0.019 a.u. Reprinted, by permission, from H. Wijnja and C. P. Schulthess, *Spectrochim. Acta A* **55**, 861 (1999), p. 866, Fig. 3. Copyright © 1999 Elsevier Science B.V.

been proposed as adsorption mechanisms, based on macroscopic observations [393–395]. However, none of these proposed mechanisms were confirmed by spectroscopic data. Figure 7.15 shows the ATR spectra of adsorbed carbonate species at the γ-Al$_2$O$_3$–solution interface in suspensions, with initial NaHCO$_3$ concentrations of 0.001 M in H$_2$O and D$_2$O. Apart from the 1027-cm^{-1} band assigned to the δOH of bayerite [γ-Al(OH)$_3$]—a product of the transformation of the aluminum oxide surface layer in water—a doublet of bands at 1390 and 1510 cm^{-1} is present in the spectra. The same bands were observed for the solution concentrations between 0.0005 and 0.02 M. Comparing with ATR spectra of solvated carbonate and bicarbonate and invoking the results of the normal coordinate analysis and adsorption studies on other substrates, these bands were assigned to carbonate adsorbed through the formation of an inner sphere complex in which carbonate has monodentate coordination. It is interesting that for this system, DRIFTS spectra of dried samples were no different than the in situ spectra. Applying the same approach, Persson et al. [396] found that acetate is adsorbed onto the bayerite-modified surface of γ-Al_2O_3 at pH < 4.3 as mononuclear outer sphere complex, in contrast with complexes formed in aqueous solution. Peak et al. [397] studied the mechanism of sulfate adsorption on goethite (α-FeOOH) and found that sulfate forms both outer and inner sphere complexes on goethite at pH < 6, while at higher pH an outer sphere complex is formed exclusively and the relative amount of outer sphere sulfate complexation increases with decreasing ionic strength. Based on the in situ ATR spectra (Fig. 2.50) Hug [387] concluded that at pH between 3 and 5 sulfate is adsorbed on hematite (α-Fe_2O_3) as a monodentate complex, but after removal of the solvent the surface species are monodentate bisulfate or bidentate sulfate.

DRIFTS spectra of two 5-μm galena powders obtained by dry grinding two different samples are shown in Fig. 7.16 [384]. Spectra of a 1:5.3 mixture of the galena with KBr were measured against KBr for each sample. Comparison with spectra of reference compounds allowed identification of the bands from basic lead sulfide at 1171, 1055, 967, 631, and 597 cm^{-1}; basic lead carbonate at 1435, 1400, 840, and 679 cm^{-1}; and the 986-cm^{-1} band of lead thiosulfate in the spectrum of sample a. For sample b, however, the absorption of the oxidation product is much weaker, exhibiting only the bands at 1121 and 984 cm^{-1} of lead thiosulfate. In addition, the XPS spectra revealed the presence of elemental sulfur in both samples [384]. Such differences in the oxidation products can be

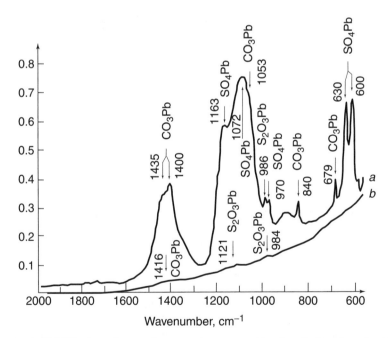

Figure 7.16. DRIFTS of two 5-μm galena samples prepared by dry grinding. Samples (a) and (b) are p- and n-type semiconductors, respectively. Reprinted, by permission, from J. M. Cases and P. de Donato, *Int. J. Mineral Process.* **33**, 49 (1991), p. 54, Fig. 2. Copyright © 1991 Elsevier Science B.V.

attributed [373] to different semiconducting properties, as samples *a* and *b* are *p*- and *n*-type semiconductors, respectively. The grinding conditions affect the nature of the surface oxidation products substantially. For sample *b*, fine grinding at pH 4 leads to negligible amounts of PbS_2O_3 [377]. An increase in pH during grinding results in the formation of more PbS_2O_3 and basic lead carbonate [375, 376]. In no instance was lead sulfate detected [384]. It is interesting that sulfoxy species (sulfates, thiosulfates, and sulfites) observed by IR spectroscopy are not detectable by XPS [398], as these species appear to decompose in the presence of atmospheric CO_2 and under UHV [398].

7.4.2. Adsorption of Oleate on Calcium Minerals

It is known [399, 400] that most Ca minerals [fluorite (CaF_2), calcite ($CaCO_3$), apatite $Ca_{10}(PO_4)_6(OH,Cl,F)_2$, scheelite $CaWO_4$, gypsum ($CaSO_4 \cdot 2H_2O$)] are floated by oleates at slightly basic or basic pH[†]. Several mechanisms of the reagent adsorption have been discussed in the literature. Since the Ca minerals are semisoluble, it was suggested [401, 402] that the adsorption results from complexation of the hydrolyzed dissolved species with the reagent followed by precipitation of the complex. The reagent can be attached to the surface by the reversible adsorption of both the molecular and ionic forms [403], the flotability being proportional to the fraction of the molecular form [404]. These processes, however, can be complicated by such phenomena as *micellization* and *precipitation* of the surfactant [361, 405, 406]. Chemisorption of oleic acid and oleate on fluorite was suggested by French et al. [407] in 1954, based on their IR spectroscopic results. During the last decade, the calcium minerals–oleate system has been studied by IR spectroscopy by Miller and co-workers [381, 408–415], Mielczarski and co-workers [416–422], and Rao and co-workers [423, 424].

On the basis of the IR data, the general consensus is that when minerals exhibit 100% flotability, several types of chemisorbed oleate as well as precipitated calcium oleate are present at the surface, in agreement with the initial hypothesis of Peck [425]. However, there is controversy concerning the assignment of the adsorption bands of adsorbed oleate and, hence, the structure of surface species. It stems from the strong dependence of the oleate–calcium coordination on the deposition and precipitation conditions–an effect well known for L monolayers and LB films of alkanoates [426]. The amount of oleate adsorbed on a fluorite slab, as measured by IR spectroscopy, is quite different than that for fine fluorite particles [412]. For the slab, there exists an adsorption saturation point, but for fine particles, the amount of adsorbed oleate increases with concentration, although the concentration dependence of the amount of oleate in the

---

[†] Froth flotation separation of minerals requires the mineral of interest to have a hydrophobic surface while the other minerals present in the pulp (a mixture of 25% m/m powdered ore with water) are hydrophilic [361, 362]. The particles of the hydrophobic mineral can be captured by the gas bubbles and raised to the pulp surface where they are collected to the froth. Flotability of the mineral depends not only on the amount of adsorbed hydrophobizing reagent (the *collector*), which in most cases does not exceed a monolayer, but also on the form in which it is adsorbed [361, 363].

supernatant (probed after sedimenting the particles) exhibits a maximum. These contradictions have been resolved by Mielszarski et al. [422] by measuring the adsorption isotherms for finely ground particles by the solution depletion method, in which the residual oleate concentration was determined after both sedimentation and centrifugation of the particles [422]. These measurements have shown that at a certain minimum concentration, oleate forms calcium oleate microcrystals and/or micelles containing calcium ions. These agglomerates are present in the filtrate, giving rise to the increased oleate surface coverage observed in the IR spectra of the particles.

In Fig. 7.17, curve a shows the DRIFTS spectrum of 10-μm fluorite conditioned in 5×10^{-6} M of sodium oleate at pH 10.0 ± 0.1, with close to a monolayer of oleate immediately after adsorption. The spectra were measured without any dissolution of the sample with KBr in order to avoid the damaging adsorbed layer and to increase the surface sensitivity to below monolayer coverage (Section 4.2). In the head group region, in addition to a band at 1468 cm^{-1} assigned to the δCH$_2$ vibrations of the carbon–hydrogen chains, there are three distinct bands at 1574, 1539, and $\sim$1415 cm^{-1}, the former two bands due to asymmetric stretching vibrations of the carboxylate group and the latter band to symmetric stretching vibrations, also of the carboxylate group. The $\sim$1570- and 1540-cm^{-1} bands in the spectra of precipitated calcium oleate [364, 427] and CaSt LB films [428] suggest the assignment of the 1574- and 1539-cm^{-1} bands to the bulk (three-dimensional) precipitates. A close examination of spectrum a in Fig. 7.17 reveals a weak broad band at 1558 cm^{-1} band,

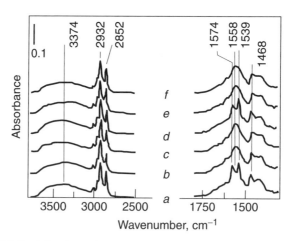

Figure 7.17. DRIFTS of fluorite with close to a monolayer of oleate (a) immediately after adsorption and after the following treatments: (b) 3 h in vacuum of 10^{-4} mbar; (c) 21 h in vacuum of 10^{-4} mbar; (d) 20 min exposed to water vapor; (e) 4 h in vacuum of 10^{-4} mbar; (f) 70 h in vacuum of 10^{-4} mbar. All spectra recorded at room temperature. Reprinted, by permission, from E. Mielczarski, Ph. de Donato, J. A. Mielczarski, J. M. Cases, O. Barres, and E. Bouquet, *J. Colloid Interface Sci.* **226**, 269 (2000), p. 275 Fig. 7. Copyright © 2000 Academic Press.

which is much more pronounced in the IR spectra reported in other works (see Ref. [414] and references therein). This band has been regarded as characteristic of chemisorbed oleate (bridging interaction) [408–429]. However, the Ca oleate aggregates formed in the pulp are characterized by a band at 1561 cm^{-1} [422], in contrast to the Ca oleate precipitates. Furthermore, as illustrated by spectra $b-f$ in Fig. 7.17, the absorption bands of carboxylate groups of the adsorbed oleate are very sensitive to the surface environment. Based on these and other arguments, the 1574- and 1539-cm^{-1} bands have been attributed to the bidentate and unidentate surface complexes of carboxylate groups [416, 421], while the 1558-cm^{-1} band is attributed to the Ca oleate aggregates formed in the pulp [422], in accordance with early spectroscopic data [425, 430].

Figure 7.17 illustrates how evacuation of the sample can influence the structure of the adsorbed film spectra. Subjecting the fluorite with oleate adsorbed to a vacuum of 0.2 Pa changes the spectrum dramatically; the doublet at 1574 and 1539 cm^{-1} transforms into a broad singlet at 1558 cm^{-1}, the $\nu_{as}CH_2$ band shifts from 2921 to 2922 cm^{-1} and broadens by about 10% [indicative of increased mobility and disorder in the adsorbed layer (Section 3.11.1)], and the 3374-cm^{-1} band of coadsorbed water decreases. These observations indicate [422] that (i) the water molecules incorporated in the oleate monolayer stabilize its order and (ii) the original ordered structure of the adsorbed layer is reversibly disturbed by water removal (curves $d-f$). As shown by Rabinovich and co-workers [431], the drying effect on the film order depends on interplay between the ordering influence of the hydrophobic interaction of the chains and the disordering influence of the thermal motion of the water. Thus, analyzing ATR spectra of LB films of dimethyldioctadecylammonium bromide (DDOAB) on silicon in air and water, a substantial decrease in the chain order down to almost full disorder upon drying was found for the film deposited with 12.5 mN · m^{-1} surface pressure, which has relatively low initial chain packing density. In contrast, for the film deposited at 25 mN · m^{-1}, the order parameter was high in both air and water, the value in water being slightly lower than in air. The decreasing order of the DDOAB film in water as compared with air was shown by Tsao et al. [432] using AFM. This effect should be taken into account extrapolating data obtained ex situ on the molecular order and orientation of adsorbed species to the in situ conditions.

Finally, the numerous IR spectroscopic studies of adsorption of oleate on Ca minerals and glasses [433, 434] offer an opportunity to compare efficiency of different IR spectroscopic techniques. Thus the ATR method for powders [414] (Section 4.2.4) appears to be most suitable for studying oleate on Ca–oxy-anion minerals (calcite, gypsum, apatite, etc.) as compared to the KBr transmission [425], KBr DRIFTS [435], and ATR with sampling technique E [413] (Section 4.1.3). The ATR method provides the shortest penetration depth, which is advantageous for the system under consideration since Ca–oxy-anion minerals partially absorb in the 1700–1500-cm^{-1} range. Also, the DRIFTS spectra of oleate species adsorbed on apatite powder show that mixing with KBr at a sample concentration higher than $\sim$10% results in a gradual change of the doublet at

$\sim$1570 and 1540 cm^{-1} into a broad band at 1560 cm^{-1} [413], which, however, can be explained by absorption saturation.

7.4.3. Structure of Adsorbed Films of Long-Chain Amines on Silicates

Numerous technologies involve the modification of silicate surfaces by the adsorption of surfactants [436, 437]. One common system is that of long-chain primary amines adsorbed on silicates. Since the 1930s this system has been studied extensively with indirect methods such as the measurement of ζ-potential, adsorption isotherms, contact angles, and flotation tests (see reviews [438–440]). Based on these data, the adsorption of amines on silicates at neutral pH has been explained mainly by the Gaudin–Fuerstenau or hemimicelle model (HM) [441–444]. This model postulates that amine cations are adsorbed in the outer Stern layer by electrostatic attraction to negatively charged surface sites. At some threshold concentration the actual concentration of amine near the quartz surface becomes higher than the critical micelle concentration (CMC) and a two-dimensional process similar to ordinary bulk (three-dimensional) micellization takes place. As a result, the surfactant aggregates at the surface and its adsorption becomes appreciable. These two-dimensional aggregates are called hemimicelles and the critical concentration is referred to as the critical hemimicelle concentration (CHC). The hemimicelles appear to hydrophobatize the silicate surface by exposing the hydrophobic chains to the solution.

However, there had not been a spectroscopic study that confirmed this HM. Moreover, this model is unable to explain certain experimental data and does not account for the thermodynamical requirement that the solution temperature be below the Krafft point for micellization to take place. In order to understand these discrepancies, adsorption of long-chain amines was studied spectroscopically on quartz and albite, and the results were compared with the indirect data [325, 372]. DRIFTS, IRRAS, and XPS of quartz and albite conditioned in solutions of dodecyl (C_{12}) and hexadecyl (C_{16}) ammonium chloride (ACl) and acetate (AAc) of different initial concentrations (C_b) were measured.

Figure 7.18 shows selected DRIFTS spectra of the quartz treated for 1 h in solutions of C_{12}–ACl and C_{12}–AAc. (The spectrum of untreated quartz is shown in Fig. 7.14.) All spectra display typical bands in the 3000–2800-cm^{-1} range due to the νCH vibrations of alkyl chains. A broad structural band centered at $\sim$3250–3000 cm^{-1} is attributed to H bonded νN$^+$H, νOH, and νNH stretching vibrations. A close inspection of these spectra reveals that twice the position of this band shifts abruptly and bathychromically at a concentration close to the CHC and when the bulk amine precipitates (vide infra). It is well known [445] that a red shift of any H bonded stretching mode indicates an increase in H bond strength. In all the spectra a negative 3745-cm^{-1} band for free surface silanols is observed, suggesting that the silanols interact with absorbed amine. This interaction is likely H bonding between a proton of the head group of amine cations and the SiOH oxygen. In addition, adsorption of the C_{12}–ACl at $C_b = C_{pr} > 2 \times 10^{-3}$ M results in appearance of a series of new bands at 3330,

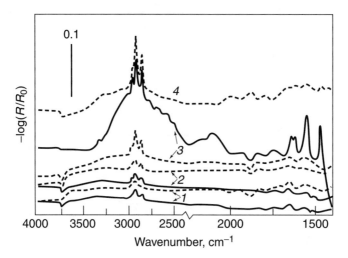

Figure 7.18. DRIFTS spectra of quartz powder (5 μm) after 1 h treatment with C_{12}–ammonium acetate (AC) (dashed lines) and C_{12}– Cl ammonium (solid lines) solutions (pH 6–7) with concentrations (1) 5×10^{-6} M, (2) 1×10^{-4} M, (3) 2×10^{-3} M, and (4) 1×10^{-2} M. Spectrum 3 for C_{12}–ammonium Cl is divided by 5. Reprinted, by permission, from I. V. Chernyshova, K. Hamunantha Rao, A. Vidyadhar, and A. V. Shchukarev, *Langmuir* **16**, 8071–8084 (2000), p. 8075, Fig. 4. Copyright © 2000 American Chemical Society.

1650–1640, 1570, and 1506 cm^{-1}, characteristic of molecular amine. Using the radioisotope data [446], the sensitivity of the DRIFTS measurements performed in the νCH range can be estimated at $\sim$0.1 ML of the C_{12} hydrocarbon chains.

Figure 7.19 shows the correlation between the surface coverage of the amines, the number of silanol groups reacted, the molecular order, the ζ-potential measured on the same sample, and the flotation recovery measured for a coarser ($-150 + 38$-μm) particle fraction. The surface coverage of the amine is proportional to the intensity of the $\nu_{as}CH_2$ band, A_n, which is normalized to the number of the methylene units in the chain. Similar to amine adsorption isotherms [447, 448], the A_n curves exhibit a break at a CHC of $\sim$1 $\times$ 10^{-4} M. The number of reacted surface silanol groups is assumed to be proportional to the peak intensity, E, of the negative 3745-cm^{-1} band. The E curves reach a plateau at a certain value of bulk amine concentration, above which the quartz particles are totally screened from interaction with the solution by the adsorbed amine. The number of "saturated" silanol groups (given by the plateau height) is higher for acetate salts. Because the maximum quantity of adsorbed amine is higher for the chloride salts (Fig. 7.19a), coadsorption of acetate ion was suspected. Another conclusion drawn from the E curves is that the surface silanols are saturated at a concentration much higher than the CHC, in agreement with other data [438, 439, 443].

It is important to note that the increase in surface coverage of the amine is accompanied by a red shift of the νCH$_2$ bands (Fig. 7.19c). The beginning of this shift coincides with the CHC. According to the correlation discussed in

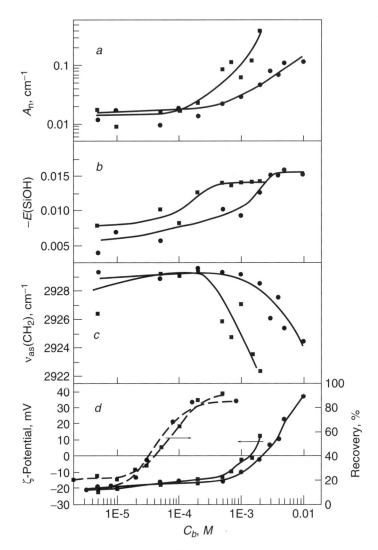

Figure 7.19. Adsorption characteristics of (●) C_{12}–ammonium Ac and (■) C_{12}–ammonium Cl on quartz powder at pH 6–7 and $t = 20 \pm 2°C$. (*a*) Normalized integrated intensity $A_n = A/11$ cm^{-1}, of $\nu_{as}CH_2$ band; (*b*) peak intensity of absorption band of silanol groups at 3745 cm^{-1}; (*c*) frequency of $\nu_{as}CH_2$ band; (*d*) ζ-potential (solid lines) and flotation recovery (dashed lines) as function of bulk amine concentration. Reprinted, by permission, from I. V. Chernyshova, K. Hamunantha Rao, A. Vidyadhar, and A. V. Shchukarev, *Langmuir* **16**, 8071–8084 (2000), p. 8078, Fig. 9. Copyright © 2000 by American Chemical Society.

Section 3.11.2, this implies that the order of the adsorbed layer increases at the CHC. The shift reaches its maximum when molecular amine precipitates at the surface. This confirms that after the break in the isotherm the molecules begin to aggregate on the surface, and their order increases.

Thus, contrary to the HM, the DRIFTS spectra indicate that the adsorbed amine is H bonded with surface silanols and, what is more important, at the CHC these H bonds become stronger, implying some qualitative change in the layer. At higher concentrations, ordinary bulk precipitation of the amine takes place, rather than the previously suggested formation of a bilayer [441, 448, 449]. However, the composition of the adsorbed layer at intermediate concentrations is not clear from the DRIFTS spectra. As revealed by XPS [325], in this concentration range, neutral and protonated amine are both present at the surface, while at lower concentrations, only the protonated form is observed. This suggests that the neutral amine appears at the surface when a critical local surface concentration of the surfactant is reached, establishing the following equilibrium:

$$
\begin{array}{c}
\text{H} \\
| \\
\text{RN}...\text{HOSi} \equiv \Leftrightarrow \text{RNH}_3^+... \ {}^-\text{OSi} \equiv \\
| \\
\text{H}
\end{array}
\qquad\qquad (A)
$$

The neutral molecules change the structure of the adsorbed layer substantially by screening the electrostatic repulsion between the head groups. This enhances the adsorption and increases the density and the order of the monolayer, which explains the decrease in νCH_2 frequencies.

It is interesting how the characteristics extracted from the DRIFTS spectra correlate with such macroscopic characteristics of adsorption as ζ-potential and flotation recovery. Figure 7.19d shows how the ζ-potential curves follow the corresponding A_n curves. After the break, the amount of amine adsorbed from the amine acetate solutions is less than from chlorine salt solutions. This correlates with the lower ζ-potentials observed for quartz, which has reacted with acetate salt solution. Because in this concentration range neutral molecules are adsorbed first in a monolayer and then as bulk precipitates, the effect of coadsorption of acetate ions does not render the surface charge more negative, and the presence of acetate ions in the double layer suppresses the formation of neutral molecules by precipitation. Another conclusion from the DRIFTS spectra is that the increase in ζ-potential is due not to adsorption of cations, as was suggested from studies of the ζ-potential curves and adsorption isotherms in the absence of spectroscopic data [441–443], but rather to precipitation of molecular amine. The ζ-potential of amine colloidal precipitates is positive and increases with increasing amine solution concentration [450]. On the other hand, the flotation results appear to be independent of the origin of the amine counterion, an effect attributed to surface *inhomogeneity*, because for flotation to occur, one highly hydrophobic patch is sufficient.

Identical correlations have been observed for albite [372], confirming the validity of the two- and three-dimensional precipitation models in interpreting the mechanism of adsorption of weak-electrolyte-type surfactants on silicates. However, the concentration at which two-dimensional precipitation for albite occurred was much lower than that for quartz. Replacing Si^{4+} with Al^{3+} in a second-nearest-neighbor site increases the proton affinity of the surface OH groups, while charge balancing Na^+ ions have the opposite effect. Since the albite surface conditioned in water is depleted of Na^+ ions, but these may not be completely substituted by protons, it is possible that the surface silanols on albite bear a higher electron density and, therefore, are stronger proton acceptors than the silanols on quartz. This shifts equilibrium (A) to the left.

In order to quantify the increasing order in the adsorbed amine film, the species adsorbed on the silicate slabs were characterized by IRRAS [325, 372, 451]. Figures 7.20 and 7.21 show the IRRAS spectra of the (002) quartz surface treated for 5 min in solutions (pH 6.5) of C_{16}–ACl (1×10^{-4} M) and C_{16}–AAc (1×10^{-4} and 1.5×10^{-4} M). Apart from the absorption in the 2900- and 1465-cm^{-1} regions due to the stretching and bending vibrations of the CH groups, respectively, the s-polarized spectra exhibit bands at 1740–1750, 1650, 1600, and 1547 cm^{-1}, which were assigned tentatively to coadsorbed (bi)carbonate, molecular amine, coadsorbed water, and contaminations. In addition, the p-polarized spectra contain the complex broad band in the 2000–4000-cm^{-1} range due to H bonded νOH and νNH vibrations. The form and the position of this band varied from spectrum to spectrum, because of the simultaneous contributions of positively and negatively going bands due to the stretching vibrations of H bonded OH and NH groups. These bands are hardly detectable in the s-polarized spectra, implying that the TDMs of the H bonded groups are oriented preferentially perpendicular to the surface.

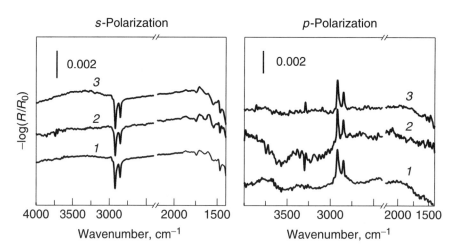

Figure 7.20. Spectra from IRRAS of polished quartz surface conditioned in solutions (pH 6.5) of (1) C_{16}–ammonium acetate (1×10^{-4} M), (2) C_{16}–ammonium chloride (1×10^{-4} M), and (3) C_{16}–ammonium acetate at 1.5×10^{-4} M. Angle of incidence is 73°. The s-polarized spectra are shown on left and p-polarized spectra on right. Spectra were measured using Perkin-Elmer 2000 spectrometer, with MCT detector, at 4 cm^{-1} resolution, by coadding 1000–2000 scans, both in s- and p-polarization.

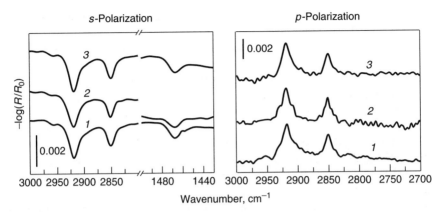

Figure 7.21. Same as in Fig. 7.20 with enlarged regions of the νCH and δCH modes and with background correction. Adapted, by permission, from I. V. Chernyshova and K. Hamunantha Rao, *J. Phys. Chem. B* **105**, 810–820 (2001), p. 817, Fig. 10. Copyright © 2001 American Chemical Society.

At a higher amine concentration of 1.5×10^{-4} *M* (Fig. 7.20, curve 3), a weak band with a substructure is noticeable in the 3000–3300-cm^{-1} region of the *s*-polarized spectrum and is assigned to precipitated three-dimensional clusters of molecular amine. The formation of the bulk amine phase was revealed by the presence of the 3330-cm^{-1} band in the DRIFT spectra characteristic of the molecular amine (Fig. 7.18). However, this band is absent in the spectrum measured by IRRAS. This was explained to be a result of two crystalline states of three-dimensional clusters precipitated at 1.5×10^{-4} *M*, one on a polished quartz surface and one on fine quartz particles. The different crystalline states could arise because of the incorporation of ionic amine into the three-dimensional clusters in the former case. The frequencies of $\nu_{as}CH_2 = 2918.3 \pm 0.1$ cm^{-1} indicate (Section 3.11.1) that the hydrocarbon chains in the films whose spectra are shown in Figs. 7.20 and 7.21 are well ordered. The film formed from 1×10^{-4} *M* solution is characterized by splitting of the δCH$_2$ band into two components, 1468.3 and 1460 cm^{-1} (shoulder) in the case of amine acetate and 1468 and 1474 cm^{-1} (shoulder) in the case of amine chloride (Fig. 7.21). This splitting can arise due to either orthorhombic or monoclinic packing of the hydrocarbon chains (Section 3.11.2). Taking into account that the (002) quartz surface consists of oxygen atoms positioned rhombically, it is more probable that the adsorbed species adopt monoclinic-like packing. This qualitative conclusion is confirmed by the biaxial symmetry of the layers (see below). However, the presence of hexagonally packed domains that are responsible for the 1468-cm^{-1} component and the wide FWHM cannot be excluded. The splitting is barely distinguishable for the amine film formed from 1.5×10^{-4} *M* solution. Instead, a broad band arises at 1467 cm^{-1} (Fig. 7.19, curve 3). This conflicts with the observed biaxial symmetry of the film (see below) but may be explained by the heterogeneous (domainlike) composition of the films, just as a single δCH$_2$ band is observed in IRRAS of hexadecan-1-ol L monolayers in the condensed state, in which orthorhombic packing dominates [452].

As pointed out in Section 3.11.4, the IRRAS method offers an intrinsic advantage in MO measurements on transparent substrates. Namely, the electric field component perpendicular to the surface (which is the most sensitive to the tilt angle) is comparable in magnitude to that parallel to the surface, unlike in ATR measurements, in which the angle of incidence exceeds the critical angle by more than 10°. This advantage has been exploited in a number of structural studies, including those of LB films [453, 454], SAMs [455–457], adsorbed surfactants [325, 372, 451], and L monolayers at the AW interface [458–460].

In the DR approach [Eq. (3.60)] [461] to estimating MO parameters from the spectra shown in Fig. 7.19, the anisotropy of the quartz substrate was neglected and a value of 1.49 was taken for the refractive index n_3 of quartz. The optical parameters of the film at the $\nu_{as}CH_2$ frequency ($n_{2\perp} = 1.49$, $n_{2\parallel} = 1.55$, $k_{2\,max} = 1.04$) were taken from Ref. [453]. Assuming the known orientation, the surface coverage was estimated from the band intensities using Eqs. (1.87). From the data obtained (Table 7.5), at a concentration of 1×10^{-4} M, the amount of amine adsorbed is on the order of one monolayer, with an average tilt angle of about 30°, which is consistent with the crystalline state of the adsorbed monolayer. For comparison, a tilt angle of 30° and similar band positions have been reported for SAMs of alkanthiols on gold [462, 463].

The biaxiality of the amine layers was calculated both from DR_{as} and $(A_{as}/A_s)^s$ values, with the help of the plots in Figs. 3.87 and 3.88. Assuming that the crystals formed by drying a drop of a high-concentrated solution of hexadecyl amine in ethanol on a KBr plate are isotropically oriented, the transmission spectrum gives $k_{as}/k_s = A_{as}/A_s = 1.5$. As seen from Table 7.5, the azimuthal angles obtained by each technique are in good agreement, indicating that the biaxial model chosen for the MO calculations is suitable. The film biaxiality as calculated from

Table 7.5. Characteristics of adsorbed layers of alkyl amines from IRRAS spectra in Figure 7.21

Spectral Characteristic	C_{16}–ACl, 1×10^{-4} M	C_{16}–AAc, 1×10^{-4} M	C_{16}–AAc, 1.5×10^{-4} M
$\nu_{as}CH_2^{s}$ [a], cm^{-1}	2918.4	2918.3	2918.4
$\nu_{s}CH_2^{s}$, cm^{-1}	2850.8	2851.0	2850.6
DR_{as}	−0.69	−0.62	−0.77
DR_s	−0.62	−0.72	−0.91
(γ, ϕ) calculated from DRs	(28°, 39°)	(30°, 52°)	(43°, 48°)
$(A_{as}/A_s)^s$	1.56	1.38	1.35
$(A_{as}/A_s)^p$	1.40	1.62	1.58
ϕ calculated from $(A_{as}/A_s)^s$	40°	52°	40°
Number of monolayers	1.04	1.16	1.26

[a]Band position in s-polarized spectrum.

the intensity ratios is rather low [$\pm 10°$ in average from the magic ($45°$) angle]. The imperfection in the surface of the substrate caused by polishing could lead to a decrease in the biaxiality.

It is interesting (Table 7.5) that the hydrocarbon chains in an amine layer of 1.26 ML formed from 1.5×10^{-4} M solution have a higher average angle of inclination ($\gamma = 43°$). The "extra" band at 3000–3300 cm^{-1} in the corresponding s-polarized spectrum (Fig. 7.20, curve 3) suggests that at this concentration, the three-dimensional precipitation has already occurred, and as a result, chaotically oriented domains are present at the surface as well as the monolayer-thick highly organized patches with the monoclinic ($\sim 30°$ tilt) subcell. Also apparent in Table 7.5 is the nonmonotonic concentration dependence of the surface coverage by C$_{12}$–AAc. This effect is analogous to that observed in the oleate–fluorite system (Section 7.3.2) and is explained by the bulk precipitation of the reagent, owing to which the actual concentration of the dissolved surfactant is much less than the net concentration.

The quantitative data are in agreement with the following observations from the polarized IRRAS. A close inspection of the vCH region in the s- and p-polarized spectra reveals no v_s^{ip}CH$_3$ band at 2870 cm^{-1} for the amine monolayer formed from 1×10^{-4} M solution, which confirms that the tilt angle is close to $30°$. As a matter of fact, since the TDM of the v_s^{ip}CH$_3$ modes is inclined by $35.5°$ away from the chain axis (Fig. 3.75), at a chain tilt angle of $35.5°$ it is perpendicular to the substrate surface. Under these conditions, the corresponding absorption band will be absent in the s-polarized spectrum and negative in the p-polarized spectrum. (The latter feature was not distinguished due to the low SNR.) By contrast, the positive v_s^{ip}CH$_3$ band is distinct in the p-polarized spectra of the amine adsorbed from 1.5×10^{-4} M solution (curve 3 in Fig. 7.21), which can be attributed to the presence of three-dimensional precipitates.

In the case of albite the hydrocarbon chains of the two-dimensional amine precipitates were found [372] to be perpendicular to the surface, which was attributed to polycrystallinity of the surface and a lower average distance between basic adsorption sites. The tails of long-chain amines and alcohols have tilt angles of $\sim 30°$ and $25°$, respectively, when adsorbed alone and an angle of $\sim 37°$ when coadsorbed from a binary 1:1 solution [451]. Thus it was concluded that the two-dimensional precipitate formed in the latter case is a solid solution of the molecules and not eutectics.

7.4.4. Interaction of Xanthate with Sulfides

The interfacial behavior of alkyl xanthates (a typical member of the O-alkyldithiocarbonate collector family) (Fig. 7.22) is of great interest in flotation studies, where these reagents have been used for the selective hydrophobization of sulfides since 1925. A variety of hypotheses have been put forward to explain this phenomenon (see Refs. [365, 369, 464–466] for review) that can be classified as either *chemical* or *electrochemical*. The former category includes adsorption [467–469], coordination [470], and replacement of surface oxidation products [471–473] or lattice ions [474, 475]. However, this hypothesis neglects

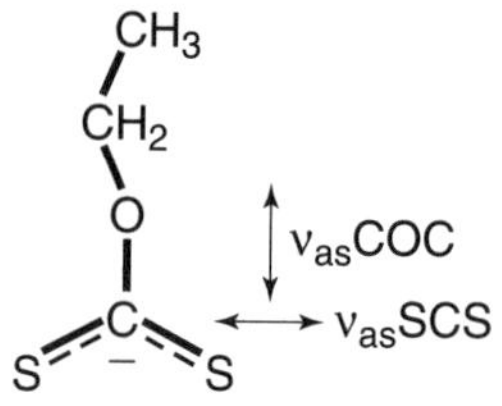

Figure 7.22. Ethyl xanthate ion.

the formation of dixanthogen [(ROCS$_2$)$_2$, the most hydrophobic derivative of xanthate] and fails to explain the poor flotabilities of highly oxidized lead and copper sulfides when the solubility product is significantly exceeded [476] and of slightly oxidized sphalerite (ZnS), which is an insulator.

The electrochemical mechanism, also called the *mixed-potential mechanism* [477], assumes charge transfer within a particle from the cathodic patch, at which oxygen is reduced, to the anodic one, at which the sulfide itself and/or xanthate anion are oxidized. This mechanism describes a broad spectrum of interfacial phenomena involving, as an intermediate step, a redox reaction in which the anodic and cathodic processes are spatially separated. Some examples of this include electrocatalytic chemisorption of xanthate and synthesis of dixanthogen and precipitation of xanthate–metal complexes (nucleation of a microphase of metal xanthate). In the latter reaction, the anodic sulfide dissolution is initiated with the ionization of surface metal atoms, and the metal ions thus produced on the surface are transferred into aqueous solution to form hydrated metal ions or metal–ion complexes associated with anions [478]. The ionization of surface metal atoms is an electrochemical oxidation, whereas the hydration or complexation of metal ions is a chemical process (an acid-based reaction).

The complex chemistry of the surface reactions in a pulp coupled with the instability of the adsorbed species makes xanthate adsorption processes quite challenging to study. A great deal of the progress in this area has been made through IR spectroscopy. The dual (chemical and electrochemical) reactivity of sulfides in the pulp has led to two types of IR spectroscopic studies: one in which the surface potential is chemically controlled and the other in which it is controlled potentiostatically. Chemical control, which is implemented by the addition of potential-determining agents to the solution, is more closely related to industrial practice. Potentiostatic control is "cleaner" and more flexible, allowing the electrochemical components of the surface reactions to be distinguished. In the present section, the results of the IR spectroscopic studies of the xanthate adsorption under chemical control will be considered, and the spectroelectrochemical studies are covered in Section 7.5.2.

The analysis of IR spectra regarding the different adsorption forms of xanthate is considerably simplified by the sensitivity of IR spectra to the coordination of the xanthate group (Fig. 7.23) [369, 479]. As the "degree of covalency" of the bond between the sulfur atom of xanthate and the heavy metal cation

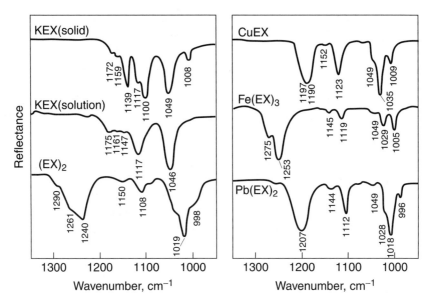

Figure 7.23. ATR spectra of reference compounds of ethyl xanthate. Reprinted, by permission, from J. O. Leppinen, C. I. Basilio, and R. H. Yoon, *Int. J. Miner. Process.* **26**, 259 (1989), p. 262, Figs. 2 and 3. Copyright © 1989 Elsevier Publishers B.V.

(S−Me) increases, the ν_{as}COC mode shifts from ~1100 cm^{-1} for a complex with K or Na to ~1200 cm^{-1} for complexes with Cu(I) [480], Au (I) [481], Ag(I) [482], Pb(II), Zn(II) [483, 484], and Cd(II) [484], in which the S−Me bond is partially covalent, and reaches a maximum (1260–1290 cm^{-1}) for bidentate complexes with Ni(II) [482], Pt(II), Co(II), Cr(II) [485], Fe(III) [486], and dixanthogen [(R−O−C−S$_2$)$_2$], in which the S−S bond is almost covalent. The band at 1000–1040 cm^{-1} in the IR spectra of metal−xanthate complexes has a large contribution from the ν_{as}SCS mode. The origin of the band at 1110–1120 cm^{-1} is more uncertain: Mielczarski [327] associated it with the complex νCOC + νSCS mode. However, as will be shown below, if a microphase of metal xanthate and dixanthogen can be easily distinguished by simulating spectra of ultrathin films of reference compounds, the assignment of the IR spectral bands of xanthate adsorbed at (sub)monolayer coverage is the subject of much debate.

The ex situ studies of xanthate adsorption under chemically controlled conditions have been conducted by transmission [487–489] and DRIFTS [375–377, 385, 386, 480, 484, 490–495] on powdered sulfides, by ATR on thin polycrystalline synthetic films of PbS [496–498] and single-crystal sphalerite [499], and by IRRAS on sulfide and metal plates [327, 329, 330, 481, 482, 500].

Persson et al. [385, 386, 484, 493] performed comparative DRIFTS studies of ethyl xanthate (EX) adsorption on fresh and oxidized natural minerals and sulfur- and metal-rich synthetic Cu$_2$S, ZnS, CdS, and PbS following different pretreatments of the absorbents. Only small amounts of Cu(I)EX were found on the chalcocite surface, which was very slightly oxidized, but relatively large amounts were detected on the oxidized synthetic surfaces. In the cases of sphalerite and metal-rich Zn(II) and Cd(II) sulfides, the ν_{as}SCS band shifted to 1058 cm^{-1} from 1033–1036 cm^{-1} for solid Zn(II) and Cd(II) xanthates. In the spectra of metal-rich synthetic ZnS and CdS, the ν_{as}COC band was at 1225 cm^{-1}, which was assumed to be characteristic of the bidentate xanthate complexes [484]. There was an additional band at 1290 cm^{-1} in the DRIFTS of EX adsorbed from water and acetone on sphalerite and from water on synthetic PbS, from which the surface oxidation products had been removed by pretreatment with ethylenediaminetetraacetic acid (EDTA). In the case of sphalerite, this band disappeared after washing the sample with water but decreased only slightly after washing with acetone. On this basis, this band was assigned to a relatively sparse packing of xanthate complexes, in which xanthate is monodentately coordinated to cations in the outermost surface layer and the ethyl group has freedom to move and rotate. Because under these conditions xanthate does not float sphalerite, it can be concluded that the surface covered by the chemisorbed complexes is not hydrophobic enough. To make sphalerite particles flotable, activation by Cu(II) ions is often used. Adsorption of EX on the activated sphalerite gave rise to bands of Cu(I)EX and ethyl dixanthogen [493], which suggests that the Cu(II) ions in the surface layer oxidize the EX ions to dixanthogen, while the Cu(I) ions formed react with the excess EX to form Cu(I)EX. In contrast to sphalerite and synthetic PbS, chemisorption on pyrite and galena was not detected by DRIFTS, the only surface species being dixanthogen (on pyrite) and lead xanthate (on galena). However, these results disagree with the in situ spectra (vide infra and Section 7.5.2). A strong additional band at 1258 cm^{-1} was observed [497, 498] in the ATR spectra of n-butyl xanthate (BX) adsorbed from water on fresh synthetic stoichiometric PbS. Differing from the value of 1264 cm^{-1} characteristic of butyl dixanthogen (see Fig. 7.42 later), this band position was interpreted in terms of chelate formation. When the reagent was adsorbed on oxidized PbS, the ν_{as}COC band was split into components at 1185 and 1196 cm^{-1}, which were assigned to xanthate complexes with nonequivalent bonds, and a component at 1270 cm^{-1}, which was assigned to adsorbed dixanthogen. Regardless of the actual origin of the extra bands observed for the synthetic sulfides, the difference between the IR spectra for natural and synthetic sulfide implies that the synthetic sulfides are different than the naturally occurring minerals.

Interesting results on the xanthate chemisorption were obtained ex situ by Ihs and co-workers for adsorption of long-chain xanthate on Au, Ag, and Ni [481, 482]. Spectral simulations were used to determine chain orientation, while XPS was used to characterize head group interactions. The experimental IRRAS spectra of octyl xanthate (OX) on gold are shown in Fig. 7.24. At low coverages, only one band at 1205 cm^{-1} due the ν_{as}COC mode is observed in the head group region, but at high coverages (spectra not shown) the spectra are very similar to the simulated spectrum of an isotropic film of bulk Au(I)OX on gold. An analogous spectral behavior was observed by Sunholm and Talonen [501] for electrocatalytic adsorption of EX on a silver electrode. The absence of the

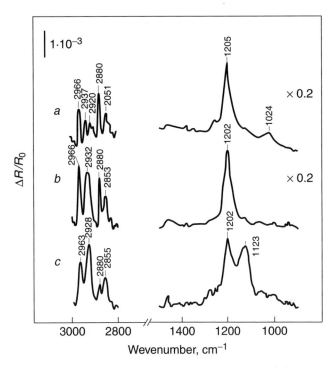

Figure 7.24. Experimental spectra from IRRAS of octyl xanthate (OX) adsorbed on gold. Concentrations and adsorption times are (a) 1 mM and 10 min, (b) 10 μM and 10 min, and (c) 1 μM and 10 min. Reprinted, by permission, from A. Ihs, K. Uvdal, and B. Liedberg, *Langmuir* **9**, 733 (1993), p. 736, Fig. 5. Copyright © 1993 American Chemical Society.

$\nu_{as}SCS$ band near $1040-1050$ cm^{-1} at the initial adsorption step was attributed to the xanthate coordination to the gold surface through both sulfur atoms, so that the TDM of the $\nu_{as}SCS$ mode is parallel to the surface (Fig. 7.22) and the IRRAS signal is forbidden (SSR for metals). This interpretation was confirmed by XPS, which revealed equivalent bonding of the adsorbed xanthate sulfurs. The tilt angle of $\sim 30°$, which was calculated from the relative intensities of the $\nu_{as}CH_2$ and ν_sCH_2 bands of the hydrocarbon chains, also agrees well with the S_2C-O axis being perpendicular to the surface. Based on the SSR for metals, Mielczarski [329] assigned the bands at 1197, 1127, and 1050 cm^{-1} in IRRAS of EX adsorbed on metallic copper to the $\nu_{as}COC$ mode of the xanthate group perpendicular to the surface.

At submonolayer coverages of OX on gold (curve a in Fig. 7.24) and silver [482] and EX on Cu [329, 482], chalcocite (Cu$_2$S) [329, 330], and silver [482, 501], an additional band was observed at ~ 1225 cm^{-1}, which disappeared at multilayer coverage or when a well-organized full monolayer was formed after washing the sample with water. The polarization dependence of the band intensity implies that the corresponding TDM is almost perpendicular to the surface. As opposite to the assignment by Persson and co-workers of the extra band at 1225 cm^{-1} in the spectra of metal-rich synthetic ZnS and CdS [484], Mielczarski and co-workers assigned this band to "(a) product(s) of surface-induced decomposition of xanthate" [330]. However, this interpretation has been debated extensively (see Section 7.5.2 for more detail).

Already in the first IR study of the xanthate adsorption by Poling and Leja [502], it was noted that the absorption band frequency of the ν_{as}COC band from bulk lead xanthate is $1210–1220$ cm^{-1}, while the lead–xanthate complex on galena at low surface coverages is characterized by a band at $1170–1180$ cm^{-1}. These authors interpreted this band behavior to be indicative of monocoordination of xanthate to lead cations in the first monolayer. According to the data of Cases and co-workers [375–377, 490], a monolayer of amyl xanthate (AX) on "preoxidized" galena is characterized by ν_{as}COC bands at 1189 and 1179 cm^{-1} and a ν_{as}SCS band at 1026 cm^{-1}. As surface coverage is increased from 1 to 6 ML, both the ν_{as}COC and ν_{as}SCS bands move gradually to 1202 cm^{-1} and 1036 cm^{-1}, respectively, compared with the values of 1220 and 1022 cm^{-1} for stoichiometric Pb(AX)$_2$. However, upon observing a similar shift for xanthate species adsorbed on oxidized galena, which are removed by washing with water, Persson et al. [493] concluded that this shift does not necessarily indicate chemisorption. In fact, such a shift can be a product of an optical effect (Section 3.3) or dipole-dipole coupling (Section 3.6). Both these viewpoints are consistent with the fact that the ν_{as}COC band of polycrystalline Pb(EX)$_2$ layers deposited from acetone on a Ge IRE [503, 504] shifts from 1190 cm^{-1} at sub-monolayer coverages to 1210 cm^{-1} for multilayers, while the ν_{as}SCS band shifts from 1035 to 1019 cm^{-1}.

To verify the optical hypothesis, spectrum simulations for the IRE($n_1 = 2.4$)–galena plate ($n_2 = 3.91$, $d_2 = 10$ μm)–xanthate film–water($n_4 = 1.26$, $k_4 = 0.036$) system in the region of $1300–900$ cm^{-1} at an angle of incidence of $36°$ were performed [478] for different thicknesses of the isotropic lead xanthate [Pb(X)$_2$] film. The complex permittivity of the film was modeled with a three-oscillator formula [Eq. (1.46)] with $\varepsilon_\infty = 2.5$; $\gamma_j = 20$ cm^{-1} ($j = 1, 2, 3$); $\nu_{01} = 1200$ cm^{-1}, $S_1 = 0.08$ (ν_{as}COC); $\nu_{02} = 1130$ cm^{-1}, $S_2 = 0.04$ (νCOC + νSCS); and $\nu_{03} = 1030$ cm^{-1}, $S_3 = 0.08$ (ν_{as}SCS), which were approximated from the optical constants extracted from the transmission spectrum of lead butyl xanthate pressed in a KBr pellet. Figure 7.25 shows that the optical effect produces splitting of the ν_{as}COC band in the unpolarized ATR spectra. With increasing film thickness from 3 to 100 nm, the spectra are unchanged. Similar results were obtained for the film at the Ge–air interface. Hence, the fact that the ν_{as}COC band shifts to higher frequencies with increasing adsorbed film thickness cannot be attributed to the optical effect, and even in the case of the cast films on Ge, the film structure changes with increasing surface coverage. However, a close inspection of the spectra reported by Persson et al. (see, e.g., Fig. 1 in Ref. [493]) reveals that the relative intensity of the ν_{as}SCS band in the DRIFTS spectra of cleaned galena conditioned with EX solution is significantly less than that of bulk Pb(EX)$_2$, which can be considered as an indicator of

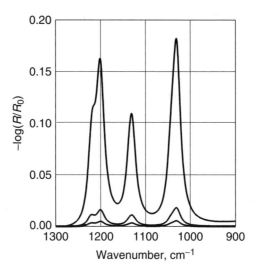

Figure 7.25. Thickness dependence of simulated ATR spectra of model organic film at galena–water interface. Simulation were performed for angle of incidence of 36°. Refractive indices for IRE, galena, and water were $n_1 = 2.4$, $n_2 = 3.91$, and $n_4 = 1.26$, respectively, and $k_4 = 0.036$. Thickness of galena layer was taken as $d_2 = 10$ μm. Optical constants of organic film were chosen to be close to those of lead xanthate but approximated by three-oscillator Maxwell–Helmholtz–Drude formula with $\varepsilon_\infty = 2.5$; $\gamma_j = 20$ cm^{-1} ($j = 1, 2, 3$); $\nu_{01} = 1200$ cm^{-1}, $S_1 = 0.08$ (ν_{as}COC); $\nu_{02} = 1130$ cm^{-1} (νCOC + νSCS); $S_2 = 0.04$; $\nu_{03} = 1030$ cm^{-1}, $S_3 = 0.08$ (ν_{as}SCS). Film thickness was 3, 10, and 100 nm from bottom curve. Reprinted, by permission, from I. V. Chernyshova, *J. Phys. Chem. B* **105**, 8185 (2001), p. 8188, Fig. 4. Copyright © 2001 American Chemical Society.

different xanthate–lead coordination in the fist adsorbed monolayer and bulk xanthate (see Section 7.5.2).

The inherent instability of the adsorbed xanthate species makes data obtained in situ more reliable. The pseudo in situ ATR method (Section 4.2.3) was used to elucidate details of the EX adsorption [505] on chalcopyrite (CuFeS$_2$), tetra-hedrite (Cu$_{12}$Sb$_4$S$_{13}$), and tennantite (Cu$_{12}$As$_4$S$_{13}$), on pyrite (FeS$_2$) in the absence and presence cyanide [506], and on galena [507] and isobutyl xanthate on pyrrho-tite (Fe$_x$S$_y$)–pentlandite [(Ni,Fe)$_9$S$_8$] mixtures in the absence [508] and presence of depressants (diethylenetriamine and sodium metabisulfite) [509]. Figure 7.26 shows the ATR spectra of 30 μm chalcopyrite powder in contact with KEX solutions of different concentrations (pH 9.5, adjusted with KOH) after condi-tioning for about 20 min. The spectra of the paste pressed against a Ge IRE were measured with 25 reflections at an angle of incidence of 45°. The reference spectrum is that of chalcopyrite in contact with water of pH 9.5. At the lowest xanthate concentration (1.9×10^{-5} M, curve 1) only the bands at 1210–1200 and $\sim$1030 cm^{-1}, which resemble those of CuEX (Fig. 7.23), are present, the surface coverage being about 0.3 ML. At higher concentrations, the bands at 1263, 1239, 1110, and 1026 cm^{-1} due to dixanthogen, the band at 1217 cm^{-1} assigned on the basis of XPS data [510] to an iron–xanthate complex, and the

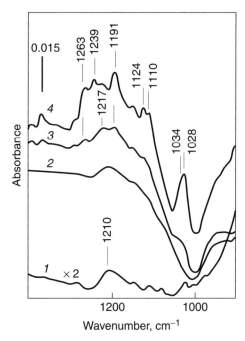

Figure 7.26. In situ ATR spectra of chalcopyrite (CuFeS$_2$) powder (<30 μm) in contact with ethyl xanthate solution pH 9.5 (adjusted with KOH) at various initial concentrations: 1, 1.9 × 10^{-5} M; 2, 5.0 × 10^{-5} M; 3, 1.7 × 10^{-4} M; 4, 7.7 × 10^{-4} M. Reference is ATR spectrum of chalcopyrite in contact with water, pH 9.5. Adapted, by permission, from J. A. Mielczarski, J. M. Cases, and O. Barres, *J. Colloid Interface Sci.* **178**, 740 (1996), p. 743, Fig. 3. Copyright © 1996 Academic Press, Inc.

broad negative-going band at ~1000 cm^{-1} due to desorbed surface oxidation products appear. At low concentrations the intensity of the absorption bands of adsorbed xanthate was found to be proportional to the negative intensity of the washed-off oxidation products, suggesting that at least part of the xanthate is adsorbed through ionic exchange. An analogous increase in the amount of surface metal–xanthate species and a decrease in the amount of surface oxidation products has been reported for galena [507], tetrahedrite, and tennantite [510]. In contrast, the in situ spectra of powdered pyrrhotite and pentlandite revealed no detectable surface oxidation products on the initial surface and only dixanthogen after immersion in xanthate solutions [511, 512]. After conditioning the minerals in the same solution, dixanthogen adsorption on pentlandite increased but decreased on pyrrhotite. This effect was explained within the framework of the mixed-potential model.

Leppinen [483] studied the adsorption of EX on nonactivated pyrite, pyrrhotite, chalchopyrite, and sphalerite and on the same minerals activated with copper sulfate. He used the particle-bed electrode (Fig. 4.51) [513, 514] under open-circuit conditions. The IR results were compared to the flotation data. Dixanthogen was found on nonactivated pyrite, pyrrhotite, and chalchopyrite. Iron xanthate was also found on pyrite at submonolayer coverages, and copper xanthate coexisted with dixanthogen on chalcopyrite. After activation, a Cu(I)EX complex was present on all the minerals studied. At acidic pH, adsorption on nonactivated surfaces increased, and neutral pH favored EX adsorption on activated minerals. Figure 7.27 shows the ATR spectra of pyrite particles after conditioning at two different KX concentrations at pH 5.0. When compared with the spectra of Fe(III)(EX)$_3$ and (EX)$_2$ (Fig. 7.23), the low-concentration spectrum can be assigned to a mixture of Fe(III)(EX)$_3$ and (EX)$_2$, while the relative content of dixanthogen increases with increasing concentration. The pH dependence of the intensity of the ν_{as}COC band in the 1230–1240-cm^{-1} region for EX on pyrite and the flotation recovery are shown in Fig. 7.28. The flotation results show the same trend as the IR signal, with a characteristic S-shape and a minimum at neutral pH, also observed by other authors [515, 516]. The intensity plot allows the sensitivity to be estimated as ~0.1 ML.

Comparison of the IR spectra measured in situ and ex situ of xanthate adsorbed on sulfides revealed a number of problems associated with ex situ studies. First, drying of the sample increases ordering of the adsorbed xanthate layer [329, 330]. Second, it was found [500] that even relatively "heavy" amyl dixanthogen evaporates rather quickly in open air, not to mention the UHV conditions (required for XPS measurements), which should also be considered when interpreting XPS data and correlating it with IR data. For example, XPS of an abraded galena surface incubated for 5 min in 10^{-3} M n-butyl xanthate (BX) at +0.20 V (SHE) [517] reveals only chemisorbed and bulk Pb(BX)$_2$, while the in situ ATR

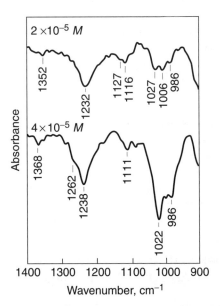

Figure 7.27. ATR spectra of pyrite particles after conditioning at two different concentrations of potassium ethyl xanthate (KEX). Reprinted, by permission, from J. O. Leppinen, *Int. J. Miner. Process.* **30**, 245 (1990), p. 251, Fig. 4. Copyright © 1990 Elsevier Publishers B.V.

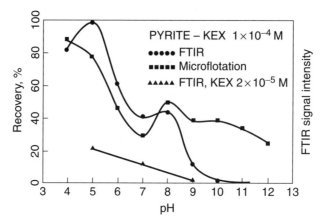

Figure 7.28. Dependence of FTIR signal intensity and microflotation recovery on pH for pyrite in potassium ethyl xanthate (KEX) solution of 1×10^{-4} M and FTIR signal intensity in KEX solution of 2×10^{-5} M. Full length of intensity axis corresponds to 1.0% absorption. Reprinted, by permission, from J. O. Leppinen, *Int. J. Miner. Process.* **30**, 245 (1990), p. 254, Fig. 7. Copyright © 1990 Elsevier Publishers B.V.

spectra recorded under the same conditions [518] show the additional presence of dixanthogen. In an analogous experiment, no dixanthogen was detected by XPS for the chalcopyrite–amyl xanthate (AX) system, which is in disagreement with the IR spectra recorded before the sample was introduced into the XPS instrument [327, 510]. The instability of dixanthogen at the solid–air(vacuum) interface could explain why dixanthogen has not been observed spectroscopically ex situ on the galena surface at concentrations used in flotation ($<10^{-4}$ M) but has only been detected by the extraction method [465, 519]. Third, as illustrated by Fig. 2.52, the dixanthogen component of the adsorbed layer is eliminated by the DRIFTS sampling procedure. This could explain the similarity of the surface composition of floated and nonfloated minerals observed by DRIFTS by Cases and co-workers [377, 490].

Table 7.4 allows comparison of the IR spectroscopic techniques in terms of the surface sensitivities to adsorbed xanthate.

Thus, in contrast to macroscopic, indirect characterization of adsorption processes, IR spectra give information about the chemical and structural identity of the adsorbed layer, which can aid in determining the mechanism of adsorption unequivocally. Examples relevant to electrochemistry are discussed in the following section.

7.5. ELECTROCHEMICAL REACTIONS AT SEMICONDUCTING ELECTRODES: COMPARISON OF DIFFERENT IN SITU TECHNIQUES

Electrochemically generated ultrathin films on semiconductors have been studied intensively in recent years in the context of flotation, electrocatalysis, solar energy storage, microelectronics, photocatalytic reactions, photodegradation of organics, and other technologies in which an interfacial layer with tailored characteristics is required [520, 521]. The nature and kinetics of the interaction of electro- or chemically-active species in solution with a semiconductor electrode (SE) are controlled by the Fermi level (the electrochemical potential of the electrons in the solid) at the semiconductor surface relative to the band edges and the distribution and density of the surface states, when the latter are present [37, 520–523]. This means that the reactivity of the electrode surface can be manipulated simply by varying its potential, without changing the chemical environment. Because of the three energetically different pathways for charge transfer (via the conduction band, the valence band or surface states), redox processes at semiconductors differ essentially from those at metals [520].

Within the last decade, STM, low-angle X-ray and neutron scattering, extended X-ray absorption fine structure (EXAFS), X-ray absorption near-edge structure (XANES), quartz microbalance (QCM) methods, and vibrational spectroscopies such as IR, Raman, or sum-frequency generation have been used for understanding the rich and important electrochemistry of semiconductors [524–527]. Among these methods, IR spectroscopy is now the most widely used [528, 529]. Interfacial processes on microelectronics-related semiconductors (Si [530–550] and GaAs [541, 551–553], and, as a model, Ge [554–557]) have been frequently studied IR spectroelectrochemically. When moderately doped, these materials are transparent over a wide spectral range ($E_{g(Si)} = 1.0$–1.1 eV, $E_{g(Ge)} = 0.67$ eV, and $E_{g(GaAs)} = 1.4$ eV), since the lattice vibrations are either at the boundary with the IR range or in the far-IR. The same is true for CdTe ($E_g \approx 2.5$ eV) used as a material for a photoexcited electrode [558, 559]. This makes "reactive" ATR spectroscopy the most appropriate method for probing their interface with an electrolyte (Section 4.1.3). Electrochemical reactions on semiconducting sulfides [330, 478, 514, 560–562] and wide-bandgap semiconducting oxides such as TiO_2 [563–569] and ZnO [570] have also been studied in situ. The sulfides are of great interest for flotation and the oxides can act as photocatalysts in reactions such as degradation of organic pollutants or solar energy converters. For these systems, in situ spectra can be obtained by ATR and IRRAS, the SE being a powder, a slab, or a thin plate (Sections 4.6.2 and 4.6.3). However, the interfacial sensitivity and the body of information provided by different techniques are different. This will be discussed using the example of the anodic oxidation of silicon and galena (natural PbS) and the oxidation of xanthate on sulfides.

7.5.1. Anodic Oxidation of Semiconductors

Silicon. The surface of silicon immersed in fluoride media is of interest for semiconductor processing and production of porous silicon (Section 5.7) [541, 542, 549, 550]. A typical current–potential curve of *p*-Si in a fluoride elec-trolyte (0.975 MNH$_4$Cl + 0.025 MNH$_4$F + 0.025 MHF, pH 2.8) measured at a rotating disk electrode at rotation rate of 3000 rpm and potential scanning speed of 5 mV $\cdot$ s^{-1} is shown in Fig. 7.29. The steep rise of the current density near

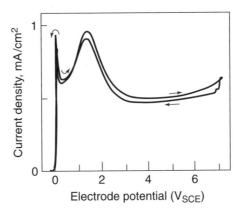

Figure 7.29. Typical current–potential curve of *p*-Si [(100) orientation, doping 10^{15} cm^{-3}] in dilute fluoride electrolyte (here 0.975 *M* NH$_4$Cl + 0.025 *M* NH$_4$F + 0.025 *M* HF, i.e., fluoride concentration 0.05 *M*, pH 2.8). Rotating disk electrode: rotation rate 3000 rpm, potential scanning speed 5 mV·s^{-1}. Reprinted, by permission, from F. Ozanam, C. da Fonseca, A. V. Rao, and J.-N Chazalviel, *Appl. Spectrosc.* **51**, 519 (1997), p. 521, Fig. 1. Copyright © 1997 Society for Applied Spectroscopy.

0 V is associated with porous silicon formation, and at more positive potentials silicon electropolishing occurs. The formation of a silicon oxide layer instead of Si–H groups of porous silicon is evidenced by the static PDIR spectra of the interface (Fig. 7.30), from which the background absorption by free carriers (vide infra) was subtracted. The splitting of the ν_{TO} and ν_{LO} bands of silicon oxide (at 1050–1070 and 1210–1230 cm^{-1}, respectively) and their average wavenumber were used to determine the quality of the oxide, as discussed in Sections. 5.1–5.3. In this example, it was revealed that the oxides formed in the second current plateau region (+4 to +6 V, SHE) are the most compact. At the current onset corresponding to the end of the second electropolishing plateau as well as at potentials lower than +4 V, the density of the oxide layer decreases. This was attributed to increasing porosity and hydration, respectively. The intensity of the ν_{TO} band in the in situ spectra of the oxides was used to estimate the oxide thickness. The oxide layer was found to increase nearly linearly with potential, by ~9 Å · V^{-1}. Bands with increasing negative intensity (Fig. 7.27) were attributed to the δNH$_4^+$, δH$_2$O, νNH, and νOH modes of the electrolyte, whose contribution in the net ATR spectrum is attenuated as the oxide layer grows (Section 3.7).

As discussed in Section 3.7, a smooth background that increases at lower wavenumbers is characteristic of free-carrier absorption, also termed *Drude absorption* (DA) (Figs. 3.44 and 3.47). The potential dependence of DA for the silicon–fluoride electrolyte interface is shown in Fig. 7.31. Free-carrier absorption is negligible at negative potentials, at which the silicon electrode is depleted. Since DA obeys the Schottky–Mott law (inset in Fig. 7.31), the flat-band potential was found to be ~+0.4 V. Weak absorption by free carriers

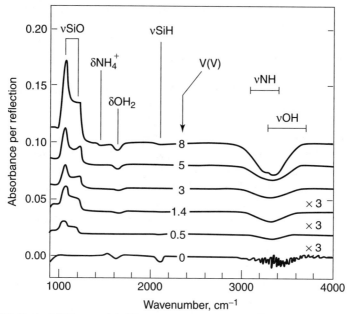

Figure 7.30. Typical ATR potential-difference spectra of p-Si electrode in same electrolyte as for Fig. 7.29. Electrode potential is indicated on each curve. Reference at −0.5 V. Baseline of spectra was incremented by arbitrary absorbance value of 0.02 for each spectrum from 0 V (no shift) to 8 V (shifted by 0.1 V). Reprinted, by permission, from F. Ozanam, C. da Fonseca, A. V. Rao, and J.-N Chazalviel, *Appl. Spectrosc.* **51**, 519 (1997), p. 522, Fig. 2. Copyright © 1997 Society for Applied Spectroscopy.

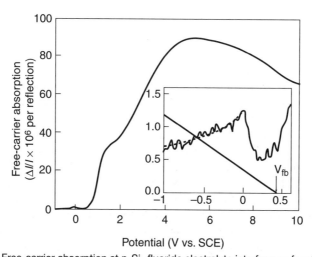

Figure 7.31. Free-carrier absorption at p-Si–fluoride electrolyte interface as function of potential determined as IR absorption in 1000–2000-cm^{-1} range, induced by potential modulation (50 mV, 1 kHz). Fluoride concentration 0.1 M, pH 4.5. Inset shows variation of $\Delta I/I$ on expanded scale in depletion region. Dashed and solid straight lines are Mott–Schottky fits for $\Delta I/I$ and $(\Delta I/I)^{-2}$, respectively; V_{fb} shows calculated value of flat-band potential. Reprinted, by permission, from F. Ozanam, C. da Fonseca, A. V. Rao, and J.-N Chazalviel, *Appl. Spectrosc.* **51**, 519 (1997), p. 523, Fig. 3. Copyright © 1997 Society for Applied Spectroscopy.

from $+0.5$ V to $\sim+1$ V indicates that the oxide layer that grows in this potential range has poor blocking capabilities. The potential modulation spectra revealed that at these potentials a part of the νOH signal is phase shifted. Since a delayed response is expected for trapping/detrapping, the low free-carrier absorption can be ascribed to trapping of the generated carriers by SiOH group traps, which act as adsorption sites. Between $+1$ and $+5$ V, the DA increases significantly, which means that free holes are accumulated at the electrode–electrolyte interface. During this accumulation, the more structurally perfect oxide form is acting as a barrier, as was inferred from the vibrational spectra.

To gain insight into the mechanism of the transition between the hydrogenated state and the oxidized state, the time-dependent ATR spectra of a (100) p-Si electrode in a dilute fluoride electrolyte were measured when the initially flat electrode was polarized in the potential range of porous-silicon formation [549, 550]. These spectra (Fig. 7.32b) show (i) an increase of the electrolyte IR absorption bands, which can be associated with the growth of a porous silicon layer with electrolyte-filled pores; (ii) the appearance and increase up to a steady-state level of a silicon oxide absorption band in the 1100-cm^{-1} region (this oxide is markedly different from that of an oxide formed at higher potentials (Fig. 7.30); and (iii) a fast decrease of the SiH$_x$ band, followed by its increase up to values much in excess of the initial level. A quantitative analysis [549] that takes into account the difference between the specific surface area A (corresponding to the porous layer) and an electrochemically active area $a < A$ indicates that the transition to the electropolishing regime occurs when the local oxide coverage of the electrochemically active part of the surface is on the order of a monolayer. Hence, silicon dissolution proceeds only partly through oxide formation–dissolution, and a parallel reaction pathway involving direct attack of silicon by fluoride species must also be present. This direct reaction path probably becomes predominant in concentrated fluoride electrolytes. As distinct from the above case of a dilute fluoride electrolyte, anodization in concentrated HF (>1 M) produces large currents. The ATR spectra of the silicon–electrolyte interface (Fig. 7.32a) show a significant increase of the νSiH band with time and no trace of oxide or of oxygen- or fluoride-backbonded SiH groups (around 2250 cm^{-1}).

IR spectroscopy has been found [537, 540, 549–551, 556, 557] to be an effective tool for studying hydrogen adsorption at electrodes. This problem is of interest since hydrogen at some of the adsorption sites is thought to act as an intermediate in the hydrogen evolution reaction [571]. The ATR spectra of the n-Ge–1 M HClO$_4$ interface [556, 557] (Fig. 3.44b) show that during the negative scan between -0.5 and -0.9 V (SCE) the Ge–H groups (a positive band at $\sim$2000 cm^{-1}) reversibly substitute the O–H groups (a negative broad band at $\sim$3200 cm^{-1}), which is in agreement with the accepted electrochemical model for Ge electrodes. Analysis of the νGeH band showed [556] that this band consists of two components: a dominant band in the 1970-cm^{-1} region attributed to GeH species and a weaker band around 2030 cm^{-1} to GeH$_2$ species. Based on dependencies of these two bands on IR light polarization, crystallographic orientation, and potential, the GeH groups on terraces and the GeH and GeH$_2$

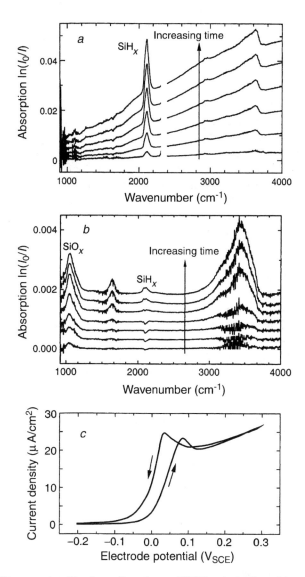

Figure 7.32. Changes in IR absorption for p-Si(100) electrode after polarization in porous-silicon formation region (*a*) in concentrated HF electrolyte (25% ethanolic HF, 10 mA $\cdot$ cm^{-2}, times of 65, 195, 325, 455, 585, and 715 s) and (*b*) in dilute fluoride electrolyte (0.01 M HF + 0.01 M NH$_4$F, potential 0 V (SCE), times of 33, 96, 159, 348, 852, 1100, and 1293 s). (*c*) Voltammogram recorded in second electrolyte (sweeping rate 5 mV $\cdot$ s^{-1}, notice electropolishing plateau at ~20 mA $\cdot$ cm^{-2}). (*a*) Spectra are dominated by fast increase in specific surface area and associated absorbance of surface SiH, whereas multiple contributions can be seen in (*b*), especially presence of submonolayer of oxide. Sloping baseline in (*a*) is due to increased light diffusion by porous Si layer. Reprinted, by permission, from J.-N. Chazalviel, A. Belaidi, M. Safi, F. Maroun, B. H. Erne, and F. Ozanam, *Electrochim. Acta* **45**, 3205 (2000), p. 3207, Fig. 1. Copyright © 2000 Elsevier Science Ltd.

groups at terrace boundaries were distinguished correlated with specific peaks in the voltammograms (Fig. 3.45*a*). The free-carrier absorption (Fig. 3.45*b*) was used for extracting additional information about the Ge–electrolyte interface (see Section 3.7).

ATR studies of the Si(111)–fluoride electrolyte interface under oscillating currents [548] revealed that the absorption of the Si–O groups oscillates with the same frequency as the current, which was interpreted in terms of a competition between oxidation of silicon and oxide etching.

Thus, the origin of the interfacial processes can be determined from the PDIR spectra, evolution of the interfacial film structure can be followed, and the kinetics of the interfacial film growth can be quantified. Applying the potential modulation method, absorptions with different phase delays can be distinguished, which provides a variety of information, including that on the origin of the interfacial traps of charge.

Galena. Galena is a natural lead sulfide (PbS)—a semiconductor with a band gap of about 0.37 eV [572]. Most galenas have *n*-type conductivity. This mineral is a major source of lead and toxic contamination of water. Synthetic lead sulfide is in IR detectors and semitransparent glasses and is considered a possible material for solar energy cells. Therefore, a large number of studies of galena oxidation have been undertaken (see Ref. [379] for review).

Under oxygen-free and collector-free conditions at slightly basic pH (8–9.5) galena can be floated in a particle bed electrode flotation cell with maximum recovery at $\sim+0.25$ V [515, 573]. However, in the presence of oxygen the collectorless flotability vanishes [363]. It is now well accepted that galena is oxidized through an electrochemical corrosion mechanism with oxygen reduction as the cathodic process [465]. Nevertheless, there is no agreement regarding the mechanism of the anodic oxidation. The following four reactions, which are all thermodynamically probable, have all been suggested as possible initial steps of the galena anodic decomposition:

(I) Formation of a metastable sulfur-rich sulfide underlayer (or elemental sulfur), yielding a monolayer of PbOH [574] and Pb(OH)$_2$ [575–577] in alkaline solutions:

$$PbS + 2H_2O + 2xh^+ \Rightarrow Pb_{1-x}S + xPb(OH)_2, \tag{Ia}$$

or injecting the lead cations (incongruent dissolution) into acidic solution [578–580] as follows:

$$PbS + 2xh^+ \Rightarrow Pb_{1-x}S + xPb^{2+}, \tag{Ib}$$

(h$^+$ denotes a hole).

(II) At pH 7–10, direct formation of sulfur and a lead (hydro)xyde monolayer through reaction with water [581–584] as follows:

$$PbS + 2H_2O + 2h^+ \Rightarrow Pb(OH)_2 + S + 2H^+, \qquad (II)$$

or lead borate, if the measurements are performed in tetraborate buffer ($Na_2O_4B_7 \times 10H_2O$), instead of lead sulfide, a lead–borate complex [585, 586].

(III) A congruent dissolution–oxidation mechanism [379, 587–591]:

$$PbS \Rightarrow Pb^{2+} + S^{2-}. \qquad (III)$$

Hydrolysis of the released ions and anodic oxidation of the hydrolyzed sulfur anion to elemental sulfur ($HS^- \Rightarrow S^0 + H^+ + 2e$ at pH 9.2) as follows:

Mechanisms I and II were suggested to explain classical electrochemical and photovoltaic data. Mechanism III has been put forward recently on the basis of general surface chemistry but is only supported by AFM and XPS data at acidic pH. The problem is that neither elemental sulfur nor the sulfur-rich layer has been clearly detected on galena that has been slightly oxidized at neutral or slightly basic pH, despite numerous attempts [373, 379, 398, 575, 576]. Nowak et al. [379, 398] have ascribed this inconsistency to the natural presence of surface contamination, which inhibits oxidation at open-circuit potentials (OCPs) and interferes in the analysis of XPS data. Buckley and Woods [592] postulated that the ultrathin surface oxidation layer relaxes as soon as the sample is removed from the solution, because of the different surface potentials in the UHV and electrochemical environments. Regarding in situ methods, Raman spectroscopy [593] is insensitive to the surface oxidation products formed at the initial stages of galena oxidation (elemental sulfur was detected only at +0.37 V to +1.24 V, SHE).

At higher potentials, based on both thermodynamic arguments and measurements of the ionic composition of the solution, the following reactions have been suggested for the second step of the galena anodic decomposition at pH 7–10. Galena first is oxidized by reaction (II) but yielding a bulk oxidation layer with $(OH)_2$ and Pb vacancies as charge carriers [581–584, 594]. As an alternative, lead hydroxide can appear simultaneously with the following:

(IV) Lead thiosulfate [595],

$$2PbS + 5H_2O \Rightarrow PbS_2O_3 + Pb(OH)_2 + 8H^+ + 8e, \qquad (IV)$$

or

(V) Anions of thiosulfate [586, 594, 596] or more thermodynamically stable sulfate [597],

$$2PbS + 7H_2O \Rightarrow 2Pb(OH)_2 + S_2O_3{}^{2-} + 10H^+ + 8e, \quad \text{(Va)}$$

$$2PbS + 2CO_3{}^{2-} + 3H_2O \Rightarrow 2PbCO_3 + S_2O_3{}^{2-} + 6H^+ + 8e, \quad \text{(Vb)}$$

$$2S^\circ + 3H_2O \Rightarrow S_2O_3{}^{2-} + 6H^+ + 4e \quad \text{(Vc)}$$

[reactions for sulfate are similar to (Va) and (Vb)].

To determine which of these possibilities is occurring, in situ ATR/FTIR spectra of the galena–borate buffer interface were measured [560]. The sample was prepared by sampling method C (Section 4.1.3): A galena plate cut along a (100) crystallographic plane was glued to a chalcogenide glass hemicylindric IRE. The working surface was wet polished successively with 1.0-, 0.3-, and 0.05-μm alumina (Buehler) and washed thoroughly with distilled water. It has been shown by impedance spectroscopy [584] that the electrochemical responses of cleaved and polished galena are not very different. The status quo of the surface was restored by holding the electrode at −0.5 V, which is close to the flat-band potential under these conditions [584, 585], for 1 h. This procedure provides [598, 504] reproducible electrochemical behavior. After cathodic reduction, the galena potential was increased stepwise from −0.5 V to +0.7 V and an ATR/FTIR spectrum (8 cm^{-1} resolution) was taken at each step, while the reference spectrum was measured at −0.5 V.

The FTIR ATR spectra were recorded at an angle of incidence of 37° with a Perkin-Elmer model 1760 FTIR spectrometer equipped with a micro-ATR unit TR-5 and MCT detector. The electrode was held for 5 min at each potential, including 2 min for collecting a 200-scan spectrum with 8 cm^{-1} resolution. Classical electrochemical equipment was employed in a three-electrode configuration. The potentials were measured against a saturated potassium chloride electrode connected via a Luggin capillary to the cell, although all potentials reported here are converted to the standard hydrogen electrode (SHE).

The oxidation current–potential ($I - V$) dependences for the galena electrode in the deaerated 0.01 and 0.05 M borate (Fig. 7.33a) are similar to those reported in other works [465, 504, 578, 598, 599]. In particular, there is a prepeak at $\sim$+0.15 and +0.10 V in the 0.01 and 0.05 M buffers, respectively. The ATR-PDIR spectra, which were measured simultaneously, are shown in Figs. 7.34 and 7.35 for 0.01 and 0.05 M borate, respectively. In the case of the 0.01 M borate (Fig. 7.34), broad bands at 1360 and 960 cm^{-1} are apparent at +0.2 V. The latter is assigned to PbSO$_3$. Since lead hydroxide, (basic) carbonate, and lead–polyborate complexes all absorb in the 1360–1440-cm^{-1} spectral range, the 1360-cm^{-1} band assignment was verified by XPS to be due to Pb(OH)$_2$ [373]. An additional narrow band arises at 982 cm^{-1} (marked by an arrow in Fig. 7.34a) at +0.3 V. At more positive potentials, this band is accompanied by a satellite at 1007–1010 cm^{-1}, while the broad band at 960 cm^{-1} splits into components

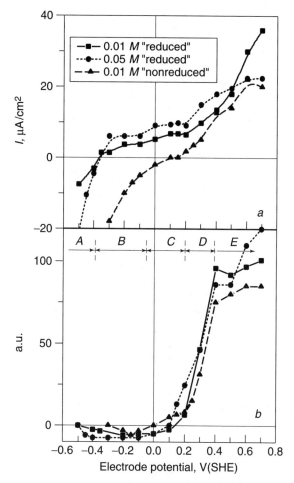

Figure 7.33. (a) Current–potential dependences for galena electrode in 0.01 M and 0.05 M deaerated borate buffer (pH 9.18), measured simultaneously with ATR/FTIR spectra shown in Figs. 7.34 and 7.35. (b) Dependence of smooth background at low wavenumbers on electrode potential, taken at 800 cm^{-1} in spectra referred to spectrum measured at initial potential of -0.5 V. Reprinted, by permission, from I. V. Chernyshova, *J. Phys. Chem. B* **105**, 8178 (2001), p. 8180, Fig. 1. Copyright © 2001 American Chemical Society.

1200, 1115–1120, and 950 cm^{-1}. The bands at 1115–1120 and 982 cm^{-1} are due to, respectively, the $\nu_{as}SO_3{}^{2-}$ and $\nu_s(SO_3{}^{2})$ modes of bulk PbS_2O_3. The pair of bands at 1220 and 1007–1010 cm^{-1} was assigned to a less stable crystallographic modification of PbS_2O_3, in which the $S_2O_3{}^{2-}$ ion coordinates Pb^{2+} by S-bridging. These features were also observed qualitatively in the 0.05 M buffer (Fig. 7.35). However, lead hydroxide appears 0.1 V lower than in the 0.01 M buffer and more is formed (a factor of 3–4 more) than in the 0.01 M borate when the amount of the sulfoxy compounds is approximately the same.

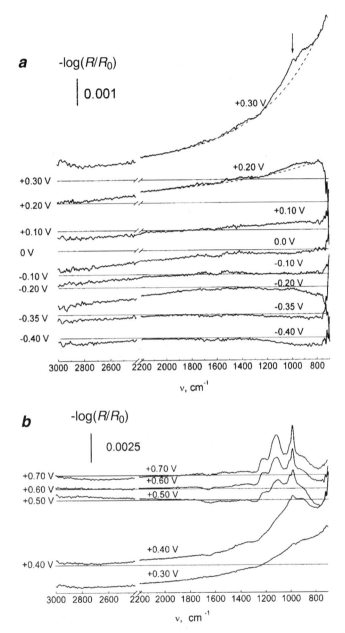

Figure 7.34. ATR unpolarized difference spectra of galena–0.01 *M* borate, pH 9.18, interface during positive scan. Smooth background at low frequencies (dashed lines) arises from free-hole absorption. Reference for spectrum marked by −0.40 V is spectrum measured at −0.50 V after reduction. Each of the other curves is spectrum measured at indicated potential after subtracting spectrum measured at preceding potential. Horizontal lines indicate zero absorbance. Reprinted, by permission, from I. V. Chernyshova, *J. Phys. Chem.* B **105**, 8178 (2001), p. 8180, Fig. 2. Copyright © 2001 American Chemical Society.

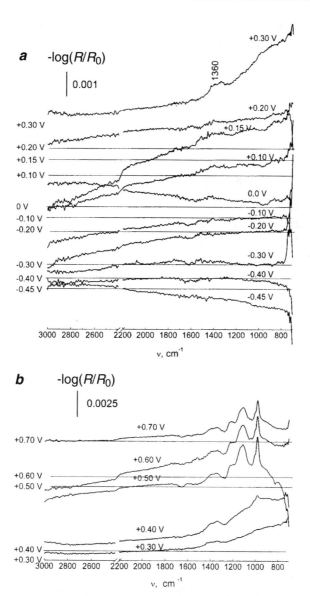

Figure 7.35. ATR unpolarized difference spectra of galena–0.05 M borate (pH 9.18) inter-face during positive scan. Each of the other curves is spectrum measured at potential indicated after subtracting spectrum measured at preceding potential. Horizontal lines indi-cate zero absorbance. Notice high-wavenumber background presenting down-going slope at $v > 1500$ cm^{-1}. The more concentrated the electrolyte, the larger this slope (compare with Fig. 7.34). In corresponding potential region, potential falls mainly across space charge region. On the basis of analysis of the potential dependence, this low-wavenumber background absorption was assigned to changing occupancy and density of both the defect levels in the surface layer and surface states. Reprinted, by permission, from I. V. Chernyshova, *J. Phys. Chem. B* **105**, 8178 (2001), p. 8181, Fig. 3. Copyright © 2001 American Chemical Society.

These results allow one to (i) rule out reactions (V) in which solvated sulfoxy ions form before their precipitation as complexes with lead and (ii) conclude that the reaction

$$PbS + 3H_2O \Rightarrow PbSO_3 + 6H^+ + 6e, \tag{VIa}$$

precedes or at least accompanies the formation of PbS_2O_3. Since formation reactions for the sulfoxy compounds and $Pb(OH)_2$ are different, reaction (IV) should also be ruled out. A possible reaction of the PbS_2O_3 formation without releasing lead cations is

$$PbS + S^0 + 3H_2O \Rightarrow PbS_2O_3 + 6H^+ + 6e, \tag{VIb}$$

which flows when elemental (active [562]) sulfur and a PbS unit are in close vicinity to each other. This requirement imposes kinetic limitations on the formation of PbS_2O_3.

However, to draw a more detailed picture of the anodic processes, additional information is needed, which may be extracted from the analysis of free-carrier absorption (Section 3.7). This absorption is marked by dashed curves in Fig. 7.34a. To follow these changes, the background absorption at 800 cm^{-1} was measured relative to that in the spectrum obtained at -0.5 V and plotted as a function of potential in Fig. 3.33b [560]. Different potential regions for 0.05 M borate are labeled by A, B, C, D, and E. The decrease in free-carrier absorption in region A is caused by subtraction of electrons from the accumulation layer. In region B, free-carrier absorption is constant and minimal, as expected for the depletion regime. Region C coincides with the prepeaks in the $I-V$ dependences (Fig. 7.33a). Here, the electrode is in the inversion regime when holes are thermally generated in sufficient concentration to promote the galena decomposition without degeneracy. In agreement with the theory, the concentration of holes and, hence, the upward band bending increase as the buffer ionic strength increases, which is balanced by a higher excess of borate anions in the outer Helmholtz plane. As a result, the negative drop of potential in the Helmholtz layer is higher, which provides a higher rate of injection of lead cations into solution. Region D spreads from $\sim+0.2$ V (the onset of the basic anodic current) to $+0.4$ V. In this region, anodic decomposition is accelerated since the Fermi level is pinned at the edge of the valence band and the sulfide behaves like a metal. Finally, in region E, the main surface reaction becomes the formation of the lead sulfoxy compounds. The rate of accumulation of holes decreases since the holes captured by sulfur now leave the PbS crystal. In parallel, $Pb(OH)_2$ is still formed supplying holes.

The fact that holes are generated and accumulated at the interface during the galena decomposition in regions C, D, and E implies the formation of a sulfur-rich layer, which is inconsistent with reaction (III). In the case of reactions (II) and (Ia), the increase in hole absorption would be accompanied from the very beginning by an increase of the absorption band of lead hydroxide, which is not observed in 0.01 M buffer. Hence, because the formation of lead hydroxide is preceded by the generation and accumulation of holes and possibly by crystallization of elemental sulfur, reaction (Ib) is the most probable under the selected conditions, while reaction

$$PbS + 2h^+ \Rightarrow Pb^{2+} + S^0 \tag{VII}$$

rather than reaction (II) appears to be responsible for the fist stage of bulk oxidation, and as noticed [515], its onset is close to the flotation edge. Lead hydroxide is formed by precipitation when the solubility product is reached at the interface. The rate of this reaction controlled by the mass transfer of lead cations into solution increases when the concentration of the electrolyte is increased.

Thus, IR spectra suggest the following interpretation of the collectorless flotation of galena in deaerated atmosphere at basic pH. As expected, the galena flotability between +0.15 and +0.35 V (maximum at +0.25 V [515]) can be attributed to the formation of the sulfur-rich layer and crystallization of elemental sulfur. The flotation is eliminated upon formation of lead–sulfoxy compounds (PbS_2O_3 and $PbSO_3$) and lead hydroxide in electrolytes of low and high ionic strengths, respectively. It can be predicted that the flotation edge will shift toward higher potentials as the ionic strength of the electrolyte decreases.

To summarize, the in situ data on the chemical identity of the interfacial species may be insufficient to determine the mechanism of the reaction on a semiconductor electrode. In this case, additional information on charge transfer can be obtained from analysis of free-carrier absorption.

7.5.2. Anodic Reactions at Sulfide Electrodes in Presence of Xanthate

At least four forms of xanthate adsorbed on sulfides have been identified using IR spectroscopy. These are a chemisorbed radical, a metal–xanthate complex, dixanthogen, and a monothiocarbonate (MTC). However, information on the conditions (potentials) under which each form arises and which specific form (or combination of forms) provides flotability of the sulfide is in many cases very confusing, primarily due to a lack of in situ spectroscopic data.

For example, Fig. 7.36a shows voltammograms for a galena electrode in 10^{-3} M solutions of methyl, ethyl, and butyl xanthates [465]. The anodic prewave has been assigned to the reversible underpotential chemisorption of xanthate [465, 504, 598] by the reaction

$$ROCS_2^- \Rightarrow (ROCS_2)_{ads} + e. \tag{7.2}$$

The resulting rise in current is consistent with the formation of lead xanthate according to the reactions

$$PbS + 2ROCS_2^- \Rightarrow Pb(ROCS_2)_2 + S^0 + 2e$$
$$[365, 464, 504, 515] \tag{7.3}$$

$$2PbS + 4ROCS_2^- + 3H_2O \Rightarrow 2Pb(ROCS_2)_2 + S_2O_3^{2-}$$
$$+ 6H^+ + 8e \quad [515] \tag{7.4}$$

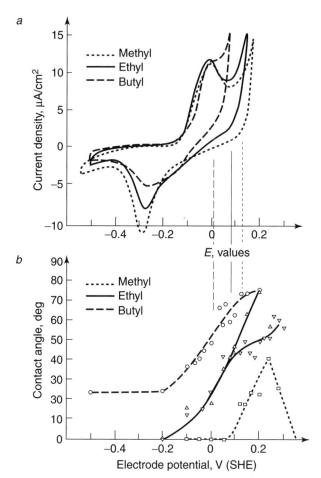

Figure 7.36. Galena electrode at 25°C in 0.05 M sodium tetraborate solution (pH 9.2) containing 1 g · L^{-1} of three potassium alkylxanthates. (a) Cyclic voltammograms at 4 mV·s^{-1}; (b) contact angles measured after holding electrode at each potential for 30 s. Vertical lines are reversible potentials of xanthate–dixanthogen couples. Reprinted, by permission, from J. R. Gardner and R. Woods, *Austr. J. Chem.* **30**. 981 (1977), p. 984, Fig. 1. Copyright © 1977 Commonwealth Scientific and Industrial Research Organization (CSIRO) Publishing.

or through a dissolution–precipitation mechanism [560, 561], in which the cations are first released by the mineral following reaction (Ib) (Section 7.5.1) and are then complexed with xanthate anions:

$$Pb^{2+} + 2ROCS_2^- \Rightarrow Pb(ROCS_2)_2. \tag{7.5}$$

In addition, electrocatalyzed synthesis of dixanthogen can contribute to the anodic current [365, 465],

$$(ROCS_2)_{ads} + ROCS_2^- \Rightarrow (ROCS_2)_2 + e \tag{7.6}$$

or

$$2ROCS_2^- \Rightarrow (ROCS_2)_2 + 2e \qquad (7.7)$$

(the reversible potentials are indicated in Fig. 7.36). At higher potentials, lead xanthate can be decomposed by the reaction

$$Pb(ROCS_2)_2 + 2H_2O \Rightarrow Pb(OH)_2 + (ROCS_2)_2 + 2H^+ + 2e, \qquad (7.8)$$

whose potential, however, is somewhat higher than the flotation edge [602]. Moreover, dixanthogen can decompose, with the concurrent formation of hydrophilic MTC and elemental sulfur:

$$(ROCS_2)_2 + 2H_2O + 4h^+ \Rightarrow (ROCSO)_2 + 4H^+ + 2S^0. \qquad (7.9)$$

Obviously, to define the reactions that occur in such a complex electrochemical system, the surface species must be identified. The first IR SEC investigation of xanthate adsorption as a function of increasing potential was performed by Leppinen et al. [513, 514] in 1988. Below, the contribution of this and subsequent work to a microscopic understanding of the anodic processes on the natural sulfides will be discussed. The spectra presented were obtained in deaerated 0.05 M sodium tetraborate buffer (pH 9.2) unless otherwise indicated. The potentials were converted in the SHE scale.

ATR on Mineral-Bed Electrodes. ATR at a mineral-bed electrode was employed to study the anodic oxidation of ethyl xanthate (EX) on chalcocite, chalcopyrite, pyrite, and galena [513, 514]. The optical scheme of the SEC cell is shown in Fig. 4.51. Prior to the addition of xanthate to the buffer, the electrode was polarized cathodically in order to remove any oxidation products that are formed during sample preparation. After a polarization period of 15 min at the selected potential, the electrode was pressed against a Ge IRE and the spectrum was measured while applying a potential less than or equal to +0.1 V. Otherwise, the spectra were recorded at open-circuit potential (OCP) just after the polarization to avoid corrosion of the Ge IRE.

A sequence of ATR spectra for a chalcocite (Cu_2S) electrode in the presence of 10^{-4} M KEX at increasing potentials is shown in Fig. 7.37a [514]. The spectra observed from -0.25 V were attributed to Cu(I)EX (Fig. 7.23). The potential dependence of the $\nu_{as}COC$ band, the flotation recovery, and the current are shown in Fig. 7.34b. After correcting for the difference in the EX concentration in the flotation test (a concentration of $1.9 \times 10^{-5} M$ implies a cathodic shift of the flotation curve by 0.043 V), the onset of the flotation is seen to coincide with the appearance of the IR signal and the first two peaks in the voltammogram. Thus the spectra suggest that the maximum flotation may be a result of the formation of multilayers of Cu(I)EX, and the current peaks can be attributed to different mechanisms of copper xanthate formation. However, a closer inspection of the spectra shown in Fig. 7.37a reveals a number of differences from the spectrum of bulk Cu(I)EX, which were ignored by the authors. At potentials below +0.05 V,

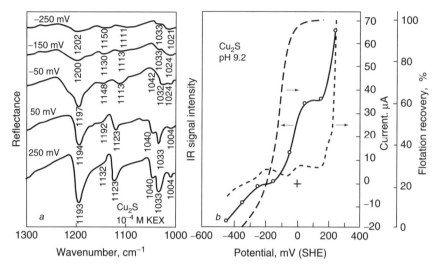

Figure 7.37. (a) FTIR spectra of potassium ethyl xanthate (KEX) adsorbed on chalcocite powder at different potentials in deaerated 0.05 M $Na_2B_4O_7$ (pH 9.2) and 10^{-4} KEX solution. Each spectrum consists of 100 scans at resolution of 4 cm^{-1}. (b) Comparison of effect of applied potential on IR signal intensity (solid line) of ethyl xanthate adsorbed on chalcocite and voltammogram for chalcocite–carbon paste electrode (short-dashed line) at pH 9.2 in presence of 10^{-4} KEX. Scan rate 1 mV · s^{-1}. Also shown are flotation results at KEX addition of 1.9×10^{-5} M (long-dashed line). Reprinted, by permission, from J. O. Leppinen, C. I. Basilio, and R. H. Yoon, *Int. J. Miner. Process.* **26**, 259 (1989), p. 263, Figs. 4 and 5. Copyright © 1989 Elsevier Science Publishers B.V.

an additional component of the ν_{as}SCS band is observed at $\sim$1024 cm^{-1} and the ν_{as}COC band at $\sim$1200 cm^{-1} is broadened. At -0.05 V, an additional shoulder at $\sim$1220 cm^{-1} is distinct in the spectrum. These features can be assigned to the chemisorption of xanthate, which is expected based on the electrochemical dependences [601].

In the case of galena (Fig. 7.38), the spectra at potentials starting from -0.25 V are practically identical to those of $Pb(EX)_2$ (Fig. 7.23), although this was over-looked by the authors [514]. As this potential is essentially below the reversible potential of $+0.053$ V for reaction (7.3), it was suggested [330] that the under-potential adsorption of xanthate takes place by ionic exchange, with residual oxidation products remaining after the electrode cleaning by cathodic polariza-tion. The spectrum intensity was shown to follow the potential dependence of the anodic current and the contact angle, although the origin of the prewave in the voltammogram and the steep increase in the contact angle (Fig. 7.33) are unclear. Note that for both chalcocite and galena, no traces of dixanthogen were found in the mineral-bed-electrode ATR spectra, in contrast to the IRRAS and ATR data (vide infra).

In situ IRRAS. ATR in Otto's geometry (Section 2.5.4) was applied to the EX–chalcocite system in 13-reflection geometry [330]. The main advantage of

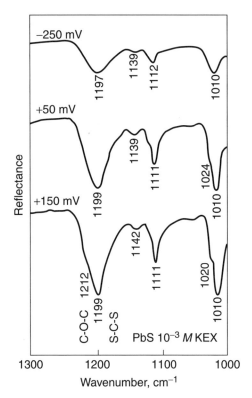

Figure 7.38. FTIR spectra of KEX adsorbed on galena at different potentials in 0.05 M Na$_2$B$_4$O$_7$ (pH 9.2) and 10^{-3} KEX solution. Reprinted, by permission, from J. O. Leppinen, C. I. Basilio, and R. H. Yoon, *Int. J. Miner. Process.* **26**, 259 (1989), p. 269, Fig. 10. Copyright © 1989 Elsevier Science Publishers B.V.

this technique for flotation studies is that the orientation of the adsorbate can be determined. The spectra (200–1000 scans per spectrum at a resolution of 4 cm^{-1}) were obtained under the optimum conditions as determined by spectral simulations. For an isotropic xanthate monolayer on chalcocite ($n = 5.10$, $k = 0.18$), these conditions included a ZnSe window, a 1-μm water interlayer, an angle of incidence of 45°, and p-polarized radiation. The average thickness of the solution layer was determined experimentally from the intensity of the water band. The SEC measurements were preceded by polishing, with no other pretreatments of the electrode. The electrode was held for 10 min at each potential.

The spectra of a chalcocite electrode polarized in 5×10^{-4} M EX are shown in Fig. 7.39b. The reference spectrum for spectrum a was taken at −0.32 V, well below the electroactive region (Fig. 7.39a). Each of the other spectra was referenced to the spectrum measured at the previous potential, which implies that the spectra only reflect (potential-induced) changes. The broad bands at 1200 and 1120 cm^{-1} at −0.02 V and the band at 1225 cm^{-1} at +0.015 V were assigned to the COC group, which is formed during decomposition of the first monolayer of

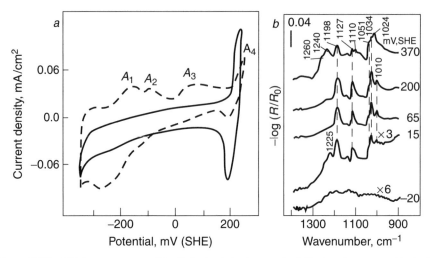

Figure 7.39. (a) Voltammograms of Cu_2S electrode in absence (solid line) and presence (dashed line) of ethyl xanthate. Concentration of xanthate solution was $5 \times 10^{-4}M$ at pH 9.2 (0.05 M borate buffer). (b) In situ reflection (thin-layer cell) spectra of Cu_2S electrode in 5×10^{-4} M xanthate solution at pH 9.2 (0.05 M borate buffer). Spectra were recorded using Bruker IFS88 FTIR spectrometer equipped with MCT detector (4 cm^{-1} resolution, 200–1000 scans). Each spectrum is referred to spectrum recorded at potential one step lower. Spectrum taken at −20 mV is referred to spectrum taken at −320 mV, which is below potential of first oxidation peak A_1. Measurements were conducted in deaerated solution. Reprinted, by permission, from J. A. Mielczarski, E. Mielczarski, J. Zachwieja, and J. M. Cases, *Langmuir* **11**, 2787 (1995), p. 2795, Figs. 6 and 7. Copyright © 1995 American Chemical Society.

xanthate. By comparison with the simulated spectra for a 1-nm isotropic Cu(I)EX film (Fig. 7.40), the other bands between 0.015 and ∼0.2 V were attributed to randomly oriented Cu(I)EX. Under these assumptions, the thickness of the layer formed at +0.015 V (Fig. 7.39*b*) was estimated to be somewhat greater than a monolayer. The typical bands of dixanthogen are pronounced at +0.37 V. The formation of both copper xanthate and dixanthogen is observed at overpotentials (relative to the thermodynamical potentials). Thus, it was concluded [330] that the FTIR investigations "do not support the previously proposed explanation that xanthate can be adsorbed on a cuprous sulfide surface at a very low potential which is more 200 mV below the potential calculated from thermodynamic data."

This viewpoint was extensively debated [601]. The counterarguments were based on the observation of a single band in the vicinity of 1200 cm^{-1} for xanthate adsorbed at low coverages on Cu [329, 481], Ag [481, 501], Ni [481], and Au [482] (Fig. 7.24), which was interpreted (convincingly) to be chemisorbed xanthate oriented with the C−O−C axis perpendicular to the surface and the TDM of the ν_{as} SCS vibration parallel to the surface. However, this explanation was rejected by Mielczarski et al. [602] based on the spectral simulations. According to the latter, an ultrathin isotropic monolayer of bulk copper xanthate on a semitransparent substrate must yield either positive or negative bands

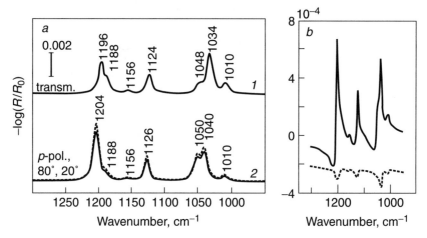

Figure 7.40. Simulated spectra of isotropic 10-Å film of cuprous ethyl xanthate complex: (*a*) (1) transmission spectrum; (2) reflection spectrum of film deposited on copper measured with *p*-polarization, at angles of incidence of 80° (solid line) and 20° (×50, dashed line). Reprinted, by permission, from J. A. Mielczarski, *J. Phys. Chem.* **97**, 2649 (1993), p. 2651, Fig. 3. Copyright © 1993 American Chemical Society (*b*) Reflection *p*- (solid line) and *s*-polarized (dashed lines) spectra of film on Cu_2S probed through ZnSe window at angles of incidence of 45° and water interlayer 1 μm thick. Adapted, by permission, from J. A. Mielczarski, E. Mielczarski, J. Zachwieja, and J. M. Cases, *Langmuir* **11**, 2787 (1995), p. 2792, Fig. 3. Copyright © 1995 American Chemical Society.

in the vicinity of 1030 cm^{-1} in the *p*-polarized reflection spectrum, and these are due to the SCS group oriented perpendicular or parallel to the surface, respectively.

The above inconsistency can be resolved by a normal-coordinate analysis on the different xanthate complexes, along with spectral simulations using aniso-tropic Fresnel formulas. Since the absorption index introduced in these formulas may be up to three times greater than that for the isotropic model (Section 3.11.3), perpendicular orientation of the COC group TDM may result in the TO−LO splitting being larger than the splitting of 1196−1204 cm^{-1} predicted for copper at an angle of incidence of 80° (Fig. 7.40*a*). Moreover, as none of the absorption bands under discussion represents a pure vibrational mode, different coordination to the surface adsorption sites may result in a redistribution of the contributions of the elementary bonds into the normal mode and a change in the force constant and TDM of the resulting mode.

In situ IRRAS with a Ge prism window was used by Laajalehto et al. [603] to study the effect of pyrite activation by copper and lead ions at pH 5, 6.5, and 9. It was found that the xanthate interaction with copper-activated pyrite resembles that of chalcopyrite, resulting in adsorption and dixanthogen formation. In similar experiments with lead-activated pyrite, only very weak absorption bands of the adsorbed collector were observed, implying that lead depresses rather than activates pyrite.

ATR on Plate-Shaped Electrodes. ATR was used to study the adsorption of *n*-butyl xanthate (BX) on thin galena and pyrite plate electrodes [478, 518]. The same approach as for the study of galena anodic oxidation (Section 7.5.1) was applied. Figure 7.41 shows the spectra obtained for the galena electrode at pH 9.2 at xanthate concentration of 8×10^{-5} *M* and deaerated conditions. The reference spectrum was measured at -0.5 V before introducing xanthate into the

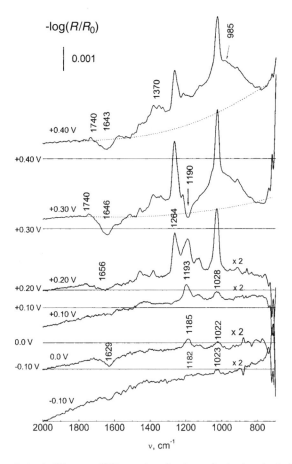

Figure 7.41. Unpolarized difference ATR spectra of galena electrode–electrolyte interface at potentials starting at -0.5 V. Electrolyte is 8×10^{-5} *M* potassium *n*-butyl xanthate solution in borate buffer (pH 9.18) at N_2 atmosphere. Spectra were obtained with Perkin-Elmer 1760X FTIR spectrometer with MCT detector. Each spectrum is average of 200 scans with 4 cm^{-1} resolution and is represented relative to spectrum measured one step before. Horizontal lines indicate zero absorbance. Additional features of spectrum baselines are upward sloping in long-wavelength part of spectra (marked with dotted lines) due to hole absorption and downward trend in short-wavelength part of spectra (>1500 cm^{-1}) at potentials from -0.1 to $+0.1$ V, attributed to recharging of surface states and defect levels. Reprinted, by permission, from I. V. Chernyshova, *J. Phys. Chem. B* **105**, 8185 (2001), p. 8187, Fig. 2. Copyright © 2001 American Chemical Society.

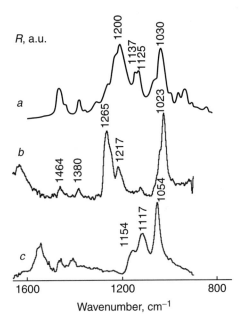

Figure 7.42. IR spectra of reference compounds: (a) DRIFTS of Pb(BX)$_2$ in KBr; (b) ATR of (BX)$_2$; (c) ATR of aqueous solution of KX. Reprinted, by permission, from I. V. Chernyshova, *J. Phys. Chem. B* **105**, 8185 (2001), p. 8187, Fig. 3. Copyright © 2001 American Chemical Society.

cell. Xanthate adsorption is seen from −0.1 V. Compared to the spectra of bulk Pb(BX)$_2$ (Fig. 7.42), the relative intensity of the ν_{as}SCS band near 1023 cm^{-1} is lower, and both the ν_{as}SCS and ν_{as}COC bands are red shifted, which can be ascribed to a somewhat different coordination of the xanthate radical to the surface than in Pb(BX)$_2$. The rate of xanthate adsorption increases, and the bands shift toward higher wavenumbers at +0.1 V. This observation can be explained by the formation of bulk lead xanthate. The bands at 1265 and 1023 cm^{-1} of dixanthogen are distinct at +0.2 V, which is close to the reversible reaction potential (7.7). Dixanthogen is still formed at potentials higher than +0.2 V, while both the galena electrode and the adsorbed xanthate film are decomposed. The decomposition is evidenced by (1) the negative band from lead xanthate at 1190 cm^{-1}, (2) the positive broad band at 1370–1400 cm^{-1} from lead hydroxide, (3) the sloping background absorption (marked by dashed lines in Fig. 7.41) attributed to holes generated by anodic decomposition of galena by reaction (Ib) (Section 7.5.1), (4) the absorption bands at 1115 and 985 cm^{-1} of lead thiosulfate and a broad band at 950 cm^{-1} of lead sulfite seen at +0.3 and +0.4 V, and (5) a weak band at 1743 cm^{-1} assigned to the carbonyl group of monothiocarbonate (ROCSO$^-$) species. For elucidating the effect of xanthate concentration, the ATR spectra were obtained in 1 × 10^{-3} *M* xanthate solution (Fig. 7.43). In this case, the xanthate adsorption is detectable at a lower potential (−0.2 V). The band at 1265 cm^{-1} from dixanthogen appears in the spectra at +0.1 V, in agreement with thermodynamic data, and thereafter the surface film composition

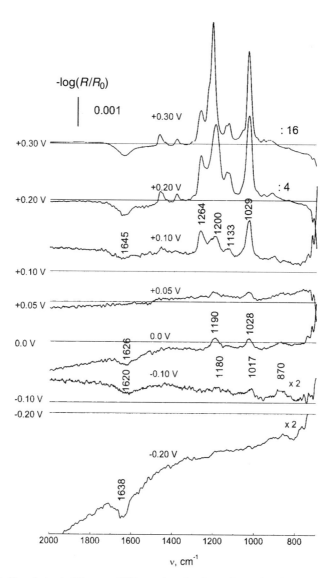

Figure 7.43. Unpolarized difference ATR spectra of galena electrode–electrolyte interface at potentials starting at −0.5 V. Electrolyte is 1×10^{-3} M potassium *n*-butyl xanthate solution in 0.05 M borate buffer (pH 9.18) at N_2 atmosphere. Other conditions as in Fig. 7.41. Reprinted, by permission, from I. V. Chernyshova, *J. Phys. Chem. B* **105**, 8185 (2001), p. 8189, Fig. 5. Copyright © 2001 American Chemical Society.

is unchanged and galena decomposition is suppressed. The electrode processes are not accompanied by the increase in free-carrier absorption that was observed in galena oxidation both in the absence of xanthate (Figs. 7.34 and 7.35) and at the low xanthate concentration (Fig. 7.41).

Thus, at both concentrations, the formation of bulk lead xanthate is preceded by chemisorption of xanthate. Dixanthogen is formed with no overpotential. However, there are principal differences in the electrode processes at low and high concentrations of the collector. At low concentrations, bulk $Pb(BX)_2$ is more likely to be formed by the precipitation mechanism (7.5), while dixanthogen forms by reaction (7.7). At higher potentials, lead xanthate transforms into lead hydroxide, against the synthesis of dixanthogen, which can be described by the overall reaction (7.8). Finally, dixanthogen decomposes into a dimer of MTC by the reaction (7.9), while galena decomposes into lead sulfite and lead thiosulfate by reactions (VI) (Section 7.5.1). At high concentrations of xanthate, the electrochemical decomposition of galena is inhibited by chemisorption (7.2), which can be interpreted within the framework of the Gerisher theory [604, 605] of the competition between redox processes and anodic decomposition reactions at the semiconductor–electrolyte interface. At the next step, lead xanthate and then lead xanthate and dixanthogen together are formed by reaction of the chemisorbed species with lead sulfide:

$$PbS + 2(ROCS_2)_{ads} \Rightarrow Pb(ROCS_2)_2 + S^0, \tag{7.10}$$

$$PbS + 4(ROCS_2)_{ads} \Rightarrow Pb(ROCS_2)_2 + (ROCS_2)_2 + S^0. \tag{7.11}$$

There are obvious parallels between the IR data obtained for the 8×10^{-5} M BX solution and the galena flotation data. According to Trahar [515], galena ground under reducing conditions starts to float at -0.1 V in 2.3×10^{-5} M EX at pH 8, reaching maximum recovery at $+0.2$ V; a rapid decrease in flotability follows at $+0.3$ V. The conclusion from this is that the maximal flotability is provided by dixanthogen, rather than chemisorbed xanthate (suggested by XPS data [365]) or lead xanthate (suggested by DRITFS data [484]). However, even in the presence of dixanthogen, flotation is suppressed by the precipitation of lead hydroxide.

In the case of pyrite, the 1260- and 1007-cm^{-1} absorption bands were observed at -0.5 V as well as the distinct band of water at 1650 cm^{-1} (Fig. 7.44). According to the DRIFTS [606] and XPS [607] data, different xanthate-related species with a common chemical formula $Fe(OH)_n X_m$ can be formed under the above-mentioned conditions. The SNR is lower in the spectra than in the case of galena, which is explained by a higher doping level of the pyrite sample. The intensity of the band near 1260 cm^{-1} corresponds to 0.5 ML at 0.0 V. At $+0.05$ V, dixanthogen is formed (the bands at 1264 and 1024 cm^{-1}), and its quantity increases as the potential is further increased. Hence, in the presence of dissolved oxygen, a monolayer of xanthate is chemisorbed on pyrite, and the high contact angle observed at potentials higher than $+0.05$ V [608] is caused by the dixanthogen formation.

Finally, there are also some trends observed in the bending band $\delta_s H_2O$ of water near 1600–1650 cm^{-1} in the in situ ATR spectra (Figs. 7.41, 7.43, and 7.44). For galena (Figs. 7.41 and 7.43), the chemisorption and precipitation of lead xanthate within the $[-0.1(-0.2)]-[+0.1(+0.05)]$-V range results in the negative band at 1630–1640 cm^{-1}, which increases steeply and shifts to 1645–1655 cm^{-1} when dixanthogen appears. Ascribing the negative intensity

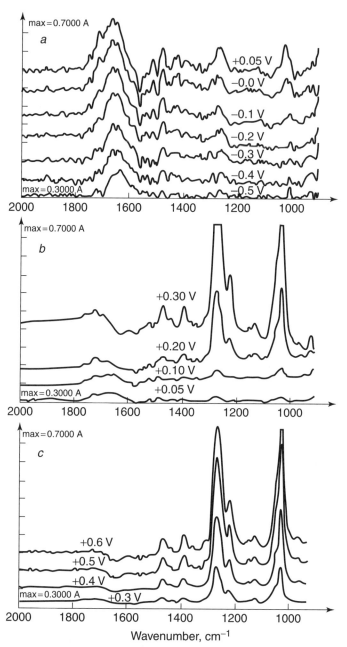

Figure 7.44. ATR spectra of surface compounds on pyrite electrode in $10^{-3}\,M$ potassium *n*-butyl xanthate solution in 0.01 *M* borate buffer (pH 9.18) at ambient atmosphere at potentials starting at -0.6 V. Reference taken at -0.6 V. Spectra obtained with Perkin-Elmer 1760X FTIR spectrometer with MCT detector. Each spectrum is average of 200 scans with 8 cm^{-1} resolution and multiplied by factor of (*a*) 300; (*b*) 50; (*c*) 10. Reprinted, by permission, from I. V. Chernyshova and V. P. Tolstoy, *Appl. Spectrosc.* **49**, 665 (1995), p. 668, Fig. 4. Copyright © 1995 Society for Applied Spectroscopy.

to the optical effect (screening of the organic film) fails to explain the sudden change of the band intensity and position (also compare with the spectra in Figs. 7.34 and 7.35). One can conclude that at least a part of these spectral changes are due to the reorganization of water molecules near the electrode surface, which is covered by hydrophobic dixanthogen. Specifically, a hydrophilic surface is covered by a water layer of a higher density than the icelike structure bordering with a hydrophobic surface (Section 3.7). Thus, increasing the hydrophobicity of a surface will give rise to a water spectrum with negative intensity. Deposition of a more hydrophobic ultrathin coating will result in the rupture of the strongest H bonds between the surface and water and the displacement of water molecules from superficial pores and defects due to a hydrophobic effect. This will cause a shift of the negative $\delta_s H_2O$ band to a higher frequency.

Several technical points may explain the discrepancies between the results obtained using different ATR techniques. First, for potentials higher than +0.1 V, the ATR spectra of mineral-bed electrodes were measured at OPC. As shown for galena [478], when the oxidizing potential is switched off, dixanthogen is instantly dispersed into solution or reduced. Another explanation [330] for the difference between the mineral-bed and plate electrode results is contamination of the solution with dixanthogen when the mineral-bed electrode was reduced prior to anodic polarization, made possible because of the proximity of the counter electrode to the working electrode (Fig. 4.51). In addition, as already mentioned, the mineral-bed electrode is difficult to polarize uniformly, so that incomplete removal of the oxidation products is a possibility.

Performance of the ATR techniques in IR SEC measurements on sulfide electrodes can now be analyzed. As seen from Table 7.4 (Section 7.3), the three techniques have comparable surface sensitivity. For absorbing sulfides (such as chalcocite), the sampling procedure, however, is easier for the ATR in Otto's geometry and on mineral-bed electrodes.

To summarize, in situ IR spectroscopy allows the adsorption of organic molecules at semiconductor electrodes to be studied with a sensitivity on the order of 0.1 ML and provides information not only on the chemical identity of the surface species and their orientation but also on the nature of the charge transfer processes at the interface and the surface hydrophobicity/hydrophilicity. Because of the instability of adsorbed xanthate, the surface composition of the electrode after decoupling from the electrochemical system differs from that existing under an applied potential.

7.6. STATIC AND DYNAMIC STUDIES OF METAL ELECTRODE–ELECTROLYTE INTERFACE: STRUCTURE OF DOUBLE LAYER

Ultrathin films at the metal–electrolyte interface can be probed in situ by a variety of methods, including surface-enhanced Raman spectroscopy (SERS) [609], X-ray absorption and diffraction methods, second-harmonic generation

(SHG) [610], total internal reflection fluorescence (TIRF) spectroscopy [611], UV/Vis spectroscopy, IR spectroscopy, ellipsometry [612], quartz crystal microbalance (QCM) [613], STM [614], and AFM (see Refs. [618–623] for review of in situ spectroscopies). From those, only the vibrational methods (SHG, SERS, and IR spectroscopy) are able to provide information on both the chemical composition and the structure of the species adsorbed. The IR SEC experiment is simpler and more accessible than SHG and SERS, making IR spectroscopy the dominant tool for studying electrochemical reactions at metallic electrodes [616, 617, 624–641].

Procedures for preparing electrode surfaces (Section 4.10), the technical aspects of measuring spectra at the metal–electrolyte interface (Section 4.6), and the problems that can arise in interpreting the resulting spectra have already been considered (Section 3.7). The contribution of IR SEC studies to an understanding of the adsorption of CO and NO and small organic molecules (methanol, ethanol, formic acid, etc.), the reduction of CO_2 on ordered noble metals, electrochemical polymerization, and the structure of the electrochemical double layer (DL) have been discussed in various recent reviews [635, 638, 641]. Below, the information that can be obtained from the IR SEC measurements is listed and two IR SEC studies of the DL structure are considered. An example in which in situ IRRAS is used to follow peptide oxidation on a Pt electrode is discussed in Section 7.8.1.

Apart from the molecular-level information on the adsorbed species, such as chemical identity and orientation, and types of adsorbate–adsorbent and adsorbate–adsorbate interactions, which can reveal the origin of the electrode process, IR spectra of nonspecifically adsorbed ions present in the DL can be used for estimating the potential of zero charge (PZC) [642]. In particular, the absorption bands (especially, ν_5) of the NO^{3-} anion in the DL will be broadened due to ion paring of nitrate anion with hydronium ions. Hence, they exhibit minimum FWHM at the PZC if the specific absorption of the ions is negligible.

In the slow-diffusion regime in the thin layer, the PDIR spectra exhibit both positive and negative bands due to species removed from the solution and generated at the surface. Provided that these species absorb at different wavenumbers, the solution-phase band can be used to quantify the extent of adsorption and therefore to establish relationships between band intensity and coverage [643–647]. Analysing the band frequency for adlayer isotopic mixtures (e.g., ^{13}CO–^{12}CO [644, 648, 649]) or mixed adsorbate (e.g., CO + NO [646, 650] and physisorbed and chemisorbed nucleic acid [651]), the character of the surface filling can be determined (islandlike or homogeneous, eutectics or solid solution). The CO adlayer spectrum can be used to determine the crystallographic orientation of the exposed surface plane (for review, see Refs. [640, 652]), the nature of the structural adsorption sites [653], and surface reconstruction [654, 655]. Time-dependent spectra (Fig. 4.55) provide information on the kinetics and dynamics of the interfacial processes [651–662], while the temporal relationship between time-dependent behavior of different interfacial species can be deduced with two-dimensional correlation analysis of the dynamic spectra [662].

The contribution of the near-surface layer of the electrolyte to the ATR and IRRAS signals (Section 3.7) offers the opportunity to study the *structure of the DL* which is one of the most important characteristics of an electrochemical system and controls the kinetics of the electrochemical processes. This is one of the original problems to which the IR SEC method has been applied [663]. Based on the potential-dependent spectra of water in the DL, the nature of the rearrangement of the water molecules and hydronium ions has been investigated (Table 3.4). IR SEC studies of the orientation of inorganic anions in the DL have been reviewed by Iwasita and Nart [630, 637]. It has been proven that electrochemically inert anions (CO_3^{2-}, SO_4^{2-}, and PO_4^{3-}), hydrogen, oxygen, and cyanide, which reversibly adsorb–desorb, undergo reorientation and a partial electron transfer at the electrode surface. Weaver and Zou [640] discussed in detail the connection between the in situ and ex situ results for adsorption at electrified interfaces. It has been shown that removal of an electrode from an electrochemical system alters significantly the binding site arrangement of adsorbed CO, which depends on the electrode potential, the potential drop within the DL, and the ion and solvent environment.

Osawa and co-workers [664] took a complex approach using both PDIR and the potential-modulation method to study the kinetics and dynamics of the charge transfer between a SAM of 4-mercaptopyridine (PySH) and an Au electrode. The experiments were performed in the SEC cell shown in Fig. 4.49 with an Au nanolayer electrode. The SEIRA optical configuration provided spectra of the first adsorbed monolayers selectively. The cell was built into the set-up sketched in Fig. 4.56. Cyclic voltammograms for a gold electrode modified with PySH in 0.1 M $HClO_4$ demonstrated no Faradaic current in the potential range between -0.1 and $+0.4$ V [versus a saturated calomel electrode (SCE)]. Nevertheless, the PDIR spectrum of the SAM changes with applied potential, as shown in Fig. 7.45. The arrows in the figure represent the directions of the spectral changes associated with the potential sweep from -0.1 to $+0.4$ V. The bipolar feature around 1470 cm^{-1} indicates that the band shifts with potential, while the bands at 1618 and 1598 cm^{-1} decrease in intensity as the potential increases, without changing their positions. The 1618- and 1470-cm^{-1} bands are assigned to totally symmetric (a_1) in-plane ring modes of the pyridine moiety. The 1598-cm^{-1} band was assigned tentatively to a b_1 mode of the protonated pyridine ring moiety. The intensity of the 1618-cm^{-1} band was found to be proportional to the applied potential. The spectral changes described were completely reversible against the potential change in the DL, indicating that the SAM is stable under potential modulation. The authors explained the different potential dependence of the modes belonging to the same (a_1) symmetry with the model of charge transfer between the surface and adsorbate [632], which predicts that the change in the potential can result in different degrees of charge transfer and consequently changes in the vibrational frequencies for vibrational modes even of the same symmetry. On the basis of the time dependence of the 1618-cm^{-1} band for a potential step from -0.1 to $+0.3$ V measured with a time resolution of 50 μm, it was concluded

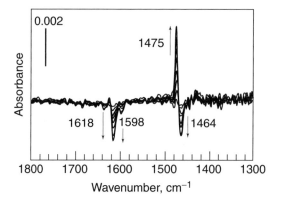

Figure 7.45. Potential-difference unpolarized ATR-SEIRA spectra of 4-mercaptopyridine (PySH) SAM on 20-nm-thick (80-nm-size particles) Au evaporated electrode in 0.1 M HClO$_4$. Reference potential was −0.1 V (SCE). Arrows show changes of peaks for positive shift of electrode potential from −0.3 to +0.4 V. Spectra were recorded using Bio-Rad FTS 60A/896 FTIR spectrometer equipped with dc-coupled MCTdetector and bandpass optical filter transmitting between 4000 and 1000 cm^{-1}. Spectrometer was operated in rapid-scanning mode and spectra were collected sequentially during potential sweep at 5 mV · s^{-1}. Sixty-four interferograms were coadded to record each spectrum, which required about 10 s. Reprinted, by permission, from K. Ataka, Y. Hara, and M. Osawa, *J. Electroanal. Chem.* **473**, 34 (1999), p. 37, Fig. 3. Copyright © 1999 Elsevier Science S.A.

that the rate of the spectral change is equivalent to or much faster than the time constant of the DL charging.

To gain additional insight into the interfacial process, the potential-modulation spectra were measured. Figure 7.46 shows typical in-phase (*IP*) and quadrature (*Q*) spectra acquired at several modulation frequencies. The modulation amplitude was set to 400 mV to achieve a SNR sufficient for kinetic analysis. The spectral features of the *IP* and *Q* spectra are identical to those of the potential-difference spectra shown in Fig. 7.43, except for the signs of the bands, which vary with the potential-modulation frequency. The *IP* (real) and *Q* (imaginary) intensities of the 1618-cm^{-1} band plotted over a complex plane were fitted by a semicircle, as is typical for ac impedance data for electron transfer-limited reactions. Analysis of this plot gave a rate of spectral change of 5.4 × 10^5 s^{-1}, which is much faster than the potential change in the double layer (1.9 × 10^3 s^{-1}), suggesting that the spectral shifts are due to charge transfer rather than to the Stark effect.

As mentioned in Section 3.7, reorientation of the electrolyte ions by changing the electric field will affect the in situ IR spectra of the electrode–electrolyte interface. Figure 7.47 shows the SNIFTIRS spectra measured in a "thin-layer" cell, with a hemispherical window at the gold surface in a dimethylacetamide (DMA) solution of [Cr(DMSO)$_6$](ClO$_4$)$_3$ [665]. The dimethylsulfoxide (DMSO) ligand has a band at 924 cm^{-1} due to the −S=O vibration. The ν_3 band of the perchlorate absorbs at 1100 cm^{-1}. Spectral features in the 1600–1650-cm^{-1} region of the spectrum were attributed to the νC=O mode of the solvent, DMA.

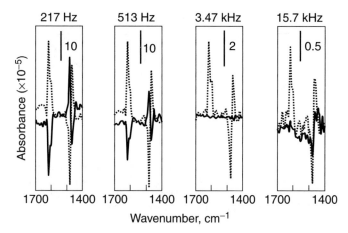

Figure 7.46. In-phase (solid) and quadrature (dashed) potential-modulated ATR-SEIRA spectra of 4-mercaptopyridine (PySH) SAM on 20-nm-thick (80-nm-size particles) Au evaporated electrode in 0.1 M HClO$_4$. Modulation frequencies are shown. Amplitude of potential modulation was 400 mV, between −0.1 and 0.3 V (SCE). Spectra were recorded using Bio-Rad FTS 60A/896 FTIR spectrometer equipped with dc-coupled MCT detector and bandpass optical filter transmitting between 4000 and 1000 cm^{-1}. Spectrometer was operated in step-scanning mode using setup shown in Fig. 4.56. Reprinted, by permission, from K. Ataka, Y. Hara, and M. Osawa, *J. Electroanal. Chem.* **473**, 34 (1999), p. 39, Fig. 6. Copyright © 1999 Elsevier Science S.A.

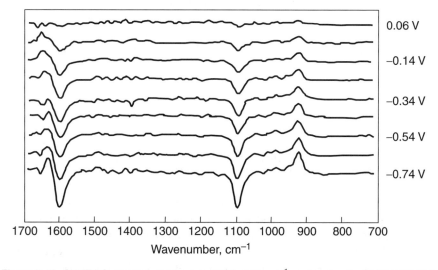

Figure 7.47. SNIFTIRS spectra obtained in 1700–700-cm^{-1} region for 2 mM [Cr(DMCO)$_6$] (ClO$_4$)$_3$ solution in dimethylacetamide at Au electrode. Reference potential was 0.16 V against ferrocene/ferrocinium reference with sample potentials at 100-mV intervals in negative direction as indicated. Reprinted, by permission, from W. R. Fawcett, A. A. Kloss, J. J. Calvente, and N. Marinkovic, *Electrochim. Acta* **44**, 881 (1998), p. 882, Fig. 1. Copyright © 1998 Elsevier Science Ltd.

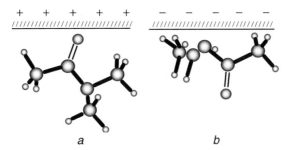

Figure 7.48. Schematic diagram indicating orientation of dimethylacetamide at positively charged metal electrode (*a*) and at negatively charged electrode (*b*). Reprinted, by permission, from W. R. Fawcett, A. A. Kloss, J. J. Calvente, and N. Marinkovic, *Electrochim. Acta* **44**, 881 (1998), p. 883, Fig. 2. Copyright © 1998 Elsevier Science Ltd.

The positive band at 924 cm^{-1} demonstrates the accumulation of Cr(DMSO)$_6$ in the DL as the electrode potential becomes more negative. At the same time, an increasingly negative band at 1100 cm^{-1} demonstrates depletion of ClO$_4^-$ in the DL. An additional feature is the bipolar band at 1637 and 1605 cm^{-1}. This was assigned to the −C=O stretching vibration in the DMA and results from reorientation of solvent dipoles. The corresponding reorientation is depicted in Fig. 7.48. At positive potentials, some fraction of the DMA molecules at the interface are oriented with the oxygen atom pointing toward the metal. In this orientation, the electrophilic metal draws electron density out of the −C=O bond, thereby weakening it. The 1605-cm^{-1} band corresponds to this orientation. At negative potentials, the solvent dipoles are oriented with the −C=O bond pointing to the solvent, and νC=O is 1637 cm^{-1}. This observation confirms spectroscopically the phenomenon predicted by thermodynamic data. Moreover, it was shown that quantitative analysis of the ClO$_4^-$ band based on the BLB law and Gouy–Chapman theory allows the ionic charge density (the surface excess of the perchlorate ion) to be estimated in the diffuse layer in the vicinity of the PZC.

7.7. THIN POLYMER FILMS, POLYMER SURFACES, AND POLYMER–SUBSTRATE INTERFACE

Thin polymer films play a central role in many important modern technologies [436], including micro- and nanotechnologies [666, 667]. IR spectroscopy has served as a routine tool for studying various chemical and physical properties of polymers for roughly half a century. The theoretical background, technical aspects of the method, and analysis of the results obtained have been covered exhaustively in numerous reviews [668–683] and monographs [684–693]. In the present section, the application of IR spectroscopy to thin polymer films, polymer surfaces, and polymer–substrate interfaces will be discussed briefly, with an emphasis on recent results that demonstrate optimal application of each IR spectroscopic method. Although in some examples the polymer films are micrometers thick and therefore do not fall into the "ultrathin film" category, they have been included here when the particular problem is of special interest.

IR spectroscopic studies of polymer films several micrometers thick have been performed in the normal-incidence transmission mode (see examples in Table 7.6 and Refs. [684, 692, 685]). By this method, the chemical and geometric structure of a single polymeric chain, the type and degree of orientation of polymeric chains and crystallites, the degree of crystallinity, and correlation of these microscopic characteristics with macroscopic physical and chemical properties of the films have all been investigated. Using two-dimensional correlation analysis (Section 3.8), microheterogeneity at the molecular level has been revealed for a series of films (see Ref. [694] for a review of early work). In particular, a significant decrease in the dynamic response of the *trans*-C—O stretching band of 5×-drawn poly(ethylene terephthalate) (PET) above the glass transition (114.5°C) was related to concurrent changes in the macroscopic mechanical properties of the PET film [695], and several secondary structures of silk fibroin were found by resolving the complex amide I band of the fibroin film [696] in the transmission experiments.

An alternative to transmission is the ATR method (Table 7.6). Its main advantage is in eliminating the interference patterns in the spectra of micrometer-thick films. The ATR method has been used to elucidate the mechanism of destruction of polyethylene [697] and natural and synthetic rubber [698, 699] under loading. It was revealed that an increase in the surface concentration of terminal groups precedes the destruction. Two polymers, poly(methyl methacrylate) (PMMA) and poly(ethylene oxide) (PEO), adsorbed onto a solid (Ge) substrate were found to be metastable [700–702]. However, stable films in which the PEO backbone is parallel to the surface can be prepared following a special protocol [703]. Numerous ATR studies have investigated changes in polymer surfaces under different physical and chemical treatments (Table 7.7). For example, it was found [704–706] that under photooxidation and corona discharge, the surface layer of unstable polypropylene becomes enriched by hydroxyl (3400 cm^{-1}) and carbonyl (1715 cm^{-1}) groups. The concentration of these functionalities decreases exponentially from the surface of the film inward. An analogous technique was used to study the oxidation of polyolefins by ozone [707]. Adsorption of organic and inorganic molecules on polymers has been thoroughly discussed by Urban [673, 686] and more recently by Chen and McCarthy [708].

Interaction between polymer substrates and between different polymers is of great importance for today's technologies, particularly in protective coatings and composites. IR spectroscopy offers a unique opportunity to study the chemical and structural composition of the interface and the adhesion forces (Table 7.8). Reviews of this topic were written by Urban [685, 689]. In most cases, the ATR method is used, as it has a higher selectivity toward the interface than the transmission and IRRAS methods. In ATR experiments of Akutin et al. [709] to study the adhesion of poly(ethylene) to a quartz IRE, it was found that H bonding is dominant between the surfaces. The ATR method was used for characterizing the protective properties of coatings in coating–substrate systems such as a water–reducible epoxy on Ge, a clear epoxy on Si, a clear epoxy on silane-treated Si, and a TiO_2 pigmented epoxy coating on an iron-coated KRS-5 under applied potential [710]. A review of recent work measuring diffusion coefficients in polymer films in real time by ATR has been written by Marand and co-workers [711], who also studied molecular interdiffusion at a poly(vinyl pyrrolidone) (PVP)–vinyl ester interface. The IR bands at 1717 and 1507 cm^{-1}, characteristic of the vinyl ester monomer, and the bands at 1669 and 1419 cm^{-1}, characteristic of the PVP, were used in a quantitative analysis. The diffusion coefficients were determined from intensity variations in these selected bands recorded as a function of time and by fitting these data to the Fickian model. A mutual diffusion coefficient was found to be on the order of 2×10^{-8} cm$^2 \cdot$ s^{-1} at 100°C, which is close to the diffusion coefficient of small molecules. This discrepancy was attributed to the limitations of the one-dimensional Fickian model.

ATR is an effective method to characterize quantitatively the depth profiles of a polymer sample. The problem of depth profiling the polymer film can be solved by taking advantage of the fact that the decay characteristics of the probing evanescent wave can be controlled by the angle of incidence, the refractive index of the IRE, and the characteristics (refractive index and thickness) of the buffer layer [712a]. The sampling depth of IRRAS also depends on the geometry of the measurements, a fact that has also found application in depth profiling [712b]. Recent developments and different depth-profiling techniques are discussed in Ref. [713].

Polymer films on metals and the transient region between the polymer and the metal can be probed by IRRAS. In particular, Tamada et al. [714], using the chamber shown in Fig. 4.42, monitored photoinduced polymerization of *N*-vinylcarbazole (NVCz) films on Ag. These polymer films can serve as hole transport layers in electroluminescent devices of flat displays. The peaks at 1641, 1455, and 752 cm^{-1} (Fig. 7.49) were assigned to the vinyl stretching vibration (termed C=C), carbazole-ring synergistic vibration (denoted by Cz–R), and C–H deformation out of the carbazole-ring plane (denoted by C–H), respectively. The kinetics of film growth and the effect of the substrate temperature on these kinetics were evaluated from the difference between the intensity of the vinyl stretching vibration in the spectrum of the polymerized film and that of the isotropic NVCz film of the same thickness. Because IRRAS of ultrathin films on metals satisfies the metal SSR, the orientation of the carbazole ring in the monolayer nearest to the metal was determined from the ratio of the intensities of the C–H and Cz–R bands [715]. It was found that the ultrathin layer in the immediate vicinity of the metal surface is anisotropic, with an average angle between the carbazole plane and the substrate surface of 60°. The ratio of the intensities of the C=C and Cz–R bands and the degree of polymerization of the film were measured as a function of the substrate temperature (Fig. 7.50), which revealed that the polymerization was optimal at 275 K.

Another approach to probe the intermediate region between a metal surface roughened by grinding and a polymer is DRIFTS [716]. To determine the

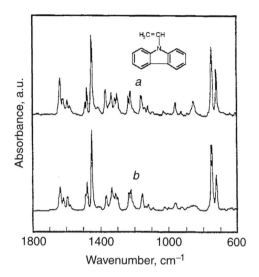

Figure 7.49. Spectra measured by IRRAS of (a) isotropic NVCz film and (b) vapor deposition polymerization (VDP) film deposited at substrate temperature of 255 K. Reflection spectrum of isotropic NVCz film was obtained by KK transformation of transmission absorption spectrum of KBr composite containing intrinsic NVCz. Reprinted, by permission, from M. Tamada, H. Koshikawa, and H. Omichi, *Thin Solid Films* **292**, 164 (1997) p. 165, Fig. 2. Copyright © 1997 Elsevier Science S.A.

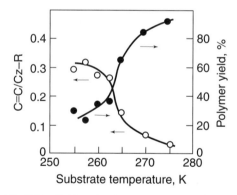

Figure 7.50. Values of C=C/Cz–R (○) and polymer yield (●) of NVCz film deposited by vapor deposition polymerization at various substrate temperatures. Reprinted, by permission, from M. Tamada, H. Koshikawa, and H. Omichi, *Thin Solid Films* **292**, 164 (1997), p. 167, Fig. 7. Copyright © 1997 Elsevier Science S.A.

compounds formed at the ethylene-vinyl acetate (EVA)–steel interface, the background spectrum of the metal surface was measured, the polymer was deposited and tiered off, and the DRIFTS spectrum was measured against the background spectrum. This spectrum (Fig. 7.51) exhibits an absorption band at 1592 cm^{-1} assigned to the carboxylate group, which indicates formation of a polymer–metal

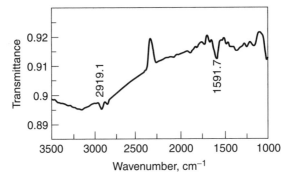

Figure 7.51. FTIR spectrum of steel surface after failure of steel–ethylene-vinyl acetate assembly. Reprinted, by permission, from S. Bistac, M. F. Vallat, and J. Schultz, *Appl. Spectrosc.* **51**, 1823 (1997), p. 1824, Fig. 2. Copyright © 1997 Society for Applied Spectroscopy.

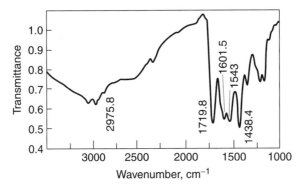

Figure 7.52. FTIR spectrum of maleic anhydride deposited on iron plate. Reprinted, by permission, from S. Bistac, M. F. Vallat, and J. Schultz, *Appl. Spectrosc.* **51**, 1823 (1997), p. 1824, Fig. 3. Copyright © 1997 Society for Applied Spectroscopy.

carboxylate complex at the EVA–steel interface. The reactive group was attributed to the anhydride groups present in the EVA. To study the mechanism of formation of this complex in more detail, IRRAS was applied to a layer of maleic anhydride on steel (Fig. 7.52), because the reaction of this compound with vinyl acetate is used in the synthesis of EVA. The band from the C=O group in the anhydride has practically disappeared from this spectrum, while a weak band has appeared at 1720 cm^{-1}, corresponding to a carboxylic acid group, which is formed by the opening of the anhydride ring. Moreover, other new peaks can be observed at 1601, 1543, and 1438 cm^{-1} that do not belong to either the metal surface or the anhydride molecule. These peaks are characteristic of carboxylate groups, formed in a chemical reaction between maleic anhydride and the iron surface.

The DRIFTS method was used for studying the surface of polymer fibers, including wood fibers [717–720] and powdered cellulose [721, 722]. A nanometer-range sensitivity was demonstrated in the detection of organic contaminants on a polymer film and on a silicon wafer surface by using Johnson's technique [723] (Section 4.1.4).

When it is necessary to characterize the heterogeneity of a polymer film or a composite film both along and across the surface, FTIR microscopy is used [724, 725] (see also Section 4.3). For example, the degradation of an acrylic polymer automotive coating that had been subjected to Florida sun for 3 years was studies with a FTIR microscope using synchrotron radiation [726].

The important information that can be provided by IR spectra is the molecular orientation in/on polymer films, which include SAMs as the specific case (Section 3.11). In the case of self-supporting anisotropic films, the linear dichroism is usually calculated from normal-incidence transmission spectra measured at two mutually perpendicular positions of the polarizer [684]. Obviously, this approach is insensitive to the modes perpendicular to the film surface. This problem is circumvented by using a combination of the normal-incidence transmission with metallic IRRAS [727], since these methods have complimentary selection rules — the modes whose TDMs are parallel or perpendicular to the surface are active in transmission or IRRAS, respectively. This technique was used to study the MO in ultrathin n-alkylacrylamide LB films [727, 728]. A strong biaxial distribution was found in these LB films in which the carbon–hydrogen chains are inclined in the dipping direction [727].

The dependence of the band intensities on molecular orientation (Section 3.11.4) and the SNR and surface selectivity advantages of PM-IRRAS (Section 4.7) allowed Gregoriou et al. [729] to study the composition and conformation of the individual components of patterned ionic assemblies 15–40 nm thick, which presented alternating bilayers of sulfonated polystyrene–poly(diallyldimethylammonium chloride) fabricated by ionic multilayer assembly on a patterned SAM on gold. Potential application of these include optical waveguides and photoresponsive systems. Evidence was found that polymeric chains on the surface in the presence of electrolyte assume a three-dimensional random coil conformation. Pezolet et al. [730] applied double (polarization and absorption) modulation to follow in situ dynamics of orientation during the deformation and relaxation processes of both stretched films of polystyrene and polyvinylmethylether blends and optically oriented copolymers containing azo-benzene side chains. It was demonstrated that this approach allows extension of the infrared linear dichroism (ILD) method toward the study of low-anisotropic polymers and rapid orientation processes. To enhance the spectral resolution, the time-dependent spectra were subjected to 2DIR spectroscopy, which provided valuable information about the relative movement of the different chemical groups of the polymers.

Many of the molecular orientation studies have been performed with ATR. In particular, orientation of PMMA LB films on Ge has been shown [731] to depend on the tacticity of the PMMA. Uniaxial orientation of the methylene and carbonyl TDMs is characteristic for atactic PMMA films, whereas the isotactic films have biaxial symmetry. In-plane orientation of polymer molecules found in thin polyimide films has been attributed [732] to cooperative effects at the film–substrate interface. The perpendicular orientation of the N–H and C=O groups in a polyamide film to the stretching direction of the film determined by ATR has been confirmed with X-ray diffraction [733].

It is impossible within the space of the present section to cover all aspects of IR spectroscopic studies of polymer films and surfaces. Samples investigated, the IR techniques used, and the main results obtained in recent studies are tabulated in Tables 7.6–7.8 in order to obtain a general idea about the wide variety of information that can be extracted with IR spectroscopy.

Table 7.6. IR spectroscopic studies of processes in polymer films

Film Material	Method[†]	Additional Information	Reference
Acrylic copolymers	Fluorescence, ATR	Structure and H bonding of dissolved water in polymers	1069
bi-Propylperylenedimide films	T, IRRAS, FT-Raman	Fabrication and optical properties	1070
Cellulose acetate membranes	ATR	Study of water	1071
Dimethyl 1,4-dihydro-2,6-dimethyl-4-(2-nitro-phenyl)-3,5-pyridinedicarboxylate	IRRAS	Evaluation of photostability	1072
Epoxy resins	ATR	Oxidative photodegradation	1073
Melamine acrylic polymer	T auto-sampling device	Moisture-enhanced photolysis under UV–50°C condition	1074
Perfluoropolymer films on Al and tetraethylorthosilicate surfaces	ATR, AFM	Anti-stiction layer deposited in micro-mirror arrays, presence of CF, CF_2, CF_3, and C–C stretching bands on Al	1075
PET	Real-time T and WAXS	Uniaxial drawing following by taut annealing	1076
PET commercially manufactured	Polarized ATR	Biaxial orientation gradients	1077
PET and PVC	ATR	Structure and dynamics of water sorbed into polymer films	1078
Poly(2-naphthol) film	ATR	Study of redox process	1079
Poly(5-amino-1-naphthol) probe	ATR	Ion exchange studies	1080
Poly(isobutylene)	Fiber-optic probe with ATR and T head	Real-time monitoring of polymerization	1081

Table 7.6. (*Continued*)

Film Material	Method[†]	Additional Information	Reference
Poly(L-glutamate) on Si wafers and quartz	T, circular dichroism, UV/Vis, ellipsometry, small-angle X-ray reflection	Ring-opening polymerization of N-carboxyanhydrides of L-glutamates, confirmation of pure α-helix conformation of grafted polypeptide layers measurements	1082
Poly(L-lysine)	2D IR of phase-resolved temperature modulation ATR spectra	Secondary structure changes	1083
Poly(methoxy ethylacrylate), PMMA, poly(2-hydroxy ethylmethacrylate), poly(vinyl methylether) (PVME)	ATR	Relationship between spectra of sorbed water and polymer structure (biocompatibility), measurements of diffusion coefficients of water vapor	1084
PMMA	In situ ATR	Plasma-polymerization, effects of reactor parameters on gas-phase and film deposition processes	1085
Poly(N-vinylcarbazole)	IRRAS	Photoinduced vapor deposition and polymerization	714, 715
Poly(phenylacetylene)	T, fiber-optic FTIR technique	Real-time monitoring of polymerization	1086
Poly(propylene) (PP)	ATR	Method for measuring diffusion of mineral oil and commercial fluorocarbon ether lubricant (Krytox)	1087
Poly(styrene-co-methacrylo-nitrile)	Flash pyrolysis and TG–FTIR	Thermal degradation mechanisms	1088
Poly(vinyl chloride) (PVC)	ATR and Raman microscopy	Investigation of diffusion and distribution of silane coupling agents	1089
Polyaniline	ATR	Base–acid transitions of different forms of polyaniline in leucomeraldine form	1090

Table 7.6. *(Continued)*

Film Material	Method†	Additional Information	Reference
Polyaniline	Electrochemical methods, in situ ATR, in situ IRRAS	Electropolymerization, structures of very first products	1091
Polyaniline free-standing films	T	Removal of incorporated H bonded solvent (*N*-methylpyrrolidone)	1092
Poly(isobutylene)	ATR	Multicomponent diffusion of methyl ethyl ketone and toluene from vapor sorption	1093
Polyisoprene film crack edge	μ-FTIR	Study of changes in molecular structure induced by crack formation, relationship between orientation and stress	1094
Polyphenylene	In situ ATR	Growth mechanism during electrooxidation of biphenyl in methylene dichloride	1095
Polypyrrole	ATR	Polymerization of pyrrole in iron-exchanged montmorillonite	1096
Polystyrene (PS) and polyethylene (PE)	IRRAS, in situ fiber-optic ATR	Plasma polymerization in plasma bulk	1097
Polysulphone ultrafiltration membranes	ATR	Characterization of clean and fouled	1098
Polyurethanes	μ-FTIR T	In-vivo degradation of polyurethanes	1099
PP	ATR	Acetone diffusion at 278–308 K, estimation of activation energy for diffusion of 98 kJ · M; technique for improving contact between film and IRE, based on controlling penetrant fluid pressure above threshold value (>230 kPa) in ATR flow cell	1100

Table 7.6. (*Continued*)

Film Material	Method[†]	Additional Information	Reference
PP/PE copolymers	TG-DTA coupled with FTIR	Degradation	1101
PS and PMMA	ATR	Water diffusion process in each type of polymer is quantified and discussed	1102
PS and PVME blends	T, AFM	Photo-oxidation	1103
PS in poly(vinylidene fluoride) and poly(vinylidene fluoride/hexafluoro-propylene)	T, ATR, SEM	Influence of ionizing radiation on film structure	1104
Silicone membranes and human stratum corneum (SC)	ATR	Morphological structure of inner and outer regions of human SC, diffusional pathlengths of 4-cyanophenol	1105a
Sulfonated poly(ether sulfone) membranes	ATR	Measurements of kinetics of water diffusion (sorption/desorption) in films with different degrees of sulfonation	1105b

[†]T — IR transmission, see Acronym listing for other abbreviations.

Table 7.7. IR spectroscopic studies of polymer surfaces[†]

Material	Method	Additional Information	Reference
Biomer(TM) surface	ATR, XPS, contact angle	Effect of toluene	1106
Contact-lens materials	ATR	Surface modification by silanization	1107
Epoxy compound	IRRAS	Polymersation on top of benzotriazole adsorbed on Cu	1108
Ethyl acrylate (EA)/ methaczylic acid (MAA) latex films	ATR	Mobility and surfactant migration	1109
Linear low-density polyethylene (LLDPE) films	μ-FTIR	Evaluating bulk-to-surface partitioning of erucamide	1110

Table 7.7. *(Continued)*

Material	Method	Additional Information	Reference
Nafion film embedded with dimethylglyoxime probe molecules	ATR	Analytical approach to detection of Ni^{2+} based on appearance of unique peak at 1572 cm^{-1} that corresponds to $\nu C=N$ in nickel dimethylglyoximate	1111a
PET uniaxially drawn	ATR	Surface structure and orientation	1111b
Poly(dimethylsiloxane) surfaces	ATR	Effect of discharge gases on microwave plasma reactions with imidazole	1112
Poly(ethylene glycol) (PEG) films	ATR	Water inside and at surface	1113
Poly(methylsilane)	T	Oxidation and transformation into poly(carbosilane) on surfaces of silicon single-crystal wafers	1114
Poly(N-vinylimidazole)	IRRAS, XPS	Oxidation on Cu	1115
Poly(octadecyl methacrylate)	ATR	Adsorption behavior of two Mytilus edulis foot proteins	1116
Poly(thienylpyrrole) thin film electrodcs	ATR	Polymerization process in acetonitrile containing different supporting salts	1117
Poly(vinyl trimethylsilane-b-dimethylsiloxane) copolymer membranes	ATR	Bulk and surface composition	1118
Poly(vinylidene fluoride) films	ATR	Structural and quantitative analysis of surface	1119
Polymer specimen surfaces with a rough relief	ATR	Recording of IR spectra using thermostatic IREs	1120
PP	ATR	Surface after plasma treatment	1121
PS	ATR	Surface migration of PS-b-PMMA block copolymer additives in PS hosts	1122
PS/poly(dimethylsiloxane)-co-polystyrene blends	ATR	Quantitative analysis of surface segregation	1123
PS/PVME blends	ATR	Study of surface enrichment	1124
PTFE	ATR, XPS	Optimization of surface modification procedures of vascular prostheses, adsorption of proteins	1125

Table 7.7. (Continued)

Material	Method	Additional Information	Reference
PVC	ATR	Microwave plasma reactions with imidazole	1126
Styrene/n-butyl acrylate latexes	Polarized ATR	Orientation of adsorbed sodium dioctylsulfosuccinate (SDOSS) surfactant	1127
Urethane acrylate coating films	ATR	Structural changes at the film–air interface upon UV curing	1128

†Polymer abbreviation as in Table 7.6, T — IR transmission, see Acromyn listing for other abbreviations.

Table 7.8. IR spectroscopic studies of interface of two media one of which is polymer†

System	Method	Additional Information	Reference
Copolymer of vinyl imidazole and vinyl trimethoxy silane films on Cu	IRRAS, SEM	Effect of copolymer composition on Cu corrosion protection at 360°C in air	1129
EA/MAA latex–substrate interface	ATR	Orientation of acid functionalities	1130
Epoxy resin–polymer laminate structures	ATR	Interfacial interactions	1131
Epoxy resins	T	Water vapor transport	1132
Glycerogelatin films	ATR	Study of diffusion of ethanol	1133
Hydrogel–hydrogel interface with PEG as adhesion promoter	SNOM	Influence of incorporated PEG on adhesion, dependence of PEG diffusion across interface on PEG molecular weight and contact time	1134
Interface between oxazoline-functionalized polymers	μ-FTIR	Interfacial reactions	1135
Latex paint	ATR	Simultaneous measurement of water diffusion, swelling, and calcium carbonate removal	1136
Latex–dioctylsulfosuccinate (SDOSS) surfactant	ATR and S^2 PAS	Quantitative analysis of SDOSS distribution at both film–air and film–substrate interfaces	1137
Long-chain diacetylene monocarboxylic acid	IRRAS	Carboxylate–counterion interactions at AW interfaces and changes in these interactions during photopolymerization	1138

Table 7.8. (*Continued*)

System	Method	Additional Information	Reference
Molten maleic anhydride copolymers–amino functionalized Si interface	ATR	Interfacial reactions	1139
PET films	ATR	Study of liquid diffusion processes in films: comparison of water with simple alcohols	1140
PMMA–Cr interface	PM-IRRAS, XPS	Detection of new species at interface and reorientation of polymer ester side chains	1141
Poly(5-amino-1,4-naphthoquinone) film	ATR	Redox process in aqueous and organic media	1142
Polyimide–metal (Au, Ag, Cu, Pd, Cr, K) interface	In situ IRRAS, NEXAFS	Vapor-phase deposited film on Pt(111) with metal overlayer from submonolayer to several monolayers thick; electron transfer from K to polyimide	1143
Poly(acrylic acid)–AlOOH–Al interface	PM-IRRAS, XPS	Conversion of carboxylic acids to monodentate carboxylate species	1144
poly(L-lactide)/poly(D-lactide) stereocomplex at AW interface	In situ PM-IRRAS, T	Orientation and molecular structure, polylactide α-helices oriented parallel to water surface	1145
Poly(L-lysine) and silica nanoparticles on a chemically modified gold surface	SPR, in situ PM-IRRAS	Characterization of films, acetone and nitromethane vapor adsorption into porous ultrathin films	1146
Poly(N-isopropyl-acrylamide)/water	Thermal μ-ATR	Quantitative study of molecular structure	1147
Poly(o-phenylenediamine) film on Pt electrode	In situ IRRAS, T	Film structure, oxidation reaction	1148
PP–silane interface	In situ ATR	Examination of water diffusion	1149
PS–poly(n-butyl acrylate) latex interfaces	ATR	Orientation and mobility of surfactants	1150
PS-polyethylene oxide (PEO) copolymers at AW interface	PM-IRRAS	Average tilt of PEO chains at surface densities intermediate between dilute regime and brush regime	1151

Table 7.8. (*Continued*)

System	Method	Additional Information	Reference
PS–PVME	ATR, rheometry	Mutual diffusion interface both below and above glass transition temperature of polystyrene	1152
Sulfonated poly(ether sulfone) membranes.	ATR	Water diffusion processes	1153

†Polymer abbreviation as in Table 7.6, T — IR transmission, see Acromyn listing for other abbreviations.

7.8. INTERFACIAL BEHAVIOR OF BIOMOLECULES AND BACTERIA

Adsorption of biomolecules and bacteria, and the function of biological membranes, is of great interest in many medical and technological areas, including molecular electronics, drug release systems, implants, chromatography, and biosensors. An understanding of biofilm–host interactions is very important in research on bacterial infections, biofouling, biodegradation, biocorrosion, bioinhibition, biomineralogy, bioleaching (hydrometallurgial dissolution of minerals), and bioflotation. IR spectroscopy has helped to gain insight into these processes [332, 333, 734–754]. This method is advantageous when information on the structure of the entire system is needed, rather than a detailed evaluation of the atomic structure, as it can provide a lower structural resolution of adsorbed biomolecules than X-ray diffraction, fluorescence spectroscopy, NMR, AFM, and STM (see Refs. [746, 750, 752] for brief reviews). The widespread use of IR spectroscopy in studies of biofilms stems from the fact that IR spectra can be obtained noninvasively, nondestructively, in real time (minimizing the effect of anisotropic fluctuations in the average signal), in situ, in vitro, in vivo, and for a wide range of biological systems.

Biomolecules are classified as weak oscillators. The optimum conditions for spectral measurements and spectral interpretation for low-absorbing ultrathin films have been discussed in Chapters 2 and 3, respectively. To date, IR spectroscopy has been used mainly in the "standard" multiple-reflection ATR geometry, usually with Ge, ZnSe, or Si IREs and at an angle of incidence of 45° (see Refs. [740, 742, 743, 745, 748, 751, 755, 756] for brief reviews); in general this provides a sufficient SNR in the amide I spectral region. Thus there remains the possibility of enhancing the surface sensitivity by adjusting the angle of incidence and/or choosing a different IRE. A strategy used in IR studies of protein affinity for different adsorption sites is the modification of the substrate surfaces by techniques such as silanization of the IRE surfaces with a variety of silanizing agents [757–759]. SAMs of silanes on Ge [760] and Si [761] were found to be stable enough to permit protein adsorption studies. By virtue of their general integrity and by proper choice of chemical functionality at the ω-terminus, alkanethiol SAMs can be prepared with prescribed surface properties such as wetting, adhesion, and binding affinity [436]. To simulate the surfaces of bones and teeth, apatite and calcium phosphate layers were deposited onto IREs [762–764], while titanium [764] and polymer [765–775] coatings were used in the ATR studies of biocompatibility. Recent improvements/modifications in the application of the ATR method to biomolecular films include the use of waveguide sensors [774, 776, 777], HATR for powders [778–780], 2DIR [781], TR S^2-FTIR [782], and μ-FTIR (Section 4.3).

The concentration of proteins in blood plasma is 60 mg $\cdot$ mL^{-1}. In order to minimize any contribution to the signal by dissolved species [743], the ATR measurements are conducted in very dilute protein solutions ($\sim$1 mg $\cdot$ mL^{-1}). However, this is not the case with coated ATR, where the thickness of the thin-film substrate controls the penetration depth (see Section 4.1.3 for more detail). For example, at a solution concentration of 60 mg $\cdot$ mL^{-1} and under the optical conditions of 45° Ge MIRE coated with a 0.4-μm-thick polymer coating, the contribution of the bulk solution was estimated to be 0.47% of the total spectra [769]. When it is necessary to follow adsorption at higher protein concentrations, special cells and measurement protocols based on internal standards are used for correcting for the bulk protein signal, which are discussed in detail by Jakobsen and Strand [743].

IRRAS has been used in ex situ measurements on metals and metal oxides for various experiments, including the determination of the optical constants of adsorbed proteins [783] and in situ for probing the air–water interfaces [332–334] and for SEC studies [752]. However, the advantages of the BML-IRRAS technique (Section 2.3.3) have not fully been exploited yet. Apart from the high sensitivity to ultrathin films on dielectrics, this technique allows investigation of biofilms at the liquid–liquid interface, provided that surface chemistry and functional groups of one of the contacting liquids can be simulated by a SAM on gold [784, 785]. The transmission method has been employed to a lesser extent and only for ex situ measurements [741, 786].

In studies of biomolecules, the in situ and in vitro data are obviously more interesting and will be considered exclusively below, with an emphasis on the type of the information that can be extracted and specific spectroscopic problems rather than on the bioscientific aspect of the work. The special requirements of sampling for the IR spectroscopic studies of biomolecular ultrathin films are discussed in detail in Refs. [332, 742–744, 748, 751] and are not repeated here.

7.8.1. Adsorption of Proteins and Model Molecules at Different Interfaces

Proteins are complex polyelectrolytes with molecular weights in the range of 10^4–10^6 au and a marginally stable structure based on highly specific sequences consisting mainly of 20 different amino acids. Fully covered protein monolayers are 2–10 nm thick [787–789].

Adsorption of Amino Acids and Peptides at Different Interfaces. To understand the elementary interactions between active sites and functional groups on the substrate and amino acid side-chain groups at the outermost surface of the protein, the adsorption of well-defined model molecules such as small peptides and amino acids on metallic and modified metallic surfaces has been studied [783]. Ihs et al. [790, 791] investigated the adsorption of glycine, L-alanine, and β-alanine on Cu and Au. A very good agreement between the experimental and simulated spectra from IRRAS indicated that glycine coordinates to dissolved copper ions and forms a layer on the Cu surface approximately 1 nm thick of anhydrous $Cu(II)(Gly)_2$. The structure of the adsorbed film depends on the Cu surface microstructure, whereby the cis form of the complex is more pronounced on $Cu(111)$ and the trans form dominates on "polycrystalline"-like Cu [790]. Trifunctional L-cysteine and 3-mercaptopropionic acid were shown to form monomolecular films on evaporated Au surfaces, bonding to Au atoms through the SH group [791]. In addition, the adsorbed 3-mercaptopropionic acid molecules were H bonded to nearest neighbors, forming lateral dimers. On the Cu surfaces, the molecules were believed to coordinate to both the surface and dissolved Cu ions. The growth and structure of the films were strongly dependent on the concentration of the molecules in solution and also on the pH of the solution.

Recognition mechanisms of a foreign surface by a living system are investigated extensively, being one of the determining features of biocompatibility. The adsorption of lysine and lysine peptides on powdered TiO_2 and $Cr(III)$ oxide–hydroxide, which model the surface of titanium and stainless steel implants, respectively, were studied in situ by HATR [779, 780]. The pH dependence of lysine adsorption was investigated with a technique called surface titration by internal reflectance spectroscopy (STIRS) [792]. It was shown that this amino acid is adsorbed electrostatically on TiO_2 at pH 5–7. Increasing crystallinity of the TiO_2 film did not affect the lysine adsorption. The maximum adsorption was found near the lysine isoelectric point of 9.8. If the number of lysine units was increased to 3–5, the carboxylate group became involved in the peptide–TiO_2 interaction. In contrast, lysine peptides and polylysine (PL) did not adsorb to the hydrophobic ZnSe surface or positively charged $Cr(III)$ oxide–hydroxide, which suggests that lysine is adsorbed mainly through electrostatic interactions. ATR studies of PL and polyglutamic (PG) acid adsorption [793] showed that for a neutral hydrophobic Si surface, the plateaus for PL and PG adsorption decrease as the degree of ionization increases, because of electrostatic repulsion between segments. On a hydrophilic Si surface, negatively charged SiOH groups increase the adsorption of PL by attraction and decrease the adsorption of PG by repulsion. Decreasing the surface charge by decreasing the pH reduces these effects. Screening by the electrolyte KBr has no effect at the neutral surface, but at the charged surface the adsorbed amounts of PL and PG increase distinctly at electrolyte concentrations higher than 0.4 *M*, again demonstrating the importance of electrostatic interactions. The molecular recognition capabilities of a novel nucleolipid amphiphile, octadecanoyl ester of

1-(2-carboxyethyl) adenine, to the complementary nucleobases, thymidine and uridine, in the LB film matrix were investigated using transmission, polarized transmission, and ATR spectroscopy [794]. The results showed that the molecular recognition between the adenine moiety in the head group of the nucleolipid amphiphile and thymidine or uridine in the subphase takes place through multiple H bonding. The thymidine-containing LB films of the nucleolipid amphiphile were found to be biaxial, while the uridine-containing LB films or the LB films that do not contain the complementary nucleobases are uniaxial, which was attributed to the steric effect caused by the methyl of the thymidine ring. The order–disorder transition processes were found to be similar in the three kinds of LB films mentioned above.

Redox processes of proteins at electrodes have been studied with the view that the electrode–electrolyte interface can be regarded as a model for biological interfaces [795]. SNIFTIRS measurements [796] indicated that L-phenylalanine is weakly chemisorbed on the Au(111) electrode at negative potentials, changes orientation at potentials close to the PZC, and is oxidized at positive potentials. SPAIRS studies of the anodic oxidation of glycine, serine, and alanine on Pt(111) [797, 798] revealed the significant role played by R-groups bonded to the α-carbon atom in the adsorption and oxidation of these amino acids. All of these molecules decompose with an accompanying irreversible adsorption of cyanide which blocks the surface active sites. In the case of glycine, glycinate anions were reversibly adsorbed as well, coordinated to the Pt(111) surface through the carboxylate group in a bridge configuration. Reversibly adsorbed serine and alanine were not detected. Instead, linear and multibonded CO and dissolved CO_2 were the oxidation products. However, it was recently shown for glycine (G) [799] that the electrochemical behavior of a single amino acid differs from that of the peptide. Figure 7.53 shows the current–potential curves and the spectra from IRRAS of the surface species appearing on a Pt electrode during an anodic scan in the presence of diglycine (G_2) at pH 13. The upward peaks at 1590, 1400, and 1320 cm^{-1} in Fig. 7.53c were attributed to the δNH, $\nu_{as}COO^-$, and νCN modes of G_2, respectively. The peak direction indicates a decrease in the G_2 concentration in the thin layer at potentials from +0.6 V to higher potentials. On the other hand, downward peaks are seen at 2343, 2850, 1744, 1448, and 1238 cm^{-1}, which are due to substances formed during the anodic oxidation of G_2. The peak at 2343 cm^{-1} is due to CO_2, and the absorption at 2850 cm^{-1} is a combination band from amino acid. Other peaks were all attributed to the COOH group of the acid form of glycine: 1744 cm^{-1}, νC=O; 1448 cm^{-1}, δ(OH in plane); 1238 cm^{-1}, νC–O. The band at 1400 cm^{-1} is negative between +0.3 and +0.6 V (Fig. 7.53b) but positive at potentials higher than +1.0 V (Fig. 7.53c), which indicates that the terminal COO$^-$ group is first accumulated at a lower potential and then subjected to oxidative decomposition. The loss of G_2 leads to the generation of glycine (1744 cm^{-1}) and CO_2 (2343 cm^{-1}). The 1744-cm^{-1} peak increases with increasing potential, which is different from the result of the anodic oxidation of glycine itself, when the decomposition was observed at potentials higher than +1.1 V [800]. This observation supports the

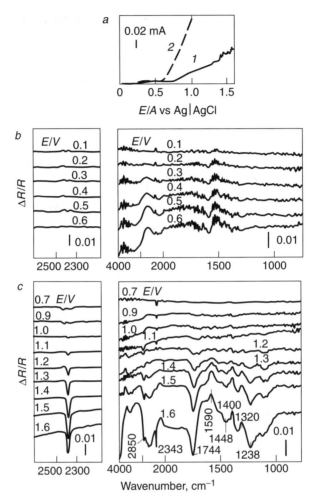

Figure 7.53. (a) Current–potential curves of Pt electrode in 0.1 *M* NaOH solution pH 13 with (1) and without (2) 50 m*M* diglycine (G$_2$) at scan rate of 0.6 mV · s^{-1}. (b, c) In situ (thin-layer configuration) spectra taken by IRRAS under same conditions as those noted in (a) were normalized to those obtained at 0 and +0.6 V, respectively. Potentials are against Ag–AgCl–saturated KCl electrode. Spectra were recorded with Shimadzu FTIR 8100M spectrometer equipped with MCT detector. Total of 50 interferometric scans were accumulated with electrode polarized at given potential. IR window was disk of CaF$_2$ and the angle of incidence was 70°. Reprinted, by permission, from K. Ogura, M. Nakayama, K. Nakaoka, and Y. Nishihata, *J. Electroanal. Chem.* **482**, 32 (2000), p. 33, Fig. 1. Copyright © 2000 Elsevier Science S.A.

assignment of the 1744-cm^{-1} peak to the acid form of glycine, whose formation is favored in the thin layer at high potentials due to simultaneous oxidation of water to O$_2$. Thus, the spectra from IRRAS show that the anodic oxidation of G$_2$ leads to formation of the acid form of glycine and the oxidative decomposition of the terminal COO$^-$ group.

Protein Structure. In the IR spectrum, the absorption bands due to the protein backbone, consisting of amide groups R−CON−R and side-chain modes, are distinguishable. The amide A band at ~3300 cm^{-1} is due to N−H stretching. The absorption bands in the 1600–1700, 1510–1580-, and 1200–1350-cm^{-1} regions are labeled amide I, II, and III (marked with a prime in the case of deuteration), respectively. Normal coordinate analysis [801, 802] has revealed that amide I is primarily (76%) the v_sC=O mode with some contribution from CN (14%) and CCN (10%) deformation. In contrast, amide II and amide III are heavily mixed modes. The amide II is an out-of-phase combination of δ^{ip}NH (43%) and v_sCN (29%) with minor contributions from δ^{ip}C=O, v_sC−C, and v_sN−C. The amide III is an in-phase combination of δ^{ip}N−H (55%) with some contributions from v_sC−C (19%), v_sC−N (15%), and δ^{ip}C=O. The amino acid side-chain absorption (due to groups such as C−H, COOH, C=O, −NH$_2$, and −NH$_3{}^+$) overlaps in many cases with amide I and II absorption.

The key feature that allows IR spectroscopy to be used to study proteins is the dependence of the amide band on the protein secondary structure (α-helix, parallel and antiparallel β-sheets, β-turns, and random). The frequency–structure correlations have been most reliably established for the amide I band (Table 7.9), although a number of exceptions to these correlations have already been reported [748, 803], and the assignment for parallel and antiparallel β-sheets is still debatable [751, 804]. Similar data for amide II bands are less well understood and hence less useful [803]. Being of lower intensity but free from interference with water (see below), the amide III band is particularly attractive for structural studies [805–812]. A number of comprehensive recent reviews contain more detailed information on the amide band assignment [748, 749, 802, 803, 811, 813–817].

The main problem in the IR spectroscopic studies of protein adsorption is that the amide bands of a protein in an aqueous environment are intrinsically broad and

Table 7.9. IR frequencies of common protein secondary structures

Structure	Amide I		Amide II	
	Frequency Range (cm^{-1})	References	Frequency Range (cm^{-1})	References
α-Helix	1654–1660	811	1290–1335	811
	1648–1660	748		
β-Sheet	1690–1698	811	1215–1250	811
	1624–1642	811		
Parallel	1642	804		
Antiparallel	1675–1695	748		
	1630–1621	804,1160		
Aggregated strands	1610–1628	804		
Turns	1660–1685	748		
Unordered	1645–1653	804	1250–1290	804
	1652–1660	748		

when several conformations coexist, the amide bands contain many overlapping contributions. To distinguish between these (to enhance the band resolution), mathematical, physicochemical, and dynamic approaches have been developed.

The *mathematical approach* involves deconvolution of the different component bands [818] and factor analysis [819–821]. The problems associated with the practical applications of these procedures are discussed in Refs. [751, 822]. The specific problem with factor analysis for the adsorbed molecules is that the calculations require the spectra of different structural conformations on the surface, which is a very difficult requirement to meet. The *physicochemical method* involves either site-directed isotope labeling and/or mutagenesis [823] or analyzing the spectral response to a change in the protein environment or state [824]. For example, if water is substituted by deuterium oxide (to acquire the spectrum, special flow cells have been designed and now are commercially available), bound water is immediately exchanged and a δD_2O band appears near 1200 cm^{-1}, while the amide band shifts downward about 10 cm^{-1}. Due to different accessibilities, the amide protons in different conformations will be exchanged to different extents. Specifically, unordered structures will undergo H/D exchange at much higher rates than regular structures, which is often used to distinguish between α-helical and random structures. However, interpretation of the band perturbations may not be straightforward if the H/D exchange is not complete [825]. In this case, one can apply a more complex, *dynamic approach* (Section 3.8). This approach can be regarded as a logical development of the physicochemical approach, when the spectra of a protein film measured after a perturbation (e.g., H/D exchange [826], change in surface pressure [827], pressure [828], and temperature [829], triggering a reaction [830, 831]) are mathematically treated. The amide groups in the different environments will have different phase delays, which can be distinguished either by electronic processing of the absorption–modulation signal (see Section 4.9.2 for more detail) or by performing two-dimensional correlation analysis of dynamic spectra. The latter approach to resolving the secondary structure of a myoglobulin film is considered in Section 3.8.

It was shown [832] that linear dichroism measurements following H/D exchange allow the tertiary protein structure to be characterized. In addition, the disappearance upon H/D exchange of the residual amide II band at 1550 cm^{-1} in the spectrum of adsorbed protein, which occurs when the protons in the protein core become accessible, can be used as an indication of the protein unfolding [833].

Interference of Water. A problem related to the above-mentioned problem of amide I band resolution is that of δH_2O absorption of water is in the amide I spectral range. This problem is common in studies of both bulk and adsorbed proteins. Typically [742, 751], if the protein is not strongly charged, the water spectrum measured under the same conditions but in the absence of the protein may simply be subtracted. Several empirical techniques have been used to select the scaling factor for this subtraction. Assuming that the water $\delta + \rho$ band at 2125 cm^{-1} (Fig. 1.4) is not affected by the presence of protein, the scaling factor can be chosen as that required to remove this band from the resulting spectrum [834]. Alternatively, the scaling factor can be obtained by fitting the 1790–1990-cm^{-1} region [835] or the water spectrum may be subtracted (added) so as to make the region from 1740 to 1990 cm^{-1} flat [751]. However, these procedures fail to exclude the absorption of hydration (bound) water, which is characterized by extinction coefficients and wavenumbers different from those of bulk water (see Sections 3.7.2 and 7.5.2 and Ref. [742]). Although it has been reported that the relative intensities of the amide I components are not affected by over- or undersubtraction of the δH_2O band [836], the accuracy of this claim is somewhat in doubt [742, 837].

The interference of water absorption is eliminated when the physicochemical and dynamic approaches are used for measuring the spectra, since a water response to a perturbation differs from that of a protein (vide supra). As an alternative, the spectrum can be rendered practically free from background absorption with a polarization modulation technique. To illustrate, Fig. 7.54 shows the spectrum from PM-IRRAS of an acetylcholinesterase (AChE) (enzyme) monolayer at the air–water interface at different surface pressures [838]. The spectra were

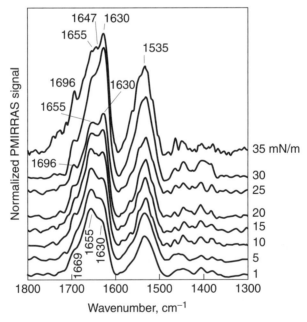

Figure 7.54. Normalized PM-IRRAS spectra of acetylcholinesterase (AChE) monolayer at AW interface at different surface pressures. AChE solution was spread at zero surface pressure. PM-IRRAS spectra recorded on Nicolet 740 spectrometer equipped with MCT detector using setup shown in Fig. 4.52. Modulation frequency was set up at 1666 cm^{-1} and 400 scans were collected for each spectrum. Angle of incidence was 75°. Reprinted, by permission, from L. Dziri, B. Desbat, and R. M. Leblanc, *J. Am. Chem. Soc.* **121**, 9618 (1999), p. 9620, Fig. 1. Copyright © 1999 American Chemical Society.

measured with the PM setup shown in Fig. 4.52. Examination of the amide I band revealed three overlapping components at 1655, 1630, and 1696 cm^{-1}. The amide II band was centered at 1535 cm^{-1}. The δCH absorption by side chains was at 1450 cm^{-1}. The 1655-cm^{-1} component was assigned to a α-helical conformation, whereas the shoulders at 1630 and 1696 cm^{-1} are due to parallel and antiparallel β-sheets, respectively (Table 7.9). This result agrees with the AChE native structure in solution. With increasing surface pressure, the relative intensities of the β-sheet bands increased and a new band at 1647 cm^{-1} associated with a second class of α-helices appeared. The redistribution of the amide band components was interpreted in terms of the protein orientation. It was concluded that the α-helices lie parallel to the water surface at low surface pressures. As the monolayer is compressed, the tilt axis of the helix became perpendicular to the surface. The same trend was observed for β-sheets. The percentages of the different conformations as a function of the surface pressure were determined from the integrated intensities of each conformation, expressed as a fraction of the total amide I band area. However, this does not allow real surface concentrations to be determined. The same approach was used to study enzyme–substrate and enzyme–lipid interactions.

The protein hydration technique has also been suggested to distinguish water absorption [742]. It requires a set of spectra for different degrees of protein hydration from the gas phase (the degree of hydration is followed by differential thermogravimetry). However, it is not clear whether these data can apply to a liquid environment.

Quantitative Analysis. Quantitative analysis of protein adsorption can be performed by radiolabeling (I^{125}) and a fluorescence technique. However, it has been revealed [839, 840] that the adsorption behavior of labeled proteins onto a surface differs from that of unlabeled ones, which may lead to misinterpretation. The possibilities of ATR in providing qualitative data have been discussed by Chittur [751] and Chittur and co-workers [841–843] and briefly reviewed by Jakobsen and Strand [743]. To obtain kinetic information on the protein adsorption, the process is followed by plotting the area under the amide I or II band as a function of time [843]. ATR measurements of protein adsorption density make use of the Sperline equation (1.114). The literature has many examples of its practical applications, including IgG adsorption on coated silicon substrates [844] and albumin adsorption on polyurethane [769, 772, 773]. In particular, a quantitative ATR study of adsorption of albumin on polyurethane in a flow cell [769] demonstrated that the adsorption density changes linearly with albumin bulk concentration. At the physiological concentration of 45 mg $\cdot$ mL^{-1}, the adsorption density was calculated to be 3.9 μg $\cdot$ cm^{-2}, which corresponds to an 80-nm film. These data contradict the Langmuir-type adsorption isotherm obtained at a low concentrations ($\sim$0.6 mg $\cdot$ mL^{-1}) by the radiolabeling technique [845]. The difference can be ascribed to the washing step in the radiolabeling measurements, which should remove any weakly bound protein. The adsorption of hemoglobin to a polystyrene thin fim was analyzed quantitatively by integrated optical waveguide ATR (IOW-ATR) spectrometry [774]. Protein adsorption densities were determined in the presence of bulk dissolved protein by measuring evanescent attenuation of propagating modes that were prism coupled into and out of the polymer film. Fitting of the Langmuir adsorption model into the isotherm data indicated that hemoglobin binds with high affinity to polystyrene and forms a complete monolayer at bulk concentrations greater than 3 μM. However, since IR spectroscopy cannot resolve IR spectra of different adsorbed proteins at the surface, the only reliable technique for quantification of the competitive adsorption of proteins onto the solid–liquid interface is the measurement of protein depletion by HPLC [846].

Molecular Orientation. The molecular orientation (MO) and the secondary structure are key characteristics of adsorbed proteins. IR spectroscopy is able to quantify both of these characteristics, although this is a rather complicated problem because of the complexity of protein structures. For a known protein secondary structure, the MO of the protein main axis with respect to the substrate can be determined most accurately by the method described in Ref. [789]. In the simplest case of an α-helix and uniaxial distribution of the amide I TDMs the anisotropic absorption indices of the protein film are expressed by Eq. (3.49), where the angle between the TDM of the amide I band and the α-helix long axis is taken to be 39° (the most commonly cited value). After measuring the absorption indices, the tilt angle of the α-helix long axis can be obtained. For bacteriorhodopsin (bR) monolayers deposited on a Ge IRE and on water (Fig. 3.85), the tilt angles were found [789a] to be 26.5° and 36°, respectively, which is in agreement with the data obtained from other methods. The problem is more complex in the case of biaxial distributions and β-sheets, which have biaxial symmetry. The problem was solved [789b, 789c] for the LB films comprised of pure α-helices and antiparallel β-sheets of poly-γ-benzyl-L-glutamate (PBG) and a model synthetic peptide K(LK)$_7$, respectively, in which the secondary structures are oriented parallel to the substrate. The anisotropic optical constants of these films on gold and CaF$_2$ were measured by metallic IRRAS and polarized transmission, respectively, and recalculated in the molecular coordinate systems of the α-helix and the β-sheet. From the values of the integrated anisotropic extinction coefficients, such useful characteristics for analytical applications as the oscillator strengths and the relative integrated intensities of the amide I and amide II modes were obtained for the two secondary structures. Moreover, it was possible to estimate for PBG the angle between the TDM of the amide I mode and the helix axis as 34°–38° [789b]. A more accurate value of 38° $\pm$ 1° was obtained for the PBG LB films transferred on a Ge IRE by fitting either the dichroic ratio of the amide I and amide II bands or the ratio of the intensities of these bands measured by the ATR method. In the special case when the sheet folds into a closed β-barrel, the uniaxial symmetry is restored, which simplifies the determination of the orientation [847].

To quantify the orientation of a pulmonary surfactant-specific protein SP-C incorporated into simple and mixed-lipid monolayers at the AW interface, Gericke et al. [848] applied the spectrum fitting procedure using the optical constants expressed by Eq. (3.56). This protein is a promoter of the spreading of surface-active phospholipids at the air–alveolar interface, which is necessary for normal breathing. It was found that the SP-C helix tilt angle changes from 24° in a 1,2-dipalmitoyl-phosphatidylchlorine (DPPC) bilayer to 70° in the mixed monolayers, whereas the chain tilt angle of DPPC decreased from 26° in pure lipid monolayers (comparable to bilayers) to 10° in the mixed monolayer films.

A less accurate (Section 3.11.3) but more widely used approach involves fitting the DR [Eq. (3.58)] or another combination of s- and p-polarized ATR or incline-angle-of-incidence transmission spectra with the formulas operating in terms of MSEF [745, 748, 751, 849]. Such formulas have been derived for a single- as well as multicomponent amide I band of α-helices (see Ref. [849] and references therein) and for the amide II bands of β-sheets [850]. In one example, ATR and transmission measurements allowed the angle of orientation of peptide nanotubes in ordered phospholipid multilayers to be determined and was found to depend on the substratum [851]. However, as discussed in Section 3.11.3, the "MSEF" formulas incorporate a systematic error, which increases with increasing tilt angle, in neglecting the absorption of the film, although this should not affect the conclusions concerning general regularities of conformational changes.

Selected Examples of In Situ Studies of Protein Adsorption. Reviews of IR spectroscopic studies on protein adsorption can be found in Refs. [751, 852]. In particular, ATR studies of the denaturation of ribonuclease A (a "hard" protein) on a hydrophilic Ge IRE surface were performed by Bentaleb and co-workers [853]. A decrease in the relative content of β-sheets and an increase in turns and unordered structures were found in the adsorbed protein. This allowed the decrease in the protein exchange rate observed by the radiolabeling technique to be correlated with the decrease in the proportion of ordered protein. A similar trend for milk-α-lactalbumin on a Ge surface was also observed [854]. The findings confirmed that the main driving force for protein adsorption is entropy gain, due to increased rotational freedom in the protein molecule. The entropy gain was shown to induce the structural rearrangement (denaturation) not only of such "soft" proteins as fibrinogen and albumin adsorbed on cellulose [765, 766] but also of such a hard protein as lysozyme adsorbed on hydrophilic silica [855].

FTIR spectroscopy has proven to be particularly useful in gaining an understanding of the *biocompatibility* phenomenon. It is believed [746, 841, 856, 857] that protein adsorption is the initial step in the interaction of blood with implanted biomaterials, followed by adhesion of cells and subsequent tissue attachment. This implies that the substrate surface characteristics influence the process, which was confirmed by ATR studies of albumin adsorption on calcium phosphate bioceramics and titanium [763] and segmented polyurethane [764], albumin and fibrinogen on acetylated and unmodified cellulose [765, 766], poly(acrylic acid)–mucin bioadhesion [767], polyurethane–blood contact surfaces [768], and other proteins on poly(ester)urethane [769], polystyrene [767, 771] and poly(octadecyl methacrylate) [771] and by IRRAS study of adsorption of proteins on Cu [858]. Another branch of IR spectroscopic studies of protein adsorption relates to *microbial adhesion* (Section 7.8.3).

7.8.2. Membranes

A cell membrane consists mainly of a lipid matrix, which serves as a cell permeability barrier, and a skeleton containing proteins and other components. The
lipid content depends on the type of the cell, with the two main classes of lipids
being phospholipids and glycolipids. Mammalian membranes can also contain
another important lipid, cholesterol. The most typical membrane structure is the
laminar one, in which phospholipids are arranged in bilayers with two hydrocarbon arrays in the internal hydrophobic region and polar groups exposed to the
external aqueous interfaces. Peripheral proteins are attached to these latter groups,
and intrinsic proteins are embedded in the hydrophobic region. The proteins are
responsible for most of the activity of biological membranes. Their conformational changes and structural reorganization is believed to depend strongly on the
interaction with the lipid bilayer [747, 748, 859].

Since the first attempt of Chapman and co-workers in 1966 [860], IR spectroscopy has become one of the most frequently used tools for elucidating lipid
properties and the mutual effects of different lipids and proteins, which are of
interest for different aspects of bioscience and biosensor design (see Refs. [333,
748, 861–864] for review). The IR methods used are transmission, ATR (MIR),
and IRRAS for model monolayer, bilayer, and multibilayer membranes and biological membranes. To perform in situ measurements on the membranes of intact
individual cells (e.g., as a function of cell membrane potential), planar miniature
waveguides can be used instead of the ATR optics [865]. PM-IRRAS has been
applied to obtain high-performance spectra of model membranes at the AW interface [866–875]. The experimental data focus mainly on the correlation between
the structure of the matrix amphiphile or phospholipid film and the structure of
the constituent species, the subphase composition, the surface pressure, and other
external conditions, as well as the interaction of such monolayers with peptides
and proteins (for reviews, see Refs. [332–334, 876, 877]).

The IR spectra of lipids can be divided into the bands that originate from the
hydrophobic acyl chains, those from the hydrophilic head groups, and those from
the interfacial region (Fig. 7.55). The general approach in IR studies of molecular assembly in lipids involves the analysis of all of these bands as a function
of various parameters, including external perturbations such as temperature, or
surface pressure, the lipid molecular structure and composition, and the type of
extraneous molecule (drag, peptide or protein). Measurements of the acyl-chain
and protein orientation are also informative. As shown in Section 3.11.1, the frequencies and widths of the stretching CH_2 or CD_2 bands of acyl chains are very
sensitive to the lipid state. This feature has shown that phospholipid monolayers
at the AW and solid–water interface can exist in many physical states, depending
on their molecular composition and structure, temperature, and environment [332,
333, 861, 878–880]. Target deuteration of the acyl chains of phospholipids has
been used to determine the depth dependence of conformational (trans–gauche)
states and microphase separation in phospholipid acyl chains [333, 881]. The
head-group bands were also found to be dependent on the phase of the lipid.
The phase transition shifts the $\nu C=O$ band from 1738 cm^{-1} (characteristic of

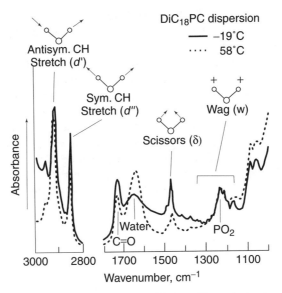

Figure 7.55. IR bands used to monitor phase behavior of long-chain phospholipids. These bands, which are marked on spectrum, represent methylene antisymmetric ($d-$) and symmetric ($d+$) C−H stretching, scissoring (δ), and wagging (w) modes. Spectra shown are of fully hydrated diC18PC the gel phase at $-19°$ (solid curves) and in liquid–crystalline phase at $58°$ (dotted curves). Types of methylene–hydrogen motion associated with bands of interest are depicted. Reprinted, by permission, from R. G. Snyder, G. L. Liang, H. L. Strauss, and R. Mendelsohn, *Biophys. J.* **71**, 3186 (1996), p. 3188, Fig. 1. Copyright © 1996 Biophysical Society.

the gel phase) to 1733 cm^{-1} (characteristic of the liquid crystalline phase) [882]. This shift was interpreted as a result of the intensity increase of the component at 1727 cm^{-1} upon a change in hydration during the transition. The head group bands have also been used to study the H bonding and interactions with organic and inorganic cations, including divalent cations, local anaesthetics, and basic peptides [748]. For example, upon hydration the $\nu_{as}PO_2$ frequency of anhydrous phosphatidylcholines decreases from ~1260 to 1220–1240 cm^{-1} [883]. Dehydration due to the incorporation of diacylglycerol and fatty acid into hydrated DPPC bilayers caused an upward shift of the νC=O band [884].

Changes in the protein structure upon binding to membranes or under membrane-mimicking conditions can be investigated by analysis of the amide bands [813, 885, 886]. This technique was used by Wu et al. [887] for in situ studies of annexin (AxV) in the Ca^{2+}–phospholipid–protein ternary complex monolayer at the AW interface. This protein, which is a member of a family of proteins that exhibit functionally relevant Ca^{2+}-dependent binding to anionic phospholipid membranes, was found to be significantly protected from thermal denaturation relative to AxV alone, Ca^{2+}–AxV, or lipid–AxV. The IRRAS data suggest that the secondary structure of AxV is strongly affected by the Ca^{2+}–membrane component of the ternary complex, whereas lipid conformational order is unchanged by the protein.

Before penetrating a cell, all external species (drug, virus, etc.) encounter the cell membrane. The effect of various drugs on phospholipid model membranes was investigated by IR spectroscopy. It was found from the example of loperamide (an amphiphilic drug from the opiate family) that the interaction of a drug with a phospholipid bilayer does not affect the wavenumber of the νCH_2 modes at temperatures far from the transition temperature, but this changes in the vicinity of the transition temperature, when the transition point and the transition width will be altered [888]. Adsorption of adriamycin, an antitumor chemical, on cardiolipin, a phospholipid specific to the inner mitochondrial membrane, was found [889] to affect both the structure of the drug and that of the lipid. The ATR DR measurements [890] showed that the peptide inhibiting the virus–cell diffusion is adsorbed with helices parallel to the membrane surface, unlike its mutants.

7.8.3. Adsorption of Biofilms

A bacterium is a unicellular microorganism. Microorganisms tend to adhere to surfaces and to form a gel layer called biofilm. Figure 7.56 illustrates that the absorption bands of all the bacterium components (proteins, RNA/DNA, polysaccharides, and lipids) and the extracellular polymeric substance (EPS) moieties (e.g., the carbonyl stretch associated with an ester linkage or a carboxylic acid salt) can be distinguished in the IR spectra [891–893]. This feature has been exploited for the identification of microorganisms [893, 894], bioprocess monitoring (when coupled with a statistical analysis) [895], and quantitative detection of the changes in the major compounds in bacteria [896] (when coupled with chemometric calibration methods) [897–899] (see Refs. [744, 749, 900, 901] for review).

Bacterial adhesion and biofilms have been studied in situ by ATR [891, 902–904]. For these measurements, special flow accessories are commercially available with a CIRCLE (Spectra-Tech, Stamford, CN) and trapezoidal (Spectra-Tech and Harrick Scientific) IRE. The latter is more suitable when the IR spectroscopic analysis is to be combined with another method, such as XPS or microscopic observation. An accessory with a trapezoidal IRE is described by Chittur et al. [841]. The use of a multichannel ATR/FTIR spectrometer specially designed for on-line examination of microbial biofilms [905] circumvents the problem of the reproducibility of the biofilm formation (see below).

Bacteria are typically enveloped in an EPS composed of proteins and extracellular polysaccharides, which connects bacterial cells with one another and with the substrate. This adhesive film imparts a uniform negative charge to the surface, masking its native properties [906–908]. Ishida and Griffiths [909] studied the adsorption of proteins (albumin and β-lactoglobulin) and polysaccharides (gum arabic and alginic acid) onto a Ge IRE by ATR. It was found that the proteins are more firmly bound to the surface than the polysaccharides. The adsorption of alginic acid was enhanced by the presence of albumin, β-lactoglobulin, and myoglobulin, while the neutral polysaccharide dextran was excluded from the

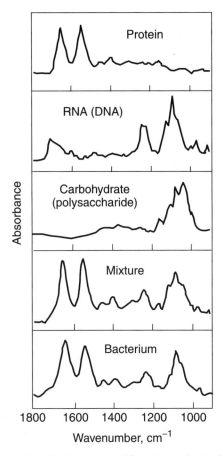

Figure 7.56. ATR FTIR spectra of bulk samples of (from top to bottom): protein (bovine serum albumin); RNA (ribonucleic acid, sodium salt yeast); carbohydrate, [D(+)-cellobiose]; mixture of appropriate proportions of protein, RNA, and carbohydrate found in bacterium, *Pseudomonas aeruginosa*. Reprinted, by permission, from P. A. Suci, J. D. Vrany, and M. W. Mittelman, *Biomaterials* **19**, 327 (1998), p. 330, Fig. 1. Copyright © 1998 Elsevier Science Ltd.

substrate conditioned by the presence of proteins but adsorbed onto the bare Ge surface [910]. The influence of divalent cations and pH on the adsorption of a polysaccharide produced by marina bacteria on germanium oxide was studied by Bhosle et al. [911]. These authors found no evidence for participation in the adhesion of H bonding to the oxide surface. Baty et al. [912] studied the adsorption of mussel adhesive proteins (MAPs) from the marine mussel *Mytilus edulis* on polystyrene (PS) and poly(octadecyl methacrylatc) (POMA) by ATR, XPS, and AFM. The ATR spectra displayed significant differences in the amide III region for the protein adsorbed on these substrates, which can be ascribed to different MAP secondary structures on PS and POMA. The amount of adsorbed polysaccharide adhesion, which was extracted from the EPS of *Hyphomonas*, on a Ge

IRE was found [913] to be larger than that of MAP and an acidic polysaccharide. The ATR spectra demonstrated that the polysaccharide adsorbed preferentially out of the EPS. The subsequent conditioning of the Ge substratum with either albumin or MAP decreased the adsorption of the adhesive polysaccharide significantly. Conditioning Ge with these proteins also decreased the adhesion of whale cells. In the context of biofouling, the adsorption of salivary proteins, albumin, and milk on six polymers was investigated by Imai and Tamaki [914]. It was shown that the amounts of proteins adsorbed depend strongly on the polymer. The increase in the contrast of the ATR solution spectrum as the thickness of the Cu coating on an IRE decreases was used [915] to study how bacteria cause copper corrosion. The slow removal of the copper film by polysaccharides supported theories that bacterial glycocalyx is important in this corrosion process [915a].

The time dependence of the area under one of the absorption bands of a bacterium allows the kinetics of the colonization of the surface to be monitored. If the absorbance is measured for the first monolayer of cells, such a dependence is called a "lateral growth curve." The experimental design and practical features to obtain a lateral growth curve, to monitor the time dependence of antimicrobial penetration in (or disappearance from) a biofilm, and to study the interaction between different reagents and a biofilm are all discussed by Suci et al. [891, 916]. To compare the effect of two antimicrobial agents, it was suggested to culture biofilms simultaneously under nearly identical conditions on two IREs directly in the chamber of a double-beam FTIR spectrometer. Mathematical analysis of the appearance/disappearance kinetics for an antibiotic in a biofilm can determine if the agent is bound to the film components. If it is, by introducing an adsorption/desorption term into Fick's equation, the density of the adsorption sites and the rates of adsorption and desorption can be adjusted to fit the model into the experimental curve. Another method for calculating the density of sites for the adsorption of an antimicrobial agent onto a biofilm is described by Nichols et al. [917].

Suci et al. [918] have demonstrated that the possibilities of ATR can be extended by combining the method with reflected differential interference contrast spectroscopy. It was shown that such an approach allows the biofilm structure, distribution, and chemistry at solid–liquid interfaces to be investigated simultaneously, the data being obtained in situ, quasi-simultaneously, nondestructively, and in real time.

Schmitt et al. [904] characterized physiological changes in bacteria (*Pseudomonas putida*) exposed to a biodegradable organic pollutant — toluene. The ATR spectra were measured in a three-channel FTIR spectrometer designed for three ATR flow cells. In channel I, the sterile medium was fed as a control. Channel II contained a biofilm and was used to obtain the signal of the resulting toluene-free biofilm. In channel III, the procedure was the same as in channel II but toluene was added in defined concentrations. In channel III, the detection of bioluminescence was achieved by fiber optics. Sterilization of the system was carried out with ethylene oxide. Figure 7.57 shows the spectra collected over 100 h for two concentrations (5 and 15 ppm) of toluene. The effect of 5 ppm

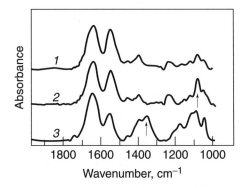

Figure 7.57. In situ ATR spectra of *P. putida* adsorbed at Ge IRE (50 × 20 × 3 mm, with 19 active reflections inside) after 100 h of adsorption: (1) biofilm without toluene; (2) 5 ppm toluene; (3) 15 ppm toluene. Arrows: increase of polysaccharide peaks at 5 ppm toluene and of carboxylic group peak at 15 ppm toluene. Spectra were obtained with multichannel ATR/FTIR spectrometer constructed on basis of RFX-30 FT IR interferometer (Laser Precision Analytical). Reprinted, by permission, from J. Schmitt, D. E. Nivens, D. C. White, and H.-C. Flemming, *Water Sci. Tech.* **32**, 149 (1995), p. 154, Fig. 5. Copyright © 1996 International Association on Water Quality (IAWQ).

toluene is manifested as changes in the polysaccharide region compared to the toluene-free film. At 15 ppm, significantly more carbonyl groups are formed and were attributed to the EPS.

Thus, ATR spectra can provide chemical and structural information about the processes in the biofilm, kinetic data on the biofilm growth, and penetration of another agent into the biofilm.

REFERENCES

1. P. Somasundaran and S. Krishnakumar, *Colloids Surf.* **123–124**, 491 (1997).

2. M. W. Urban, *Vibrational Spectroscopy of Molecules and Macromolecules on Surfaces*, Wiley, New York, 1993.

3. M. L. Hair, *Infrared Spectroscopy in Surface Chemistry*, Marcel Dekker, New York, 1967.

4. L. Little, *Infrared Spectra of Adsorbed Species*, Academic, London, New York, 1966.

5. A. V. Kiselev and V. I. Lygin, *Infra-red Spectra of Surface Compounds*, Wiley, New York, 1975.

6. A. A. Davydov, *Infrared Spectroscopy of Adsorbed Species on the Surface of Transition Metal Oxides*, Wiley, New York, 1990.

7. P. R. Griffiths, K. W. Van Every, and N. A. Wright, *Fresenius Z. Anal. Chem.* **324**, 571 (1986).

8. A. W. Adamson, *Physical Chemistry of Surfaces*, 4th ed., Wiley, New York, 1982.

9. A. M. Buswell, K. Krebs, and W. H. Rodebush, *J. Am. Chem. Soc.* **59**, 2603 (1937).

10. A. N. Terenin, *Zhurnal phyzicheskoi khimii* **14**, 1362 (1940).

11. A. N. Terenin and N. G. Iaroslavskii, *Izvestia AN USR, ser. Phyz.* **2**, 303 (1945).

12. A. N. Sidorov, *Doklady AN USSR* **95**, 1235 (1954).

13. E. A. Paukshtis, *IR Spectroscopy in Heterogeneous Acid-Base Catalysis*, Nauka, Novosibirsk, 1992, (in Russian).

14. R. P. Eischens, *Science* **146**, 486 (1964).

15. V. Lorenzelli, *Chim. Anal.* **48**, 637 (1966).

16. R. Hardeveld, *Chem. Weekbe* **62**, 261 (1966).

17. K. Klier, *Catal. Rev.* **1**, 207 (1967).

18. D. J. C. Yates, *J. Solid State Chem.* **12**, 282 (1975).

19. M. Hair, *J. Non.-Cryst. Solids* **19**, 299 (1975).

20. H. Ishida and J. L. Koenig, *Am. Lab.* **10**, 33 (1978).

21. L. Kubelkova, H. Hoger, A. Riva, and F. Trifira, *Zeolites* **3**, 244 (1983).

22. J. W. Strojek, J. Mielczarski, and P. Nowak, *Adv. Colloid Interface Sci.* **19**, 309 (1983).

23. V. E. Henrich, *Rep. Progr. Phys.* **48**, 1481 (1985).

24. E. Knozinger, P. Hoffman, and R. Echterhoff, *Microchim. Acta* **2**, 27 (1988).

25. A. M. Bradshaw and E. Schweizer, in R. J. H. Clark and R. E. Hester (Eds.), *Spectroscopy of Surfaces*, Wiley, New York, 1988, p. 413.

26. Y. J. Chabal, *Surf. Sci. Reports* **8**, 211 (1988).

27. N. Sheppard, *Spectrochim Acta (Spec. Suppl.)*, 239 (1990).

28. F. Bensebaa and T. H. Ellis, *Prog. Surf. Sci.* **50**, 173 (1995).

29. G. Busca, *J. Mol. Struct.* **218**, 363 (1990).

30. N. Sheppard, *Opt. Spectrosc.* **83**, 539 (1997).

31. A. A. Tsyganenko, *Opt. Spectrosc.* **83**, 547 (1997).

32. H. G. Karge, *Micropor. Mesopor. Mater.* **22**, 547 (1998).

33. H. Knozinger and S. Huber, *J. Chem. Soc. Farad. Trans.* **94**, 2047 (1998).

34. H. S. Han, C. H. Kim, and K. Kim, *Appl. Spectrosc.* **52**, 1047 (1998).

35. M. Che and A. J. Tench, in D. D. Eley, H. Pines, and P. B. Weiz (Eds.), *Advances in Catalysis*, Vol. 32, Academic, New York, 1983

36. P. A. Thiel, *Acc. Chem. Res.* **24**, 31 (1991).

37. S. R. Morrison, *The Chemical Physics of Surfaces*, 2nd ed., Plenum, New York, 1990.

38. H. Knozinger, in P. Schuster, G. Zundel, and C. Sandorfy (Eds.), *The Hydrogen Bond*, North-Holland, Amsterdam, 1976, Chapter 27.

39. P. A. Thiel, *Acc. Chem. Res.* **24**, 31 (1991).

40. E. F. Vansant, P. Van Der Voort, and K. C. Vrancken, *Characterisation and Chemical Modification of the Silica Surface*, Elsevier, Amsterdam, 1995.

41. C. Mortra and G. Magnacca, *Catal. Today* **27**, 497 (1996).

42. W. Stumm, *Chemistry of the Solid-Water Interface*, Wiley, New York, 1992, p. 14.

43. J. B. Peri, *J. Phys. Chem.* **69**, 231 (1965).

44. J. B. Peri and R. B. Hannan, *J. Phys. Chem.* **64**, 1526 (1960).

45. A. A. Tsyganenko and V. N. Filimonov, *J. Mol. Struct.* **19**, 579 (1973).

46. H. Knozinger and P. Ratnasamy, *Catal. Rev. Sci. Eng.* **4**, 31 (1978).

47. A. A. Tsyganenko and P. P. Mardilovich, *J. Chem. Soc. Faraday Trans.* **92**, 4843 (1996).

48. X. Liu and R. E. Truitt, *J. Am. Chem. Soc.* **119**, 9856 (1997).

49. M. I. Zaki and H. Knozinger, *Mater. Chem. Phys.* **17**, 201 (1987).

50. R. A. Santen, *Catalysis: An Integrated Approach*, Elsevier, Amsterdam, 1999.

51. C. Mortra, A. Zacchina, S. Coluccia, and A. Chiorino, *J. Chem. Soc. Faraday Trans. I* **77**, 1544 (1977).

52. S. M. Riseman, F. E. Massoth, G. M. Dhar, and E. M. Eyring, *J. Chem. Phys.* **86**, 1760 (1982).

53. N. Takagi, J. Yoshinobu, M. Kawai, *Phys. Rev. Lett.* **73**, 292 (1994).

54. J. Happel, *Isotopic Assessment of Heterogeneous Catalysis*, Academic, New York, 1986.

55. P. Biloen, J. N. Helle, F. G. A. van Den Berg, and W. M. H. Sachtler, *J. Catal.* **81**, 450 (1983).

56. S. S. C. Chuang, M. A. Brundage, M. W. Bbalakos, and G. Srinivas, *Appl. Spectrosc.* **49**, 1151 (1995).

57. Y. Kiroda, S. Kittaka, S. Takahara, T. Yamaguchi, and M.-C. Bellissent-Funel, *J. Phys. Chem. B* **103**, 11064 (1999).

58. M. Foster, M. Furse, and D. Passno, *Surf. Sci.* **502**, 102 (2002).

59. J. Chalmers and P. Griffiths (Eds.), *Handbook of Vibrational Spectroscopy*, Vol. 4, Wiley, New York, 2002, Chapter 9.

60. T. E. Madey, J. T. Sandstrow, Jr., and R. J. H. Voorhoeve, *Treatise Solid State Chem.* **6B**, 1 (1976).

61. N. Sheppard and T. T. Nguen, *Adv. Infrared Raman Spectrosc.* **5**, 67 (1978).

62. E. A. Paukshtis and E. N. Yurchenko, *Russ. Chem. Rev.* **52**, 426 (1983).

63. J. Mink, *Acta Phys. Hung.* **61**, 71 (1987).

64. J. T. Yates, Jr., P. Gelin, and T. Brebe, *Prepr. Amer. Chem. Soc. Div. Pet. Chem.* **29**, 908 (1984).

65. J. B. Peri, *Catal. Sci. Technol.* **5**, 171 (1984).

66. P. J. Collins, *Catalysis* **7**, 105 (1985).

67. K. Tamaru, in J. Anderson and M. Boudart (Eds.), *Catalysis: Science and Technology*, Vol. 9, Springer-Verlag, Berlin, 1991, p. 87.

68. J. A. Lercher, *Proc. SPIE-Int. Soc. Opt. Eng.* **1575**, 125 (1992).

69. G. A. Ozin, C. L. Bowes, and M. R. Steele, *Mater. Res. Soc. Symp. Proc.* **277**, 105 (1992).

70. H. Knozinger, in E. Umbach and H.-J. Freund (Eds.), *Adsorption on Ordered Surfaces of Ionic Solids and Thin Films*, Springer Series in Surface Science K33, Springer-Verlag, Berlin, 1993, p. 257.

71. J. P. Blitz and S. M. Augustine, *Spectroscopy* **9**, 28 (1994).

72. G. Coudurier and F. Lefevre, in B. Jmelik and J. C. Vedrinr (Eds.), *Catalyst Characterization Physical Techniques for Solid Materials*, Plenum, New York, 1994, p. 11.

73. S. F. Parker, A. Amorelli, Y. D. Amos, C. Hughes, N. Rorter, and J. R. Walta, *J. Chem. Soc. Faraday Trans.* **91**, 517 (1995).

74. J. E. Wachs, *Colloids Surf. A.* **105**, 143 (1995).

75. Q. Xin and X. Gao, *Prog. Nat. Sci.* **5**, 279 (1995).

76. J. E. Wachs, *Catal. Today* **27**, 437 (1996).

77. O. Busca, *Catal. Today* **27**, 323 (1996).

78. O. Busca, *Catal. Today* **27**, 457 (1996).

79. C. Marterra and G. Magnacca, *Catal. Today* **27**, 497 (1996).

80. G. Leofanti, G. Tozzola, M. Podovan, G. Petrini, S. Bordiga, and A. Zecchina, *Catal. Today* **34**, 307 (1997).

81. R. Cracium, D. J. Miller, N. Dulamita, and J. E. Jackson, *Prog. Catal.* **5**, 55 (1996).

82. V. A. Matyshak and O. V. Krylov, *Catal. Today* **25**, 1 (1995).

83. J. A. Lercher, Ch. Grundling, and G. Eder-Mirth, *Catal. Today* **27**, 353 (1996).

84. J. A. Lercher, V. Veefkind, and K. Fajerwerg, *Vib. Spectrosc.* **19**, 107 (1999).

85. G. N. Zhizhin, A. A. Sigarev, and V. A. Yakovlev, *Mikrochim. Acta* **14**(Suppl.), 669 (1997).

86. E. V. Avieva, G. Beitel, L. A. Kuzik, A. A. Sigarev, V. A. Yakovlev, G. N. Zhizhin, A. F. G. van der Meer, and M. J. van der Wiel, *Appl. Spectrosc.* **51**, 584 (1997).

87. E. V. Avieva, G. Beitel, L. A. Kuzik, A. A. Sigarev, V. A. Yakovlev, G. N. Zhizhin, A. F. G. van der Meer, and M. J. van der Wiel, *J. Mol. Struct.* **449**, 119 (1998).

88. Y. Ishino, R. T. Graf, and H. Ishida, *Proc. SPIE* **553**, 480 (1985).

89. R. J. Bell, R. W. Alexander, C. A. Ward, and I. L. Tyler, *Surf. Sci.* **48**, 253 (1975).

90. J. Lauterbach, R. W. Boyle, M. Schick, W. J. Mitchell, B. Meng, and W. H. Weinberg, *Surf. Sci.* **350**, 32 (1996). Erratum: *Surf. Sci.* **366**, 228 (1996).

91. H. J. Krebs and H. Luth, *Appl. Phys.* **14**, 337 (1977).

92. R. K. Sharma, W. A. Brown, and D. A. King, *Surf. Sci.* **414**, 68 (1998).

93. M. Sushchikh, J. Lauterbach, and W. H. Weinberg, *Surf. Sci.* **393**, 135 (1997).

94. R. G. Greenler, *J. Chem. Phys.* **44**, 310 (1966).

95. R. G. Greenler, *J. Chem. Phys.* **50**, 1963 (1969).

96. M. J. D. Low and J. C. McManus, *Chem. Commun.* **1**, 166 (1967).

97. P. Hollins and J. Pritchard, *Progr. Surf. Sci.* **19**, 275 (1985).

98. A. Horn, in R. J. H. Clark and R. E. Hester (Eds.), *Spectroscopy of Surfaces*, Wiley, New York, 1998, p. 273.

99. H. Knozinger, *Acta Cient. Venez.* **24**, 76 (1973).

100. R. Greenler, in T. S. Jayadevaian and R. Vanselow (Eds.), *Surface Science: Recent Progress and Perspectives*, CRC Press, Cleveland, 1974, p. 136.

101. R. Greenler, *Crit. Rev. Solid State Sci.* **4**, 415 (1974).

102. H. G. Tompkins, in A. W. Czanderna (Ed.), *Methods of Surface Analysis*, Elsevier, New York, 1975, p. 447.

103. I. Ratajezykowa, *Wiad. Chem.* **29**, 7 (1975).

104. J. Pritchard, *Dechema-Monogr.* **78**, 231 (1975).

105. J. Pritchard and T. Catterick, *Experimental Methods in Catalytic Research* **3**, 281 (1976).

106. H. G. Tompkins, *Appl. Spectrosc.* **4**, 377 (1976).

107. M. J. Dignam and J. Fedyk, *Appl. Spectrosc. Rev.* **14**, 249 (1978).

108. P. Hollins and J. Pritchard, in R. F. Willis (Ed.), *Vibrational Spectroscopy of Adsorbates*, Springer-Verlag, Berlin, 1980, p. 171.

109. M. H. Dunken and R. Stepanowitz, *Z. Chem.* **23**, 353 (1983).

110. M. Handke, M. Milosevic, and N. Harrick, *Vib. Spectrosc.* **1**, 251 (1991).

111. M. A. Chesters, *J. Electron Spectrosc. Related Phenom. A*, **38**, 123 (1986).

112. M. A. Chesters, *Vacuum* **38**, 428 (1988).

113. A. M. Bradshaw, *NATO ASI Ser. C* **496**, 207 (1993).

114. B. N. J. Persson and A. I. Volokitin, *J. Electron Spectrosc. Related Phenom.* **64–65**, 23 (1993).

115. R. Raval, *Surf. Sci.* **331–333**(Pt. A), 1 (1995).

116. N. Takao, *Hyomen Kagaku* **15**, 152 (1994).

117. R. G. Greenler and R. K. Brandt, *Colloid Surf. A Physicochem. Inginer. Aspects* **104**, 19 (1995).

118. T. Takenaka and J. Unemura, *Stud. Interface Sci.* **4**, 145 (1996).

119. M. Kawai, *Bull. Mater. Sci.* **20**, 769 (1997).

120. R. J. Lippert, B. D. Lamp, and M. D. Porter, in F. M. Mirabella (Ed.), *Modern Techniques in Applied Molecular Spectroscopy*, Wiley, New York, 1998, pp. 83–126.

121. R. Raval, A. J. Roberts, J. Williams, T. S. Nunney, and M. Surman, *Proc. SPIE-Int. Soc. Opt. Eng.* **3153**, 110 (1997).

122. F. M. Hoffmann, *Surf. Sci. Rep.* **3**, 107 (1983).

123. D. L. Allara, *Crit. Rev. Surf. Chem.* **2**, 91 (1993).

124. F. M. Hoffmann and M. D. Weisel, *J. Vac. Sci. Technol.* **11**, 1957 (1993).

125. P. Hollins, *Vacuum* **45**, 705 (1994).

126. P. Hollins, *Surf. Sci. Rep.* **16**, 51 (1992).

127. K. Horn and J. Pritchard, *Surf. Sci.* **55**, 701 (1976).

128. A. I. Volokitin, *Phys. Chem. Mech. Surf.* **5**, 1064 (1990).

129. E. Borquet and H.-L. Doi, *J. Electron Spectrosc. Related Phenom.* **54–55**, 573 (1990).

130. E. Borquet, J. Dvorak, and H.-L. Doi, *Proc. SPIE-Int. Soc. Opt. Eng.* **2125**, 12 (1994).

131. A. J. Volokitin and B. N. J. Persson, *J. Exp. Theor. Phys. (USA)* **81**, 545 (1995).

132. J. P. Camplin, J. C. Cook, and E. M. McCash, *J. Chem. Soc. Faraday Trans.* **31**, 3563 (1995).

133. J. C. Cook and E. M. McCash, *Surf. Sci.* **356**, L445 (1996).

134. A. G. Malshukov, *Solid State Commun.* **79**, 81 (1991).

135. C. J. Hirschmugl, G. P. Williams, F. M. Hoffmann, and Y. J. Chabal, *Phys. Rev. Lett.* **65**(4), 480 (1990).

136. S. Carter, S. J. Culik, and J. M. Bowman, *J. Chem. Phys.* **107**, 10458 (1997).

137. A. G. Yodh and H. W. K. Tom, *Phys. Rev. B Condens. Matter* **45**, 14302 (1992).

138. W. Erley, *J. Electron Spectrosc. Related Phenom.* **44**, 65 (1987).

139. K. Horn, M. Hussain, and J. Pritchard, *Surf. Sci.* **63**, 244 (1977).

140. J. Pitchard, T. Catterick, and R. K. Gupta, *Surf. Sci.* **53**, 1 (1975).

141. R. A. Shiglishi and D. A. King, *Surf. Sci.* **58**, 379 (1976).

142. R. A. Shiglishi and D. A. King, *Surf. Sci.* **75**, L397 (1978).

143. K. Horn and J. Pritchard, *J. Phys. (France)* **38**, 164 (1977).

144. R. A. Shiglishi and D. A. King, *Surf. Sci.* **75**, L397 (1978).

145. D. Hoge, M. Tueshaus, P. Gardner, and A. Bradshaw, *Stud. Surf. Sci. Catal (Struct. React. Surf.)* **48**, 493 (1988).

146. B. N. Persson and R. Ryberg, *Phys. Rev. B Condens. Matter* **40**, 10273 (1989).

147. R. Ryberg, *Phys. Rev* **40**, 8567 (1989).

148. R. Ryberg, *Phys. Rev. B* **40**, 5849 (1989).

149. L. F. Sutcu, J. L. Wrang, and H. W. White, *Phys. Rev. B Condens. Matter* **41**, 2164 (1990).

150. J. D. Beckerie, R. R. Cavanagh, M. P. Cassara, E. J. Heilvell, and J. C. Stephenson, *J. Chem. Phys.* **95**, 5403 (1991).

151. H. W. White, L. F. Suta, and J. L. Wragg, *Proc. SPIE-Int. Soc. Opt. Eng.* **1636**, 27 (1992).

152. R. Ryberg, *Phys. Rev. B Condens. Matter* **44**, 13160 (1991).

153. R. R. Cavanagh, J. D. Beckerle, M. P. Casassa, E. J. Hellwell, and J. C. Stephenson, *Surf. Sci.* **269–270**, 113 (1992).

154. J. A. Rodriquez, C. M. Tsuong, and D. W. Goodman, *J. Chem. Phys.* **96**, 7814 (1992).

155. D. Jakob, *Phys. Rev. Lett.* **79**, 2919 (1997).

156. P. Gardner, R. Martin, M. Tueshaus, and A. M. Bradshaw, *J. Electron Spectrosc. Related Phenom.* **54–55**, 619 (1991).

157. K. Fukui, H. Miyauchi, and Y. Jwasawa, *J. Phys. Chem.* **100**, 18795 (1996).

158. R. K. Sharma, W. A. Brown, and D. A. King, *Chem. Phys. Lett.* **291**, 1 (1998).

159. D. K. Lambert and R. G. Tobin, *Surf. Sci.* **232**, 149 (1990).

160. J. Xu, and J. T. Yates, Jr., *Surf. Sci.* **327**, 193 (1995).

161. J. Xu, P. N. Henriksen, and J. T. Yates, Jr., *Langmuir* **10**, 3663 (1994).

162. J. S. Luo, R. G. Tobin, D. K. Lambert, F. T. Wagner, and T. E. Maylan, *J. Electron Spectrosc. Related Phenom.* **54–55**, 469 (1991).

163. H. Moritz and H. Lueth, *Vacuum* **41**, 63 (1990).

164. M. K. Ainsworth, M. R. S. McCoustra, M. A. Chesters, N. Sheppard, and C. De La Cruz, *Surf. Sci.* **437**, 9 (1999).

165. H. C. Eckstrom and W. H. Smith, *J. Opt. Soc. Am.* **57**, 1132 (1967).

166. E. F. McCoy and R. St. Smart, *Surf. Sci.* **39**, 109 (1973).

167. J. G. Chen, W. Erley, and H. Ibach, *Surf. Sci.* **223**, L891 (1989).

168. J. Miragliotta, R. S. Polizotti, P. Rabinowitz, Z. D. Cameron, and R. B. Hall, *Appl. Phys. A Solids Surf.* **A51**, 221 (1990).

169. M. Kawa and Y. Yoshinoby, *Springer Ser. Solid State Sci.* **121**, 78 (1996).

170. V. M. Bermudez, *J. Vac. Sci. Technol. A* **9**, 3169 (1991).

171. A. Banadara, S. Dobashi, J. Kubota, C. Hirose, and S. S. Kano, *Surf. Sci.* **387**, 312 (1997).

172. S. Haq, J. G. Love, and D. A. King, *Surf. Sci.* **275**, 170 (1992).

173. J. Yoshinibu and M. Kawai, *J. Electron Spectrosc. Related Phenom.* **64–65**, 207 (1993).

174. J. G. Chen, W. Erley, and H. Ibach, *Surf. Sci.* **227**, 79 (1990).

175. T. Yuzawa, T. Higashi, J. Kubota, J. Kondo, K. Domen, and C. Hirose, *Surf. Sci.* **325**, 223 (1995).

176. H. E. Sanders, P. Gardner, D. A. King, and M. A. Morris, *Surf. Sci.* **304**, 159 (1994).

177. R. Raval, S. Haq, M. A. Harrison, G. Blyholder, and D. A. King, *Chem. Phys. Lett.* **167**, 391 (1990).

178. R. Raval, M. A. Harrison, and D. A. King, *Surf. Sci.* **211–212**, 61 (1989).

179. X. Xu, P. Chen, and D. W. Goodman, *J. Phys. Chem.* **98**, 9242 (1994).

180. J. Szany, W. K. Kuhn, and D. W. Goodman, *J. Phys. Chem.* **98**, 2978 (1994).

181. J. Szany and D. W. Goodman, *J. Phys. Chem.* **98**, 2972 (1994).

182. K. Wolter, O. Seiferth, J. Libuda, H. Kuhlenbeck, M. Baeumer, and H.-J. Freund, *Chem. Phys. Lett.* **277**, 513 (1997).

183. X. Xu, P. Chen, and D. W. Goodman, *J. Phys. Chem.* **98**, 9242 (1994).

184. K. J. Lyons, J. Xie, W. J. Mitchell, and W. H. Weinberg, *Surf. Sci.* **325**, 85 (1995).

185. M. Schik, J. Lauterbach, and W. H. Weinberg, *J. Vac. Sci. Technol. A* **14**(Pt. 2), 1448 (1996).

186. D. Reinalda and V. Ponec, *Surf. Sci.* **91**, 113 (1980).

187. H. Langsam-Wing and D. W. Goodmen, *Catal. Lett.* **5**, 353 (1990).

188. R. L. Toomes and D. A. King, *Surf. Sci.* **349**, 19 (1996).

189. J. W. He, W. K. Kuhn, and D. W. Goodman, *Surf. Sci.* **262**, 351 (1992).

190. W. K. Kuhn, J. W. He, and D. W. Goodman, *J. Phys. Chem.* **98**, 259 (1994).

191. W. K. Kuhn, J. W. He, and D. W. Goodman, *J. Phys. Chem.* **98**, 264 (1994).

192. M. Kaltchev and W. T. Tysoe, *Catal. Lett.* **53**, 145 (1998).

193. J. T. Yates, R. G. Greenler, I. Rotajczykova, and D. A. King, *Surf. Sci.* **36**, 739 (1973).

194. C. H. F. Peden, D. W. Goodman, M. D. Weisel, and F. M. Hoffman, *Surf. Sci.* **253**, 44 (1991).

195. F. M. Hoffman, M. D. Weisel, and C. H. F. Peden, *J. Electron Spectrosc. Related Phenom.* **54–55**, 779 (1990).

196. P. M. Parlett and M. A. Chesters, *Surf. Sci.* **357–358**, 791 (1996).

197. M. D. Weisel, J. L. Robbins, and F. M. Hoffman, *J. Phys. Chem.* **97**, 9441 (1993).

198. F. M. Hoffman, M. D. Weisel, and C. H. F. Peden, *Surf. Sci.* **253**, 59 (1991).

199. B. E. Hayden and A. M. Bradshaw, *Surf. Sci.* **125**, 787 (1983).

200. B. N. J. Persson and R. Ryberg, *Phys. Rev. B* **24**, 6954 (1981).

201. B. N. J. Persson, E. M. Hoffmann, and R. Ryberg, *Phys. Rev. B* **34**, 2266 (1986).

202. D. C. Langreth and M. Persson, *Phys. Rev. B* **43**, 1353 (1991).

203. A. Grossman, W. Erley, and H. Ibach, *Surf. Sci. Lett.* **330**, L646 (1995).

204. P. Gardner, M. Tueshaus, R. Martin, and A. M. Bradshaw, *Surf. Sci.* **240**, 112 (1990).

205. T. T. Magkoev, M. Song, K. Fukutani, and Y. Murata, *Surf. Sci.* **330**, L669 (1995).

206. D. Loffreda, D. Simon, and P. Sautet, *Chem. Phys. Lett.* **291**, 15 (1998).

207. W. A. Brown, P. Gardner, M. P. Jigato, and D. A. King, *J. Chem. Phys.* **102**, 7277 (1995).

208. J. G. Chen, W. Erley, and H. Ibach, *Surf. Sci.* **224**, 215 (1989).

209. P. Jakob, *Surf. Sci.* **427–428**, 309 (1999).

210. W. Erley, *Surf. Sci.* **205**, L771 (1988).

211. S. Aminpirooz, A. Schmalz, L. Becker, and J. Haas, *Phys. Rev. B* **45**, 6337 (1992).

212. M. C. Asensio, D. P. Woodruff, A. W. Robinson, K.-M. Schlinder, P. Gardener, D. Ricken, A. M. Bradshaw, I. C. Cunesa, and A. R. Gonzalez-Elipe, *Chem. Phys. Lett.* **192**, 259 (1992).

213. L. D. Mapledoram, A. Wander, and D. A. King, *Chem. Phys. Lett.* **208**, 409 (1993).

214. D. J. Oakes, M. R. Cosestra, and M. A. Chesters, *Faraday Discuss.* **96**, 325 (1993).

215. J. Yoshinobu, H. Ogasawa, and M. Kawai, *Phys. Rev. Lett.* **75**, 2176 (1995).

216. K. Watanabe, K. Sawabe, and Y. Matsumoto, *Phys. Rev. Lett.* **76**, 1751 (1996).

217. S. Jchihara, J. Kondo, K. Domen, and C. Hirose, *Surf. Sci.* **357–358**, 634 (1996).

218. J. Kubota, S. Jchihara, J. Kondo, K. Domen, and C. Hirose, *Langmuir* **12**, 1926 (1996).

219. J. Kubota, S. Jchihara, J. Kondo, K. Domen, and C. Hirose, *Surf. Sci.* **357–358**, 634 (1996).

220. F. Zaera, H. Hoffman, and P. R. Griffiths, *J. Electron Spectrosc. Related Phenom.* **54–55**, 705 (1990).

221. F. Zaera and N. Bernstein, *J. Am. Chem. Soc.* **116**, 4881 (1994).

222. C. French and I. Harrison, *Surf. Sci.* **387**, 11 (1997).

223. N. R. Gleason, C. J. Jenks, C. R. French, B. E. Bent, and F. Zaera, *Surf. Sci.* **405**, 238 (1998).

224. J. Fah and M. Trenary, *Langmuir* **10**, 3649 (1994).

225. J. Fan and M. Trenary, *Surf. Sci.* **282**, 76 (1993).

226. W. G. Golden, H. Hunziker, and M. S. de Vries, *J. Phys. Chem.* **98**, 1739 (1994).

227. T. Iwasita, J. L. Rodriquez, and X. H. Xia, *Langmuir* **16**, 5478 (2000).

228. M. J. Hoestler, R. G. Nuzzo, and G. S. Girolami, *J. Am. Chem. Soc.* **116**, 11608 (1994).

229. W. L. Manner, A. R. Bishop, G. S. Girolami, and R. G. Nuzzo, *J. Phys. Chem. B* **102**, 8816 (1998).

230. M. Kovar, R. V. Kasza, K. Griffiths, P. R. Norton, G. P. Williams, and D. Van Campen, *Surf. Rev. Lett.* **5**, 589 (1998).

231. A. Quick, V. M. Browne, S. G. Fox, and P. Hollins, *Surf. Sci.* **221**, 48 (1989).

232. M. E. Brubaker and M. Trenary, *J. Chem. Phys.* **85**, 6100 (1986).

233. H. E. Sanders, P. Gardner, and D. A. King, *Surf. Sci.* **331–333**(Pt. B), 1496 (1995).

234. T. Kratochwil, M. Wittman, and J. Kueppers, *J. Electron Spectrosc. Related Phenom.* **64–65**, 607 (1993).

235. A. M. Coats, E. Cooper, and R. Raval, *Surf. Sci.* **307–309**, 89 (1994).

236. E. Zahidi, M. Castonguay, and P. McBreen, *J. Am. Chem. Soc.* **116**, 5847 (1994).

237. E. Zahidi, M. Castonguay, and P. McBreen, *Chem. Phys. Lett.* **236**, 122 (1995).

238. E. Cooper and R. Raval, *Surf. Sci.* **331–333**(Pt. A), 94 (1995).

239. E. Zahidi, M. Castonguay, and P. H. McBreen, *Chem. Phys. Lett.* **236**, 122 (1995).

240. E. Cooper, A. Coate, and R. Raval, *J. Chem. Soc. Faraday Trans.* **91**, 3703 (1995).

241. T. S. Nanney, J. J. Birtill, and R. Raval, *Surf. Sci.* **427–428**, 282 (1999).

242. K. Horn and J. Pritchard, *Surf. Sci.* **52**, 437 (1975).

243. J. Kubota, J. Kondo, K. Domen, and C. Hirose, *J. Electron Spectrosc. Related Phenom.* **64–65**, 137 (1993).

244. C. L. A. Lamout, B. N. J. Persson, and G. P. Williams, *Chem. Phys. Lett.* **243**, 429 (1995).

245. E. M. McCash, *Vacuum* **40**, 423 (1990).

246. J. Kubota, J. Kondo, K. Domen, and C. Hirose, *Hyomen Kagaru* **15**, 329 (1994).

247. S. L. Silva, T. M. Pham, and F. M. Leibsle, *Surf. Sci.* **452**, 79 (2000).

248. S. C. Street and A. J. Gellman, *J. Phys. Chem.* **100**, 8338 (1996).

249. A. Peremans, F. Maseri, J. Darville, and J. M. Gilles, *J. Vac. Sci. Technol. A* **8**, 3224 (1990).

250. K. C. Lin, R. A. Tobin, and P. Dumas, *J. Vac. Sci. Technol. A* **13**(Pt. 2), 1579 (1995).

251. K. C. Lin and L. East, *Phys. Rev. B Condens. Matter* **49**, 17273 (1994).

252. J. Kubota, J. N. Kondo, K. Domen, and C. Hirose, *J. Phys. Chem.* **98**, 7653 (1994).

253. M. Kaltchev, A. W. Thompson, and W. T. Tysoe, *Surf. Sci.* **391**, 145 (1997).

254. W. Kai, Z. Runsheng, J. Jintend, L. Sihua, and W. Sicheneng, *J. Electron Spectrosc. Related Phenom.* **76**, 201 (1995).

255. K. Kretzschnar, J. K. Sass, P. Hoffman, A. Ortega, A. M. Bradshaw, and S. Holloway, *Chem. Phys. Lett.* **78**, 410 (1981).

256. K. Kretzschnar, J. K. Sass, A. M. Bradschaw, and S. Holloway, *Surf. Sci.* **115**, 183 (1982).

257. P. M. Parlett and M. A. Chesters, *Surf. Sci.* **357–358**, 791 (1996).

258. C. A. Mins, M. D. Weisel, F. M. Hoffman, J. H. Sinfeld, and J. M. White, *J. Phys. Chem.* **97**, 12656 (1993).

259. L. M. Jlharco, A. R. Garcia, and J. L. da Silva, *Surf. Sci.* **392**, L27 (1997).

260. J. Kiss and F. Solymosi, *J. Catal.* **179**, 277 (1998).

261. D. M. Riffe and A. J. Sievers, *Phys. Rev. B* **41**, 6, 3406 (1990).

262. J. B. Restorff and H. D. Drew, *Surf. Sci.* **88**, 399 (1979).

263. J. N. Kondo, T. Higashi, H. Yamamoto, M. Haar, and K. Domen, *Surf. Sci.* **349**, 294 (1996).

264. W. A. Brown, P. Gardner, and D. A. King, *Surf. Sci.* **330**, 41 (1995).

265. G. J. Szulczewski and J. M. White, *Surf. Sci.* **406**, 194 (1998).

266. N. Osaka, M. Akita, and K. Jtoh, *J. Phys. Chem. B* **102**, 6817 (1998).

267. S. Serghini-Monim, K. Griffiths, P. K. Norton, and R. J. Puddephatt, *J. Am. Chem. Soc.* **117**, 14 (1995).

268. R. Li and K. P. Baluer, *Surf. Sci.* **331–333**(Pt. A), 100 (1995).

269. M. A. Chesters and D. A. Slater, *Proc. SPIE-Int. Soc. Opt. Eng.* **2089**, 16 (1993).

270. P. Hollins, A. A. Davis, D. A. Slater, M. A. Chesters, E. C. Hargreaves, P. M. Perlett, J. C. Wenger, and M. Surman, *J. Chem. Soc. Faraday Trans.* **92**, 879 (1996).

271. Ph. Parent, C. Laffon, G. Tourillon, and A. Gassuto, *J. Phys. Chem.* **99**, 5058 (1995).

272. M. Akita, N. Osaka, and K. Iton, *Surf. Sci.* **405**, 172 (1998).

273. J. Wang and B. E. Koel, *J. Phys. Chem. A* **102**, 8573 (1998).

274. P. Hollins, A. A. Davis, D. A. Slater, M. A. Chesters, E. C. Hargreaves, P. M. Parlett, J. C. Wanger, and M. Surman, *J. Chem. Soc. Faraday Trans.* **92**, 879 (1996).

275. S. Fujii, N. Osaka, M. Akita, and K. Itoh, *J. Phys. Chem.* **99**, 6994 (1995).

276. M. R. Peng and J. E. Reutt-Robey, *Surf. Sci.* **336**, L755 (1995).

277. H. Masuda, *J. Jpn. Inst. Metals* **62**, 961 (1998).

278. T. G. Koch, N. S. Holmes, T. B. Roddis, and J. R. Sodeau, *J. Chem. Soc. Faraday Trans.* **92**, 4787 (1996).

279. K. Uvdal, L. Persson, and B. Liedberg, *Langmuir* **11**, 1252 (1995).

280. G. Kurth and T. Bein, *Langmuir* **11**, 578 (1995).

281. A. J. Maeland, R. Rittenhouse, W. Lahar, and P. V. Romano, *Thin Solid Films* **21**, 67 (1974).

282. G. N. Kuznetsova and V. P. Tolstoy, *Zhurnal Prikladnoi Spektroskopii* **27**, 490 (1977).

283. J. Chatelet, H. H. Claassen, D. M. Gruen, I. Sheft, and R. B. Wright, *Appl. Spectrosc* **29**, 185 (1975).

284. R. Guillamet, M. Lenglet, and F. Adam, *Solid State Commun.* **81**, 633 (1992).

285. Th. Scherubl and L. K. Thomas, *Appl. Spectrosc.* **51**, 844 (1997).

286. B. C. Trasferetti, C. U. Davanzo, N. C. da Cruz, and M. A. B. de Maraes, *Appl. Spectrosc.* **54**, 687 (2000).

287. J. Itoh, T. Sasaki, and T. Ishikawa, *Zairyo to Kankyo* **46**, 777 (1997).

288. T. Takamura, H. Kihara-Morishita, and U. Moriyama, *Thin Solid Films* **6**, R17 (1970).

289. D. Hinze and W. Zeimbrock, *Thin Solid Films* **35**, 175 (1976).

290. T. Wadayama, T. Hihara, A. Hatta, and W. Suetaka, *Appl. Surf. Sci.* **48–49**, 409 (1991).

291. B. Mihailova, V. Engstrom, J. Hedlund, A. Holmgren, and J. Sterte, *Microporous Mesoporous Matter.* **32**, 297 (1999).

292. R. M. Almeida, T. A. Guiton, and C. G. Pantano, *J. Non-Cryst. Solids* **119**, 238 (1990).

293. V. P. Tolstoy, G. N. Kuznetsova, S. I. Koltsov, and V. B. Aleskovskii, *Zhural prikladnoi khimii* **53**, 2353 (1980).

294. G. Poling *J. Electrochem Soc.* **116**, 958 (1969).

295. A. Pandey, N. S. S. Murty, and S. M. Patel, *Process Control Quality* **11**, 363 (2000).

296. (a) S. Zakipour and C. Leygraf, *Brit. Corros. J.* **27**, 295 (1992); (b) T. Aastrup and C. Leygraf, *J. Electrochem Soc.* **144**, 2986 (1997).

297. J. Itoh, T. Sasaki, T. Ohtsuka, and M. Osawa, *J. Electroanal. Chem.* **473**, 256 (1999).

298. P. A. Thiel and T. E. Madey, *Surf. Sci. Rep.* **7**, 211 (1989).

299. F. Mansfeld (Ed.), *Corrosion Mechanisms*, Marcel Dekker, New York, 1987.

300. P. R. Roberge, *Handbook of Corrosion Engineering*, McGraw-Hill, New York, 2000.

301. J. C. Scully, *The Fundamentals of Corrosion*, Pergamon, Oxford, 1990.

302. L. H. Little, *Australs Corros. Eng.* **14**, 17 (1970).

303. S. M. Thibalt and J. M. Talbot, *C. R. Acad. Sci.* **C272**, 805 (1971).

304. S. Friberg and H. Muller, *Ann. Univ. Ferrara Ser. 5 Suppl.* **5**, 93 (1971).

305. N. Harrick, *Internal Reflection Spectroscopy*, Wiley, New York, 1967.

306. Y. J. Chabal, in F. M. Mirabella (Ed.), *Internal Reflection Spectroscopy*, Marcel Dekker, New York, 1993, pp. 191–232.

307. L. H. Sharke, *Proc. Chem. Soc.* **1961**, 461 (1961).

308. S. L. Grigorovich, V. I. Lygin, and V. A. Fedorov, *Kollodnyi zhurnal* **33**, 345 (1972).

309. S. L. Grigorovich, A. V. Kiselev, V. I. Lygin, and V. A. Fedorov, *Zhurnal phyzicheskoi khimii* **46**, 2870 (1972).

310. T. Takenaka, K. Naogarami, H. Goton, and R. Goton, *J. Colloid Interface Sci.* **44**, 249 (1973).

311. A. M. Iasynich, H. B. Mark, and C. N. Giles, *J. Phys. Chem.* **80**, 839 (1976).

312. V. A. Bernshtein, U. D. Varfolomeev, and V. V. Nikitin, *Physica tverdogo tela* **13**, 693 (1971).

313. V. A. Bershtein and V. V. Nikitin, *Proc. 10th Int. Cong. Glas. Kyoto* **9**, 105 (1974).

314. G. L. Haller and R. W. Rice, *J. Phys. Chem.* **74**, 4386 (1970).

315. L. M. Struck and M. P. D'Evelyn, *J. Vac. Sci. Technol. A* **11**, 1992 (1993).

316. L. Ley, B. F. Mantel, K. Matura, M. Stammler, K. Janischowsky, and J. Ristein, *Surf. Sci.* **427–428**, 245 (1999).

317. M. McGonigal, J. N. Russell, P. E. Pehsson, H. G. Maguire, and J. E. Butler, *J. Appl. Phys.* **77**, 4049 (1995).

318. R. T. Yang, M. J. D. Low, G. L. Haller, and J. Penn *J. Coll. Interf. Sci.* **44**, 249 (1973).

319. V. M. Zolotarev, A. F. Perveev, G. A. Muranova, and T. G. Arkatova, *Zhurnal prikladnoi spektroskopii* **16**, 331 (1972).

320. A. F. Perveev, V. M. Zolotarev, P. P. Egorov, and G. A. Muranova, *Optika i spektroskopia USSR* **32**, 607 (1972).

321. A. F. Perveev and G. A. Muranova, *Sov. J. Opt. Tehnol.* **2**, 73 (1973).

322. D. B. Parry and J. M. Harris, *Appl. Spectrosc.* **42**, 997 (1988).

323. T. R. E. Sinpson, D. A. Russel, I. Chambrier, M. J. Cook, A. B. Horn, and S. C. Thorpe, *Sensors Activators B* **29**, 353 (1995).

324. A. Couzis and E. Gulari, in P. M. Holland and D. N. Rubingh (Eds.), *Mixed Surfactant Systems*, ACS Symposium Series 501, American Chemical Society, Washington, DC, 1992, pp. 354–365.

325. I. V. Chernyshova, H. K. Rao, A. Vidyadhar, and A. V. Shchukarev, *Langmuir* **16**, 8071 (2000).

326. A. Udagawa, T. Matsui, and S. Tanaka, *Appl. Spectrosc.* **40**, 794 (1986).

327. J. A. Mielczarski, E. Mielczarski, and J. M. Cases, *J. Colloid Interface Sci.* **188**, 150 (1997).

328. J. A. Mielczarski and R. H. Yoon, *J. Phys. Chem.* **93**, 2034 (1989).

329. J. A. Mielczarski, *J. Phys. Chem.* **97**, 2649 (1993).

330. J. A. Mielczarski, E. Mielczarski, J. Zachwieja, and J. M. Cases, *Langmuir* **11**, 2787 (1995).

331. J. A. Mielczarski and E. Mielczarski, *J. Phys. Chem. B* **103**, 5852 (1999).

332. R. A. Dluhy, S. M. Stephens, S. Widayati, and A. D. Williams, *Spectrochim. Acta A* **51**, 1413 (1995).

333. (a) R. Mendelsohn and C. R. Flach, in J. Chalmers and P. Griffiths (Eds.), *Handbook of Vibrational Spectroscopy*, Vol. 2, Wiley, New York, 2002, p. 1028; (b) R. Mendelsohn, J. W. Brauner, and A. Gericke, *Annu. Rev. Phys. Chem.* **46**, 305 (1995).

334. D. Blaudez, T. Buffeteau, B. Desbat, and J. M. Turlet, *Current Opinion Colloid Interface Sci.* **4**, 265 (1999).

335. J. Heidberg, E. Kamphoff, O. Schonecas, H. Stein, and H. Weiss, *Ber. Bunsenges. Phys. Chem.* **94**, 118 (1990).

336. J. Heidberg, E. Kamphoff, O. Schonekas, H. Stein, and H. Weiss, *Ber. Bunsenges. Phys. Chem.* **94**, 112 (1990).

337. J. Heidberg, E. Kampshoff, R. Kuchnemuth, and O. Schonekas, *Surf. Sci.* **272**, 306 (1992).

338. O. Berg and R. Dissekamp, *Surf. Sci.* **277**, 8 (1992).

339. J. Heidberg, E. Kampshoff, R. Kuhnemuth, O. Schonekas, H. Stein, and H. Weiss, *Surf. Sci.* **226**, L43 (1990).

340. O. Berg and G. Ewing, *J. Vac. Sci. Technol. A* **8**(Pt. 2), 2653 (1990).

341. C. Noda, H. H. Richardson, and G. E. Ewing, *J. Chem. Phys.* **92**, 2099 (1990).

342. H.-C. Chang and G. E. Ewing, *Chem. Phys.* **139**, 55 (1989).

343. D. J. Dai and G. E. Ewing, *J. Electron Spectrosc. Related Phenom.* **64–65**, 101 (1994).

344. D. K. Lambert, G. P. M. Poppe, and C. M. J. Wijers, *J. Chem. Phys.* **103**, 6206 (1995).

345. J. Heidberg, E. Kampshoff, R. Kuehnemuth, M. Suhren, and H. Weiss, *Surf. Sci.* **269**, 128 (1992).

346. J. Heidberg, E. Kampshoff, R. Kuehnemuth, and O. Schoenekaes, *Can. J. Chem.* **72**, 795 (1994).

347. J. Heidberg, N. Gushanskaya, O. Shoenekaes, and R. Shwarte, *Surf. Sci.* **331–333** (Pt. B), 1473 (1995).

348. J. Heidberg and W. K. Haser, *J. Electron Spectrosc. Related Phenom.* **54–55**(Spec. Issue), 971 (1990).

349. O. Berg, G. E. Ewing, A. W. Meredith, and A. J. Stone, *J. Chem. Phys.* **104**, 6843 (1996).

350. H. Weiss, *Surf. Sci.* **331–333**, 1453 (1995).

351. J. Heidberg, M. Kandel, D. Meine, and V. Wildt, *Surf. Sci.* **331–333**(Pt. B), 1467 (1995).

352. A. Allouche, F. Cora, and C. Girardet, *Chem. Phys.* **201**, 59 (1995).

353. J. Heinberg and D. Meine, *Surf. Sci. Lett.* **279**, L175 (1992).

354. J. E. Jaffe and A. C. Hess, *J. Chem. Phys.* **104**, 3348 (1996).

355. L. N. Kuraeva, V. M. Zolotarev, Yu. O. Lisizyn, and S. S. Kachk, *Kolloidnyi zhurnal* **1**, 138 (1979).

356. C. M. Aubuchon, B. S. Davison, A. M. Nishimura, and N. J. Tro, *J. Phys. Chem.* **98**, 240 (1994).

357. G. E. Ewing, *Springer Ser., Surf. Sci.* **33**, 57 (1993).

358. M. Foster and G. E. Ewing, *Surf. Sci.* **427/428**, 102 (1999).

359. W. Stumm and J. J. Morgan, *Aquatic Chemistry*, 3rd ed., Wiley, New York, 1996.

360. A. R. Hind, S. K. Bhargava, and A. McKinnon, *Adv. Colloid Interface Sci.* **93**, 91 (2001).

361. J. Leja, *Surface Chemistry of Froth Flotation*, Plenum, New York, 1982.

362. M. H. Jones and J. T. Woodcock (Eds.), *Principles of Mineral Flotation: The Wark Symposium*, Australasian Institute of Mining and Metallurgy, Victoria, 1984.

363. V. A. Chanturia and V. E. Vigdergaus, *Electrochemistry of Sulfides*, Nauka, Moscow, 1993 (in Russian).

364. E. W. Giesekke, *Int. J. Mineral Process.* **11**, 19 (1983).

365. R. Woods, in J. O'M. Bockris, B. E. Conway, and R. E. White (Eds.), *Modern Aspects of Electrochemistry*, No. 29, Plenum, New York, 1996, p. 401.

366. T. Mallouk and D. J. Harrison, *Interfacial Design and Chemical Sensing*, ACS Symposium Series 561, American Chemical Society, Washington, DC, 1994.

367. Ch. P. Huang, Ch. R. O'Melia, and J. J. Morgan, *Aquatic Chemistry: Interfacial and Interspecies Processes*, Advances in Chemistry Series 244, American Chemical Society, Washington, DC, 1995.

368. I. N. Plaksin and V. I. Solnyshkin, *Infrared Spectroscopy of Surface Layers on Minerals*, Nauka, Moscow, 1966.

369. G. W. Poling, in M. C. Fuerstenau (Ed.), *Flotation: A. M. Gaudin Memorial Volume*, Vol. 1, American Institute of Mining, Metallurgical and Petroleum Engineers, New York, 1976, p. 334.

370. C. A. Prestige, J. Ralson, and R. St. C. Smart, *Int. J. Miner. Process.* **38**, 205 (1993).

371. J. L. Cecile, in K. S. E. Forssberg (Ed.), *Developments in Mineral Processing, Vol. 6: Flotation of Sulphide Minerals*, Elsevier, Amsterdam, 1985, p. 61.

372. I. V. Chernyshova, K. H. Rao, A. Vidyadhar, and A. V. Shchukarev, *Langmuir* **17**, 775 (2001).

373. I. V. Chernyshova and A. I. Andreev, *Appl. Surf. Sci.* **108**, 225 (1997).

374. K. Laajalehto, P. Nowak, A. Pomianowski, and E. Suoninen, *Colloids Surf.* **57**, 319 (1991).

375. J. M. Cases, M. Kongolo, P. de Donato, L. Michot, and R. Erre, *Int. J. Mineral Process.* **28**, 313 (1990).

376. J. M. Cases, M. Kongolo, P. de Donato, L. Michot, and R. Erre, *Int. J. Mineral Process.* **30**, 35 (1990).

377. M. Kongolo, J. M. Cases, P. de Donato, L. Michot, and R. Erre, *Int. J. Mineral Process.* **30**, 195 (1990).

378. R. K. Rath, S. Subramanian, and T. Pradeep, *J. Colloid Interface Sci.* **229**, 82 (2000).

379. P. Nowak, K. Laajalehto, and I. Kartio, *Colloids Surf.* **161**, 447 (2000).

380. C. M. Coretsky, D. A. Sverjensky, J. W. Salisbury, and D. M. D'Aria, *Geochim. Cosmochim. Acta.* **61**, 2193 (1997).

381. E. V. Kalinkina, A. M. Kalinkin, W. Forsling, and V. N. Makarov, *Int. J. Mineral Process.* **61**, 273 (2001).

382. E. V. Kalinkina, A. M. Kalinkin, W. Forsling, and V. N. Makarov, *Int. J. Mineral Process.* **61**, 289 (2001).

383. E. Mako, R. L. Frost, J. Kristof, and E. Horvath, *J. Colloid Interface Sci.* **244**, 359 (2001).

384. J. M. Cases and P. de Donato, *Int. J. Mineral Process.* **33**, 49 (1991).

385. P. Persson and I. Persson, *J. Chem. Soc. Faraday Trans.* **87**, 2779 (1991).

386. P. Persson and I. Persson, *Colloids Surf.* **58**, 161 (1991).

387. S. J. Hug, *J. Colloid Interface Sci.* **188**, 415 (1997).

388. J. D. Russell, E. Paterson, A. R. Fraser, and V. C. Farmer, *J. Chem. Soc. Faraday I* **71**, 1623 (1975).

389. R. L. Parfitt and J. D. Russell, *J. Soil Sci.* **28**, 297 (1977).

390. M. I. Tejedor-Tejedor and M. A. Anderson, *Langmuir* **6**, 602 (1990).

391. H. Wijnja and C. P. Schulthess, *Soil Sci. Soc. Am. J.* **65**, 324 (2001).

392. H. Wijnja and C. P. Schulthess, *Spectrochim. Acta A* **55**, 861 (1999).

393. G. Sposito, *The Surface Chemistry of Soils*, Oxford University Press, New York, 1984, pp. 138–141.

394. C. P. Schulthess and J. F. McCarthy, *Soil Sci. Soc. Am.* **54**, 688 (1990).

395. A. Van Green, A. P. Robertson, and J. O. Leckie, *Geochim. Cosmochim. Acta* **58**, 2073 (1994).

396. P. Persson, M. Karlsson, and L.-O. Ohman, *Geochim. Cosmochim. Acta* **62**, 3657 (1998).

397. D. Peak, R. G. Ford, and D. L. Sparks, *J. Colloid Interface Sci.* **218**, 289 (1999).

398. P. Nowak and K. Laajalehto, *Colloids Surf.* **157**, 101 (2000).

399. A. Gaudin, *Flotation*, 2nd ed., McGraw-Hill, New York, 1957.

400. M. A. Eigeles, *Basis of Flotation of Non-Sulfide Ores*, Nedra, Moscow, 1964.

401. I. A. Kakovski and E. I. Silina, in O. S. Bogdanov and I. A. Kakovski (Eds.) *Theoretical Studies of Flotation Process*, Mekhanobr, Leningrad, 1955, p. 53 (in Russian).

402. C. Du-Rietz, in E. Öhman (Ed.), *Progress in Mineral Dressing*: *Transactions of International Mineral Dressing Congress*, Stockholm, Almqvist & Wiksell, 1957, p. 34.

403. M. Cook and I. Nixon, *J. Phys. Colloid. Chem.* **54**, 445 (1950).

404. E. Shafrin and U. Tsisman, in *Monomolecular Layers*, Inostrannaia Liretatura, Moscow, 1956, p. 152.

405. H. S. Hanna and P. Somasundaran in M. C. Fuerstenau (Ed.), *Flotation: A. M. Gaudin Memorial Volume, Vol. 1*, American Institute of Mining, Metallurgical, and Petroleum Engineers, New York, 1976, p. 197.

406. K. P. Ananthapadmanabhan and P. Somasundaran, *Colloids Surf.* **13**, 151 (1985).

407. R. O. French, M. E. Wadsworth, M. A. Cook, and I. B. Culler, *J. Phys. Chem.* **58**, 805 (1954).

408. J. J. Kellar, W. M. Cross, and J. D. Miller, *Sep. Sci. Technol.* **25**, 2133 (1990).

409. W.-H. Jang and J. D. Miller, *Langmuir* **9**, 3159 (1993).

410. M. L. Free, W.-H. Jang, and J. D. Miller, *Colloids Surf.* **93**, 127 (1994).

411. J. D. Miller, W.-H. Jang, and J. J. Kellar, *Langmuir* **11**, 3272 (1995).

412. M. L. Free and J. D. Miller, *Int. J. Miner. Process.* **48**, 197 (1996).

413. Y. Lu, J. Drelich, and J. D. Miller, *J. Colloid Interface Sci.* **202**, 462 (1998).

414. C. A. Young and J. D. Miller, *Int. J. Miner. Process.* **58**, 331 (2000).

415. C. A. Young and J. D. Miller, *Minerals Metallurg. Process.* **18**, 38 (2001).

416. J. A. Mielczarski, J. A. Cases, P. Tekely, and D. Canet, *Langmuir* **9**, 3357 (1993).

417. J. A. Mielczarski and E. Mielczarski, *J. Phys. Chem.* **99**, 2649 (1993).

418. J. A. Mielczarski, J. A. Cases, E. Bouquet, O. Barres, and J. F. Delon, *Langmuir* **9**, 2370 (1993).

419. J. A. Mielczarski and E. Mielczarski, *J. Phys. Chem.* **99**, 3206 (1995).

420. J. A. Mielczarski, E. Mielczarski, and J. A. Cases, *Langmuir* **14**, 1739 (1998).

421. J. A. Mielczarski, E. Mielczarski, and J. A. Cases, *Langmuir* **15**, 500 (1999).

422. E. Mielczarski, Ph. de Donato, J. A. Mielczarski, J. M. Cases, O. Barres, and E. Bouquet, *J. Colloid Interface Sci.* **226**, 269 (2000).

423. K. Hanumantha Rao, J. M. Cases, P. deDonato, and K. S. E. Forssberg, *J. Colloid Interface Sci.* **145**, 314 (1991).

424. K. Hanumantha Rao, J. M. Cases, and K. S. E. Forssberg, *J. Colloid Interface Sci.* **145**, 330 (1991).

425. A. S. Peck, *Infrared Studies of Oleic Acid and Sodium Oleate Adsorption on Fluorite, Barite, and Calcite.* U. S. Bureau of Mines Report of Investigation No. 6202, 1963.

426. D. K. Schwartz, *Surf. Sci. Reports* **27**, 241 (1997).

427. D. H. Lee and R. A. Condrate, *J. Materials Sci.* **34**, 139 (1999).

428. F. Kimura, J. Umemura, and T. Takenaka, *Langmuir* **2**, 96 (1986).

429. W. Q. Gong, A. Parentich, L. H. Little, and L. J. Warren, *Langmuir* **8**, 118 (1992).

430. V. M. Lovell, L. A. Goold, and N. P. Finkelstein, *Int. J. Miner. Process.* **1**, 183 (1974).

431. Ya. I. Rabinovich, D. A. Guzonas, and R.-H. Yoon, *J. Colloid Interface Sci.* **155**, 221 (1993).

432. Y. Tsao, S. X. Yang, D. F. Evans, and H. Wennerstrom, *Langmuir* **7**, 3154 (1991).

433. D. H. Lee and R. A. Condrate, Sr., *J. Materials Sci.* **34**, 139 (1999).

434. D. H. Lee and R. A. Condrate, Sr., *J. Materials Sci.* **35**, 4961 (2000).

435. B. M. Antti and E. Forssberg, *Miner. Eng.* **2**, 217 (1989).

436. A. Ulman, *An Introduction to Ultrathin Organic Films: From Langmuir-Blodgett to Self-Assembly*, Academic, Boston, 1991, pp. 101–236.

437. J. C. Riviere and S. Mihra (Eds.), *Handbook of Surface and Interface Analysis*, Marcel Dekker, New York, 1998.

438. R. W. Smith and J. L. Scott, *Mineral Process. Extractive Metallurgy Rev.* **7**, 81 (1990).

439. R. W. Smith and S. Akhtar, in M. C. Fuerstenau (Ed.), *Flotation: A. M. Gaudin Memorial Volume*, American Institute of Mining, Metallurgical, and Petroleum Engineers, New York, 1976, Chapter 5, pp. 87–116.

440. R. W. Smith, in P. Somasundaran and B. M. Moudgil (Eds.), *Reagents in Mineral Technology*, Marcel Dekker, New York, 1988, pp. 219–256.

441. A. M. Gaudin and D. W. Fuerstenau, *Trans. Soc. Min. Eng. AIME.* **202**, 958 (1955).

442. P. Somasundaran and D. W. Fuerstenau, *J. Phys. Chem.* **70**, 90 (1966).

443. B. E. Novich and T. A. Ring, *Langmuir* **1**, 701 (1985).

444. D. W. Fuerstenau and H. M. Jang, *Langmuir* **7**, 3138 (1991).

445. G. C. Pimentel and A. L. Mc Clellan, *The Hydrogen Bond*, Freeman, San Francisco, 1960.

446. P. L. deBruyn, *Trans AIME* **202**, 291 (1955).

447. H. Schubert, *Freiberger Forschungshefte A* **514**, 3 (1972).

448. J. M. Cases and B. Mataftschiev, *Surf. Sci.* **9**, 57 (1968).

449. J. M. Cases and F. Villieras, *Langmuir* **8**, 1251 (1992).

450. J. S. Laskowski, R. M. Vurdela, and Q. Liu, in K. S. E. Forssberg (Ed.), *Developments in Mineral Processing*, Proceedings of the XVI International Mineral Processing Congress, Stockholm, Sweden, Elsevier, Amsterdam, 1988, pp. 703–715.

451. I. V. Chernyshova and H. K. Rao, *Langmuir* **17**, 2711 (2001).

452. A. Gericke, J. Simon-Kutscher, and H. Huhnerfuss, *Langmuir* **9**, 3115 (1993).

453. D. Blaudez, T. Buffeteau, B. Desbat, P. Fournier, A.-M. Ritcey, and M. Pezolet, *J. Phys. Chem. B* **102**, 99 (1998).

454. T. Hasegawa, J. Umemura, and T. Takenaka, *J. Phys. Chem.* **97**, 9009 (1993).

455. H. Hoffmann, U. Mayer, and A. Krischanitz, *Langmuir* **11**, 1304 (1995).

456. A. N. Parikh and D. L. Allara, *J. Phys. Chem.* **96**, 927 (1992).

457. D. L. Allara, A. N. Parikh, and F. Rondelez, *Langmuir* **11**, 2357 (1995).

458. A. Gericke, A. V. Michailov, and H. Huhnerfuss, *Vib. Spectrosc.* **4**, 335 (1993).

459. C. R. Flach, A. Gericke, and R. Mendelsohn, *J. Phys. Chem. B* **101**, 58 (1997).

460. Y. Ren, C. W. Meuse, S. L. Hsu, and H. D. Stidham, *J. Phys. Chem.* **98**, 8424 (1994).

461. I. V. Chernyshova and H. K. Rao, *J. Phys. Chem. B* **105**, 810 (2001).

462. K. Sinniah, J. Cheng, S. Terrettas, J. E. Reuttrobey, and C. J. Miller, *J. Phys. Chem.* **99**, 14500 (1995).

463. H. Hoffman, U. Mayer, H. Brunner, and A. Krischanitz, *Vib. Spectrosc.* **8**, 151 (1995).

464. R. Woods, in M. C. Fuerstenau (Ed.), *Flotation: A. M. Gaudin Memorial Volume, Vol. 1*, American Institute of Mining, Metallurgical, and Petroleum Engineers, New York, 1976, p. 298.

465. R. Woods, in M. H. Jones and J. T. Wooscock (Eds.), *Principles of Mineral Flotation*, Australasian Institute of Mining and Metallurgy, Parkville, Victoria, 1984, p. 91.

466. Th. Healy, in M. H. Jones and J. T. Wooscock (Eds.), *Principles of Mineral Flotation*, Australasian Institute of Mining and Metallurgy, Parkville, Victoria, 1984, p. 43.

467. K. L. Sutherland and J. W. Wark, *Principles of Flotation*, Australasian Institute of Mining and Metallurgy, Melboure, 1955.

468. Z. X. Sun, *Surface Reactions in Aqueous Metal Sulfide Systems*, Ph.D. Thesis, Lulea University of Technology, Lulea, 1991.

469. W. Forsling and Z. X. Sun, *Int. J. Miner. Process.* **51**, 81 (1997).

470. G. Winter, *Rev. Inorg. Chem.* **2**, 253 (1980).

471. V. I. Klassen and V. A. Mokrousov, *Introduction into the Theory of Flotation*, Gosgortekhizdat, Moscow, 1959 (in Russian).

472. K. L. Sutherland and J. W. Wark, *Principles of Flotation*, Australasian Institute of Mining and Metallurgy, Melbourne, 1955.

473. V. A. Konev, *Flotation of Sulfides*, Nedra, Moscow, 1985 (in Russian).

474. D. A. Shvedov, *Gorno-obogatitelnyi Zhurnal* **6**, 24 (1936).

475. D. A. Shvedov and I. N. Shorsher, *Gorno-obogatitelnyi zhurnal* **8**, 9 (1937).

476. V. A. Glembotski and E. A. Anfimova, *Flotation of Oxidized Ores of Non-Ferrous Metals*, Nedra, Moscow, 1966 (in Russian).

477. S. G. Salamy and J. C. Nixon, *Aust. J. chem.* **7**, 146 (1954).

478. I. V. Chernyshova, *J. Phys. Chem. B* **105**, 8185 (2001).

479. N. B. Colthup and L. P. Powell, *Spectrochim. Acta* **43A**, 317 (1987).

480. P. Persson, Ph.D. Thesis, *On the Adsorption of Alkylxanthate Ions on Sulfide Mineral and Synthetic Metal Sulfides*, Uppsala University, Uppsala, 1990.

481. A. Ihs, K. Uvdal, and B. Liedberg, *Langmuir* **9**, 733 (1993).

482. A. Ihs, Ph.D. Thesis, *Organic Molecules on Metal Surfaces: Structural Studies Using Infrared and X-ray Photoelectron Spectroscopy*, Linkoping University, Linkoping, 1993.

483. J. O. Leppinen, *Int. J. Miner. Process.* **30**, 245 (1990).

484. P. Persson, B. Malmensten, and I. Persson, *J. Chem. Soc. Faraday Trans.* **87**, 2769 (1991).

485. A. T. Pilipenko and I. V. Mel'nikova, *Heorganicheskaya khimial* **15**, 1186 (1970) (in Russian).

486. X.-H. Wang, *J. Colloid Interface Sci.* **171**, 413 (1995).

487. R. G. Greenler, *J. Phys. Chem.* **66**, 879 (1962).

488. A. M. Kongolo, J. M. Cases, A. Burneau, and J. J. Predali, in M. J. Jones and R. Oblatt (Eds.), *Reagents in Mineral Industry*, Institute of Mining and Metallurgy, London, 1984, p. 79.

489. M. C. Fuerstenau, M. C. Kuhn, and D. A. Elgillani, *Trans. AIME* **241**, 148 (1968).

490. Ph. de Donato, J. M. Cases, M. Kongolo, L. J. Michot, and A. Burneau, *Colloids Surf.* **44**, 207 (1990).

491. P. Persson and I. Persson, *Colloids Surf.* **58**, 161 (1991).

492. A. Etahiri, B. Humbert, K. El Kacemi, B. Marouf, and J. Bessiere, *Int J Miner Process* **52**, 49 (1997).

493. I. Persson, P. Persson, M. Valli, S. Fozo, and B. Malmensten, *Int. J. Miner. Process.* **33**, 67 (1991).

494. J. M. Cases, Ph. de Donato, M. Kongolo, and L. J. Michot, in *XVIII International Mineral Processing Congress, 23–28 May 1993, Sydney, Australia*, Autralasian Institute of Mining and Metallurgy, Parkville, Victoria, 1993, p. 663.

495. C. A. Prestige, J. Ralson, and R. St. C. Smart, *Int. J. Miner. Process.* **38**, 205 (1993).

496. K. Laajalehto, P. Nowak, A. Pomianowski, and E. Suoninen, *Colloids Surf.* **57**, 319 (1991).

497. V. A. Konev and E. D. Kriveleva, *Int. J. Miner. Process.* **28**, 189 (1990).

498. E. D. Kriveleva and V. A. Konev, *Tsvetnye metally* **1**, 115 (1990).

499. M. L. Larsson, A. Holmgren, and W. Forsling, *Langmuir* **16**, 8129 (2000).

500. J. A. Mielczarski, E. Mielczarski, and J. A. Cases, *Langmuir* **12**, 6521 (1996).

501. G. Sunholm and P. Talonen, *J. Electroanal. Chem.* **380**, 261 (1995).

502. G. W. Poling and J. Leja, *J. Phys. Chem.* **67**, 2121 (1963).

503. J. Mielczarski, P. Nowak, and J. W. Strojek, *Polish J. Chem.* **54**, 279 (1980).

504. J. Lekki and T. Chmielewski, *Fyzykochem. Probl. Metallurgii* **21**, 127 (1989).

505. J. A. Mielczarski, J. M. Cases, and O. Barres, *J. Colloid Interface Sci.* **178**, 740 (1996).

506. C. A. Prestige, J. Ralson, and R. St. C. Smart, *Int. J. Miner. Process.* **38**, 205 (1993).

507. J. Mielczarski, P. Nowak, and J. Strojek, *Polish J. Chem.* **54**, 279 (1980).

508. V. Bozkurt, Z. Xu, and J. A. Finch, *Int. J. Miner. Process.* **52**, 203 (1998).

509. V. Bozkurt, Z. Xu, and J. A. Finch, *Can. Metallurg. Quart.* **38**, 105 (1999).

510. J. A. Mielczarski, J. A. Cases, M. Alnot, and J. J. Ehrhardt, *Langmuir* **12**, 2531 (1996).

511. Z. Xu and J. A. Finch, *Trans. IMM* **105**, C197 (1996).

512. V. Bozkurt, Z. Xu, and J. A. Finch, *Int. J. Miner. Process.* **52**, 203 (1998).

513. J. O. Leppinen, C. I. Basilio, and R. H. Yoon, in P. E. Richardson and R. Woods (Eds.), *Electrochemistry in Mineral and Metal Processing*, Vol. 2, American Electrochemical Society, Pennington, NJ, 1988, p. 49.

514. J. O. Leppinen, C. I. Basilio, and R. H. Yoon, *Int. J. Miner. Process.* **26**, 259 (1989).

515. W. J. Trahar, in M. H. Jones and J. T. Woodcock (Eds.), *Principles of Mineral Flotation*, Australasian Institute of Mining and Metallurgy, Parkville, Victoria, 1984, p. 117.

516. D. W. Fuerstenau and R. K. Mishra, in H. H. Jones (Ed.), *Complex Sulphide Ores (Rome)*, Institute of Mining and Metallurgy, London, 1981, 271.

517. A. V. Shchukarev, I. M. Kravets, and A. N. Buckle, *Int. J. Miner. Process.* **44**, 99 (1994).

518. I. V. Chernyshova and V. P. Tolstoy, *Appl. Spectrosc.* **49**, 665 (1995).

519. A. Golikov, *Tsvetnye metally. (N.Y.)* **2**, 19 (1961).

520. H. O. Finklea (Ed.), *Semiconductor Electrodes (Studies in Physical and Theoretical Chemistry)*, Elsevier, Amsterdam, 1988.

521. A. J. Nozik and R. Memming, *J. Phys. Chem.* **100**, 13061 (1996).

522. F. F. Volkenstein, *Electronic Processes on the Semiconductor Surface*, Nauka, Moscow, 1987 (in Russian).

523. V. A. Myamlin and Yu. V. Pleskov, *Electrochemistry of Semiconductors*, Plemun, New York, 1967.

524. K. E. R. England, J. M. Charnock, R. A. D. Pattrick, and D. J. Vaugan, *Mineralogical Magazine* **63**, 559 (1999).

525. P. A. Christensen and A. Hamnett, *Techniques and Mechanisms in Electrochemistry*, Blackie, London, 1994.

526. X. Itaya, *Prog. Surf. Sci.* **58**, 121 (1998).

527. C. A. Melendres and A. Tadjeddine (Eds.), *Synchrotron Techniques in Interfacial Electrochemistry*, Kluwer, Dordrecht, 1993.

528. (a) J. N. Chazalviel, B. H. Erne, F. Maroun, and F. Ozanam, *J. Electroanal. Chem.* **502**, 180 (2001); (b) J. N. Chazalviel, B. H. Erne, F. Maroun, and F. Ozanam, *J. Electroanal. Chem.* **509**, 108 (2001).

529. J.-N. Chazalviel, S. Fellah, and F. Ozanam, *J. Electroanal. Chem.* **524–525**, 137 (2002).

530. A. V. Rao, J.-N. Chazalviel, and F. Ozanam, *J. Appl. Phys.* **60**, 696 (1986).

531. A. V. Rao and J.-N. Chazalviel, *J. Electrochem Soc.* **134**, 2777 (1987).

532. L. M. Peter, D. J. Blackwood, and S. Pons, *J. Electroanal. Chem.* **294**, 111 (1990).

533. J.-N. Chazalviel, V. M. Dubin, K. C. Mandal, and F. Ozanam, *Appl. Spectrosc.* **47**, 1411 (1993).

534. J. Rappich, H. J. Lewerenz, and H. Gerischer, *J. Electrochem. Soc.* **140**, L187 (1993).

535. Q. Fan and L. M. Ng, *J. Electrochem. Soc.* **141**, 3369 (1994).

536. V. M. Dubin, F. Ozanam, and J.-N. Chazalviel, *Vib. Spectrosc.* **8**, 159 (1995).

537. J. Rappich, M. Aggour, S. Rauscher, H. J. Lewerenz, and H. Jungblut, *Surf. Sci.* **335**, 160 (1995).

538. J. Rappich and H. J. Lewerenz, *Thin Solid Films* **276**, 25 (1996).

539. J. Rappich and H. J. Lewerenz, *Electrochim. Acta* **41**, 675 (1996).

540. F. Ozanam, A. Djebri, and J.-N. Chazalviel, *Electrochim. Acta* **41**, 687 (1996).

541. C. da Fonseca, F. Ozanam, and J.-N. Chazalviel, *Surf. Sci.* **365**, 1 (1996).

542. F. Ozanam, C. da Fonseca, A. Venkateswara Rao, and J.-N. Chazalviel, *Appl. Spectrosc.* **51**, 519 (1997).

543. C. da Fonseca, F. Ozanam, and J.-N. Chazalviel, *Microchim. Acta* **14**(Suppl.), 811 (1997).

544. J.-N. Chazalviel and F. Ozanam, *J. Appl. Phys.* **81**, 7684 (1997).

545. S. Cattarin, J.-N. Chazalviel, C. da Fonseca, F. Ozanam, L. M. Peter, G. Schichthorl, and J. Stumper, *J. Electrochem. Soc.* **145**, 498 (1998).

546. P. Allongue, C. H. de Villeneure, J. Pinson, F. Ozanam, J.-N. Chazalviel, and X. Wallart, *Electrochim. Acta* **43**, 2791 (1998).

547. J.-N. Chazalviel, C. da Fonseca, and F. Ozanam, *J. Electrochem. Soc.* **145**, 964 (1998).

548. S. Rauscher, O. Nast, H. Junglut, and H. J. Lewerenz, *Proc. Electrochem. Soc.* **97–35**, 439 (1998).

549. A. Belaidi, M. Safi, F. Ozanam, J.-N. Chazalviel, and O. Gorochov, *J. Electrochem. Soc.* **146**, 2659 (1999).

550. J.-N. Chazalviel, A. Belaidi, M. Safi, F. Maroun, B. H. Erne, and F. Ozanam, *Electrochim. Acta* **45**, 3205 (2000).

551. B. H. Erne, F. Ozanam, and J.-N. Chazalviel. *Phys. Rev. Lett.* **80**, 4337 (1998).

552. B. H. Erne, M. Shchakovsky, F. Ozanam, and J.-N. Chazalviel, *J. Electrochem. Soc.* **145**, 447 (1998).

553. B. H. Erne, F. Ozanam, M. Shchakovsky, D. Vanmaekelbergh, and J.-N. Chazalviel. *J. Phys. Chem. B* **104**, 5961 (2000).

554. K. C. Mandal, F. Ozanam, and J.-N. Chazalviel, *J. Electroanal. Chem.* **336**, 153 (1992).

555. J.-N. Chazalviel, K. C. Mandal, and F. Ozanam, *SPIE Conf. Proc.* **1575**, 40 (1992).

556. F. Maroun, F. Ozanam, and J.-N. Chazalviel. *J. Phys. Chem B* **103**, 5280 (1999).

557. F. Maroun, F. Ozanam, and J.-N. Chazalviel, *Surf. Sci.* **427–428**, 184 (1999).

558. B. A. Blajeni, M. A. Habib, I. Taniguchi, and J. O'M. Bockris, *J. Electroanal. Chem.* **157**, 399 (1983).

559. Q. Fan and L. M. Ng, *J. Electroanal. Chem.* **398**, 151 (1995).

560. I. V. Chernyshova, *J. Phys. Chem. B* **105**, 8178 (2001).

561. I. V. Chernyshova, *Russ. J. Electrochem.* **37**, 679 (2000).

562. G. H. Kelsall, Q. Yin, D. J. Vaughan, K. E. R. England, and N. P. Braudon, *J. Electroanal. Chem.* **471**, 116 (1999).

563. L. Kavan, P. Krtil, and M. Gratzel, *J. Electroanal. Chem.* **373**, 123 (1994).

564. K. Shaw, P. Christensen, and A. Hamnett, *Electrochim. Acta* **41**, 719 (1996).

565. P. Krtil, L. Kavan, and I. Hoskovcova, *J. Appl. Electrochem.* **26**, 523 (1996).

566. J. Klima, K. Kratochilova, and J. Ludvik, *J. Electroanal. Chem.* **427**, 57 (1997).

567. P. Krtil, L. Kavan, and K. Macounova, *J. Electroanal. Chem.* **433**, 187 (1997).

568. P. Christensen, J. Eameaim, and A. Hamnett, *Phys. Chem. Chem. Phys.* **1**, 5315 (1999).

569. C. L. P. S. Zanta, A. R. deAndrade, and J. F. C. Boodts, *J. Appl. Electrochem.* **30**, 467 (2000).

570. J. Lee and Y. Tak, *J. Ind. Eng. Chem.* **5**, 87 (1999).

571. J. O'M. Bockris and A. K. N. Reddy, *Modern Electrochemistry, Vol. 2*, Plenum, New York, 1970, pp. 1231–1250.

572. D. J. Vaughan and J. R. Craig, *Mineral Chemistry of Metal Sulfides*, Cambridge University Press, Cambridge, 1978.

573. J. R. Gardner and R. Woods, *Austral. J. Chem.* **26**, 1635 (1973).

574. E. Ahlberg and A. E. Broo, *Int. J. Miner. Process.* **33**, 135 (1991).

575. A. N. Buckley and R. Woods, *Appl. Surf. Sci.* **17**, 1984, 401.

576. A. N. Buckley and G. W. Walker, in K. S. E. Forssberg (Ed.), *Developments in Mineral Processing*, Proceedings of the XVI International Mineral Processing Congress, Stockholm, Sweden, Elsevier, Amsterdam, 1988, p. 589.

577. A. N. Buckley and R. Woods, *J. Electroanal. Chem.* **370**, 295 (1994).

578. P. E. Richardson and E. E. Maust, in M. C. Fuerstenau (Ed.), *Flotation: A. M. Gaudin Memorial Volume*, Vol. 1, American Institute of Mining, Metallurgical, and Petroleum Engineers, New York, 1976, p. 364.

579. P. E. Richardson, Y.-Q. Li, and R.-H. Yoon, *XVIII International Mineral Processing Congress, Sydney, Australia, May 23–28, 1993*, Australasian Institute of Mining and Metallurgy, Sidney, 1993, p. 757.

580. I. J. Kartio, K. Laajalehto, E. J. Suoninen, A. N. Buckley, and R. Woods, *Colloids Surf.* **133**, 303 (1998).

581. D. Schuhmann, in S. H. Castro Flores and J. A. Moisan (Eds.), *Flotation*, Proceedings 2nd Lat.-Amer. Congress, Conception, Aug. 19–23, 1985, Elsevier, Amsterdam, 1988, p. 65.

582. E. Ndzebet, D. Schuhmann, and P. Vanel, *Electrochim. Acta* **39**, 745 (1994).

583. D. Schuhmann, *New J. Chem.* **17**, 551 (1993).

584. Th. Peuporte and D. Schuhmann, *J. Electroanal. Chem.* **385**, 9 (1995).

585. S. Fletcher and M. D. Horne, *Int. J. Miner. Process.* **33**, 145 (1991).

586. R. Woods, *Electrochemistry of Sulfide Mineral Flotation, Lectures*, School of Science Griffith University, Australia, 2000.

587. B. S. Kim, R. A. Hayers, C. A. Prestige, J. Ralson, and R. St. Smart, *Langmuir* **11**, 2554 (1995).

588. G. Wittstock, I. Kartio, D. Hirsch, S. Kunze, and R. Szargan, *Langmuir* **12**, 5709 (1996).

589. Z. Sun, W. Forsling, L. Rönngren, and S. Sjöberg, *Int. J. Miner. Process.* **33**, 83 (1991).

590. Z. Sun, W. Forsling, L. Rönngren, S. Sjöberg, and P. W. Schindler, *Colloids Surf.* **59**, 243 (1991).

591. W. Forsling and Z. Sun, *Int. J. Miner. Process.* **51**, 81 (1997).

592. A. N. Buckley and R. Woods, *J. Appl. Electrochem.* **26**, 899 (1996).

593. S. B. Turcotte, R. E. Benner, A. M. Riley, J. Li, M. E. Wadeworth, and D. M. Bodily, *J. Electroanal. Chem.* **347**, 195 (1993).

594. P. O. Lam-Thi, M. Lamache, and D. Bauer, *Electrochim. Acta* **29**, 217 (1984).

595. R. L. Paul, M. L. Nicol, J. W. Diggle, and A. P. Saunders, *Electrochem. Acta* **23**, 625 (1978).

596. D. Schuhmmann, A. M. Guinard-Baticle, P. Vanel, and A. Talib, *J. Electrochem. Soc.* **134**, 1128 (1987).

597. J. R. Gardner and R. Woods, *J. Electroanal. Chem.* **100**, 447 (1979).

598. T. Chmielewski and J. Lekki, *Miner. Eng.* **2**, 387 (1989).

599. P. E. Richardson and C. S. O'Dell, *J. Electrochem. Soc.* **132**, 1350 (1985).

600. P. J. Guy and W. J. Trahar, *Int. J. Miner. Process.* **12**, 15 (1984).

601. R. Woods and R.-H. Yoon, *Langmuir* **13**, 876 (1997).

602. J. A. Mielczarski, *Langmuir* **13**, 878 (1997).

603. K. Laajalehto, J. Leppinen, I. Kartio, and T. Laiho, *Colloids. Surf. A* **154**, 193 (1999).

604. H. Gerischer *Semiconductors Electrochemistry: Physical Chemistry*, Academic, New York, 1970, p. 463.

605. H. Gerischer, *Surf. Sci.* **13**, 265 (1969).

606. X.-H. Wang, *J. Colloid Interface Sci.* **171**, 413 (1995).

607. E. Suoninen and K. Laajalehto, in *XVIII International Mineral Processing Congress, Sydney, Australia, May 23–28, 1993*, Australasian Institute of Mining and Metallurgy, Parkville, Victoria, 1993, p. 625.

608. J. R. Gardner and R. Woods, *Austr. J. Chem.* **30**, 981 (1977).

609. A. G. Brolo, D. E. Irish, and B. D. Smith, *J. Mol. Struct.* **405**, 29 (1997).

610. G. Lupke, *Surf. Sci. Rep.* **35**, 75 (1999).

611. A. N. Asanov and L. L. Larina, in M. J. Allen (Ed.), *Charge and Field Effects in Biosystems — 3 (3rd International Symposium)*, Birkhauser, Boston, 1992, p. 13.

612. A. A. Tidblad, J. Martensson, *Electrochim. Acta* **42**, 389 (1997).

613. M. R. Deakin and D. A. Buttry, *Anal. Chem.* **61**, 1147A (1989).

614. K. Itaya, *Progr. Surf. Sci.* **58**, 121 (1998).

615. H. Abruna (Ed.), *Electrochemical Interfaces: Modern Techniques for In-Situ Interface Characterization*, VCH, New York, 1991.

616. D. Aurbach, B. Markovsky, M. D. Levi, E. Levi, A. Schechter, M. Moshkovich, and Y. Cohen, *J. Power Sources* **82**, 95 (1999).

617. J. Lipkowski, *Can. J. Chem.* **77**, 1163 (1999).

618. P. A. Christensen and A. Hammett, *Techniques in Electrochemistry*, Blackie, London, 1994.

619. A. Wieckowski (Ed.), *Interfacial Electrochemistry: Theory, Experiment and Applications*, Marcel Dekker, New York, 1999.

620. J. O'M. Bockris and S. U. M. Khan, *Surface Electrochemistry*, Plenum, New York, 1993.

621. J. Lipkowski and P. N. Ross (Eds.), *Adsorption of Molecules at Metal Electrodes*, Vol. 1, VCH, New York, 1992.

622. J. E. Pemberton and S. D. Garvey, in P. Vanysek (Ed.), *Modern Techniques in Electroanalysis*, Chemical Analysis Series, Vol. 139, Wiley, New York, 1996, p. 59.

623. H. D. Abruna, J. H. White, M. J. Albarelli, G. M. Bommarito, M. G. Bedzyk, and M. McMillan, *J. Phys. Chem.* **92**, 7045 (1988).

624. N. M. Markovic and P. N. Ross, *Surf. Sci. Rep.* **45**, 121 (2002).

625. N. M. Markovic and P. N. Ross, *Electrochim. Acta* **45**, 4101 (2000).

626. P. A. Christensen and A. Hammett, *Compr. Chem. Kinet* **29**, 1 (1989).

627. J. McQuillan, *Chem. N. Z.* **54**, 61 (1990).

628. S. C. Chang and M. J. Weaver, *J. Phys. Chem.* **95**, 5391 (1991).

629. S. M. Stole, D. D. Popenoe, and M. P. Porter, in H. D. Abruna (Ed.), *Electrochemical Interfaces: Modern Techniques for In-Situ Interface Characterization*, VCH, New York, 1991, p. 339.

630. T. Iwasita and F. C. Nart, in H. Gerischer and C. W. Tobias (Eds.), *Advances in Electrochemical Science and Engineering*, Vol. 4, VCH Publishers, New York, 1991, p. 123.

631. B. Beden, J.-M. Leger, and C. Lamy, in J. O. M. Bockris, B. E. Conway, and R. E. White (Eds.), *Modern Aspects of Electrochemistry*, Vol. 22, Plenum, New York, 1992, p. 97.

632. R. J. Nichols, in J. Lipkowski and P. N. Ross (Eds.), *Adsorption of Molecules at Metal Electrodes*, VCH, New York, 1992, p. 347.

633. M. J. Weaver, N. Kizhakevarian, X. Jiang, J. Villegas, C. Stuhlman, A. Tolia, and X. Gao, *J. Electron Spectrosc. Related Phenom.* **64–65**, 351 (1993).

634. K. Keiji, *Hyomen Kagaku* **11**, 8 (1994).

635. J. E. Pemberton and S. D. Garvey, in P. Vanysek (Ed.), *Modern Techniques in Electroanalysis*, Wiley, New York, 1996, pp. 59–106.

636. C. Korzeniewski, *Crit. Rev. Anal. Chem.* **27**, 81 (1997).

637. T. Iwasita and F. C. Nart, *Prog. Surf. Sci.* **55**, 271 (1997).

638. M. I. S. Lopes and L. Proenca, *Port. Electrochim. Acta* **15**, 81 (1997).

639. B. Orel, *Acta Chim. Slov.* **44**, 397 (1997).

640. M. J. Weaver and S. Zou, in R. J. H. Clark and R. E. Hester (Eds.), *Spectroscopy for Surface Science*, Wiley, New York, 1998, pp. 219–271.

641. P. A. Christensen and A. Hammett, *Electrochim. Acta* **45**, 2443 (2000).

642. N. Marinkovic, J. J. Calvente, A. Kloss, Z. Kovaccova, and W. R. Fawcett, *J. Electroanal. Acta* **467**, 325 (1999).

643. D. S. Corrigan and M. J. Weaver, *J. Electroanal. Chem.* **239**, 55 (1988).

644. M. W. Severson, C. Stuhlmann, I. Villegas, and M. J. Weaver, *J. Chem. Phys.* **103**, 9832 (1995).

645. C. Tang, S. Zou, M. W. Severson, and M. J. Weaver, *J. Phys. Chem. B* **102**, 8546 (1998).

646. C. Tang, S. Zou, M. W. Severson, and M. J. Weaver, *J. Phys. Chem. B* **102**, 8796 (1998).

647. H.-Q. Li, S. G. Roscoe, and J. Lipkowski, *J. Electroanal. Chem.* **478**, 67 (1999).

648. C. S. Kim, W. J. Tornquist, and C. Korzeniewski, *J. Chem. Phys.* **101**, 9113 (1994).

649. S. C. Chang and M. Weaver, *J. Phys. Chem.* **92**, 4582 (1990).

650. C. Tang, S. Zou, M. W. Severson, and M. J. Weaver, *Surf. Sci.* **412/413**, 344 (1998).

651. K. Ataka and M. Osawa, *J. Electroanal. Chem.* **460**, 188 (1999).

652. S. C. Chang, J. D. Roth, Y. Ho, and M. J. Weaver, *J. Electron Spectrosc. Related Phenom.* **54/55**, 1185 (1990).

653. C. S. Kim and C. Korzeniewski, *Anal. Chem.* **69**, 2349 (1997).

654. H. Ogasawara, J. Inukai, and M. Ito, *Chem. Phys. Lett.* **198**, 389 (1992).

655. S. Zou, R. Gomez, and M. J. Weaver, *Surf. Sci.* **399**, 270 (1998).

656. S. I. Yanger and D. W. Vidrine, *Appl. Spectrosc.* **40**, 174 (1986).

657. J. Daschbach, D. Heisler, and B. S. Pons, *Appl. Spectrosc.* **40**, 489 (1986).

658. K. Kitamura, M. Takahashi, and M. Ito, *J. Phys. Chem.* **92**, 3320 (1988).

659. M. Osawa, K. Yoshii, K. Ataka, and T. Yotsuyanagi, *Langmuir* **10**, 640 (1994).

660. M. Osawa and K. Yoshii, *Appl. Spectrosc.* **51**, 512 (1997).

661. M. Osawa, K. Yoshii, Y. Hibino, T. Nakano, and I. Noda, *J. Electroanal. Chem.* **426**, 11 (1997).

662. H. Noda, K. Ataka, L. Wan, and M. Osawa, *Surf. Sci.* **427–428**, 190 (1999).

663. (a) S. Pons, T. Davidson, and A. Bewick, *J. Electroanal. Chem.* **140**, 211 (1982); (b) S. Pons, *J. Electroanal. Chem.* **150**, 495 (1983).

664. K. Ataka, Y. Hara, and M. Osawa, *J. Electroanal. Chem.* **473**, 34 (1999).

665. W. R. Fawcett, A. A. Kloss, J. J. Calvente, and N. Marinkovic, *Electrochim. Acta* **44**, 881 (1998).

666. T. B. Dudrovsky, Z. Hou, P. Troeve, and N. L. Abott, *Anal. Chem.* **71**, 327 (1999).

667. T. Kratzmuller, D. Appelhans, and H.-G. Braun, *Adv. Mater.* **11**, 555 (1999).

668. G. Leukroth, *Gummi, Asbest, Kunstst* **28**, 1118 (1970).

669. H. G. Tompkins, *Thin Solid Films* **119**, 337 (1984).

670. M. Ito, *Shikizai Kyokaishi (Jap.)* **57**, 200 (1984).

671. T. Nguyen, *Prog. Org. Coat.* **13**, 1 (1985).

672. J. O. Koenig, *Polym. Mater. Sci. Eng.* **64**, 27 (1991).

673. M. W. Urban, *Adv. Chem. Ser.* **236**, 3 (1993).

674. H. W. Siesler, *Adv. Chem. Ser.* **236**, 41 (1993).

675. M. Sargent and J. L. Koenig, *Adv. Chem. Ser.* **236**, 191 (1993).

676. K. A. B. Lee, *Appl. Spectrosc. Rev.* **28**(3), 231 (1993).

677. G. Miller, *Appl. Spectrosc.* **47**(2), 222 (1993).

678. M. B. Mitchell, *Adv. Chem. Ser.* **236**, 351 (1993).

679. J. M. Chalmers and N. J. Everall, in B. J. Hunt and M. I. James (Eds.), *Polymer Characterisation*, Blackie Academic and Professional, Glasgow, 1993, pp. 69–114.

680. R. Meier, *Trends Polym. Sci.* **2**, 53 (1994).

681. V. A. Berschtein and V. A. Ryzhov, *Ads. Polym. Sci.* **114**, 43 (1994).

682. J. Yarwood, *Spectrosc. Eur.* **8**, 8 (1996).

683. J. M. Chalmers, N. J. Everall, and S. Ellison, *Micron* **27**, 315 (1996).

684. R. Zbinden, *Infrared Spectroscopy of High Polymers*, Academic, New York, 1964.

685. H. W. Siesler and K. Holland-Moritz (Eds.), *Infrared and Raman Spectroscopy of Polymers*, Marcell Dekker, New York, 1980.

686. H. Ishida (Ed.), *Infrared Fourier Transform Characterization of Polymers*, Plenum, New York, 1987.

687. A. Garton, *IR Spectroscopy of Polymer Blends, Composites, and Surfaces*, Oxford, New York, 1992.

688. M. Urban, *Vibrational Spectroscopy of Molecules and Macromolecules on Surfaces*, Wiley, New York, 1993.

689. D. I. Bower and W. F. Maddams, *The Vibrational Spectroscopy of Polymers*, Cambridge University Press, Cambridge, 1992.

690. M. W. Urban and C. D. Craver (Eds.), *Surface-Property Relations in Polymers, Spectroscopy and Performance*, American Chemical Society, Washington, DC, 1993.

691. M. Urban, *Attenuated Total Reflectance Spectroscopy of Polymers*, American Chemical Society, Washington, DC, 1996.

692. A. H. Kuptsov and G. N. Zhizhin (Eds.), *Handbook of FT Raman and IR Spectra of Polymers*, Elsevier, Oxford, 1998.

693. J. L. Koenig, *Spectroscopy of Polymers*, 2nd ed., Elsevier, Amsterdam, 1999.

694. I. Noda, A. E. Dowrey, and C. Marcott, *Appl. Spectrosc.* **47**, 1317 (1993).

695. M. Sonoyama, K. Shoda, G. Katagiri, H. Ishida, T. Nakano, S. Shimada, T. Yokoyama, and H. Toriumi, *Appl. Spectrosc.* **51**, 598 (1997).

696. M. Sonoyama, M. Miyizawa, G. Katagiri, and H. Ishida, *Appl. Spectrosc.* **51**, 545 (1997).

697. V. I. Vettegren, I. I. Novak, and A. E. Chmel, *Vysokomolekularnye soedinenia* **17B**, 605 (1975).

698. V. I. Dyrda, V. I. Vettegren, and V. P. Nadutyi, *Kauchuk i resina* **10**, 30 (1974).

699. V. I. Dyrda, V. I. Vettegren, and V. P. Nadutyi, *Kauchut i resina (Russ.)* **4**, 26 (1976).

700. P. Franz and S. Granik, *Macromolecules* **27**, 2553 (1994).

701. H. M. Shneider, S. Granick, and S. Smith, *Macromolecules* **27**, 4714 (1994).

702. H. M. Shneider, S. Granick, and S. Smith, *Macromolecules* **27**, 4721 (1994).

703. E. P. Enriquez and S. Granick, *Colloids Surf. A* **113**, 11 (1996).

704. D. J. Carlson and D. M. Wiles, *Polym. Lett.* **8**, 419 1970.

705. D. J. Carlson and D. M. Wiles, *Can. J. Chem.* **48**, 2397 (1970).

706. D. J. Carlson and D. M. Wiles, *Macromolecules* **4**, 174 (1971).

707. U. A. Pentin, B. I. Tarasevich, and B. S. Elzefon, *Zhurn. Fizicheskoi Chimii (Russ.)* **46**, 2116 (1972).

708. W. Chen and T. J. McCarthy, *Macromolecules* **31**, 3648 (1998).

709. M. S. Akutin, A. N. Shabadash, M. L. Korber, I. O. Stalnova, and B. V. Alekseev, *Vysokomolekularnye soedinenia* **16A**, 659 (1074).

710. T. Nguyen, E. Byrd, D. Bentz, and C. Lin, *Prog. Organic Coatings* **27**, 181 (1996).

711. C. M. Laot, E. Marand, and H. T. Oyama, *Polymer* **40**, 1095 (1999).

712. (a) E. Marand and L. M. Smart, *Appl. Spectrosc.* **49**, 513 (1995); (b) G. Boven, R. H. G. Brinkhuis, E. J. Vorenkamp, and A. J. Schouten, *Macromolecules* **24**, 967 (1991).

713. (a) J. Chalmers and P. Griffiths (Eds.), *Handbook of Vibrational Spectroscopy*, Vol. 2, Wiley, New York, 2002, Chapter 9; (b) S. Ekgasit and H. Ishida, *Appl. Spectrosc.* **51**, 1488 (1997).

714. M. Tamada, H. Koshikawa, and H. Omichi, *Thin Solid Films* **292**, 164 (1997).

715. M. Tamada, H. Koshikawa, and H. Omichi, *Thin Solid Films* **292**, 113 (1997).

716. S. Bistac, M. F. Vallat, and J. Schultz, *Appl. Spectrosc.* **51**, 1823 (1997).

717. S. Moon and J. Jang, *J. Appl. Polym. Sci.* **68**, 1117 (1998).

718. M. Kazayawoko, J. J. Balatinecz, and R. T. Woodhams, *J. Appl. Polym. Sci.* **66**, 1163 (1997).

719. I. I. Salame and T. J. Bandosz, *Ind. Eng. Chem. Res.* **39**, 301 (2000).

720. A. Valadez-Gonzalez, J. M. Cervantes-Uc, R. Olayo, and P. J. Herrera-Franco, *Composites B* **30**, 321 (1999).

721. L. P. Ilharco, A. P. Garcia, J. L. da Silva, and L. F. V. Ferreira, *Langmuir* **13**, 4126 (1997).

722. S. Kokot, L. Marahusin, and D. P. Schweinsberg, *Microchim. Acta* **14**(Suppl.), 201 (1997).

723. N. T. M. Johnson, *Off. Dig. Oil Colour Chem. Assoc.* **32**, 1067 (1960).

724. J. M. Chalmers, N. J. Everall, and S. Ellison, *Micron* **27**, 315 (1996).

725. I. Karamancheva, V. Stefov, B. Soptrajanov, G. Danev, E. Spasova, and J. Assa, *Vib. Spectrosc.* **19**, 369 (1999).

726. D. L. Wetzel and R. O. Carter, *AIP Conf. Proc.* **430**, 567 (1998).

727. T. Miyashita and T. Suwa, *Langmuir* **10**, 3387 (1994).

728. T. Miyashita and T. Suwa, *Thin Solid Films* **284–285**, 330 (1996).

729. V. G. Gregoriou, R. Hapanowicz, S. L. Clark, and P. T. Hammond, *Appl. Spectrosc.* **51**, 470 (1997).

730. M. Pezolet, C. Pellerin, R. E. Prud'homme, and T. Buffeteau, *Vib. Spectrosc.* **18**, 103 (1998).

731. H. Li, Z. Wang, B. Zhao, H. Xiong, X. Zhang, and J. Shen, *Langmuir* **14**, 423 (1998).

732. T. Fujiama, *J. Spectrosc. Soc. Jpn.* **25**, 255 (1976).

733. F. Druschke, H. W. Siester, G. Spielgies, and H. Tender, *Polym. Eng. Sci.* **17**, 93 (1977).

734. J. Chalmers and P. Griffiths (Eds.), *Handbook of Vibrational Spectroscopy*, Vol. 5, Wiley, New York, 2002.

735. G. Chen, R. J. Palmer, and D. C. White, *Biodegradation* **8**, 189 (1997).

736. J. M. Anderson, A. Hiltner, M. J. Wiggins, M. A. Schubert, T. O. Collier, W. J. Kao, and A. B. Mathur, *Polymer Int.*, **46**, 163 (1998).

737. M. Fletcher, *Methods Microbiol.* **22**, 251 (1990).

738. A.A. Christy, Y. Ozaki, and V.G. Gregoriou, *Modern Fourier Transform Infrared Spectroscopy*, Elsevirer, Amsterdam, 2001, Chapter 8.

739. (a) P. K. Sharma and K. H. Rao, *Miner. Metallurg. Proc.* **16**, 35 (1999); (b) P. K. Sharma, K. H. Rao, K. S. E. Forssberg, and K. A. Natarajan, *Int. J. Miner. Process.* **62**, 3 (2001).

740. S. Krimm and J. Bandekar, *Adv. Prot. Chem.* **38**, 181 (1986).

741. E. Goormaghtigh and J.-M. Ruysschaert, in R. Brasseur (Ed.), *Molecular Description of Biological Membranes by Computer-Aided Conformational Analysis*, CRC Press, Boca Raton, FL, 1990, pp. 285–329.

742. U. P. Fringeli, in F. M. Mirabella, Jr. (Ed.), *Internal Reflection Spectroscopy, Theory and Application*, Marcel Dekker, New York, 1993, p. 255.

743. R. J. Jakobsen and S. W. Strand, in F. M. Mirabella, Jr. (Ed.), *Internal Reflection Spectroscopy, Theory and Application*, Marcel Dekker, New York, 1993, p. 107.

744. J. Twardowski and P. Anzenbacher, *Raman and IR Spectroscopy in Biology and Biochemistry*, Ellis Horwood, London, and Polish Scientific Publishers, Warsaw, 1994.

745. E. Goormaghtigh, V. Cabiaux, and J.-M. Ruysschaert, in H. J. Hilderson and G. B. Ralson (Eds.), *Subcellular Biochemistry*, Vol. 23: *Physicochemical Methods in the Study of Biomembranes*, Plenum, New York, 1994, p. 329.

746. J. L. Brash and T. A. Horbett (Eds.), *Proteins at Interfaces: Fundamentals and Applications*, ACS Symposium Series 602, American Chemical Society, Washington, DC, 1995.

747. H. H. Mantsch and D. Chapmen (Eds.), *Infrared Spectroscopy of Biomolecules*, Wiley-Liss, New York, 1996.

748. L. K. Tamm and S. A. Tatulian, *Q. Rev. Biophys.* **30**, 365 (1997).

749. H.-U. Gremlich and B. Yan (Eds.), *Infrared and Raman Spectroscopy of Biological Materials*, Marcel Dekker, New York, 2001.

750. K. K. Chittur, *Biomaterials* **19**, 301 (1998).

751. K. K. Chittur, *Biomaterials* **19**, 357 (1998).

752. S. G. Roscoe, in J. O'M. Bockris, B. E. Conway, and R. E. White (Eds.), *Modern Aspects of Electrochemistry*, No. 29, Plenum, New York, 1996, p. 319.

753. B. Schrader (Ed.), *Infrared and Raman Spectroscopy: Methods and Applications*, VCH, Weinheim, 1995, Chapter 4.7.

754. W. G. Charaklis and K. C. Marshall (Eds.), *Physical and Chemical Properties of Biofilms*, Wiley-Interscience, New York, 1990.

755. D. R. Scheuing (Ed.), *Fourier Transform Infrared Spectroscopy in Colloid and Interface Science*, ACS Symposium Series 447, American Chemical Society, Washington, DC, 1991, p. 13.

756. W. Plieth, W. Kozlowski, and T. Twomey, in J. Lipkowski and P. N. Ross (Eds.), *Adsorption of Molecules on Metal Electrodes*, VCH, New York, 1992, p. 239.

757. P. Hofer and U. P. Fringeli, *Biophys. Struct. Mech.* **6**, 67 (1979).

758. C. E. Giacomelli, M. G. E. G. Bremer, and W. Norde, *J. Colloid Interface Sci.* **220**, 13 (1999).

759. S. R. Ge, K. Kojio, and T. Kajiyama, *J. Biomat. Sci. Polym. (Ed.)*, **9**, 131 (1998).

760. S. S. Cheng, K. K. Chittur, C. N. Sukenik, L. A. Culp, and K. Lewandowska, *J. Colloid Interface. Sci.* **162**, 135 (1994).

761. S. R. Ge, K. Kojio, A. Takahara, and T. Kajiyama, *J. Biomater. Sci. — Polymer Edition* **9**, 131 (1998).

762. J. L. Ong, K. K. Chittur, and L. C. Lucas, *J. Biomed. Mater. Res.* **28**, 1337 (1994).

763. E. Ruckenstein, S. Gourisankar, and R. E. Baier, *J. Colloid Interface Sci.* **96**, 245 (1983).

764. H. Zeng, K. K. Chittur, and W. Lacefield, *Biomaterials* **20**, 377 (1999).

765. T. A. Giroux and S. L. Cooper, *J. Colloid Interface Sci.* **146**, 179 (1991).

766. M. Muller, C. Werner, K. Grundke, K. J. Eichhorn, and H. J. Jacobasch, *Macromol. Symp.* **103**, 55 (1996).

767. M. Muller, C. Werner, K. Grundke, K. J. Eichhorn, and H. J. Jacobasch, *Mikrochim. Acta* **14**(Suppl.), 671 (1997).

768. L. H. Marsh, M. Coke, P. W. Dettmar, R. J. Ewen, M. Havler, T. G. Nevell, J. D. Smart, J. R. Smith, B. Timmins, J. Tsibouklis, and C. Alexander, *J. Biomed. Mater. Res.* **61**, 641 (2002).

769. K. Grundke, H. J. Jacobasch, F. Simon, and S. Schneider, in K. L. Mittal (Ed.), *Polymer Surface Modification: Relevance to Adhesion*, VSP (International Science Publishers), Utrecht und Tokio, 1995, pp. 431–454.

770. J. S. Jeon, R. P. Sperline, and S. Raghavan, *Appl. Spectrosc.* **46**, 1644 (1992).

771. R. Kellner, G. Gidaly, and F. Unger, *Adv. Biomater.* **3**, 423 (1982).

772. A. M. Baty, P. A. Suci, B. J. Tyler, and G. G. Greesey, *J. Colloid Interface Sci.* **177**, 307 (1996).

773. W. G. Pitt and S. L. Cooper, *J. Biomed. Mater. Res.* **22**, 359 (1988).

774. F.-N. Fu, M. P. Fuller, and B. R. Singh, *Appl. Spectrosc.* **47**, 98 (1993).

775. S. S. Saavedra and W. M. Reichert, *Langmuir* **7**, 995 (1991).

776. N. I. Afanasieva, *Macromol. Symp.* **141**, 117 (1999).

777. M. S. Braiman and S. E. Plunkett, *Appl. Spectrosc.* **51**, 592 (1997).

778. H.-S. Liu, Y.-C. Wang, and W.-Y. Chen, *Colloids Surf. B* **5**, 25 (1995).

779. A. D. Roddick-Lanzilotta, P. A. Connor, and A. J. McQuillan, *Langmuir* **14**, 6479 (1998).

780. A. D. Roddick-Lanzilotta and A. J. McQuillan, *J. Colloid Interface Sci.* **217**, 194 (1999).

781. A. Nabet and M. Pesolet, *Appl. Spectrosc.* **51**, 466 (1997).

782. K. Ataka and M. Osawa, *J. Electroanal. Chem.* **460**, 188 (1999).

783. P. Tengvall, I. Lundström, and B. Liedberg, *Biomaterials* **19**, 407 (1998).

784. Y. Cheng and R. M. Corn, *J. Phys. Chem.* **103**, 8726 (1999).

785. Y. Cheng, L. Murtomaki, and R. M. Corn, *J. Electroanal. Chem.* **483**, 88 (2000).

786. J. Wang, *Thin Solid Films* **379**, 224 (2000).

787. W. J. Dillman and I. F. Miller, *J. Colloid Interf. Sci.* **44**, 221 (1973).

788. J. L. Brash and D. J. Lyman, *J. Biomed. Mater. Res.* **3**, 175 (1969).

789. (a) D. Blaudez, F. Boucher, T. Buffeteau, B. Desbat, M. Grandbois, and C. Salesse, *Appl. Spectrosc.* **53**, 1299 (1999); (b) T. Buffeteau, E. Le Calvez, S. Castano, B. Desbat, D. Blaudez, and J. Dufourcq, *J. Phys. Chem. B* **104**, 4537 (2000); (c) T. Buffeteau, E. Le Calvez, B. Desbat, I. Pelletier, and M. Pezolet, *J. Phys. Chem. B* **105**, 1464 (2001).

790. A. Ihs, B. Lindberg, K. Uvdal, C. Törnkvist, P. Bodö, and I. Lundström, *J. Colloid Interf. Sci.* **140**, 192 (1990).

791. A. Ihs and B. Lindberg, *J. Colloid Interf. Sci.* **144**, 282 (1990).

792. K. D. Dobson, P. A. Connor, and A. J. McQuillan, *Langmuir* **13**, 2614 (1997).

793. E. Killmann and M. Reiner, *Tenside Surfactants Detergents* **33**, 220 (1996).

794. J. Huang and Y. Liang, *Thin Solid Films* **325**, 210 (1998).

795. E. F. Bowden, F. M. Hawkridge, and H. N. Blount, in S. Srinivasan, Yu. A. Chiznadzhiev, J. O. M. Bockris, B. Conway, and E. Yeager (Eds.), *Comprehensive Treatise of Electrochemistry*, Plenum, New York, 1985.

796. H.-Q. Li, A. Chen, S. G. Roscoe, and J. Lipkowski, *J. Electroanal. Chem.* **500**, 299 (2001).

797. (a) F. Huerta, E. Morallon, F. Cases. A. Rodes, J. L. Vazquez, and A. Aldaz, *J. Electroanal. Chem.* **421**, 179 (1997); (b) F. Huerta, E. Morallon, F. Cases. A. Rodes, J. L. Vazquez, and A. Aldaz, *J. Electroanal. Chem.* **431**, 269 (1997).

798. F. Huerta, E. Morallon, J. L. Vazquez, J. M. Perez, and A. Aldaz, *J. Electroanal. Chem.* **445**, 155 (1998).

799. K. Ogura, M. Nakayama, K. Nakaoka, and Y. Nishihata, *J. Electroanal. Chem.* **482**, 32 (2000).

800. K. Ogura, M. Kobayashi, M. Nakayama, and Y. Miho, *J. Electroanal. Chem.* **463**, 218 (1999).

801. S. Krimm and J. Bandekar, *Adv. Prot. Chem.* **38**, 181 (1986).

802. J. Bandekar, *Biochim. Biophys. Acta* **1120**, 123 (1992).

803. M. Jackson and H. H. Mantsch, *Crit. Rev. Biochem. Mol. Biol.* **30**, 95 (1995).

804. S. Seshadri, R. Khurana, and A. L. Fink, *Methods Enzymol.* **309**, 559 (1999).

805. F.-N. Fu, D. B. DeOliveira, W. R. Trumble, H. K. Sarkar, and B. R. Singh, *Appl. Spectrosc.* **48**, 1432 (1994).

806. B. R. Singh, M. P. Fuller, and G. Schiavo, *Biophys. Chem.* **46**, 155 (1990).

807. B. R. Singh, D. B. DeOliveira, F.-N. Fu, and M. P. Fuller, *SPIE Biomol. Spectr. III* **1890**, 47 (1993).

808. B. R. Singh, F.-N. Fu, and D. N. Ledoux, *Struct. Biol.* **1**, 358 (1994).

809. K. Griebenow and A. M. Klebanov, *Biotechnol. Bioeng.* **53**, 351 (1997).

810. K. Griebenow and A. M. Klebanov, *J. Am. Chem. Soc.* **118**, 11695 (1996).

811. K. Griebenow, A. M. Santos, and K. G. Carrasquillo, *Internet J. Vibrational Spectrosc.* **3**, Edition1, available on-line: http://www.ijvs.com//volume3/edition1/section3.htrr.

812. G. Anderle and R. Mendelsohn, *Biophys. J.* **52**, 69 (1987).

813. M. S. Braiman and K. L. Rothschild, *Ann. Rev. Biophys. Biophys. Chem.* **17**, 541 (1988).

814. W. Surewicz and H. H. Mantsch, *Biochim. Biophys. Acta* **952**, 115 (1988).

815. J. L. R. Arrondo, A. Miga, J. Castresana, and F. M. Goni, *Prog. Biophys. Mol. Biol.* **59**, 23 (1993).

816. P. I. Haris and D. Chapman, *Biopolym. (Peptide Sci.)* **37**, 251 (1995).

817. W. K. Surewicz and H. H. Mantsch, *Spectroscopic Methods Determining Protein Structure in Solution*, VCH, New York, 1996.

818. (a) H. H. Mantsch, D. J. Moffat, and H. L. Casal, *J. Mol. Struct.* **173**, 285 (1988); (b) G. Vedantham, H. G. Sparks, S. U. Sane, S. Tzannis, and T. M. Przybycien, *Anal. Biochem.* **285**, 33 (2000).

819. (a) D. C. Lee, P. I. Haris, D. Chapman, and R. C. Mitchell, *Biochemistry* **29**, 9185 (1990); (b) F. Dousseau and M. Pezolet, *Biochemistry* **29**, 8771 (1990).

820. V. Acha, J. M. Ruysschaert, and E. Goormaghtigh, *Anal. Chim. Acta* **435**, 215 (2001).

821. E. R. Malinowski, *Factor Analysis in Chemistry*, 2nd ed., Wiley, New York, 1991.

822. (a) L. A. Forato, R. Bernardes, and L. A. Colnago, *Quimica Nova* **21**, 146 (1998); (b) S. Wi, P. Pancoska, and T. A. Keiderling, *Biospectrosc.* **4**, 93 (1998).

823. (a) C. F. C. Ludlam, I. T. Arkin, X.-M. Liu, M. S. Rothman, P. Path, S. Aimoto, S. O. Smith, D. M. Engelman, and K. J. Rothschild, *Biophys. J.* **70**, 1728 (1996); (b) K. Gerwert, *Biol. Chem.* **380**, 931 (1999).

824. (a) R. M. Nyquist, D. Heitbrink, C. Bolwien, T. A. Wells, R. B. Gennis, J. Heberle, *Febs Lett.* **505**, 63 (2001); (b) S. Servagent-Noinville, M. Revault, H. Quiquampoix, M. H. Baron, *J. Colloid Interface Sci.* **221**, 273 (2000); (c) E. Kauffmann, N. C. Darnton, R. H. Austin, and K. Gerwert, *Proc. National Acad. Sci. USA* **98**, 6646 (2001).

825. S. N. Timasheff, H. Susi, and L. Stevens, *J. Biol. Chem.* **242**, 5467 (1967).

826. (a) N. L. Sefara, N. P. Magtoto, and H. H. Richardson, *Appl. Spectrosc.* **51**, 536 (1997); (b) S. Lecomte, C. Hilleriteau, J. P. Forgerit, M. Revault, M. H. Baron, P. Hildebrandt, T. Soulimane, *Chembiochem.* **2**, 180 (2001).

827. D. L. Elmore and R. A. Dluhy, *Colloids Surf. A* **171**, 225 (2000).

828. (a) L. Smeller, K. Heremans, *Vib. Spectrosc.* **19**, 375 (1999); (b) L. Smeller, P. Rubens, J. Frank, J. Fidy, and K. Heremans, *Vib. Spectrosc.* **22**, 119 (2000).

829. Y. Ozaki, Y. Liu, and I. Noda, *Appl. Spectrosc.* **51**, 526 (1997).

830. P. G. H. Kosters, A. H. B. de Vries, R. P. Kooyman, *Appl. Spectrosc.* **54**, 1659 (2000).

831. (a) B. Czarnik-Matusewicz, K. Murayama, Y. Q. Wu, and Y. Ozaki, *J. Phys. Chem. B* **104**, 7803 (2000); (b) J. F. Halsall, M. Kalaji, and A. L. Neal, *Biofouling* **16**, 105 (2000).

832. (a) E. Goormaghtigh, L. Vigneron, G. A. Scarborough, and J.-M. Ruysschaert, *J. Biol. Chem.* **269**, 27409 (1994); (b) L. Vigneron, J.-M. Ruysschaert, and E. Goormaghtigh, *J. Biol. Chem.* **270**, 17685 (1995); (c) H. H. J. De Jongh, E. Goormaghtigh, and J.-M. Ruysschaert, *Biochemistry* **134**, 172 (1995).

833. A. H. Clark, D. H. P. Saunderson, and A. Suggett, *Int. J. Pept. Protein Res.* **17**, 353 (1981).

834. F. Dousseau, M. Therrien, and M. Pezolet, *Appl. Spectrosc.* **43**, 538 (1989).

835. J. R. Powell, F. M. Wasacz, and R. J. Jakobsen, *Appl. Spectrosc.* **40**, 339 (1986).

836. P. W. Holloway and H. H. Mantsch, *Biochemistry* **28**, 931 (1989).

837. M. Jackson and H. Mantsch, *Appl. Spectrosc.* **46**, 699 (1992).

838. L. Dziri, B. Desbat, and R. M. Leblanc, *J. Am. Chem. Soc.* **121**, 9618 (1999).

839. H. G. W. Lensen, D. Bargeman, P. Bergveld, and J. Feijen, *J. Colloid Interface Sci.* **99**, 1 (1984).

840. G. Desmet, D. Thomas, and J. L. Boitieux, *J. Chromatogr.* **376**, 199 (1986).

841. K. K. Chittur, D. J. Fink, T. B. Huston, R. M. Gendreau, R. J. Jakobsen, and R. I. Leininger, in J. L. Brash and T. A. Horbett (Eds.), *Proteins at Interfaces*, American Chemical Society, Washington, DC, 1987, p. 362.

842. K. K. Chittur, D. J. Fink, R. I. Leininger, and T. B. Huston, *J. Colloid Interface Sci.* **111**, 419 (1986).

843. D. J. Fink, T. B. Huston, K. K. Chittur, and R. M. Gendreau, *Anal. Biochem.* **165**, 147 (1987).

844. J. Buijs and W. Norde, *Langmuir* **12**, 1605 (1996).

845. D. J. Lyman and K. Knutson, in E. P. Goldberg and A. Nakajima (Eds.) *Biomedical Polymers*, Academic, New York, 1980, p. 1.

846. H.-S. Liu and Y.-C. Wang, *Colloids Surf. B* **5**, 35 (1995).

847. N. A. Rodionova, S. A. Tatulian, T. Surrey, F. Jahnig, and L. K. Tamm, *Biochemistry* **34**, 1921 (1995).

848. A. Gericke, C. R. Flach, and R. Mendelsohn, *Biophys. J.* **73**, 492 (1997).

849. D. Marsh, *Biophys. J.* **77**, 2630 (1999).

850. D. Marsh, *Biophys. J.* **72**, 2710 (1997).

851. H. S. Kim, J. D. Hartgerink, and M. R. Ghadiri, *J. Am. Chem. Soc.* **120**, 4417 (1998).

852. V. F. Kalinsky, *Appl. Spectrosc. Rev.* **31**, 193 (1996).

853. A. Bentaleb, A. Abele, Y. Haikel, P. Schaaf, and J. C. Voegel, *Langmuir* **14**, 6493 (1998).

854. A. Bentaleb, A. Abele, Y. Haikel, P. Schaaf, and J. C. Voegel, *Langmuir* **15**, 4930 (1999).

855. A. Ball and R. A. Jones, *Langmuir* **11**, 3542 (1995).

856. R. E. Baier and R. C. Dutton, *J. Biomed. Res.* **3**, 191 (1969).

857. T. A. Horbett, in S. L. Cooper and N. A. Peppas (Eds.), *Biomaterials: Interface Phenomena and Applications*, American Chemical Society, Washington DC, 1982, p. 233.

858. M. Trojanowicz, *Anal. Lett.* **33**, 1387 (2000).

859. K. M. Merz and B. Roux (Eds.), *Biological Membranes: A Molecular Perspective from Computation and Experiment*, Birkhauser, Boston, 1996, pp. 145–174.

860. D. Chapman, *J. Am. Oil Chem.* **42**, 353 (1965).

861. R. A. Dluhy, S. M. Stephens, S. Widayati, and A. D. Williams, *Spectrochim. Acta A* **51**, 1413 (1995).

862. E. Goormaghtigh and J.-M. Ruysschaert, in R. Brasseur (Ed.), *Molecular Description of Biological Membranes by Computer-Aided Conformational Analysis*, CRC Press, Boca Raton, FL, 1990, p. 285.

863. P. H. Axelsen and M. J. Citra, *Prog. Biophys. Molec. Biol.* **66**, 227 (1996).

864. M. Jackson and H. Mantsch, *Spectrochim. Acta Rev.* **15**, 53 (1993).

865. M. S. Braiman and S. E. Plunkett, *Appl. Spectrosc.* **51**, 592 (1997).

866. W.-P. Ulrich and H. Vogel, *Biophys. J.* **76**, 1639 (1999).

867. S. E. Taylor, B. Desbat, and G. Schwarz, *Biophys. Chem.* **87**, 63 (2000).

868. M. Grandbois, B. Desbat, and C. Salesse, *Langmuir* **15**, 6594 (1999).

869. M. Grandbois, B. Desbat, and C. Salesse, *Biophys. Chem.* **88**, 127 (2000).

870. M. Laguerre, J. Dufourcq, S. Castano, and B. Desbat, *Biochim. Biophys. Acta* **1416**, 176 (1999).

871. S. Castano, B. Desbat, and J. Dufourcq, *Biochim. Biophys. Acta* **1463**, 65 (2000).

872. I. Vergne and B. Desbat, *Biochim. Biophys. Acta* **1467**, 113 (2000).

873. E. Bellet-Amalric, D. Blaudez, B. Desbat, F. Graner, F. Gauthier, and A. Renault, *Biochim. Biophys. Acta* **1467**, 131 (2000).

874. L. Dziri, B. Desbat, and R. M. Leblanc, *J. Am. Chem. Soc.* **121**, 9618 (1999).

875. I. Estrela-Lopis, G. Brezesinski, and H. Mohwald, *Biophys. J.* **80**, 749 (2001).

876. D. Blaudez, T. Buffeteau, and J. C. Cornut, *Thin Solid Films* **242**, 146 (1994).

877. D. Blaudez, J. M. Turlet, and B. Desbat, *J. Chem. Soc. Faraday Trans.* **92**, 525 (1996).

878. D. Marsh, *Chem. Phys. Lipids* **57**, 109 (1991).

879. R. A. Dluhy, Z. Ping, K. Faucher, and J. M. Brockman, *Thin Solid Films* **327–329**, 308 (1998).

880. D. Blaudez, T. Buffeteau, B. Desbat, and J. M. Turlet, *Current Opinion Colloid Interface Sci.* **4**, 265 (1999).

881. R. Mendelsohn and R. G. Snyder, in K. M. Merz and B. Roux (Eds.), *Biological Membranes: A Molecular Perspective from Computation and Experiment*, Birkhauser, Boston, 1996, pp. 145–174.

882. H. H. Mantsch and R. N. Mc Elhaney, *Chem. Phys. Lipids* **57**, 213 (1991).

883. I. Ueda, J.-S. Chiou, P. R. Krishna, and H. Kamaya, *Biochim. Biophys. Acta* **1190**, 421 (1994).

884. F. Lopez-Garcia, J. Villalain, J. C. Gomez-Fernandez, and P. J. Quinn, *Biophys. J.* **66**, 19991 (1994).

885. P. I. Haris, D. Chapman, and G. Benga, *Eur. J. Biochem.* **233**, 659 (1995).

886. T. Heimburg and D. Marsh, *Biophys. J.* **65**, 2408 (1993).

887. F. Wu, C. R. Flach, B. A. Seaton, T. R. Mealy, and R. Mendelsohn, *Biochemistry* **38**, 792 (1999).

888. E. Hantz, A. Cao, and R. S. Phadke, *Chem. Phys. Lipids* **51**, 75 (1989).

889. E. Goormaghtigh, R. Brausseur, P. Huart, and J. M. Ruysschaert, *Biochem.* **26**, 1789 (1987).

890. I. Ben-Efraim, Y. Kliger, C. Hermesh, and Y. Shai, *J. Mol. Biol.* **285**, 609 (1999).

891. P. A. Suci, J. D. Vrany, and M. W. Mittelman, *Biomaterials* **19**, 327 (1998).

892. D. Naumann, D. Helm, and H. Labischinski, *Nature* **351**, 81 (1991).

893. D. Helm, H. Labischinski, G. Schallehn, and D. Naumann, *J. Gen. Microbiol.* **137**, 69 (1991).

894. V. D. Kuznetsov, E. D. Zagreba, N. A. Bashkhatova, D. Y. Pavlovisha, and M. K. Grube, *Microbiology* **61**, 903 (1992).

895. K. C. Schuster, F. Mertens, and J. R. Gapes, *Vib. Spectrosc.* **19**, 467 (1999).

896. A. A. Kamlev, M. Ristic, L. P. Antonyuk, A. V. Chernyshev, and V. V. Ignatov, *J. Mol. Struct.* **408–409**, 201 (1997).

897. E. D. Zagreba, V. Savenkov, and M. Ginovska, *Microbial Conversion — Fundamental Aspects*, Zinatne, Riga, Latvia, 1990, p. 139.

898. M. Grube, J. Zagreba, and M. Fomina, *Vib. Spectrosc.* **19**, 301 (1999).

899. B. K. Bjorn, W. G. Wade, and R. Goodacre, *Appl. Spectrosc.* **52**, 823 (1998).

900. D. Naumann, D. Helm, H. Labischinski, and P. Giesbrecht, in W. H. Nelson (Ed.), *Modern Techniques for Rapid Microbiological Analysis*, VCH, New York, 1991, p. 43.

901. N. M. Tsyganenko, A. A. Tsyganenko, C. Picand, J. Travert, and G. Novel, in J. C. Mervin, S. Turrell, and J. P. Huvenne (Eds.), *Spectroscopy for Biological Molecules*, Kluwer, Dordrecht, 1995, p. 513.

902. P. Nichols, M. Henson, J. Guckert, and D. C. White, *J. Microb. Meth.* **4**, 79 (1985).

903. J. Schmitt and H.-C. Flemming, *Int. Biodeterioration Biodegradation* **41**, 1 (1998).

904. J. Schmitt, D. Nivens, D. C. White, and H.-C. Flemming, *Water Sci. Tech.* **32**, 149 (1995).

905. D. E. Nivens, J. Schmit, J. Sniatecki, T. Anderson, J. Q. Chambers, and D. C. White, *Appl. Spectrosc.* **47**, 668 (1993).

906. J. W. Costerton, T. J. Marrie, and K. J. Cheng, in D. C. Savage and M. Fletcher (Eds.), *Bacterial Adhesion: Mechanisms and Physiological Significance*, Plenum, New York, 1985, p. 3.

907. D. E. Nivens, J. Q. Chambers, T. R. Anderson, A. Tunlid, J. Smith, and D. C. White, *J. Microbiol. Meth.* **17**, 199 (1993).

908. K. C. Marshall, *ASM News* **58**, 202 (1992).

909. K. P. Ishida and P. R. Griffuths, in D. R. Scheuing (Ed.), *Fourier Transform Infrared Spectroscopy in Colloid and Interface Science*, ACS Symposium Series 447, American Chemical Society, Washington, DC, 1991, p. 208.

910. K. P. Ishida and P. R. Griffuths, *J. Colloid Interface Sci.* **160**, 190 (1993).

911. N. Bhosle, P. A. Suci, A. M. Baty, R. M. Weiner, and G. G. Geesey, *J. Colloid Interface Sci.* **205**, 89 (1998).

912. A. M. Baty, P. A. Suci, B. J. Tyler, and G. G. Geesey, *J. Colloid Interface Sci.* **177**, 307 (1996).

913. B. Frolund, P. A. Suci, S. Langille, R. M. Weiner, and G. G. Geesey, *Biofouling* **10**, 17 (1996).

914. Y. Imai and Y. Tamaki, *J. Prosth. Dentistry* **82**, 348 (1999).

915. (a) T. Iwaoka, P. R. Griffiths, J. T. Kitasako, and G. G. Gees, *Appl. Spectrosc.* **7**, 1062 (1986); (b) P. J. Bremer and G. G. Geesey, *Appl. Environ. Microbiol.* **57**, 1956 (1991); (c) G. G. Geesey and P. J. Bremer, *Mar. Technol. Soc. J.* **24**, 36 (1990); (d) J. G. Jolley, G. G. Geesey, M. R. Hankins, R. B. Wright, and P. L. Wichlacz, *Appl. Spectrosc.* **43**, 1062 (1989).

916. P. A. Suci, M. W. Mittelman, F. P. Yu, and G. G. Geesey, *Antimicrob. Agents Chemother.* **38**, 2125 (1994).

917. W. W. Nichols, M. J. Evans, M. P. E. Slack, and H. L. Walmsley, *J. Gen. Microbiol.* **135**, 1291 (1989).

918. P. A. Suci, K. J. Siedlecki, R. J. J. Palmer, D. C. White, and G. G. Geesey, *Appl. Environ. Microbiol.* **63**, 4600 (1997).

919. X. Liu and R. E. Truitt, *J. Amer. Chem. Soc.* **119**, 9856 (1997).

920. G. Turnes Palomino, S. Bordiga, A. Zecchina, G. L. Marra, and C. Lamberti, *J. Phys. Chem. B* **104**, 8641 (2000).

921. W. Zhang, P. G. Smirniotis, M. Gangoda, and R. N. Bose, *J. Phys. Chem. B* **104**, 4122 (2000).

922. F.-W. Schuetre, F. Roessner, J. Meusinger, and H. Papp, *Stud. Surf. Sci. Catal.* **112**, 127 (1997).

923. T. Mori, Y. Kuroda, Y. Yoshikawa, M. Nagao, and S. Kittaka, *Langmuir* **18**, 1595 (2002).

924. G. Muller, E. Bodis, J. Kornatowski, and J. A. Lercher, *Phys. Chem. Chem. Phys.* **1**, 571 (1999).

925. M. Trombetta, G. Busca, M. Lenarda, L. Storaro, R. Ganzerla, L. Piovesan, A. J. Lopez, M. Alcantara-Rodriguez, and E. Rodriguez-Castellon, *Appl. Catal. A* **193**, 55 (2000).

926. J. Billingham, C. Breen, and J. Yarwood, *Clay Minerals* **31**, 513 (1996).

927. G. Martra, *Appl. Catal. A* **200**, 275 (2000).

928. W. Daniell, U. Schubert, R. Glockler, A. Meyer, K. Noweck, and H. Knozinger, *Appl. Catal. A* **196**, 247 (2000).

929. J. Datka, B. Gil, and J. Weglarski, *Micropor. Mesopor. Mater.* **21**, 75 (1998).

930. P. M. Kumar, S. Badrinarayanan, and M. Sastry, *Thin Solid Films* **358**, 122 (2000).

931. S. Kotrel, J. H. Lunsford, and H. Knozinger, *J. Phys. Chem. B* **105**, 3917 (2001).

932. B. H. Li and R. D. Gonzalez, *Catal. Today* **46**, 55 (1998).

933. W. Daniell, N.-Y. Topsoe, and H. Knozinger, *Langmuir* **17** 6233 (2001).

934. H. Wiame, C. Cellier, and P. Grange, *J. Catal.* **190**, 406 (2000).

935. C. Peuker, *J. Mol. Struct.* **349**, 317 (1995).

936. A. M. Turek, I. E. Wachs, and E. DeCanio, *J. Phys. Chem.* **96**, 5000 (1992).

937. R. R. Ivlieva and V. Yu. Borovkov, *Kinet. Catal.* **33**, 2790 (1992).

938. M. Daturi, C. Binet, S. Bernal, J. A. Perez Omit, and J. C. Lavalley, *J. Chem. Soc. Faraday Trans.* **94**, 1143 (1998).

939. N. Essayem, A. Holmqvist, P. Y. Gayraud, J. C. Vedrine, and Y. Ben Taarit, *J. Catal.* **197**, 273 (2001).

940. I. A. Beta, H. Bohlig, B. Hunger, *Thermochim. Acta* **361**, 61 (2000).

941. N. I. E. Shewring, T. G. J. Jones, G. Maitland, and J. Jarwood, *J. Colloid Interface Sci.* **176**, 308 (1995).

942. Y. Kuroda, S. Kittaka, S. Takahara, T. Yamaguchi, and M.-C. Bellissent-Funel, *J. Phys. Chem. B* **103**, 11064 (1999).

943. A. Paulidou and R. M. Nix, *Surf. Sci.* **470**, L104 (2000).

944. P. Van Der Voort, M. G. White, M. B. Mitchell, A. A. Verberckmoes, and E. F. Vansant, *Spectrochim. Acta, A* **53**, 2181 (1997).

945. R. Nakamura, K. Ueda, and S. Sato, *Langmuir* **17**, 2298 (2001).

946. B. Lee, J. N. Kondo, K. Domen, and F. Wakabayashi, *J. Mol. Catal. A* **137**, 269 (1999).

947. J. E. Bateman, R. D. Eagling, B. R. Horrocks, and A. Houlton, *J. Phys. Chem. B* **104**, 5557 (2000).

948. K. Domen, T. Fujino, A. Wada, C. Hirose, and S. S. Kano, *Micropor. Mesopor. Mater.* **21**, 673 (1998).

949. W. Lin and H. Frei, *J. Am. Chem. Soc.* **124**, 9292 (2002).

950. M. Maciejewski, P. Fabrizioli, J. D. Grunwaldt, O. S. Beckert, and A. Baiker, *Phys. Chem. Chem. Phys.* **3**, 3846 (2001).

951. N. Y. Topsoe and H. Topsoe, *J. Mol. Catal. A* **141**, 95 (1999).

952. V. Yu. Borovkov, M. Jiang, and Y. Fu, *J. Phys. Chem.* **103**, 5010 (1999).

953. K. Hadjiivanov and H. Knozinger, *J. Catal.* **191**, 480 (2000).

954. M. Katoh, T. Yamazaki, and S. Ozawa, *J. Collid Interface Sci.* **203**, 447 (1998).

955. V. S. Kamble and N. M. Gupta, *Phys. Chem. Chem. Phys.* **2**, 2661 (2000).

956. B. S. Shete, V. S. Kamble, N. M. Gupta, and V. B. Kartha, *Phys. Chem. Chem. Phys.* **1**, 191 (1999).

957. K. Hadjiivanov, B. M. Reddy, and H. Knozinger, *Appl. Catal. A* **188**, 355 (1999).

958. K. Hadjiivanov, P. Massiani, and H. Knozinger, *Phys. Chem. Chem. Phys.* **1**, 3831(1999).

959. K. Pokrovski, K. T. Jung, and A. T. Bell, *Langmuir* **17**, 4297 (2001).

960. C. Laberty, C. Marquez-Alvarez, C. Drouet, P. Alphonse, and C. Mirodatos, *J. Catal.* **198**, 266 (2001).

961. K. Hadjiivanov and H. Knozinger, *Chem. Phys. Lett.* **303**, 513 (1999).

962. O. Seiferth, K. Wolter, B. D. Dillmann, G. Klivenyi, H. J. Freund, D. Scarano, and A. Zecchina, *Surf. Sci.* **421**, 176 (1999).

963. V. B. Kazansky, V. Y. Borovkov, A. I. Serykh, and M. Bulow, *Phys. Chem. Chem. Phys.* **1**, 3701 (1999).

964. V. B. Kazansky, *J. Mol. Catal. A* **141**, 83 (1999).

965. V. B. Kazansky, V. Yu. Borovkov, A. Serich, and H. G. Karge, *Micropor. Mesopor. Mater.* **22**, 251 (1998).

966. T. Ando, M. Jshi, M. Kamo, and Y. Sato, *Diamond Relat. Mater.* **4**, 607 (1995).

967. K. Hadjiivanov and H. Knozinger, *Catal. Lett.* **58**, 21 (1999).

968. G. H. Smudde, Jr., T. L. Slager, C. G. Coe, J. E. MacDongal, and S. J. Weigel, *Appl. Spectrosc.* **49**, 1747 (1995).

969. R. F. Lobo, K. M. Bulanin, and M. O. Bulanin, *J. Phys. Chem. B* **104**, 1269 (2000).

970. M. A. Centeno, I. Carrizosa, and J. A. Odriozola, *Appl. Spectrosc.* **53**, 800 (1999).

971. K. Hadjiivanov and H. Knozinger, *Phys. Chem. Chem. Phys.* **2**, 2803 (2000).

972. K. Hadjiivanov, V. Avreyska, D. Klissurski, and T. Marinova, *Langmuir* **18**, 1619 (2002), 1619.

973. P. E. Fanning and M. A. Vannice, *J. Catal.* **207**, 166 (2002).

974. C. Borensen, U. Kirchner, V. Scheer, R. Vogt, and R. Zellner, *J. Phys. Chem. A* **104**, 5036 (2000).

975. C. Li, K. Domen, K. Maruya, and T. Onhishi, *J. Am. Chem. Soc.* **111**, 7683 (1989).

976. K. M. Bulanin, J. C. Lavalley, J. Lamotte, L. Mariey, N. M. Tsyganenko, and A. A. Tsyganenko, *J. Phys. Chem. B* **102**, 6809 (1998).

977. O. V. Manoilova, J. C. Lavalley, N. M. Tsyganenko, and A. A. Tsyganenko, *Langmuir* **14**, 5813 (1998).

978. A. Travert, O. V. Manoilova, A. A. Tsyganenko, F. Mauge, and J. C. Lavalley, *J. Phys. Chem. B* **106**, 1350 (2002).

979. M. El-Maazawi, A. N. Finken, A. B. Nair, and V. H. Grassian, *J. Catal.* **191**, 138 (2000).

980. M. B. Mitchell, V. R. Chakravarthy, and M. G. White, *Langmuir* **10**, 4523 (1994).

981. K. Krauss, A. Drochner, M. Fehlings, J. Kunert, and H. Vogel, *J. Mol. Catal. A* **162**, 405 (2000).

982. J. P. Blitz, *Coloid Surf.* **63**, 11 (1992).

983. G. Hubner, G. Rauhut, H. Stoll, and E. Roduner, *Phys. Chem. Chem. Phys.* **4**, 3112 (2002).

984. A. Simon, L. Delmotte, J. M. Chezeau, A. Janin, and J.-C. Lavalley, *Phys. Chem. Chem. Phys.* **1**, 1659 (1999).

985. B. E. Hayden, A. King, and M. A. Newton, *J. Phys. Chem. B* **103**, 203 (1999).

986. M. Trombetta, A. G. Alejandre, and G. Busca, *Appl. Catal. A* **198**, 81 (2000).

987. A. V. Ivanov, G. W. Graham, and M. Shelef, *Appl. Catal. B* **21**, 243 (1999).

988. N. Nesterenko, E. Lima, P. Graffin, L. C. de Menorval, M. Lasperas, D. Tichit, and F. Fajula, *New J. Chem.* **23**, 665 (1999).

989. C. Morterra, G. Magnacca, and V. Bolis, *Catal. Today* **70**, 43 (2001).

990. E. Lehmann, C. Chmelik, H. Scheidt, S. Vasenkov, B. Staudte, J. Karger, F. Kremer, G. Zadrozna, and J. Kornatowski, *J. Am. Chem. Soc.* **124**, 8690 (2002).

991. F. Bauer, E. Geidel, W. Geyer, and C. Peuker, *Microp. Mesop. Mater.* **29**, 109 (1999).

992. L. J. Burcham, G. T. Deo, X. T. Gao, and I. E. Wachs, *Top. Catal.* **11**, 85 (2000).

993. I. Shimizu, H. Okabayashi, K. Taga, E. Nishio, and C. J. O'Connor, *Vibr. Spectrosc.* **14**, 113 (1997).

994. S. C. Ringwald and J. E. Pemberton, *Environm. Sci. Technol.* **34**, 259 (2000).

995. C. Paze, B. Sazak, A. Zecchina, and J. Dwyer, *J. Phys. Chem. B* **103**, 9978 (1999).

996. S. Huber and H. Knozinger, *J. Mol. Catal. A* **141**, 117 (1999).

997. V. A. Self and P. A. Sermon, *Rev. Sci. Instrum.* **67**, 2096 (1996).

998. S. S. Deshmukh, V. I. Kovalchuk, V. Y. Borovkov, and J. L. d'Itri, *J. Phys. Chem. B* **104**, 1277 (2000).

999. S. Vasenkov and H. Frei, *J. Phys. Chem. A* **104**, 4327 (2000).

1000. F. Schuth, *J. Phys. Chem.* **96**, 7493 (1992).

1001. J. R. Combes, L. D. White, and C. P. Tripp, *Langmuir* **15**, 7870 (1999).

1002. E. Wackelgard, *J. Phys.: Condens. Mater.* **8**, 4289 (1996).

1003. L. Harris and J. Piper, *J. Opt. Soc. Am.* **52**, 223 (1962).

1004. T. S. Eriksson, A. Hjortsberg, G. A. Niklasson, and C. G. Granquist, *Appl. Opt.* **20**, 2742 (1981).

1005. F. P. Mertens, *Surf. Sci.* **71**, 161 (1978).

1006. L. A. Nimon and G. K. Korpi, *Plating* **59**, 431 (1972).

1007. O. P. Konovalova, O. Y. Rusakova, and I. I. Shaganov, *Sov. J. Opt. Technol.* **55**, 402 (1988).

1008. B. C. Trasferetti, C. U. Davanzo, N. C. da Cruz, and M. A. B. de Moraes, *Appl. Spectrosc.* **54**, 687 (2000).

1009. T. Schram, J. DeLaet, and H. Terryn, *J. Electrochem. Soc.* **145**, 2733 (1998).

1010. V. P. Tolstoy and G. N. Kuznetsova, *Pribory i tekhnika experimenta* **3**, 268 (1976).

1011. S. Tibault, *Thin Solid Films* **35**, L33 (1976).

1012. H. Kihara-Morishita, T. Takamura, and T. Takeda, *Thin Solid Films* **6**, R29 (1970).

1013. T. Takamura and H. Kihara-Morishita, *J. Electrochem. Soc.* **122**, 386 (1975).

1014. T. Takamura and H. Kihara-Morishita, *Thin Solid Films* **24**, 57 (1974).

1015. R. J. Thibeau, *J. Electrochem. Soc.* **127**, 1913 (1980).

1016. Th. Scheruebl and K. Thomas, *Fresenius J. Anal. Chem.* **349**, 216 (1994).

1017. C. Arroyave, F. A. Lopez, and M. Morcillo, *Corros. Sci.* **37**, 1751 (1995).

1018. A. Raman and B. Kuban, *Corrosion USA* **44**, 483 (1988).

1019. A. Pilon, K. C. Cole, and D. Noel, *Proc. SPIE–Int. Soc. Opt. Eng.* **1145**, 209 (1989).

1020. F. P. Mertens, *Corrosion (USA)* **34**(10), 359 (1978).

1021. M. D. Porter, D. L. Allara, and T. B. Bright, *Proc. SPIE–Soc. Photo-Opt. Instrum. Eng.* **553**, 488 (1985).

1022. D. K. Offesen, *J. Electrochem. Soc.* **132**, 2250 (1985).

1023. Y. Gonzaler, M. C. Lafont, N. Pebere, G. Chatainier, J. Pebere, G. Chatainier, and J. Roy, *Corros. Sci.* **37**, 1823 (1995).

1024. J. M. Morgan, J. S. McNatt, M. J. Shepard, N. Farkas, and R. D. Ramsier, *J. Appl. Phys.* **91**, 9375 (2002).

1025. M. R. Peng and Y. E. Rentt-Robey, *Surf. Sci.* **336**, L755 (1995).

1026. M. Wadsak, M. Schreiner, T. Aastrup, and C. Leygraf, *Surf. Sci.* **454–456**, 246 (2000).

1027. T. Aastrup and C. Leygraf, *J. Electrochem. Soc.* **144**, 2986 (1997).

1028. G. W. Poling, *J. Electrochem. Soc.* **116**, 958 (1969).

1029. F. J. Boerio and L. Armogan, *Appl. Spectrosc.* **32**, 509 (1978).

1030. R. G. Greenler, R. R. Rahn, and J. P. Schwartz, *J. Catal.* **23**, 42 (1971).

1031. A. Surca, B. Orel, and B. Pihlar, *J. Sol-Gel Sci. Technol.* **8**, 743 (1997).

1032. D. Persson and C. Leygraf, *J. Electrochem. Soc.* **139**, 2243 (1992).

1033. (a) M. Le Calvar and M. Lenglet, *Stud. Surf. Sci. Catal.* **48**, 567 (1988); (b) M. Lenglet, F. F. Delaunay, and B. Lefez, *Mater. Sci. Forum* **251-2**, 267 (1997).

1034. B. Lefez, S. Jouen, J. Kasperek, and B. Hannoyer, *Appl. Spectrosc.* **55**, 935 (2001).

1035. (a) L. Legrand, S. Savoye, A. Chausse, and R. Messina, *Electrochim. Acta* **46**, 111 (2000); (b) L. Legrand, G. Sagon, S. Lecomte, A. Chausse, and R. Messina, *Corrosion Sci.* **43**, 1739 (2001); (c) L. Legrand, M. Abdelmoula, A. Gehin, A. Chausse, and J. M. R. Genin, *Electrochim. Acta* **46**, 1815 (2001).

1036. L. J. Simpson, *Electrochim. Acta* **43**, 2543 (1998).

1037. B. L. Frey, D. G. Hanken, and R. M. Corn, *Langmuir* **9**, 1815 (1993).

1038. G. A. Swallow and G. C. Allen, *Oxide Metals* **17**, 41 (1982).

1039. K. Honda, T. Atake, and Y. Saito, *Appl. Spectrosc.* **46**, 464 (1992).

1040. Th. Scherubl and L. K. Thouns, *Appl. Spectrosc.* **51**, 844 (1997).

1041. B. X. Huang, P. Tornatore, and Y.-S. Li, *Electrochim. Acta* **46**, 671 (2001).

1042. B. G. Frederick, G. Apai, and T. N. Rhodin, *J. Electron Spectrosc. Related Phenom.* **54–55**, 415 (1990).

1043. S. Thibault and J. Talbot, *Ann. Univ. Ferrara Sez. 5 Suppl.* **5**, 75 (1971).

1044. (a) K. Aramaki, *Corrosion Sci.* **41**, 1715 (1999); (b) K. Nozawa and K. Aramaki, *Corrosion Sci.* **41**, 57 (1999); (c) K. Aramaki, *Corrosion Sci.* **42**, 1975 (2000).

1045. V. Subramanian and W. J. van Ooij, *Corrosion* **54**, 204 (1998).

1046. G. Grundmeier and M. Stratmann, *Materials And Corrosion* (Werkstoffe Und Korrosion) **49**, 150 (1998).

1047. J. Jang and E. K. Kim, *J. Appl. Polymer Sci.* **71**, 585 (1999).

1048. S. Rajendran, B. V. Apparao, N. Palaniswamy, V. Periasamy, and G. Karthikeyan, *Corrosion Sci.* **43**, 1345 (2001).

1049. Y. Gonzalez, M. C. Lafont, N. Pebere, G. Chatainier, J. Roy, and T. Bouissou, *Corrosion Sci.* **37**, 1823 (1995).

1050. S. Manov, F. Noli, A. M. Lamazouere, and L. Aries, *J. Appl. Electrochem.* **29**, 995 (1999).

1051. K. Aramaki, *Corrosion Sci.* **43**, 1985 (2001).

1052. (a) S. Thibault and J. Talbot, *Bull. Soc. Chim. France* **4**, 1348 (1972); (b) S. Thibault and J. Talbot, *5 Congr. Eur. Corrs. Paris* **1**, 193 (1973).

1053. H. Tobe, N. Morito, K. Monma, and W. Suetaka, *J. Jpn. Inst. Metals* **38**, 770 (1974).

1054. N. Morito and W. Suetaka, *J. Jpn. Inst. Metals* **36**, 1131 (1972).

1055. S. Yoshida and H. Ishida, *J. Chem. Phys.* **78**, 6960 (1983).

1056. J. O. Nilsson, C. Tornkvist, and B. Lieberg, *Appl. Surf. Sci.* **37**, 306 (1989).

1057. S. Yoshida and H. Ishida, *Appl. Surf. Sci.* **89**, 39 (1995).

1058. R. L. Opila, H. W. Krautter, B. R. Zegarski, L. H. Dubois, and G. Wender, *J. Electrochem. Soc.* **142**, 4074 (1995).

1059. W. Yan, H. C. Lin, and C. N. Cao, *Electrochim. Acta* **45**, 2815 (2000).

1060. R. Tremont, H. De Jesus-Cardona, J. Garcia-Orozco, R. J. Castro, and C. R. Cabrera, *J. Appl. Electrochem.* **30**, 737 (2000).

1061. F. E. Varela, B. M. Rosales, G. P. Cicileo, and J. R. Vilche, *Corrosion Sci.* **40**, 1915 (1998).

1062. K. Hyuncheol and J. Jyongsik, *Polymer* **39**, 4065 (1998).

1063. S. Yoshida and H. Ishida, *Appl. Surf. Sci.* **20**, 497 (1983).

1064. S. Yoshida and H. Ishida, *J. Chem. Phys.* **78**, 6960 (1983).

1065. H. G. Tompkins and S. P. Sharma, *Surf. Interface Anal.* **4**, 261 (1982).

1066. H. Ishida and S. Yoshida, *Appl. Surf. Sci.* **89**, 39 (1995).

1067. W. N. Richmond, P. W. Faguy, and S. C. Weibel, *J. Electroanal. Chem.* **448**, 237 (1998).

1068. G. P. Cicileo, B. M. Rosales, F. E. Varela, and J. R. Vilche, *Corrosion Sci.* **41**, 1359 (1999).

1069. M. Thouvenin, I. Linossier, O. Sire, J. J. Peron, and K. Vallee-Rehel, *Macromol.* **35**, 489 (2002).

1070. S. Rodrigues-Llorente, R. Aroca, and J. Duff, *J. Mater. Chem.* **8**, 2175 (1998).

1071. D. Murphy and M. N. De Pinho, *J. Membr. Sci.* **106**, 245 (1995).

1072. R. Teraoka, M. Otsuka, and Y. Matsuda, *Int. J. Pharm.* **184**, 35 (1999).

1073. G. Zhang, W. G. Pitt, and N. L. Owen, *J. Appl. Polym. Sci.* **54**, 419 (1994).

1074. T. Nguyen, J. Martin, E. Byrd, and N. Embree, *Polym. Degrad. Stabil.* **77**, 1 (2002).

1075. K.-K. Lee, N.-G. Cha, J.-S. Kim, J.-G. Park, and H.-J. Shin, *Thin Solid Films* **377–378**, 727 (2000).

1076. A. C. Middleton, R. A. Duckett, I. M. Ward, A. Mahendrasingam, and C. Martin, *J. Appl. Polym. Sci.* **79**, 1825 (2001).

1077. N. Everall, D. MacKerron, and D. Winter, *Polymer* **43**, 4217 (2002).

1078. C. Sammon, C. Mura, J. Yarwood, N. Everall, R. Swart, and D. Hodge, *J. Phys. Chem. B* **102**, 3402 (1998).

1079. M.-C. Pham and P.-C. Lacaze, *J. Electrochem. Soc.* **141**, 156 (1994).

1080. C. Barbero, O. Haas, and M. C. Pham, *J. Electrochem. Soc.* **142**, 1829 (1995).

1081. J. E. Puskas, A. J. Michel, and L. B. Brister, *Kautschuk Gummi Kunststoffe* **53**, 587 (2000).

1082. R. H. Wieringa, E. A. Siesling, P. F. M. Geurts, P. J. Werkman, E. J. Vorenkamp, V. Erb, M. Stamm, A. J. Schouten, *Langmuir* **17**, 6477 (2001).

1083. M. Muller, R. Buchet, and U. P. Fringeli, *J. Phys. Chem.* **100**, 10810 (1996).

1084. H. Kitano, K. Ichikawa, M. Fukuda, A. Mochizuki, and M. Tanaka, *J. Colloid Interface Sci.* **242**, 133 (2001).

1085. Y. V. Pan, E. Z. Barrios, and D. D. Denton, *J. Polym. Sci. A* **36**, 587 (1998).

1086. M. Hofmann, J. E. Puskas, and K. Weiss, *Eur. Polym. J.* **38**, 19 (2002).

1087. Y. Yi, K. Nerbonne, and J. Pellegrino, *Appl. Spectrosc.* **56**, 509 (2002)

1088. M. Yang, T. Tsukame, H. Saitoh, and Y. Shibasaki, *Polym. Degrad. Stabil.* **67**, 479 (2000).

1089. P. Eaton, P. Holmes, and J. Yarwood, *Appl. Spectrosc.* **54**, 508 (2000).

1090. Z. Ping, G. E. Nauer, H. Neugebauer, and J. Theiner, *J. Electroanal. Chem.* **420**, 301 (1997).

1091. A. Zimmermann and L. Dunsch, *J. Mol. Struct.* **410**, 165 (1997).

1092. H. Kim and J. Jang, *J. Appl. Polym. Sci.* **68**, 775 (1998).

1093. S. U. Hong, T. A. Barbari, and J. M. Sloan, *J. Polym. Sci. B* **36**, 337 (1998).

1094. J. Glime and J. L. Koenig, *Rubber Chem. Technol.* **73**, 47 (2000).

1095. M.-C. Pham, *J. Electroanal. Chem.* **277**, 327 (1990).

1096. P. W. Faguy, R. A. Lucas, and W. Ma, *Colloid Surf. A* **105**, 105 (1995).

1097. J. Meichsner, *Contributions Plasma Phys.* **39**, 427 (1999).

1098. A. Pihlajamaki, P. Vaisanen, and M. Nystrom, *Colloid Surf. A* **138**, 323 (1998).

1099. S. J. McCarthy, G. F. Meijs, N. Mitchell, P. A. Gunatillake, G. Heath, A. Brandwood, K. Schindhelm, *Biomaterials* **18**, 1387 (1997).

1100. Y. Yi, K. Nerbonne, and J. Pellegrino, *J. Polym. Sci. B* **38**, 1773 (2000).

1101. J.-P. Gibert, J.-M. Cuesta, A. Bergeret, and A. L. Crespy, *Polym. Degrad. Stabil.* **67**, 437 (2000).

1102. I. Linossier, F. Gaillard, M. Romand, and J. F. Feller, *J. Appl. Polym. Sci.* **66**, 2465 (1997).

1103. B. Mailhot, S. Morlat, and J.-L. Gardette, *Polymer* **41**, 1981 (2000).

1104. C. Aymes-Chodur, N. Betz, M. C. Porte-Durrieu, C. Baquey, and A. Le Moel, *Nucl. Instrum. Meth. B* **151**, 377 (1999).

1105. (a) M. A. Pellett, A. C. Watkinson, J. Hadgraft, and K. R. Brain, *Int. J. Pharmaceutics* **154**, 217 (1997); (b) S. Hajatdoost and J. Yarwood, *J. Chem. Soc. Faraday Trans.* **93**, 1613 (1997).

1106. N. Nurdin, P. Francois, and P. Descouts, *J. Biomat. Sci. Polym. Edit.* **7**, 49 (1995).

1107. C. G. L. Khoo, J. B. Lando, and H. Ishida, *J. Polym. Sci. B* **28**, 213 (1990).

1108. D. Qinpin and M. Min, *Spectrosc. Lett.* **28**, 43 (1995).

1109. T. A. Thorstenson, L. K. Tebelius, and M. W. Urban, *Polym. Mater. Sci. Eng.* **68**, 203 (1993).

1110. N. B. Joshi, and D. E. Hirt, *Appl. Spectrosc.* **53**, 11 (1999).

1111. (a) T. Ponnuswamy and O. Chyan, *Anal. Sci.* **18**, 449 (2002); (b) D. J. Walls, *Appl. Spectrosc.* **45**, 1193 (1991).

1112. H. Kim and M. W. Urban, *Langmuir* **15**, 3499 (1999).

1113. M. Ide, D. Yoshikawa, D. Maeda, and H. Kitano, *Langmuir* **15**, 926 (1999).

1114. M. Scarlete, S. Brienne, I. S. Butler, and J. F. Harrod, *Chem. Mater.* **6**, 977 (1994).

1115. J. E. Hansen, B. I. Rickett, J. H. Payer, and H. Ishida, *J. Polym. Sci. B* **34**, 611 (1996).

1116. P. A. Suci and G. G. Geesey, *Colloid Surf. B* **22**, 159 (2001).

1117. Z. Ping and G. E Nauer, *J. Electroanal. Chem.* **416**, 157 (1996).

1118. B. Zelei, T. Szekely, N. K. Gladkova, and S. G. Durgaryan, *Spectrochim. Acta A*, **44A**, 1117 (1988).

1119. K. J. Kuhn, B. Hahn, V. Percec, and M. W. Urban, *Appl. Spectrosc.* **41**, 843 (1987).

1120. E. V. Kober and A. E. Chmel, *J. Appl. Spectrosc.* **64**, 140 (1997).

1121. M. Rochery, T. M. Lam, and J. S. Crighton, *Macromol. Symp.* **119**, 277 (1997).

1122. H. Lee and L. A. Archer, *Polymer* **43**, 2721 (2002).

1123. J. X. Chen and J. A. Gardella, *Appl. Spectrosc.* **52**, 361 (1998).

1124. J. M. G. Cowie, B. G. Devlin, and I. McEwen, *J. Polymer*, **34**, 501 (1993).

1125. C. Werner and H. J. Jacobash, *Int. J. Artificial Organs* **22**, 160 (1999).

1126. B. R. Schmitt, H. Kim, and M. W. Urban, *Polym. Mater. Sci. Eng.* **75**, 71 (1996).

1127. B. J. Niu and M. W. Urban, *Polym. Mater. Sci. Eng.* **73**, 325 (1995).

1128. K. Yukiyasu and M. W. Urban, *Progr. Organic Coatings* **35**, 247 (1999).

1129. H. Kim and J. Jang, *Polymer* **39**, 4065 (1998).

1130. L. K. Tebelius and M. W. Urban, *Polym. Mater. Sci. Eng.* **71**, 482 (1994).

1131. J. K. F. Tait, G. Davies, R. McIntyre, and J. Yarwood, *Vib. Spectrosc.* **15**, 79 (1997).

1132. S. Cotugno, D. Larobina, G. Mensitieri, P. Musto, and G. Ragosta, *Polymer* **42**, 6431 (2001).

1133. A. M. Tralhao, A. C. Watkinson, and N. A. Armstrong, *Pharm. Research* **12**, 572 (1995).

1134. J. J. Sahlin and N. A. Peppas, *J. Biomater. Sci. -Polymer Edition* **8**, 421 (1997).

1135. R. Schafer, J. Kressler, and R. Mulhaupt, *Acta Polym.* **47**, 170 (1996).

1136. C. M. Balik and J. R. Xu, *J. Appl. Polym. Sci.* **52**, 975 (1994).

1137. B. J. Niu and M. W. Urban, *J. Appl. Polym. Sci.* **70**, 1321 (1998).

1138. C. Ohe, H. Ando, N. Sato, Y. Urai, M. Yamamoto, and K. Itoh, *J. Phys. Chem. B* **103**, 435 (1999).

1139. T. Bayer, K. J. Eichhorn, K. Grundke, and H. J. Jacobasch, *Macromol. Chem. Phys.* **200**, 852 (1999).

1140. C. Sammon, J. Yarwood, N. Everall, R. Smart, and D. Hodge, *Polymer* **41**, 2521 (2000).

1141. R. Tannenbaum, C. Hakanson, A. Zeno, and M. Tirrell, *Langmuir* **18**, 5592 (2002).

1142. B. Piro, E. A. Bazzaoui, M.-C. Pham, P. Novak, and O. Haas, *Electrochim. Acta* **44**, 1953 (1999).

1143. T. Strunskus, M. Grunze, G. Kochendoerfer, and Ch. Wöll, *Langmuir* **12**, 2712 (1996).

1144. M. R. Alexander, G. Beamson, C. J. Blomfield, G. Leggett, and T. M. Duc, *J. Electron Spectrosc. Related Phenom.* **121**, 19 (2001).

1145. H. Bourque, I. Laurin, M. Pezolet, J. M. Klass, R. B. Lennox, G. R. Brown, *Langmuir* **17**, 5842 (2001).

1146. C. R. Evans, T. A. Spurlin, and B. L. Frey, *Anal. Chem.* **74**, 1157 (2002).

1147. S.-Y. Lin, K.-S. Chen, and R.-C. Liang, *Polymer* **40**, 2619 (1999).

1148. X. Q. Lin and H. Q. Zhang, *Elecrochim. Acta* **41**, 2019 (1996).

1149. S. H. McKnight and J. W. Gillespie Jr., *J. Appl. Pol. Sci.* **64**, 1971 (1997).

1150. L. K. Tebelius and M. W. Urban, *Polym. Mater. Sci. Eng.* **71**, 154 (1994).

1151. C. M. Yam, L. Zheng, M. Salmain, C. M. Pradier, P. Marcus, G. Jaouen, *Colloid Surf. B* **21**, 317 (2001).

1152. S. Vaudreuil, H. Qiu, S. Kaliaguine, M. Grmela, and M. Bousmina, *Macromol. Symp.* **158**, 155 (2000).

1153. M. R. Pereira and J. Yarwood, *J. Chem. Soc. Faraday Trans.* **92**, 2737 (1996).

APPENDIX

As discussed in Section 3.1, IR spectra of ultrathin films of strong oscillators measured with p-polarization at grazing angles of incidence exhibit the band(s) that are absent in the spectrum of the absorption coefficient of the parent material. Therefore, the traditional interpretation based on comparison with the normal-transmission spectra is incorrect in this case. Among other means to distinguish this effect (Section 3.2.3), one can use Table A.1, in which either the frequencies (ν_{LO}) of the most intensive absorption bands in the metallic IRRAS spectra of ultrathin films or the parameters of the elementary oscillators [Eq. (1.46)] are collected for a number of inorganic and organic compounds. Using the data from this table, it should be kept in mind that the position of the ν_{LO} band in the experimental spectra of ultrathin films is affected by porosity/inhomogeneity and crystallinity of the film (Section 3.9.3). (In the table KK refers to the Kramers–Kronig transform and DA to dispersion analysis.) Table A.2 briefly describes optical and physical properties of some IR transparent materials employed as immersion media or materials for IREs and substrates in IR spectroscopic measurements of ultrathin films.

Table A.1. Characteristics of most intensive bands in IR spectra of ultrathin films with strong LO–TO splitting

Compound	Crystal Type, Orientation of Electric Field Relative to Crystal Axes		ε_∞	ν_{LO}, cm^{-1}	ν_{TO}, cm^{-1}	S	γ_{LO}, cm^{-1}	γ_{TO}, cm^{-1}	Comments	References
					I. Oxides					
AlO			2.9		866	0.48		57	IRRAS of 28-nm film	1
α-Al$_2$O$_3$	Corundum	$E\perp c$		900	635				Raman	2, 3
		$E\|c$		871	583					
γ-Al$_2$O$_3$				960					Thin film on Al	4
BaO	Cubic			425	132					5
BeO	Wurzite	$E\perp c$		1085	684				Raman	6, 7
		$E\|c$		1095	725					
		$E\perp c$	2.99		680	4.66		12.8		8
		$E\|c$	2.95		724	3.99		11.6		
CaO	Cubic			575	286				Reflectance	9
CdO	Cubic			380	270					10
Co$_3$O$_4$	Spinel			684					Thin film	11

Material	Structure	Polarization		Frequencies		Intensity		Notes	Ref
α-Cr₂O₃	Corundum	E⊥c	6.2	420, 446, 602, 766, 602, 759	417, 444, 532, 613, 538, 613				12
		E∥c	6.1						
Cr₂O₃				733				Oxidation film on stainless steel	13
Cu₂O	Cubic, $O_h^4(P_n3_m)$		6.5	635	609	0.7			14
	Cubic, $O_h^4(P_n3_m)$		6.9	638	613	0.57	18.3		8
CuO	Monoclinic, C_{2h}^6 (C2/c)	E∥z		600	421			Thin film	15
		E∥x, y		624	603			Raman and IR	16
α-Fe₂O₃	Corundum	E⊥c		230, 368, 494, 662, 414, 662, 655	227, 286, 437, 524, 299, 526	1.1, 12, 2.9, 1.1, 11.5, 2.2	4, 8, 20, 25, 15, 30		17
Fe₂O₃				655				Oxidation film on stainless steel	13

Table A.1. (*Continued*)

Compound	Crystal Type, Orientation of Electric Field Relative to Crystal Axes	ε_∞	TO–LO Splitting of Most Intensive Bands in IR Region or DA Data					Comments	References
			ν_{LO}, cm^{-1}	ν_{TO}, cm^{-1}	S	γ_{LO}, cm^{-1}	γ_{TO}, cm^{-1}		
GeO$_2$	Rutile		755	455	7.78		15.9		18
H$_2$O	Liquid	1.56		1641	0.013		80	ATR	19
In$_2$O$_3$	Cubic, T_h^7		513	489				Reflectance of samples pressed from powder and KK	20
MgO	Cubic	3.0	725	401	6.6		7.6		21
		2.964		396	6.8		7.6		8
				643	0.043		90		
MnO		4.95	570	270	15.7		25.7	Reflectance and DA	22
MoO$_3$	Orthorhombic, $P_{bnm} \equiv D_{2h}^{16}$		974	818					23
α-SiO$_2$ $\quad E \perp c$		2.356		1227	0.009		135		24
				1163	0.01		7.0		
				1072	0.67		7.6		
				797	0.11		7.2		
				697	0.18		8.4		
				450	0.82		4.1		
				394	0.33		2.8		
$E \| c$		2.383		1220	0.011		183		
				1080	0.67		7.5		

Material	Structure	Pol.	ε					Comments	Ref.
SiO	Amorphous		2.1		778	0.10	7.8	DA of IRRAS of 1-nm film within MIS structure,	1
					539	0.006	21.6		
					509	0.05	7.1		
					495	0.66	4.5		
					364	0.68	5.1		
SnO_2		$E \perp c$		757	605				25, 26
		$E \parallel c$		703	465				
SrO	Cubic			487	227				5, 27
TiO_2	Amorphous			821	1191	0.049	49	IRRAS and KK for PECVD films on Al	28
					1101	0.31	46		
					790	0.17	96		
TiO_2	Rutile	$E \perp c$	6	831	183	81.5	35		29
				458	388	1.08	23		
					500	2.00	22		
		$E \parallel c$	7.2		167	163			
TiO_2	Rutile	$E \perp c$	6	831	508		50	Product fit	30
		$E \parallel c$	7.8	458	388		38		
TiO_2	Rutile	$E \perp c$		831	508		27	Raman	31
				458	388		76		
		$E \parallel c$		811	167				
ThO_2	Fluorite			568	281				32

Table A.1. (*Continued*)

Compound	Crystal Type, Orientation of Electric Field Relative to Crystal Axes	ε_∞	ν_{LO}, cm^{-1}	ν_{TO}, cm^{-1}	S	γ_{LO}, cm^{-1}	γ_{TO}, cm^{-1}	Comments	References
UO_2	Fluorite		555	281				Raman	33
V_2O_5	$E\|a$		952	765	1.78		40		34
	$E\|b$		1037.7	975	0.6		2		
	$E\|c$		842.4	505	8.21		19		
V_2O_5	Orthorhombic, terminal $\nu(V{-}O)$ bridging $\nu(V{-}O{-}V)$		1035 895	1016 795					35
WO_3			970	~700				ATR of sputtered WO_3 film	36
Y_2O_3			620 535 486 456 412 359 315	555 490 461 415 371 335 303					8
ZnO			585	413				Reflectance	9
ZrO_2	Fluorite		680	354					2

TO–LO Splitting of Most Intensive Bands in IR Region or DA Data

II. Antimonides, Arsenides, Carbides, Fluorides, Hydrides, Nitrides, Phosphides, and Sulfides

Material	Structure	Pol.	n						Comment	Ref.
AlSb	Zincblende			340	319					37
AlAs	Zincblende			402	364					38
SiC	Zincblende		$6.35 + 0.36i$		799	2.82	39		Laser abladed 108-nm film on Si	39
α-SiC	Wurtzite	$E \perp c$		970	797				Raman	40
		$E \| c$		964	788					
β-SiC	Zincblende		6.7	969	793.6	3.3	8.5		Sum fit of transmission and reflectance of films on Si	41, 42
	Zincblende			973	797				Thin film	43
CaF$_2$	Fluorite		2.045	473	257					44, 45
KF	Cubic			326	190					46
LiF	Cubic		1.96	306	503	6.8	18.4			21
						0.11	91			
MgF$_2$	Rutile	$E \perp c$		617	450					47–49
		$E \| c$		625	399					
NaF	Cubic			418	244					46
SrF$_2$	Fluorite			366	217					44
BaF$_2$	Fluorite			319	184					44

Table A.1. (*Continued*)

Compound	Crystal Type, Orientation of Electric Field Relative to Crystal Axes	ε_∞	TO–LO Splitting of Most Intensive Bands in IR Region or DA Data					Comments	References
			ν_{LO}, cm^{-1}	ν_{TO}, cm^{-1}	S	γ_{LO}, cm^{-1}	γ_{TO}, cm^{-1}		
PbF$_2$	Fluorite		340					IRRAS of film on Al	50
LaF$_3$	$E \perp c$		457	356					51, 52
	$E \parallel c$		468	323					
KMgF$_3$	Cubic perovskite		551	458					53
KMnF$_3$			483	399					53
CsNiF$_3$	Hexagonal, D_{6h}^4 $E \parallel c$		457	332			16	Reflectance	54
AlN	Wurzite $E \perp c$		895	672					55, 56
	$E \parallel c$		888	659					
		$4.25 + 0.02i$		662		3.39	17	Laser abladed 813-nm film	39
			887					IRRAS of 1-nm MOCVD film on NiAl	57
			818						
BN	Hexagonal $E \perp c$	4.95	778	767				Thin film	58
			1610	1367					
	$E \parallel c$		828	783					
			1595	1510					
	Zincblende		1305	1055					59

	Structure	n	Frequency		Intensity	p-Polarized transmission at 60° of thin film		Reference
GaN	Hexagonal		735			14		60
	Cubic		739			11		
InN		5.8	570				Reflectance of thin film	61
Si_3N_4			1100				Thin film	62
a-SiN:H		3.68		852	1.88	138	DA	1
				465	0.65	126		
a-Si:H		11.2		2016	0.017	97	DA	1
AlON	Spinel	3.14		346	1.0	55		8
				395	0.4	51		
			551	495	3.33	99		
			718	634	1.32	95		
			869	738	0.16	65		
			969	920	0.03	46		
			920–946				IRRAS of 0.8-nm MOCVD film on NiAl	57
BP	Zincblende		829	799				59
GaP	Zincblende		403	367			Raman	37, 63
InP	Zincblende		345	304				37
FeS_2	Cubic, T_h^6		439	422				64, 65
			421	412				
			411	401				
			350	348				
			294	293				

Table A.1. (Continued)

Compound	Crystal Type, Orientation of Electric Field Relative to Crystal Axes	ε_∞	ν_{LO}, cm^{-1}	ν_{TO}, cm^{-1}	S	γ_{LO}, cm^{-1}	γ_{TO}, cm^{-1}	Comments	References
			TO–LO Splitting of Most Intensive Bands in IR Region or DA Data						
PbS	Cubic	18.96	212	71	150				66
ZnS	Zincblende		352	282					67, 68
CdIn$_2$S$_4$		6.5	338	304	0.59		9.1		69
III. Binary Oxides (Salts of Metal-Oxide Acids)†									
CoAlO$_4$	Spinel		768	650					70
CoCr$_2$O$_4$			699	608	0.03		26.8		71
CuCr$_2$O$_4$	Tetragonal $14_{1/amd}$		656	576	0.067		73.2		71
FeCr$_2$O$_4$			693	608				Reflectance of pressed pellets, KK	72
MgCr$_2$O$_4$			725	633	0.045		53.2		71
MnCr$_2$O$_4$			683	602				Reflectance of pressed pellets, KK	72
NiCr$_2$O$_4$			692	610				Reflectance of pressed pellets, KK	72
ZnCr$_2$O$_4$			706	633				Reflectance of pressed pellets, KK	72

Material	Structure	Orientation							Method	Ref
$CaCuO_2$				598						73
$La_2CuO_{4-\delta}$	Tetragonal	$E \perp c$	3	418, 700, 351, 456, 579	359, 669, 314, 354, 516				Sum fit of reflectance and KK	74
		$E \parallel c$	3.6							
$Nd_2CuO_{4-\delta}$	Tetragonal	$E \perp c$	7	409, 582	347, 516		40, 65	12, 36	Sum fit of reflectance and KK	74
$LaFeO_3$	Orthorhombic, P_{nma}		2.65	566	541	0.133	33.7	32.6		75
$CoFe_2O_4$	Spinel			680					IRRAS of thin film	11
$Y_3Fe_5O_{12}$	Garnet		2.89	712	590		25	18		76
Al_2MgO_4				800	670					8
$CaMoO_4$	Scheelite	$E \perp c$		910	790					77, 78
		$E \parallel c$		898	450					
$PbMoO_4$	Scheelite	$E \perp c$		886	744					78
		$E \parallel c$		865	745					
$Rb_{0.3}MoO_3$		$E \perp b$		473	290		8.7	38	Reflectance	79

Table A.1. (*Continued*)

Compound	Crystal Type, Orientation of Electric Field Relative to Crystal Axes	ε_∞	ν_{LO}, cm^{-1}	ν_{TO}, cm^{-1}	S	γ_{LO}, cm^{-1}	γ_{TO}, cm^{-1}	Comments	References
LiNbO$_3$	$E \perp c$	5.06		152	22		2		80
				236	0.8		3		
				265	5.5		3		
				322	2.2		3.5		
				363	2.3		12		
				431	0.18		5.2		
				586	3.3		20.5		
				670	0.2		31.5		
	$E \| c$	5.48		248	16		5.2		
				274	1		4		
				307	0.16		8		
				628	2.55		21		
				692	0.13		34		
NaNbO$_3$			876	535					81
KNbO$_3$			826	521					81
LaNiO$_3$	Rhombohedral		831	370.9	5.297	507	181	Reflectance and DA	75
BaTiO$_3$			717	487					81
	$E \perp c$		706	482		22	21		8
	$E \| c$		729	507		34	45		
CaTiO$_3$			866	610					81
PbTiO$_3$			750	500					81

Material	Structure	Orientation						Method	Ref.
$SrTiO_3$				795	545				81
$CaWO_4$	Scheelite	$E \perp c$	3.65	147.5	143	1.5	10		77
				248	202	3.5	10		
				368	309	1.06	117		
				905	793	0.93	9.5		
		$E \| c$	3.75	182	180	0.3	11		
				327	237	4.65	18		
				448	435	0.22	11		
				893	778	0.97	54		
					360	0.1	36		
$CdWO_4$	Wolframite			~900				Reflectance and KK	82
$PbWO_4$	Scheelite	$E \perp c$		869	756				78, 83
		$E \| c$		866	764				
$SrZrO_3$	Orthorhombic			704	519				84

IV. Ternary Oxides

Material	Structure	Orientation						Method	Ref.
$SrLaAlO_4$	Tetragonal, $14/mmm\text{-}D_{uh}^{17}$	$E \| a$	4.2	582.2	447.4	6.3	6.3	Reflectance and KK	85
		$E \| c$	4.2	557.8	347.8	26.6	79.6		
$Ca_{0.86}Sr_{0.14}CuO_2$				405	320		40	Reflectance and KK	86
$LaSrGaO_4$	Tetragonal	$E \| a$	4.54	502	327		10.4	Reflectance and KK	87
		$E \| c$	4.02	505	399		30		
$CoFeMnO_4$	Spinel			687				IRRAS of thin film	11
$La_{0.5}Sr_{1.5}MnO_4$	Perovskite	$E \| a$			367		53	Reflectance and KK	88
		b			374		50		
		$E \| c$							

681

Table A.1. (Continued)

Compound	Crystal Type, Orientation of Electric Field Relative to Crystal Axes	ε_∞	TO–LO Splitting of Most Intensive Bands in IR Region or DA Data					Comments	References
			ν_{LO}, cm^{-1}	ν_{TO}, cm^{-1}	S	γ_{LO}, cm^{-1}	γ_{TO}, cm^{-1}		

V. Carbonates, Chlorates, Germanates, Nitrites, Phosphates, Phosphites, Silicates, and Sulfates

Compound	Orientation	ε_∞	ν_{LO}, cm^{-1}	ν_{TO}, cm^{-1}	S	γ_{LO}, cm^{-1}	γ_{TO}, cm^{-1}	Comments	References
CaCO$_3$			1545	1407				Reflectance	89
			1549	1407				Raman	90
			890	872					
K$_2$CO$_3$			1492	1400				Reflectance	89
Li$_2$CO$_3$			1606	1420				Reflectance	89
Na$_2$CO$_3$			1543	1419				Reflectance	89
SrCO$_3$			1538	1445				Reflectance	89
NaClO$_3$			1030	988				IRRAS	9
CuGeO$_3$	$E\|y$	3.0	805	720		35	9		91
	$E\|x$	3.5	860	777		35	20		
AgNO$_3$			1410	1316				Reflectance	89
Ba(NO$_3$)$_2$		2.37		729	0.006		1.7		92
				817	0.006		2.7		
				1352	0.23		12.5		
				1412	0.017		10		

Compound	Mineral	Orientation	Value	ν_a	ν_b	Frequencies	Intensity	Angle	Method	Ref.
$Ca(NO_3)_2$				1470	1360				Reflectance	89
$CsNO_3$				1421	1330				Reflectance	89
KNO_3				1445	1375				Reflectance	89
$LiNO_3$				1480	1345				Reflectance	89
$NaNO_3$				1459	1350				Reflectance	89
$Pb(NO_3)_2$			2.7			723 806 1316 1318	0.003 0.006 0.32 0.015	32.5	Reflectance	76
$RbNO_3$				1440	1340				Reflectance	89
$Sr(NO_3)_2$				1460	1360				Reflectance	89
$TlNO_3$				1418	1280				Reflectance	89
$NaNO_2$				1458	1355				IRRAS	9
$AlPO_4$	α-Betonite	$E\|c$	1252	1108						93
$Ca_5(PO_3)_3F$	Apatite	$E\perp c$	2.66 2.67			300 575 598	1.68 0.066 0.132	10 9 13		92
		$E\|c$				1043 1091 300 559 1030	0.36 0.03 0.56 0.443 0.45	9 12.5 8		

Table A.1. (Continued)

Compound	Crystal Type, Orientation of Electric Field Relative to Crystal Axes	ε_∞	TO–LO Splitting of Most Intensive Bands in IR Region or DA Data					Comments	References
			ν_{LO}, cm^{-1}	ν_{TO}, cm^{-1}	S	γ_{LO}, cm^{-1}	γ_{TO}, cm^{-1}		
ZrSiO$_4$	$E\|c$		1106	989					94
	$E\perp c$		1034	885					
α-KAl(SO$_4$)$_2$ · 12H$_2$O	Cubic (Pa3) (111)		1134	1093				Reflectance and KK	95
γ-NaAl(SO$_4$)$_2$ · 12H$_2$O	Cubic (Pa3) (111)		1146	1102				Reflectance and KK	95
β-CsFe(SO$_4$) · 12H$_2$O	Cubic (Pa3) (210)		1129	1090				Reflectance and KK	95
VI. Organic Films									
Meta-aniline	Orthorombic, Pbc2$_1$ $M\|b$		1527	1509				96	
Hexamethylene-tetra-amine C$_6$H$_{12}$N$_4$		2.53	675.5	672	0.029		1.7		97
			815	811.5	0.022		2.6		
			1020	1006	0.071		3.1		
			1243	1237	0.024		3.7		

684

Polyimide	2.82		1717	0.069	19.6		98
CO adsorbed at CH$_2$Cl$_2$–Pt interface	2.56		2000	0.07	60	DA of in situ ATR spectra	19
Poly(methyl methacrylate)		1737	1727			ν C=O, KK	99
Perfluoropoly-ether	1.7		740	0.013	25		100–102
			805	0.008	30		
			980	0.24	20		
			995	0.006	20		
			1060	0.008	30		
			1125	0.050	50		
			1180	0.025	50		
			1240	0.081	50		
			1305	0.076	50		

†In alphabetical order of anions

Table A.2. Optical and physical properties of IR materials

Material	Composition	Transparency Range (cm^{-1})	Refractive Index	Wavenumber (cm^{-1})	Additional Information
α-Al$_2$O$_3$	Sapphire	1780–50,000	1.60	1870	Hard, inert crystal
AMTIR	Se, As, and Ge glass	725–11,000	2.5	1000	Relatively hard, good with acids, incompatible with alkalis
As$_{0.22}$S$_{0.33}$Br$_{0.44}$	As–S–Br liquid glass	900–4000	2.10	1250	$T_{\text{melt}} = -50°C$, used as IRE material for liquid ATR, toxic
BaF$_2$		740–50,000	1.42	1000	Weakly soluble in water and organic solvents
CaF$_2$ (Irtran-3)		1120–50,000	1.41	2500	Withstands high pressure
CdTe (Irtran-6)		360–20,000	2.67	1000	Brittle, easily cracked
Diamond	Carbon	2500–4500; 33–1660	2.40	1000	Pressure and scratch resistant, very hard
GaAs		630–4900	3.27	1000	Hard, brittle
Ge		830–5500	4.00	1000	Hard, brittle
IKS-24 (Russian trade mark)	As$_x$Se$_y$I	600–10,000	2.38	1400	Brittle, toxic
IKS-35 (Russian trade mark)	As$_x$Se$_y$I glass	560–7000	2.37	1250	Melting point $\approx$ 90°C, toxic
KRS-5	Mixture of TlBr and TlI	400–20,000	2.37	1000	Temperature sensitive, toxic
Si		1500–8300	3.42	1000	Hard, brittle
SiO$_2$	Fused silica	2900–50,000	1.43	4000	
TiO$_2$	Rutile	1400–25,000	$n_{\parallel} = 2.32$	2000	
ZnS (Irtran-2)		950–17,000	2.20	1000	Comparable to ZnSe but slightly harder and more chemical resistant
ZnSe (Irtran-4)		650–20,000	2.40	—	Most popular ATR material
ZrO$_2$–(12%)Y$_2$O$_3$	Zirconia	2000–40,000	2.04	2340	Chemically inert

REFERENCES

1. R. Brendel and D. Bormann, *J. Appl. Phys.* **71**, 1 (1992).
2. S. Shin and M. Ishigame, *Phys. Rev. B* **34**, 8875 (1986).
3. S. P. S. Porto and R. S. Krishnan, *J. Chem. Phys.* **47**, 1009 (1967).
4. K. Nishikida and R. W. Hannach, *Appl. Spectrosc.* **46**, 999 (1992).
5. M. Galtier, A. Montaner, and G. Vidal, *J. Phys. Chem. Solids* **33**, 2295 (1972).
6. E. Loh, *Phys. Rev.* **166**, 673 (1967).
7. C. A. Arguello, D. L. Rousseau, and S. P. S. Porto, *Phys. Rev.* **181**, 1351 (1969).
8. E. D. Palik (Ed.), *Handbook of Optical Constants of Solids III*, Academic, New York, 1998.
9. J. B. Bates and M. H. Brooker, *J. Phys. Chem. Solids* **32**, 2403 (1971).
10. Z. V. Popovic, G. Stanisic, D. Stojanovic, and R. Kostic, *Phys. Stat. Solidi (B)* **165**, K109 (1991).
11. M. Lenglet and B. Lefez, *Solid State Commun.* **98**, 8,689 (1996).
12. D. R. Renneke and D. W. Lynch, *Phys. Rev.* **138**, A530 (1965).
13. D. K. Offesen, *J. Electrochem. Soc.* **132**, 2250 (1985).
14. P. Dawson, M. M. Harqreave, and G. R. Wilkinson, *J. Phys. Chem. Solids* **34**, 2201 (1973).
15. B. Lefez, R. Souchet, K. Kartouni, and M. Lenglet, *Thin Solid Films* **268**, 45 (1995).
16. S. Guda, D. Peebles, and T. J. Wieting, *Phys. Rev. B* **43**, 13092 (1991).
17. S. Onari, T. Arai, and K. Kudo, *Phys. Rev. B* **16**, 1717 (1977).
18. D. M. Roessler and W. A. Albers, *J. Phys. Chem. Solids* **33**, 293 (1972).
19. T. Burgi, *Phys. Chem. Chem. Phys.* **3**, 2124 (2001).
20. H. Sobota, H. Neuman, G. Kuhn, and V. Riede, *Cryst. Res. Technol.* **25**, 1, 61 (1990).
21. J. R. Jasperse, A. Kahan, and J. N. Plendle, *Phys. Rev.* **146**, 256 (1966).
22. S. Mochizuki, *J. Phys. Condens. Matter.* **1**, 51, 10351 (1989).
23. K. Edo, *J. Solid State Chem.* **95**, 64 (1991).
24. W. Spitzer and D. Kleinman, *Phys. Rev.* **121**, 1324 (1961).
25. R. Summit, *J. Appl. Phys.* **39**, 3762 (1967).
26. J. F. Scott, *J. Chem. Phys.* **53**, 852 (1970).
27. J. L. Jacobson and E. R. Nixon, *J. Phys. Chem. Solids* **29**, 967 (1968).
28. B. C. Trasferetti, C. U. Davanzo, N. C. da Cruz, and M. A. B. de Moraes, *Appl. Spectrosc.* **54**, 687 (2000).
29. W. G. Spitzer, R. C. Miller, D. A. Kleiman, and L. E. Howarth, *Phys. Rev.* **126**, 1710 (1962).
30. (a) F. Gervais and B. Piriou, *J. Phys.C* **7**, 2374 (1974); (b) F. Gervais and B. Piriou, *Phys. Rev. B* **10**, 1642 (1974).
31. G. R. Wilkinson, in A. Anderson (Ed.), *Raman Effect*, Vol. 2, Marcel Dekker, New York, 1973, Chapter 1.
32. P. G. Axe and G. D. Pettit, *Phys. Rev.* **151**, 676 (1966).
33. P. G. Marlowe and J. P. Russell, *Phil. Mag.* **14**, 409 (1966).

34. P. Clauws and J. Vennik, *Phys. Stat. Solidi (B)* **76**, 707 (1976).

35. A. Surca, B. Orel, G. Drazic, and B. Pihlar, *J. Electrochem. Soc.* **146**, 232 (1999).

36. T. A. Taylor and H. H. Patterson, *Appl. Spectrosc.* **48**, 674 (1994).

37. A. Mooradian and G. B. Wright, *Solid State Commun.* **4**, 431 (1966).

38. M. Ilegems and G. L. Pearson, *Phys. Rev. B* **1**, 1576 (1970).

39. H. Hobert, H. H. Dunken, J. Meischien, and H. Stafast, *Vib. Spectrosc.* **19**, 205 (1999).

40. D. W. Feldman, J. H. Parker, W. J. Choyee, and L. Patrick, *Phys. Rev.* **170**, 698 (1968).

41. W. G. Spitzer, D. A. Kleiman, C. F. Frosch, and D. J. Walsh, in J. R. O'Connor and J. Smiltens (Eds.), *Silicon Carbide, A High Temperature Semiconductor*, Pergamon, New York, 1960.

42. W. G. Spitzer, D. A. Kleiman, and D. J. Walsh, *Phys. Rev.* **113**, 133 (1959).

43. T. Zorba, D. I. Siabkas, and C. C. Katsidis, *Microelectron. Eng.* **28**, 229 (1995).

44. W. Kaiser, W. G. Spitzer, R. H. Kaiser, and L. E. Howarth, *Phys. Rev.* **127**, 1950 (1962).

45. I. Richman, *J. Chem. Phys.* **41**, 2836 (1966).

46. S. S. Mitra, in S. Nudelman and S. S. Mitra (Eds.), *Optical Properties of Solids*, Plenum Press, New York, 1969, Chapter 14.

47. S. P. S. Porto, P. A. Fleury, and T. C. Damen, *Phys. Rev.* **154**, 522 (1967).

48. J. Giordano and C. Benoit, *J. Phys. C* **21**, 2749 (1988).

49. C. Benoit and J. Giordano, *J. Phys. C* **21**, 5209 (1988).

50. A. Lehmann, W. Heerdegeen, G. Schirmer, H. Mutschke, W. Richter, E. Hacker, and R. Dohle, *Phys. Stat. Solidi (A)* **119**, 683 (1990).

51. R. P. Bauman and S. P. S. Porto, *Phys. Rev.* **161**, 842 (1967).

52. R. P. Lownders, J. F. Parrish, and C. H. Perry, *Phys. Rev.* **182**, 913 (1969).

53. C. H. Perry and E. F. Young, *J. Appl. Phys.* **38**, 4616 (1967).

54. N. Primeau, S. Jandl, M. Banville and A. Caille, *Solid State Commun.* **75**, 2, 121 (1990).

55. J. A. Sanjurjo, E. Lopex-Cruz, P. Vogl, and M. Cardona, *Phys. Rev. B* **28**, 4579 (1983).

56. A. T. Collins, E. C. Lightowlers, and P. J. Dean, *Phys. Rev.* **158**, 833 (1967).

57. V. M. Bermudez, *Thin Solid Films* **347**, 195 (1999).

58. R. Geick, C. H. Perry, and G. Rupprecht, *Phys. Rev.* **146**, 543 (1966).

59. P. J. Gielisse, S. S. Mitra, J. N. Plendl, R. D. Griffis, L. C. Mansur, R. Marshall, and E. A. Pascoe, *Phys. Rev.* **155**, 1039 (1967).

60. M. Giehler, M. Ramsteiner, O. Brandt, H. Yang, and K. H. Ploog, *Appl. Phys. Lett.* **67**, 6, 733 (1995).

61. T. Inushima, T. Shiraishi, and V. Yu. Davydov, *Solid State Commun.* **110**, 491 (1999).

62. M. Firon, C. Bonnelle, and A. Mayeux, *J. Vac. Sci. Technol. A* **14**, 2488 (1996).

63. A. S. Barker, *Phys. Rev.* **165**, 917 (1968).

64. H. D. Lutz, G. Schneider, and G. Kliche, *J. Phys. Chem. Solids* **46**, 437 (1985).

65. H. Vogt, T. Chattopadhyay, and H. J. Stolz, *J. Phys. Chem. Solids* **44**, 869 (1983).

66. R. Geick, *Phys. Lett.* **10**, 51 (1964).

67. A. Manabe, A. Mitsuishi, and H. Yoshinga, *Jap. J. Appl. Phys.* **6**, 593 (1967).

68. W. G. Nilsen, *Phys. Rev. B* **182**, 838 (1969).

69. H. Neuman, W. Kissinger, F. Levy, H. Sobota, and V. Riede, *Cryst. Res. Technol.* **24**, 1165 (1989).

70. H. Shirai, Y. Morioka, and J. Nakagawa, *J. Phys. Soc. Jap.* **51**, 592 (1982).

71. H. D. Lutz, B. Muller, and H. J. Steiner, *J.Solid State Chem.* **90**, 54 (1991).

72. M. M. T. Anki and B. Lefez, *Appl. Opt.* **35**, 1399 (1996).

73. S. Lupi, P. Galvani, M. Capizzi, P. Maseli, W. Sadovski, and E. Walker, *Phys. Rev. B* **45**, 12470 (1992).

74. J. G. Zhang, G. W. Lehman, and P. C. Eklund, *Phys. Rev. B* **45**, 4660 (1992).

75. N. E. Massa, H. Falcon, H. Salva, and R. E. Carbonio, *Phys. Rev. B* **56**, 10178 (1997).

76. V. M. Zolotarev, V. N. Morosov, and E. V. Smirnova, *Optical Constants of Natural and Technical Media*, Khimia, Leningrad, 1984, 216pp.

77. A. S. Barker, *Phys. Rev. A* **135**, 742 (1964).

78. P. Tarte and M. Liegeois-Duyckaerts, *Spectrochim. Acta* **28A**, 2029 (1972).

79. S. Jangl, M. Banville, C. Pepin, J. Marens, and C. Schlenker, *Phys. Rev. B* **40**, 12487 (1989).

80. A. S. Barker and J. R. Lowdon, *Phys. Rev.* **158**, 433 (1964).

81. W. Zong, R. D. King-Smit, and D. Vanderbilt, *Phys. Rev. Lett.* **72**, 3618 (1994).

82. J. Gabrusenocs, A. Veispals, A. Von Czarnovski, and K.-H. Meiwes-Broer, *Electrochim. Acta* **46**, 2229 (2001).

83. J. M. Stencel, E. Siberman, and J. Springer, *Phys. Rev. B* **12**, 5435 (1976).

84. K. Wakamura, *Solid State Ionics* **145**, 315 (2001).

85. G. Komba, E. Buihadera, and A. Pajaczkovska, *Phys. Status Solidi (A)* **168**, 317 (1998).

86. G. Burns, M. K. Crawford, F. M. Dacol, E. M. McCarron III, and T. M. Shaw, *Phys. Rev. B* **40**, 6717 (1989).

87. A. W. MacConnell, T. Timusk, A. Dabkovski, and H. A. Dabkovska, *Physica C* **292**, 233 (1997).

88. J. H. Jung, H. J. Lee, T. W. Noh, and Y. Moritimo, *J. Phys: Condens. Mat.* **12**, 9799 (2000).

89. J. B. Bates and M. H. Brooker, *Chem. Phys. Lett.* **21**, 349 (1973).

90. S. P. S. Porto, J. A. Giordmaine, and T. C. Damen, *Phys. Rev.* **147**, 608 (1966).

91. S. D. Devic, M. J. Constantinovic, Z. V. Popovic, G. Dhalenne, and A. Revcolevschi, *J. Phys. Cond. Mater.* **48**, L745 (1994).

92. L. C. Kravits, J. D. Kingsley, and E. L. Elkin, *J. Chem. Phys.* **49**, 4600 (1968).

93. A. Goullet, T. Bretagnon, J. Camassel and J. Pascual, *Phys. Scr.* **42**, 478 (1990).

94. P. Dawson, M. M. Hargreave, and G. R. Wilkinson, *J. Phys. C* **4**, 240 (1971).

95. V. Ivanovski, V. M. Petrusevski, and B. Soptrajanov, *Vib. Spectrosc.* **19**, 425 (1999).

96. M. M. Szostak, N. Le Calve, F. Romaino, and B. Pasquier, *Chem. Phys.* **187**, 3, 373 (1994).

97. G. N. Zhizhin, M. A. Mockaleva, and V. A. Yakovlev, *Phizika tverdogo tela* **17**, 2217, 1975.

98. P. A. Kawka and R. O. Buckius, *Int. J. Thermophys.* **22**, 517 (2001).

99. K. Yamamoto and H. Ishida, *Vib. Spectrosc.* **8**, 1 (1994).

100. V. J. Novotny, I. Hussla, J.-M. Turlet, and M. R. Philpot, *J. Chem. Phys.* **90**, 5861 (1989).

101. J. Pacansky, R. J. Waltman, and M. Maier, *J. Phys. Chem.* **91**, 1225 (1987).

102. J. Pacansky, C. D. England, and R. J. Waltman, *Appl. Spectrosc.* **40**, 8 (1986).

INDEX

Abnormal absorption, 219
Absorbance, 36, 433
  integral, 20–21, 58, 182, 438, 441, 447
Absorption, 5, 9
  optical theory of, *see* Maxwell theory;
      Dispersion model
  physical models of, 10–11
Absorption band
  broadening of, 205, 225, 256
    inhomogeneous, 172
  complex, resolution of components, *see*
      Principal component analysis;
      Correlation analysis
  distortions of, *see* Spectrum distortions
  intensity of, 35, 249. *See also* Absorption
      depth
  narrowing of, 187, 596
  at ν_{LO}, 157, 533
  at ν_{TO}, 157
  shift of, *see* Band repulsion
    blue, 62, 142, 161–163, 178, 181–185,
        202–203, 226, 246, 255
    red, 15, 62, 107, 163, 185, 192, 195, 202,
        205, 206, 220–230, 250
  splitting of, 191, 198, 205, 224, 245, 257,
      263–265, 278, 559. *See also*
      Berreman effect
Absorption coefficient, *see* Decay constant
Adsorption density/isotherm, 58–59, 522, 545,
    596, 621. *See also* Quantitative analysis
    (linearity range)
Absorption depth, 34
Absorption/extinction cross section, 66, 224
Absorption edge, 12
Absorption factor, 35
Absorption index, *see* Extinction coefficient
Absorption spectrum, xxv, 7, 22, 41, 65. *See
    also* Absorption

Ac advantage, 376
Acceptor states, 187, 525
Accumulation layer, 187, 206, 497, 574, 582
Acetaldehyde, 532
Acetone, 132, 191, 205, 232, 522, 564, 566,
    608, 612
Acetonitrile, 115, 191, 205–209, 522, 528, 532,
    610
Acetylcholinesterase, 620–621
Acroleine, 532
Acrylic acid, 532
Acrylic polymer, 605, 606
Acrylonitrile, 532
Active/inactive medium, 148, 149, 151
Adenine, 616
Adhesion, 122, 323, 329, 344, 392, 514, 601,
    611, 614, 623. *See also* Bacterium
    adhesion
Adriamycin, 626
Adsorption, 514–515
  from gaseous phase, 514–547. *See also* Cell,
      temperature- and
      pressure-programmed
  from solution, 547–570, 613–616, 621–629.
      See also Spectroelectrochemical
      (SEC) experiments; Cell, solid–liquid
      interface
  in pores, *see* Porous film; Substrate,
      particles, porous
Adsorption isotherm, 373, 521, 543, 552, 555
Aerosil, 121, 329
Ag, 19, 86, 113, 122, 123, 126–127, 192, 195,
    199, 211, 230–238, 241, 242, 258, 316,
    328, 359, 367, 371, 384, 391, 392, 516,
    532, 534, 564–565, 588, 602, 612
AgBr, 318
AgCl, 173, 318, 330, 393, 543
AgNO$_3$, 682

691